ANTIFUNGAL COMPOUNDS

VOLUME 2

Interactions in Biological and Ecological Systems

Edited by

MALCOLM R. SIEGEL
Department of Plant Pathology
University of Kentucky
Lexington, Kentucky

HUGH D. SISLER
Department of Botany
University of Maryland
College Park, Maryland

Senior Editor of Volume 2: HUGH D. SISLER

MARCEL DEKKER, INC. New York and Basel

Library of Congress Cataloging in Publication Data (Revised)

Main entry under title:

Antifungal compounds.

Includes bibliographical references and indexes.
CONTENTS: v. 1. Discovery development, and uses. v. 2. Interactions in 76-46827 biological and ecological systems.
1. Fungicides. I. Siegel, Malcolm R. II. Sisler, Hugh Delane, 1922-
SB951.3.A57 628.9 76-46827
ISBN *0-8247-6558-3*

MARCEL DEKKER, INC.
270 Madison Avenue, New York, New York 10016

Current printing (last digit):
10 9 8 7 6 5 4 3 2

PRINTED IN THE UNITED STATES OF AMERICA

PREFACE

The need for fungicides is created by fungal activities detrimental to the welfare of mankind. As the bases for such destructive activities are identified, compounds are sought to control the fungi involved. Fungi had been identified as the basis of many problems a half century or more ago, but at that time very few fungicides had been found to control them. A great number of compounds have since been discovered, new methods of application have been developed, and new concepts on modes of action have evolved. At no time in the past has interest in fungicides been greater than during the current period of transition from the protective to the systemic fungicide.

A number of years have elapsed since a comprehensive coverage of fungicides appeared in two volumes edited by D. C. Torgeson (Fungicides: An Advanced Treatise, 1967). Since then, there have been marked advances in the development of systemic compounds, and a rising awareness of the toxicological hazards and the environmental impact of fungicides has evolved. With the advent of more selective fungicides, fungal resistance has emerged as a major problem. In this rapidly changing field, it is the purpose of this book and its companion volume to summarize and evaluate recent developments, to integrate these with significant developments of the past, and to attempt some projections into the future. We believe the various contributors to the volumes have achieved a reasonable measure of success for these goals.

The overall organization of the two books follows a biological—biochemical—ecological approach rather than one based on chemical class. However, individual authors have utilized the latter approach where it was most appropriate for their particular contribution.

The first of the two volumes focuses on the discovery, development, and use of fungicides and the problems associated therewith in plant pathology, medicine, wood preservation, and industry. This volume considers fungicides according to their effects and their fate as they interact with biological, biochemical, and ecological systems.

We are grateful to the many authors whose efforts have made these volumes possible. We wish to thank Carolyn Siegel, Patricia Sisler, Libbie Jones, Debby Owen, and Jeanne Kelleher for generous assistance in the preparation of the manuscripts.

Malcolm R. Siegel
Hugh D. Sisler

CONTENTS

CONTRIBUTORS

H. M. DEKHUIJZEN*
Institute for Organic Chemistry TNO, Central Organization for Applied Scientific Research in the Netherlands, Utrecht, The Netherlands

J. DEKKER
Laboratory of Phytopathology, Agricultural University, Wageningen, The Netherlands

L. V. EDGINGTON
Department of Environmental Biology, University of Guelph, Guelph, Ontario, Canada

LAWRENCE FISHBEIN
Chemistry Division, National Center for Toxicological Research, Jefferson, Arkansas

S. G. GEORGOPOULOS†
Department of Biology, Nuclear Research Center "Demokritos," Athens, Greece

A. KAARS SIJPESTEIJN
Institute for Organic Chemistry TNO, Central Organization for Applied Scientific Research in the Netherlands, Utrecht, The Netherlands

KAZUO KAKIKI
Fungicide Laboratory, Institute of Physical and Chemical Research, Wako-Shi, Saitama, Japan

DONALD D. KAUFMAN
Pesticide Degradation Laboratory, Agricultural Environmental Quality Institute, Agricultural Research Service, United States Department of Agriculture, Beltsville, Maryland

---

*Present address: Center for Agrobiological Research, Wageningen, The Netherlands

†Present address: Department of Plant Pathology, Agricultural College of Athens, Votanicos, Athens, Greece

JOSEPH KUĆ
Department of Plant Pathology, University of Kentucky, Lexington, Kentucky

H. LYR
Institute of Plant Protection Research, Kleinmachnow, Academy of Agricultural Sciences of the German Democratic Republic

TOMOMASA MISATO
Fungicide Laboratory, Institute of Physical and Chemical Research, Wako-Shi, Saitama, Japan

GAYLORD D. PAULSON
Metabolism and Radiation Research Laboratory, United States Department of Agriculture, Agricultural Research Service, Fargo, North Dakota

CAROL A. PETERSON*
Department of Environmental Biology, University of Guelph, Guelph, Ontario, Canada

NANCY N. RAGSDALE
Department of Botany, University of Maryland, College Park, Maryland

DONALD V. RICHMOND
Department of Agriculture and Horticulture, Long Ashton Research Station, University of Bristol, Bristol, England

LOUIS SHAIN
Department of Plant Pathology, University of Kentucky, Lexington, Kentucky

MALCOLM R. SIEGEL
Department of Plant Pathology, University of Kentucky, Lexington, Kentucky

J. W. VONK
Institute for Organic Chemistry TNO, Central Organization for Applied Scientific Research in the Netherlands, Utrecht, The Netherlands

DAVID WOODCOCK
Department of Agriculture and Horticulture, Long Ashton Research Station, University of Bristol, Bristol, England

---

*Present address: Department of Biology, University of Waterloo, Waterloo, Ontario, Canada

CONTENTS OF VOLUME 1

Chapter 1

SOIL-FUNGICIDE INTERACTIONS

Donald D. Kaufman

Pesticide Degradation Laboratory
Agricultural Environmental Quality Institute
Agricultural Research Service
United States Department of Agriculture
Beltsville, Maryland

I. INTRODUCTION

Organic pesticides occur in detectable amounts in many parts of the environment. An understanding of the fate and behavior of biologically active substances in the total environment is a necessity for the chemicals to be used safely, and for new products to be developed which will not produce adverse effects. Fungicides are an important part of our chemical arsenal for controlling agricultural pests. Although they were among the earliest agricultural chemicals developed for use, they currently rank third in total sales. Common and chemical names of fungicides are listed in Table 1.

Fungicides enter into the soil from a variety of treatments. Excess solution from spray applications to the aerial parts of plants may drip from leaves on to the soil surface or may reach the soil surface directly. In preplanting, plant, or other treatments, fungicides can be applied directly to the soil either in solution, as dusts, or in granular formations. The ultimate behavior of a given pesticide in soil depends primarily on the physical, chemical, and biological properties of both the soil and chemical. Sorption, diffusion, volatilization, leaching, runoff, microbial and chemical degradation, photodegradation, and plant uptake are all significant processes affecting fungicide behavior in soil. The interactions of these various parameters determine the effectiveness of the chemical applied to soil and its residual life in soil.

According to Munnecke [1], much of the early research on soil fungicides was empirical. A transition from empirical to experimental research began about 1950 [1, 2] when research on chemical disinfestation was initiated. Despite the view of McNew [3] who believed that research on fungicides in general had the disadvantage of scientists working in a "vacuum," it is interesting to note the significant contributions of research on fungicides in soils to our overall understanding of the behavior of all pesticides in soil. As a disciple of plant pathology and soil microbiology who has worked on the soil microbiology of pesticides (primarily herbicides and insecticides), the author cannot help but be impressed with the contribution of early soil fungicide and fumigant research to our understanding of the behavior of all pesticides in soil. The preponderance of soil fungicide behavior studies, however, have dealt with the basic problems of efficacy. Since the publication of Rachel Carson's Silent Spring [4] increased awareness of our environment dictates that we not only understand the behavior (efficacy) of chemicals released into the environment, but that we also aim for a total concept of the ultimate fate of the chemical. The development and enactment of the U.S. Environmental Protection Agencies Guidelines for Pesticide Registration now requires extensive investigations of the fate and behavior of all chemicals ultimately scheduled for release into the environment. The prominence of herbicide and insecticide development at that critical point in time has resulted in a vast amount of scientific effort being put into understanding both the environmental fate and behavior of those chemicals in soil. With few exceptions, very little attention by comparison has been devoted to fungicide research. While much of the information necessary for continued registration of fungicides must have been developed, very little has appeared in the literature.

Several reviews have been published on the environmental behavior of fungicides [5-17]. The very excellent review by Goring [5], which discusses the physical aspects of soil in relation to the action of soil fungicides,

TABLE 1

Common and Chemical Names of Some Important Fungicides

Common name or designation	Chemical name	Other designations
ACNQ	2-Amino-3-chloro-1,4-naphthoquinone	
Actinomycin D	$C_{62}H_{86}N_{12}O_{16}$	
Anilazine	2,4-Dichloro-6-(o-chloroanilino)-s-triazine	Dyrene, Kemate
Aureomycin	$C_{22}H_{23}ClN_2O_8$	Chlortetracycline
BAS-307	2-Chlorobenzanilide	
BAS-3050	2-Methylbenzanilide	Mebenil
BAS-3191	2,5-Dimethyl-3-furanilide	Furcarbanil
Benomyl	Methyl 1-(butylcarbamoyl)-2-benzimidazolecarbamate	Benlate
Binapacryl	2-sec-Butyl-4,6-dinitrophenyl-3-methyl-2-butenoate	Endosan, Morocide, NIA-9044
Captafol	cis-N[(1,1,2,2-Tetrachloroethyl)thio]-4-cyclohexene-1,2-dicarboximide	Difolatan
Captan	N-[(Trichloromethyl)thio]-4-cyclohexene-1,2-dicarboximide	Orthocide
Carbon disulfide	Carbon disulfide	
Carboxin	5,6-Dihydro-2-methyl-1,4-oxathiin-3-carboxanilide	Vitavax, DCMO
Ceresan L	Methylmercuric 2,3-dihydroxypropylmercaptide (2.89%) + methylmercuric acetate (0.62%)	

TABLE 1 (Cont.)

Common name or designation	Chemical name	Other designations
Ceresan M	N-(Ethylmercuri)-p-toluenesulfonanilide	
Chloramphenicol	D(-)-Threo-2,2-dichloro-N-[β-hydroxy-α-(hydroxymethyl)-p-nitrophenethyl]acetamide	Chloromycetin
Chloranil	Tetrachloro-p-benzoquinone	Spergon
Chloroneb	1,4-Dichloro-2,5-dimethoxybenzene	Demosan, Thersan SP
Chloropicrin	Trichloronitromethane	
Chlorothalonil	Tetrachloroisophthalonitrile	Bravo, Daconil, Termil
Cycloheximide	3-[2-(3,5-Dimethyl-2-oxocyclohexyl)-2-hydroxyethyl]-glutarimide	Acti-dione
Cycloheximide oxime	Cycloheximide oxime	
1,3-D	1,3-Dichloropropene and related chlorinated C_3 hydrocarbons	DD, Telone
D-198	2,5-Dimethyl-1,4-benzoquinone monoxime	
Dazomet	Tetrahydro-3,5-dimethyl-2H-1,3,5-thiadiazine-2-thione	DMTT, Mylone
DBCP	1,2-Dibromo-3-chloropropane	Fumagon, Fumazon, Nemagon
DCNA	2,6-Dichloro-4-nitroaniline	Botran, dichloran

Diazoben	Sodium p-(dimethylamino)benzenediazosulfonate	Dexon, BAY-22555
Dichlone	2,3-Dichloro-1,4-naphthoquinone	Phygon
Dinocap	2-(1-Methylheptyl)-4,6-dinitrophenyl crotonate	Karathane, Mildex, DNOCP
DNBP	2-sec-Butyl-4,6-dinitrophenol	Dinoseb
Dodine	n-Dodecylguanidine acetate	Cyprex
Ethylene dibromide	1,2-Dibromoethane	EDB, Dowfume
ETMT	5-Ethoxy-3-(trichloromethyl)-1,2,4-thiadiazole	Terrazole
Ferbam	Ferric dimethyldithiocarbamate	Fermate
Griseofulvin	(2S-trans)-7-chloro-2^1,4,6-trimethoxy-6^1-methylspiro-[benzofuran-2(3H),1^1-[2]cychohexene]-3,4^1-dione	
Hexachlorobenzene	Hexachlorobenzene	HCB
Hexachlorophene	2,2^1-Methylenebis[3,4,6-trichlorophenol]	NABAC
Maneb	Manganous ethylenebis[dithiocarbamate]	Manzate
Metham	Sodium methyldithiocarbamate	SMDC, Vapam
Methylarsinic sulfide	Methylarsinic sulfide	Rhizoctol
Methyl bromide	Bromomethane	MBR
Methyl isothiocyanate	Methyl isothiocyanate	MIT
Nabam	Disodium ethylenebis[dithiocarbamate]	Parzate
NIA-5961	1-Chloro-2-nitropropane	Lanstan

TABLE 1 (Cont.)

Common name or designation	Chemical name	Other designations
NIA-9102	Mixture of 5.2 parts by weight of ammoniates of [ethylene-bis(dithiocarbamate)]zinc with 1 part by weight ethylene-bis[dithiocarbamic acid], bimolecular and trimolecular cyclic anhydrousulfides and disulfides	Polyram, metiram
Oxycarboxin	5,6-Dihydro-2-methyl-1,4-oxathiin-3-carboxanilide 4,4-dioxide	Plantvax
Oxythioquinox	Cyclic S,S-(6-methyl-2,3-quinoxalinediyl)dithiocarbonate	Morestan
Panogen 15	Cyano(methylmercuri)guanidine	MMDD
Patulin	4-Hydroxy-4H-furo[3,2-C]pyran-2(6H)-one	Clavacin, clavatin, expansine
PCNB	Pentachloronitrobenzene	Terraclor, quintozene
PCP	Pentachlorophenol	DOWCIDE 7
PCP-(Na)	Sodium pentachlorophenate	
PMA	(Acetato)phenylmercury	
Streptomycin sulfate	2,4-Diguanidino-3,5,6-trihydroxy-cyclohexyl-5-deoxy-2-O-[2-deoxy-2-(methylamino)-α-glucopyranosyl]-3-formyl-L-lyxopentanofuranoside sulfate	Agri-mycin, Agri-strep

Streptothricin	$C_{19}H_{34}N_8O_8$	
Subtilin	$C_{148}H_{234}N_{38}O_{41}S_5$(diacetate composition)	
TCNA	2, 3, 5, 6-Tetrachloro-4-nitroanisole	
TCNB	1, 2, 4, 5-Tetrachloro-3-nitrobenzene	Tecnazene
Thiophanate-methyl	Dimethyl [1, 2-phenylenebis(iminocarbonothioyl)]-bis[carbamate]	Topsin, Mildothane
Thiram	Bis(dimethylthiocarbamoyl) disulfide	TMTD, Arasan
TPTH	Triphenyltin hydroxide	Du-Ter, fentin hydroxide
Triarimol	α-(2, 4-Dichlorophenyl)-α-phenyl-5-pyrimidinemethanol	EL 273
Triphenyltin acetate	Triphenyltin acetate	TPTA, Brestan
Zinc ion-maneb complex	Coordination products of zinc ion and manganous ethylenebis[dithiocarbamate]	Dithane M-45, mancozeb
Zineb	Zinc ethylenebis[dithiocarbamate]	
Ziram	Zinc dimethyldithiocarbamate	Cuman

describes much of the current thought on soil-pesticide interactions and was particularly helpful, along with the discussions of Munnecke [1] in the preparation of this chapter. The reader is referred to these and other reviews [6-17] for more detailed, elaborate discussions of soil fungicide behavior. In addition to discussing these aspects briefly, the bulk of this chapter will be directed to a discussion of current literature dealing with the fate of fungicides in the soil environment.

II. BEHAVIOR AND FATE OF FUNGICIDES IN SOIL

A. Methods Used in Investigating Soil Fungicide Behavior

Munnecke [1] summarized methods used to assess the persistence of fungicides in soils including: chemical analysis [18-27]; response of spores or mycelium of test fungi buried in treated soil [28-34]; response of spores in agar media to extracts from fungicide treated soil [35], or to diffusates directly from soil [36]; infestation of agar media in treated soil [37]; disease reaction of susceptible plants in infested soil treated with fungicides [38-41]; response of fungi to vapors from treated soil [23, 29, 42]; and combination of one or more of the above [23, 37, 38, 43, 44].

Chemical analyses are the most qualitative and quantitative in terms of actual persistence of the active ingredient and the characterization of degradation rates and products. In view of our continuing concern for environmental pollution, these methods are the most preferred. No single technique, however, is suited for all purposes. While chemical assays indicate actual extractable toxicant, they are not necessarily indicative of actual residual toxicity. Bioassays more closely measure actual fungitoxicity. Rigid standardization of the quantity and genetic stability of the test material is necessary for reproducible results. Viability, number, and age of the assay organism are critical aspects of the bioassay method.

B. Sorption

Sorption is perhaps the most important single factor affecting the behavior of pesticides in the soil environment. Adsorption to soil constituents will affect the rate of volatilization, diffusion, or leaching, as well as the availability of the chemical to microbial or chemical degradation, or uptake by plants or other organisms. Burchfield [10], Kreutzer [45], Goring [5, 11], and Munnecke [1] have discussed the relevance of adsorption to fungicide and/or fumigant behavior in soil. The reviews of Goring [5, 11] were particularly helpful in the preparation of this section. For more detailed discussions of the sorption phenomenon the reader is referred to these reviews.

The concentration of pesticide in soil water or air, and the length of time it is maintained, determine the extent to which the organism is controlled and how long control will be maintained. This, in turn, depends on the physical, chemical, and biological properties of the toxicant and the soil, and their interactions. Sorption is chemical or physical. Anionic or cationic compounds are chemically sorbed, whereas nonionic compounds are physically sorbed. Hydrogen bonding is intermediate between physical and chemical sorption.

Physical sorption generally involves several layers and low binding strengths. Chemical sorption involves high binding strength, and while several layers may be present, only the first layer is chemically bonded to the surface. The Langmuir isotherm equations generally apply to compounds adsorbed in monomolecular layers. S-Shaped isotherms have been obtained from fumigants in soils, and the Brunauer, Emmett, and Teller (BET) equations have been applied to these cases, based on the assumption that the divergence from linearity is due to multimolecular adsorption resulting from action of van der Waals forces [46]. Although Burchfield [10] stated that the basic theory is unsound, the BET equations have been widely used.

Considerable effort has been expended to determine the relative importance of clay and organic matter in sorption. The task is difficult because both clay and organic matter sorb chemically and physically. Hartley [47] suggested that clay had "preoccupied" most researchers minds, and that it was not important with large uncharged molecules. It is now generally accepted that organic matter is most important, except in dry soils where clay content and specific surfaces are the principal factors regulating sorptive capacity.

Soil pH and moisture are critical factors affecting sorption of some pesticides. The pH of soil as well as the charge on clays affects the dissociation of pesticides and thus is very important in soil processes. Frissel [48] concluded that all compounds are adsorbed strongly at low pH, anionic substances are adsorbed negatively at slightly basic conditions, and nonionic compounds are moderately adsorbed. According to Crafts [49], cationic compounds are immediately affected by the base-exchange complex in moist soil in a manner similar to the rapid immobilization of ammonia by soil. A similar effect occurs with quaternary nitrogen compounds. Although they may be precipitated as insoluble compounds of soil bases, nonpolar and anionic compounds are not so affected by soil pH. The pH effect was demonstrated strikingly by Harris and Warren [50] with 2-*sec*-butyl-4,6-dinitrophenol, dinoseb (DNBP), which was strongly adsorbed at low pH but only slightly adsorbed at high pH (Fig. 1). Munnecke and Martin [51] observed a similar effect in three soils treated with methyl isothiocyanate (MIT). Although the slope for the three soils varied, the amount of MIT released increased linearly as the soil pH increased.

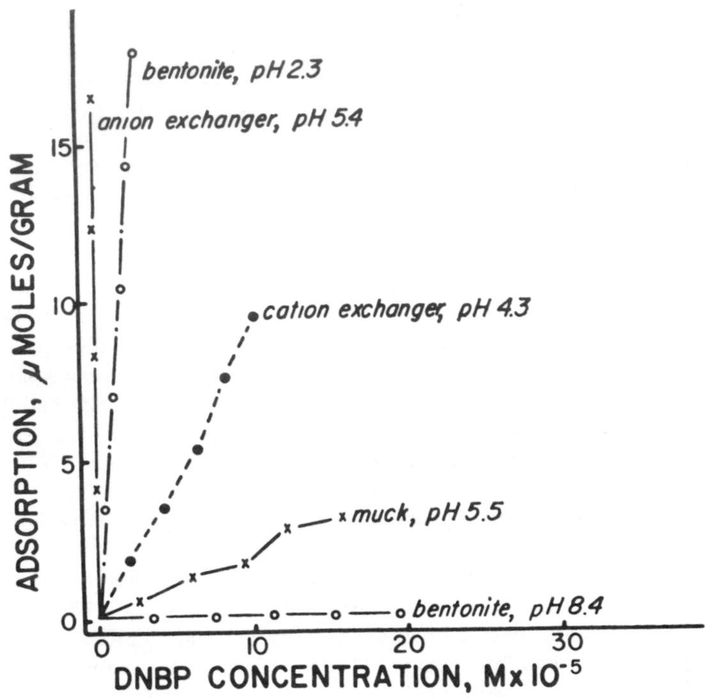

FIG. 1. Adsorption of dinoseb. (Reproduced from Ref. 50 by permission of the Weed Science Society of America.)

Sorption of some pesticides is more complete in dry soils than in wet soils. Sorption of the herbicide S-ethyl dipropylthiocarbamate (EPTC) on dry montmorillonite was attributed to coordination of cations with the carbonyl group or nitrogen, and bonding of the hydrogen on the methylene groups to the clay surfaces. Although stable complexes were formed in a humid atmosphere, EPTC was completely displaced when the EPTC-clay complex was added to water. Goring [5] expressed doubt that any nonionic toxicant can compete successfully with water for the surfaces of clay minerals, which accounts for observations in which toxicants sorbed by dry soils or clays are rapidly released by the addition of water.

Wade [52, 53], Call [54, 55], and Jurinak and Volman [46] observed that ethylene dibromide (EDB) was rapidly sorbed by dry soils, but the amount sorbed dropped off sharply as moisture contents increased. Call [55] examined EDB sorption on 20 different soils held at moisture contents corresponding to field capacity. The data were expressed as sorption coefficients and correlations with surface area, organic matter content, moisture content, and clay content were obtained. Clay content and specific surface area were the principal factors regulating sorptive capacity of dry soils, whereas organic matter was critical at field capacity. Leistra [56] obtained similar results with cis- and trans-1,3-dichloropropene (1,3-D; DD), but also noted effects of temperature and isomeric form. The amount adsorbed at 2°C was about three times greater than that adsorbed at 20°C. The trans-isomer was more strongly adsorbed than the cis-isomer.

C. Diffusion and Volatilization

Diffusion and volatilization are important factors affecting the distribution and persistence of some pesticides in soil. The significance of these processes becomes most apparent when pesticides are found in air or on soils at sites far from those originally treated. Although the majority of organic chemicals used in pest control have vapor pressures that are low by the standards of practical organic chemistry, the ability of these chemicals to diffuse and to volatilize or pass into vapor phase enables their movement from the application site. Compounds with high vapor pressures penetrate field soils best, if the gas is not allowed to escape into the air. The vapor pressure of methyl bromide is 1,380 mmHg at 20°C as opposed to 20 mmHg for chloropicrin [57]. It is useless to apply methyl bromide without covering the soil. Tarping frequently enhances the effectiveness of compounds with lower vapor pressures such as chloropicrin and MIT.

Goring [5] suggested that Fick's first law of diffusion [58] was probably applicable to all diffusion processes in soil, including movement of pesticides through air, water, and organic matter. Simply stated, the rate of movement of a pesticide is directly proportional to the concentration of the toxicant and its diffusion coefficient. The diffusion coefficient is a quantitative way of expressing the difficulty diffusing molecules have in moving through different transfer media such as air, water, or organic matter.

Goring [5] stressed the significance of water/air ratios and organic matter/water ratios. Pesticides with water/air ratios under 10,000 should diffuse primarily through air, whereas those with ratios over 30,000 should diffuse primarily through water. Methyl bromide and carbon disulfide have water/air ratios under 10 and move so rapidly they are lost within hours to days when introduced into soil. Compounds such as chloropicrin, EDB, 1,3-D, and MIT with ratios of 10 to 100 disappear within a few weeks. 1-Chloro-2-nitropropane, 1,2-dibromo-3-chloropropane (DBCP), and 2-chloro-6-(trichloromethyl)pyridine with ratios of 100-2,000 could remain in the soil for months if not decomposed by other mechanisms.

The partition coefficient suggested by Goring [5] is important in the process of volatilization. Usually the amount of pesticide applied is much less than that required to saturate the soil-water phase. Dry soil surfaces may adsorb pesticides strongly, but at normal soil moisture content, volatilization of pesticides takes place from a dilute aqueous solution at the exposed soil surface. An estimate of potential volatility can be obtained from the ratio of water solubility to vapor pressure, since this indicates what portion of the pesticide is in the vapor phase. This ratio is useful as a guide, but it is affected by adsorption, which decreases the amount of material present in the vapor phase.

The diffusion rate of a gas is influenced by molecular weight, temperature, presence of codiffusing gases, continuity of air spaces, and distribution of fumigant between air, water, and solid phases of soil. This latter distribution is influenced by temperature, moisture, air space, clay, and organic matter [1]. Diffusion is more important than movement in water in affecting soil penetration by methyl bromide and carbon disulfide. Both air diffusion and water flow are important with water soluble compounds such as metham and dazomet, whose activity depends upon volatile breakdown products [1]. In either case, however, soil moisture content affects diffusion greatly because the physical blocking of free air space by water hinders diffusion.

Application of volatile pesticides to soil surfaces generally results in their rapid dissipation. Losses from soil surfaces have even been reported for triazines [59] and ureas [60] which have very low vapor pressures. Thus incorporation into the soil is required for most effective use of such compounds as methyl bromide, carbon disulfide, chloropicrin, and chloroneb. Compounds such as metham, dazomet, or even the fumigant DBCP can be applied to the soil surface if it is then incorporated or leached into the soil [61].

There are many methods for applying volatile compounds to soil and these have been reviewed extensively [5, 11, 62]. The basic pattern of diffusion, once incorporated or injected, is illustrated in Fig. 2. The area under each curve represents the dose an organism would receive at that distance from the injection point.

Volatile compounds readily diffuse through soil, but care must be taken to avoid letting the chemical dissipate too rapidly to effect the desired level of control. Tarping or special injection techniques are thus necessary for most efficient use of volatile materials. Nonvolatile substances require thorough mixing with the soil for effective control. Goring [5] suggested that semivolatile compounds may have certain advantages. They need not be thoroughly mixed with the soil, yet they are more efficient than volatile compounds because of their greater tendency to stay in the treated zone. 2-Chloro-6-(trichloromethyl)pyridine applied in a band with ammonium fertilizer will control _Nitrosomonas_ in the band but not in the bulk of the soil [63, 64]. Chloroneb on cottonseed should diffuse into the soil and control _Rhizoctonia_ around the hypocotyl [5].

Pentachloronitrobenzene is applied in the seed row [65-67] or by seed treatment to control _Rhizoctonia_. It has sufficient vapor pressure that it would diffuse appreciably through soil [5]. Although Munnecke et al. [23] were unable to demonstrate the presence of fungitoxic vapors in soil treated

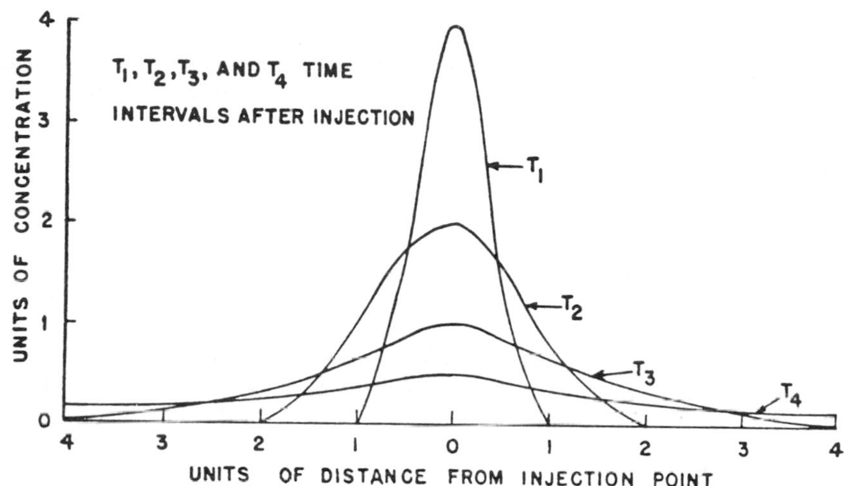

FIG. 2. Relation between distance from injection point and concentration of fumigant in soil at various time intervals after injection. (Reproduced from Ref. 5 with permission from Annual Review of Phytopathology. Copyright 1967 by Annual Reviews, Inc. All rights reserved.)

with PCNB, Richardson and Munnecke [40] observed fungitoxic concentrations of PCNB in soil air with the aid of a more sensitive bioassay method. While it is questionable whether methylmercuric dicyandiamide or some decomposition product may diffuse into the soil zone around seeds [23, 68], its redistribution on seed because of its volatility is well known [69]. Whatever diffuses is highly toxic to *Pythium* but not to *Rhizoctonia* [23].

D. Leaching

Toxicants are moved into soil by rainfall or irrigation. The movement of the toxicant is governed by three processes: desorption, diffusion through water, and hydrodynamic dispersion [70, 71]. Solubility of the toxicant in soil solution has also been considered important [1, 5, 47]. Desorption is proportional to the amount of toxicant sorbed [70, 71]. Diffusion through the water obeys Fick's law [1]. Hydrodynamic dispersion [71] is caused by water percolating downward through the centers of the pores more rapidly than along the sides. The net result of these three processes is increased spread of the toxicant throughout the soil. Thomas [71] discussed in detail the distribution patterns for solutes moved through soil by water. Figure 3 shows typical patterns for a toxicant placed on the soil surface and leached with increasing amounts of water. The amounts of water required to move concentration peaks to selected depths are directly related to the organic matter/water ratios of the toxicants being leached [5].

Movement of water in soil is upward as well as downward. If it is neither too easy nor too difficult to leach the toxicant, this alternating movement tends to distribute it evenly throughout the profile.

Although considerable information is available on vapor-phase movement of volatile fungicides in soil, very little has been published on the movement of nonvolatile fungicides. Early efforts with some soil fungicides focused on soil drenches which carried the assumption of some fungicide transport. Munnecke [72] examined the movement of 13 fungicides in large diameter soil columns and found that water-soluble compounds such as nabam and Panogen are greatly affected by leaching. Less soluble materials such as thiram, captan, ferbam, and PCNB were less easily leached, and their movement was affected to a greater degree than could be explained solely on

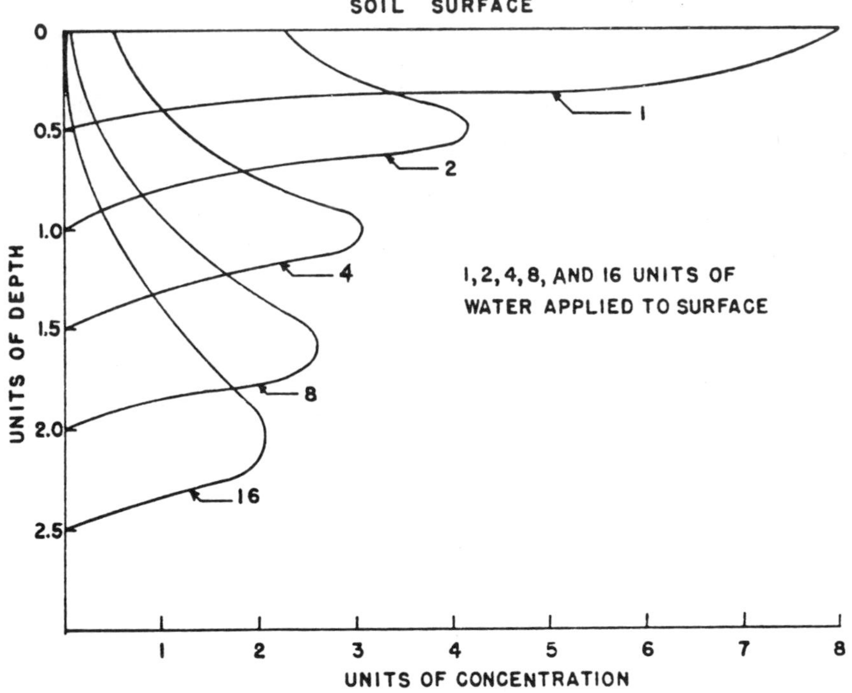

FIG. 3. Relation between amounts of water applied and concentration of toxicant at various depths, assuming application of toxicant to soil surface. (Reproduced from Ref. 5 with permission from Annual Review of Phytopathology. Copyright 1967 by Annual Reviews. Inc. All rights reserved.)

on the basis of water solubility. He also concluded that penetration was greater for solution formulations than for suspensions; for finer fungicide particles than for coarse particles; for coarser-textured soils; and for initially dry soils.

The soil thin-layer chromatographic (soil TLC) technique has proved useful in studies of relative mobility of pesticides [73-75]. Water moves in the soil layer by unsaturated flow [76], resembling typical water flow under field conditions. Helling et al. [77] evaluated 33 fungicides in terms of leaching and diffusion characteristics with a fungal bioassay developed for use with soil TLC. The results of this investigation are summarized in Table 2. Cycloheximide, cycloheximide oxime, Ceresan L, diazoben, and oxycarboxin were found to be relatively mobile compounds. Chloranil, chloroneb, DCNA, dichlone, dodine, hexachlorophene, oxythioquinox, PCNB, TCNA, ETMT, and zineb were immobile. The behavior of dodine was similar to that of the herbicides paraquat and diquat. All three are organic cations and are strongly adsorbed to soil because of the soils cation exchange capacity.

The movement of a number of compounds on soil TLC plates were characterized by several inhibition zones. Some of these spots were presumably due to fungitoxic degradation products formed either on the soil TLC plate, or more likely in the original samples. Ceresan L contains two active ingredients, methylmercuric acetate and a methylmercuric mercaptide, whose leaching characteristics are completely different. It was reasoned that the methylmercury cation of the methylmercuric acetate would be more strongly adsorbed by soil ion exchange and therefore be the immobile fraction. The dithiocarbamates do yield many fungitoxic by-products [78, 79].

The mobility order nabam > maneb > zineb was determined by both bioassay and autoradiography. Movement of these fungicides in five soils was inversely related to soil organic matter content [77].

One must also consider the inherent persistence of each chemical in soil in evaluating whether any mobile fungicide might pollute groundwater. The half-life of cycloheximide, the most mobile fungicide is very short.

The relative mobility in soil TLC of the fungicides evaluated was similar to that reported by other more cumbersome methods for nabam [72, 80], thiram [72, 80], captan [72, 80], zineb [72], PCNB [80], PMA [81], Panogen [23], benomyl [82], ACNQ [83], dichlone [5, 34, 72], chloranil [5], diazoben [84], chloroneb [85], DCNA [86], PCP [34], and dinocap [33, 34].

E. Photolysis

The action of sunlight may chemically alter and destroy pesticides in the environment. The rate of breakdown and the potential of the products to contaminate the environment influence the usefulness of the parent compound. Since environmental photochemistry is complicated by the possible interaction

TABLE 2

Fungicide Mobility[a] in Hagerstown Silty Clay Loam Soil TLC Plates [77]

Fungicide	Mobility Class[b]
ACNQ	1
Anilazine	1
Benomyl	2
Binapacryl	3,$\underline{1}$[c]
Captan	$\underline{3}$,1
Ceresan L	$\underline{4}$,1
Ceresan M	1
Chloranil	1
Chloroneb	1
Cycloheximide	4
Cycloheximide Oxime	5
D-198	3
DCNA	1
Diazoben	$\underline{4}$,1
Dichlone	1
Dinocap	$\underline{2}$,1
Dodine	1
E-275	$\underline{1}$,1
ETMT	1
Ferbam	5,$\underline{3}$,3,1
Hexachlorophene	1
Maneb	$\underline{2}$,1
Nabam	5,4,$\underline{2}$,1
Oxycarboxin	4
Oxythioquinox	1
Panogen 15	4,$\underline{3}$,3,1

TABLE 2 (Cont.)

Fungicide	Mobility Class[b]
PCNB	1
PCP	3
PCP-Na	3
TCNA	1
Thiram	5,4,2,1
Zineb	1
Ziram	5,3,2,2,1

[a]Mean based on individual averages for each assay organism. Assay organisms included *Aspergillus fumigatus*, *Diplodia zeae*, *Fusarium moniliforme* (two isolates), *F. roseum*, *Helminthosporium sativum*, *Penicillium chrysogenum*, *P. rugulosum*, *Rhizoctonia solani*, *Trichoderma viride*, *Chlorella sorokiniana*.

[b]Mobility classes:
1 = immobile = R_f 0 to 0.09
2 = low mobility = R_f 0.10 to 0.34
3 = intermediate = R_f 0.35 to 0.64
4 = mobile = R_f 0.65 to 0.89
5 = very mobile = R_f 0.90 to 1.0

[c]Fungicides with multiple numbers indicate presence of multiple active ingredients. Underlined number indicates principal ingredient toxic to most assay organisms.

of the reacting molecule with many environmental components, laboratory experiments have been necessary to obtain comparative data for determining the effect of factors that might modify products and rates of reaction.

There are two fundamental laws of photochemistry: (1) the Grotthus-Draper law which states that only the light adsorbed by a molecule is responsible for reaction; and (2) the Stark-Einstein law which states that the absorption of light by a molecule is a one-quantum process, so that the sum of the primary quantum yields, must be unity [87]. Determination of quantum yields requires elaborate experimental techniques. Major attention has been directed recently to atmospheric photochemistry, where consideration can be given to reactions of organic molecules at low concentrations in the vapor phase [88].

"Sensitized" processes, where light is absorbed by a "sensitizer" molecule which adopts an excited electronic configuration, are believed important in the environment. This excited electronic state possesses a relatively long life time during which the molecule can transfer energy to a molecule of another species which does not absorb light in the same region as the sensitizer molecule. Thus, a molecule which does not absorb light in the visible region of the spectrum may undergo reaction in the presence of a sensitizer absorbing at longer wave lengths [89-92].

Solvent, concentration, and aeration are important factors affecting rates of photolysis. Temperature seldom influences rates of photochemical reaction. Water is the predominant solvent for environmental studies of photolysis. Organic solvents are frequently limited by their ability to transmit light, or by the fact that they themselves may participate in the chemical reactions. Many complex polymeric substances may be formed in the presence of high concentrations. First-order rate laws apply to decomposition rates by photolysis in dilute aqueous solutions. The presence or absence of oxygen also influences the course and products of reaction. Photooxidation reactions predominate in the presence of oxygen rather than nitrogen [93].

Except when pesticides enter into aquatic systems, their primary points of exposure are either in the vapor phase or on leaf or soil surfaces. Photochemical studies have been performed on thin layers of pure material deposited on glass plates in attempts to emulate environmental surface photochemistry. Silica-gel coated glass plates or thin layers of soil have been used. Few studies have been reported on soil surfaces but the rates of photolysis appear to be slower than on silica-gel [94]. Changes in absorption spectra occur, however, with pesticides deposited on surfaces [95]. While such spectral changes are a general phenomenon and are dependent upon the nature of the compound [96], care must be taken in extrapolating photochemical information from one system to another.

Photodecomposition appears to be a major factor in the degradation of some fungicides under certain conditions. Exposure of aqueous diazoben solutions to sunlight destroys its fungistatic activity and its ability to control seedling diseases when used as a soil treatment [97]. Prolonged exposure of treated seeds to sunlight reduces protection against damping-off. Diazoben, bright yellow in aqueous solution, becomes colorless on exposure to light. Subsequently in an oxygen atmosphere, a pink color develops with time, becoming purple and gradually dark brown. The amorphous nature of the residue when the purple solution was lyophilized indicates polymerization of reaction products.

These observations and those of others [98, 99] with diazo compounds similar to diazoben, suggest a two-stage decomposition: (1) a dissociation in aqueous solution, forming the diazonium ion; and (2) absorbed light catalyzes a reduction of the diazonium groups resulting in the evolution of nitrogen. Subsequent oxidation is followed by polymerization of the reaction products. Polymerization of phenolic materials is not uncommon in nature. The environmental significance of such polymers derived from pesticide products is not known. Despite its susceptibility to photodecomposition, once in the soil diazoben appears to be fairly persistent [100].

Other fungicides also appear sensitive to light. Exposure of chloranil to ultraviolet resulted in a two-step reduction [101]. In the presence of ultraviolet, dichlone reacted with pyridine nucleotides and underwent reduction [101]. Solutions of dichlone bleach upon standing in glass bottles in light [102], and the effectiveness of fungicidal seed treatment is greatly reduced upon exposure to the sun. Even in subdued daylight, benzene solutions of dichlone underwent photochemical replacement of the chlorines with phenyl groups via free-radical reactions [103].

Extensive studies have been made of PCP photolysis in aqueous systems. It provides a good example of the importance of irradiation conditions to photolytic pathways. Using a low-pressure mercury lamp at approximately room temperature in each instance, Mitchell [104] observed that PCP remained unchanged when irradiated on filter paper, while Crosby and Hamadmad [105] demonstrated that only 2,3,5,6-tetrachlorophenol was formed in hexane solution, whereas tetrachlorophenol, chloranil, and a large portion of humic acid polymer appeared in water at slightly acidic pH. Furthermore, irradiation of PCP by sunlight in water led to a loss of aromaticity and formation of alicyclic compounds. Kuwahara and associates [106-108] found tetrachlororesorcinol, chloranilic acid, and other oxidized products after irradiation in water at slightly alkaline pH. Octachlorodibenzo-p-dioxin was also detected under these conditions [109].

Pentachloronitrobenzene appeared to be more resistant to photodecomposition. Ultraviolet irradiation of PCNB solutions in various solvents decomposed the compound only very slowly [105]. Reductive dechlorination was the primary reaction observed with PCNB, pentachlorobenzene, and PCP. The nitro group of PCNB was also removed to form pentachlorobenzene, but corresponding nitroanilines were not detected.

Although it is obvious from these results that photodecomposition is a significant factor affecting stability of some fungicides in aqueous solutions and on exposed seeds, considerably more information is needed to fully assess the impact of this process on fungicides on soil, or particularly on plant surfaces where disease control may occur.

F. Chemical and Microbial Degradation

An important part of the soil-pesticide interaction involves the rate of decomposition of the pesticide. Pesticide decomposition occurs by chemical reactions with soil constituents, or as a result of biochemical interactions with soil microorganisms. The rate at which pesticides decompose in soil is characteristic of both the chemical and the soil. Whether chemical or biochemical processes are most important in decomposition of a specific pesticide is probably most dependent upon the chemical characteristics of the pesticides, but may also be dependent upon specific soil characteristics. Very astute investigations are needed to accurately determine whether degradation is actually biological or chemical. Occasionally, due to apparent similarities, chemical degradative mechanisms may be misinterpreted as biological mechanisms by the more casual investigator.

Soil factors known to affect pesticide decomposition include temperature, aeration, microbial population, pH, organic matter, clay, cation exchange capacity, and moisture. Soil chemical and physical processes, and microbial populations may be similarly affected by these factors. The influence of environmental factors on physical processes such as leaching, diffusion, volatilization, or adsorption may parallel those effects observed on microbial populations. In a milieu such as soil, the nature, number, rate, and complexity of chemical interactions can be expected to increase or decrease under many of the same conditions causing increased or decreased microbial activity. Increased moisture increased dazomet decomposition [54], but metham decomposition increases only with decreased moisture [110]. The rate of decomposition of both chemicals increases with increased temperature [51, 110]. Solubility, sorption by organic matter, and catalysis at clay surfaces are factors believed to affect these interactions [21, 111]. Microbial decomposition and activity are generally enhanced by increasing temperature and moisture [5, 60, 112-118]. The activity of some microorganisms, however, may be enhanced by decreased soil moisture contents.

Decomposition of EDB [119], 1,3-D [120], metham [10, 111], and methyl isothiocyanate [21] are accelerated by added organic matter. Organic matter content is generally considered a good indicator of microbiological activity, and decomposition also seems to be increased with increased organic matter [5]. Soil pH is known to affect microbial populations. Metham decomposition is accelerated by metallic cations such as copper and iron [120].

Chemical reactions most likely to occur are decomposition in water or hydrolysis [5, 10, 25, 32, 72, 110, 111, 120-124], and nucleophilic substitution by active groups of organic matter [10, 11, 16, 21, 111, 119, 125-128]. The role of free radicals (high-energy hydroxyl groups) in decomposition of pesticides in soil has recently been examined [129, 130]. The dithiocarbamates tend to hydrolyze or decompose readily in water [111, 169]. Burchfield [10] suggested that dichlone, chloranil, anilazine, PCNB, captan, nitrohalobenzenes, and compounds with allylic halogens participated in substitution reactions. Others were described by Moje [16]. Since the concentration of toxicant is usually low relative to the substrate with which it is reacting, first-order kinetics are believed applicable whether the reaction occurring is hydrolysis, or uni- or bimolecular substitution [5, 10, 16].

Microbial degradation accounts for much of the loss of chemicals from soil [5, 131-133]. The kinetics of decomposition typical for enrichment cultures were established in early studies with (2,4-dichlorophenoxy)acetic acid (2,4-D) [134]. Such results encouraged isolation of soil microorganisms specifically responsible for decomposing select pesticides. While many organisms capable of degrading specific pesticides have been isolated and characterized, some chemicals failed to support microbial growth or enrichment during the degradation process. Other chemicals appeared persistent or otherwise resistant to microbial degradation. Hence the concepts of microbial fallability [135, 136] and cometabolism [137, 138] were introduced. While some microorganisms thrive and enrich on easily degraded compounds, some pesticides are initially unsuitable as energy sources for microorganisms. The more persistent pesticides are less efficiently degraded by the myriad enzymatic reactions generated by soil microorganisms.

First-order kinetics apply when the concentration of the pesticide is low relative to the biological activity in the soil [60, 113, 118, 139-141]. Michaelis-Menten kinetics seem to apply when pesticide concentration increases and the rate of decomposition changes from being proportional to being independent of concentration [139, 142].

1. Aliphatic Fungicides

Ethylene dibromide decomposed very slowly in stored soil, with some 9% still remaining after 172 days [119]. Very little evidence for microbial involvement was obtained [143]. Loss presumably occurs primarily by diffusion. A thorough investigation of the sorption of EDB on various soils and the thermodynamics of adsorption by calcium montmorillonite, calcium kaolinite, and other soils has been described [46, 54, 55, 119, 144-147]. Sorption coefficients were tabulated for soils of different organic, moisture, and clay contents. Temperature coefficients of sorption and the heat of sorption were tabulated [55] and the mechanism of sorption in moist soils investigated [54]. The significance of these findings were discussed in the preceding section on adsorption.

Methyl bromide is a gas and is quite volatile. It is applied as a gas beneath a cover and its activity is dependent upon gaseous diffusion. The diffusion rate is influenced by molecular weight, temperature, continuity of air spaces, presence of codiffusing gases, and distribution of fumigant between air, water, and solid phases of soil. The latter distribution is influenced by temperature, moisture, air space, and organic matter [11] as discussed in previous sections.

Little is known about methyl bromide degradation in soil. Presumably, much of it is dissipated primarily by sorption and diffusion. Under conditions of fumigation, methyl bromide absorbed by wheat underwent decomposition with the formation of inorganic bromides [148, 149]. Some free methanol was produced by hydrolysis. Methyl bromide was found to be active in methylating the histidine and lysine of wheat protein [148, 149].

Cis- and *trans*-1,3-dichloropropene were hydrolyzed in wet soil to the corresponding allyl alcohol [121]. When measured by optical density, the growth of *Bacillus subtilis* and *Arthrobacter globiformis* in the presence of 100-1,000 ppm DD, 1,2-dichloropropane, 2,2,2-trichloro-1-ethoxyethanol, 1,3-dichloropropane, 2,3-dichloropropene, and allyl alcohol surpassed that of controls [150], suggesting the possibility that these materials could serve as additional substrates. No degradation products were characterized, however.

Allyl alcohol was first discovered as a promising treatment for controlling damping-off pathogens of pine seedlings [151]. Overman and Burgis [152] observed that while allyl alcohol controlled species of *Pythium* and *Rhizoctonia*, it stimulated growth of *Trichoderma viride*. Several strains of *T. viride* capable of metabolizing allyl alcohol were subsequently isolated, and its rapid detoxification in soil was conclusively demonstrated [153]. A number of bacteria, including *Pseudomonas fluorescens*, *P. putida*, and *Nocardia corallina* were also capable of utilizing allyl alcohol for growth [154]. Although no degradation products have been described, the work of Legator and Racusen [155] may be relevant. They demonstrated that the toxicity of allyl alcohol to *R. solani* was attributable to its in vivo conversion to the aldehyde acrolein, which reacts readily with thiols.

The fungicide dodine is a relatively simple molecule compared with the complex molecular structures of many other pesticides. Although its fate in soil has not been established, in plants the major degradation product is creatine. This is accounted for by β-oxidation of the aliphatic group followed by *N*-methylation [156].

2. Antibiotics

The antifungal antibiotic cycloheximide lost practically all of its activity after 11 days in nonsterile soil, whereas only 70% was lost in the same period in an autoclaved soil [157]. Patulin lost activity only slowly in a sterile black prairie loam soil, but was rapidly degraded in nonsterile soil.

Griseofulvin was degraded rapidly in a fresh garden loam soil [158-160] and a *Pseudomonas* sp. was isolated which could utilize griseofulvin as a sole carbon source. Three different monodemethylgriseofulvins were demonstrated during the enzymatic destruction of griseofulvin [161]. 4-Demethylgriseofulvin was produced by *Microsporum canis*, while *Botrytis allii* produced 1-demethylgriseofulvin (griseofulvic acid), and *Cercospora melonis* produced 6-demethylgriseofulvin.

Waksman and Woodruff [162] attributed the inactivation of actinomycin in soil to both adsorption on soil particles as well as microbial degradation. A direct relationship between inoculum size and the extent of actinomycin D decomposition by *Achromobacter* sp. was subsequently observed [163]. Streptomycin was inactivated both by strong adsorption to soil particles [158] and by microbial degradation [164]. Nissen [165], however, failed to demonstrate decomposition of either aureomycin or streptomycin by using a carbon dioxide measurement technique.

Simon and Gottlieb [166] stressed the importance of the chemical nature of the antibiotic since basic antibiotics such as streptomycin, streptothricin, and subtilin are effectively inactivated by adsorption on colloidal complexes such as clays and soil organic matter. While base-exchange studies indicated a high degree of irreversibility, this was only possible with other large cations. Thus it would seem likely that basic antibiotics such as streptomycin may not play an active biological role in soil. The neutral antibiotic chloramphenicol does not appear to be inactivated to anywhere near the same extent [167]. The soil microflora, notably Bacillus subtilis possessed a marked ability to metabolize chloramphenicol. p-Nitrophenylserinol and dichloroacetic acid were isolated from broth cultures of Proteus vulgaris and B. subtilis containing chloramphenicol [168].

3. Dithiocarbamates

The dithiocarbamates include both fungicidal and herbicidal compounds. Thiram, ferbam, nabam, zineb, maneb, NIA 9102, a zinc ion-maneb complex, and ziram are fungicidal, whereas 2-chloroallyl diethyldithiocarbamate (CDEC) is a herbicidal dithiocarbamate. Metham is a broad-spectrum pesticide, but is used primarily as a fungicide and herbicide.

Recent reviews of the chemistry and mode of action of dithiocarbamate fungicides suggest chemical degradation of these compounds into organic compounds including ethylenediamine, carbon disulfide, ethylenethiourea (ETU), ethylenethiuram disulfide, ethylenethiuram monosulfide, isothiocyanates, metallic sulfides, elemental sulfur, and carbonyl sulfide [23, 35, 40, 78, 169-184]. Limited microbial degradation has been observed [185-199], and the formation of conjugates has been reported [189-190]. The toxicological properties of these degradation products have been partially evaluated [191-195].

Nabam, maneb, zineb, NIA 9102, and the zinc ion-maneb complex have recently come under close scrutiny since they are known to degrade to ethylenethiourea, a known carcinogen. The appearance of low levels of ETU in certain foodstuffs has provided impetus for a flurry of research activity directed at assessing the environmental fate and behavior of these fungicides and their degradation products, particularly ETU. Although most of the results of current investigations involving these materials have not been published, some information is available.

Ethylenethiourea residues have been detected in apples [196], tomato foliage [197], potatoes [198], and kale and lettuce [199]. Its degradation in plants [200] and soil, air, and water [197, 201-204] has been characterized. These investigations generally concur in that ETU degrades under most environmental conditions. Research in our laboratory [203] suggests that in most soils ETU degrades by formation of ethyleneurea, which is subsequently degraded to other products including CO_2. Degradation may occur by photolytic, chemical, or biological mechanisms.

Metham decomposes chemically in soil to methyl isothiocyanate which is the primary toxicant [23, 25, 110, 111, 205, 206]. The chemical reactions of metham and its decomposition products have been examined under varying soil conditions [25]. Under alkaline conditions metham decomposes to methyl isothiocyanate and elemental sulfur. Carbon disulfide, hydrogen sulfide, methylamine, methyl isothiocyanate, and N,N-dimethylthiuram disulfide were formed under acid conditions. Several of these products interact with one another to form additional products. Microbial involvement in the biodegradation of metham in soil does not appear to be significant. The ultimate fate of these reaction products in the environment is not known.

Thiram has been used both as an accelerator for vulcanization and as a fungicide for seeds. It also has been used as an antiseptic spray and a bacteriostat in soap. Thiram degradation occurred more rapidly in nonsterile than in sterile soil [207]. The degradation of two unidentified divalent sulfur degradation products of thiram also occurred more rapidly in nonsterile than sterile soil. The possibility of fungicidal degradation products of thiram has been suggested [7]. The observations of Richardson [35] indicated that the protection to plants lasted longer than did the fungicide. Munnecke and Mickail [208] also observed with bioassay procedures that the soil retained fungitoxicity for a long period. Maeda and Tonomura [209] isolated a pseudomonad capable of utilizing thiram as a source of carbon, nitrogen, and sulfur. Dithiocarbamate (DDC), dimethylamine (DMA), formaldehyde, elemental sulfur, and methionine were detected in the culture as degradation products of thiram (see Fig. 4).

Raghu et al. [207] also isolated a *Pseudomonas* sp. capable of degrading thiram. The isolate of Maeda and Tonomura [209] was capable of utilizing thiram as sources of carbon, nitrogen, and sulfur, whereas the isolate of Raghu et al. [207] was incapable of growing in a mineral salts medium containing thiram as a sole carbon source. Thus, it appears that the latter isolate metabolized thiram by the phenomenon of cometabolism. Sijpesteijn and Vonk [210] suggested that thiram could degrade into DMA and CS_2. The formation of DMA in thiram-treated soils [211, 212] and isolated microbial cultures [209, 212] has been demonstrated. Although CS_2 had been indicated as one of the degradation products, very little has been reported about its presence in soil. Gas chromatographic analysis of the headspace gas of flasks containing nonsterile thiram-treated soil revealed the presence of CS_2 [212]. No CS_2 was isolated from sterile soil. The presence of CS_2 in the headspace gas of *Pseudomonas* inoculated medium containing thiram was also demonstrated [212]. Sisler and Cox [186] observed the evolution of CS_2 by *Fusarium* spores treated with thiram.

Dimethylamine is volatile under alkaline conditions but may also be metabolized by bacteria [210]. Ayanaba et al. [211] observed the formation of dimethylnitrosamine (DMNA) in flooded thiram-treated soil in the presence of nitrate or nitrite. They also demonstrated the formation of DMNA with added DMA and nitrite. Both DMA and DMNA disappeared with time from the soil, however. The potential formation of nitrosamines in the environment has been a growing concern, for reasons of carcinogenicity.

A possible degradation pathway for thiram in soil was proposed by Raghu et al. [212] based on their results and those of others (see Fig. 4). Thiram is degraded to the dithiocarbamate ion which can readily form heavy metal complexes with Cu, Zn, Fe, etc. [212]. The microbial formation of the amino- and ketobutyric acids has been indicated [210]. Maeda and Tonomura [209] demonstrated the formation of elemental sulfur, methionine, and formaldehyde in culture filtrates of *Pseudomonas* treated with thiram.

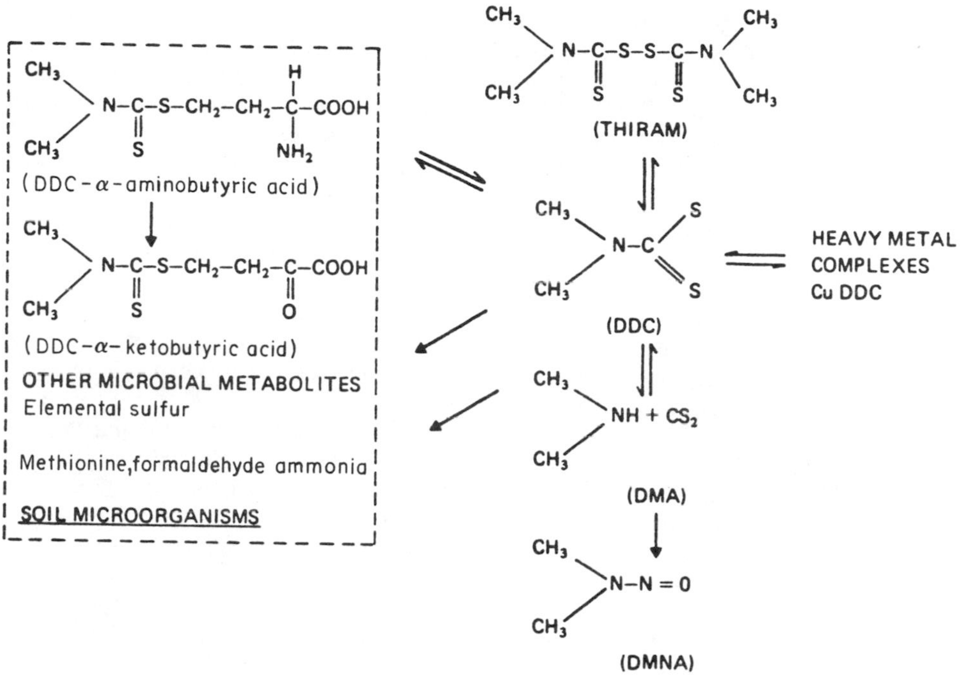

FIG. 4. Proposed degradation pathway of thiram in soil [212].

4. Organometallic Compounds

Although nickel-containing fungicides have been tested since 1908 [213] essentially nothing is known about their fate and behavior in soil.

Trialkyl and triaryl derivatives of tin have been tested and used as fungicides on numerous crops [214-218]. The degradation of tri[phenyl-^{14}C]tin acetate (TPTA) was followed in two soil types over a period of 9 weeks [219]. Degradation and cleavage of the phenyl rings resulted in evolution of 26-28% of the [^{14}C]TPTA applied as $^{14}CO_2$. No other degradation products were identified. Preliminary investigations in our own laboratory with tri[phenyl-^{14}C]tin hydroxide suggest a slow conversion of the [^{14}C]phenyl groups to $^{14}CO_2$ [220].

Recent incidents of mercury poisoning in certain Scandinavian countries, Japan, and North America have resulted in restricted use or banning of these materials. The major concern has been for the presence of alkylated mercury (methyl-, dimethyl-,) and elemental mercury which is biologically alkylated. Alkylated mercury products have shown some potential for biomagnification. The first step in the biomagnification process apparently occurs in the sediment layers of lakes and streams where metallic mercury is oxidized to divalent mercury [221]. Alkylation of divalent mercury to methyl- and dimethylmercury is a biological reaction. Putrescent homogenates of *Xiphophorus helleri* and *Gadus colja* alkylate Hg^{2+} to dimethylmercury, presumably through the intermediate formation of methylmercury [222]. The demonstration that the methyl group was transferred from methylcobalamine to Hg^{2+} by the action of an anaerobic bacterium isolated from sediment provided direct evidence for a microbial methylation process [223].

Of the approximately 20 different mercury-containing fungicides, eight contain alkylmercury radicals that could be released into the environment during decomposition. All 20 could eventually yield alkylated mercury through decomposition to free mercury with subsequent alkylation of the free mercury. This latter process would presumably require a much longer time than the direct release of alkylated mercury.

Very little is known about the persistence of mercury fungicides in our environment, the rate of elemental mercury formation, or the rate of its subsequent alkylation. Semesan and Panogen are degraded by soil microorganisms, but decomposition products have not been identified [123]. The degradation of (acetato)phenylmercury (PMA) has been examined by several investigators [68, 224, 225]. (Acetato)phenylmercury persists in soil up to 6 months, and can be lost as metallic mercury vapors [68]. Diphenylmercury is also one of the major metabolites of PMA [224]. Other microorganisms convert phenylmercury to metallic mercury [225].

5. Phthalimides

Captan, captafol, and folpet react readily with thiols such as cysteine and glutathione at pH levels in excess of 4.0-5.0. Phthalimide, thiophosgene, dithiophosgene, tetrahydrophthalimide, carbonyl sulfide, and hydrogen sulfide are degradation products obtained from reactions of captan and folpet with microbial cell components [226-230]. The trichloromethylthio groups and thiophosgene were believed to be intermediates in the formation of 2-thioxo-4-thiazolidine carboxylic acid, another degradation product. Although neither captan or folpet appear capable of reacting with other than sulfhydryl groups, the trichloromethylthio groups or phosgene released during this reaction is apparently capable of reacting with amino, hydroxyl, sulfhydryl, and possibly other groups [226, 227]. Histidine and serine react with thiophosgene producing UV-absorbing compounds.

Saccharomyces cerevisiae degraded captan to thiophosgene and di(thiophosgene) [228]. Captan also reacts with nabam, zineb, and ferbam to form bis(dimethylthiocarbamyl) trithiocarbonate [231].

Munnecke [232] reported that a 65-day half-life for captan could be achieved equally well in nonsterile as in sterile soil. Burchfield [18], however, reported a half-life of only 3 to 4 days at pH 7. Domsch [233] determined the half-life to be >50 days.

6. Polychlorinated Benzenes

Dichloronitroaniline (DCNA) is widely used for the postharvest control of *Rhizopus* rot on sweet potato roots and on fruits of cherries, nectarines, and peaches. It is also used on a number of vegetable crops for the control of foliage diseases caused by species of *Botrytis* and species of *Sclerotinia* [234]. Groves and Chough [235] observed the degradation of [^{14}C]DCNA in soil and an isolated bacterial culture. While a substantial amount of [^{14}C]DCNA was evolved as $^{14}CO_2$, no other products were identified.

Microorganisms rapidly metabolized DCNA to a number of unidentified compounds in soil under flooded conditions [236]. Conversion of DCNA in soil to unidentified products was nearly complete within 30-40 days in each soil type. One product appeared similar to a suspected condensation product of 2,6-dichloro-*p*-phenylenediamine (DCPD) characterized by Van Alfen [237] as having the composition $C_{12}H_6N_4Cl_4$. *Escherichia coli* B, *Pseudomonas cepacia*, and an unidentified bacterium metabolized DCNA in liquid cultures [234]. *E. coli* B and *P. cepacia* both converted DCNA to DCPD and 4-amino-3,5-dichloroacetanilide (ADCAA). The latter, but no detectable amounts of DCPD were produced by the unknown bacterium. At least four other unidentified metabolites were produced by the bacteria.

Dichloronitroaniline is only slightly water soluble, leaches slowly, is strongly adsorbed in clay soils, and is not readily extractable from treated soils [238].

Mycelia of *Rhizoctonia solani* converted chloroneb to 2,5-dichloro-4-methoxyphenol (DCMP) apparently by cleavage of the methoxy linkage or dealkylation [239]. Chloroneb was metabolized to an unidentified product by *Neurospora crassa* [240]. Of 23 organisms grown in the presence of chloroneb, 13 demethylated the fungicide to produce DCMP [241]. A *Fusarium* sp. was most active, with the ability to convert 50% of the chloroneb to DCMP in 5 days. There were eight organisms, particularly *Trichoderma viride* and *Mucor ramannianus*, capable of converting DCMP back to chloroneb. *Cephalosporium gramineum*, *Rhizoctonia solani*, *M. ramannianus*, and *Fusarium* could both methylate and demethylate DCMP to produce chloroneb and 2,5-dichlorohydroquinone, respectively. The microbial methylation of DCMP to chloroneb, may account in part for the relative stability and long-term effectiveness of chloroneb in soil [241].

The half-life of chloroneb in soil was approximately 3 to 6 months when incorporated into the soil 2-3 inches (5-.7.5 cm) below the surface at a rate of 2 lb active per acre (2.25 kg/ha) [242]. About 90% of the residual activity recovered from treated soil was intact chloroneb. The remainder of the activity consisted of polar compounds which could not be identified as 2,5-dichloro-4-methoxyphenol, 2,5-dichlorohydroquinone, or 2,5-dichloroquinone.

Microorganisms have a rather widespread ability to convert PCNB to pentachloroaniline [243]. Reduction was favored by flooding [244] and the resulting pentachloroaniline (PCA) was reported to be stable in both moist and submerged soils. There is some evidence that pentachlorophenol is also a metabolite of PCNB. The formation of PCA extends the activity of PCNB in soil, since PCA was also slightly toxic to Rhizoctonia solani. An additional factor which may account for extended activity of PCNB in soils is that some soil organisms are capable of oxidizing PCA to PCNB [245]. Pentachlorothioanisole (PCTA) has also been detected as a degradation product of PCNB [246].

Among the compounds identified in extracts of PCNB-treated greenhouse soils [247] were PCNB, PCA, PCTA, tetrachloronitrobenzene (TCNB), tetrachloroaniline (TCA), tetrachlorothioanisole (TCTA), hexachlorobenzene (HCB), and pentachlorobenzene (QCB); HCB, QCB, and TCNB are impurities in technical PCNB [247]. Tetrachloroanisole and TCTA are presumably degradation products of PCA and PCTA, respectively. Pentachloronitrobenzene was rapidly lost from soils made anaerobic by flooding or exclusion of oxygen [248]. Losses were mainly due to microbial degradation. Loss rates were inversely related to the content of organic matter. Organic amendments increased the rate of PCNB loss under flooded conditions, but not at field capacity. Degradation in flooded soils was retarded by free oxygen or combined oxygen in nitrate form, but not by MnO_2 or Fe_2O_3. Rate of loss varied directly with temperature from 15 to 35°C. Pentachloroaniline and possibly PCTA were found to be degradation products of PCNB.

Some loss of PCNB occurs through volatilization from aerated soils [249]. Kaufman [246] observed that HCB, PCA, and PCTA losses could also occur by volatilization from aerated soils. Half-lives of 4.7 to 9.7 months were found for PCNB added to three soil types incubated in flasks at 25°C [250].

Azobenzene formation in soil has been reported for a number of aniline-based pesticides. Azobenzenes are formed through the condensation of two aromatic amine molecules. Concern has been expressed for azobenzenes since some are potentially carcinogenic compounds [251]. Treatment of different soils with PCNB at a level of 1,000 ppm and incubation for 24 days at 30°C did not result in the formation of polychloroazobenzenes, although PCA was formed [252]. This negative finding agrees with previous work which demonstrated that azobenzenes are only formed from substituted anilines having at least one free ortho position.

Pentachlorophenol has many pesticidal uses. An isolate of Cephaloascus fragrans adapted to concentrations of PCP and some degradation was observed with Trichoderma virgatum, but no metabolites were identified [253]. Other investigations indicated that the disappearance of PCP from soils was more rapid in watered than unwatered soils and that decomposition proceeded more rapidly in soils high in organic matter than in those low in organic matter [254]. Leaching was not important in PCP disappearance.

Information regarding degradation of PCP emanates largely from its use as a wood preservative or as a herbicide in rice paddies. The work by Lyr [255] and others [253, 256, 257] indicates that wood-rotting and wood-staining fungi do degrade PCP. The role of a lactase type of polyphenol oxidase in PCP degradation has been suggested [255, 258].

Until recently, PCP was one of the major herbicides used in rice paddy fields. Although photodecomposition is believed to be the major degradation process in paddy fields, a substantial portion of PCP applied to the soil surface in flooded paddy fields is known to infiltrate into the soil with percolating water [259]. Based on the results of experiments with soil sterilization, soil temperature, and PCP degradation products it is believed that PCP degradation proceeds by both biological and chemical means [259].

Hayashi and Watanabe [260] isolated seven bacterial species which degraded PCP. The methyl ether of PCP and the dimethyl ether of tetrachlorohydroquinone were isolated from the culture media of these bacteria by Suzuki and Nose [261]. Niki et al. [262] detected methyl ethers of three tetrachlorophenols in extracts of PCP-treated soil. The isolation of tetra- and trichlorophenols from PCP-treated soils has also been reported [259, 263, 264]. Kuwatsuka and Igarashi [263, 264] detected three tetrachlorophenols and four trichlorophenols in PCP-treated soils. Kuwatsuka [259] detected 2,3,4,5-, 2,3,4,6-, and 2,3,5,6-tetrachlorophenols and 2,3,6-, 2,4,6-, 2,3,5-, and 2,3,4- and/or 2,4,5-trichlorophenol in PCP-treated soils. The major products were 2,3,4,5-tetrachlorophenol and 2,3,6- and 2,4,6-trichlorophenols. The other phenols were present in only trace amounts. No di- or monochlorophenols were reported. However, it would seem likely that both mono- and dichlorophenols would be further degraded, perhaps at a more rapid rate than the parent phenols, and thus present few residues problems.

Kuwatsuka [259] discussed the effect of numerous soil factors on PCP degradation in soil. Temperature, aeration, and organic matter were closely related to the rate of PCP dissipation from soil. Cation-exchange capacity and soil pH were only partially related to PCP degradation, whereas soil texture, clay content, degree of base saturation, and free iron oxides were not closely related to the rate of PCP degradation. At higher temperatures PCP disappeared more rapidly from soils under anaerobic (flooded) conditions than under aerobic conditions. The half-life of PCP at an initial soil concentration of 100 ppm was 10 to 40 days at 30°C under flooded conditions, whereas almost 100% was detected in aerobic soil after two months. Pentachlorophenol degraded more rapidly in high-organic-matter soils than in soils low in organic matter.

Hexachlorobenzene is a seed fungicide which has been used for bunt control in wheat since 1945 [265]. By comparison with other pesticides, its total agricultural use is relatively small, i.e., only 6,800 kg were used for seed fungicide purposes in 1971 [266]. Until recently, HCB was also an impurity in the herbicide dacthal (dimethyl tetrachloroterephthalate) and the fungicide PCNB [267]. While apparently no major environmental residue problems have resulted from its use as a seed protectant [267], recent surveys have demonstrated that HCB may be an important contaminant in some niches of our environment.

A major route of HCB into the environment is apparently as a by-product from the manufacture of perchloroethylene, chlorine gas, carbon tetrachloride, trichloroethylene, and other chlorinated hydrocarbons. In 1972 it was estimated that between 1.1 and 2.2 million kg of HCB was produced from these industrial processes [268]. Recently found HCB in Louisiana cattle was apparently related to airborne industrial emissions [269], whereas residues in sheep from Texas and California were traced to pesticides contaminated with HCB [267]. Residues of HCB have also been found elsewhere in the environment [266].

Very little is known about the fate and behavior of HCB in the environment. Adsorption to soil particles or volatilization into the atmosphere may be the principal mechanisms of HCB loss. No loss of HCB occurred at any rate or storage condition from soil treated with 0.1-100 ppm HCB and stored under sterile or nonsterile aerobic and anaerobic conditions for 1 year in covered containers that retarded volatilization [266]. Concentrations of 45.2, 24.4, 7.9, 4.7, and 3.4% of that originally determined at day 1 (5.6 ppm) were found after 0.5, 1, 6.5, 12, and 19 months, respectively, in turf-grass plots maintained under greenhouse conditions [270]. In this study, HCB was believed to have volatilized rapidly from both plant and soil surfaces. Concentrations of HCB at the depths of 2-4 cm did not change significantly over a 19-month period. No degradation products were observed in either investigation.

The derivatives formed when yeast cells were exposed to chlorothalonil resembled those formed by the reaction of glutathione with chlorothalonil in vitro [271]. Coenzyme A and 2-mercaptoethanol also readily formed derivatives with chlorothalonil in vitro. Infrared spectrography and chromatography of the four derivatives obtained, indicated that one to four of the halogens had been substituted by 2-mercaptoethanol [271]. Chlorothalonil also readily degraded in fresh rumen contents to two unidentified metabolites [272].

Several tetrachloronitrobenzenes ("TCNBs") exist. According to Kuchar et al. [273], 2,3,4,5-TCNB is an impurity in PCNB. The TCNB found in the extracts of PCNB treated soil by de Vos et al. [247] had the same gas chromatographic retention time as 2,3,5,6-TCNB, which is applied in smokes as a fungicide for lettuce. Isomer comparisons were not possible in their work, however, since no pure 2,3,4,5-TCNB was available. Reduction of the nitro group to form the corresponding tetrachloroaniline was observed in rabbit gut [274-276] with 2,3,4,5-TCNB, 2,3,5,6-TCNB, and 2,3,4,6-TCNB [277]. Tetrachloroaniline, a mercapturic acid, sulfate, glucuronide, and 4-amino-2,3,5,6-tetrachlorophenol were also detected in the urine of rabbits dosed with 2,3,5,6-TCNB [274-276].

7. Quinoid

Exposure of chloranil to ultraviolet irradiation resulted in a two-step reduction [101]. When chloranil was added to cultures of *Aspergillus niger*, *Neurospora crassa*, or *Mucor* sp. and submitted to electron spin resonance (ESR) spectroscopy, free radicals accumulated rapidly. Yeast suspensions and extracts also exhibited an accumulation and decay of a free radical which corresponded to the semiquinone [278].

Both chloranil and dichlone can participate in substitution reactions involving a reactive chlorine atom and functional groups present on amino acids, peptides, or proteins [279]. Burchfield [18] observed that dichlone degraded much more rapidly in a silt loam soil than in an aqueous medium, suggesting that similar reactions may well influence the stability of these compounds in an environment as rich in organic matter as some soils.

8. Systemic Fungicides

Although benomyl has been introduced as a systemic fungicide, it is also known to have nematicidal and miticidal activities. Hydrolysis of glucuronide and/or sulfate conjugates formed in rats yields methyl 5-hydroxy-2-benzimidazolecarbamate. Whether hydroxylation occurs before or after loss of the butylcarbamoyl group [280] has not been determined. Benomyl degrades rapidly in aqueous solutions, forming the methyl ester of 2-benzimidazolecarbamic acid (BCM or MBC) [281]. This product is also quite toxic to microorganisms. Because benomyl is so unstable, it has been proposed that MBC is the actual toxicant in plant tissues removed from the point of benomyl application [281]. A more polar, but unidentified product of MBC has also been isolated.

Benomyl was degraded to MBC and 2-aminobenzimidazole in bare and turf-field soils [282]. The half-life of the total benzimidazole-containing residues was about 3 to 6 months on turf, and about 6 to 12 months on bare soil. The major portions of the residues were found in the top 4 inches (10 cm) of soil.

Thiophanate-methyl is a broad-spectrum systemic fungicide. In soil it underwent rapid conversion to MBC [283]. The rate of conversion was reduced in steam-treated soil. The rate of conversion was more than four times faster in soil at pH 7.4 than in soil at pH 5.6. Soil incubated with [2-^{14}C(ring)]MBC and [methyl-^{14}C]MBC released less than 1 and 16%, respectively, of the applied radioactivity as $^{14}CO_2$. The difference in $^{14}CO_2$ evolution from the two labeling patterns was attributed to the hydrolysis of the carbamate moiety of MBC yielding 2-aminobenzimidazole. Nearly all of the extractable ^{14}C chromatographed as MBC.

The systemic fungicide carboxin was degraded rapidly in soil [284]. Although hydrolysis of carboxin would produce aniline, this mode of degradation has not been observed in either soil or aqueous systems. The major route of degradation in both systems appears to be via the formation of the sulfoxide and subsequently the sulfone. The sulfone was observed only in aqueous systems. The first product, the sulfoxide, is much less active than the parent material, carboxin. Complete conversion of carboxin to its sulfoxide in soil required approximately 14 days. Acid pH ranges of 2 to 4 favored a more rapidoxidation of carboxin in aqueous systems.

Wallnofer et al. [285] observed that cultures of Rhizopus japonicus could convert carboxin into carboxin sulfoxide and carboxin sulfone (oxycarboxin). Both the sulfoxide and sulfone were produced when the cultures were maintained under aerobic conditions. When R. japonicus was cultured under limited anaerobic conditions, the sulfoxide, but no sulfone was produced. A third metabolite was produced, however, under the partial anaerobic conditions. This metabolite had an identical UV maximum as a compound observed under similar conditions and had been recognized as a substituted anilide [286].

Oxycarboxin, in addition to being a metabolite of carboxin, is itself a systemic fungicide. Oxycarboxin degraded rapidly and extensively in hydroponic solutions [287]. The degradation was concomitant with bacterial growth and an increase in pH. The products were found to be translocated into bean plants. Degradation products identified included 2-(vinylsulfonyl)-acetanilide, 4-phenyl-3-thiomorpholinone-4,4-dioxide, and 2-(2-hydroxyethylsulfonyl)acetanilide.

The Rhizopus japonicus which metabolized carboxin [284] was also capable of metabolically converting three other systemic fungicides. 2-Methylbenzanilide (BAS-3050) and 2-chlorobenzanilide (BAS-307) were metabolized to their corresponding p-hydroxyanilino derivatives, 2-methyl-4'-hydroxybenzanilide, and 2-chloro-4'-hydroxybenzanilide, respectively [288]. R. japonicus and related fungi converted 2,3-dimethyl-3-furancarboxylic acid anilide (BAS 3191) into two metabolites, 2-hydroxymethyl-5-methyl-3-furancarboxylic acid anilide and 2-methyl-5-hydroxymethyl-3-furancarboxylic acid anilide [289]. R. nigricans, R. peka, and two strains of Mucor sp. were also capable of BAS-3191 metabolism. Since these materials are useful as seed treatments for food crops, their fate and behavior in soil is of interest.

9. Miscellaneous

Dazomet is used principally as a soil sterilant, being effective against many weed seeds, nematodes, and soil fungi. Activation and degradation of dazomet in soil is dependent on moisture and is a function of time and temperature [1, 5, 290]. Although it behaves in soil quite similarly to metham, it is slightly more persistent, lasting from several days to 4-5 weeks. The principal toxicant of dazomet is actually its degradation product methyl isothiocyanate. Other degradation products include formaldehyde and methylamino methyldithiocarbamate which further degrades to methylamine, hydrogen sulfide, and methyl isothiocyanate. As with metham degradation products, many of the dazomet products may react with one another and form additional products. Methylamine and hydrogen sulfide react with formaldehyde and form (methylamino)methanol, trimethylamine, and 1,3,5-trithiocyclohexane, which eventually degrades to CO_2, NH_3, SO_2, and H_2O. Methyl isothiocyanate reacts with ammonia, amines, or sulfhydryl groups.

Methyl isothiocyanate also will react with water and form CO_2, H_2S, and methylamine.

Diazoben appears to be fairly persistent in soil, despite its high susceptibility to photodecomposition [100]. Diazoben remained at relatively high concentrations 1 year after soil was treated in the greenhouse. Analysis of a silt loam treated in the field in May, sampled in October, and then used for growing several winter pea crops in the greenhouse revealed that large amounts of diazoben remained. Soil from the same field sampled in the subsequent April (after fall plowing and wintering) still contained substantial amounts of diazoben. Rhizoctonia decomposed diazoben in the presence of reduced diphosphopyridine nucleotide (DPNH) [291].

Karanth and Vasantharajan [292] found diazoben to be much less persistent even when applied at higher rates [100-200 ppm). Less than 1% of the diazoben remained in the soil after 60 days. Addition of organic nutrients such as glucose or straw facilitated a more rapid degradation. They believed that a higher incubation temperature (30°C) and moisture content may have accounted for the more rapid degradation. No degradation products were characterized.

Anilazine is an s-triazine fungicide. Burchfield [18] observed the stability of four organic fungicides in a silt loam to decrease in the order of captan > dichlone > 1-fluoro-2,4-dinitrobenzene > anilazine. Based on colorimetric analyses of reactive halogen, one-half of the anilazine added to moist soil disappeared within half a day whereas 3 to 4 days were required for the deterioration of the same proportion of captan. Fungicide decomposition was much slower in air-dried soil. Stability in soil could not be correlated with hydrolysis rate in aqueous buffer solutions. Captan survived seven times longer than anilazine in moist soil, whereas in aqueous buffer at pH 7 anilazine was approximately 200 times more stable than captan. This phenomenon is not too surprising in view of results obtained with s-triazine herbicides such as atrazine (2-chloro-4-[ethylamino]-6-[isopropylamino]-s-triazine). Atrazine is fairly stable in aqueous solutions, but is readily dehalogenated in soil [293, 294].

III. EFFECTS OF FUNGICIDES ON SOIL MICROORGANISMS

Pesticides are used deliberately to alter the ecology, that is, elimination or restriction of undesirable species in favor of species considered desirable by man. The ubiquitous nature of so many biological and biochemical processes make it unlikely that even highly specific pesticides will not affect some other nontarget organisms. Thus, it behooves man to determine what ecological changes pesticides produce, which changes are temporary or permanent, and decide which are acceptable or unacceptable. The ultimate problem is not perpetuation of a specific community for abstract reasons, but rather maximizing productivity while minimizing environmental pollution.

In a very broad sense, soil is a living entity unto itself. It contains structural material; it consumes oxygen and nutritive materials; it respires CO_2; and it is composed of functional units which contribute processes and products to the well-being of the whole. The intricate relations of the delicately balanced systems existing in soil are extensively described in microbiology [295], biochemistry [296], and soil texts. A short but excellent discussion of the subject is given by Audus [297] in his treatment of the effects of herbicides on soil microorganisms. Other reviews are also available [298-302]. Careful examination of the reviews reveals that with few exceptions concentrations of pesticides that result from normal rates of application for pest control have no long-lasting effects on microorganisms. It is interesting to note that many of these reviews deal with the effects of herbicides or insecticides on soil microbial processes. While it is recognized that fungicides are generally effective in eradicating or otherwise suppressing soil-borne pathogens, very little information is available concerning the effect of fungicides on other soil microorganisms or processes. It would seem probable that fungicides might indeed similarly effect other soil processes. Primary attention will be given in the ensuing discussion to the effects of fungicides on soil organisms or processes other than the pathogen they are designed to control.

Domsch [303] suggested that the ecological significance of any temporary change in the composition of microbial populations can only be judged when the role of groups of microorganisms within the ecosystem is properly identified. Interruptions of the energy flow and inhibition of functions within the ecosystem are important criteria for the secondary effects of pesticides. Most microorganisms are enzymatically well equipped. The elimination or reduction of pesticide sensitive members generally triggers a sequence of replacements within the population, resulting in the preservation of the soil's metabolic integrity. This, therefore, should be one of the major criteria in the selection and use of pesticides whose ultimate fate rests in the soil environment. Pesticides which permanently alter or destroy the soil metabolic integrity should be avoided or severely restricted in their usage.

When considering the effects of pesticides on microorganisms it is important to consider just where in the environment the effect may be occurring. Chemical and physical properties of both the environment and the pesticide may interact to limit any ecological effect of a given pesticide. Pesticides strongly adsorbed to soil particles may remain concentrated in a limited portion of the soil profile, thus representing a concern in only a small confined area. Similarly, specific placement of a pesticide in the soil such as in band application, or in the furrow, or seed treatments may limit the influence of pesticides to the immediate location.

Interpreting potentially significant effects of fungicides on soil microbial populations in terms of numbers is difficult. Bollen [304] reviewed the literature citing effects of fungicides and fumigants on soil microbial populations. Various investigators observed that dazomet [305], nabam [305],

EDB [304], and DD [306] decreased bacterial populations. Actinomycete populations were stimulated by metham [307], but suppressed by captan [307], methylmercury oxinate [305], EDB [128], methyl bromide [306], and metham [306]. Algae were inhibited by allyl alcohol [307].

No fungicides were described as immediately stimulating soil fungal populations under the same conditions, but CS_2 [308], chloropicrin [308], DD [308], EDB [308], metham [308], methylmercury oxinate [305], captan [307], nabam [307], allyl alcohol [307], and dazomet [305] suppressed soil fungi. Conflicting reports on the effects of a few individual compounds are obvious. Whether or not these effects are due to soil type difference, methods used in characterizing microbial populations, or both, can only be speculated. In most cases it is not known whether populations increased as a result of stimulation by the fungicide or as a result of reduced competition by other inhibited segments of the population. Population increases following some treatments are known to be the result of proliferation of only a few species. *Trichoderma viride* is often prominent in the fungal population developing after soil fumigation. This organism, rather than the direct fungicidal action of the chemical applied, may destroy certain pathogens such as *Armillaria mellea*, *Phytophthora*, *Pythium*, and *Rhizoctonia* [128, 309]. The frequency with which *T. viride* has been reported as a dominant recolonizer of fumigated soils may result as much from its high growth rate as from its tolerance [309]. Similar results may occur with other chemicals and organisms. For these reasons, and those suggested by Domsch [303], an examination of the soils metabolic integrity may be more meaningful.

Oxygen and/or substrate consumption, respiration, enzyme activity, ammonification, nitrification, denitrification, nodulation, sulfur oxidation, etc., are some of the soil processes which may more significantly reflect the effects of pesticides on soil microorganisms. Roa [310] examined the effects of the fumigants EDB, methyl bromide, 1,3-D, and metham at field rates and at concentrations ten times greater in laboratory investigations. Carbon dioxide evolution from the decomposition of native organic matter was increased by all compounds except metham, which also retarded decomposition of added dextrose. Chandra and Bollen [311] observed soil respiration was retarded for 28 days by dazomet and nabam. The effect lessened at 42 days, while at 56 days the treated soils evolved more CO_2 than did the control. Although the use of fumigants to retard biological degradation and consequent subsidence of muck soils has been suggested, Stotzky et al. [312] demonstrated that dazomet, EDB, and several other fumigants were not effective on Rifle peat. Carbon dioxide evolution from soil was only slightly influenced by benomyl even at high concentrations [313]. Even with 200 ppm of benomyl in soil treated with 0.2% cellulose or chitin, maximum CO_2 evolution occurred only a few days later than in the controls, though an increased lag phase for cellulose decomposition was found. Hofer et al. [314] observed some increase in CO_2 evolution without added substrate and a small decrease with added starch during 3 days incubation with benomyl. Diazoben was found to retard the breakdown of glucose and paddy straw added to soil [292]. Inhibition of CO_2 production occurred during the early stages of glucose decomposition, and with straw throughout the 60-day incubation period.

Captan, methylarsinic sulfide, mercuric chloride, diazoben, and PCNB were examined for their effect on oxygen consumption by soil [315]. Captan, methylarsinic sulfide, and mercuric chloride exerted only a temporary effect on soil respiration, whereas diazoben and PCNB had no measurable effect [315]. The effect of captan on the oxidative decomposition of glucose, aesculin, chitin, and tannin was also studied in soil [316]. The primary action of captan was a strong inhibition of initial oxygen uptake in soil treated with glucose, aesculin, and chitin. The decomposition of the four model substrates was in no case completely inhibited. As a rule, the inhibition was longer the less accessible the substrate or its breakdown products were to microbial attack.

The processes involved within the nitrogen cycle are extremely important to some soils and appear to be fairly sensitive to effects by pesticides. Ethylenedibromide, methylbromide, 1,3-D, and metham slightly increased ammonia production from the native organic matter, but temporarily depressed the ammonification of added peptone [310]. Jones [317] also noted that over a wide range of concentrations EDB stimulated ammonification. Similar results were obtained by Munnecke and Ferguson [318] for metham, methyl bromide, and chloropicrin. Captan, thiram, and a 2.5% (w/w) mercury fungicide (Verdasan) did not affect ammonification when applied at low rates, but there was a marked increase in ammonium nitrogen with increased fungicide concentrations [319]. Dubey and Rodriguez [320] observed that ammonification was inhibited by maneb and anilazine only at high concentrations of the active ingredient. A tenfold increase in the amount of ammonium nitrogen followed fumigation with chloropicrin [321].

Prasad, Rajale, and Lakhdive [322] reviewed the nitrification-retarding properties of agricultural chemicals, including both fungicides and fumigants. They noted that fungicides are effective in retarding nitrification even when applied at normal dosage rates. Fumigants applied at field rates have a greater inhibitory effect upon nitrification, presumably because they come into direct contact with a large portion of the soil biota. Associated with a decrease in nitrification in such treated soils is an increase in the amount of ammonium nitrogen. At very low concentrations captan, thiram, and Verdasan stimulated nitrification [322] or did not affect the process. At higher concentrations these fungicides led to a progressive decrease in nitrate production. Maneb and anilazine also inhibited nitrification [320]. Significantly more nitrite and nitrate ammonium was found in benomyl-treated soil than in untreated soil [313]. Hofer et al. [314], however, found a more or less strong inhibition of nitrification in a benomyl-treated humiferous sandy soil. Van Faassen [313] observed that the oxidation of nitrite to nitrate with isolated cultures of *Nitrosomonas* and *Nitrobacter* was more sensitive to benomyl than the oxidation of ammonium to nitrite.

Nitrification was retarded for a few weeks by EDB, methyl bromide, 1,3-D, and metham [17]. Similar results were obtained by Munnecke and Ferguson [32] for methyl bromide, metham, and chloropicrin. Nitrification of ammonium sulfate and ammonium hydroxide was completely inhibited for 30 days by dazomet and nabam; when added 30 days after fungicide treatment the nitrification of the two ammonium compounds was depressed by only 50% [311]. Ferbam, ziram, and zineb at 50 ppm were temporarily toxic to nitrification in soil [323]. A 7- to 8-week lag period in nitrification was caused by 1,3-D and Wolcott et al. [324] discussed the significance of such prolonged inhibition periods in altering seasonal distribution of ammonium and nitrate nitrogen in the field. Nodulating bacteria for alfalfa were less sensitive than those for clover [325]. These results indicate the advisability of testing fungicides proposed for use on leguminous seeds for their interaction with inoculants.

Microorganisms play an important part in the mineralization of many soil nutrients. The application of benomyl, captan, thiram, and PCNB generally increased amounts of exchangable Mn, Na, and Zn [326]. In thiram-treated soils there were significant increases in Na, K, and Zn. At high rates benomyl decreased Cu, but increased Zn at the lower rate.

The accurate measurement of specific soil enzyme reactions has only begun recently. Although numerous pesticides are known to inhibit specific enzymic reactions, their effects on soil enzymes are still not known. The dehydrogenase activity of a soil was not significantly affected by 20 or 200 parts per million of benomyl in soil, either in the absence of additional substrate or with simultaneous addition of glucose [313]. Some soil microorganisms are capable of producing compounds which affect soil aggregation. Diazoben was observed to reduce the proportion of water-stable aggregates in soil [327] but there was a recovery in the aggregate content with the disappearance of the fungicide.

Another more recent aspect of the effect of pesticides on soil microorganisms is their effect on the biodegradation or persistence of other pesticides. Pesticides are rarely applied individually at present, and some applications may contain as many as seven or eight different pesticides. Interactions may also result from serial applications during a single season. Numerous incidences are now known where the biodegradation of one pesticide is affected by the presence of the other. Although fungicides might logically be immediately suspect of such effects, very little is actually known about their interactions with other pesticides. More information appears to be available regarding interactions between herbicices or herbicides and insecticides.

Kaufman [328] examined combinations of one or two of the fungicides captan, Ceresan-M, Panogen, and PCNB with each of the insecticides carbaryl, 1,1,1-trichloro-2,2-bis(p-chlorophenyl)ethane (DDT), heptachlor, and toxaphene, for their individual and combined effects on the behavior and persistence of the herbicide chlorpropham in soil. The pesticides were selected as combinations likely to occur in cotton-cropped soils. Carbaryl,

PCNB, and heptachlor increased the persistence of chlorpropham in soil when applied either alone or in combination with other pesticides. Combinations of DDT and captan also increased soil persistence of chlorpropham, although neither pesticide alone affected chlorpropham persistence. The results obtained with PCNB, heptachlor, or DDT-captan combinations, however, were not as striking as those obtained with carbaryl. Subsequent investigations have described the mechanism of these interactions and some of the parameters which affect it [329, 330]. Carbaryl functions as a competitive inhibitor of the chlorpropham-hydrolyzing enzyme produced by soil microorganisms. Similar results were also obtained by Walker [331] who observed that the half-life of chlorpropham in soil was increased from 11.5 to 18.5 days in soil treated with PCNB. The half-life of CDEC was also increased slightly by PCNB.

REFERENCES

1. D. E. Munnecke, in Fungicides (D. C. Torgeson, ed.), Academic Press, New York, 1967, p. 510.
2. W. H. Fuchs, in Handbuch der Pflanzenkrankheiten (O. Appel, ed.), Vol. 6, Parey, Berlin and Hamburg, 1952, p. 144.
3. G. L. McNew, Agr. Chem., 8, 44, 109 (1953).
4. R. Carson, Silent Spring, Houghton Mifflin Co., Boston, 1962.
5. C. A. I. Goring, Ann. Rev. Phytopathol., 5, 285 (1967).
6. D. Woodcock, in Fungicides (D. C. Torgeson, ed.), Academic Press, New York, 1967, p. 613.
7. D. Woodcock, in Soil Biochemistry (A. D. McLaren and J. Skujins, eds.), Vol. 2, Marcel Dekker, New York, 1971, p. 337.
8. A. G. Newhall, Bot. Rev., 21, 189 (1955).
9. N. M. Hall and L. F. L. Clegg, Proc. Soc. Appl. Bacteriol., 12, 105 (1949).
10. H. P. Burchfield, in Plant Pathology (J. G. Horsfall and A. E. Dimond, eds.), Vol. 3, Academic Press, New York, 1960, p. 477.
11. C. A. I. Goring, Advan. Pestic. Control Res., 5, 47 (1962).
12. K. H. Domsch, Ann. Rev. Phytopathol., 2, 293 (1964).
13. W. A. Kreutzer, in Plant Pathology (J. G. Horsfall and A. E. Dimond, eds.), Vol. 3, Academic Press, New York, 1960, p. 431.
14. J. B. Kendrick, Jr. and G. A. Zentmyer, Advan. Pestic. Control Res., 1, 219 (1957).
15. J. P. Martin, Residue Rev., 4, 96 (1963).
16. W. Moje, Advan. Pestic. Control Res., 3, 181 (1960).
17. O. Vaartaja, Bot. Rev., 30, 1 (1964).
18. H. P. Burchfield, Contrib. Boyce Thompson Inst., 20, 205 (1959).
19. J. T. Hughes, Rept. Glasshouse Crops Res. Inst., 1952, p. 108.
20. J. J. Jurinak and T. S. Inouye, Soil Sci. Soc. Amer. Proc., 27, 602 (1963).

21. K. Kotter, J. Willenbrink, and K. Junkmann, Z. Pflanzenkrankh. Pflanzenschutz, 68, 407 (1961).
22. W. Moje, D. E. Munnecke, and L. T. Richardson, Nature, 202, 831 (1964).
23. D. E. Munnecke, K. H. Domsch, and J. W. Eckert, Phytopathology, 52, 1298 (1962).
24. N. J. Turner, Ph.D. Thesis, Oregon State University, Corvallis, 1962.
25. N. J. Turner and M. E. Corden, Phytopathology, 53, 1388 (1963).
26. J. Willenbrink, E. Schulze, and K. Junkmann, Z. Pflanzenkrankh. Pflanzenschutz, 68, 92 (1961).
27. M. G. Ashley, B. L. Leigh, and L. S. Lloyd, J. Sci. Food Agr., 14, 153 (1963).
28. S. H. F. Chinn and R. J. Ledingham, Phytopathology, 52, 1041 (1962).
29. M. E. Corden and R. A. Young, Phytopathology, 52, 503 (1962).
30. K. H. Domsch and P. Schicke, Nachrbl. Deut. Pflanzenschutzdienst (Stuttgart), 12, 121 (1960).
31. L. J. Klatz, T. A. DeWolfe, and R. C. Baines, Plant Disease Reptr., 43, 1174 (1959).
32. D. E. Munnecke and J. Ferguson, Phytopathology, 43, 375 (1953).
33. A. G. Newhall, Plant Disease Reptr., 42, 677 (1958).
34. G. A. Zentmyer, Phytopathology, 45, 398 (1955).
35. L. T. Richardson, Can. J. Bot., 32, 335 (1954).
36. D. E. Munnecke, Phytopathology, 48, 61 (1958).
37. C. L. Maurer, R. Baker, D. J. Phillips, and L. Danielson, Phytopathology, 52, 957 (1962).
38. K. H. Domsch, Plant Soil, 10, 114 (1958).
39. L. T. Richardson, Plant Disease Reptr., 44, 104 (1960).
40. L. T. Richardson and D. E. Munnecke, Phytopathology, 54, 836 (1964).
41. J. H. Reinhart, Plant Disease Reptr., 44, 648 (1960).
42. H. Jacks and H. C. Smith, New Zealand J. Sci. Technol., A33, 69 (1952).
43. K. H. Domsch, Phytopathol. Z., 25, 311 (1956).
44. K. H. Domsch, Plant Soil, 10, 122 (1958).
45. W. A. Kreutzer, in Plant Pathology (J. G. Horsfall and A. E. Dimond, eds.), Vol. 3, Academic Press, New York, 1960, p. 431.
46. J. J. Jurinak and D. H. Volman, Soil Sci., 83, 487 (1957).
47. G. S. Hartley, in The Physiology and Biochemistry of Herbicides (L. J. Audus, ed.), Academic Press, New York, 1964, p. 11.
48. M. J. Frissel, Verslag. Landbouwk. Onderzoek., 67, 1 (1961).
49. A. S. Crafts, Proc. Western Weed Control Conf., 18, 43 (1962).
50. C. I. Harris and G. F. Warren, Weeds, 12, 120 (1964).
51. D. E. Munnecke and J. P. Martin, Phytopathology, 54, 941 (1964).
52. P. Wade, J. Sci. Food Agr., 5, 184 (1954).
53. P. Wade, J. Sci. Food Agr., 6, 1 (1955).
54. F. Call, J. Sci. Food Agr., 8, 630 (1957).
55. F. Call, J. Sci. Food Agr., 8, 137 (1957).

56. M. Leistra, J. Agr. Food Chem., 18, 1124 (1970).
57. C. R. Youngson, R. G. Baker, and C. A. I. Goring, J. Agr. Food Chem., 10, 21 (1962).
58. J. Crank, The Mathematics of Diffusion, Oxford University Press, London, 1965.
59. P. C. Kearney, T. J. Sheets, and J. W. Smith, Weeds, 12, 83 (1964).
60. G. D. Hill, J. W. McGrahen, H. M. Baker, D. W. Finnerty, and C. W. Bingeman, Agron. J., 47, 93 (1955).
61. C. R. Youngson, C. A. I. Goring, and R. L. Noveroske, Down to Earth, 23 (1967).
62. J. E. Peachey and M. R. Chapman, "Chemical Control of Plant Nematodes," Commonwealth Bur. Helminthology, Gt. Brit. Tech. Commun., 36, 1966, 119 pp.
63. C. A. I. Goring, Soil Sci., 93, 431 (1962).
64. G. O. Turner and C. A. I. Goring, Down to Earth, 22, 19 (1966).
65. D. K. Bell and J. H. Owen, Plant Disease Reptr., 47, 1016 (1963).
66. L. S. Bird, Phytopathology, 55, 497 (1965).
67. C. R. Maier, Plant Disease Reptr., 45, 276 (1961).
68. K. Kimura and V. L. Miller, J. Agr. Food Chem., 12, 253 (1964).
69. O. Linstrom, J. Agr. Food Chem., 8, 217 (1960).
70. R. S. Adams, Jr., Soil Sci. Soc. Amer. Proc., 30, 689 (1966).
71. G. W. Thomas, J. Agr. Food Chem., 11, 201 (1963).
72. D. E. Munnecke, Phytopathology, 51, 593 (1961).
73. C. S. Helling and B. C. Turner, Science, 162, 562 (1968).
74. C. S. Helling, Soil Sci. Soc. Amer. Proc., 35, 737 (1971).
75. R. J. Stipes and D. R. Oderwald, Phytopathology, 60, 1018 (1970).
76. C. S. Helling, Soil Sci. Soc. Amer. Proc., 35, 732 (1971).
77. C. S. Helling, D. G. Dennison, and D. D. Kaufman, Phytopathology, 64, 1091 (1974).
78. R. A. Ludwig and G. D. Thorn, Advan. Pest Control Res., 3, 219 (1960).
79. R. G. Owens, in Fungicides (D. C. Torgeson, ed.), Vol. 2, Academic Press, New York, 1969, p. 147.
80. J. B. Kendrick, Jr. and J. T. Middleton, Plant Disease Reptr., 38, 350 (1954).
81. S. Aomine, H. Kawasaki, and K. Inoue, Soil Sci. Plant Nutr., 13, 186 (1967).
82. R. E. Pitblado and L. V. Edgington, Phytopathology, 62, 513 (1972).
83. H. Hagimoto, Weed Res., 9, 296 (1969).
84. R. D. Raabe and J. H. Hurlimann, Phytopathology, 55, 1072 (1965).
85. R. C. Rhodes, I. J. Belasco, and H. L. Pease, J. Agr. Food Chem., 18, 542 (1970).
86. K. Groves and K. S. Chough, J. Agr. Food Chem., 18, 1127 (1970).
87. J. C. Calvert and J. N. Pitts, Jr., Photochemistry, Wiley, New York, 1966.

88. A. Hecht and J. H. Seinfeld, Environ. Sci. Technol., 6, 47 (1972).
89. J. R. Plimmer and V. I. Klingebiel, Science, 174, 407 (1971).
90. J. R. Plimmer, P. C. Kearney, D. D. Kaufman, and F. S. Guardia, J. Agr. Food Chem., 15, 966 (1967).
91. J. D. Rosen and M. Siewierski, J. Agr. Food Chem., 18, 943 (1970).
92. C. S. Foote, Science, 162, 693 (1968).
93. J. R. Plimmer, V. I. Klingebiel, and B. E. Hummer, Science, 167, 67 (1970).
94. J. R. Plimmer, Abstr. 168th Meet., Amer. Chem. Soc., Atlantic City, 1974, No. 29.
95. J. R. Plimmer, Pesticide Chem., 6, 47 (1972).
96. P. A. Leermakers, H. T. Thomas, L. D. Weis, and F. C. Jones, J. Amer. Chem. Soc., 88, 5075 (1966).
97. F. J. Hills and L. D. Leach, Phytopathology, 52, 51 (1962).
98. K. H. Saunders, The Aromatic Diazo Compounds and Their Technical Application, Edward Arnold and Co., London, 1969, 442 pp.
99. K. Venkataraman, The Chemistry of Synthetic Dyes, Vols. 1 and 2, Academic Press, Inc., New York, 1952.
100. R. Alconero, Phytopathology, 56, 869 (1966).
101. G. Zweig and J. E. Hitt, Abstr. 155th Meet., Amer. Chem. Soc., San Francisco, 1968, A-21.
102. D. J. Neadle and R. J. Pollitt, J. Chem. Soc., 1969c, 2127.
103. E. R. White, W. W. Kilgore, and G. Mallett, J. Agr. Food Chem., 17, 585 (1969).
104. L. C. Mitchell, J. Assoc. Offic. Anal. Chemists, 44, 643 (1961).
105. D. G. Crosby and N. Hamadmad, J. Agr. Food Chem., 19, 1171 (1971).
106. M. Kuwahara, N. Kato, and K. Munakata, Agr. Biol. Chem. (Tokyo), 30, 232, 239 (1966).
107. K. Munakata and M. Kuwahara, Residue Rev., 25, 13 (1969).
108. M. Kuwahara, N. Shindo, and K. Munakata, Nippon Nigie Kagakn Kaishi, 44, 169 (1970).
109. D. G. Crosby, A. S. Wong, J. R. Plimmer, and E. A. Woolson, Science, 173, 748 (1971).
110. N. J. Turner, M. E. Corden, and R. A. Young, Phytopathology, 52, 756 (1962).
111. R. A. Gray, Phytopathology, 52, 734 (1962).
112. R. J. Aldrich, J. Agr. Food Chem., 1, 257 (1953).
113. P. Burschell and V. H. Freed, Weeds, 7, 157 (1960).
114. C. A. Edwards, Soils Fertilizers, 27, 451 (1964).
115. C. A. Edwards, Residue Rev., 13, 83 (1966).
116. E. P. Lichtenstein and K. R. Schulz, J. Econ. Entomol., 53, 192 (1960).

117. E. P. Lichtenstein and K. R. Schulz, J. Econ. Entomol., 57, 618 (1964).
118. P. H. Schuldt, H. P. Burchfield, and H. Bluestone, Phytopathology, 47, 534 (1957).
119. W. J. Hanson and R. W. Nex, Soil Sci., 76, 209 (1953).
120. M. G. Ashley and B. L. Leigh, J. Sci. Food Agr., 14, 148 (1963).
121. C. E. Castro and N. O. Belser, J. Agr. Food Chem., 14, 69 (1966).
122. D. E. Munnecke, Phytopathology, 48, 581 (1959).
123. W. C. Spanis, D. E. Munnecke, and R. A. Solberg, Phytopathology, 52, 455 (1962).
124. D. C. Torgeson, D. M. Yoder, and J. B. Johnson, Phytopathology, 47, 536 (1957).
125. H. P. Burchfield and P. H. Schuldt, J. Agr. Food Chem., 6, 106 (1958).
126. H. P. Burchfield and E. E. Storrs, Contrib. Boyce Thompson Inst., 18, 395 (1956).
127. W. Moje, J. Agr. Food Chem., 7, 702 (1959).
128. W. Moje, J. P. Martin, and R. C. Baines, J. Agr. Food Chem., 5, 32 (1957).
129. D. D. Kaufman, J. R. Plimmer, P. C. Kearney, J. Blake, and F. S. Guardia, Weed Sci., 16, 266 (1968).
130. J. R. Plimmer, P. C. Kearney, D. D. Kaufman, and F. S. Guardia, J. Agr. Food Chem., 15, 996 (1967).
131. D. Woodcock, Ann. Rev. Phytopathol., 2, 321 (1964).
132. D. D. Kaufman, in Pesticides in Soil and Water (W. D. Guenzi, ed.), Soil Science Society of America, Madison, Wis., 1974, p. 133.
133. D. D. Kaufman and P. C. Kearney, Residue Rev., 32, 235 (1970).
134. L. J. Audus, in Herbicides and the Soil (E. K. Woodford and G. R. Sagar, eds.), Blackwell, Oxford, 1960, p. 1.
135. M. Alexander, Soil Sci. Soc. Amer. Proc., 29, 1 (1965).
136. M. Alexander and B. K. Lustigman, J. Agr. Food Chem., 14, 410 (1966).
137. R. S. Horvath, J. Agr. Food Chem., 19, 291 (1971).
138. R. S. Horvath and M. Alexander, Can. J. Microbiol., 16, 1131 (1970).
139. J. W. Hamaker, Advan. Chem. Ser., 60, 122 (1966).
140. T. J. Sheets, J. Agr. Food Chem., 12, 30 (1964).
141. T. J. Sheets and C. I. Harris, Residue Rev., 11, 119 (1965).
142. J. W. Hamaker, C. R. Youngson, and C. A. I. Goring, Down to Earth, 23, 2 (1967).
143. P. Wade, J. Sci. Food Agr., 5, 288 (1954).
144. F. Call, J. Sci. Food Agr., 8, 143 (1957).
145. J. J. Jurinak, Soil Sci., 21, 599 (1957).
146. J. J. Jurinak, J. Agr. Food Chem., 5, 598 (1957).
147. J. J. Jurinak and D. H. Volman, Soil Sci., 86, 6 (1958).
148. R. G. Bridges, J. Sci. Food Agr., 6, 261 (1955).

149. F. P. Winteringham, W. A. Harrison, R. G. Bridges, and R. M. Bridges, J. Sci. Food Agr., 6, 251 (1955).
150. J. Altman and S. Lawlor, J. Appl. Bact., 29, 260 (1966).
151. R. M. Lindgen and B. W. Henry, Plant Disease Reptr., 33, 228 (1940).
152. A. J. Overman and D. S. Burgis, Phytopathology, 46, 532 (1956).
153. H. L. Jensen, Tidsskr. Planteavl., 65, 185 (1961).
154. H. L. Jensen, Nature (London), 183, 903 (1959).
155. M. Legator and D. Racusen, J. Bacteriol., 77, 120 (1959).
156. A. N. Curry, J. Agr. Food Chem., 10, 13 (1962).
157. D. Gottlieb, P. Siminoff, and M. M. Martin, Phytopathology, 42, 493 (1952).
158. E. G. Jefferys, J. Gen. Microbiol., 7, 295 (1952).
159. J. M. Wright, Ann. Appl. Biol., 43, 288 (1955).
160. J. M. Wright and J. F. Grove, Ann. Appl. Biol., 45, 36 (1957).
161. B. Boothroyd, E. J. Napier, and G. A. Somerfield, Biochem. J., 80, 34 (1961).
162. S. A. Waksman and H. B. Woodruff, J. Bacteriol., 40, 581 (1940).
163. E. Katz and P. Pienta, Science, 126, 402 (1957).
164. D. Pramer and R. L. Starkey, Science, 113, 127 (1951).
165. T. V. Nissen, Nature, 174, 226 (1954).
166. P. Simon and D. Gottlieb, Phytopathology, 41, 420 (1951).
167. D. Gottlieb and P. Siminoff, Phytopathology, 42, 91 (1952).
168. G. N. Smith, C. S. Worel, and B. L. Lilligren, Science, 110, 297 (1949).
169. C. M. Menzie, Metabolism of Pesticides, U.S. Fish. Wildlife Serv., Spec. Sci. Rept. 127, Washington, D.C., 1969.
170. A. L. Morehart and D. F. Crossan, Studies on the Ethylenebisdithiocarbamate Fungicides, Delaware Univ. Agr. Expt. Sta., Bull. No. 357, Newark, 1965.
171. R. G. Owens, Ann. N.Y. Acad. Sci., 160, 114 (1969).
172. A. Kaars Sijpesteijn and J. Kaslander, Outlook Agr., 4, 119 (1964).
173. G. Czeglédi-Jankó, J. Chromatogr., 31, 89 (1967).
174. H. M. Dekhuijzen, Mededel. Landbouwhogeschool Opzoekingssta. Staat Gent., 26, 1542 (1961).
175. H. M. Dekhuijzen, Nature (London), 191, 198 (1961).
176. H. M. Dekhuijzen, Plant Pathol. (Netherlands), 70, 1 (1964).
177. J. R. Iley and J. G. A. Fiskell, Soil Crop Sci. Soc. Florida Proc., 23 (1963).
178. L. E. Lopatecki and W. Newton, Can. J. Bot., 30, 131 (1952).
179. R. A. Ludwig and G. D. Thorn, Plant Disease Reptr., 37, 1953.
180. R. A. Ludwig, G. D. Thorn, and D. M. Miller, Can. J. Bot., 32, 48 (1954).
181. W. Moje, D. E. Munnecke, and L. T. Richardson, Nature (London), 202, 831 (1964).

182. S. Rich and J. G. Horsfall, Amer. J. Bot., 37, 643 (1950).
183. J. W. Vonk and A. Kaars Sijpesteijn, Ann. Appl. Biol., 65, 489 (1970).
184. J. W. Vonk and A. Kaars Sijpesteijn, Pestic. Biochem. Physiol., 1, 163 (1971).
185. C. E. Cox, H. D. Sisler, and R. A. Spurr, Science, 114, 643 (1951).
186. H. D. Sisler and C. E. Cox, Phytopathology, 41, 565 (1951).
187. H. D. Sisler and C. E. Cox, Amer. J. Bot., 41, 338 (1954).
188. R. M. Weed, S. E. A. McCallan, and L. P. Miller, Contrib. Boyce Thompson Inst., 17, 299 (1953).
189. A. Kaars Sijpesteijn and G. J. M. Van Der Kerk, Ann. Rev. Phytopathol., 3, 127 (1965).
190. A. Kaars Sijpesteijn, J. Kaslander, and G. J. M. Van Der Kerk, Biochem. Biophys. Acta, 62, 587 (1962).
191. R. Engst and W. Schnaak, Z. Lebensmitteluntersuch. Forsch., 134, 216 (1967).
192. R. Engst, W. Schnaak, and H. J. Lewerenz, Z. Lebensmitteluntersuch. Forsch., 138, 91 (1971).
193. S. L. Graham and W. H. Hansen, Bull. Environ. Contam. Toxicol., 7 (1972).
194. J. Seifter and W. J. Ehrich, J. Pharmacol. Exptl. Therap., 92, 303 (1948).
195. R. B. Smith, Jr., J. K. Finnegan, P. S. Larson, P. F. Sohyoun, M. L. Dreyfuss, and H. B. Haay, J. Pharmacol. Exptl. Therap., 109, 159 (1953).
196. W. H. Newsome, Abstr. 164th Meet., Amer. Chem. Soc., Div. Pestic. Chem., New York, 1972, No. 43.
197. C. H. Blazquez and W. A. Plummer, Abstr. 164th Meet., Amer. Chem. Soc., Div. Pestic. Chem., New York, 1972, No. 44.
198. R. F. Cook and B. C. Leppert, Abstr. 164th Meet., Amer. Chem. Soc., Div. Pestic. Chem., New York, 1972, No. 50.
199. G. Yip, J. H. Onley, and S. F. Howard, J. Assoc. Offic. Anal. Chemists, 54, 1373 (1971).
200. D. S. Frear, U.S.D.A. Metabolism Radiation Res. Lab., Fargo, N. Dakota, personal communication, 1973.
201. J. Hylin, Abstr. 164th Meet., Amer. Chem. Soc., Div. Pestic. Chem., New York, 1972, No. 49.
202. R. D. Ross and D. G. Crosby, Abstr. 164th Meet., Amer. Chem. Soc., Div. Pestic. Chem., New York, 1972, No. 47.
203. D. D. Kaufman and C. L. Fletcher, Abstr. 165th Meet., Amer. Chem. Soc., Div. Pestic. Chem., New York, 1973, No. 1.
204. P. A. Cruickshank and H. C. Janow, Abstr. 164th Meet., Amer. Chem. Soc., Div. Pestic. Chem., New York, 1972, No. 46.
205. J. T. Hughes, Ann. Rept. Glasshouse Crops Res. Inst., 1960, p. 108.
206. G. A. Llyd, J. Sci. Food Agr., 13, 309 (1962).
207. K. Raghu, N. B. K. Murthy, and R. Kumarasamy, Proc. Delhi Atomic Energy Symp. Use of Plant Productivity, 1974.

208. D. E. Munnecke and K. Y. Mickail, Phytopathology, 57, 969 (1967).
209. K. Maeda and K. Tonomura, Kogyo Gijutsuin Hakko Kenkyusho Kenkyu Hokoku, 33, 1 (1968).
210. A. Kaars Sijpesteijn and J. W. Vonk, Mededel, Rijksfac. Landbouwwetenschap. Gent, 35, 799 (1970).
211. A. Ayanaba, W. Verstraete, and M. Alexander, Soil Sci. Soc. Amer. Proc., 37, 565 (1973).
212. K. Raghu, N. B. K. Murthy, R. Kumarasamy, R. S. Rao, and P. V. Sane, in Proc. Joint FAO/IAEA Division of Atomic Energy in Food and Agriculture, Vienna, 1975, p. 137.
213. The International Nickel Company, Inc., Report 1CB-39, New York, 1964.
214. V. Estienne and G. L. Hannebert, Agricultura (Louvain), 7, 473 (1959).
215. K. Hartel, Agr. Chem., 13, 69 (1958).
216. H. E. Hirschland and C. K. Banks, Advan. Chem. Series, 23, 204 (1959).
217. A. J. Pieters, Jr. and X. V. Philips-Duphar, Tin Its Uses, 5, 3 (1962).
218. V. L. Miller and C. J. Gould, Plant Disease Reptr., 47, 408 (1963).
219. A. Suss and Ch. Eben, personal communication, 1976.
220. D. D. Kaufman, U.S.D.A. Pesticide Degradation Lab., Beltsville, Md., unpublished research, 1973.
221. A. Jernelov, in Chemical Fallout (M. W. Miller and G. G. Berg, eds.), Charles C. Thomas, Springfield, Ill., 1969, p. 68.
222. S. Jensen and A. Jernelov, Nature (London), 223, 753 (1969).
223. J. M. Wood, F. S. Kennedy, and G. C. Rosen, Nature (London), 220, 173 (1968).
224. F. Matsumura, Y. Gotoh, and G. M. Boush, Science, 173, 49 (1971).
225. K. Tonomura, K. Meada, F. Futai, T. Nakagami and M. Yamada, Nature (London), 217, 644 (1968).
226. R. J. Lukens and H. Sisler, Phytopathology, 48, 235 (1958).
227. R. J. Lukens and H. Sisler, Science, 127, 650 (1958).
228. R. J. Lukens, Phytopathology, 53, 881 (1963).
229. M. R. Seigel and H. D. Sisler, Phytopathology, 58, 1123 (1968).
230. M. R. Seigel and H. D. Sisler, Phytopathology, 58, 1129 (1968).
231. R. J. Lukens, Phytopathology, 49, 339 (1959).
232. D. E. Munnecke, Phytopathology, 48, 581 (1958).
233. K. H. Domsch, Z. Pflanzenkrankh. Pflanzenschutz, 65, 651 (1958).
234. N. K. Van Alfen and T. Kosuge, J. Agr. Food Chem., 22, 221 (1974).
235. K. Groves and K. S. Chough, J. Agr. Food Chem., 18, 1127 (1970).
236. C. H. Wang and F. E. Broadbent, J. Environ. Qual., 2, 511 (1973).
237. N. K. Van Alfen, Ph.D. Thesis, University of California, Davis, 1972.
238. K. Groves, Abstr. 149th Meet., Amer. Chem. Soc., Detroit, 1965, No. 19A.
239. W. K. Hock and H. D. Sisler, J. Agr. Food Chem., 17, 123 (1969).

240. R. G. Owens, Develop. Ind. Microbiol., 1, 187 (1960).
241. M. V. Wiese and J. M. Vargas, Jr., Pestic. Biochem. Physiol., 3, 214 (1973).
242. R. C. Rhodes, H. L. Pease, and R. K. Brantley, J. Agr. Food Chem., 19, 744 (1971).
243. C. I. Chacko, J. L. Lockwood, and M. Zabik, Science, 154, 893 (1966).
244. W. H. Ko and J. D. Farley, Phytopathology, 59, 64 (1969).
245. D. D. Kaufman, U.S. Dept. Agr. Pesticide Degradation Lab., Beltsville, Md., unpublished data, 1976.
246. D. D. Kaufman, in Pesticides in the Soil: Ecology, Degradation and Movement, International Symposium on Pesticides in Soil, Michigan State University, East Lansing, 1970, p. 73.
247. R. H. de Vos, M. C. ten Noever de Brauw, and P. D. A. Olthof, Bull. Environ. Contam. Toxicol., 11, 567 (1974).
248. C. H. Wang and F. E. Broadbent, J. Environ. Quality, 2, 511 (1973).
249. J. C. Caseley, Bull. Environ. Contam. Toxicol., 3, 180 (1968).
250. C. H. Wang and F. E. Broadbent, Soil Sci. Soc. Amer. Proc., 36, 742 (1972).
251. J. H. Weisburger and E. K. Weisburger, Chem. Eng. News, 44, 124 (1966).
252. H. R. Buser and H. P. Bosshardt, Pestic. Sci., 6, 35 (1975).
253. A. J. Cserjesi, Can. J. Microbiol., 13, 1243 (1967).
254. H. C. Young and J. C. Carroll, Agron. J., 43, 504 (1951).
255. H. Lyr, Phytopathol. Z., 47, 73 (1963).
256. G. G. Duncan and F. J. Deverall, Appl. Microbiol., 12, 57 (1964).
257. R. S. Ingils and P. C. Stevenson, Res. Eng., 18, 4 (1963).
258. S. Rich and J. G. Horsfall, Proc. Natl. Acad. Sci. U.S., 40, 139 (1954).
259. S. Kuwatsuka, in Environmental Toxicology of Pesticides (F. Matsumuro, G. M. Boush, and T. Misato, eds.), Academic Press, New York, 1972, p. 385.
260. H. Hayashi and I. Watanabe, Abstr. Ann. Meet. Soc. Soil Manure Japan, 16, 44 (1970).
261. T. Suzuki and K. Nose, Abstr. 10th Meet. Weed Soc. Japan, 10, 114 (1971).
262. Y. Niki, I. Watanabe, A. Ide, and H. Watanabe, Abstr. Meet. Agr. Chem. Soc. Japan, 1971, p. 387.
263. S. Kuwatsuka and M. Igarishi, Meet. Soc. Soil Manure, Kanazawa, Chubu Bunch, 1971.
264. S. Kuwatsuka and M. Igarishi, Abstr. Ann. Meet. Soc. Soil Manure Japan, 17, 24 (1971).
265. Anonymous, 1969 Evaluations of Some Pesticide Residues in Food, FAO/WHO, Rome, 1970, p. 161.
266. A. R. Isensee, E. R. Holden, E. A. Woolson, and G. E. Jones, J. Agr. Food Chem., 24, 1210 (1976).

267. Environmental Protection Agency, Report on Environmental Contamination from Hexachlorobenzene, Office of Toxic Substances, Environmental Protection Agency, Washington, D.C., 1973.
268. C. E. Mumma and E. W. Lawless, E.P.A. Report No. 560-3-75-003, Environmental Protection Agency, Washington, D.C., 1975.
269. Anonymous, Chem. Week, April, 1973.
270. M. L. Beall, Jr., J. Environ. Qual., 5, 367 (1976).
271. P. G. Vincent and H. D. Sisler, Physiologia Plantarum, 21, 1249 (1968).
272. W. H. Gutenmann and D. W. Lisk, J. Dairy Sci., 49, 1272 (1966).
273. E. J. Kuchar, F. O. Geenty, W. P. Griffith, and R. J. Thomas, J. Agr. Food Chem., 17, 1237 (1969).
274. H. G. Bray, Z. Hybs, S. P. James, and W. V. Thorpe, Biochem. J., 52, XVII (1952).
275. H. G. Bray, Z. Hybs, S. P. James, and W. V. Thorpe, Biochem. J., 53, 266 (1953).
276. H. G. Bray, Z. Hybs, and W. V. Thorpe, Biochem. J., 49, lxv (1951).
277. J. Betts, S. P. James, and W. V. Thorpe, Biochem. J., 61, 611 (1955).
278. J. R. Rowlands and E. M. Gause, Abstr. 156th Meet., Amer. Chem. Soc., Atlantic City, 1968, No. 73.
279. R. G. Owens, Contrib. Boyce Thompson Inst., 17, 221 (1953).
280. J. A. Gardiner, R. K. Brantley, and H. Sherman, J. Agr. Food Chem., 16, 1050 (1968).
281. G. P. Clemons and H. D. Sisler, Phytopathology, 59, 705 (1969).
282. F. J. Baude, H. L. Pease, and R. F. Holt, J. Agr. Food Chem., 22, 1974.
283. J. R. Fleeker, H. M. Lacy, I. R. Schultz, and E. C. Honkom, J. Agr. Food Chem., 22, 592 (1974).
284. W. T. Chin, G. M. Stone, and A. E. Smith, J. Agr. Food Chem., 18, 731 (1970).
285. P. R. Wallnofer, M. Koniger, S. Safe, and O. Hutzinger, Intern. J. Environ. Anal. Chem., 2, 37 (1972).
286. P. R. Wallnofer, Naturwissenschaften, 55, 351 (1968).
287. J. A. Ross, B. G. Tweedy, L. C. Newby, and J. J. Bates, Abstr. 164th Meet., Amer. Chem. Soc., Div. Pestic. Chem., New York, 1972, No. 23.
288. P. R. Wallnofer, S. Safe, and O. Hutzinger, Pestic. Biochem. Physiol., 1, 458 (1972).
289. P. R. Wallnofer, M. Koniger, S. Safe, and O. Hutzinger, J. Agr. Food Chem., 20, 20 (1972).
290. D. C. Torgeson, D. M. Yoder, and J. B. Johnson, Phytopathology, 47, 536 (1957).
291. W. J. Tolmsoff, Phytopathology, 52, 755 (1962).
292. N. G. K. Karanth and V. N. Vasantharajan, Soil Biol. Biochem., 5, 679 (1973).

293. C. I. Harris, Abstr. 149th Meet., Amer. Chem. Soc., Detroit, 1965, No. 49.
294. C. I. Harris, J. Agr. Food Chem., 15, 157 (1967).
295. M. Alexander, Introduction to Soil Microbiology, Wiley, New York, 1961.
296. Soil Biochemistry; Vol. 1 (A. D. McLaren and G. M. Peterson, eds.), 1967; Vol. 2 (A. D. McLaren and J. Skujins, eds.), 1971; Vols. 3 and 4 (E. A. Paul and A. D. McLaren, eds.), 1975; Marcel Dekker, New York.
297. L. J. Audus, in The Physiology and Biochemistry of Herbicides (L. J. Audus, ed.), Academic Press, New York, 1964, p. 163.
298. W. W. Fletcher, in Herbicides and the Soil (F. K. Woodward and G. R. Sagar, eds.), Blackwell, Oxford, 1960, p. 20.
299. W. W. Fletcher, Proc. Brit. Weed Control Conf., 8, 896 (1966).
300. W. W. Fletcher, Landbouwk. Tijdschr., 78, 274 (1966).
301. A. S. Newman and C. E. Downing, J. Agr. Food Chem., 6, 352 (1958).
302. P. C. Kearney, C. I. Harris, D. D. Kaufman, and T. J. Sheets, Advan. Pest Control Res., 6, 1 (1965).
303. K. H. Domsch, in Pesticides in the Soil: Ecology, Degradation and Movement, International Symposium on Pesticides in Soil, Michigan State University, East Lansing, 1970.
304. W. B. Bollen, Ann. Rev. Microbiol., 15, 69 (1961).
305. M. E. Corden and R. A. Young, Soil Sci., 99, 272 (1965).
306. W. B. Bollen, H. E. Morrison, and H. H. Crowell, J. Econ. Entomol., 47, 303, 307 (1954).
307. K. H. Domsch, Z. Pflanzenkrankh. Pflanzenschutz, 66, 17 (1959).
308. J. P. Martin, R. C. Baines, and J. O. Erwin, Soil Sci. Soc. Amer. Proc., 21, 163 (1957).
309. J. B. Saksena, Brit. Mycol. Soc. Trans., 43, 111 (1960).
310. D. P. Roa, M. A. Thesis, Oregon State University, Corvallis, 1959.
311. P. Chandra and W. B. Bollen, Soil Sci., 92, 387 (1961).
312. G. Stotzky, W. P. Martin, and J. L. Mortensen, Soil Sci. Soc. Amer. Proc., 20, 392 (1956).
313. H. G. Van Faassen, Soil Biol. Biochem., 6, 131 (1974).
314. I. Hofer, T. Beck, and P. Wallnofer, Z. Pflanzenkrankh. Pflanzenpathol. Pflanzenschutz, 7, 398 (1971).
315. K. H. Domsch, Phytopathol. Z., 49, 291 (1964).
316. K. H. Domsch, Phytopathol. Z., 52, 1 (1965).
317. L. W. Jones, Utah State Agr. Coll. Agr. Expt. Sta. Bull., No. 390, 1956.
318. D. E. Munnecke and J. Ferguson, Plant Disease Reptr., 44, 552 (1960).
319. M. Wainwright and G. J. F. Pugh, Soil Biol. Biochem., 5, 577 (1975).
320. H. D. Dubey and R. L. Rodriguez, Soil Sci. Soc. Amer. Proc., 34, 435 (1970).

321. J. P. Winfree and R. S. Cox, Plant Disease Reptr., 42, 807 (1958).
322. R. Prasad, G. B. Rajale, and B. A. Lakhdive, Advan. Agron., 23, 337 (1971).
323. H. A. Wilson, West Virginia Univ. Agr. Expt. Sta. Bull., No. 366T, 1954.
324. A. R. Wolcott, F. Maciak, L. N. Shepard, and R. E. Lucas, Down to Earth, 16, 10 (1960).
325. A. W. Hofer, Soil Sci., 86, 282 (1958).
326. M. Wainwright and G. J. F. Pugh, Soil Biol. Biochem., 6, 263 (1974).
327. N. G. K. Karanth and V. N. Vasantharajan, Current Sci. (India), 40, 394 (1971).
328. D. D. Kaufman, Pesticide Chem., 6, 175 (1972).
329. J. Blake and D. D. Kaufman, Pestic. Biochem. Physiol., 5, 305 (1975).
330. D. D. Kaufman, P. C. Kearney, D. W. Von Endt, and D. E. Miller, J. Agr. Food Chem., 18, 513 (1970).
331. A. Walker, Hort. Res., 10, 45 (1970).

Chapter 2

SYSTEMIC FUNGICIDES: THEORY, UPTAKE, AND TRANSLOCATION

L. V. Edgington and Carol A. Peterson*

Department of Environmental Biology
University of Guelph
Guelph, Ontario, Canada

---

*Present address: Department of Biology, University of Waterloo, Waterloo, Ontario, Canada.

I. INTRODUCTION

Our discussion of fungicides systemic in plants will primarily be oriented toward the factors controlling uptake and translocation. We will neither discuss the practical uses nor attempt to tabulate the voluminous literature on systemic fungicides. Excellent reviews of systemic fungicides by Crowdy in 1972 [1] and Erwin in 1973 [2] have already been published. Systemic fungicides are also discussed in Chaps. 6, 7, 8, and 9 of Vol. 1. We shall consider first the theoretical aspects of transport in terms of the apoplast and symplast, and develop a hypothesis to explain translocation of many systemic fungicides and herbicides. Second, we will consider the uptake of fungicides by various organs of the plant (i.e., leaves, stems, roots, and seeds). Third, we will discuss translocation of fungicides within plants and finally, we shall attempt to characterize the important prerequisites for systemicity.

II. THEORY OF APOPLASTIC AND SYMPLASTIC MOVEMENT

It is customary to view the plant in terms of various organs, tissues, or cells, but it is more meaningful to divide the plant into three basic parts when discussing translocation. The three parts are the apoplast, symplast, and intercellular air spaces. The apoplast and symplast, originally introduced by Münch [3], are defined by Crafts [4] as follows: The symplast is "the continuum of interconnected protoplasts of the plant" and the apoplast is "the continuum of non-living cell-wall material that surrounds and contains the symplast." The apoplast, symplast, and air-space systems of the plant are shown in Figure 1. Movement of substances within the plant will differ greatly, depending on whether they are located in the symplast, apoplast, or air spaces. In this chapter, we will discuss the translocation of water and solutes in the apoplast and symplast but will not consider the movement of gasses in the air spaces.

The location of a substance within the plant depends on the site of origin of the substance and also on its ability to traverse the plasmalemma, since this semipermeable membrane represents the boundary between the apoplast and the symplast. A chemical which is synthesized in the cytoplasm and

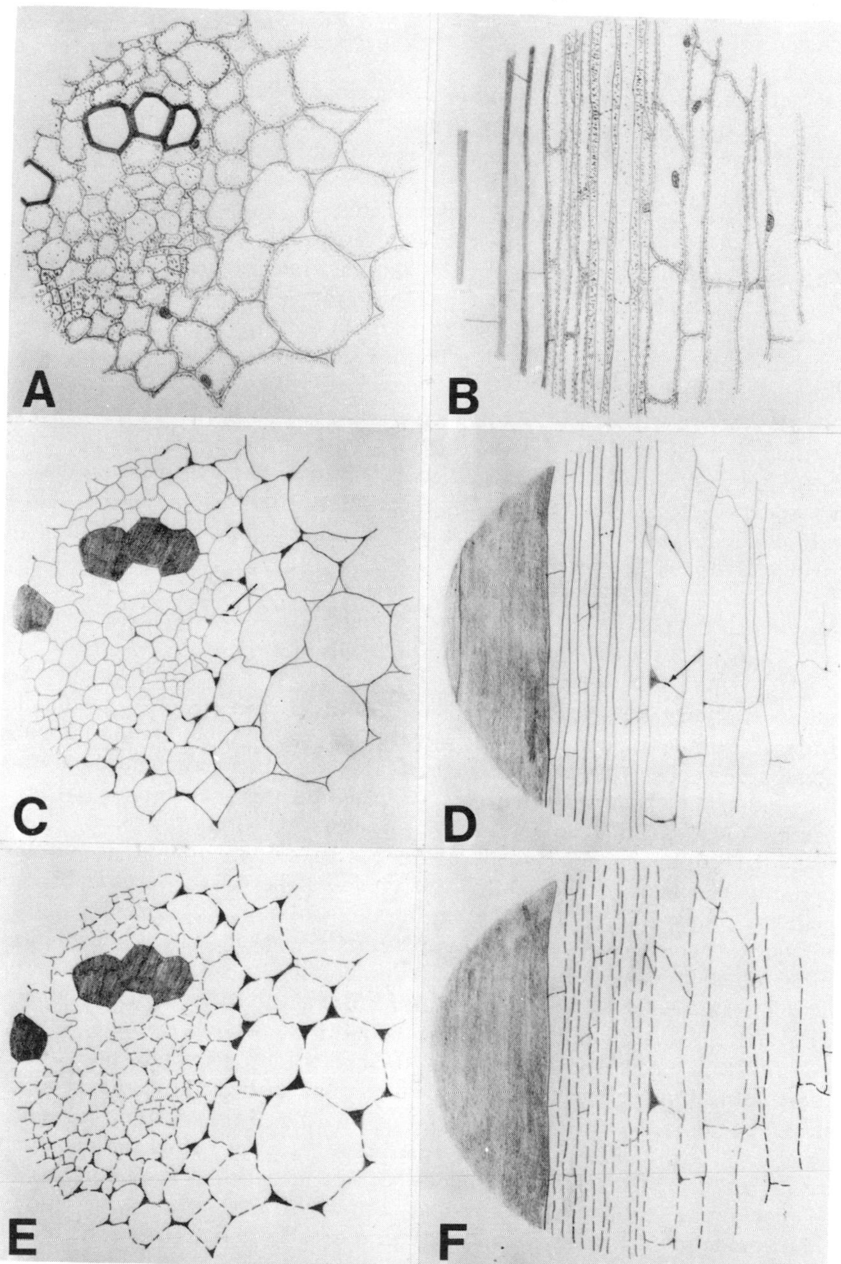

FIG. 1. The apoplast and symplast in tomato root showing part of the cortex and stele: A, C, and E are cross sections; B, D, and F are longitudinal sections; A and B are camera lucida drawings of the tissue. Cytoplasmic content is indicated by stippling. Conducting xylem elements are recognized by their thick walls and lack of cytoplasmic content. Conducting sieve tube elements have thin walls and are depicted as having little cytoplasmic content distributed uniformly through the cell. Dashed end walls indicate sieve plates. C and D show the apoplast as grey. The Casparian band of the endodermis is black (see arrows); E and F show the symplast in white.

does not penetrate the plasmalemma will necessarily remain in the symplast. Substances originating outside the plant body will initially enter the apoplast and will remain there if they are incapable of penetrating the plasmalemma. Substances which can penetrate the plasmalemma will be distributed partially in the apoplast and partially in the symplast. Some substances are taken up by the symplast and are retained there. If such a substance is introduced into the plant, it enters the apoplast initially but rapidly becomes concentrated in the symplast. It is possible for various chemicals to be located in the symplast, in the apoplast, or in both regions. Specific examples of chemicals which fall into these categories will be given in the following discussion.

With this background, let us now consider the properties of the apoplast and the symplast with respect to translocation. In the following discussion we will take advantage of the large body of information available on the translocation of herbicides since the foundations already laid in the field of herbicide transport can be used as a basis for an understanding of fungicide transport.

A. Apoplastic Transport

The apoplast of the plant consists of the relatively inert regions of the plant which lie outside the bounds of the plasmalemma. It includes the interconnected cell wall system and other associated structures such as the cuticle. Since the apoplast forms a continuous system throughout the plant, it is possible for a chemical to be very well distributed within the plant by being transported in the apoplast. All substances which enter the plant must do so via the apoplast. Movement of substances dissolved in water is brought about by a mass flow of water within the cell walls. In its primary state, the cell wall consists of cellulose microfibrils embedded in a hydrated matrix, the major constituents of the matrix being hemicelluloses, pectic substances, and proteins. Some of the interfibrillar spaces are filled with water and it is through these channels that water moving in the wall is thought to flow most freely. Particles 40 Å in diameter are able to penetrate the cell wall system of *Helixine* parenchyma tissue [5, 6] indicating that channels of this size do occur in the wall. At one time it was thought that the major pathway of water through the plant leaf and root parenchyma included the symplast as well as the apoplast but now much indirect evidence is accumulating for the apoplast as the major route of water flow [7-14].

Movement of water and solutes within the walls is not controlled by metabolic forces but rather by means of structural modifications in the wall. Water-impermeable substances such as suberin, cutin, and lignin may be deposited between the microfibrils and, if present in large amounts, greatly diminish the permeability of the wall to water. Structural barriers which prevent free movement of water in the apoplast are the Casparian band in the endodermis of the root, suberin layers in the exodermis of the root, the cuticle of leaves and other aerial organs, the bark which may be present on stems and roots, and the mestome sheath which occurs in the bundle sheath in the leaves of some grasses. These barriers will be discussed later.

Since the cell wall is negatively charged, positively charged molecules moving in the walls tend to be adsorbed and move less freely than neutral or negatively charged molecules. Relatively few studies have been done on the movement of apoplastic chemicals in the parenchyma. In addition to research work with dyes and metals already cited, Crafts [15] has studied the movement of various herbicides, one of which (monuron) was thought to be apoplastic, in blocks of potato tuber tissue. Monuron displayed an unrestricted movement throughout the tissue with an eventual accumulation of the chemical at the edges of the cube which were the evaporating surfaces. Movement of the herbicide was thought to occur via the walls although, as we will see later, this was not necessarily the case. Transport of substances within parenchyma tissue need only occur for short distances, since the majority of cells are only a short distance away from ramifications of the vascular system. Most of the research on translocation in the apoplast has been concerned with translocation within the xylem.

The conducting elements of the xylem represent a highly modified part of the apoplast (see Fig. 1). Here, very rapid movement of water and dissolved substances is made possible by the structural modifications of the conducting elements, i.e., the tracheids and vessels. In the first place, the whole interior of the cell is available for transport. Resistance to water flow in the lumen is negligible and the tortuous route of water movement in the walls of living cells is avoided. Cells of the conducting elements of the xylem are elongated with end walls variously modified to diminish resistance to water flow. In many vessels, the end walls are completely removed.

The direction of water movement in the whole plant is illustrated in Figure 2. Water moves along a gradient of water potential. Under normal conditions, soil provides the source of water for the plant and since water is lost from the aerial parts, the direction of water movement is upward in the stem and distal in the leaves. Rates of movement vary from 0 to 100 m/hr. Rates of transport are extremely variable due to variations in water supply and environmental conditions. Reversal of the direction of the transpiration stream occurs only under very unusual circumstances, such as the immersion of a leaf in water, which allows the leaf to become a source of water for the rest of the plant. Since evaporation of water is the major force controlling its movement, and aerial organs vary in their transpiration rates, it follows that certain organs will receive more water and dissolved substances than others from the transpiration stream. Leaves generally transpire more than the flowers and fruits and, therefore, receive the majority of the water and dissolved substances from the xylem.

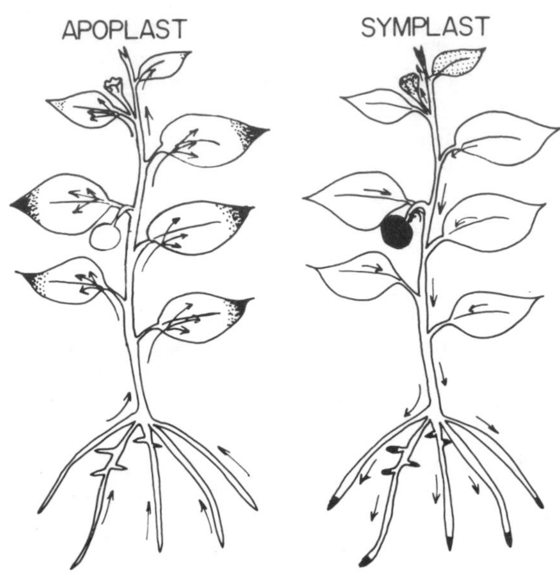

FIG. 2. Translocation patterns in the apoplast and symplast. Arrows indicate direction of transport, black indicates areas of accumulation of systemic chemicals. Stippling indicates areas of lesser accumulation. Redrawn from Ref. 88.

B. Symplastic Transport

The symplast is located within the confines of the plasmalemma and, like the apoplast, forms a continuous system through which substances may move. The protoplasts of individual cells are interconnected by means of numerous plasmodesmata. Like the cytoplasm of the cells, the plasmodesmata are bounded by the plasmalemma. Thus the plasmalemma is continuous from one cell to another. Although the internal structure of the plasmodesmata is controversial at the present time [16], it is clear that these connections between cells must function in the intercellular transport of substances. This is demonstrated by the transport of ions and dyes in the symplast of parenchyma tissue without leakage into the apoplast [17-20]. Viruses have also been found in the plasmodesmata between infected cells [21].

At the present time, we have no clear idea of the precise pathway of transport within the cell or the forces which motivate the transport. The symplast is much more complex than the apoplast chemically and the possibilities for binding and degradation of applied chemicals are increased.

Rapid, long-distance transport in the symplast occurs in the sieve tubes of the phloem (see Fig. 1). Again, we find that the cells involved in translocation are highly modified. They are elongated and have perforated end walls called sieve plates. Chloroplasts and mitochondria are present, but often poorly developed. The nucleus, ribosomes, and tonoplast are absent from the sieve tube in its mature state. The central plasm is bounded by a semipermeable plasmalemma and is connected to surrounding cells by means of plasmodesmata. Thus the contents of the sieve tube are part of the symplast. Esau [22] gives more information on phloem.

Transport of substances within the phloem is thought to occur by a mass flow mechanism by many workers [23]. Whether or not one agrees with any of the several postulated mechanisms of transport, it is possible to make several useful generalizations about transport in the phloem. The direction of transport is from areas known as "sources" to areas known as "sinks." Sources are those parts of the plant which manufacture more food than they require and are able to export it, while sinks are areas which require more food than they manufacture and are able to import it. The usual sources are mature leaves and the sinks are the roots, buds, shoot apices, flowers, and fruits. The symplastic pattern of movement in the plant is illustrated in Figure 2. All substances transported in the phloem move in the same direction, although rates of movement may differ. The controlling factor in determining the direction of flow within the plant is not simply the gross nutritional status of the plant part, but the ability of certain areas to load or unload substances into and out of the sieve tubes. For example, when a mature leaf becomes shaded, it does not begin importing food to satisfy its respiratory demands but continues to export sugar and other products of its own degeneration. In the case of storage roots, the root has more than enough carbohydrate to satisfy its own requirements but still continues to import more. These apparent anomalies are probably explained by the compartmentalization of nutrients within the cells. The major solute transported in the phloem of most plants is sucrose. Rates of transport may vary from 0 to 100 cm/hr, 20 to 30 cm/hr being typical rates.

C. Ambimobile Transport

Figure 2 shows that in many areas of the plant, the direction of transport of substances in the symplast is opposite to the direction of transport in the apoplast. This observation has allowed investigators to test the mode of movement of a given chemical by applying it to a localized area of the plant and observing its direction of movement. The mature leaf is commonly used for such a test. When a droplet of radioactive solution to be tested is placed on the leaf, the distribution of the radioactivity after a time is revealed by autoradiography. An apoplastic chemical will show a wedge-shaped acropetal distribution in contrast to a symplastic chemical which will show transport in a basipetal direction. In order to reach a definite conclusion regarding the movement of a specific chemical, it is of course necessary to confirm that the chemical which was transported was the same one that was applied initially. The urea and triazine herbicides give typical apoplastic patterns of transport, while the phenoxy herbicides give typical symplastic patterns of transport. Most groups of herbicides give a transport pattern which is a combination of apoplastic and symplastic types. These herbicides apparently move both in the apoplast and the symplast and are termed "ambimobile." More information on herbicide translocation can be obtained from Ref. 24. Many naturally occurring substances also show ambimobile transport within the plant [25].

D. Euapoplastic and Pseudoapoplastic Chemicals

In the past, it has been assumed that exogenously applied chemicals showed an apoplastic pattern of movement because they were incapable of penetrating the plasmalemma of the cell. However, there are indications that many of these so-called apoplastic chemicals do in fact enter the symplast.

1. Effects of Apoplastic Chemicals on the Symplast

Herbicides from the triazine and urea groups (all classified as apoplastic on the basis of their transport patterns) are inhibitors of photosynthesis [24]. In order to have this effect, they would have to penetrate not only the plasmalemma of the cytoplasm but also the membrane of the chloroplast. Galley and Foerster [26] have pointed out that apoplastic insecticides must be present in the symplast since they are toxic to aphids feeding in the phloem. They have demonstrated that small amounts of radioactivity are present in aphids and their honeydew when the aphids fed on leaves of plants treated with the apoplastic fungicide ethirimol (^{14}C-labeled). Fungicides containing the benzimidazole moiety are known to have cytokinin-like effects on plants [27, 28].

2. Metabolism of Apoplastic Chemicals

Certain metabolic products of applied chemicals indicate the involvement of symplastic enzymes and sugars. It has been reported that thiophanate-methyl and MBC are partially metabolized to CO_2 [29, 30]. Glycoside formation has been reported for the fungicides ethirimol [31], chloroneb [32], and methyl benzimidazol-2-yl carbamate (MBC) [33], as well as amino acid complex formation for the herbicide atrazine [34].

3. Efficient Uptake of Apoplastic Pesticides by Roots

A chemical which is confined to the apoplast should be blocked in its entry into the conducting system of the root by the Casparian band of the endodermis. Yet, a characteristic of apoplastic pesticides in general is that they are effectively transported to the rest of the plant following a root treatment. In discussing how monuron could enter roots, Donaldson et al. [35] state,

"... the monuron molecule, being relatively non-polar, may be able to 'dissolve' in the lipid of the cell membrane and passively penetrate the cell. Once inside the cell, it could move through the cortex to the stele and from there to the top of the plant via the transpiration stream. This would explain how monuron is able to bypass the Casparian strip of the endodermis." Shone and Wood [36] also conclude that in the case of the apoplastic herbicides, "... it is inevitable that this latter group of compounds must cross the endodermis via the plasmodesmata which have been shown to link the cortex and stelar tissues ... implying that these compounds can enter the protoplasts of the cortical cells."

4. Basipetal Transport of Apoplastic Fungicides

Basipetal movement from mature leaves is an indication of symplastic transport. Small amounts of basipetal movement have been reported for systemic fungicides [29, 30, 37-40]. An interesting relationship between the concentration of the applied chemical and the amount of basipetal transport achieved has been noted in the case of MBC [30]. When the concentration of [^{14}C]MBC was increased by adding unlabeled MBC to the treatment solution, more [^{14}C]MBC was transported out of a mature leaf.

Despite the above indications that pesticides displaying an apoplastic pattern of movement do enter the symplast, the lack of significant basipetal transport from mature leaves, coupled with a lack of success in extracting significant quantities of these pesticides from the symplast [30, 41] has reinforced the view that only small amounts, if any, of the pesticide actually enter the symplast. Ashton and Crafts [42] describe the penetration and movement of an apoplastic herbicide into a leaf as follows: "... it may penetrate the cuticle and then part into the aqueous phase of the apoplast system where it is free to diffuse into the inner leaf structure. If not phloem-mobile, it will remain in the apoplast and move with the transpiration water throughout the acropetal portion of the leaf with a tendency to accumulate around the edge. ..."

On the other hand, the failure to detect significant quantities of pesticide in the symplast may have been due to the removal of the pesticide from the symplast during rinsing steps in the procedures. Monuron [35, 43], ethirimol [36, 44, 45], and several other herbicides [36, 44] are known to diffuse rapidly out of cells and tissues, and in all these studies there is evidence that at least some of the pesticide was initially present in the symplast. If other apoplastic chemicals also diffuse rapidly out of cells, it is not surprising that they are not recovered from the symplast by conventional extraction techniques which involve rinsing the tissue prior to analysis. Boulware and Camper [41] found that the apoplastic herbicides chlorobromuron and fluometuron were not retained by isolated cells or protoplasts after three rinses of five minutes duration each. Confusing penetration into the cells with retention by the cells, they state, "Since fluometuron and chlorobromuron were not absorbed to any degree, there was no measureable chemical binding or transport into the protoplasts or cells." In order to explain the observation of others that fluometuron inhibits photosynthesis after a root treatment, they make the unlikely suggestion: "It appears that the herbicide must be taken up by the xylem to be distributed within the cell to the chloroplasts, rather than directly taken up by the leaf." Solel et al. [30] were likewise unable to detect any [^{14}C]MBC in isolated cucumber mesophyll cells following treatment with [^{14}C]MBC, despite the fact that the chemical showed a small amount of phloem mobility. The isolated cells treated with MBC were rinsed four times before analysis and thus a rapid diffusion of MBC out of the cells could have occurred during rinsing.

In order to determine whether or not apoplastic pesticides can penetrate into the symplast, we have used a method whereby the volume of tissue penetrated is calculated by measuring the depletion of pesticide from the external medium [46]. In the absence of metabolic breakdown of the chemical by the cell, the removal of chemical from the medium is a measure of the uptake of chemical by the tissue. One can also assume that the chemical is not concentrated within the cell, or within a localized area in the cell, if the chemical is free to diffuse out again following transfer to a fresh medium [35, 43]. Using this method, it was possible to calculate the volume of tissue penetrated by various chemicals without removing the tissue from the treatment solution.

The results of this study are presented diagrammatically in Figure 3. The fungicides MBC, ethirimol, carboxin, and oxycarboxin and the herbicides atrazine and diuron moved rapidly into discs (1 mm thick) cut from potato tubers, occupying the entire tissue volume. When the fluorescent dye trisodium,3-hydroxy-5,8,10-pyrenetrisulfonate (PTS), which is unable to penetrate into the symplast [7, 47], was tested in a similar way, only about 6% of the tissue volume was filled. When the cells of the tissue were killed by repeated freezing and thawing, the PTS in the external solution moved into the entire tissue volume. When the tissues treated with the pesticides were killed, the pesticide concentration in the tissue temporarily increased, then returned to the original concentration before freezing. The reason for this temporary change in concentration of pesticide in the tissue is not clear at present. It may be an artifact caused by freezing the tissue in the treatment solution. The ease of diffusion of the fungicides and herbicides out of the unfrozen tissue indicates that the chemicals were not bound (and therefore not concentrated) in specific areas within the tissue. Our results indicate that the fungicides and herbicides studied did penetrate into the entire cell volume and thus they entered the symplast. The pattern of uptake of (2,4-dichlorophenoxy)acetic acid (2,4-D), a known symplastic herbicide, was in contrast to the other fungicides and herbicides. It was rapidly taken up by the tissue and continued to be taken up over the entire 90-min treatment period with the result that it was concentrated about twelvefold within the tissue. After freezing, the 2,4-D was released from the tissue and at equilibrium was at equal concentration in the tissue and in

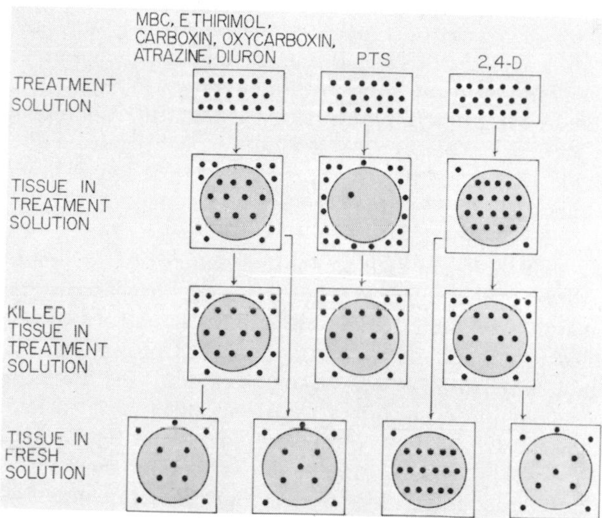

FIG. 3. Diagrammatic summary of results of dilution experiments in which potato tuber tissue (gray) was added to a treatment solution (white) containing the test substance (black dots). The density of the dots represents the concentration of the test substance.

the external solution. Unlike the "apoplastic" chemicals, very little 2,4-D diffused out of the living tissue after it was transferred to a fresh medium.

Can we generalize these results obtained with potato tuber tissue to other parts of the plant? Collander [48] states for nonelectrolytes, "Although different plant cells do vary in their permeability properties, this variation cannot obscure the fundamental fact, already stressed by Overton, that there is nevertheless a very striking similarity between the permeability of all protoplasts, especially if we disregard a few extreme cases (Beggiatoa, Oscillatoria, diatoms)." We have preliminary evidence that tissue from carrot root and celery petiole gives the same results as potato tissue in the experiment with atrazine described above. Thus it seems safe to say that lipophilic pesticides will enter all plant protoplasts.

We now come to the conclusion that the apparent apoplastic transport of a given chemical can occur when the chemical either does not penetrate into the symplast or is not retained by the symplast after it has penetrated. We suggest that the reason for the apoplastic distribution of chemicals in the latter case is a consequence of the more rapid rate of movement of the transpiration stream in the apoplast as compared to the assimilate transport in the symplast. If a chemical is freely transferred between the two systems which are moving in opposite directions, its observed movement will occur in the direction of the most rapid stream. Is it possible to obtain some symplastic transport of a fungicide by increasing its concentration in the plant? This would lead to an increased fungicide concentration within the symplast and might allow some of the chemical to be transported for a significant distance in the symplast before it was all transferred to the transpiration stream. This approach would probably not be successful, since Solel et al. [30] found that the increases in concentration of the treatment solution necessary to achieve even slight symplastic transport of MBC were too high to be economically feasible.

We propose that the term "euapoplastic" be used to describe chemicals which are confined to the apoplast, and the term "pseudoapoplastic" be used to describe chemicals which display an apoplastic pattern of transport but are not confined to the apoplast. At the present time, we would describe MBC, ethirimol, carboxin, oxycarboxin, atrazine, and diuron as pseudoapoplastic. One would expect other realtively nonpolar pesticides which enter the plant readily via the roots to be included in this group. Further work will be necessary to determine how many other chemicals should be classified as pseudoapoplastic. The transport of both euapoplastic and pseudoapoplastic chemicals can still be described as apoplastic since their movement is controlled by the transpiration stream.

E. Accumulation by the Symplast

From the discussion in the previous section, it is clear that penetration into the symplast is not sufficient to ensure significant symplastic transport. It appears that the retention of a chemical within the symplast is necessary. This idea was expressed earlier by Pickering [49] when his histoautoradiographic studies of [^{14}C]monuron penetration into leaves showed radioactivity uniformly distributed within the tissue. Unfortunately, he felt that the amount of radioactivity in the tissue was too low to draw a definite conclusion regarding its distribution.

How can a chemical foreign to the plant be retained in a mobile form by the symplast? One idea, known as the weak acid hypothesis, has been described by Crisp [50]. According to this theory, weak acids which are in the undissociated nonpolar state at low pH are able to diffuse through the plasmalemma due to their lipid solubility. Once inside the plasmalemma, the higher pH of the cytoplasm causes the molecules to dissociate and the polar anions are unable to diffuse back through the plasmalemma. In this way, the continued inward movement of the molecules and the absence of their outward movement leads to an accumulation of the anions within the cytoplasm. Further inward movement of the anions into the vacuole would be largely prevented by the tonoplast membrane. In addition, the lower pH in the vacuole would favor accumulation by the cytoplasm. We would predict from this theory that the site of accumulation within the cell would be in the cytoplasm of cells with acidic vacuoles. Swanson and Baur [51] found that the uptake of picloram by potato tuber tissue was strongly dependent on the pH of the bathing medium and state, "These data confirm earlier work with 2,4-D and other weak acid growth regulators and suggest that picloram is absorbed primarily as the undissociated molecule. In the high pH range, where maximum dissociation would be anticipated (pK of picloram = 3.58), uptake was negligible."

In a survey of various herbicides, Jacob et al. [52] looked for molecular features associated with phloem mobility. They found no correlation between a herbicide's phloem mobility and its water solubility, its breakdown in plants, or the presence of halogen, amino, monosubstituted amino, disubstituted amino, nitro, carbamate, thiocarbamate, urea, methyl or methoxy groups in the molecule. They did find a good correlation between phloem mobility and the presence of a free carboxyl group. They also state, "The necessity of carboxyl groups for phloem mobility of phenoxyacetic acids is accentuated by the fact that phenyl compounds without this group are immobile or xylem mobile." Earlier, Mitchell et al. [53] had found that carbamates containing a lactic acid group were translocated basipetally (i.e., symplastically) following leaf application in grasses while related carbamates without the lactic acid group were not. Furthermore, they induced basipetal movement in plants by substituting a lactic acid group for the isopropyl group of isopropyl N-phenylcarbamate and isopropyl N-3-chlorophenyl carbamate. Crisp [50] has proposed a scheme whereby a group could be added to the parent molecule of a pesticide so that after penetration of the cuticle and ester hydrolysis by cuticular apoplastic esterases, a moiety such as malonic or succinic acid would remain on the parent molecule. Then the chemical, being a weak acid, would penetrate into the symplast where the organic acid would be removed by β-oxidation. It was assumed that the parent molecule would be unable to leak out of the symplast but we now know that if it were pseudoapoplastic, it would leak out very readily. Therefore in a scheme of this sort, it would be essential that the removal of the acid group proceed at a rate slow enough to allow transport of the pesticide-acid complex to the desired site of action before the acid group is cleaved off. Even then, the pesticide molecules would tend to be depleted from the symplast rapidly and carried away by the transpiration stream.

We have found that the ambimobile nematocide oxamyl was not concentrated at all in our potato disc test system. The major difference between oxamyl and the other pseudoapoplastic pesticides studied was that the rate of penetration of oxamyl into the tissue and also its rate of efflux out of the tissue was relatively slow [140]. The reason for the partial symplastic transport of this chemical following a leaf application could be that once inside the symplast it has time to be transported due to its relatively slow leakage out. The tendency for the chemical moving in the transpiration stream to be concentrated at leaf tips and margins means that a reservoir of concentrated pesticide would be available in the leaf. The possible importance of the leaf tip as a region for transfer of chemicals from the xylem to the phloem has already been pointed out by Price [54].

Carrier mechanisms, whereby specific molecules or ions are moved across membranes by complexing with specific sites on protein molecules, are used to explain the uptake and retention of naturally occurring substances by the symplast but as Crowdy [55] has stated, "A plant is unlikely to develop a specific carrier mechanism for a foreign chemical completely unrelated to its natural products."

Does the accumulation and retention of a chemical in the symplast mean that it will necessarily be phloem mobile? Unfortunately, we still do not have a sufficiently detailed understanding of the biochemical steps and anatomical routes followed by assimilates, as they move from parenchyma cells into the sieve tubes, to be able to make any predictions about the pathway followed by phloem-mobile pesticides.

III. UPTAKE OF FUNGICIDES BY THE PLANT

The uptake of pesticides is most easily discussed in terms of the site of entry, for example, leaf, stem, root, and seed coat.

A. Uptake By Leaves

Most systemic fungicides have been developed commercially for use as foliar sprays. Therefore the transcuticular movement of fungicides is a vital factor for success. If we accept the role of the cuticle as reducing water loss from the plant surface, then we accept the fact that pesticides in aqueous sprays may have a low penetration efficiency. The cuticle consists of a lipoidal layer enveloping the above-ground portions of plants. The outermost portion of the cuticle consists of waxes which are made up of long-chain hydrocarbons (21-35 carbons) in various states of oxidation while the inner portion, or cutin, is a mixture of C_{16} to C_{18} chains interconnected by ester, peroxide, and ether linkages to form polymers. Waxes are thought to be the most important barrier limiting the movement of pesticides through the cuticle. Evidence to support this contention is based on movement through an isolated cuticle with waxes removed. For these studies, the intact plant leaf or fruit was dipped in chloroform or dichloromethane to remove the wax. These solvents might possibly also alter the cutin. Rather dramatic effects have been obtained by wax removal. Research with *Malus* [56] and *Populus* [57] shows a four to fivefold increase in transcuticular movement of herbicides by dipping leaves 10 sec in chloroform. Using the same technique, Drandarevski and Mayer [58] increased uptake of several systemic fungicides through epidermal sections of *Bryophyllum* and *Tulipa*. This indicates the waxes are a significant barrier to penetration. Penetration, or transcuticular movement to be more precise, is apparently not dependent on the thickness of the cuticle but upon qualitative differences between plant species [59].

With almost all pesticides tested, uptake is more rapid through the lower, or abaxial cuticle than through the upper or adaxial cuticle. An interesting exception is the report [57] that the ethyl ester of 2,4-D penetrated either surface of *Populus* leaves equally well, whereas the more polar dimethylamine form of 2,4-D penetrated the abaxial surface more efficiently. In this context, most systemic fungicides penetrate the abaxial surface more rapidly [60], which may indicate that their polar properties are favoring abaxial uptake. Since the abaxial surfaces of leaves generally have many more stomata than the adaxial surfaces [55] (see Fig. 4), it is tempting to attribute the differential uptake of pesticides to differences in stomatal frequency. While this may be partially correct, other differences between the two surfaces are also worthy of consideration. Norris and Bukovac [61] noted the lower cuticle of *Malus* contained less embedded waxes than the upper cuticle. Other differences of the abaxial cuticles could be due to the lower light intensity received during formation which might affect the oxidation and thus polarity.

The arguments for stomatal penetration also warrant consideration. A pesticide could pass through the stoma into the substomatal chamber. However, this cavity is still enclosed by a cuticle [61] (see Fig. 4D). The high relative humidity and possibly the absence of waxes in the substomatal cavity could favor penetration of the pesticide as compared to the exposed cuticle [47, 62]. The penetration of the stomata may be as a liquid from the spray droplet and would be dependent on the surface tension and contact angle. Dybing and Currier [47] reported that a surfactant is essential to allow water to move into leaves via stomata. Prasad et al. [62] and Norris and Bukovac [63] have noted that penetration of stomata by pesticides was especially enhanced with the surfactants X-77 (nonionic surfactant of Colloidal Products Corp., Sausalito, Calif.) and Vatsol OT (sodium dioctyl sulfosuccinate) which lowered the surface tension to about 30 dyn/cm^2. Evidence indicated that pesticide uptake was by stomata and they assumed that even though the substomatal chamber is lined with a cuticle (see Fig. 4D), albeit thin, the relative humidity in the chamber would be approaching 99%. Similarly then, pesticide uptake would be influenced by any factor influencing stomatal opening, e.g., light [64], relative humidity [57], temperature [65], and accumulation of photosynthesis products. For example, on a summer day, the stomata may close at midday and pesticide uptake via stomata would be nil. Later in the day the stomata may reopen, but the spray droplets have dried and consequently the uptake will still be negligible. Rewetting may give some subsequent penetration, although once the pesticide has dried the benefits of rewetting are never very large [62].

Another possibility is the uptake of pesticides through stomata in the form of gasses. Small [66] recently found pyroxychlor uptake was through the stomata as a gas. Since most systemic fungicides are not very volatile, this is the exception rather than the rule.

Transcuticular movement of fungicides had been studied by using intact leaves, epidermal strips or isolated cuticles. Intact leaves do not permit

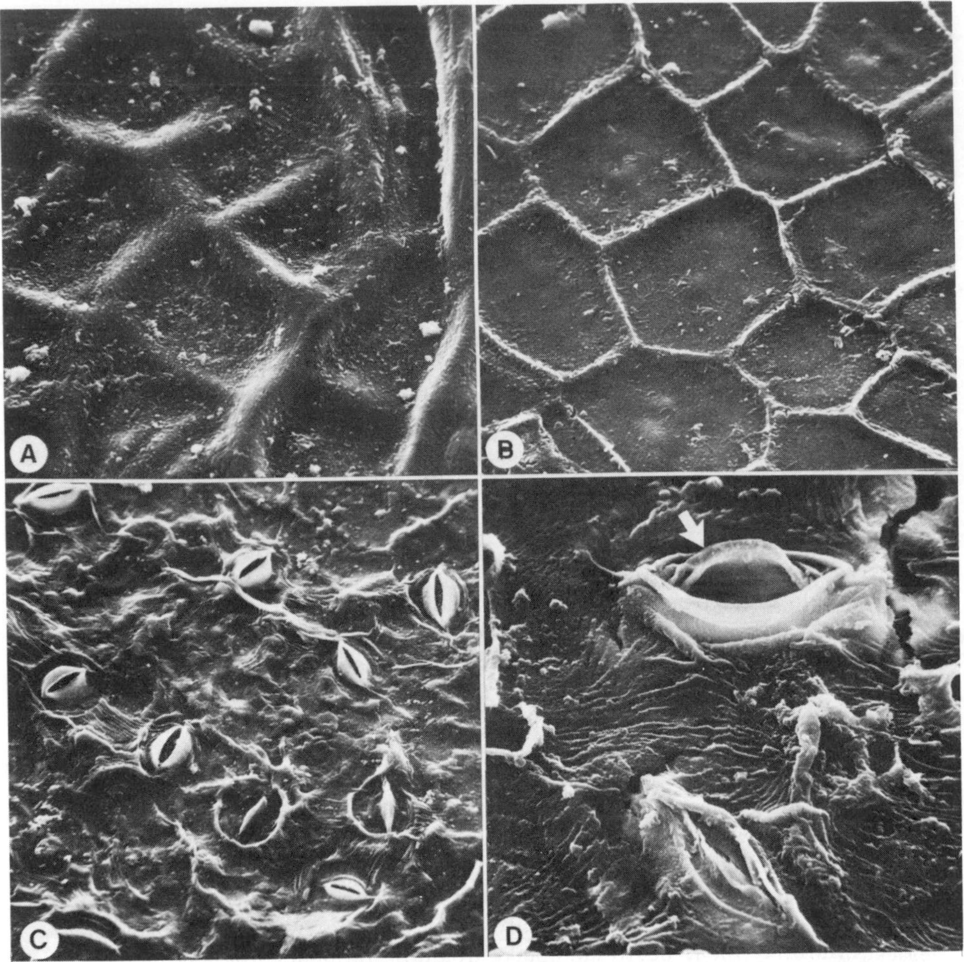

FIG. 4. Scanning electron micrograph of isolated cuticles of apple leaf: A. Adaxial cuticle viewed from outer side showing absence of stomatal pores and impression of epidermal cells; B. Adaxial cuticle viewed from inner side; C. Abaxial cuticle, outer side showing stomatal pores; D. Abaxial cuticle, inner side showing cuticle lining of substomatal chamber (see arrow). Photographs courtesy of H. Dahman and F. J. Schwinn, Ciba-Geigy AG, Basel, Switzerland.

the separation of transcuticular movement from subsequent entrance and translocation within the leaf but the cuticle is in a natural state of relative humidity at both inner and outer surfaces and movement of chemicals into the internal portion of the leaf is "normal." Drandarevski and Mayer [58] have used epidermal strips but the plants from which one can successfully remove the intact epidermis are limited. This did enable them to study the effect of light on stomatal opening, and subsequent fungicide penetration. Isolated cuticles allow the singular study of transcuticular movement. They can be used either as a membrane with a solution on both sides [63] or mounted on aluminum supports on an agar medium with fungicide penetration being bioassayed with sensitive fungi [60, 67]. Inherent deficiencies of research with isolated cuticles may be caused by possible changes in the cuticle in the process of isolation from the leaf or from the unnaturally high degree of hydration of cuticles. Regardless of the technique used, several conclusions are justified. Protective fungicides penetrate inefficiently, particularly through the adaxial surface [58, 60]. The efficiency or percentage of penetration by systemic fungicides through cuticles is extremely low. For example, benomyl penetrated the adaxial surface of cucumber leaves to a maximum of about 5% [68, 69]. Small changes in molecular structure greatly influence efficiency of transcuticular movement of fungicides. The addition of the butyl carbamoyl group to the 2 position of carbendazim (or MBC) increased uptake through isolated cuticles [60] and epidermal strips [58] about threefold. With intact cucumber leaves, Upham and Delp [68] had similar results with 0.8% for MBC vs. 5% for benomyl.

While the percent penetration in data reported for isolated cuticles and epidermal strips appears much higher than for intact leaves, this is misleading. These experiments are for uptake during 24 hr, and if we estimate that spray droplets dry and uptake ceases in 1-3 hr, then the results are quite comparable.

Solubility of fungicides in the spray solution appears important. As the concentration of fungicides exceeds the solubility in water the percentage of fungicide penetrating decreases [58, 67]. However, the total amount penetrating increases somewhat with increase in concentration. Often by altering pH and thus solubility, uptake can be enhanced or decreased [60, 70].

The formulation of systemic fungicides as emulsifiable concentrates (EC), colloidal suspensions (cols), or oil dispersible (OD) forms has been shown to enhance cuticular penetration [58, 67, 71, 72]. This dramatic shift in formulation of systemics, in contrast to protective fungicides which were all applied as wettable powders, has played a critical role in the successful commercial development of systemics. This applies primarily to foliar or dormant bark applications. For example, recent formulations of oxycarboxin, triforine, tridemorph, and pyroxychlor are EC, while pyracarbolid and benodanil are OD, and oil is added to benomyl [73, 74] and thiophanate-methyl sprays. The EC formulations are probably more efficient because they permit a rapid liquid-liquid partitioning between the organic and aqueous phases in the spray droplet. The OD formulations are apparently different in their modus operandi. The fungicide may not be very soluble in oil or water. The oil seems to diffuse into or through the cuticle and enhances penetration of the fungicide and other chemicals applied. The possible role of the emulsifier acting as a "cosolvent" in the OD formulations is intriguing.

The OD and EC liquid formulations have evolved as a very important economic saving of two- to threefold in the dosage of fungicide required per hectare [72].

B. Uptake by Stems

The stems of herbaceous plants have epidermal and cuticular layers similar to those of leaves described earlier. Penetration of fungicides into herbaceous stems is analogous to penetration into leaves as discussed in the previous section.

In woody plants, the outermost layer of the stem is the periderm or outer bark, which is composed of the cells of the phellum, phellogen, and phelloderm. At maturity, the phellum cell walls contain suberin and wax lamellae and are impervious to water [75]. The cells of the phellum (and phelloderm) arise from the meristematic phellogen. The phellogen may form at different levels within the stem in various species of plants. According to Norris [76], "The relative success of bark applications on species can be partially related to the site of development of new phelloderm. In *Pyrus* this originates shallow in or just beneath the epidermis, thus only limited dead bark will be present. In *Prunus*, the phelloderm is initiated more deeply in stem tissues and as a result greater thickness of dead periderm and phloem are usually present in *Prunus*." The cells of the phellum are closely packed and have no air spaces between them, but the lenticels in the bark are composed of loosely packed cells which do have air spaces between them. The cell walls in the lenticel may or may not be suberized. Considerable variation in the arrangement and suberization of the lenticel cells exists between species. As yet, no experimental work has correlated the type of lenticel with the penetration of pesticides. Smalley et al. [77] have observed toxicity patterns indicating that benomyl penetrated the bark of young elm twigs through the lenticels.

Some plants are more amenable to bark treatment than others. Success has been achieved in treating citrus and pine trees with insecticides, but elm tree bark is not easily penetrated. In general, it has proved difficult to treat the mature bark of thick-barked trees. Insecticides have been used in most experimental work involving bark penetration of pesticides. This work has been reviewed by Norris [78]. The advantages of trunk application of insecticides have been summarized by Norris as follows: "In addition to restricting the chemical to the specific tree, applications to the basal trunk offer several other major advantages. The dosages of insecticides required for a certain level of pest control in a tree applied via the trunk frequently are 1/10 to 1/5th those required in the soil .... Banding of the circumference of the basal trunk with chemical generally will yield the most uniform distribution of the pesticide in the tree .... Such a band treatment could produce adequate pest control in conifers that branch only inches above the soil surface .... The physical ease of placing a systemic on the basal trunk commonly surpasses that of making soil treatment to the root systems, especially of large trees .... The chemical applied to the trunk also is not as dependent upon moisture for uptake into the plant as with soil treatments. Greater accuracies in dosage are possible via the trunk than through the soil .... This route allows a more accurate timing of the necessary concentration in the tree." The same line of reasoning can be used with respect to fungicides.

As proved to be the case with MBC and thiabendazole (TBZ) applications to cotton stems [71], addition of oil to the formulation increases the effectiveness of the fungicide treatment. Hearn and Childs [74] found reduced severity of sour orange scab in the third flush of leaves after spraying orange trees with benomyl in Volck oil (product of Chevron Chemical Co., Richmond, Calif.). This indicates to us that the benomyl must have penetrated bark as this fungicide primarily moves only acropetally. Ramsdell and Ogawa [79] found that addition of Volck oil to MBC increased its penetration into almond twigs. We have found that benomyl in oil painted on elm bark will penetrate to the wood when the bark is 1 mm or less in thickness [80]. Application of fungicides as dormant sprays may be advantageous in bark treatments since the tree bark is not shielded by the leaves at that time and is less susceptible to phytotoxicity than the foliage. The reason for the increased penetration of pesticides in oil has not been established. The oil could either change the structure of the suberin and wax layers which present a barrier to entry; dissolve through these layers, carrying with it the fungicide; or provide a nondrying source of fungicide.

Coppel and Norris [81] obtained good penetration of the insecticide Bidrin through the bark of white pine trees when a Bidrin-resin powder was mixed with Estab (product of Velsicol Chemical Co., Woodstock, Ill.) plastic and painted on the trunk. They state, "The addition of the resin powder produced final formulations that were more porous in their solidified state. This characteristic was important in minimizing the phytotoxic effects of treatment, and this conclusion was further supported by the severe bark injury caused by ... wax materials ... which produced a homogeneous seal of the bark in the treatment area. ..." They obtained insect control in the upper crown for a period of 1 year after treatment. Although we have been unsuccessful in using Estab to obtain penetration of benomyl through the bark of elm trees, the success of Coppel and Norris might be repeated using other fungicides or other types of trees.

In the case of thick-barked mature trees, pesticide application to the trunk can be achieved by a variety of injection techniques. In some cases extensive bark and cambium injury resulted from the injection [82] and the trees became susceptible to infection. Nevertheless, trunk injection techniques have proven successful in many other cases. Further information on injection methods is presented in Volume 1, Chapters 7 and 8.

C. Uptake by Roots

Plant roots function as absorbing organs for water and soil minerals. Large quantities of water and dissolved substances may be taken up by roots and transported to the top of the plant. Thus roots are potentially advantageous sites for the uptake of fungicides. In the field, fungicides may be applied to plant roots as a seed treatment, in row application during seeding or as a drench later on in the growing season. The problems of distribution, adsorption, and breakdown of fungicides in the soil have been discussed in the preceding chapter. For the present discussion, we will assume that the fungicide is available to the plant roots in the soil solution or in nutrient culture.

Roots of different species of plants vary in their structure and permeability to water and solutes. An individual root also shows variation in permeability along its length. The region of highest absorption occurs in young areas of the root. In the older areas of many plants, the root epidermis becomes suberized or a bark begins to form. Both of these events lead to a progressive loss of permeability. In monocot roots, which generally lack secondary growth and suberization, the root is more uniformly permeable to water along its length. Despite the fact that the older areas of some roots are less permeable to water than younger areas, the large volume of root which exists in the suberized condition in long-lived plants, such as trees, may mean that the major part of water is taken up by the plant through the older areas of the root. In the following discussion, we will consider the uptake of fungicides in a young area of the root as shown in Figure 1. Unlike the aerial plant parts, the epidermis of the young root is not covered with a well-developed cuticle. Therefore, fungicides in solution have relatively easy access to the wall system (apoplast) of the root. Many inorganic ions are actively taken up by plant roots in processes requiring metabolic energy. Alternatively, it is possible for a chemical to be taken up passively. Indications of active uptake of a chemical are: (1) its accumulation by the roots against a concentration gradient; (2) its continued accumulation by the roots over a period of several hours; (3) inhibition of its uptake by low temperatures, respiratory inhibitors, competitive inhibitors, and anaerobic conditions; and (4) a nonlinear relationship between its rate of uptake by the roots and its external concentration. The opposite features indicate a passive mechanism of uptake.

Donaldson et al. [35] have published a very thorough comparative study of the uptake of the symplastic herbicide 2,4-D and the pseudoapoplastic herbicide monuron by barley roots. While we do not have any such complete studies of fungicide uptake by roots, the available evidence indicates a passive mechanism of uptake for those studied. Shone and Wood [36] concluded that ethirimol was taken up passively by barley roots since its uptake was not inhibited by low temperatures and it did not become concentrated within the root. Uptake of ethirimol was complete within 5 min and the level of ethirimol in the roots remained fairly constant until the end of the experiment (15 min). Sumida et al. [83] concluded that 3-(3',5'-dichlorophenyl)-5,5-dimethyloxazolidine-2,4-dione (DDOD) was taken up passively by bean roots since the fungicide was not concentrated within the root and the amount of DDOD taken up was proportional to the amount provided in the nutrient solution. We found that MBC was not concentrated in onion roots [84]. It is possible for a chemical to enter the root passively and then become bound or metabolized to other products within the root. In these cases, it may appear that the chemical is taken up against a concentration gradient although uptake is really passive. Snel and Edgington [85] found that carboxin and oxycarboxin were more concentrated in bean roots than in the external solution, but did not conclude that active uptake had occurred since most of the carboxin in the root was metabolized. Shephard [45] has shown that different species of plants display considerable variation in their uptake of the pyrimidines ethirimol and dimethirimol. Roots of woody plants such as apple and vine tended to concentrate the fungicides (see Figure 5-H), while the roots of herbaceous plants such as wheat and cucumber did not (see Figure 5-C). The increased uptake of the fungicides in woody plant roots can be partially explained by binding reactions within the roots.

Penetration into the root could conceivably occur via the apoplast up to the endodermis. Here, the Casparian band, a deposition of suberin and lignin in the tangential and radial walls of the endodermal cells (as well as in the middle lamella connecting the cells) is thought to impede the further inward movement of substances in the apoplast. It is known that the Casparian band presents a barrier to the movement of certain dyes [84, 86-88] and metal ions [12, 89-91]. It is not known whether or not this structure is also a barrier for the movement of fungicides. It has been suggested that the gaps in the Casparian band, formed during early stages of lateral root initiation, could function as pathways of entry for substances moving in the apoplast [92, 93]. However, it has been demonstrated that these gaps are not necessary for the penetration of MBC into the conducting elements of the root [84]. The MBC either diffuses through the Casparian band or penetrates the plasmalemma of the endodermal cell and moves through the symplast in order to reach the root xylem. Both these pathways have been suggested to explain the penetration of "apoplastic" herbicides into roots [94]. Donaldson et al. [35] first pointed out that monuron is taken up by barley roots in amounts indicating that it had penetrated the entire root volume, instead of approximately 14% of its volume which would be the expected result if the chemical had only penetrated the apoplast without binding. The fact that 95% of the monuron diffused out of the root quickly when it was transferred to water was evidence that the monuron was not bound within the root and thus was not concentrated at any particular site within the root. They concluded that most of the root volume is accessible to monuron and that monuron may move through the symplast of the endodermal cells. Shone and Wood [36] have demonstrated a rapid uptake of the "apoplastic" herbicides diuron, simazine, and the fungicide ethirimol into barley roots. Regarding elution following transfer of the treated roots to water, they state, "The rate of loss to the water was comparable with the rate of uptake suggesting that elution may readily remove material from the vacuoles." Absolute values of chemical eluted out in water were not given and therefore it is not possible to calculate the volume of the cell accessible to the chemicals by diffusion. Some concentrating of the chemicals within the cell did occur which the authors attributed to binding since the uptake of the herbicides was not sensitive to temperature. They postulated that the "apoplastic" herbicides studied entered the protoplasts of the cortical cells outside the endodermis and crossed the endodermis via the plasmodesmata because the tertiary thickenings of the endodermal walls described by Clarkson et al. [95] would impede direct penetration into the plasmalemma of the endodermis. It should be noted, however, that the tertiary thickenings do not occur throughout the entire root system. In younger areas of the root, the endodermis is in the primary state of development as shown in Figure 1. Thus if a chemical is capable of moving easily into and out of cells (as seems to be the case for the pseudoapoplastic pesticides studied), the chemical may then move from the apoplast of the cortex into the protoplast of the endodermal cell and out into the apoplast within the stele without necessarily going through the plasmodesmata. Also, the alternate route of entry into the stele through the Casparian band itself cannot be overlooked at the present time. Shone and Wood [36] suggest that hydroxyatrazine and ethirimol are largely located in the free space of the root since increasing the concentration of calcium ions in the treatment solution decreased the amount of protonated pesticide in the xylem sap. Why the occupation of negatively charged sites in the root by calcium ions should inhibit the movement of hydroxyatrazine or ethirimol across the root and into the xylem is not clear. Shephard [45] has studied the uptake of four pyrimidine analogues including ethirimol and dimethirimol into plants roots. Although the time intervals used were considerably longer than those used by Shone and Wood, the results of the uptake of ethirimol by wheat and cucumber roots agree well with Shone and Wood's results with barley. Shephard also demonstrated that all the ethirimol taken up by wheat roots elutes out within about 30 min after transfer to fresh solution. Uptake and elution patterns for ethirimol were more complex in the roots of the woody plants, apple, and vine. More work will be necessary to interpret the absorption of pyrimidine fungicides by these woody plants.

Using potato tuber tissue as a test system, we found that ethirimol penetrated the entire volume of the tissue and diffused freely out again, indicating that ethirimol penetrated into the protoplasts of the cells. Perhaps the rate of penetration will prove to be a critical factor in the rate of uptake of a fungicide by the plant via the roots. Some fungicides may not penetrate membranes quickly enough to keep up with the water of the transpiration stream which must move through membranes of the endodermal cells to bypass the Casparian band. The results of both Shone and Wood [36] and Shephard [45] show that the roots impede the movement of ethirimol into the transpiration stream.

Verloop [96] has shown that for stable aromatic sulfones, the amount of chemical available to the rest of the plant following a root treatment was positively correlated with the lipophilicity of the chemical. The same general result was obtained by Shone and Wood [36] for the pesticides studied. In order for chemicals to be taken up by the rest of the plant from a root treatment, they must traverse a lipid barrier, either in the form of the plasmalemma of the endodermal cell or the Casparian band itself. Therefore, it is not surprising that the lipid solubility of a chemical will be an important factor in its ability to move through plant roots.

At the other extreme, chemicals which are very lipid soluble are not readily released into the transpiration stream but are retained by the root lipids. Describing the optimum lipophilic-hydrophilic balance for movement into the transpiration stream, Crowdy et al. [97] state, "With broad bean, compounds with a hexane-water partition coefficient greater than 6 tend to be retained in the roots; a value of 0.8 is satisfactory but 0.3 may be advantageous." They also found that the partition coefficients for olive oil-water were approximately 30 times higher than for hexane-water. Therefore, a value of 6 obtained with hexane-water would be equivalent to a value of 180 in olive oil-water. This is higher than the values for the successful apoplastic pesticides studied by Shone and Wood [36]. The highest olive oil-water partition coefficient in the latter work was 50 for atrazine. Unfortunately, Verloop did not specify the solvents used in determining the lipophilic-hydrophilic balance in his work and therefore his results cannot be compared with those of Crowdy et al. and Shone and Wood.

Once inside the stele, pseudoapoplastic herbicides and fungicides move into the xylem elements and are transported to the aerial parts of the plant. Since the chemicals are easily eluted from living cells [35, 43, 44], they would enter the apoplast without difficulty even if they originally entered the stele via the protoplasts of the endodermal cells. Symplastic herbicides such as 2,4-D are generally not translocated in quantity to the top of the plant following a root treatment. Crafts [98] has suggested that this is because of strong adsorption which does not allow the molecules to enter the stele of the root. Another explanation would be that the molecules are not released into the apoplast once they are inside the stele but instead are transported toward the growing root tip in the phloem.

Fuchs et al. [99] found that applying equimolar amounts of the MBC-forming fungicides, i.e., benomyl, thiophanate-methyl, and NF-48, to plant roots did not result in equimolar concentrations of fungicide in the plant tops 2 days later. Approximate concentrations estimated from their graph (Ref. 99, Fig. 7) for bean plants were NF-48 16, benomyl 11, thiophanate-methyl 5 $\mu g/g$ fresh wt. The authors also noted an important difference in the distribution of the fungitoxicant in the foliage after cessation of treatment with the three parent fungicides. While fungicide was present only in the foliage which existed at the time of treatment in the case of 2-(3-methoxy-carbonyl-thioureido)aniline (NF-48), some fungicide was also present in the newly developed foliage in the case of the benomyl and thiophanate-methyl treatments. Furthermore, the depletion of fungicide from the central areas of leaves was much more pronounced in the case of the NF-48 treatment. The efficacy of long-term disease control from the various fungicidal treatments was benomyl > thiophanate-methyl > NF-48, which reflected the amounts of fungicide in the new growth and central areas of the leaves. They postulated that the differences in distribution of the fungitoxicant in the plants were due to a differential binding of the parent compounds in the roots and a gradual release of MBC to the rest of the plant. They suggested that the *N*-1 substituted form of the molecule might be the bound form. Treatment of the roots with heated fungicide, which resulted in a conversion of the parent chemical to MBC, gave results similar to the treatment with NF-48. In a later paper [100], it was also reported that disease control with benomyl and thiophanate-methyl was superior to disease control by heated samples of the fungicides. However, it can also be seen (Ref. 100, Table 1) that there was considerable variation in the location of fungicide after short-term treatment with the heated fungicides. The reason for this variation is not clear if the heating process did successfully convert the fungicides to MBC. Fuchs et al. [99, 100] did not recover benomyl or thiophanate-methyl from the foliage following a root treatment. This is in contrast to the results of Buchenauer et al. [101] who found thiophanate-methyl in the foliage of cotton plants following a root treatment. It would be desirable to have data on fungicide binding to the root as Fuchs et al. have postulated.

The binding of the pyrimidine fungicides to the roots of woody plants has already been mentioned [45]. This binding has resulted in poor translocation of the fungicides to the remainder of the plant following a root treatment.

In conclusion, we must point out the correlation between the type of transport exhibited by a chemical and its uptake by the root system. Substances which show an apoplastic pattern of transport within the plant are passively taken up by the roots. Euapoplastic substances such as the dye PTS are not readily transported to plant tops from a root treatment; whereas pseudoapoplastic substances are readily transported to the plant top. Symplastic substances such as 2,4-D are actively taken up by roots but are not always readily transported to the top of the plant. The amount translocated is a function of the applied concentration with proportionally more of the substance being transported upward as the concentration is increased. As new fungicides are developed, their behavior in root uptake will undoubtedly be correlated with their transport patterns in the plant.

D. Uptake by Seeds

While there are numerous indications of slight penetration by seeds of protective fungicides, the failure of protective fungicides like the dithiocarbamates to control seed decay organisms within seed is mute evidence of inefficient uptake. Organic mercury fungicides were partially successful in reducing pathogens such as Cochliobolus within the seed coat in cereals. Perhaps the greatest success has been Maude's results [102] obtained with soaking seed in 0.2% thiram for 24 hr. While many fungicides have been evaluated, few could successfully traverse the 2 mm, plus or minus, into the embryo. The discovery by von Schmeling and Kulka that loose smut of barley could be controlled by treating seed with carboxin was a milestone in seed treatment [103].

An obvious question is how much fungicide penetrates directly through the seed coat as compared with penetration into the emerging radicle. Thapliyal and Sinclair [104] studied this problem by removing the seed coat at varying times after treating seed and bioassaying the fungicide which had penetrated. Unfortunately the data are not plotted on a standard curve to give the actual percentage of fungicide in tissues. They did note that extracts from inside seed treated with benomyl and oxycarboxin reached maximum levels in 36-58 hr, whereas chloroneb gradually increased during the 96 hr tested. This seems to agree with data of Richardson [105] on chloroneb. He found thiram to be synergistic with chloroneb for the control of Pythium damping-off of peas and concluded that thiram prevents seed decay, thus enhancing uptake of chloroneb by the young seedling. Chloroneb seems to be taken up more slowly than other systemic fungicides. For example, in a very thorough time study Briggs et al. [106] recovered carboxin inside barley seeds in 6 hr, the concentration reaching a maximum after about 3 days. Concentrations of nonmetabolized carboxin were greater in the scutella and roots than in the shoots after 3 days. Subsequently, rapid degradation of carboxin masked uptake effects.

Perhaps the most unique approach to uptake of pesticides by seed treatment is the use of organic solvents. This method is still in the embryonic stage of development. The pesticide can be dissolved in organic solvents, e.g., dichloromethane, chloroform, or acetone. Seeds soaked in the solvent mixture take up the pesticide linearly with time [107]. Upon removal of seeds, the solvent evaporates leaving the pesticide within the seed. Tao et al. [107] even obtained control of internal seedborne fungi with pentachloronitrobenzene (PCNB), a nonsystemic fungicide. Economics of solvents, labor costs, and flammability of solvents will be important factors influencing the ultimate use of this unique method.

Except for the last method, the uptake of chemicals by seeds does not appear to have unique requisites for systemicity. In other words, if a fungicide is systemic by application to roots or leaves, it will be systemic when applied to seeds. Many fungicides are phytotoxic when applied to seed because the concentration is too high in the germinating seedling. Thus it is very difficult, if not impossible, to apply enough fungicide to a seed to control foliar diseases on the plant for the growing season. Ethirimol applied to barley seed at 3.3 g/kg seed will give good, but not complete, control of powdery mildew for the growing season [108, 109]. The amazing triazole, 4-n-butyl-1,2,4-triazole (RH-124), applied to seed gives control of wheat leaf rust for most of the growing season (about 70 days), but must be followed by one or two sprays of mancozeb about midseason for best control and yield benefit [110]. Obviously, the fungicide taken up from seed treatment will be diluted by plant growth, subjected to degradation, and even if the fungicide remains intact, it may soon be ineffective because it will be transported to leaf margins. The fungicide remaining in the soil around the seed is also subject to degradation and adsorption. For example, Graham-Bryce [108] reports less than 5% of ethirimol applied to seed is taken up and translocated to foliage. Consequently, we cannot generally expect to control more than seedborne fungi and seedling diseases with seed treatment.

IV. TRANSPORT OF FUNGICIDES

A. Apoplastic Transport

As noted earlier, systemic fungicides moving acropetally in the direction of the transpiration stream are really pseudoapoplastic. Very few differences exist in transport pattern between the various commercial systemic fungicides displaying apoplastic transport.

The results with carboxin [85], benomyl [111-113] and ethirimol [31] all show a fairly similar rapid movement to the margins of leaves. Results with ethirimol (see Fig. 5), adapted from research with Shephard while at our laboratory and partially reported [31, 45], typify the pattern of transport. Their movement is to the transpiring organs as initially reported [112, 114]. Since fruits have few if any stomata, transpiration by the fruit is negligible and residues have not caused serious problems. Once the fungicide has entered a leaf there is no basipetal movement down the petiole and then acropetally to enter the new leaves developed later. This is clearly shown in the time-course study in Figure 5. Only the leaves of cucumber and apple present at the time of treatment (A and D) show any radioactivity 13 (C and F) or even 27 days (G) after treatment. As with all autoradiographs, some caution must be taken in interpreting the ^{14}C-patterns. For example, the total ^{14}C in apple plants decreased with time from 2,500,000 to 480,000 to 430,000 to 210,000 dpm for 2, 6, 13, and 27 days, respectively, after treatment. Of the ^{14}C recovered only 75, 58, 48, and 22%, respectively, was still ethirimol. Thus one must consider both total activity and degradation in diagnosing patterns.

There is probably a difference between plant species in the rate of transport of fungicides. Certainly pyrimidines, as Shephard reported [45], move

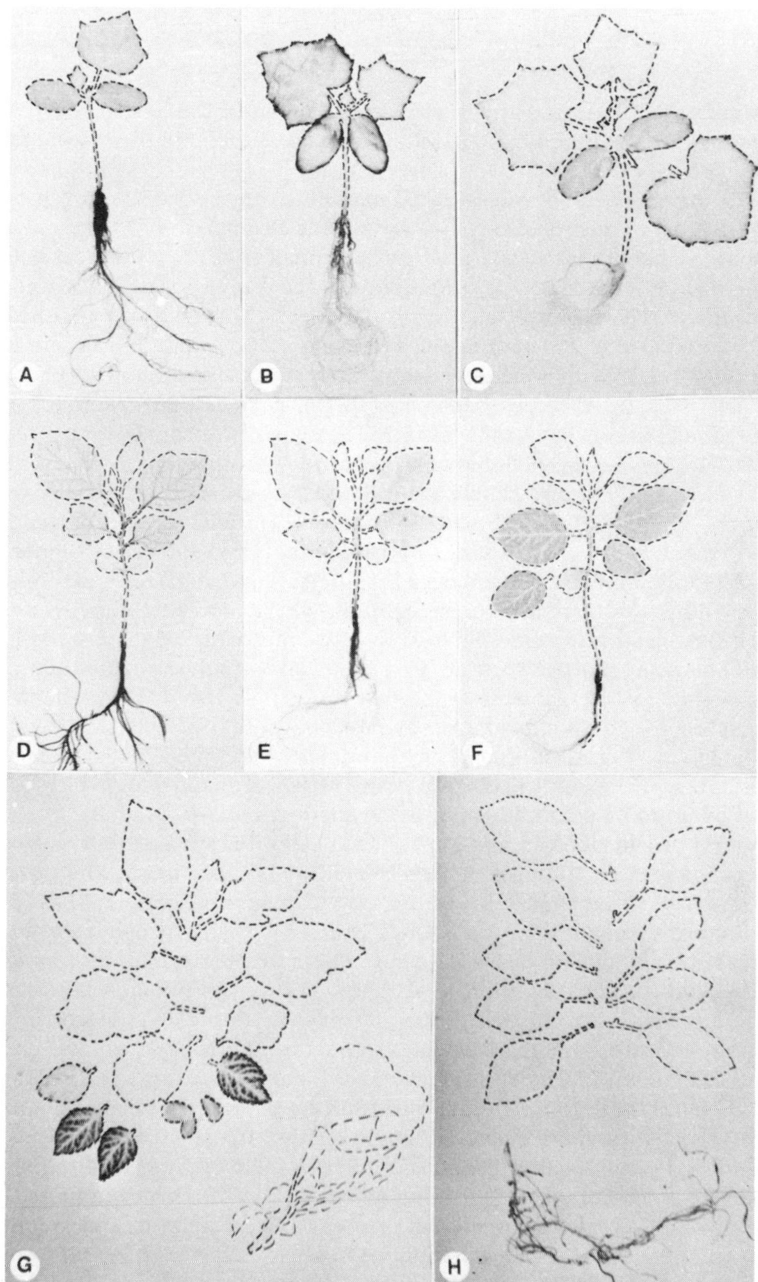

FIG. 5. Distribution of ^{14}C from a single root treatment of [^{14}C]ethirimol in cucumber and apple plants. The dotted line outlines the perimeter of the plant and the ^{14}C appears as darkened areas. A, B, and C are cucumber leaves at 2, 6, and 13 days, respectively, after treatment; D, E, F, and G are apple leaves at 2, 6, 13, and 27 days, respectively, after treatment of a 4-5 leaf apple seedling; H is an apple leaf 6 days after treatment of a 6-month-old seedling. Radioactivity indicates rapid movement and accumulation at leaf margins in cucumbers in 6 days while in young apple seedlings ^{14}C is retained in the vascular system, gradually moving out into the lamina (intercostal areas) at 27 days. In 6-month-old apple seedlings movement out of the roots is negligible at 6 days after root treatment.

much more slowly in the woody plant apple, than in the herbaceous plant cucumber (Fig. 5). It would be interesting to see if this difference is true for other fungicides.

Some differences between related fungicides in the rate of acropetal transport for a given plant species have been reported. For example, amongst the benzimidazoles, thiabendazole has generally been reported as slower moving than MBC [2]. Wang et al. [114] found that thiabendazole was apparently bound to plant constituents in cotton. However, it is difficult to interpret data on translocation per se because of degradation in plants as noted for Figure 5. The rate of degradation seems to vary with plants. For example, Noguchi et al. [29] observed differences in conversion of thiophanate-methyl between grape and French bean. Rouchaud et al. [33] found only 50% of benomyl remaining as MBC after a 2-months treatment of melons, whereas Siegel and Zabbia [113] found 78% as MBC after 52 days in dwarf pea plants. Ben-Aziz and Aharonson [116] found that thiabendazole was degraded three to four times as rapidly as MBC in pepper plants. Consequently, the studies of Rombouts [115] with a metabolically stable sulfone compound are exceptionally meaningful since there are no biases resulting from breakdown. He found the hydrophilic-lipophilic balance of the fungicides controlled their acropetal movement. The more hydrophilic, the more rapid was the translocation to leaf margins. Now that we know that many systemic pesticides are really pseudoapoplastic, this is a very logical phenomenon. The more lipophilic chemicals would partition into and remain in the living protoplasts instead of being freely transported.

Triforine is an example epitomizing the difficulty in studying acropetal transport. Drandarevski and Fuchs [117] reported the half-life as less than 10 days for tomato but up to 30 or 40 days for barley, bean, and pea. From the extensive research at Wye College by Carter et al. [118, 119] it appears the trichloroformamido group ($CCl_3CHNHCHO$) is the toxophore of triforine, whereas the remainder of the molecule has a dramatic influence on systemic activity against powdery mildew of wheat. Either an alkyloxy or alkylamino side-chain was necessary for systemicity. Whether these groups affected metabolic stability or the hydrophilic-lipophilic balance is not known.

In a very complete study of compounds related to oxathiins, which ten Haken and Dunn [120] call "*cis*-crotonanilides," no relationship could be established between the chemical's hydrophilic-lipophilic balance and its systemicity. The most interesting analogue reported was WL-24708 which is like pyracarbolid except for an ethyl group in the 6 position instead of a methyl group. This change did not cause a loss in fungitoxicity but did cause an apparent loss of systemicity in plants. Why is this so? The more recent research of White and Thorn [121] shows that the ethyl analogue is an even more potent inhibitor of the succinic acid dehydrogenase system than pyracarbolid itself. Did this change really prevent systemic movement in plants as concluded by ten Haken and Dunn or was the ethyl analogue more easily metabolized by the plant and therefore gave no systemic control of disease?

In several reports there are suggestions that translocation of systemic fungicides is influenced by diseased tissue. Upham and Delp [68] demonstrated high radioactivity in infected areas of bean plants treated with [^{14}C]benomyl as a soil drench. Rombouts [115] also noted selective movement to regions infected with powdery mildew. This may be construed as due to active uptake by fungi but is probably because of higher rates of transpiration by diseased tissue. It is certainly a fortuitous fact regardless of the driving force. Disease may also influence translocation negatively. Benomyl applied to apple leaves before leaf fall gave complete inhibition of perithecial formation and ascospore discharge [122] but spring application to fallen leaves only reduced discharge of ascospores by 93% [123]. This might be due to differences in spray coverage but it may also result from poorer diffusion of benomyl in dead tissue than in living tissue. A systemic fungicide is not necessarily any different than a protectant fungicide in its ability to diffuse in dead tissue.

B. Symplastic Transport

The development of systemic fungicides which show evidence of symplastic transport is now in its early stages. As discussed earlier, occasionally a very small amount of a pseudoapoplastic substance can be transported basipetally. Certain other chemicals also move basipetally to a limited extent. Kamimura et al. [124] found that 3-hydroxy-5-methylisoxazole (F-319) may show evidence of ambimobile transport. When [^{14}C]F-319 was applied to one rootlet of a cucumber plant, ^{14}C moved to another (untreated) one and was exuded into the medium. In order to display this pattern of movement, the chemical would first have to ascend in the transpiration stream, be transferred into the symplast, and be transported basipetally again. They also found that treatment of a cucumber leaf blade with [^{14}C]F-319 resulted in radioactivity being present in the petiole and stem below the leaf blade as well as in the upper parts of the plant. This distribution is explained by an initial basipetal transport in the phloem followed by transfer to the xylem. The authors conclude, "This distribution pattern may indicate that movement in the treated leaf was mainly in the apoplast and minimally in the symplast." Further work will be necessary to determine whether the translocated ^{14}C was in the form of F-319 itself or in one of its breakdown products, since they also showed that considerable breakdown of the chemical occurred within the time interval used in the transport studies.

The fungicide pyroxychlor, 2-chloro-6-methoxy-4-(trichloromethyl)pyridine, formerly known as Dowco 269, has been reported to control or delay the onset of diseases in the root when applied to the shoot [125-127]. Recently, Small [128] demonstrated that the fungicide itself is transported since he was able to extract it from the roots of tomato plants after a foliar treatment. After 24 hr, 14% of the translocated pyroxychlor was recovered from the roots. Pyroxychlor displays a much stronger symplastic movement than F-319. When applied to tomato roots, pyroxychlor was concentrated 50-fold by the roots but only 0.4% of the amount taken up was transported to the shoot. This behavior is typical of symplastic compounds such as 2,4-D. The mechanism leading to accumulation and retention of pyroxychlor by the plant symplast is not known at present.

Since it appears that one successful symplastic chemical has already been developed and more are likely to follow, it is of interest to discuss the probable transport behaviors of symplastic fungicides. Figure 2 shows that the fungicide would be expected to accumulate in certain locations within the plant; including the roots, newly developed foliage, flowers, and fruits. The distribution of the applied chemical to each sink area is dependent on the relative strengths of the sinks. Thus a foliar spray given at a time of rapid fruit growth would result in decreased transport to the roots. In the case of chemicals like the herbicide 2,4-D which are rapidly immobilized in the plant, it is necessary to apply the chemical when the sink areas of the plant are growing rapidly in order to obtain significant transport of chemical to the sink. The distribution of the applied chemical will also depend on the proximity of the treated area to the sink. The closer the source is to the sink, the greater is the proportion of the total chemical exported by that source to the sink. Consequently, timing of sprays will be a much more important consideration for symplastic than for pseudoapoplastic fungicides. Symplastic chemicals are transported away from mature leaves to which they are applied and therefore may fail to protect them. Hoitink and Schmitthenner [127] found that foliar applications of pyroxychlor were ineffective in controlling a foliar disease and state, "Possibly, Dowco 269 is not active for foliage diseases or is translocated to roots out of upper plant parts too rapidly to be effective. ... The volatile properties of Dowco 269 also may reduce its effectiveness as a foliar spray." Further work will be needed to determine whether this lack of protection of a sprayed leaf is a common occurrence for all strongly symplastic fungicides or whether pyroxychlor is an exception due to its volatility or its metabolic breakdown to nonfungicidal products in the leaf.

Several features of symplastic transport are useful in disease control. A chemical applied as a foliar spray is transported basipetally and, in many cases, application of a foliar spray is the most convenient treatment for root diseases. Distribution of the chemical to the new foliage means that new growth will be protected, a feature lacking for apoplastic fungicides unless they are continuously supplied to the roots. Symplastic transport of fungicides to flowers and fruits also means that these organs will be protected from disease. At the same time, fungicide accumulation in organs such as fruits and tubers may pose residue problems. Soil, root, or seed treatments will be less effective than foliar sprays since transport in the symplast is in the downward direction in the roots.

V. PREREQUISITES FOR SYSTEMICITY

Perhaps the term "systemic fungicide" has been maligned from an etymological viewpoint. We would define the term as "a fungicide which is taken into and transported within the plant." Few chemicals are entirely systemic moving throughout the plant "system." They are either moving to the transpiring organs or to metabolic sinks. Also, our present "systemic fungicides" are mostly fungistats. Some may act indirectly by altering the host metabolism and as such are neither systemic fungistats nor systemic fungicides but are systemic chemotherapeutants. What are the requisites that convey "systemicity" to a fungicide or fungistat? This is a complex question which does not have a simple answer.

Since 1965, which might be considered the year of genesis of synthetic systemic fungicides, industry has discovered systemic activity amongst structures of quite diverse chemical nature, e.g., oxathiins, thiazoles, benzimidazoles, pyridines, pyrimidines, triazoles, piperazines, and several nonaromatic types. Obviously, published information is handicapped because most information on structures which failed to prove systemic are in the record books of industry. You cannot test thousands of pyrimidines, as Shephard has reported [45], without developing an understanding of requisites for systemic control of plant disease. On the other hand, researchers in industry often know the structures which fail to control disease but do not, or cannot take the time to determine why the particular structure fails. Since failure to systemically control disease could be due to numerous factors such as photosensitivity, soil absorption, rapid metabolism by plants as well as lack of movement into and within the plant, the reason for the negative results is not always clear. Even though handicapped by lack of access to industry's data, we shall attempt to briefly outline in general terms the requisites for systemicity.

A. Penetration

First of all, the fungicide must gain entrance into the plant via the leaf, stem, root, or seed. The leaf cuticle obviously prevents many fungicides from entering and in fact most protective fungicides would be phytotoxic if they did traverse the cuticle. To accomplish penetration, the fungicide must be relatively lipid soluble. However, it must still be dissolved in a solvent (not particulate), and the most commonly used solvent is water. Since altering lipid solubility of a fungicide concomitantly alters its fungitoxicity, we should seek alternate means of penetrating through the cuticle. A logical consideration is to add alkyl or other lipophilic groups which can be autolytically, as in benomyl, and/or enzymatically cleaved, as in thiophanate-methyl [129], after traversing the cuticle. This choice also has serious deficiencies since a large percentage of systemic fungicides remain on plant leaf surfaces to act as a protectant. We cannot add lipophilic groups if they destroy fungitoxicity or we will sacrifice the dual role as a protective fungicide on the leaf surface. Fortunately, thiophanate-methyl on leaf surfaces is changed to MBC both by pH above 7 and by the ultraviolet in sunlight [130]. Thus this fungicide can function as a protectant and a systemic fungicide. Alternatively, rather than chemical structure, formulations of the fungicides as emulsifiable concentrates or oil dispersions has proved effective in enhancing uptake [67, 72, 131].

Herbaceous fruits and seeds have cuticles similar to those of the leaves and thus similar requisites. However, if the stem is woody unique formulations will be required. More research is needed to define this requisite.

The root also presents an important barrier to systemic movement. The fungicide must have water solubility to be readily available for uptake. On the other hand, roots with a suberized epidermis and/or hypodermis require some lipid solubility of the fungicide for penetration. Generally, most systemic fungicides are capable of movement whether applied to roots, stems, or leaves.

B. Movement Within Plants

An important barrier in the apoplast of the root is the Casparian band of the endodermis. As discussed earlier, our present systemic fungicides can passively diffuse through cell membranes and thus by-pass the Casparian band. As Verloop [96] showed several years ago, if metabolically stable chemicals are used a definite relationship exists between membrane permeability and oil-water partitioning. Often this relationship is obscured by metabolic instability of the fungicide within plant cells [96, 130] which we shall return to in a moment.

1. Xylem Transport

Having gained entrance into the plant, the fungicide must enter the xylem or phloem of the vascular system to be transported long distances. To be transported within the xylem, the fungicide must be actively secreted or passively eluted into the apoplast or xylem. Data on existing fungicides favor the latter process. The report of Shone and Wood [36] indicates an olive oil-water partitioning of 50 or less appears desirable for xylem transport.

2. Phloem Transport

If the fungicide is to be transported in the phloem, can we specify an oil-water partitioning coefficient? Not if we consider the fact that the pseudo-apoplastic fungicides can pass through the endodermal plasmalemma membrane. Then the same oil-water partitioning coefficient may be required for entrance into phloem cells. Thus it is not the penetration of the phloem plasmalemma but rather the accumulation against a concentration gradient or "loading" that is unique. The structural requisites for phloem loading are quite vague.

It appears that there are two possible approaches to the design of symplastic chemicals. One would be to formulate the chemical as a weak acid so that it would be taken up strongly by the symplast and retained there. This would lead to a predominantly symplastic distribution of the chemical in the plant. The other approach would be to formulate the chemical so that it would penetrate the plant membrane slowly and be retained in the symplast long enough to be translocated within that system. Because the chemical would leak out eventually, such a chemical would be ambimobile. It would seem that for overall protection of the plant, an ambimobile chemical would be the most useful type. Most problems of fungicide accumulation in specific areas of the plant, as well as their concomitant depletion from other areas, would be avoided. The tendency for accumulation in the roots due to transport of the chemical in the symplast would be negated by movement upward in the transpiration stream. Accumulation in young foliage would be negated by growth of the region and later by export of the fungicide when the leaves become mature. The tendency for accumulation in the margins and tips of leaves due to movement in the apoplast would be negated by export in the symplast. The exception to this rule would be in the case of fungicide accumulation in organs such as fruits due to symplastic transport. Since these organs are not serving as a water source for the plant, the fungicide would not be removed from them by the transpiration stream.

Fungicide movement within plants is also dependent on freedom from binding. Binding to xylem walls is obvious for large cationic fungicides, e.g., alkyl quaternary ammoniums [132, 133]. While the pyrimidine fungicides appear to be loosely bound in vascular tissues of woody plants (Figure 5), conclusions must be taken with caution. Since it appears pyrimidines as well as other systemic fungicides are really pseudoapoplastic, these fungicides could be retained in other cells such as woody parenchyma, adjacent to xylem vessels, instead of being bound by hydrogen bonding to lignin in xylem walls. Further research is needed to prove binding to lignin in xylem vessel walls.

C. Selective Toxicity

If we accept the pseudoapoplastic nature of current systemic fungicides, it follows that selective toxicity to the fungus as compared with the host plant must be required. This selective toxicity may vary with plant species. For example, carboxin is much more toxic to the succinic acid dehydrogenase system of the smut fungus, *Ustilago maydis* than to Pinto bean (*Phaseolus*) [134]. On the other hand, Mathre [134] also noted that

2,4-dimethylthiazole-5-carboxanilide (G 696), a related fungicide, was more toxic to bean than to smut. In practice, G 696 has not been developed, probably because of phytotoxicity. Mathre believes part of the selective action is because of differences in permeability of mitochondrial membranes.

Concomitant with selective toxicity between host plant and fungi has come selective toxicity between fungi. This selectivity was never very evident with protective fungicides but is very striking with systemic fungicides reaching an extreme with RH-124 (Indar) which controls only leaf rust of wheat caused by Puccinia recondita [110]. Perhaps we should develop fungicides selective for some process unique to fungi. In fact, we may have already empirically achieved this. Research with triarimol [135, 136], triforine [137], and possibly several other systemic fungicides [138] indicates interference with sterol synthesis and metabolism. These fungicides are especially toxic to obligate parasites such as powdery mildew and rusts which have haustoria. Possibly fungi utilize sterols in haustorial membranes in a process unique to fungi [139].

D. Metabolic Stability

Again, based on the pseudoapoplastic nature of systemic fungicides, the fungicide within a plant cell will be subject to degradation by many enzymes. To be successful, the fungicide must resist degradation. For example, as ten Haken and Dunn [120] concluded after examining many fungicides related to carboxin, "It is felt that the hitherto rare combination of systemic and fungicidal properties in the group of what may be called 'ciscrotonanilides' is related to their metabolic stability. ..."

In conclusion, the successful systemic fungicide (or fungistat) must have the chemical structure permitting entrance into the plant, movement through the plant, and most important, entrance into the host plant cells where it will selectively inhibit the pathogen without adversely affecting the host plant. Lastly, it must last long enough, without being degraded, to maintain a healthy plant.

REFERENCES

1. S. H. Crowdy, in Systemic Fungicides (R. Marsh, ed.), Wiley, New York, 1972, Chap. 5.
2. D. C. Erwin, Ann. Rev. Phytopathology, 11, 389 (1973).
3. E. Münch, Die Stoffbewegung in der Pflanze, Gustav Fischer, Jena, Germany, 1930, p. 73.
4. A. S. Crafts, Translocation in Plants, Holt, Rinehart and Winston, New York, 1961, p. 142.
5. S. Strugger and E. Peveling, Deut. Bot. Ges. Ber., 74, 300 (1961).
6. D. F. Gaff, D. C. Chambers, and K. Marcus, Australian J. Biol. Sci., 17, 581 (1964).

7. S. Strugger, Praktikum der Zell- und Gewebephysiologie der Pflanze, Springer, Berlin, 1949, p. 198.
8. M. Hülsbruch, in Encyclopedia of Plant Physiology, Vol. III (W. Ruhland, ed.), Springer, Berlin, 1956, p. 522.
9. P. E. Weatherley, in The Water Relations of Plants (A. J. Rutter and F. H. Whitehead, eds.), Blackwell, London, 1963.
10. M. T. Tyree, Can. J. Bot., 46, 317 (1968).
11. M. T. Tyree, J. Exptl. Bot., 20, 341 (1969).
12. T. W. Tanton and S. H. Crowdy, J. Exptl. Bot., 23, 600 (1972).
13. T. W. Tanton and S. H. Crowdy, J. Exptl. Bot., 23, 619 (1972).
14. J. S. Boyer, Planta (Berlin), 117, 187 (1974).
15. A. S. Crafts, Translocation in Plants, Holt, Rinehart and Winston, New York, 1961, p. 131.
16. A. W. Robards, Ann. Rev. Plant Physiol., 26, 13 (1975).
17. W. Schumacher, Jahrb. Wiss. Bot., 77, 685 (1933).
18. W. H. Arisz, Koninkl. Ned. Akad. Wetenschap. Proc. Ser. C, 59, 454 (1956).
19. W. H. Arisz, Protoplasma, 52, 309 (1960).
20. L. Littlefield and C. Forsberg, Physiol. Plantarum, 18, 291 (1965).
21. K. Esau, J. Cronshaw, and L. L. Hoefert, J. Cell Biol., 32, 71 (1967).
22. K. Esau, in Encyclopedia of Plant Anatomy (K. Linsbauer, founder), Part 2, Vol. 5, G. Borntraeger, Berlin, 1969.
23. A. S. Crafts and C. E. Crisp, Phloem Transport in Plants, Freeman, San Francisco, 1971.
24. F. M. Ashton and A. S. Crafts, Mode of Action of Herbicides, Wiley, New York, 1973.
25. O. Biddulph, S. Biddulph, R. Cory, and H. Koontz, Plant Physiol., 33, 293 (1958).
26. D. J. Galley and L. A. Foerster, Proc. 7th Brit. Insectic. Fungic. Conf., Brighton, England, 1973, Vol. 1, p. 171.
27. G. von Schruft, Pflanzenkrankh., 5, 280 (1971).
28. K. G. M. Skene, J. Hort. Sci., 47, 179 (1972).
29. T. Noguchi, K. Ohkuma, and S. Kosaka, Herbicides, Fungicides, Formulation Chemistry, in Proc. 2nd Intern. IUPAC Congr. Pestic. Chem., Tel Aviv, Israel, 1972, Vol. 5, p. 263.
30. Z. Solel, J. M. Schooley, and L. V. Edgington, Pestic. Sci., 4, 713 (1973).
31. B. D. Cavell, R. J. Hemingway, and G. Teal, Proc. 7th Brit. Insectic. Fungic. Conf., Brighton, England, 1973, Vol. 2, p. 431.
32. G. D. Thorn, Pestic. Biochem. Physiol., 3, 137 (1973).
33. J. P. Rouchaud, J. R. Decallonne, and J. A. Meyer, Phytopathology, 64, 1513 (1974).
34. G. L. Lamoureux, R. H. Shimabukuro, H. R. Swanson, and D. S. Frear, J. Agr. Food Chem., 18, 81 (1970).

35. T. W. Donaldson, D. E. Bayer, and O. A. Leonard, Plant Physiol., 52, 638 (1973).
36. M. G. T. Shone and A. V. Wood, J. Exptl. Bot., 25, 390 (1974).
37. B. T. Kirk, J. B. Sinclair, and E. N. Lambremont, Phytopathology, 59, 1473 (1969).
38. Z. Solel, Phytopathology, 60, 1186 (1970).
39. C. Intrieri and K. Ryugo, Plant Disease Reptr., 56, 590 (1972).
40. H. Buchenauer, D. C. Erwin, and N. T. Keen, Phytopathology, 63, 85 (1973).
41. M. A. Boulware and N. D. Camper, Weed Sci., 21, 145 (1973).
42. F. M. Ashton and A. S. Crafts, Mode of Action of Herbicides, Wiley, New York, 1973, p. 37.
43. G. Zweig and E. Greenberg, Biochem. Biophys. Acta, 79, 226 (1964).
44. M. G. T. Shone, B. O. Bartlett, and A. V. Wood, J. Exptl. Bot., 25, 401 (1974).
45. M. C. Shephard, Proc. 7th Brit. Insectic. Fungic. Conf., Brighton, England, 1973, Vol. 3, p. 841.
46. C. A. Peterson and L. V. Edgington, Pestic. Sci., 7, 483 (1976).
47. D. C. Dybing and H. B. Currier, Plant Physiol., 36, 169 (1961).
48. R. Collander, in Plant Physiology (F. C. Steward, ed.), Vol. 2, Academic Press, New York, 1959, Chap. 1.
49. E. Pickering, Ph.D. Thesis, University of California, Davis, 1965.
50. C. E. Crisp, Insecticides, in Proc. 2nd Intern. IUPAC Congr. Pestic. Chem., Tel Aviv, Israel, 1972, Vol. 1, p. 211.
51. C. R. Swanson and J. R. Baur, Weed Sci., 17, 311 (1969).
52. F. Jacob, S. Neumann, and U. Strobel, Proc. Res. Inst. Pomology, Skierniewice, Poland, 1973, Ser. E, p. 315.
53. J. W. Mitchell, I. R. Schneider, and H. G. Gauch, Science, 131, 1863 (1960).
54. C. E. Price, Proc. 7th Brit. Insectic. Fungic. Conf., Brighton, England, 1973, Vol. 1, p. 161.
55. S. H. Crowdy, Proc. 7th Brit. Insectic. Fungic. Conf., Brighton, England, 1973, Vol. 3, p. 831.
56. R. F. Norris and M. J. Bukovac, Pestic. Sci., 3, 705 (1972).
57. M. P. Sharma and W. H. Vanden Born, Weed Sci., 18, 57 (1970).
58. C. A. Drandarevski and E. Mayer, Mededel. Rijksfak. Landbouwwetenschap. Gent, 39, 1127 (1974).
59. R. F. Norris, Amer. J. Bot., 61, 74 (1974).
60. Z. Solel and L. V. Edgington, Phytopathology, 63, 505 (1973).
61. R. F. Norris and M. J. Bukovac, Amer. J. Bot., 55, 975 (1968).
62. R. Prasad, C. L. Foy, and A. S. Crafts, Weeds, 15, 149 (1967).
63. R. F. Norris and M. J. Bukovac, Physiol. Plantarum, 22, 701 (1969).
64. D. W. Greene and M. J. Bukovac, J. Amer. Soc. Hort. Sci., 96, 240 (1971).
65. D. W. Greene and M. J. Bukovac, Amer. J. Bot., 61, 100 (1974).

66. L. W. Small, private communication, 1975.
67. L. Edgington, H. Buchenauer, and F. Grossmann, Pestic. Sci., 4, 747 (1973).
68. P. M. Upham and C. J. Delp, Phytopathology, 63, 814 (1973).
69. C. J. Delp, private communication, 1975.
70. H. Buchenauer and D. C. Erwin, Phytopathology, 61, 433 (1971).
71. D. C. Erwin, R. A. Khan, and H. Buchenauer, Phytopathology, 64, 485 (1974).
72. A. I. Zaki, D. C. Erwin, and S. Tsai, Phytopathology, 64, 490 (1974).
73. T. Wicks, Plant Disease Reptr., 57, 560 (1973).
74. C. J. Hearn and J. F. L. Childs, Plant Disease Reptr., 53, 203 (1969).
75. K. Esau, Plant Anatomy, 2nd ed., Wiley, New York, 1967, p. 338.
76. D. M. Norris, Jr., Bull. Entomol. Soc. Amer., 11, 187 (1965).
77. E. B. Smalley, C. J. Meyers, R. N. Johnson, B. C. Fluke, and R. Vieau, Phytopathology, 63, 1239 (1973).
78. D. M. Norris, Jr., Ann. Rev. Entomol., 12, 127 (1967).
79. D. C. Ramsdell and J. M. Ogawa, Phytopathology, 63, 830 (1973).
80. C. A. Peterson and L. V. Edgington, Proc. 36th Can. Phytopathol. Soc., p. 27, 1970.
81. H. C. Coppel and D. M. Norris, Jr., J. Econ. Entomol., 59, 928 (1966).
82. H. E. Thompson, Bull. Entomol. Soc. Amer., 11, 198 (1965).
83. S. Sumida, R. Yoshihara, and J. Miyamoto, Agr. Biol. Chem., 37, 2781 (1973).
84. C. A. Peterson and L. V. Edgington, Phytopathology, 65, 1254 (1975).
85. M. Snel and L. V. Edgington, Phytopathology, 60, 1708 (1970).
86. J. H. Priestley and E. E. North, New Phytologist, XXI, 113 (1922).
87. H. Mager, Planta (Berlin), 19, 534 (1933).
88. C. A. Peterson and L. V. Edgington, in Systemic Fungicides (H. Lyr and C. Polter, eds.), Akademie-Verlag, Berlin, 1975, p. 287.
89. J. de Rufz de Lavison, Rev. Gen. Bot., 22, 225 (1910).
90. A. W. Robards and M. E. Robb, Science, 178, 980 (1972).
91. G. Nagahashi, W. W. Thomson, and R. T. Leonard, Science, 183, 670 (1974).
92. I. Karas and M. E. McCully, Protoplasma, 77, 243 (1973).
93. E. B. Dumbroff and D. R. Peirson, Can. J. Bot., 49, 35 (1971).
94. A. S. Crafts and S. Yamaguchi, Amer. J. Bot., 47, 248 (1960).
95. D. T. Clarkson, A. W. Robards, and J. Sanderson, Planta (Berlin), 96, 292 (1971).
96. A. Verloop, Phillips Tech. Rev., 28, 93 (1967).
97. S. H. Crowdy, J. F. Grove, and P. McCloskey, Biochem. J., 72, 241 (1959).
98. A. S. Crafts, Translocation in Plants, Holt, Rinehart and Winston, New York, 1961, p. 72.

99. A. Fuchs, G. A. van den Berg and L. C. Davidse, Pestic. Biochem. Physiol., 2, 191 (1972).
100. A. Fuchs, D. L. Fernandes, and F. W. de Vries, Neth. J. Plant Pathol., 80, 7 (1974).
101. H. Buchenauer, D. C. Erwin, and N. T. Keen, Phytopathology, 63, 1091 (1973).
102. R. B. Maude, A. S. Vizor, and C. G. Shuring, Ann. Appl. Biol., 64, 245 (1969).
103. B. von Schmeling and M. Kulka, Science, 152, 659 (1966).
104. P. N. Thapliyal and J. B. Sinclair, Phytopathology, 60, 1373 (1970).
105. L. T. Richardson, Plant Disease Reptr., 57, 3 (1973).
106. D. E. Briggs, R. H. Waring, and M. Hackett, Pestic. Sci., 5, 599 (1974).
107. K. Tao, A. A. Khan, G. E. Harman, and C. J. Eckenrode, J. Amer. Soc. Hort. Sci., 99, 217 (1974).
108. I. J. Graham-Bryce, Proc. 7th Brit. Insectic. Fungic. Conf., Brighton, England, 1973, Vol. 3, p. 921.
109. L. V. Edgington, E. Reinbergs, and M. C. Shephard, Can. J. Plant Sci., 52, 693 (1972).
110. J. B. Rowell, Nematicide Fungicide Tests, 28, 203 (1972).
111. C. A. Peterson and L. V. Edgington, Phytopathology, 60, 475 (1970).
112. C. A. Peterson and L. V. Edgington, Phytopathology, 61, 91 (1971).
113. M. R. Siegel and A. J. Zabbia, Jr., Phytopathology, 62, 630 (1972).
114. M. C. Wang, D. C. Erwin, J. J. Sims, N. T. Keen, and D. E. Borum, Pestic. Biochem. Physiol., 1, 188 (1971).
115. J. E. Rombouts, Mededel. Rijksfac. Landbouwwetenschap. Gent, 36, 63 (1971).
116. A. Ben-Aziz and N. Aharonson, Pestic. Biochem. Physiol., 4, 120 (1975).
117. C. A. Drandarevski and A. Fuchs, Mededel. Rijksfac. Landbouwwetenschap. Gent, 38, 1525 (1973).
118. G. A. Carter, L. A. Summers, and R. L. Wain, Ann. Appl. Biol., 70, 233 (1972).
119. G. A. Carter, K. Chamberlain, and R. L. Wain, Ann. Appl. Biol., 75, 49 (1973).
120. P. ten Haken and C. L. Dunn, Proc. 6th Brit. Insectic. Fungic. Conf., Brighton, England, 1973, Vol. 2, p. 453.
121. G. A. White and G. D. Thorn, Pestic. Biochem. Physiol., in press (1975).
122. S. R. Connor and J. W. Heuberger, Plant Disease Reptr., 52, 654 (1968).
123. P. M. Miller, Plant Disease Reptr., 54, 27 (1970).
124. S. Kamimura, M. Nishikawa, H. Saeki, and Y. Takahi, Phytopathology, 64, 1273 (1974).
125. J. F. Knauss, Plant Disease Reptr., 58, 1100 (1974).

126. R. L. Noveroske, Phytopathology, 65, 22 (1975).
127. H. A. J. Hoitink and A. F. Schmitthenner, Phytopathology, 65, 69 (1975).
128. L. W. Small, Proc. Amer. Phytopathol. Soc., 1974, Vol. 1, p. 144.
129. A. Matta and I. A. Gentile, Mededel. Rijksfac. Landbouwwetenschap., Gent, 36, 1151 (1971).
130. H. Buchenauer, L. V. Edgington, and F. Grossmann, Pestic. Sci., 4, 343 (1973).
131. M. Y. Chin, L. V. Edgington, G. C. A. Bruin, and E. Reinbergs, Can. J. Plant Sci., 55, 911 (1975).
132. L. V. Edgington and A. E. Dimond, Phytopathology, 54, 1193 (1964).
133. M. Salerno and L. V. Edgington, Phytopathology, 53, 605 (1963).
134. D. E. Mathre, Pestic. Biochem. Physiol., 1, 216 (1971).
135. N. N. Ragsdale, Biochim. Biophys. Acta, 380, 81 (1975).
136. N. N. Ragsdale and H. D. Sisler, Pestic. Biochem. Physiol., 3, 20 (1973).
137. J. L. Sherald, N. N. Ragsdale, and H. D. Sisler, Pestic. Sci., 4, 719 (1973).
138. T. Kato, S. Tanaka, M. Ufda, and Y. Kawase, Agr. Biol. Chem., 39, 169 (1975).
139. K. Schluter and H. C. Weltzien, Meded. Fak. Landbouwwetenschap., Gent, 36, 1159 (1971).
140. C. A. Peterson, P. P. Q. DeWildt, and L. V. Edgington, Pestic. Biochem. Physiol., in press (1977).

Chapter 3

BIOLOGICAL CONVERSION OF FUNGICIDES IN PLANTS AND MICROORGANISMS

A. Kaars Sijpesteijn, H. M. Dekhuijzen*, and J. W. Vonk

Institute for Organic Chemistry TNO
Central Organization for Applied Scientific Research
in the Netherlands
Utrecht, The Netherlands

*Present address: Center for Agrobiological Research, Wageningen, The Netherlands.

I. INTRODUCTION

This chapter intends to evaluate the biological breakdown of fungicides by plants and microorganisms. Previous reviews have covered portions of this subject [1-7].

Since 1967, many new fungicides have been introduced and a variety of investigations have been carried out on their persistence in plants, microbial cultures, soil, and animals. Fungicides introduced before that time have now been submitted to such studies as a consequence of the new, more stringent regulations for the protection of food and the environment. In this review, only the limited studies available in the literature can be discussed. Considerably more knowledge is undoubtedly present in reports of industrial laboratories to various authorities. The data we will present show how poor and fragmentary our picture still is of a subject which deserves far more attention and study. The data to be presented also demonstrate how far removed we are from the ideal situation where plant-protecting chemicals, after having done their job, should be completely degraded by chemical or biochemical processes.

Purely chemical breakdown will be treated in Chapter 5 but such processes obviously cannot always be separated from biological conversions. Sometimes the first stages of degradation take place spontaneously in water as occurs in the breakdown of the bisdithiocarbamates. In this case the resulting products do, however, again undergo microbial transformation. Conversely, thiram will be subject to biological reduction in soil, and the resulting dialkyldithiocarbamic acid under acidic condition splits spontaneously into dimethylamine and CS_2; the former product will undoubtedly serve again as a substrate for microorganisms.

For unknown ages all organic matter was of natural origin and has been subject to rapid and complete remineralization by microorganisms after life had ceased. However, there were also exceptions, especially under strictly anaerobic conditions or at extreme pH values or very low temperatures of the environment. Under such extreme conditions products like oil, coal, and methane accumulated and not only materials like wood, but even whole animals were conserved.

Heterotrophic microorganisms that contain a wide variety of enzymes capable of decomposing this natural organic matter are especially equipped for providing themselves with the energy and organic molecules necessary for growth and reproduction. Plants, being autotrophic organisms, are less well equipped for such a task.

How great are the problems that foreign organic molecules, synthesized by man, impose upon microorganisms to solve? Clearly these unnatural compounds can only undergo metabolic breakdown if the specificity of the microbial enzymes involved is wide enough to accept such compounds as a substrate. In plants, a lack of uptake by the tissues is also prohibitive for further breakdown of many compounds. Sometimes, however, not the parent compound but a metabolite is taken up by plants. Microorganisms are less selective in this respect.

Certain foreign compounds can serve so adequately as a source of carbon and energy for microorganisms that growth and multiplication will occur. For fungicides this has so far been shown to happen in a few cases, as, for instance, the growth of a species of Beyerinckia on biphenyl, of a species of Pseudomonas on pentachlorophenol, and of a species of Nocardia on carboxin as sole sources of carbon. Such odd substrates will probably only be utilized in nature in case of shortage of natural substrates as often happens in soil. If, moreover, a sufficient nitrogen source is available, microbial development may take place, thus accelerating the rate of breakdown of the substrate involved.

Next to this type of utilization of a foreign compound as a source of carbon for growth, we can distinguish other transformations which proceed only with the oxidation of a normal substrate. This type of microbial degradation is called "cometabolism." The definition given by Horvath [8] is, "... any oxidation of substances without utilization of the energy derived from the oxidation to support microbial growth." Transformations like the oxidation of a sulfide to a sulfoxide or an amino group to a nitro group belong to this category. The phenomenon of cometabolism appears to occur widely in microbial metabolism.

The rate of the cometabolic type of breakdown, in contrast to the former type where the fungicide serves as sole source of carbon for growth, is speeded up by the addition of natural oxidizable substrate. By the principle of cometabolism many foreign compounds first thought to be recalcitrant to microbial attack were shown later to be degradable to a certain extent.

As we will see, the conversion in plants and by microorganisms of the same compound often follow a different pattern. Yet, on the whole, all biological conversions fungicides undergo can be roughly grouped as oxidations, reductions, hydrolysis, and conjugate formation.

Examples of oxidation are the oxidative demethylation of chloroneb, hydroxylation reactions of mebenil and other compounds, the formation of a sulfoxide and a sulfone from carboxin, as well as of an N-oxide from tridemorph.

Reduction of aromatic $-NO_2$ to $-NH_2$ is well-known, but the formation of benzene from phenylmercuric acetate is also regarded as a reduction process.

Hydrolysis takes place with carboxin, pyrazophos, edifenphos (Hinosan), and other products.

Conjugate formation is encountered in the formation of a glucoside from the demethylated product of chloroneb, the formation of an anisole from pentachlorophenol, of an alanine derivative and a glucoside from dimethyldithiocarbamate and of a thiomethyl derivative from pentachloronitrobenzene (PCNB). In case of conjugate formation, under certain circumstances a reverse reaction may again give rise to the parent fungicide.

The survey of the literature presented in this chapter will show that the changes taking place are often only minor, and that in fact the skeleton, especially of aromatics and heterocyclics, remains largely intact even over prolonged periods. This would stress the future need for better degradable fungicides.

Even when minor changes do occur, there is nearly always a considerable loss of fungitoxicity. The conversion of dexon into a diamine and of carboxin into an active sulfone may seem to be rare exceptions in this respect. Certain microbial transformations in soil may even be reversible depending on environmental conditions. Examples are the reduction of PCNB and dichloronitroaniline (DCNA). Whereas certain fungicides like hexachlorobenzene (HCB) are completely resistant to breakdown, other compounds produce low-molecular breakdown products. For example, CO_2 or glycine can be assimilated into natural constituents of plants or microbes. Thus the label of [^{14}C]DCNA turns up in various fractions of treated plants.

The rate of breakdown is often expressed as half-life value of the parent compound. It should be noted that half-life values give a very poor indication of the time required for the total disappearance of a compound, since the shape of the disappearance curve may be very different for different compounds. Moreover, in plants the rate will depend on the species, the part of the plant studied, its age, the site of application, environmental conditions, and the concentration of the compound. In the case of microorganisms, pH value, temperature, availability of oxygen, and of sources of carbon and nitrogen will also be of importance.

For a qualitative and quantitative study of the products of transformation, the use of appropriately labeled fungicide is in most cases imperative. A quantitative balance sheet of the end product can then be made but so far this has been done for only a few compounds.

It is of much interest to know the pathway of breakdown of fungicides along with the various intermediates formed, as well as the enzyme systems involved. This knowledge is, however, still very scanty.

The following survey will present the current knowledge on the breakdown of agricultural fungicides. The compounds are classified according to structural relationship. The literature received up to November 1, 1976 has been covered.

II. DITHIOCARBAMATES

Dithiocarbamate fungicides are extensively used in agriculture [9] and their metabolism has been studied in detail. It is appropriate to distinguish clearly between two groups of compounds which, although related, show very different chemical properties [9, 10].

A. Dialkyldithiocarbamates

This group comprises the dialkyldithiocarbamates proper and their oxidation product, tetramethylthiuram disulfide. The soluble sodium salt of dimethyldithiocarbamate (sodium-DDC) (1) is often used in laboratory studies, but the insoluble iron salt (ferbam) and zinc salt (ziram) find practical application. Investigations on the fate of ziram and ferbam are exceedingly scarce

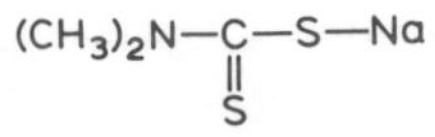

$$(CH_3)_2N{-}\underset{\underset{S}{\|}}{C}{-}S{-}Na$$

(1) sodium-DDC

but their metabolites are not expected to differ from those of sodium-DDC. Tetramethylthiuram disulfide, thiram (2), is also used as a fungicide.

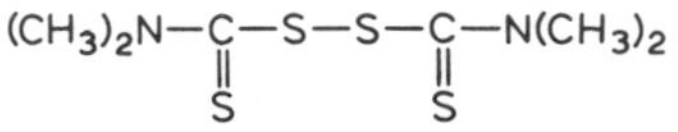

$$(CH_3)_2N{-}\underset{\underset{S}{\|}}{C}{-}S{-}S{-}\underset{\underset{S}{\|}}{C}{-}N(CH_3)_2$$

(2) thiram

Thiram can be reduced to dimethyldithiocarbamate. At low pH dimethyldithiocarbamic acid is formed which gives dimethylamine and carbon disulfide:

$$(CH_3)_2N{-}\underset{\underset{S}{\|}}{C}{-}SH \longrightarrow (CH_3)_2NH + CS_2$$

1. Metabolism in Plants

Dimethyldithiocarbamates are slightly systemic. When taken up by plants, they are rapidly converted into three fungitoxic conjugates. One of these has been characterized as the glucoside of dimethyldithiocarbamic acid, DDC-glucoside (3) [1, 2, 11, 12].

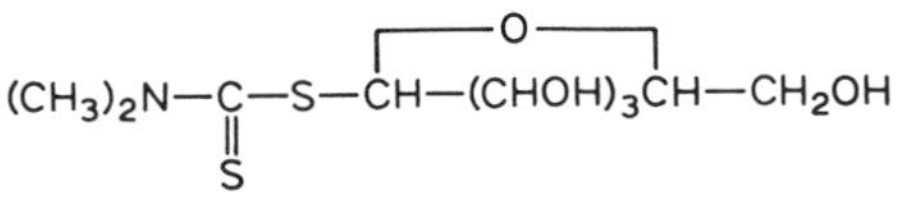

$$(CH_3)_2N{-}\underset{\underset{S}{\|}}{C}{-}S{-}\overbrace{CH{-}(CHOH)_3CH}^{O}{-}CH_2OH$$

(3) DDC-glucoside

The second fungitoxic conjugate appeared to be the L-alanine derivative of dimethyldithiocarbamic acid, DDC-alanine (4) [13], whereas the nature of the third compound remained unknown [11]. It is, however, certain that it is also a DDC conjugate [1, 2, 11-15]. A fourth, nonfungitoxic compound formed nonenzymatically from DDC-alanine by loss of dimethylamine appeared to be thiazolidine-2-thione-4-carboxylic acid, TTCA (5) [11, 16].

$(CH_3)_2N-\underset{\underset{S}{\|}}{C}-S-CH_2-\overset{H}{\underset{NH_2}{C}}-COOH$

(4) DDC-alanine

(5) TTCA

Massaux applied [^{35}S]thiram to begonia and cucumber leaves and detected the same fungitoxic compounds [17]. No other compounds were found. Hylin and Chin [18], however, studied the fate of [^{35}S] and [^{14}C]dimethyldithiocarbamate after application to Carica papaya and found four unidentified compounds, in addition to DDC-glucoside, DDC-alanine, carbon disulfide, and dimethylamine.

When sodium-DDC was fed to the roots of tomato plants, a fungitoxic compound with an R_f value equal to that of DDC-alanine was found in tomato-stem exudate [19], whereas various fungitoxic substances could be detected in sap expressed from leaves of tobacco plants treated with sodium-DDC [20].

Demethylation, which often occurs with synthetic compounds, does not take place with DDC-alanine or DDC-glucoside.

Most of the reactions in plants and those in microorganisms are interconvertible, as shown in Figure 1 [16]; thiram undergoes exactly the same reactions after reduction to DDC-ions [21].

Although the degradation products of dimethyldithiocarbamates in plants are now well known, there are still a number of unsolved problems. Dekhuijzen [16, 21] tried to make a complete balance sheet for the amount of DDC-alanine taken up by roots of cucumber plants and the total amount of DDC-alanine and conversion products present in the plants after various periods of time; DDC-alanine disappeared slowly during 3 weeks, whereas the amounts of DDC-glucoside and of TTCA increased. After 3 weeks, 90%

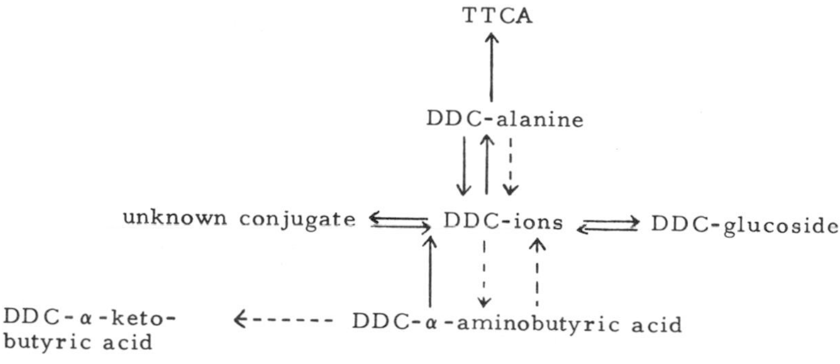

FIG. 1. Pathways of the conversion of DDC ions and their transformation products in higher plants and fungi. The solid lines indicate transformations occurring in higher plants, the dotted lines those occurring in fungi.

of the DDC-alanine taken up by the plant had disappeared, whereas only 25% could be recovered as the original compound plus that of the conversion products and TTCA. The identity of the nonrecovered part is unknown but there may have been decomposition into volatile carbon disulfide or hydrogen sulfide as well as unknown decomposition products. Breakdown products of DDC-alanine may also have been incorporated into proteins. This is indicated by the work of Hylin and Chin [18] who could recover an appreciable portion of ^{35}S from dimethyldithiocarbamate in the sulfur amino acids. Thorn and Richardson [22] showed complex formation between dithiocarbamate, copper ions, and protein. Moreover, Raghu et al. [23] found that sulfur from [^{35}S]thiram found its way into sulfur-containing compounds such as proteins in corn plants.

2. Metabolism by Microorganisms

It is interesting that fungi and bacteria do not convert sodium-DDC into DDC-alanine but into DDC-α-aminobutyric acid (6) and its corresponding α-keto acid, DDC-α-ketobutyric acid (7) [1, 11, 24-26].

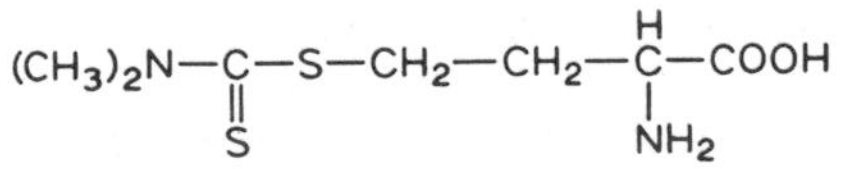

$$(CH_3)_2N-\underset{\underset{S}{\|}}{C}-S-CH_2-CH_2-\overset{H}{\underset{NH_2}{C}}-COOH$$

(6) DDC-α-aminobutyric acid

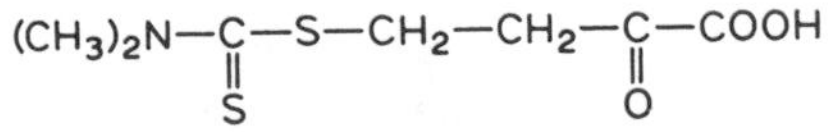

$$(CH_3)_2N-\underset{\underset{S}{\|}}{C}-S-CH_2-CH_2-\underset{\underset{O}{\|}}{C}-COOH$$

(7) DDC-α-ketobutyric acid

This was found for heavy suspensions of bakers yeast, several species of filamentous fungi, and bacteria in the presence of a source of carbon. The proportion of the amino acid conjugate increases with increasing amounts of a nitrogen source, inorganic or organic. It has been suggested [2] that the α-aminobutyric acid conjugate is formed by the enzyme system which normally couples cysteine and O-succinyl homoserine or O-acetyl homoserine to give rise to cystathionine. Similarly, the formation of DDC-alanine in plants may result from the reaction of the DDC-ion with activated L-serine.

Kaars Sijpesteijn and Vonk [26] speculate that thiram is degraded spontaneously into CS_2 and dimethylamine under slightly acid conditions. Sisler and Cox [27] showed the evolution of CS_2 by Fusarium spores treated with thiram. Pseudomonas sp. isolated from thiram-enriched soil was able to break down thiram with the evolution of CS_2 and the formation of dimethylamine [23]. Recently, Ayanaba et al. [28] reported the formation of dimethylamine in thiram-treated soils. It is not known whether these reactions are spontaneous, following acid formation by the microorganisms or whether they are also enzymatically enhanced. Ayanaba et al. [28] reported that under flooded conditions the carcinogenic compound dimethylnitrosoamine was produced in a thiram-treated soil of pH 3.8 in the presence of nitrate or nitrite. They also showed the formation of dimethylnitrosoamine from dimethylamine and nitrite. Both compounds gradually disappeared from these soils.

B. Bisdithiocarbamates

This group comprises the ethylenebisdithiocarbamates nabam, zineb, maneb, mancozeb, and Metiram, as well as propineb which is a propylene derivative. They are used against a wide range of pathogens.

Chemically the bisdithiocarbamates are related to the dialkyldithiocarbamates, both being derived from an amine and CS_2. However, an essential difference stems from the fact that the former are derived from a primary amine and hence have a reactive hydrogen on the nitrogen which is lacking in the latter. This chemical difference is reflected in the difference in the biological behavior of the two groups [9, 29, 30]. They also differ in stability, the bisdithiocarbamates being much less stable than the dialkyldithiocarbamates.

1. Nabam

Nabam, disodium ethylenebis(dithiocarbamate), (8) may be regarded as the parent compound of the ethylenebisdithiocarbamates. It is not used in agricultural practice in view of its phytotoxicity and unreliability in the field.

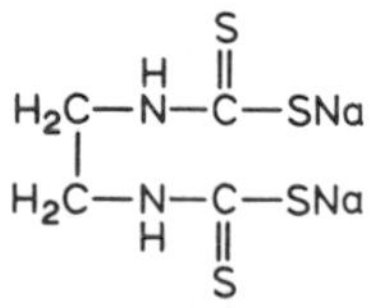

```
           S
      H    ||
H2C—N—C—SNa
  |
H2C—N—C—SNa
      H    ||
           S
```

(8) nabam

Dissolved in water, this compound is extremely unstable, giving the following products: ethylenethiourea, ETU (9); "ethylenethiuram monosulfide,"

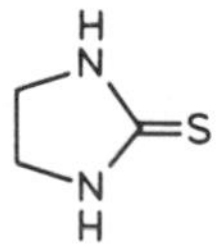

```
      H
    ┌─N
    |    \
    |     C=S
    |    /
    └─N
      H
```

(9) ETU

the structure of which was recently revised to 5,6-dihydro-3H-imidazo[2,1-c]-1,2,4-dithiazole-3-thione, DIDT (10) [31]; polymeric ethylenethiuram disulfide (11); ethylenediamine (12); ethylenediisothiocyanate (13); and elemental sulfur, CS_2 and H_2S [30, 32]. Vonk [33], working with very dilute solutions of [^{14}C]nabam, found only ETU and DIDT as main degradation products containing the ethylene moiety.

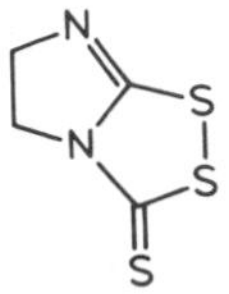

(10) DIDT

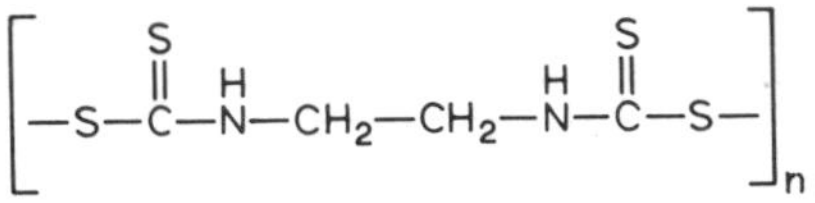

(11) polymeric ethylenethiuram disulfide

H_2C-NH_2
H_2C-NH_2

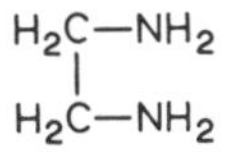

(12) ethylenediamine

$H_2C-N=C=S$
$H_2C-N=C=S$

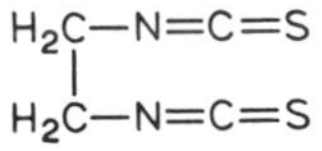

(13) ethylenediisothiocyanate

Cucumber plants which had been placed for 2 days with the roots in dilute [ethylene-^{14}C]nabam solution contained, immediately after the uptake period, mainly ETU [33, 34] and smaller amounts of 2-imidazoline (14), ethyleneurea (15), and unidentified material [33] (see Fig. 2). The ETU

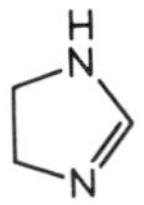

(14) 2-imidazoline

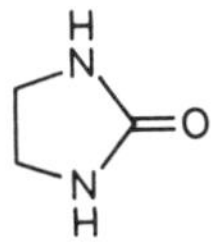

(15) ethyleneurea

found in these plants was formed in the nabam solution and selectively taken up, and converted rapidly in the plants into 2-imidazoline and ethyleneurea,

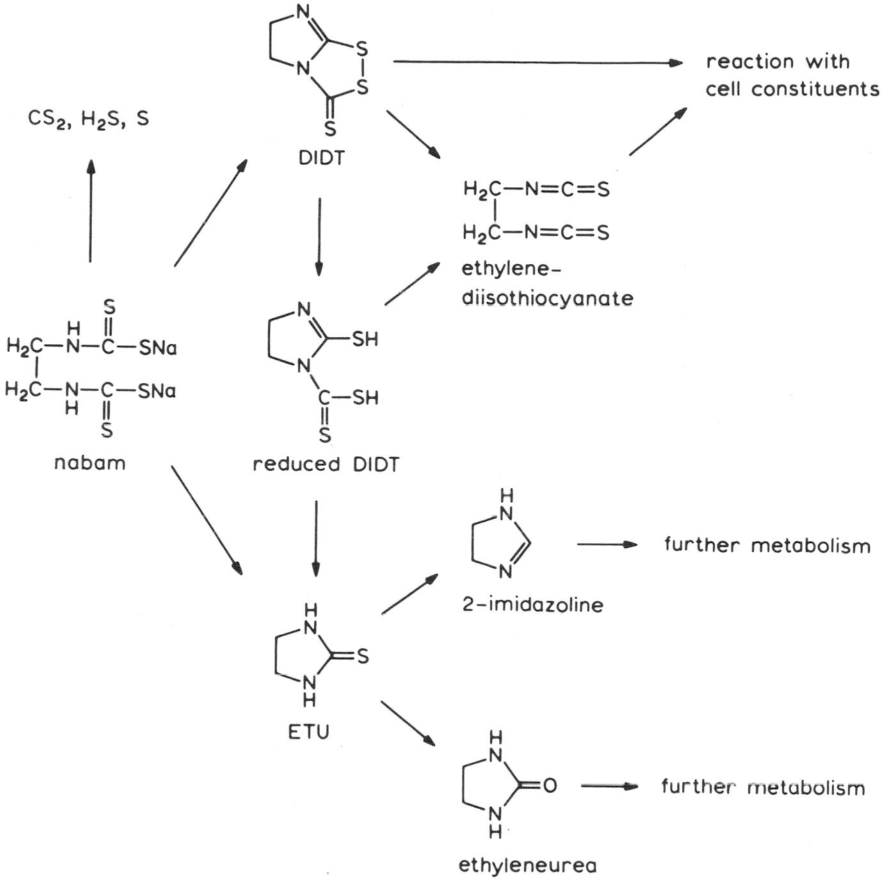

FIG. 2. Transformations involved in metabolism of nabam by plants and microorganisms.

which were degraded further. The unidentified, mostly polar, material may partly have resulted from reaction of nabam or DIDT with plant constituents. Of the radioactivity taken up, 19% had disappeared after 19 days, presumably as expired CO_2. The distribution of radioactivity among different plant parts and compounds is shown in Figure 3.

2-Imidazoline and ethyleneurea were also detected as metabolites in wheat plants and cucumber plants after root treatment with ETU [33, 35].

After leaf application of [^{14}C]nabam to cucumber plants the following compounds were present after 2 weeks: ETU, DIDT, 2-imidazoline, a small amount of ethyleneurea, and unidentified material [33, 36].

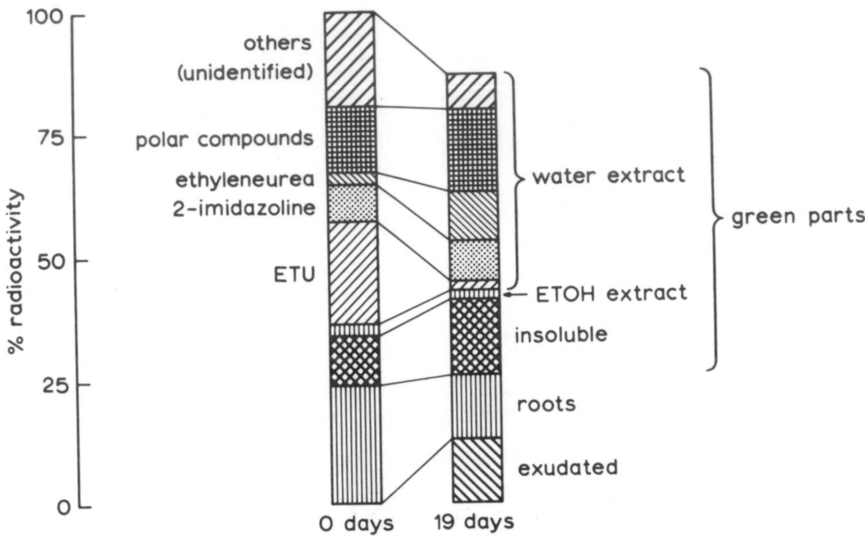

FIG. 3. Block diagram of the distribution of radioactivity among the different parts, fractions, and compounds in cucumber seedlings, 0 and 19 days after root treatment with [^{14}C]nabam.

Lyman and Lacoste [37] applied [^{14}C]ETU to the soil around tomato and potato plants. Small amounts of label were found in leaves, but only negligible residues of ETU were found in the edible parts.

Work with [^{14}C]ETU showed that ethyleneurea was a major metabolite of ETU in corn, lettuce, pepper, and tomato plants [229]. Only minor amounts of $^{14}CO_2$ were evolved.

Formation of 2-imidazoline and ethyleneurea from ETU may also occur photochemically in the presence of sensitizers [33, 38, 39].

In soil ETU and DIDT are formed from nabam [26] in addition to CS_2, H_2S, and COS [40, 41]. Ethylene diamine, a possible intermediate, is subject to very rapid microbial degradation [42]. Ethylenethiourea itself is also degraded in soil [37, 43, 44], 200 ppm being converted into ethyleneurea within 8 days. A slower but steady conversion of ETU occurred in autoclaved soil. In nonsterile, but not in autoclaved soil, further degradation into CO_2 occurred. Other metabolites of ETU in soil were hydantoin (16) and 1-(2'-imidazolin-2'-yl)2-imidazolidinethione, Jaffé's base (17) [43].

(16) hydantoin

(17) Jaffé's base

Interestingly, ETU appeared resistant to attack by pure cultures of bacteria and fungi irrespective of the presence of a source of carbon and nitrogen and by soil-enrichment cultures [26, 33, 43]. The DIDT was converted into ETU by bacteria and fungi, as well as by crude enzyme preparations from such organisms. The mechanism of the enzymic conversion was studied. The DIDT was reduced by NADH in the presence of flavins with subsequent release of CS_2 to form ETU [33] (see Fig. 2).

2. Zineb and Maneb

Because the zinc and manganous ethylenebisdithiocarbamates (18, 19) are less soluble in water than nabam, the rate of decomposition of these bisdithiocarbamates is slower than that of the latter compound. In an aqueous

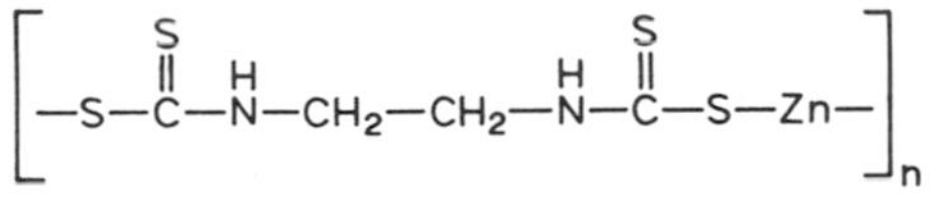

$$\left[-S-\overset{\overset{S}{\|}}{C}-\overset{H}{N}-CH_2-CH_2-\overset{H}{N}-\overset{\overset{S}{\|}}{C}-S-Zn-\right]_n$$

(18) zineb

$$\left[-S-\overset{\overset{S}{\|}}{C}-\overset{H}{N}-CH_2-CH_2-\overset{H}{N}-\overset{\overset{S}{\|}}{C}-S-Mn-\right]_n$$

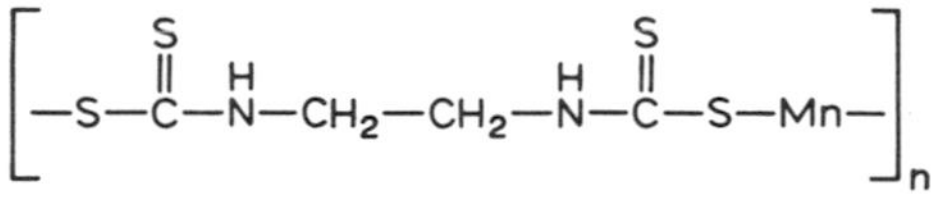

(19) maneb

medium the same decomposition products as reported for nabam were also obtained from maneb and zineb [32]. Formulations of these fungicides stored at elevated temperature and high humidity show an increase in ETU content [45]. Maneb is somewhat less stable than zineb [46].

Only a few data are available on the fate of zineb and maneb on plants. Sato and Tomizawa [47] found ETU and an unidentified compound in cucumber plants after leaf application of [^{35}S]zineb. DIDT was found on plants treated with zineb [48, 49]. Ethylenediisothiocyanate, ETU, DIDT, and sulfur were observed on fruit of tomato plants treated with zineb and maneb [50]. The detection of the highly reactive compound ethylenediisothiocyanate as a residue may be due to its formation from DIDT during the extraction with nonpolar solvents [33].

The presence of ETU on lettuce and kale treated with zineb and maneb, respectively, has been demonstrated [33, 51]. Residues of ETU were highest directly after the treatment and disappeared rapidly. They were probably due to the occurrence of ETU in the formulations.

The formation of metabolites from [^{14}C]zineb after application to the leaves of lettuce plants was studied by Vonk [230] and the following conversion products were found: ETU, DIDT, ethyleneurea, 2-imidazoline, and an unknown compound. A fairly high percentage of the radioactivity present after 21 days could not be identified.

Newsome [231, 232] determined residues of DIDT, ETU, ethylenediisothiocyanate, and ethylenediamine on tomatoes after application of zineb and maneb. Generally only small amounts were detectable 14 days after treatment. On beans higher amounts of ETU, DIDT, and ethylenediamine were found [232].

After soil treatment with [^{14}C]nabam, zineb, or maneb radioactivity was readily absorbed by soybean plants and translocated throughout the plant. Ethyleneurea was a major degradation product, but a number of unknown metabolites were formed either in soil or in plants [233].

3. Propineb, zinc 1,2-propylenebis(dithiocarbamate), Antracol

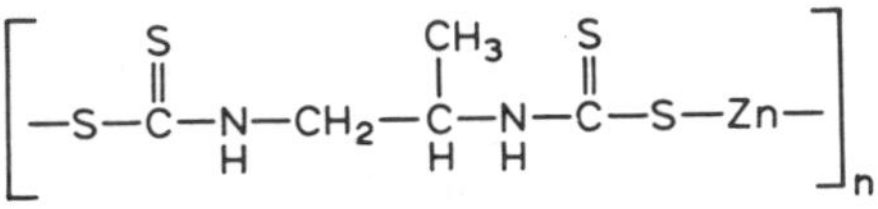

$$\left[-S-\overset{S}{\overset{\|}{C}}-\underset{H}{N}-CH_2-\underset{H}{\overset{CH_3}{\overset{|}{C}}}-\underset{H}{N}-\overset{S}{\overset{\|}{C}}-S-Zn- \right]_n$$

(<u>20</u>) propineb

After application of [^{14}C]propineb to apples and grapes, a gradually increasing amount of radioactivity was found in the apple pulp and grape juice. The breakdown of propineb starts with the formation of oligomers [52].

4. Mancozeb

Mancozeb (Dithane M-45) is the coordination product of zinc ion and manganous ethylenebis(dithiocarbamate) [53]. Formulations of mancozeb are more stable on storage than those of zineb and maneb [45]. Degradation of mancozeb slurries in water yielded zinc and manganese ions, thiosulfate ion, ETU, DIDT, and sulfur [53].

Metabolism in plants was studied with ^{3}H-, ^{35}S-, and ^{14}C-labeled material [53]. The following metabolites were detected after 2 weeks: ethyleneurea, ethylene diamine, DIDT, 2-imidazoline, ETU, 1-(2'-imidazolin-2'-yl)-2-imidazolidinethione, elemental sulfur, and sulfate ion. Potato plants sprayed in the field with [^{14}C]mancozeb contained only very small amounts of ETU and ethyleneurea in the tubers. A substantial part of the label was present in starch. It has been suggested that mancozeb is ultimately degraded into glycine via ethylene diamine [53].

Newsome [231] detected residues of DIDT, ETU, and ethylenediisothiocyanate on tomatoes of plants treated with mancozeb. Mancozeb is degraded in soil [44], where it gives rise to the evolution of CO_2 [37].

5. Metiram, Polyram

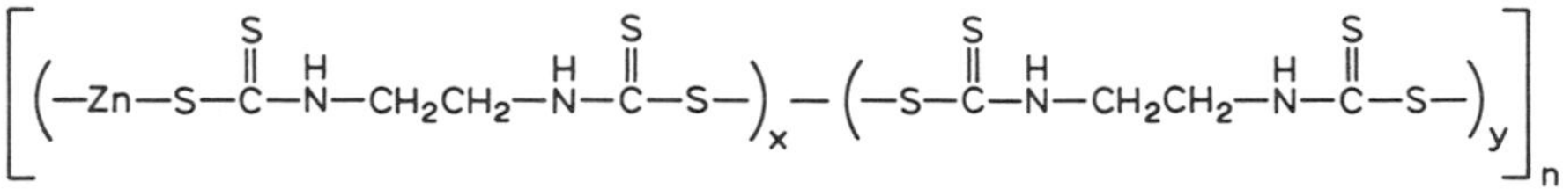

$$\left[\left(-Zn-S-\overset{S}{\overset{\|}{C}}-\overset{H}{N}-CH_2CH_2-\overset{H}{N}-\overset{S}{\overset{\|}{C}}-S-\right)_x-\left(-S-\overset{S}{\overset{\|}{C}}-\overset{H}{N}-CH_2CH_2-\overset{H}{N}-\overset{S}{\overset{\|}{C}}-S-\right)_y\right]_n$$

(21) metiram

Very little information is available on this compound, which does not accumulate in orchard soil [54]. After treatment of tomato plants, residues of DIDT and ETU were detected on the fruits [231].

Several authors have commented on the formation of ETU from bis(dithiocarbamate) residues during food processing [231, 234]. This subject is, however, beyond the scope of this chapter.

III. TRICHLOROMETHYLSULFENYL FUNGICIDES

Captan, N-(trichloromethylthio)tetrahydrophthalimide (22) and the closely related analogs folpet, N-(trichloromethylthio)phthalimide (23) and captafol, N-(1,1,2,2-tetrachloroethylthio)tetrahydrophthalimide (24) are widely used as protectant fungicides although some systemic activity has been reported [25].

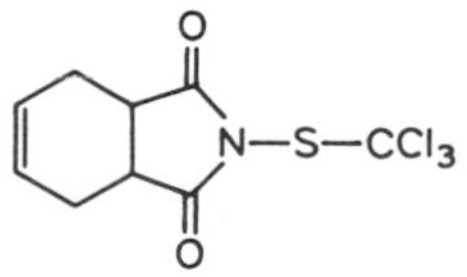

N—S—CCl_3

(22) captan

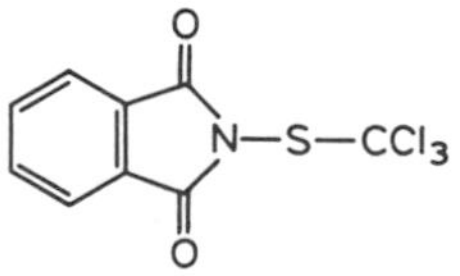

N—S—CCl_3

(23) folpet

N—S—CCl_2—CCl_2H

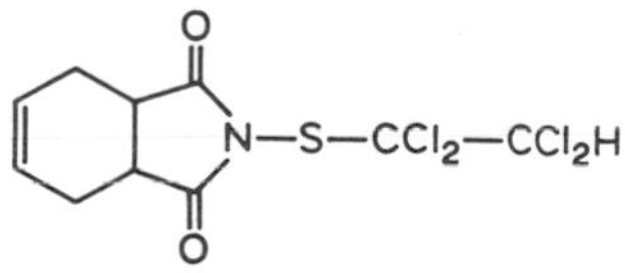

(24) captafol

Wallen and Hoffman [55] detected captan in pea seedlings after root treatment. Small amounts of [^{35}S]captan applied to the roots of broad bean plants were recovered in the leaves [56]. Captan decomposes rapidly in plants but it does not lose its fungicidal activity at the same rate, suggesting that a metabolite may be a fungicidal agent [55].

Captan is a biological alkylating agent which also reacts in vitro with thiols to yield a number of compounds, whose structure seems to indicate that the primary reaction is a transfer of the $SCCl_3$ moiety to the S atom of the thiol [57, 58]. When taken up by fungal spores, captan reacts with both soluble and insoluble thiols; the former reaction detoxifies the accumulated fungicide. Most of the label from [^{35}S]captan was recovered in oxidized glutathione (GSSG), suggesting exchange of sulfur, and in a product tentatively identified as a thiazolidine derivative of glutathione (25) [59, 60]. Another compound containing a thiazolidine ring was formed in vitro after reaction of captan with cysteine [61]. From these results [59, 61], it has been postulated that when $SCCl_3$-containing compounds react with thiols, the fungicide splits, yielding the free imide (see Fig. 4, reaction 1) and the cellular thiol which is oxidized to its disulfide, i.e., glutathione (see Fig. 4, reaction 2). The liberated trichloromethylthio $SCCl_3$ moiety can be transferred directly to thiol sites or undergo dechlorination to yield thiophosgene (see Fig. 4, reaction 3). Thiophosgene is highly reactive and can undergo rapid hydrolysis in aqueous solutions to yield carbonyl sulfide, hydrochloric acid, and hydrogen sulfide (see Fig. 4, reaction 4) or it may react with low- and high-molecular weight thiols and other groups to form various addition products (see Fig. 4, reaction 5) [60, 66].

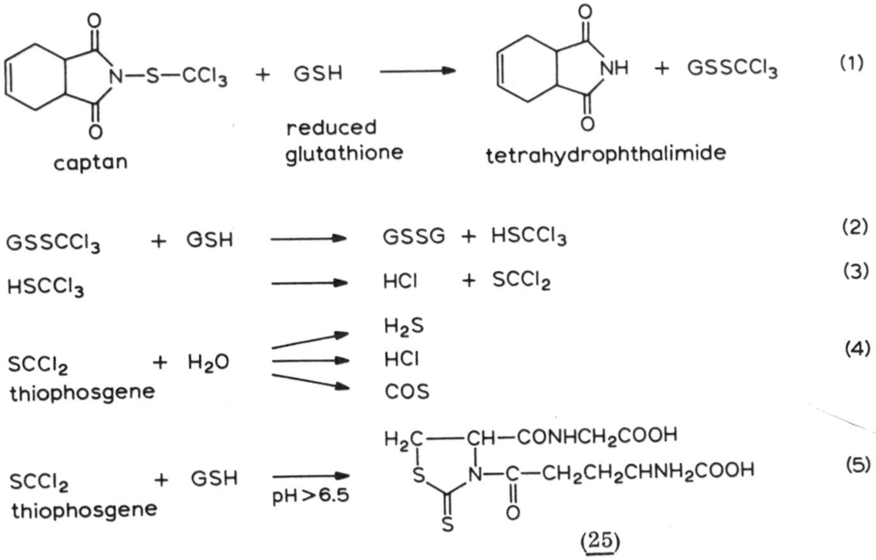

$$\text{captan (N-S-CCl}_3\text{)} + \underset{\text{reduced glutathione}}{\text{GSH}} \longrightarrow \underset{\text{tetrahydrophthalimide}}{\text{NH}} + \text{GSSCCl}_3 \qquad (1)$$

$$\text{GSSCCl}_3 + \text{GSH} \longrightarrow \text{GSSG} + \text{HSCCl}_3 \qquad (2)$$

$$\text{HSCCl}_3 \longrightarrow \text{HCl} + \text{SCCl}_2 \qquad (3)$$

$$\underset{\text{thiophosgene}}{\text{SCCl}_2} + \text{H}_2\text{O} \longrightarrow \text{H}_2\text{S},\ \text{HCl},\ \text{COS} \qquad (4)$$

$$\underset{\text{thiophosgene}}{\text{SCCl}_2} + \text{GSH} \xrightarrow[\text{pH}>6.5]{} \text{H}_2\text{C—CH—CONHCH}_2\text{COOH};\ \text{S—C(=S)—N—C(=O)—CH}_2\text{CH}_2\text{CHNH}_2\text{COOH}\ \ \underline{(25)} \qquad (5)$$

FIG. 4. Proposed pathway for the metabolism of captan in microorganisms.

Similar reactions have been reported to occur with folpet [62-66]. The direct transfer of the $SCCl_3$ moiety to thiol proteins in vitro was also demonstrated.

IV. BENZIMIDAZOLE DERIVATIVES INCLUDING THIOPHANATES

A. Benomyl and Carbendazim

Since the introduction of benomyl, methyl-(1-butylcarbamoyl)-benzimidazole-2-yl carbamate (26), this compound [67] has attracted much attention due to its excellent systemic properties in plants and broad antifungal spectrum.

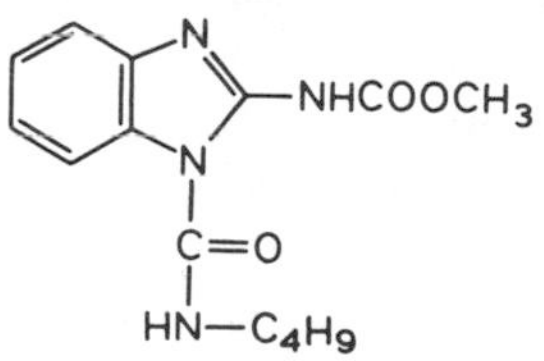

N
—NHCOOCH$_3$
N
|
C=O
|
HN—C$_4$H$_9$

(26) benomyl

In aqueous solution or in organic solvents benomyl rapidly loses the butylcarbamoyl group [68-70], giving methyl benzimidazole-2-yl carbamate, MBC (27). Thus the green parts of plants treated via the roots with benomyl

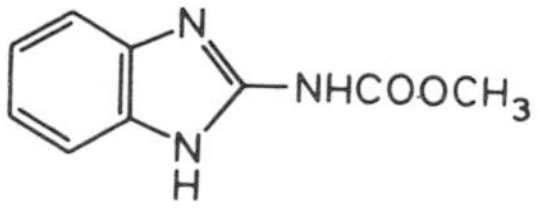

N
—NHCOOCH$_3$
N
H

(27) MBC (carbendazim)

contained only MBC and no benomyl [71-73]. In spray suspensions, however, benomyl remains largely intact and constitutes a major part of the spray deposit on various leaf surfaces for at least 3 weeks [74]. Generally, MBC is regarded as the active principle of benomyl. It is also used as a fungicide (carbendazim), but its systemic properties are inferior to those of benomyl after leaf application [75].

When benomyl is applied to roots of plants, MBC taken up in the plants is rather resistant to degradation [76, 77]. Immediately after root application of [2-^{14}C]benomyl to peas, almost all radioactivity could be extracted in the form of MBC [78]. After 52 days, 78% of the label was present as MBC and 3% in the metabolite 2-aminobenzimidazole (28).

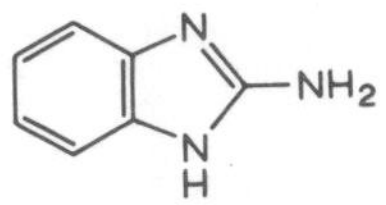

(28) 2-aminobenzimidazole

The remaining label was present in unidentified material found in the water-soluble extract or in the nonextractable plant residue. In strawberry plants, metabolism of MBC was more rapid than in peas. Considerable amounts of 2-aminobenzimidazole and unidentified compounds were present after 88 days of continuous feeding of [2-^{14}C]MBC to the roots [79]. Evidence was obtained that O- and N-conjugates were present in the water-soluble fraction and that a portion of the unidentified products were associated with hemicellulose. Since there was no indication of label in the products of intermediary metabolism, the benzimidazole nucleus is apparently highly stable. However, Solel et al. [80] found that during a period of 14 days 1.56% of ^{14}C-ring-labeled MBC applied to cucumber leaves was metabolized and respired as CO_2.

A large part of MBC taken up was loosely bound to debris of pepper and tomato plants; it could be solubilized with methanol-HCl [81]. No bound MBC was detected after treatment of leaves with this compound or with benomyl. The rate of disappearance of MBC was 13% per 10 days; small amounts of 2-aminobenzimidazole were detected [81].

After leaf application of ^{14}C-ring-labeled benomyl to several crops in the field, benomyl itself constituted a major component of the total residue for extended periods [74]. No residues other than MBC were detected on the leaves, whereas the total label gradually disappeared during a 3-week period. One week after application to bean plants, about 50% of the label could not be extracted.

Rouchaud et al. [235] detected two products resulting from cleavage of the benzimidazole nucleus, 2-aminobenzonitrile and aniline in melon plants after 2 months. Benzimidazole was found as well.

As might be expected from its behavior in aqueous solution, benomyl disappears rapidly in soil, giving rise to MBC [82, 83]. Carbendazim is much more resistant to degradation in soil where it is slowly converted into 2-aminobenzimidazole [82]. The half-life of the total benzimidazoles in benomyl-treated soils is 0.5-1 year, whereas total label of ^{14}C-ring-labeled benomyl disappeared after 1-2 years. Fleeker et al. [84] and Siegel [236] found that $^{14}CO_2$ was formed from [^{14}C]carbendazim. Carbon dioxide was released faster from methyl-labeled than from ring-labeled carbendazim, indicating faster splitting of the methylcarbamate group [84].

Type of soil, temperature, and sterilization influence the rate of disappearance of carbendazim [83]; thus it seems probable that microorganisms are involved in the breakdown.

Little is known about degradation of benomyl and carbendazim by microorganisms. Helweg [85] claimed to have isolated from benomyl-enriched soil some bacteria and fungi, which grew very slowly on benomyl as a sole source of carbon and nitrogen and which degraded this fungicide to nonfungitoxic compounds. An Achromobacter isolate was reported to degrade benomyl [86]. Mixed cultures of carbendazim-degrading bacteria were obtained by Fuchs on a medium with benomyl as a sole source of carbon [87]. Working with ring-labeled carbendazim, he observed the formation of 2-aminobenzimidazole and of an unknown compound. From the cultures, $^{14}CO_2$ evolved probably as a result of ring splitting of 2-aminobenzimidazole. Recently the formation of the 5-hydroxy derivative of carbendazim (29) by

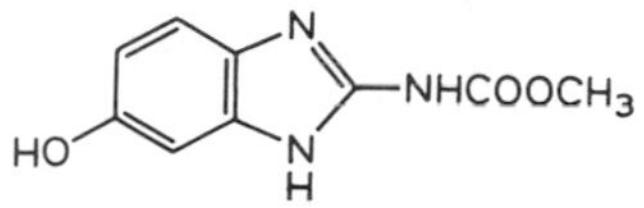

(29) 5-hydroxy MBC

fungi was observed independently by Japanese and Dutch workers [88, 89]. This compound was already known as a metabolite of benomyl in animals [90, 91].

B. Thiabendazole

The systemic fungicide thiabendazole (30), originally developed as an anthelmintic, has not obtained the same degree of practical application as benomyl, although it is almost as active against a broad spectrum of fungi [92, 93].

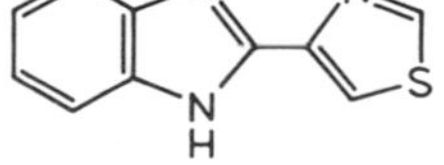

(30) thiabendazole

Erwin, Wang, and co-workers [94, 95] have studied the fate of ^{14}C-benzene-ring-labeled thiabendazole in cotton plants treated via the roots. After 3 days 60% of the label was present as the free parent compound; the remainder appeared to be bound in higher-molecular-weight compounds which have not yet been identified. They may be conjugates of thiabendazole metabolites [95]. In acetone extracts of soy bean plants treated with thiabendazole only the unaltered compound was found [96]. The rate of disappearance from pepper leaves was faster for thiabendazole than for carbendazim, i.e., 75% of the thiabendazole disappeared in 10 days [81].

On plants exposed to sunlight, small amounts of benzimidazole and benzimidazole-2-carboxamide formed from thiabendazole apparently as a consequence of photodecomposition [237].

Thiabendazole is also less persistent in soil than carbendazim. The fungistatic activity of 50 ppm in soil disappeared in 12 to 18 weeks [97]. Disappearance in a sandy loam was reported by Erwin et al. [98] and Aharonson and Kafkafi [238]. There is no information on the breakdown of this compound by pure cultures of microorganisms.

C. Fuberidazole

Fuberidazole (31) is used as a seed dressing; no information is available on breakdown by plants or microorganisms. According to Frank [99], the furan ring of fuberidazole is split open in certain animals to yield a γ-hydroxybutyric acid derivative, but it is unknown whether this reaction also takes place in plants or microbes.

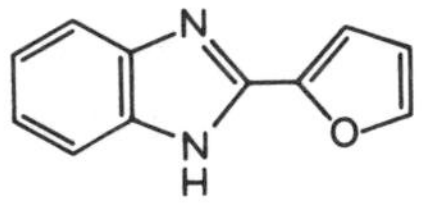

N
N
H
O

(31) fuberidazole

D. Cypendazole

Cypendazole, Bay Dam 18654 (32), has only recently been introduced as a systemic fungicide. Experiments reported by Kirkpatrick and Sinclair [100], show that it breaks down in water into MBC. This fact is not surprising in view of the structural similarity with benomyl.

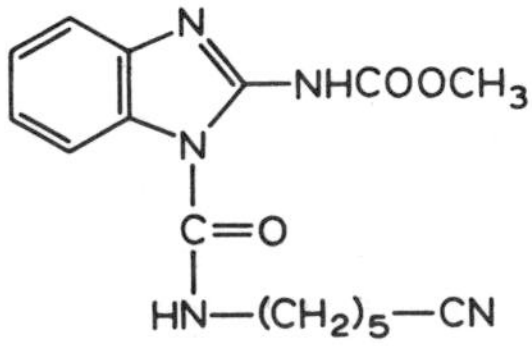

N
$-NHCOOCH_3$
N
C=O
$HN-(CH_2)_5-CN$

(32) cypendazole

E. Thiophanate-methyl and Thiophanate

Thiophanate-methyl (33) and thiophanate (34) do not belong structurally to the benzimidazoles. However, studies on their transformation and their antifungal spectrum have established that they act via the alkyl benzimidazole

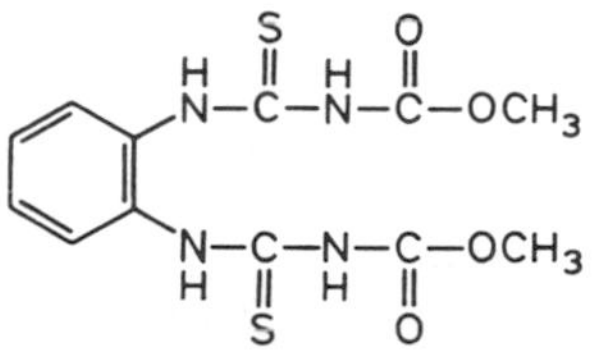

(33) thiophanate-methyl

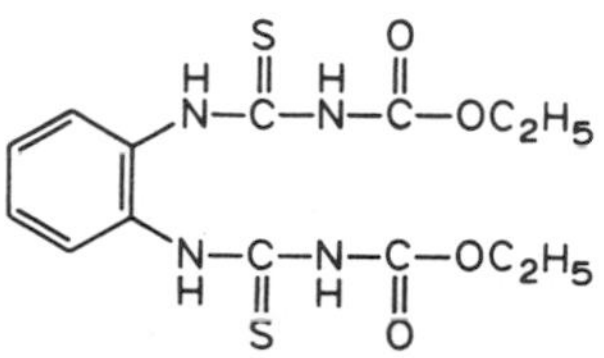

(34) thiophanate

carbamates, viz., MBC (carbendazim) and its ethyl analog, respectively. We will discuss here only the fate of thiophanate-methyl, that of thiophanate being essentially the same.

In water or sterile nutrient solution, thiophanate-methyl changes slowly to MBC [101, 102]. The rate of this conversion decreases considerably when the pH of the solution is lowered. Transformation of thiophanate-methyl into MBC occurred also upon irradiation of solutions by UV or sunlight [103] or on glass plates exposed to outdoor conditions [104]; the formation of small amounts of dimethyl 4,4'-o-phenylenebisallophanate (35) was reported [104].

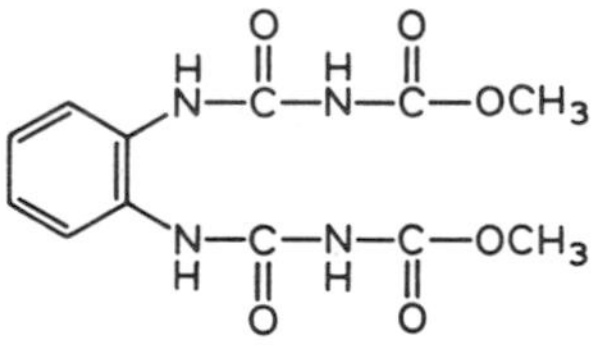

(35) dimethyl
4,4'-o-phenylenebisallophanate

The metabolism of thiophanate-methyl applied to leaves of apple and grape plants has been studied with the ^{14}C-ring-labeled compound [104]. The half-life of the parent compound was about 15 days. MBC [105] and smaller amounts of dimethyl-4,4'-o-phenylenebisallophanate were identified as metabolites. After 14 days about 10% of the applied activity was present as MBC. The formation of MBC from thiophanate-methyl on leaves is probably photochemical [103].

After root application of thiophanate-methyl to several plants, a rapid conversion to MBC occurred [73, 102, 106]. In cotton plants after 6 days, nearly all of the thiophanate-methyl was converted into MBC [106]. Apparently in plant tissues this transformation is more rapid than in water and catalyzing factors seem to be present. Active conversion of thiophanate-methyl into MBC by mycelium of several fungi was also reported [88, 101]. In this respect the observation that thiophanate-methyl is converted rapidly into MBC in the presence of slices of plant tissue [107], chloroform extracts [105], and homogenates of plants [108] is interesting. Recent work has revealed that quinones formed as intermediates upon oxidation of o-diphenols can catalyse the conversion [108].

Thiophanate-methyl is unstable is soil. Studies with variously ^{14}C- and ^{35}S-labeled compounds revealed that after 1 week only 1% of the parent compound was left [109]. MBC was the major metabolite. Its concentration increased at first but later declined. From the chromatographic data presented, it may be deduced that 2-aminobenzimidazole is also formed. These results were confirmed by Fleeker et al. [84]. The rate of conversion of thiophanate-methyl was four times faster in soil at pH 7.4 than in soil at pH 5.6. The rate was reduced by steam treatment [84]. This suggests that microbial activity might be a factor in these degradations. The degradation of the MBC formed occurred faster at higher temperature [109].

With regard to transformations of the thiophanates by pure cultures of microorganisms, little is known other than that several species of fungi can accelerate the conversion of thiophanate-methyl into MBC [88, 101].

F. Other MBC Generators

Besides cypendazole and the thiophanates, several compounds have been reported to lead to carbendazim, e.g., methyl 2,1,4-benzothiadiazin-3-yl carbamate (36) [110] and methyl quinoxaline-2-yl carbamate 1-oxide (37) [111].

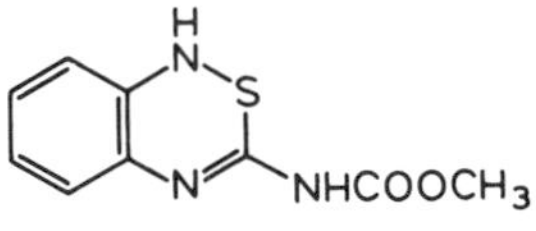

(36) methyl 2,1,4-benzothiadiazin-3-yl carbamate

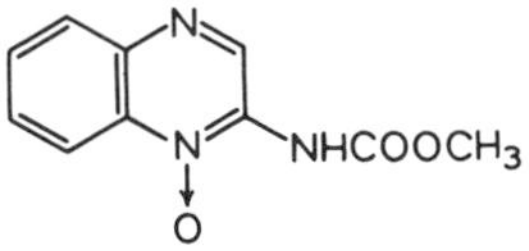

(37) methyl quinoxaline-2-yl carbamate 1-oxide

V. DINOCAP

Dinocap or karathane (38) is widely used as a fungicide against powdery mildews of apple and other plants. Little is known about the fate of this compound on plants, in soil, or in microbial cultures. It is not persistent

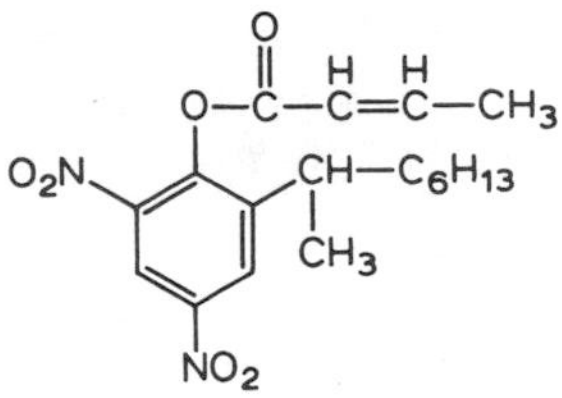

(38) dinocap

in or on cucumber plants where its residue in fruit decreased from 1.48 to 0.01 ppm in 5 days, and in leaves from 14.95 to 1.15 ppm in 7 days [112]. It is to be expected that microorganisms reduce the nitro groups to amino groups and hydrolyze the compound.

In the related herbicidal compound dinoseb (39), the isobutyl side chain is oxidized to isobutyric acid on apples [225]; reduction of the NO_2 group at the 2 position also appears to take place.

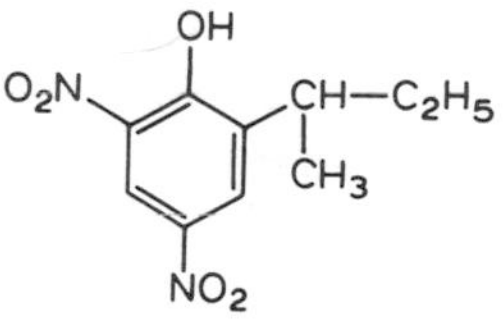

(39) dinoseb

VI. SUBSTITUTED BENZANILIDES

A. Oxathiins (Carboxin and Oxycarboxin)

1. Metabolism in Plants

Carboxin, 2,3-dihydro-6-methyl-5-phenylcarbamoyl-1,4-oxathiin (40), is a systemic fungicide which is used as a seed dressing to control smut infections of barley and wheat.

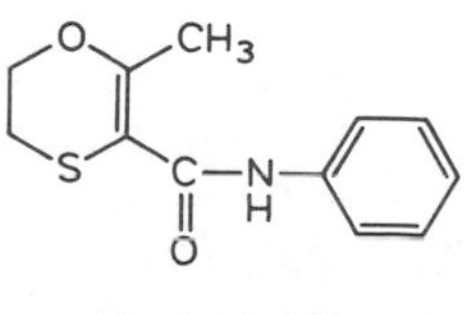

(40) carboxin,
Vitavax

The oxathiins are readily translocated and converted in plants. Several authors have observed the oxidation of carboxin into the nonfungitoxic sulfoxide (41) and the fungitoxic sulfone (42) [113-118]. The sulfoxide was

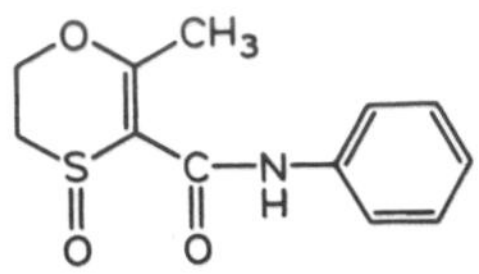

(41) carboxin sulfoxide

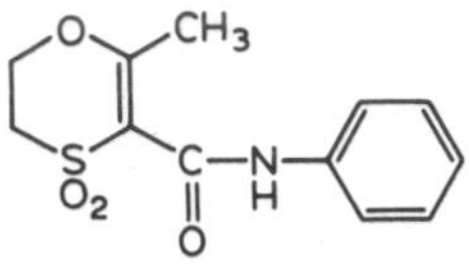

(42) oxycarboxin, Plantvax
carboxin sulfone

found to be the main product (90-92%) in acetone extracts of wheat plants treated with [^{14}C]carboxin. The acetone-soluble fraction contained about 8-10% of the sulfone, whereas only traces of carboxin could be detected [115]. Recently, however, the p-hydroxy phenyl derivative (43) was identified as the major methanol-soluble metabolite of carboxin in barley seedlings

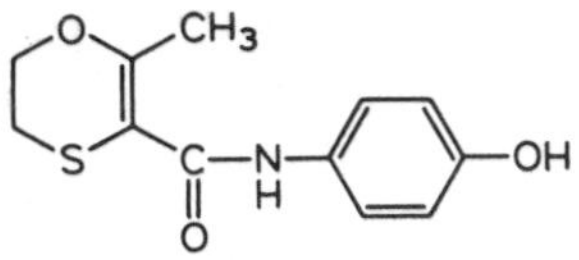

(43) p-hydroxycarboxin

and mature barley plants [118]. Only small quantities of sulfoxide could be detected. The authors doubt whether this latter product is the result of an enzymic reaction since they also found that carboxin is oxidized to its sulfoxide when allowed to stand in air at room temperature for a few days.

Hydrolysis of carboxin to aniline was found to occur in bean plants [119], but it was not detected in vivo in wheat or barley plants [115]. Yet, it seems likely that aniline may be formed as a result of hydrolytic cleavage of the carboxamide linkage of carboxin or of its oxidation products because an aryl-acyl amidase has been isolated from barley which hydrolyzed a number of carboxamide-containing compounds [120].

Time course studies over a period of 35 days showed an increasing amount of labeled chemicals in the acetone-insoluble fractions after feeding of [^{14}C]carboxin or oxycarboxin to bean plants, indicating a binding of aniline to plant polymers. This result agrees with the observation of Chin et al. [115] and Briggs et al. [118] that, as carboxin-treated wheat and barley plants approached maturity, the total percentage of extractable residues decreased and the percentage of nonextractable residues (anilide-lignin complexes) increased. Recently, it was found that upon hydrolysis the lignin fraction yielded p-hydroxycarboxin indicating a binding of this phenol to lignin [118].

2. Metabolism in Microorganisms

In soil, nonlabeled carboxin is rapidly oxidized to sulfoxides, but only traces of sulfone were detected. Hydrolysis of carboxin did not take place [121]. However, Wallnöfer and Engelhardt [122] described a hydrolytic degradation of carboxin by cultures of Bacillus sphaericus, whereas oxycarboxin was not attacked by this microorganism. Engelhardt et al. [123, 124] demonstrated the hydrolysis of carboxin by a crude enzyme preparation of B. sphaericus which resulted in the liberation of aniline. It seems likely that aniline is further broken down in soil because soil bacteria of the species Nocardia can grow on a medium with carboxin as the sole source of carbon and nitrogen [125].

Isolated mitochondria from sensitive as well as insensitive fungi were able to oxidize carboxin to the sulfoxide [126]. Further oxidation by isolated fungal mitochondria occurred only to a limited extent. Most probably flavin-containing enzymes are involved in the oxidation of carboxin.

The formation of a sulfoxide by fungi was also observed for methionine [127].

B. Mebenil, BAS 3191, and Pyracarbolid

Mebenil, 2-methylbenzanilide (44), and 2-chlorobenzanilide (45) are new systemic fungicides for control of oat smut (Ustilago) and Rhizoctonia [128]. Nothing is known about the fate of these compounds in higher plants, but it is known that soil fungi belonging to the Phycomycetes, which are not sensitive to these compounds, are active in transforming anilide fungicides. Rhizopus japonicus converts mebenil and 2-chlorobenzanilide into the corresponding p-hydroxybenzanilide derivatives (46, 47). The 2-methyl-4'-hydroxy derivative is less fungitoxic than the parent compound against Ustilago and Rhizoctonia solani [128, 129].

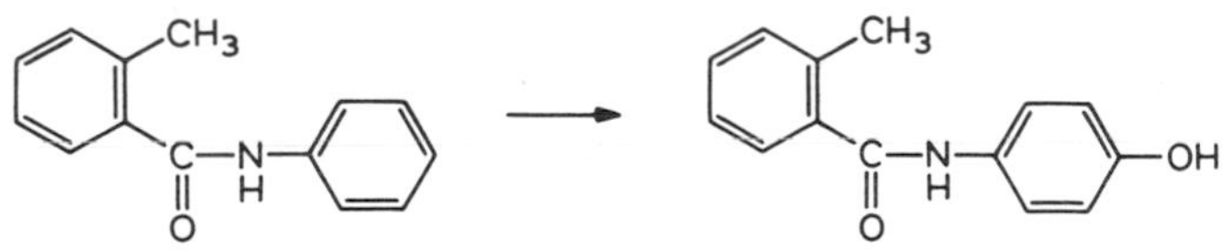

(44) mebenil

(46) 2-methyl-4-hydroxybenzanilide

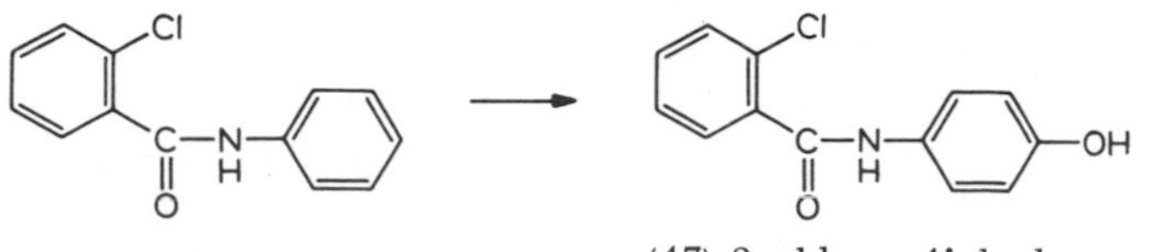

(45) 2-chlorobenzanilide

(47) 2-chloro-4'-hydroxybenzanilide

R. japonicus and two species of *Mucor* rapidly hydroxylated the methyl groups of BAS 3191, 2,5-dimethyl-3-furancarboxylic acid anilide (48) into the corresponding hydroxymethyl derivatives (49, 50) [130]. The metabolites were not further degraded. The mechanism of conversion of BAS 3191

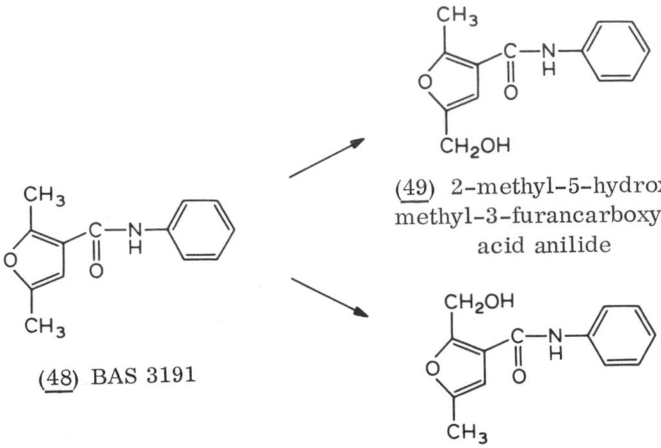

(48) BAS 3191

(49) 2-methyl-5-hydroxymethyl-3-furancarboxylic acid anilide

(50) 2-hydroxymethyl-5-methyl-3-furancarboxylic acid anilide

seems in contrast with the hydroxylation of the anilide group of mebenil by *R. japonicus* and it is rather surprising that such a conversion would not occur with BAS 3191. In this connection, it is worthwhile to mention phenyl ring hydroxylation and subsequent glucosidation in plants of monuron, a herbicide which contains an anilide group [131]. Moreover, microorganisms grew rapidly on urea herbicides carrying an anilide group. The side chain was hydrolyzed yielding the corresponding anilines [132].

Oeser et al. [239] reported that pyracarbolid is metabolized in plants to three compounds. One was tentatively identified as 2,6-dihydroxy-3-phenylaminocarbonylhex-2-ene, the result of ring splitting of the dihydropyran ring.

VII. ORGANIC PHOSPHATE ESTERS

A number of organic phosphorus fungicides have been developed and used in practice. Kitazin (*O*-*O*-diisopropyl *S*-benzylphosphorothioate), edifenphos (Hinosan, *O*-ethyl *S*,*S*-diphenylphosphorodithioate), and Inezin (*S*-benzyl *O*-ethyl phenylphosphonothioate) are all active against rice diseases, while pyrazophos (Curamil, Hoe 2873, *O*-*O*-diethyl *O*-(6-ethoxycarbonyl-5-methyl-pyrazole [1, 5-α]pyrimid-2-yl) phosphorothioate) is active against *Pyricularia oryzae* and mildew diseases [133-135].

A. Pyrazophos

Pyrazophos (51) is converted into the oxygen analog PO-pyrazophos (52), by *Pyricularia oryzae* [136], a type of reaction which is well known for organophosphorus insecticides. The oxygen analog is hydrolyzed to 2-hydroxy-5-methyl-6-ethoxycarbonylpyrazolo-[1, 5-α]pyrimidine (PP) (53) and water-soluble, nonidentified metabolites. Both known metabolites are toxic to a number of fungi. It has been suggested that the breakdown might be due to activities of microsomal mixed-function oxidases and hydrolytic enzymes.

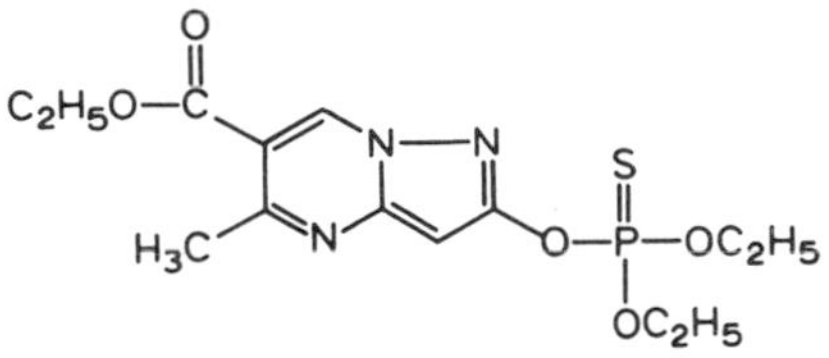

(51) pyrazophos

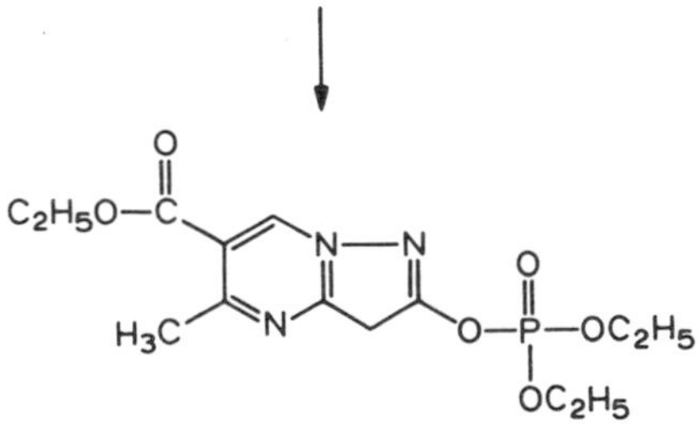

(52) PO-pyrazophos

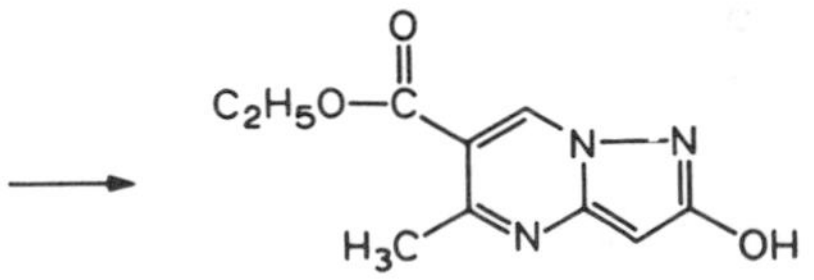

(53) PP

In cucumber the β-glucoside of PP was detected as a metabolite on the leaves. Possibly, PO-pyrazophos was present in small amounts in wheat leaves between the 4th and 10th day after treatment [240].

B. Kitazin P

[^{32}P, ^{35}S]Kitazin P (54) is easily absorbed through roots and leaf sheath of rice plants and translocated to other parts. The compound is converted into a number of polar and nonpolar metabolites [137, 138]. The main metabolite is O-O-diisopropyl hydrogen phosphorothioate which is further degraded yielding finally phosphoric acid (see Fig. 5).

The sequence in which the different compounds are formed is not exactly known, but it has been suggested that O,O-diisopropyl hydrogen phosphorothioate is desulfurated into diisopropyl phosphate, which is further hydrolyzed into monoisopropyl dihydrogen phosphate. S-Methylation of O-O-diisopropyl hydrogen phosphorothioate in rice plants and soil has also been demonstrated.

Gas chromatographic analysis showed also the occurrence of a cleavage of the P-S bond of Kitazin P yielding finally dibenzyl disulfide, and the conversion into the P-S analog of Kitazin P [138].

Metabolism of Kitazin P in *Pyricularia oryzae* presumably follows the same pathways as in plants. O,O-Diisopropyl phosphorothioate is stable for about 48 hr in *P. oryzae*. Part of Kitazin P is hydroxylated at the *m*-position of the benzyl group by *P. oryzae* [139]. This reaction was also found to occur in strains of *P. oryzae* resistant to Kitazin P. In various soils the compounds b, e, f, g, h, and i (Fig. 5) were detected as metabolites. Final breakdown products were H_2SO_4 and H_3PO_4 [241].

C. Hinosan

Edifenphos (Hinosan) is absorbed by roots and leaves of rice plants [55]. The compound remains mainly in the treated area and is less rapidly translocated and metabolized than Kitazin P [140-142]. O-Ethyl S-phenyl phosphorothioic acid and S,S-diphenyl phosphorodithioic acid are the main metabolites of Hinosan in rice plants (see Fig. 6). A time course study showed a further breakdown yielding finally ethyl phosphoric acid and phosphoric acid. Furthermore, application of [^{35}S]Hinosan gave rise to the formation of benzenesulfonic acid and sulfuric acid. So far the complete metabolic pathway of Hinosan has not been worked out.

The metabolism of Hinosan by *P. oryzae* has been studied by Uesugi and Tomizawa [143]. The compound undergoes the same reactions as in plants, including a hydrolysis of the P-S linkage and ethyl ester binding. Evidence supports the view that mycelium of *P. oryzae*, in contrast to plants, hydroxylates part of Hinosan at the *p*-position of one of the phenyl groups to give monohydroxy-Hinosan.

Tomizawa et al. [241] investigated the fate of Hinosan in various types of soils. Metabolites b, c, e, f, and h (see Fig. 6) were observed, as well as S,S,S-triphenylphosphorothioate and O,O-diethyl-S-phenylphosphorothioate.

D. Inezin

The metabolism of Inezin (56) in plants has not been thoroughly investigated but it is known that it is converted into S-benzyl hydrogen phenylphosphonothioate and phenylphosphonothioic acid in rice plants (see Fig. 7) [144].

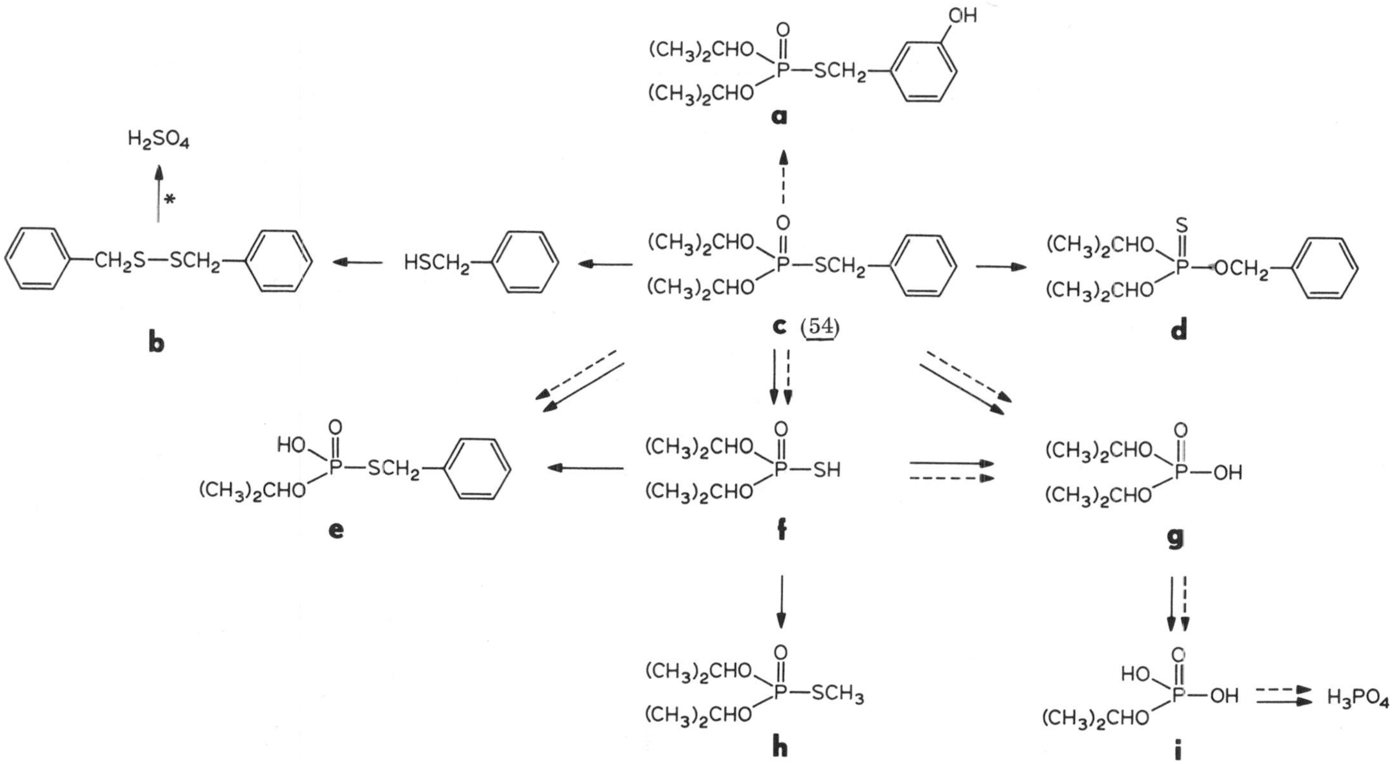

FIG. 5. Pathways of the metabolism of Kitazin P (54) in rice plants and in Pyricularia oryzae. The solid lines indicate pathways in rice plants, dotted lines those in P. oryzae, and lines marked with * indicate postulated pathways [137, 138]. a, S-m-hydroxybenzyl O-O-diisopropyl phosphorothioate; b, dibenzyl disulfide; c, Kitazin P; d, O,O-diisopropyl O-benzyl phosphorothioate; e, S-benzyl O-isopropyl phosphorothioate; f, O,O-diisopropyl hydrogen phosphorothioate; g, diisopropyl hydrogen phosphate; h, O,O-diisopropyl S-methyl phosphorothioate; i, O,O-monoisopropyl dihydrogen phosphate.

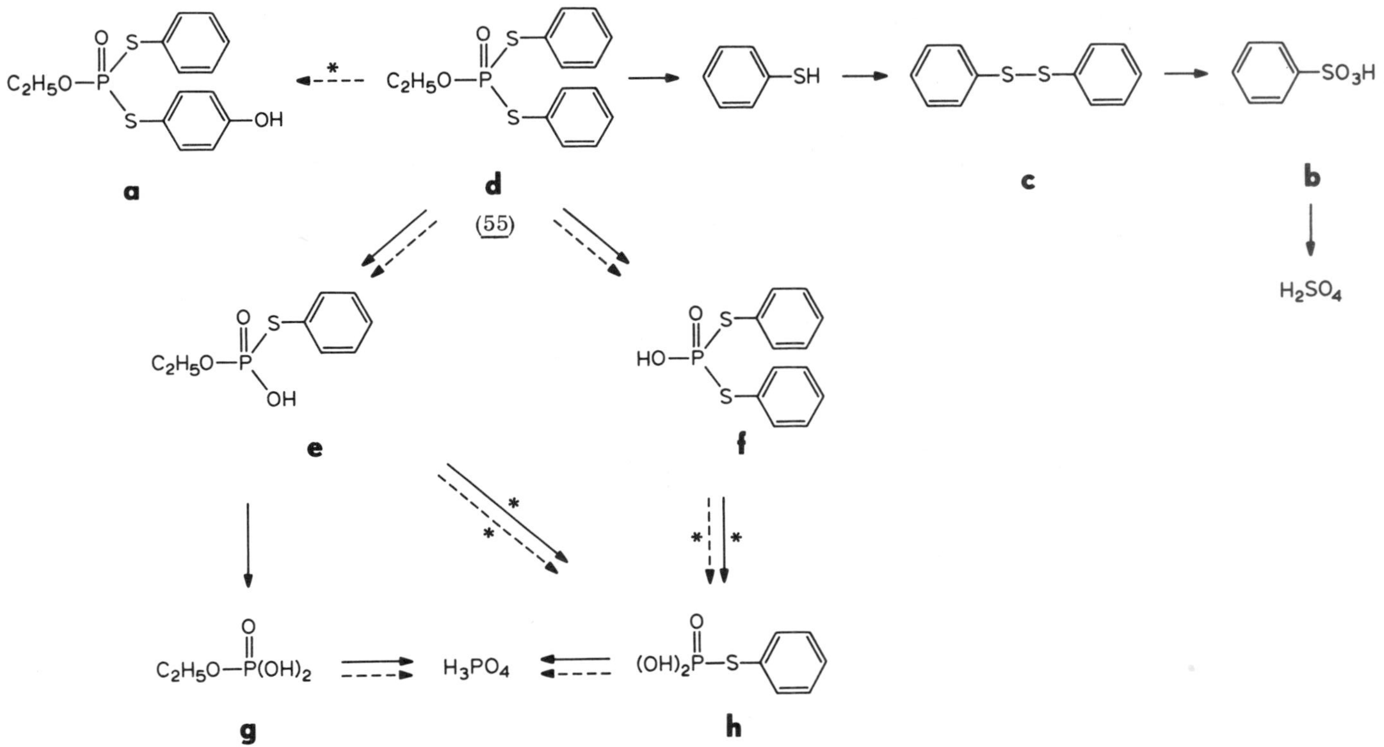

FIG. 6. Pathways for the metabolism of Hinosan (55) in rice plants and in *Pyricularia oryzae*. Solid lines indicate pathways in rice plants, dotted lines those in *P. oryzae*, and lines marked with * indicate postulated pathways [140–143]. a, monohydroxy Hinosan; b, benzene sulfonic acid; c, diphenyl disulfide; d, Hinosan; e, *O*-ethyl *S*-phenyl hydrogen phosphorothioate; f, *S*,*S*-diphenyl hydrogen phosphorodithioate; g, ethyl phosphoric acid; h, *S*-phenyl dihydrogen phosphorothioate.

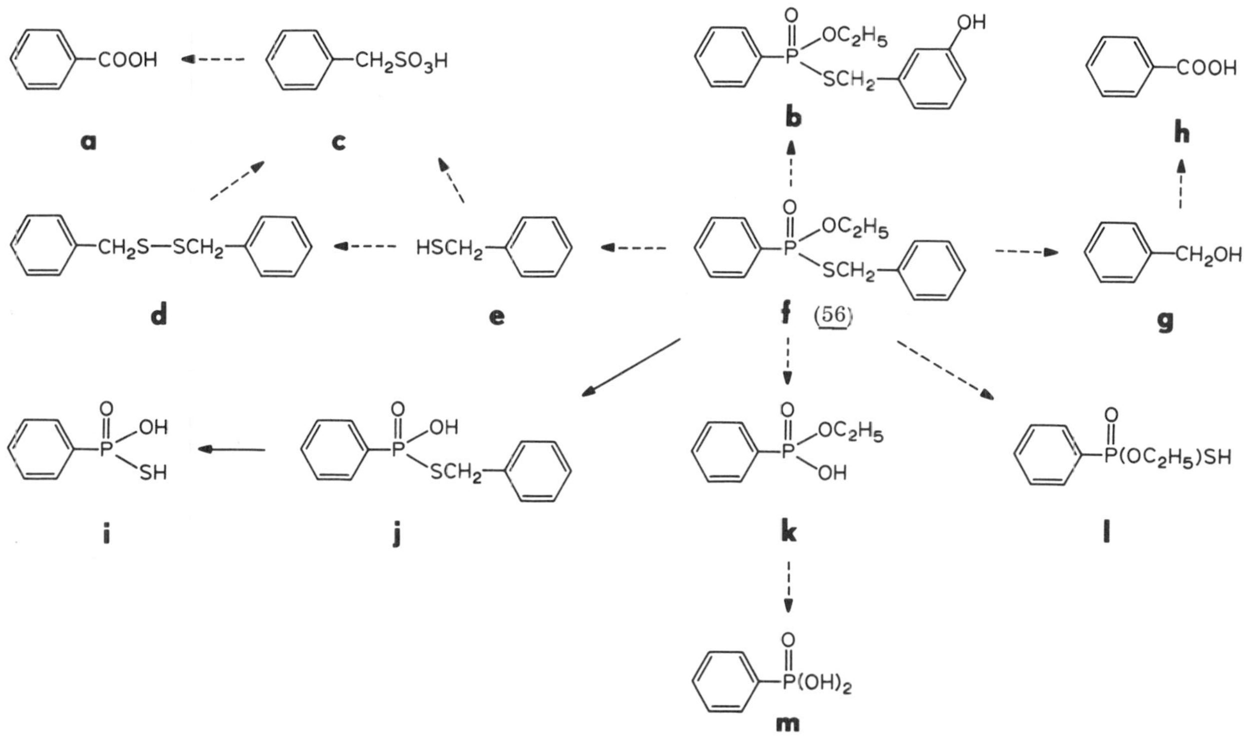

FIG. 7. Pathways for the metabolism of Inezin (56) in rice plants and in *Pyricularia oryzae*. Solid lines indicate pathways in the plants, dotted lines those in the fungus [144]. a, benzoic acid; b, *O*-ethyl *S*-*m*-hydroxybenzyl phenylphosphonothioate; c, toluene α-sulfonic acid; d, dibenzyl disulfide; e, benzylmercaptan; f, Inezin; g, benzylalcohol; h, benzoic acid; i, phenylphosphonothioic acid; j, *S*-benzyl hydrogen phenylphosphonothioate; k, *O*-ethyl hydrogen phenylphosphonate; l, *O*-ethyl hydrogen phenylphosphonothioate; m, phenyl phosphonic acid.

The fate of Inezin in mycelium of P. oryzae has been studied in more detail. The main metabolites which were found in the culture medium were identified as ethyl hydrogen phosphonate and O-ethyl hydrogen phenylphosphonothioate [144]. The latter was found to be more stable than the former which was further degraded to phenylphosphonic acid. Application of ^{14}C-benzyl-labeled Inezin revealed the formation of benzyl alcohol and benzyl sulfide which finally yielded benzoic acid.

Similar to the fate of Kitazin P in P. oryzae, the benzyl group of Inezin is hydroxylated at the meta position.

VIII. PYRIMIDINES

A. Ethirimol and Dimethirimol

Pyrimidine fungicides are specifically active against powdery mildews. Dimethirimol, Milcurb (57) and ethirimol, Milstem (58) are translocated from the roots to the parts above ground of barley and cucumber plants [145-147]. The degradation of these compounds in plants has been partly elucidated (see Fig. 8). Metabolism of both compounds follows a pattern similar to that observed in animals [146, 147]. The three routes of metabolism observed are N-dealkylation, hydroxylation of the butyl group, and conjugate formation. In the case of plants, the hydroxylation appears to be a relatively minor pathway in contrast to conjugate formation. The conjugates formed are glucosides and probably phosphates, while in animals ethirimol glucuronide was the only conjugate identified.

N-Demethylation of dimethirimol occurs very rapidly to give the N-monomethyl derivative, an active fungicide in cucumber plants. The loss of the second N-methyl group occurs more slowly and gives the relatively inactive 2-amino derivative, which is also formed from ethirimol in barley plants.

Recently two new metabolites of ethirimol (59) and (60) were reported in cereals [242].

B. 6-Azauracil

6-Azauracil (61) is another pyrimidine derivative which is systemically active against powdery mildew and a number of other diseases. The compound has no practical application. In plants as well as in fungi, 6-azauracil is converted to 6-azauridine (62) and then to the monophosphate (63) [148] which is the actual fungitoxic compound. Strains of Cladosporium cucumerinum resistant to azauracil converted this compound into the uridine derivative and the monophosphate to a lesser degree than the susceptible strain [149-151].

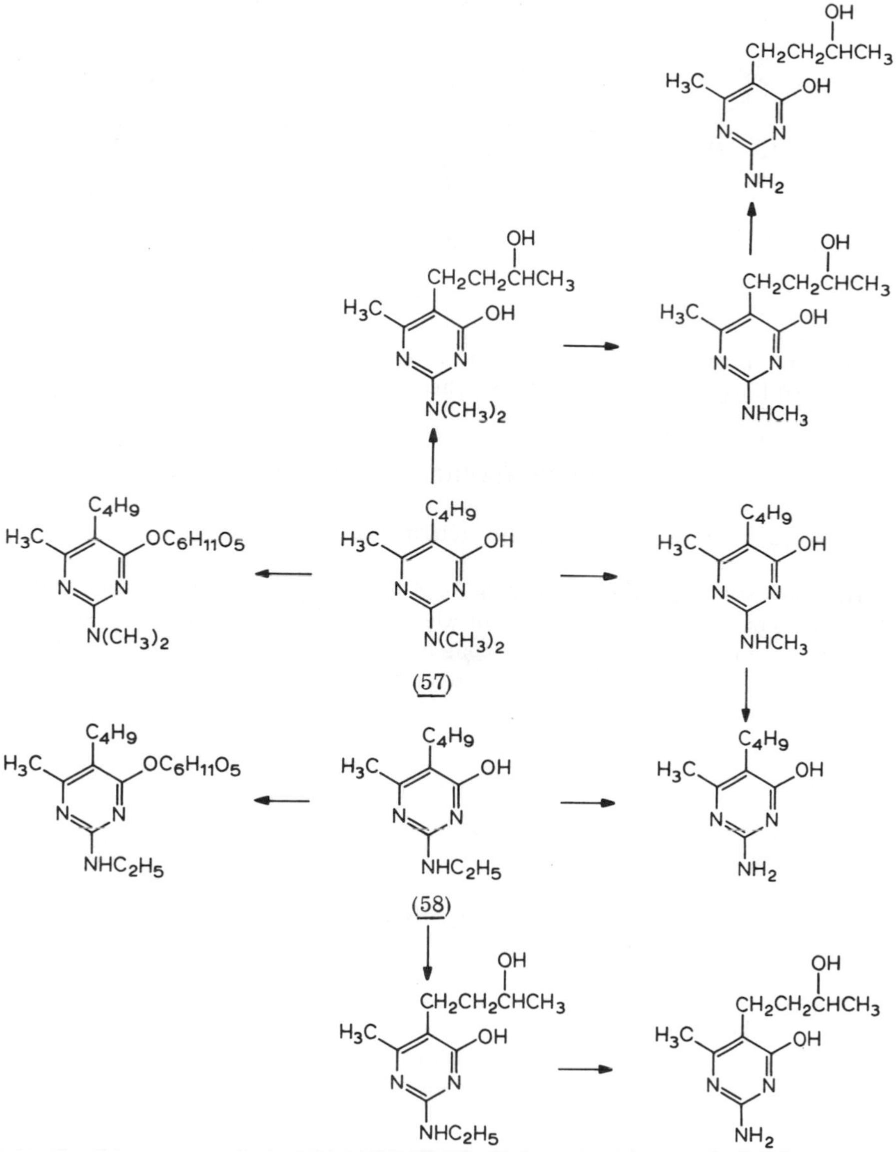

(57)

(58)

FIG. 8. Pathways for the metabolism of dimethirimol (57) and ethirimol (58) in plants.

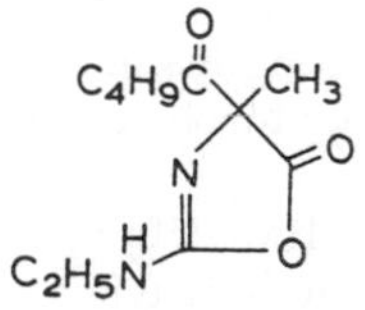

(59)

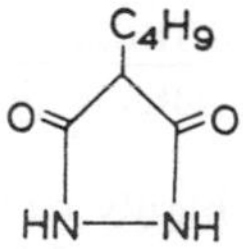

(60)

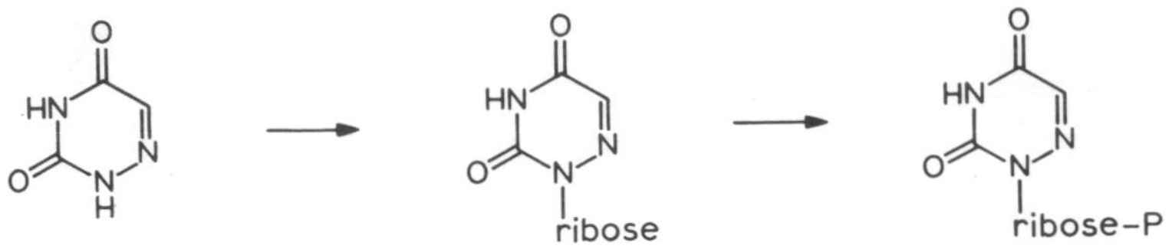

(61) 6-azauracil (62) 6-azauridine (63)

Matolcsy and Poonawalla [152] assume that, analogous to the fate of the dithiocarbamates, 2-thio-6-azauracil is converted into the alanine derivative and the glucoside by bean and wheat plants.

IX. POLYCHLORINATED AROMATICS

A. Hexachlorobenzene

Hexachlorobenzene or HCB (64) is used as a seed disinfectant, especially for the control of Tilletia tritici of wheat, on a limited basis. It also occurs as a technical impurity in quintozene and is extremely persistent in soil

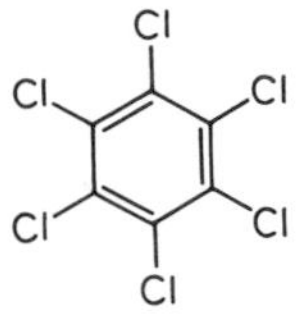

(64) HCB

[153, 154], as indicated by half-life values which amount to several years. It is rather volatile. Different plant species and parts show great differences in the quantity taken up from soil; a definite accumulation was observed in carrots and grass roots [154].

B. Quintozene

Quintozene, pentachloronitrobenzene, PCNB, or Terraclor (65) is used against *Botrytis* and *Rhizoctonia.* It contains as technical impurities 2,3,5,6-tetrachloronitrobenzene (66), pentachlorobenzene (67), and hexachlorobenzene [153].

In soil, PCNB is highly persistent. Half-life values of 4.7, 7.6, and 9.9 months have been reported for different soil types, the loss being mainly

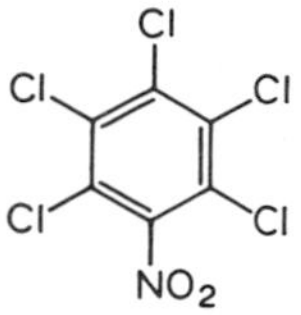

(65) PCNB

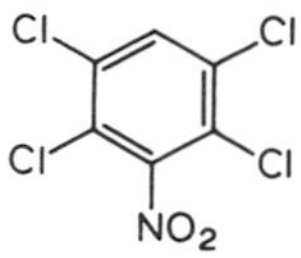

(66) tetrachloronitrobenzene

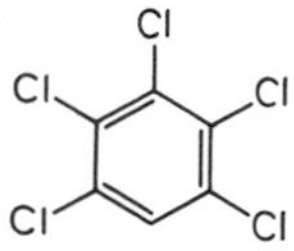

(67) pentachlorobenzene

due to volatilization [155]. In laboratory experiments, pentachlorobenzene (67), pentachloroaniline, PCA (68), and methylthiopentachlorobenzene, MTPCB (69), appear to accumulate in PCNB-treated soil [153]. Earlier Ko and Farley [156] reported that PCNB is rather stable in moist soil, but

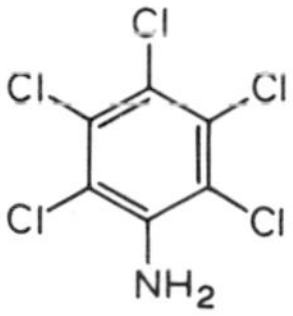

(68) PCA

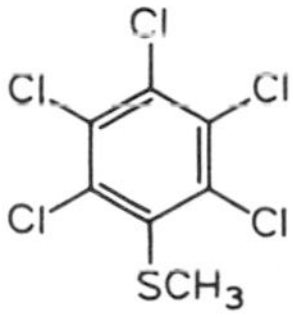

(69) MTPCB

disappears within a few weeks from submerged unsterilized soil giving partly pentachloroaniline. In soil amended with 1% saccharose, the rate of breakdown was greatly increased [157].

Pure culture studies with fungi and actinomycetes under aerobic growth conditions revealed a rather general ability to convert low concentrations of PCNB to PCA and MTPCB [158, 159].

The reduction of PCNB to PCA obviously is an unspecific biological process in which PCNB acts as a hydrogen acceptor. This process may be reversible under less reducing conditions leading again to PCNB (see DCNA, Section E). The mechanism of the formation of MTPCB is not yet clear. Possibly the $-SCH_3$ moiety is derived from methionine, the only $-SCH_3$ compound abundantly present in biological material, but experimental proof is still lacking.

In potatoes grown in soil treated with PCNB, it was present particularly in the peels. Moreover, conversion products were detected one of which was pentachloroaniline; others remained unidentified [160]. Kuchar et al. [161] presented evidence that these may have been unchanged HCB (an impurity present in the PCNB) and MTPCB. They studied the compounds present in young cotton plants grown for 1 or 2 weeks in soil containing 300 ppm PCNB. They discovered in the green parts of these plants, as well as in similarly treated young corn and soybean plants, large amounts of PCNB in addition to small amounts of the PCNB impurities (PCB, TCNB, and HCB) and the metabolites PCA and MTPCB. These metabolites detected in the plants may well have been taken up as such from soil where they are formed.

C. Tecnazene

Tecnazene, 2, 3, 5, 6-tetrachloronitrobenzene or TCNB (66) is used to control Fusarium of potato tubers. It disappears more rapidly from soil than PCNB. This is attributed to its greater volatility, the rate of biological degradation being about the same as that of PCNB [157]. There is no information on its metabolic fate in the presence of microbial cultures or plants, but one may expect conversions similar to those of quintozene.

D. Pentachlorophenol

Pentachlorophenol or PCP (72) which is strongly phytotoxic is used in wood preservation. Its activity sometimes declines due to biological conversions.

Engel et al. isolated from wood shavings treated with tetrachlorophenol (70) and used in poultry cages, the compound tetrachloroanisole (71). This compound was apparently produced by fungi and caused a musty taste in the poultry products [162]. In pure cultures, fungi also caused the methylation of tetrachlorophenol [163]. In a similar way, a pure culture of Trichoderma virgatum growing in malt extract was shown to methylate PCP (72) to pentachloroanisole (73) [164]. From an aerobic enrichment culture with PCP as

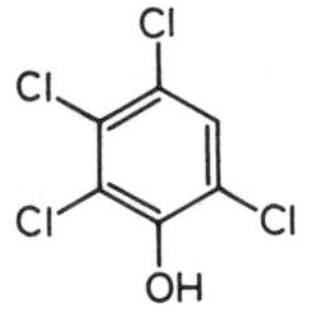

(70) tetrachlorophenol

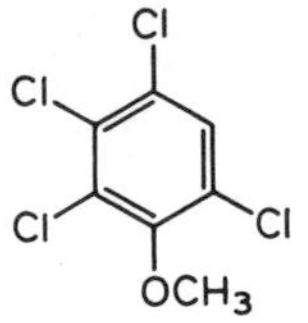

(71) tetrachloroanisole

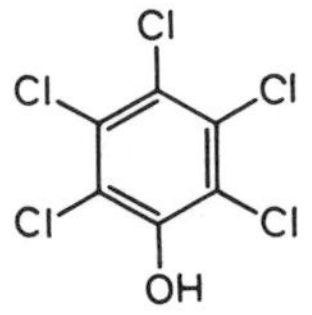

(72) pentachlorophenol

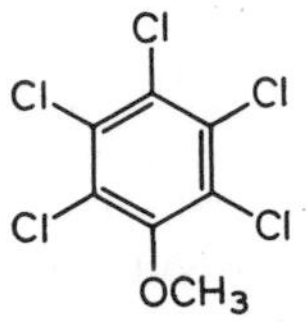

(73) pentachloroanisole

sole organic substance, Chu and Kirsch [165] obtained a coryniform bacterium which could break down PCP completely; 73% of the [^{14}C]pentachlorophenol appeared as $^{14}CO_2$ when exposed for 24 hr to washed cell suspensions in buffer. Similar results were obtained by Watanabe [166] who reported that in soil perfused with 40 ppm PCP the five chlorine atoms were liberated completely after about 3 weeks. From this soil, Pseudomonas-like bacteria were isolated which were able to grow aerobically on PCP as a sole source of carbon.

E. Dichloran

Dichloran, 2,6-dichloro-4-nitroanilin, DCNA, or Botran (74), is used as a soil fungicide and for leaf treatment against Botrytis and Rhizopus. It is taken up by the roots of many plants.

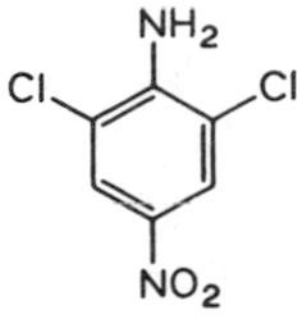

NH2, Cl, Cl, NO2

(74) DCNA

Half-life values of 30, 16.2, and 13.6 months were reported for different soil types [155]. However, in flooded soil treated with 1% glucose, only 7% of the parent compound remained after 3 days incubation [167]. No $^{14}CO_2$ was detected in 9 days, but 4-amino-3,5-dichloroacetanilide, ADCAA (75), accumulated as a major product. Some of the intermediate 2,6-dichloro-p-phenylenediamine, DCPD (76) was also present.

Pure cultures of E. coli rapidly reduced ^{14}C-DCNA to DCPD as the major metabolite, whereas for Pseudomonas cepacia ADCAA was the main product [168]. Apart from these two compounds, four minor metabolites

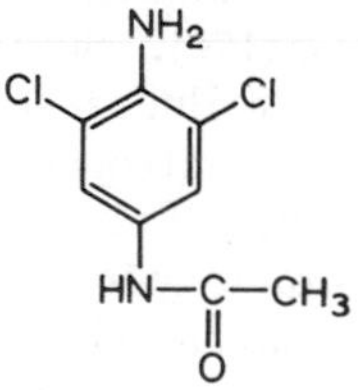

NH2, Cl, Cl, HN—C—CH3, O

(75) 4-amino-3,5-dichloroacetanilide

NH2, Cl, Cl, NH2

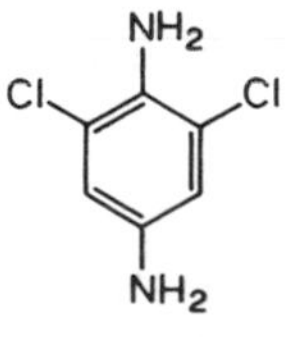

(76) 2,6-dichloro-p-phenylenediamine

were also detected in the cultures. Under anaerobic conditions the reduction to DCPD by *E. coli* was greatly accelerated. It is of interest that further breakdown was observed for a rod-shaped bacterium isolated from soil repeatedly treated with DCNA. In nutrient solution, it completely transformed 20 ppm of [^{14}C]DCNA within 5 days, 25-50% of the label being recovered as $^{14}CO_2$ [169]. No further details are known.

From a Hoagland solution, roots of tomato and lettuce seedlings rapidly took up ^{14}C-DCNA [170]. Within 11 days it was degraded in leaves and stems. Since polar metabolites were formed and no DCPD was detected, ring opening is suggested. The $^{14}CO_2$ was not measured in these experiments. A carbohydrate fraction of the plants showed radioactivity but the amino acid fraction did not.

Soybean plants are reported to metabolize DCNA to a compound tentatively identified as 4-amino-3,5-dichloromalonanilic acid (77) [171]. This compound was also recovered from bean seedlings which had taken up [^{14}C]DCNA from a nutrient solution for 2 days. The metabolite appeared

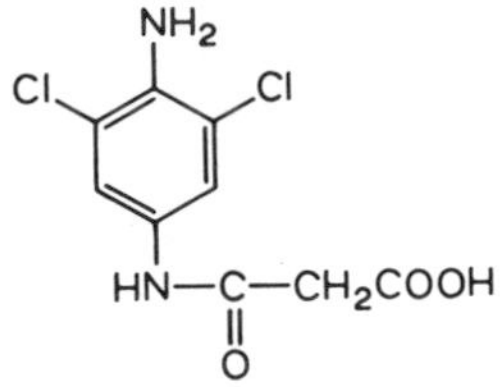

(77) 4-amino-3,5-dichloro-
malonanilic acid

to be unstable, only 1% being recovered after 51 days [169]. Only a trace of $^{14}CO_2$ was formed, the main radioactivity being recovered in the form of natural constituents.

It is not known whether the metabolites formed in soil are taken up by plant roots.

Another consideration is that the reduction of DCNA to DCPD may well be reversible. The same was suggested above for the reduction of PCNB to PCA. Thus, for instance, 4-chloroaniline is oxidized by *Fusarium oxysporum* to the corresponding nitroderivative [172]. Similarly, 2,4-dinitrophenol is not only reduced by this fungus in a glucose-containing medium, but one of the reduction products, 2-amino-4-nitrophenol, was also found to be reoxidized by this organism in absence of glucose [173]. Obviously, in the first case 2,4-dinitrophenol acts as a hydrogen acceptor, whereas in the latter the amino derivative acts as an oxidizable compound.

F. Chloroneb

Chloroneb, 1,4-dichloro-2,5-dimethoxybenzene (78) [174] is effective in soil against Rhizoctonia and Pythium and controls seedling diseases of cotton, beans, and soybeans. It has a slight systemic activity as it accumulates in the roots and the lower parts of plants [175].

In soil, ^{14}C-ring-labeled chloroneb has a half-life of 3 to 6 months [175]; the products formed are not known. Rhizoctonia solani demethylates the compound to nonfungitoxic 2,5-dichloro-4-methoxyphenol, DCMP (79) (see Fig. 9), whereas Sclerotinia rolfsii and Saccharomyces pastorianus do not [176]. Wiese and Vargas observed interconversion of these two compounds by various fungi under conditions of active growth to produce a mixture [177]. Methylation may occur by a similar mechanism as in the case of pentachlorophenol. The ability to methylate and demethylate differed for the various organisms studied. The authors assume that microbial synthesis of chloroneb from DCMP, its principal degradation product, accounts in part for the relative stability and long-term effectiveness of the fungicide in soil. Also 2,5-dichlorohydroquinone (80) was formed. No experiments on breakdown by bacteria have been reported.

Twelve days after uptake of ^{14}C-ring-labeled chloroneb, bean plants contained equal amounts of chloroneb and DCMP to a total of about 95%.

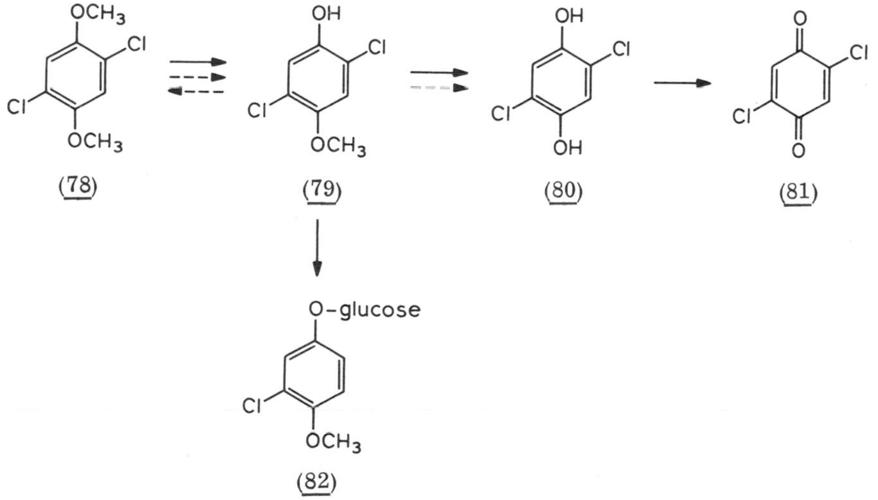

FIG. 9. Pathways for the metabolism of chloroneb (78) in plants and fungi. Solid lines indicate pathways in the plants, the dotted lines those in microorganisms.

Trace amounts of 2,5-dichloroquinone (81) and 2,5-dichlorohydroquinone (80) as well as an unknown compound were also found [175]. Thorn identified the latter as the β-D-glucoside of 2,5-dichloro-4-methoxyphenol (82) [178]. After 2 days exposure of ^{14}C-ring-labeled chloroneb to the roots of bean plants, 30% of the extractable radioactivity was recovered as this glucoside. Rhodes et al. [175] missed this product, primarily because they applied acid hydrolysis. The formation of glucosides from phenols in plant tissue is a well-known phenomenon.

G. Chlorothalonil

Chlorothalonil, 2,4,5,6-tetrachloroisophthalonitril or Daconil (83) is a broad-spectrum fungicide applied to soil and leaves to control *Rhizoctonia* and *Botrytis*. It is an alkylating agent which reacts with both low- and high-molecular-weight cellular thiols [226, 227]. Reduced glutathione reacts in vivo and in vitro with the fungicide, forming substituted chlorothalonil-glutathione derivatives.

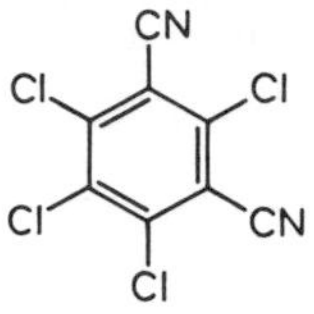

(83) chlorothalonil

H. Chloranil

Chloranil, tetrachloro-*p*-benzoquinone or Spergon (84) is a broad-spectrum fungicide used as a seed disinfectant. It is unstable to light. No information is available on its biological breakdown.

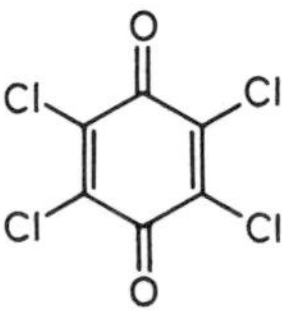

(84) chloranil

I. Dichlone

Dichlone, 2,3-dichloronaphthoquinone or Phygon (85), like chloranil, is unstable to light. It is used as a seed disinfectant as well as on fruits and vegetables.

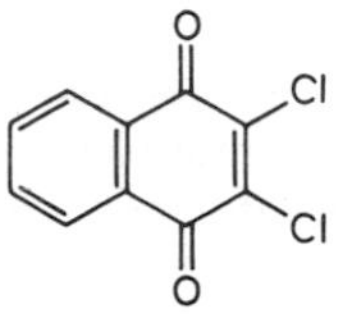

(85) dichlone

Photoconversion products discovered in crop extract will be described in Chapter 5. This fungicide has been reported to form reaction products with both thiol and amino groups in fungal cells [228].

X. ORGANOMETALLIC COMPOUNDS

A. Organotin Compounds

Triphenyltin acetate, fentinacetate or Brestan (86), and triphenyltin hydroxide, fentinhydroxide or Du-Ter are nonsystemic fungicides used to control *Phytophthora infestans* of potato and *Cercospora beticola* of sugar beet.

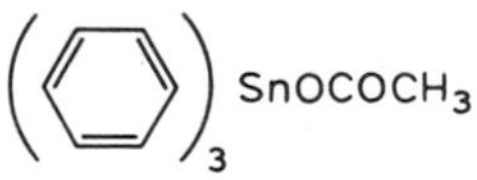

$(C_6H_5)_3SnOCOCH_3$

(86) triphenyltin acetate

In a recent study, it was reported that degradation of triphenyltin chloride on leaves of sugar beet proceeds via di- and monophenyltin compounds to inorganic tin [179]. This degradation may be a spontaneous rather than a biological conversion.

Studies on degradation in soil were made with 5 and 10 ppm of the acetate in which the aromatic carbon atoms were ^{14}C-labeled [180, 181].

Treated soil was placed in columns and evolution of $^{14}CO_2$ was monitored continuously by passing a slow stream of air through the column and absorbing the CO_2. Results as measured by evolution of $^{14}CO_2$ would indicate a half-life for triphenyltinacetate of approximately 140 days. However, its actual half-life may be much shorter since CO_2 need not be a primary product of degradation. Analogously with the microbial breakdown of phenylmercuric acetate [182] and triphenyllead acetate [183], benzene, known to undergo microbial conversion to CO_2, may be the first product split from the fungicide. Since heat-sterilized soil did not yield CO_2, microorganisms are involved in the breakdown of fentin to CO_2. Monophenyltin is supposed to accumulate as a rather stable intermediate breakdown product.

Isolated from fentin acetate treated soil in shake culture, two fungi and one gram-negative rod-shaped bacterium slowly produced $^{14}CO_2$ from the labeled compound added to nutrient solution.

B. Organomercury Compounds

The organomercury fungicides, mainly phenylmercury, methylmercury, and methoxyethylmercury salts, are mostly used as seed and bulb disinfectants. Methylmercury compounds have been withdrawn from the market in some countries due to environmental problems.

Only a few data are available concerning the metabolism of organomercurials in plants. According to Tomizawa [184], the label of [^{203}Hg]phenylmercuric acetate, PMA (87), is quickly absorbed through roots, stems, and leaves of rice plants and translocated into other tissues. In wheat and

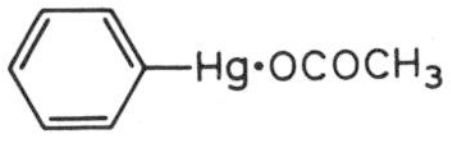

—Hg·OCOCH$_3$

(87) PMA

soybean plants, PMA was decomposed into inorganic mercury after root or leaf application [185]. Also methylmercuric chloride was transformed into inorganic mercury in wheat plants [185], whereas degradation of PMA and methyl mercuric chloride occurred in the presence of tissue of wheat roots [186].

Using a biological assay technique, Spanis et al. [187] found that methylmercury dicyandiamide and Semesan (a phenylmercury derivative) were inactivated by soil microorganisms. Phenylmercuric acetate was slowly degraded in soil with metallic mercury being formed, probably by microbial activity. Only small amounts of Hg^{2+} were found [188]. The degradation of methylmercury in soil was much slower than that of PMA. The persistence of the latter compound in the biosphere is well-known, although bacteria have been isolated which convert this compound into metallic mercury and methane [189, 190].

Some bacterial isolates resistant to PMA have been shown to degrade PMA into metallic mercury and benzene [191] or to convert PMA into diphenylmercury [192]. Tonomura and co-workers [182, 193-196] have performed an extensive study of the mechanism underlying the enzymatic conversion of organomercurials into mercury and the corresponding alkane in which NAD and NADP play a role.

XI. MISCELLANEOUS COMPOUNDS

A. Dexon

Dexon, sodium p-(dimethylamino)benzenediazosulfonate (88) is used as a soil fungicide against the Phycomycetes *Pythium*, *Aphanomyces*, and *Phytophthora*. It shows slight systemic activity and is subject to photochemical decomposition.

The rate of disappearance of Dexon from soil is rapid initially but slows down with decreasing concentration. Even after one year, substantial residues were still present [197, 198]. The products formed have not been described. A culture of *Pseudomonas fragi* was isolated from Dexon-treated soil [199]. It was unable to use Dexon as a source of carbon or nitrogen for growth but accumulated N,N-dimethyl-p-phenylene diamine (89) in a glucose mineral salts medium treated with 100 ppm Dexon. Remarkably, this diamine is as toxic to *Pythium aphanidermatum* as Dexon itself [199]. A reductase responsible for the formation of the diamine has been isolated from the bacterium [243].

$$(CH_3)_2N-C_6H_4-N{=}N-SO_2-ONa \longrightarrow (CH_3)_2N-C_6H_4-NH_2$$

(88) Dexon

(89) N,N,dimethyl-p-phenylene diamine

Four unidentified compounds were detected in a root extract of sugarbeet seedlings, treated through their roots with Dexon labeled with ^{14}C in the methyl groups [200]. Since demethylation may well take place, similar experiments with ring-labeled Dexon might be of interest.

B. Biphenyl

Biphenyl or diphenyl (90) is not used on plants or in soil but to impregnate citrus fruit wrappings.

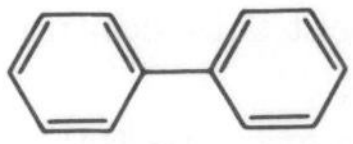

(90) biphenyl

Several bacteria were reported to use biphenyl as a sole source of carbon for growth, however, their pathways of breakdown appear to differ. In a culture of a gram-negative bacterium, 2,3-dihydroxybiphenyl was detected which was further degraded by certain cell fractions via α-hydroxy-β-phenylmuconic semialdehyde to phenylpyruvate [201]. A species of *Beyerinckia* showed the same behavior [202].

Pseudomonas putida also is able to utilize biphenyl as a source of carbon but oxidizes biphenyl via 2,3-dihydro-2,3-dihydroxybiphenyl, and 2,3-dihydroxybiphenyl to benzoic acid and 4-hydroxy-2-oxovalerate or 2-hydroxypenta-2,4-dienoate [203-205]. In this case, a *meta* cleavage takes place between the C_1 and C_2.

C. Dodine

Dodine, *n*-dodecylguanidine monoacetate, Cyprex (91) is used against apple scab; it has slight curative effect.

In river mud or soil dodine is only very slowly decomposed [206]. From soil, species of *Flavobacterium* and *Achromobacter* were isolated which, after an initial lag phase, could utilize dodine as a source of carbon. Ungerminated macroconidia of *Fusarium solani* very rapidly detoxify dodine. In 96 min, two-thirds of the [^{14}C]dodine absorbed was released as a different, unidentified compound [207].

The fate of [^{14}C]dodine (labeled in the guanidine carbon) which was painted throughout the season on leaves of an apple tree was studied by Curry [208]. The stability of dodine is very low; several weeks after treatment, various other labeled compounds were detected, including guanidine (92) or perhaps guanidine monosubstituted with a small group. After hydrolysis, labeled creatine (93) was found.

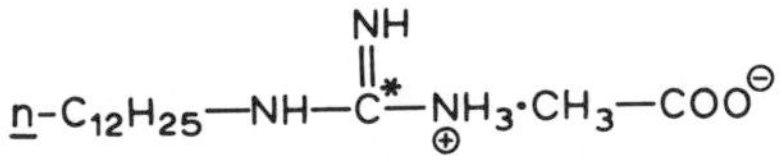

$$n\text{-}C_{12}H_{25}—NH—\overset{\overset{NH}{\|}}{C^*}—\underset{\oplus}{N}H_3\cdot CH_3—COO^{\ominus}$$

(91) dodine

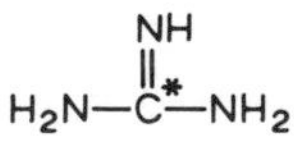

$$H_2N—\overset{\overset{NH}{\|}}{C^*}—NH_2$$

(92) guanidine

$$HOOC—CH_2—\underset{\underset{CH_3}{|}}{N}—\overset{\overset{NH}{\|}}{C^*}—NH_2$$

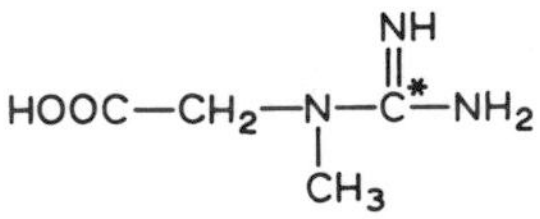

(93) creatine

D. Sklex

Sklex, 3-(3'-5'-dichlorophenyl)-5,5-dimethyloxazolidine-2,4-dione, DDOD (94), is a systemic fungicide with limited use against Sclerotinia and Botrytis.

An extensive study was made of its decomposition by plants, soil, and light [209] using DDOD ^{3}H-labeled in the benzene ring or ^{14}C-labeled at the 4-carbonyl carbon of the oxazolidine ring. In nonsterile soil, 35% of the DDOD was decomposed in 3 weeks at 25° C; 30% N-3,5-dichlorophenyl-N-α-hydroxyisobutylcarbamic acid, DHCA (95); 5% α-hydroxyisobutyryl-3,5-dichloroanilide, HDA (96) and traces of 3,5-dichloroaniline had been formed.

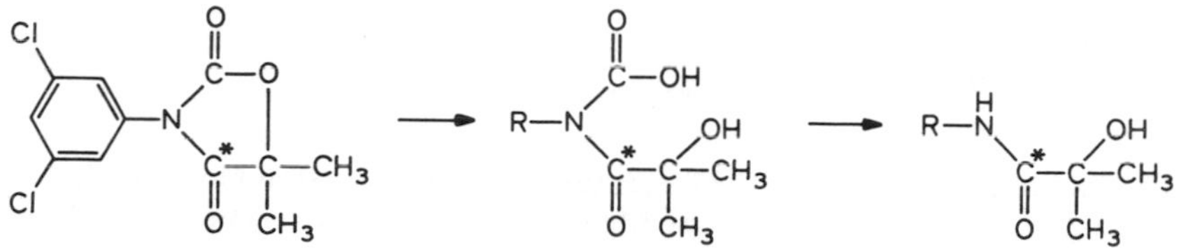

(94) sklex, DDOD → (95) DHCA → (96) HDA

The first product resulted from a hydrolytic cleavage of the oxazolidine ring, whereas the second product must have been obtained through subsequent decarboxylation. It is remarkable that these two compounds were also formed in sterilized distilled water, but not in sterile soil. Therefore, the authors conclude that microbial conversions occur in soil. DHCA and HCA were also present in bean plants root-treated with a solution of DDOD. Since these compounds were not found in bean plants injected with DDOD, it can be assumed that they have been taken up as such from the treatment solution. In grape plants, HDA and possibly a 4-hydroxy analogue of sklex were detected after injection.

E. Tridemorph

Tridemorph, N-tridecyl-2,6-dimethylmorpholine or Calixin (97) finds application in the control of powdery mildew in cereals and leaf spot in bananas. It is systemic and has both protective and curative activity.

In soil treated with 5 ppm of the compound, ^{14}C-labeled at position 2 and 6, 80% conversion occurred after 30 weeks at room temperature [210]. Degradation is initiated by the formation of tridemorph-N-oxide (98). $^{14}CO_2$ was evolved at a constant rate, 20% of the label evolving as $^{14}CO_2$ during the indicated period. 2,6-Dimethylmorpholine (99) was also detected.

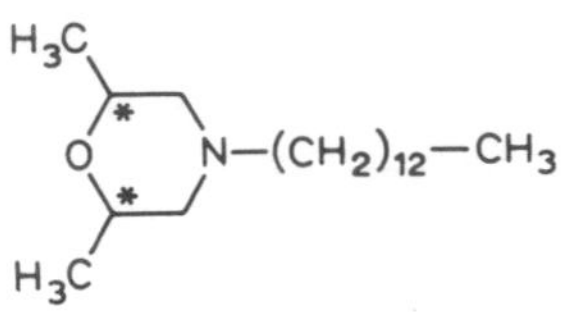

(97) tridemorph

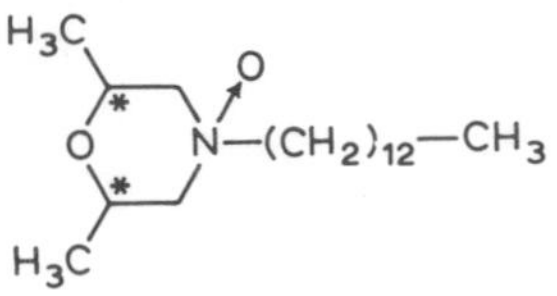

(98) tridemorph-N-oxide

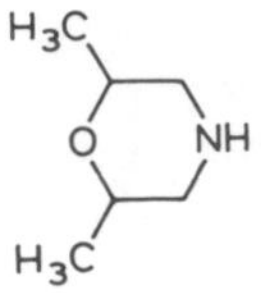

(99) 2,6-dimethylmorpholine

In cereals treated with tridemorph, the content drops from 20 ppm to less than 0.05 ppm in 48 days [211]. No conversion products have been identified so far, although indications were obtained for the formation of hydroxylated metabolites [244].

F. Triforine

The systemic fungicide N,N'-bis-(1-formamido-2,2,2-trichloroethyl)-piperazine (100) (Fig. 10) is particularly effective against powdery mildew and rusts [212,213]. After treatment, ^{3}H-piperazine- and ^{14}C-side-chain-labeled triforine are rapidly translocated to the parts above ground and accumulate in the leaves present at the time of treatment [214, 215, 219]. Triforine decomposes more rapidly in dilute aqueous solution than in plants. In water, the 1-formamido trichloroethyl groups were successively converted into glyoxyl groups, yielding N-monoglyoxyl piperazine (101), N,N'-diformylpiperazine (102), N,N'-diglyoxylpiperazine (103), and finally piperazine (104) (Fig. 10) [219]. In different plant species triforine was metabolized at different rates; in cucumber plants breakdown occurred more rapidly than in barley plants [216-220]. According to the same authors [216-219], triforine is metabolized in plants to four nonfungitoxic metabolites, one of which was identified as piperazine. The three unknown metabolites might be identical with the degradation products in aqueous solution. Since ring cleavage of the piperazine ring was not found to occur in barley plants, piperazine may be the final terminal residue of triforine in plants [217].

G. Prothiocarb

Prothiocarb, S-ethyl N-(3-dimethylaminopropyl)thiocarbamate (105), is a new systemic compound exerting a fungistatic effect on a variety of species of *Phycomycetes* [221]. When [propyl-^{14}C]prothiocarb was applied to the roots of tomato plants via a nutrient solution or a soil drench, a large amount of label, consisting mainly of unchanged parent compound, was rapidly translocated to the shoots. In addition to the main conversion

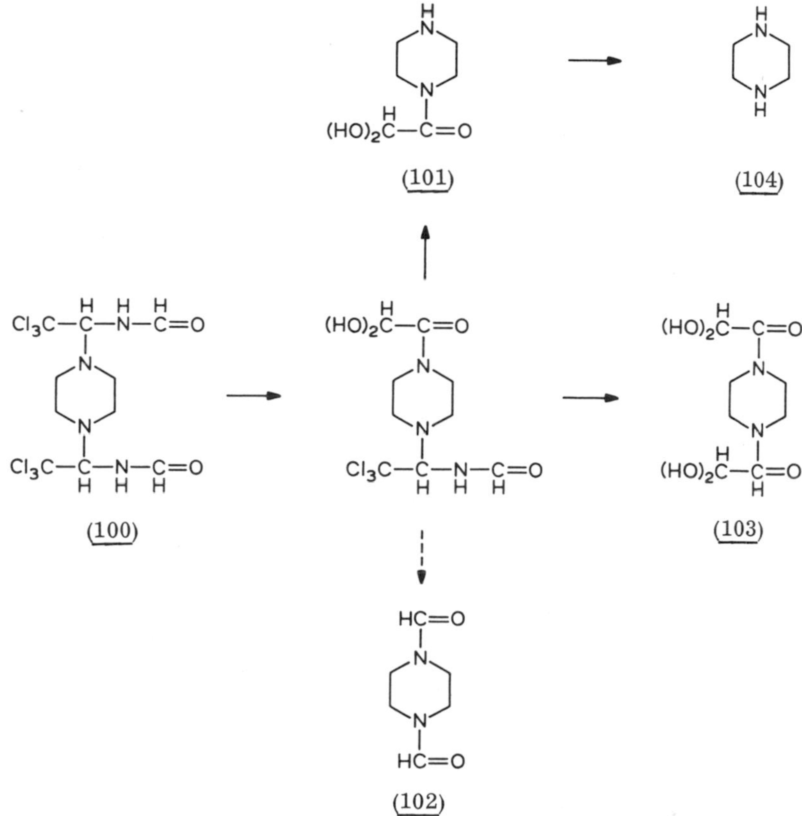

FIG. 10. Proposed pathway for the metabolism of triforine (100) in plants.

product, N,N-dimethylpropane-1,3-diamine (106), small amounts of 1-methyl-tetrahydro-1,3-diazine-2-one and bis-(3-dimethylaminopropyl)-urea were detected. In the roots and in the nutrient solution, N,N-dimethylpropane-1,3-diamine represented the major fraction [245].

Prothiocarb is degraded in soil, the formation of $^{14}CO_2$ from the labeled product being observed. The degradation occurs more readily in an alkaline soil. The half-life of the fungicide was 59 and 144 days in a sandy and a sandy loam soil, respectively. N,N-Dimethylpropane-1,3-diamine (106), diethyldisulfide, N-demethylprothiocarb (107), and 1-methyltetrahydro-1,3-diazine-2-one were tentatively identified as major metabolites [222, 245].

$(CH_3)_2N(CH_2)_3NH{-}C({=}O){-}SC_2H_5$

(105) prothiocarb

$(CH_3)_2N(CH_2)_3NH_2$

(106) N-N-dimethylpropane-1,3-diamine

$H_2N(CH_2)_3NH{-}C({=}O){-}SC_2H_5$

(107) N-demethylprothiocarb

H. Denmert

Denmert (108), labeled with ^{14}C in the methylene group attached to the phenyl ring, was gradually decomposed to yield a number of products when applied to plants or soils. The identified products were largely the same in either environment. Three primary reactions occurred: (1) oxidation of the S atoms; (2) cleavage of the dithiocarbonimidate linkage; and (3) oxidation of the methylene group [246].

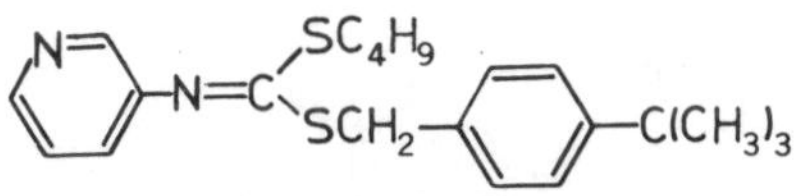

(108) Denmert

I. Blasticidin S

Blasticidin S (109) is a systemic antibiotic produced by *Streptococcus griseochromogenes*; it is used against the rice blast pathogen.

Many fungi are able to transform this compound. Breakdown by *Aspergillus fumigatus* was studied in detail. In aqueous solution, washed suspensions completely transformed blasticidin S within a few hours into a product which could be identified as deaminohydroxyblasticidin S (110) [223]. In shake culture in nutrient medium three minor products could be identified as well. Other workers [224] reported similar breakdown of Blasticidin S in cultures of bacteria and fungi.

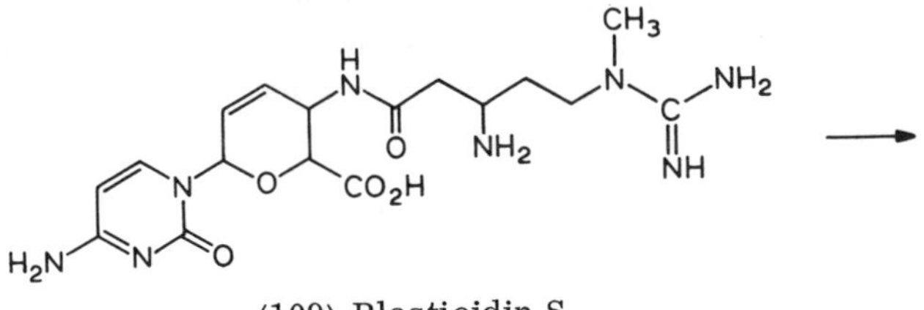

(109) Blasticidin S

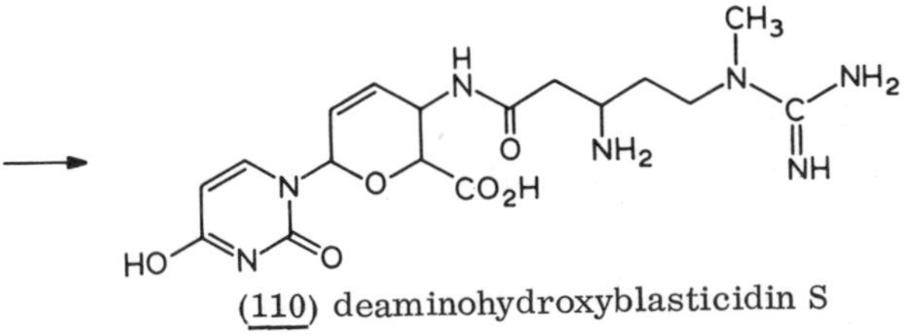

(110) deaminohydroxyblasticidin S

In plants the compound penetrates only through wounds and is then translocated upwards. Amongst the products formed, cytomycin and a trace of (110) were detected.

REFERENCES

1. A. Kaars Sijpesteijn and G. J. M. van der Kerk, Ann. Rev. Phytopathology, 3, 127 (1965).
2. A. Kaars Sijpesteijn, World Rev. Pest Control, 8, 138 (1969).
3. J. E. Casida and L. Lykken, Ann. Rev. Plant Physiol., 20, 607 (1969).
4. J. L. Garraway, Pestic. Sci., 3, 449 (1972).
5. A. Kaars Sijpesteijn, in Systemic Fungicides (R. W. Marsh, ed.), Longman, London, 1972, Chap. 7, pp. 132-155.
6. D. G. Crosby, Ann. Rev. Plant Physiol., 24, 467 (1973).
7. C. S. Helling, P. C. Kearney, and M. Alexander, in Advances in Agronomy (N. C. Brady, ed.), Vol. 23, Academic Press, New York, 1971, pp. 147-240.
8. R. S. Horvath, Bacteriol. Rev., 36, 146 (1972).
9. G. D. Thorn and R. A. Ludwig, The Dithiocarbamates and Related Compounds, Elsevier, Amsterdam, 1962.
10. G. J. M. van der Kerk, Mededel. Landbouwhogeschool Gent, 21, 305 (1956).
11. J. Kaslander, Ph.D. Thesis, University of Utrecht, 1966.
12. J. Kaslander, A. Kaars Sijpesteijn, and G. J. M. van der Kerk, Biochem. Biophys. Acta, 52, 396 (1961).
13. J. Kaslander, A. Kaars Sijpestcijn, and G. J. M. van der Kerk, Biochem. Biophys. Acta, 60, 417 (1962).
14. H. M. Dekhuijzen, Nature, 191, 198 (1961).
15. H. M. Dekhuijzen, Mededel. Landbouwhogeschool. Gent, 26, 1542 (1961).
16. H. M. Dekhuijzen, Neth. J. Plant Pathol., 70, Suppl. 1, 1 (1964).
17. F. Massaux, Mededel. Landbouwhogeschool. Gent, 28, 590 (1963).
18. J. W. Hylin and B. H. Chin, 6th Intern. Congr. Plant Protection, Vienna, 1967, p. 614.
19. G. D. Thorn and W. H. Minshall, Can. J. Bot., 42, 1405 (1964).
20. R. Corbaz, Phytopathol. Z., 44, 101 (1962).
21. H. M. Dekhuijzen, J. W. Vonk, and A. Kaars Sijpesteijn, in Pesticide Terminal Residues (A. S. Tahori, ed.), Butterworths, London, 1971, pp. 233-241.
22. G. D. Thorn and L. T. Richardson, Can. J. Bot., 40, 25 (1962).
23. K. Raghu, N. B. K. Murthy, R. Kumarasamy, R. Roa Sudha, and P. V. Sane, private communication, 1974.
24. A. Kaars Sijpesteijn, J. Kaslander, and G. J. M. van der Kerk, Biochem. Biophys. Acta, 62, 587 (1962).

25. A. Kaars Sijpesteijn, H. M. Dekhuijzen, J. Kaslander, C. W. Pluijgers, and G. J. M. van der Kerk, Mededel. Landbouwhogeschool Gent, 28, 597 (1963).
26. A. Kaars Sijpesteijn and J. W. Vonk, Mededel. Landbouwhogeschool Gent, 35, 799 (1970).
27. H. D. Sisler and C. E. Cox, Phytopathology, 41, 565 (1951).
28. A. Ayanaba, W. Verstraete, and M. Alexander, Soil Sci. Soc. Amer. Proc., 37, 565 (1973).
29. H. L. Klopping, Ph.D. Thesis, University of Utrecht, 1951.
30. H. L. Klopping and G. J. M. van der Kerk, Rec. Trav. Chim., 70, 949 (1951).
31. C. W. Pluijgers, J. W. Vonk, and G. D. Thorn, Tetrahedron Letters, 1971, p. 1317.
32. R. Engst and W. Schnaak, Z. Lebensmitteluntersuch. Forsch., 143, 99 (1970).
33. J. W. Vonk, Ph.D. Thesis, University of Utrecht, 1975.
34. J. W. Vonk and A. Kaars Sijpesteijn, Ann. Appl. Biol., 65, 489 (1970).
35. J. W. Vonk and A. Kaars Sijpesteijn, Pestic. Biochem. Physiol., 1, 163 (1971).
36. A. Kaars Sijpesteijn and J. W. Vonk, in Pesticides (Suppl. 3 of Environ. Quality Safety), F. Coulston and F. Korte, eds., Thieme Verlag, Stuttgart, 1975, p. 57.
37. W. R. Lyman and R. J. Lacoste, in Pesticides (Suppl. 3 of Environ. Quality Safety), F. Coulston and F. Korte, eds., Thieme Verlag, Stuttgart, 1975, p. 67.
38. P. A. Cruickshank and M. C. Jarow, J. Agr. Food Chem., 21, 333 (1973).
39. R. D. Ross and D. G. Crosby, J. Agr. Food Chem., 21, 335 (1973).
40. D. E. Munnecke, Ann. Rev. Phytopathology, 10, 375 (1972).
41. W. Moje, D. E. Munnecke, and L. T. Richardson, Nature, 202, 831 (1964).
42. L. E. Den Dooren de Jong, Ph.D. Thesis, University of Delft, 1926.
43. D. D. Kaufman and C. L. Fletcher, 2nd Intern. Congr. Plant Pathology, Minneapolis 1973, Abstr. No. 1018.
44. C. H. Blazquez, J. Agr. Food Chem., 21, 330 (1973).
45. W. R. Bontoyan and J. B. Looker, J. Agr. Food Chem., 21, 338 (1973).
46. G. Czeglédi-Jankó, J. Chromatog., 31, 89 (1967).
47. T. Sato and C. Tomizawa, Bull. Natl. Inst. Agr. Sci., Ser. C., 12, 181 (1960).
48. R. A. Ludwig, G. D. Thorn, and D. M. Miller, Can. J. Bot., 32, 48 (1954).
49. T. Hirai, Ann. Phytopathol. Soc. Japan, 27, 122 (1962).
50. R. Engst, W. Schnaak, and H. Rattba, Nachrbl. Deut. Pflanzenschutzdienst (Berlin), 22, 26 (1968).
51. G. Yip, J. H. Onley, and S. F. Howard, J. Assoc. Official Anal. Chemists, 54, 1373 (1971).
52. K. Vogeler, Ph. Dreze, A. Rap, H. Steffan, and H. Ullemeyer, in Pesticides (Suppl. 3 of Environ. Quality Safety), F. Coulston and F. Korte, eds., Thieme Verlag, Stuttgart, 1975, p. 67.

53. W. R. Lyman, in Pesticide Terminal Residues (A. S. Tahori, ed.), Butterworths, London, pp. 243-256.
54. R. J. Kuhr, A. C. Davis, and J. B. Bourke, Bull. Environ. Contam. Toxicol., 11, 224 (1974).
55. V. R. Wallen and I. Hoffman, Phytopathology, 49, 680 (1959).
56. E. Somers and D. V. Richmond, Nature, 194, 1194 (1962).
57. R. G. Owens and G. Blaak, Contrib. Boyce Thompson Inst., 20, 475 (1960).
58. H. D. Sisler, Conn. Agr. Expt. Stat. Bull., No. 663, 1963, p. 116.
59. D. V. Richmond and E. Somers, Ann. Appl. Biol., 57, 231 (1966).
60. D. V. Richmond and E. Somers, Ann. Appl. Biol., 62, 35 (1968).
61. R. J. Lukens and H. D. Sisler, Phytopathology, 48, 179 (1958).
62. M. R. Siegel and H. D. Sisler, Phytopathology, 58, 1123 (1968).
63. M. R. Siegel and H. D. Sisler, Phytopathology, 58, 1129 (1968).
64. M. R. Siegel, J. Agr. Food Chem., 18, 819 (1970).
65. M. R. Siegel, J. Agr. Food Chem., 18, 823 (1970).
66. M. R. Siegel, Pestic. Biochem. Phys., 1, 225 (1971).
67. C. J. Delp and H. L. Klöpping, Plant Disease Reptr., 52, 95 (1968).
68. G. P. Clemons and H. D. Sisler, Phytopathology, 59, 705 (1969).
69. W. W. Kilgore and E. R. White, Bull. Environ. Contam. Toxicol., 5, 67 (1970).
70. M. Chiba and F. Doornbos, Bull. Environ. Contam. Toxicol., 11, 273 (1974).
71. J. J. Sims, H. Mee, and D. C. Erwin, Phytopathology, 59, 1775 (1969).
72. C. A. Peterson and L. V. Edgington, J. Agr. Food Chem., 17, 898 (1969).
73. A. Fuchs, G. A. van den Berg, and L. C. Davidse, Pestic. Biochem. Physiol., 2, 192 (1972).
74. F. J. Baude, J. A. Gardiner, and J. C. Y. Han, J. Agr. Food Chem., 21, 1084 (1973).
75. P. M. Upham and C. J. Delp, Phytopathology, 63, 814 (1973).
76. D. Netzer and I. Dishon, Phytoparasitica, 1, 33 (1973).
77. C. A. Peterson and L. V. Edgington, Phytopathology, 60, 475 (1970).
78. M. R. Siegel and A. J. Zabbia, Jr., Phytopathology, 62, 630 (1972).
79. M. R. Siegel, Phytopathology, 63, 890 (1973).
80. Z. Solel, J. M. Schooley, and L. V. Edgington, Pestic. Sci., 4, 713 (1973).
81. A. Ben-Aziz and N. Aharonson, Pestic. Biochem. Physiol., 4, 120 (1974).
82. F. J. Baude, H. L. Pease, and R. F. Holt, J. Agr. Food Chem., 22, 413 (1974).
83. A. Helweg, Tidsskr. Planteavl., 77, 232 (1973).
84. J. R. Fleeker, H. Morgan Lacy, I. R. Schultz, and E. C. Houkom, J. Agr. Food Chem., 22, 592 (1974).
85. A. Helweg, Soil Biol. Biochem., 4, 377 (1972).
86. R. E. Weeks and H. G. Hedrick, Bacteriol. Proc., Abstr. A, 78, 14 (1971).

87. A. Fuchs and G. J. Bollen, in Systemfungizide (H. Lyr and C. Polter, eds.), Akademie-Verlag, Berlin, 1975, p. 121.
88. Y. Yasuda, S. Hashimoto, Y. Soeda, and T. Noguchi, Ann. Phytopathol. Soc. Japan, 39, 49 (1973).
89. L. C. Davidse, Pestic. Sci., 6, 538 (1976).
90. J. A. Gardiner, R. K. Brantley, and H. Sherman, J. Agr. Food Chem., 16, 1050 (1968).
91. J. A. Gardiner, J. J. Kirkland, H. L. Klöpping, and H. Sherman, J. Agr. Food Chem., 22, 419 (1974).
92. L. V. Edgington, K. L. Khew, and G. L. Barron, Phytopathology, 61, 32 (1971).
93. R. W. Marsh (ed.), Systemic Fungicides, Longman, London, 1972, Chaps. 7 and 10.
94. D. C. Erwin, M. C. Wang, and J. J. Sims, Phytopathology, 60, 1291 (1970).
95. M. C. Wang, D. C. Erwin, J. J. Sims, N. T. Keen, and D. E. Borum, Pestic. Biochem. Physiol., 1, 188 (1971).
96. L. E. Gray and J. B. Sinclair, Phytopathology, 61, 523 (1971).
97. R. B. Hine, D. L. Johnson, and C. J. Wenger, Phytopathology, 59, 798 (1969).
98. D. C. Erwin, J. J. Sims, D. E. Borum, and J. R. Childers, Phytopathology, 61, 964 (1971).
99. A. Frank, Acta Farmacol. Toxicol., 29, Suppl. 2, 1 (1971).
100. B. L. Kirkpatrick and J. B. Sinclair, Phytopathology, 63, 1532 (1973).
101. J. W. Vonk and A. Kaars Sijpesteijn, Pestic. Sci., 2, 160 (1971).
102. T. Noguchi, K. Ohkuma, and S. Kosaka, in Pesticide Terminal Residues (A. S. Tahori, ed.), Butterworths, London, 1971, pp. 257-270.
103. H. Buchenauer, L. V. Edgington, and F. Grossmann, Pestic. Sci., 4, 343 (1973).
104. Y. Soeda, S. Kosaka, and T. Noguchi, Agr. Biol. Chem. (Tokyo), 36, 931 (1972).
105. Y. Soeda, S. Kosaka, and T. Noguchi, Agr. Biol. Chem. (Tokyo), 36, 817 (1972).
106. H. Buchenauer, D. C. Erwin, and N. T. Keen, Phytopathology, 63, 1091 (1973).
107. A. Matta and I. A. Gentile, Mededel. Rijksfac. Landbouwwetenschap. Gent, 36, 1151 (1971).
108. J. W. Vonk and B. Mihanovic, Neth. J. Plant Pathol., Suppl., in press (1977).
109. T. Noguchi, in Environmental Toxicology of Pesticides (F. Matsumura, G. Boush, and T. Misato, eds.), Academic Press, New York, 1972, p. 607.
110. J. A. Styles and R. Garner, Mutation Res., 26, 177 (1974).

111. R. A. Burrell, J. M. Cox, and E. G. Savins, J. Chem. Soc. Perkin, I, 1973, p. 2707.
112. H. Su, W. W. Yu, and T. J. Wang, Nung Yeh Yen Chiu, 20, 84 (1971) (in Chinese); Chem. Abstr. (English transl.), 75, 97372 (1971).
113. A. I. Allam and J. B. Sinclair, Phytopathology, 59, 1548 (1969).
114. M. Snel and L. V. Edgington, Phytopathology, 60, 1708 (1970).
115. W. T. Chin, G. M. Stone, and A. E. Smith, J. Agr. Food Chem., 18, 709 (1970).
116. W. T. Chin, G. M. Stone, and A. E. Smith, in Pesticide Terminal Residues (A. S. Tahori, ed.), Butterworths, London, 1971, pp. 271-279.
117. P. Leroux and M. Gredt, Phytiat. Phytopharm., 21, 45 (1972).
118. D. E. Briggs, R. H. Waring, and A. M. Hackett, Pestic. Sci., 5, 599 (1974).
119. L. Newby and B. G. Tweedy, Phytopathology, 60, 6 (1970).
120. H. Lyr, G. Ritter, and C. Polter, in Systemfungizide (H. Lyr and C. Polter, eds.), Akademie-Verlag, Berlin, 1975.
121. W. T. Chin, G. M. Stone, and A. E. Smith, J. Agr. Food Chem., 18, 731 (1970).
122. P. R. Wallnöfer and G. Engelhardt, Arch. Mikrobiol., 80, 315 (1971).
123. G. Engelhardt, P. R. Wallnöfer, and R. Plapp, Appl. Microbiol., 22, 284 (1971).
124. G. Engelhardt, P. R. Wallnöfer, and R. Plapp, Appl. Microbiol., 26, 709 (1973).
125. R. Bochoffer, O. Oltmans, and F. Lingens, Arch. Mikrobiol., 90, 141 (1973).
126. H. Lyr, G. Ritter, and L. Banasiak, Z. Allgem. Mikrobiol., 14, 313 (1974).
127. M. Uyeda, S. Takeobu, and M. Hongo, Agr. Biol. Chem. (Tokyo), 38, 1797 (1974).
128. E. H. Pommer, Mededel. Landbouwhogeschool Gent, 33, 1019 (1968).
129. P. R. Wallnöfer, S. Safe, and O. Hutzinger, Pestic. Biochem. Physiol., 1, 458 (1971).
130. P. R. Wallnöfer, M. Koniger, S. Safe, and O. Hutzinger, J. Agr. Food Chem., 20, 20 (1972).
131. S. S. Lee and S. C. Fang, Weed Res., 13, 59 (1973).
132. G. W. McClure, Weed Science, 22, 323 (1974).
133. Y. Uesugi, Japan. Pestic. Inform., 2, 11-14 (1970).
134. A. H. Huisman, Mededel. Landbouwhogeschool Gent, 37, 511 (1972).
135. M. A. de Waard, Neth. J. Plant Pathol., 78, 186 (1972).
136. M. A. de Waard, Mededel. Landbouwhogeschool Wageningen, 74, 14 (1974).
137. H. Yamamoto, C. Tomizawa, Y. Uesugi, and T. Murai, Agr. Biol. Chem., 37, 1553 (1973).
138. T. Masuda and J. Kanazawa, Agr. Biol. Chem., 37, 2931 (1973).

139. C. Tomizawa and Y. Uesugi, Agr. Biol. Chem., 36, 294 (1972).
140. K. Ishizuka, I. Takase, K. E. Tan, and S. Mitsui, Agr. Biol. Chem., 37, 1307 (1973).
141. I. Takase, K. E. Tan, and K. Ishizuka, Agr. Biol. Chem., 37, 1563 (1973).
142. I. Ueyama, Y. Uesugi, C. Tomizawa, and T. Murai, Agr. Biol. Chem., 37, 1543 (1973).
143. Y. Uesugi and C. Tomizawa, Agr. Biol. Chem., 35, 941 (1971).
144. Y. Uesugi and C. Tomizawa, Agr. Biol. Chem., 36, 313 (1971).
145. R. S. Elias, M. C. Shephard, B. K. Snell, and J. Stubbs, Nature, 219, 5159 (1968).
146. A. Calderbank, Acta Phytopathol. Acad. Sci. Hung., 6, 355 (1971).
147. P. Slade, B. D. Cavell, R. J. Hemingway, and M. J. Sampson, 2nd Intern. Congr. Pestic. Chem., Tel Aviv, 1971, p. 5.
148. J. Dekker and A. J. P. Oort, Phytopathology, 54, 815 (1964).
149. J. Dekker, Neth. J. Plant Pathol., 74, 127 (1968).
150. H. M. Dekhuijzen and J. Dekker, Pestic. Biochem. Physiol., 1, 11 (1971).
151. H. M. Dekhuijzen and J. Dekker, Acta Phytopathol. Acad. Sci. Hung., 6, 339 (1971).
152. G. Matolcsy and N. Poonawalla, Acta Phytopathol. Acad. Sci. Hung., 2, 109 (1967).
153. J. Beck and K. E. Hansen, Pestic. Sci., 5, 41 (1974).
154. J. H. Smelt and M. Leistra, Agr. Environ., 1, 65 (1974).
155. C. H. Wang and F. E. Broadbent, Soil Sci. Soc. Amer. Proc., 36, 742 (1972).
156. W. H. Ko and J. D. Farley, Phytopathology, 59, 64 (1969).
157. J. C. Caseley, Bull. Environ. Contam. Toxicol., 3, 180 (1968).
158. C. I. Chacko, J. L. Lockwood, and M. Zabik, Science, 154, 893 (1966).
159. T. Nakanishi and H. Oku, Phytopathology, 59, 1761 (1969).
160. S. Gorbach and U. Wagner, J. Agr. Food Chem., 15, 654 (1967).
161. E. J. Kuchar, F. O. Geenty, W. P. Griffith, and R. J. Thomas, J. Agr. Food Chem., 17, 1237 (1969).
162. C. Engel, A. P. de Groot, and C. Weurman, Science, 154, 270 (1966).
163. R. F. Curtis, C. Dennis, J. M. Gee, M. G. Gee, N. M. Griffith, D. G. Land, J. L. Peel, and D. Robinson, J. Sci. Food Agr., 85, 811 (1974).
164. A. J. Cserjesi and E. L. Johnson, Can. J. Microbiol., 18, 45 (1972).
165. J. P. Chu and E. J. Kirsch, Appl. Microbiol., 23, 1033 (1972).
166. I. Watanabe, Soil Sci. Plant Nutr. (Tokyo), 19, 109 (1973).
167. N. K. van Alfen and T. Kosuge, J. Agr. Food Chem., 24, 584 (1976).
168. N. K. van Alfen and T. Kosuge, J. Agr. Food Chem., 22, 221 (1974).
169. K. Groves and K. S. Chough, J. Agr. Food Chem., 18, 1127 (1970).

170. A. J. Lemin, J. Agr. Food Chem., 13, 557 (1965).
171. U.S. Dept. of Agriculture, 1967 Progress Report on Pesticides and Related Activities, Washington, D.C., 1968, p. 194.
172. D. D. Kaufman, J. R. Plimmer, and U. I. Klingebiel, J. Agr. Food Chem., 21, 127 (1973).
173. C. Madhosingh, Can. J. Microbiol., 7, 553 (1961).
174. B. T. Kirk, J. B. Sinclair, and E. N. Lambremont, Phytopathology, 59, 1473 (1969).
175. R. C. Rhodes, H. L. Pease, and R. K. Brantley, J. Agr. Food Chem., 19, 745 (1971).
176. W. K. Hock and H. D. Sisler, J. Agr. Food Chem., 17, 123 (1969).
177. M. V. Wiese and J. M. Vargas, Pestic. Biochem. Physiol., 3, 214 (1973).
178. G. D. Thorn, Pestic. Biochem. Physiol., 3, 137 (1973).
179. K. D. Freytag, Pestic. Sci., 5, 731 (1974).
180. R. D. Barnes, A. T. Bull, and R. C. Poller, Chem. Ind. (London), 1971, p. 204.
181. R. D. Barnes, A. T. Bull, and R. C. Poller, Pestic. Sci., 4, 305 (1973).
182. K. Furukawa, T. Suzuki, and K. Tonomura, Agr. Biol. Chem. (Tokyo), 33, 128 (1969).
183. B. Williams, L. G. Dring, and R. T. Williams, Biochem. J., 127, 24P (1972).
184. C. Tomizawa, J. Food Hyg. Soc. Japan, 7, 26 (1966).
185. M. Takeda, K. Isobe, T. Nigo, H. Tanabe, and I. Kawashiro, J. Food Hyg. Soc. Japan, 12, 152 (1971).
186. M. Takeda and K. Isobe, J. Food Hyg. Soc. Japan, 12, 160 (1971).
187. W. C. Spanis, D. E. Munnecke, and R. A. Solberg, Phytopathology, 52, 455 (1962).
188. Y. Kimura and V. L. Miller, J. Agr. Food Chem., 12, 253 (1964).
189. W. J. Spranger, J. L. Spigarelli, J. M. Rose, and H. P. Miller, Science, 180, 192 (1973).
190. W. J. Spranger, J. L. Spigarelli, J. M. Rose, R. S. Flippin, and H. P. Miller, Appl. Microbiol., 25, 488 (1973).
191. J. D. Nelson, W. Blair, F. E. Brinckman, R. R. Colwell, and W. P. Iverson, Appl. Microbiol., 26, 321 (1973).
192. F. Matsumura, Y. Gotoh, and G. M. Boush, Science, 173, 49 (1971).
193. K. Tonomura, K. Maeda, F. Futai, T. Nakagami, and M. Yamada, Nature, 217, 644 (1968).
194. K. Tonomura and F. Kanzaki, Biochim. Biophys. Acta, 184, 227 (1969).
195. K. Furukawa, T. Suzuki, and K. Tonomura, Agr. Biol. Chem. (Tokyo), 33, 128 (1969).
196. K. Furukawa and K. Tonomura, Agr. Biol. Chem. (Tokyo), 35, 604 (1971).

197. R. Alconero and D. J. Hagedorn, Phytopathology, 58, 34 (1968).
198. J. E. Mitchell and D. J. Hagedorn, Phytopathology, 61, 978 (1971).
199. N. G. K. Karanth, S. G. Bhat, C. S. Vaidyanathan, and V. N. Vasantharajan, Appl. Microbiol., 27, 43 (1974).
200. F. J. Hills, Phytopathology, 52, 389 (1962).
201. D. Lunt and W. C. Evans, Biochem. J., 118, 54P (1970).
202. D. T. Gibson, R. L. Roberts, M. C. Wells, and V. M. Kobal, Biochem. Biophys. Res. Commun., 50, 211 (1973).
203. D. Catelani, C. Sorlini, and V. Treccani, Experientia, 27, 1173 (1971).
204. D. Catelani, A. Colombi, C. Sorlini, and V. Treccani, Biochem. J., 134, 1063 (1973).
205. D. Catelani and A. Colombi, Biochem. J., 143, 431 (1974).
206. M. C. Goldberg and R. L. Wershaw, U.S. Geol. Survey Progr. Paper D235, Washington, D.C., 1969.
207. J. A. Bartz and J. E. Mitchell, Phytopathology, 60, 350 (1970).
208. A. N. Curry, J. Agr. Food Chem., 10, 13 (1962).
209. S. Sumida, R. Yoshihara, and J. Miyamoto, Agr. Biol. Chem. (Tokyo), 37, 2781 (1973).
210. S. Otto and N. Drescher, Proc. 7th Brit. Insectic. Fungic. Conf., Brighton, England, 1973, p. 56.
211. Calixin, Documentation BASF, Landwirtschaftliche Versuchsstation, Limburgerhof, FRG, gpe/II-1976.
212. A. Fuchs, S. Duma, and J. Voros, Neth. J. Plant Pathol., 77, 42 (1971).
213. C. Ebenebe, V. v. Bruchhausen, and F. Grossmann, Pestic. Sci., 5, 17 (1974).
214. J. D. Gilpatrick and J. Bourke, Mededel. Landbouwhogeschool Gent, 38, 1537 (1973).
215. C. A. Drandarevski and A. Fuchs, Mededel. Landbouwhogeschool Gent, 38, 1525 (1973).
216. A. Fuchs, M. Viets-Verweij, J. Voros, and F. W. de Vries, Acta Phytopathol. Hung., 6, 347 (1971).
217. A. Fuchs, M. Viets-Verweij, and F. W. de Vries, Phytopathol. Z., 75, 111 (1972).
218. V. v. Bruchhausen and M. Stiasni, Pestic. Sci., 4, 767 (1973).
219. A. Fuchs and W. Ost, Arch. Environ. Contam. Toxicol., 4, 30 (1976).
220. A. Fuchs, F. W. de Vries, and A. M. J. Aalbers, Pestic. Sci., 7, 115 (1976).
221. M. G. Bastiaansen, E. A. Pieroh, and E. Aalbers, Mededel. Rijksfac. Landbouwwetenschap. Gent, 39, 1019 (1974).
222. I. Iwan and D. Goller, in Pesticides (Suppl. 3 of Environ. Quality Safety), F. Coulston and F. Korte, eds., Thieme Verlag, Stuttgart, 1975, p. 319.
223. H. Seto, N. Otake, and H. Yonehara, Agr. Biol. Chem. (Tokyo), 30, 877 (1966).

224. I. Yamaguchi, K. Takagi, and T. Misato, Agr. Biol. Chem. (Tokyo), 36, 1719 (1972).
225. D. R. Hawkins and V. H. Saggers, Pestic. Sci., 5, 497 (1974).
226. P. G. Vincent and H. D. Sisler, Physiol. Plantarum, 21, 1249 (1968).
227. R. W. Tillman, M. R. Siegel, and J. W. Long, Pestic. Biochem. Physiol., 3, 160 (1973).
228. R. G. Owens and G. Blaak, Contrib. Boyce Thompson Inst., 20, 475 (1960).
229. R. E. Hoagland and D. S. Frear, J. Agr. Food Chem., 24, 129 (1976).
230. J. W. Vonk, Mededel. Rijksfac. Landbouwwetenschap. Gent, 41, 1883 (1976).
231. W. H. Newsome, J. Agr. Food Chem., 24, 999 (1976).
232. W. H. Newsome, J. B. Shields, and B. C. Villeneuve, J. Agr. Food Chem., 23, 756 (1975).
233. R. G. Nash, J. Agr. Food Chem., 24, 596 (1976).
234. R. Engst, Environmental significance of ETU, to be published by IUPAC.
235. J. P. Rouchaud, J. R. Decallonne, and J. A. Meyer, Phytopathology, 64, 1513 (1974).
236. M. R. Siegel, Phytopathology, 65, 219 (1975).
237. T. A. Jacob, J. R. Carlin, R. W. Walker, F. J. Wolf, and W. J. A. VandenHeuvel, J. Agr. Food Chem., 23, 704 (1975).
238. N. Aharonson and U. Kafkafi, J. Agr. Food Chem., 23, 720 (1975).
239. H. Oeser, S. Gorbach, and E. Boerner, in Pesticides (Suppl. 3 of Environ. Quality Safety), F. Coulston and F. Korte, eds., Thieme Verlag, Stuttgart, 1975, p. 557.
240. S. Gorbach, W. Thier, H. M. Kellner, E. F. Schulze, K. Kuenzler, and H. Fisher, in Pesticides (Suppl. 3 of Environ. Quality Safety), F. Coulston and F. Korte, eds., Thieme Verlag, Stuttgart, 1975, p. 840.
241. C. Tomizawa, Y. Uesugi, I. Ueyama, and H. Yamamoto, J. Environ. Sci. Health, B 11, 231 (1976).
242. R. J. Hemingway, in The Persistence of Insecticides and Herbicides (K. I. Beynon, ed.), British Crop Protection Council, Monograph No. 17, Droitwich, 1976, p. 55.
243. N. G. K. Karanth, S. G. Bhat, C. S. Vaidyanathan, and V. N. Vasantharajan, Pestic. Biochem. Physiol., 6, 20 (1976).
244. R. H. Waring and M. S. Wolfe, Pestic. Science, 6, 169 (1975).
245. J. Iwan, B. Hertel, and J. Radquweit, Neth. J. Plant Pathol., Suppl., in press (1977).
246. H. Ohkawa, R. Shibaike, Y. Okihara, M. Morikawa, and J. Miyamoto, Agr. Biol. Chem., 40, 943 (1976).

Chapter 4

BIOLOGICAL CONVERSIONS OF FUNGICIDES IN ANIMALS

Gaylord D. Paulson

Metabolism and Radiation Research Laboratory
United States Department of Agriculture
Agricultural Research Service
Fargo, North Dakota

I. INTRODUCTION

With the increased development and use of pesticides, a growing concern for possible harmful effects of these compounds in the environment has developed. Because animals are likely to be exposed to fungicides either directly or indirectly through contaminated feedstuffs, this concern is justified, especially since these compounds, or their metabolites, or both, may contaminate foods used for human consumption. Consequently, the number of studies and reports dealing with the metabolic fate of fungicides in animals and man has increased. This chapter is a summation of the available literature concerning the metabolic transformations of fungicides (exceptions cited below) in animals.

Publications by Torgeson [1], Thomson [2], and others have documented the fact that a tremendous variety of compounds (elemental, inorganic, and organic) have been or are being used as fungicides. Because of time and space limitations and because of the nature of most fungicides used in modern agricultural practices, this review will be restricted to organic fungicides [1]. Because comprehensive reviews about the fate of mercury in the environment, including that in mercury-containing fungicides, have been published recently [3-5], those compounds will not be discussed in this chapter.

The organic fungicides are, in themselves, a very diverse group of compounds that are used on a wide variety of crops or products and with varied cultural practices. Thus, these compounds could be classified in many ways; however, in this review, the general classification scheme used by Torgeson [1] is followed. Information emphasized in this review includes (1) the rates and routes of elimination of the fungicide or their metabolites, or both; (2) the isolation and identification of fungicide metabolites in milk, tissues, eggs, urine, and feces; and (3) the quantitation of the parent compounds, their metabolites, or both, in the various fractions. Special emphasis is placed on the critical evaluation of the supportive evidence for the structure of metabolites that have been reported.

A literature search of Chemical Abstracts and Pesticide Abstracts was conducted in an attempt to obtain all pertinent information published from 1964 through 1974. Additional information from older literature was also included when it was relevant to the compounds being discussed. Although a sincere effort was made to include all available literature concerning the animal metabolism of the organic fungicides, some reports were probably overlooked. Any such errors were oversights and not intentional omissions or deletions.

II. THIOCARBAMATE AND RELATED FUNGICIDES

The thiocarbamate fungicides that have been studied in animals were reportedly absorbed from the gut to varying degrees and metabolized to a number of products. However, the degree of metabolism and the nature of the metabolites vary considerably with structural types [6].

Vekshtein and Khitsenko [7] administered a single oral dose of ziram to rats and then killed the animals at 0.5, 16, 24, 48, 72, and 144 hr after dosing. Ziram was detected in the blood for 2 days after dosing and in the intestine, liver, and kidneys until the end of the experiment. However, no ziram was detected in the urine of these animals. Large amounts of tetramethylthiuram disulfide and smaller amounts of the dimethylamine salt of dimethyldithiocarbamic acid were excreted in the urine from the 2nd until the 6th day of the experiment (see Table 1). The expiratory air from these animals contained carbon disulfide and dimethylamine. The body tissues contained tetramethylthiuram disulfide, tetramethylthiourea, the dimethylamine salt of dimethyldithiocarbamic acid, carbon disulfide, and dimethylamine (see Table 1). The concentration of ziram in the tissues decreased and the metabolites generally increased with time after dosing; the largest buildup of ziram metabolites was in the spleen and lungs. Vekshtein and Khitsenko [7] also found that ziram was converted to a variety of similar metabolites when incubated with gastric juice in vitro (see Table 1). More recent studies confirmed that [^{35}S]ziram was metabolized when given orally to rats [8, 9].

An interesting study by Eisenbrand et al. [10] showed that the carcinogen, dimethylnitrosamine, was formed when ziram was incubated with sodium nitrite under conditions similar to those existing in the stomach of the rat. The same compound was formed in vivo when ziram (10^{-4} moles) and $NaNO_2$ (4×10^{-3} moles) were given simultaneously by stomach tube. The average yield of dimethylnitrosamine from ziram during the 15-min experimental period was about 0.9% of the theoretical value. Ferbam and DPTD also yielded dimethylnitrosamine under similar conditions; Eisenbrand et al. [10] concluded that residual dithiocarbamate fungicides in the human diet represent a potential starting material for the formation of carcinogenic nitrosamines when they are ingested with nitrite.

Approximately 50% of the radioactivity in a single oral dose of [^{35}S]ferbam or [^{14}C]ferbam was absorbed from the gut of rats within 24 hr after dosing [11]. When rats were given the [^{35}S]ferbam, 18% of the ^{35}S was eliminated in the expiratory air, 23% in the urine, and 1% in the bile within 24 hr. When rats were given [^{14}C]ferbam, less than 0.1% of the ^{14}C was eliminated in the expiratory air, 43% in the urine, and 1.4% in the bile. The bile and urine did not contain any of the parent compound but contained a number of metabolites that were not identified. When [^{14}C]ferbam was given to pregnant rats, the fetuses contained ^{14}C, and when [^{14}C]ferbam was given to lactating rats, the milk contained ^{14}C. The nature of the radiolabeled compound(s) in the tissues and milk was not determined.

Tetraethylthiuram disulfide (TTD) has been used to treat chronic alcoholism and probably for this reason its animal metabolism has been more thoroughly investigated than that of other thiuram disulfides. Many of the early studies with this compound have been summarized [6, 12, 13]. In general, these studies indicated that TTD was absorbed from the gut [14-17] and that a large percentage of the sulfur was excreted in the urine [15]. However, significant amounts of sulfur from TTD were retained in the body for several days after dosing [15]. The parent compound, TTD, was not detected in the urine of humans and animals [14, 17] but its reduction product, diethyldithiocarbamate, was observed in rabbit urine [17, 18], rabbit

TABLE 1

The Metabolism of Thiocarbamate and Related Fungicides in Animals

Parent compound and name	Metabolites			
	Proposed structure	Source	Evidence for structure	Reference
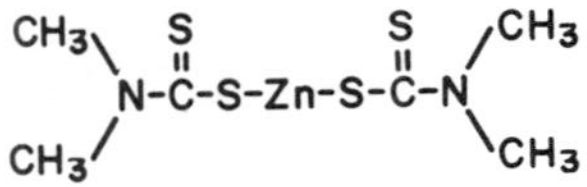 $(CH_3)_2N\text{-}C(=S)\text{-}S\text{-}Zn\text{-}S\text{-}C(=S)\text{-}N(CH_3)_2$ Ziram Zinc dimethyldithiocarbamate	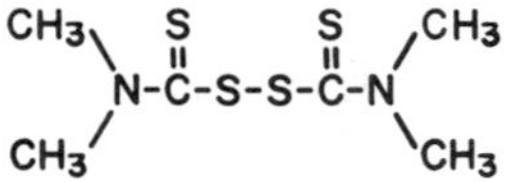 $(CH_3)_2N\text{-}C(=S)\text{-}S\text{-}S\text{-}C(=S)\text{-}N(CH_3)_2$	Rat gastric juice in vitro, tissues, and urine	Comparative TLC[a] UV spectroscopy	7
	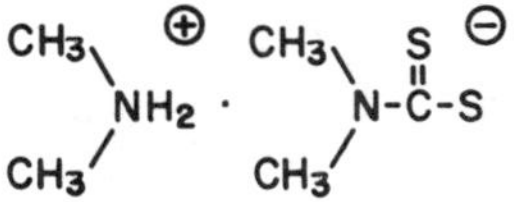 $(CH_3)_2NH_2^{\oplus} \cdot (CH_3)_2N\text{-}C(=S)\text{-}S^{\ominus}$	Rat gastric juice in vitro, tissues, and urine	Comparative TLC and UV spectroscopy	7
	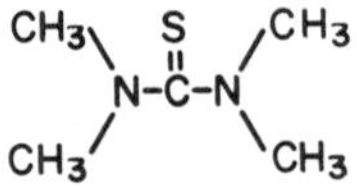 $(CH_3)_2N\text{-}C(=S)\text{-}N(CH_3)_2$	Rat gastric juice in vitro, tissues, and urine	Comparative TLC and UV spectroscopy	7
	CS_2	Rat tissues, urine, and expiratory air	Colorimetric assay	7
	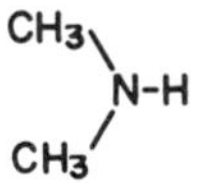 $(CH_3)_2N\text{-}H$	Rat tissues and urine	Comparative TLC and UV spectroscopy	7
		Expiratory air from rat	Derivative formation	7

	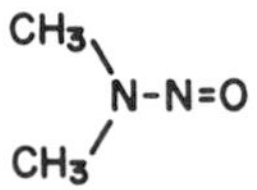 $(CH_3)_2N-N=O$	Rat stomach contents	UV spectroscopy, GLC,[b] MS[c]	10
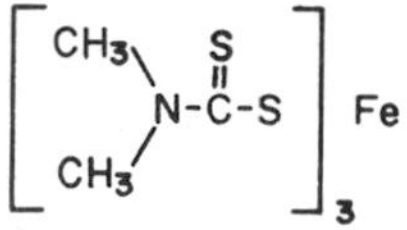 $[(CH_3)_2N-C(=S)-S]_3Fe$ Ferbam Ferric dimethyl-dithiocarbamate	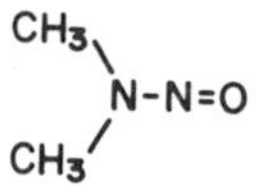 $(CH_3)_2N-N=O$	Rat stomach contents	UV spectroscopy, GLC, MS	10
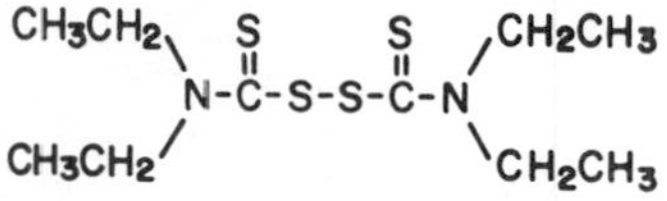 $(CH_3CH_2)_2N-C(=S)-S-S-C(=S)-N(CH_2CH_3)_2$ TTD Tetraethylthiuram disulfide	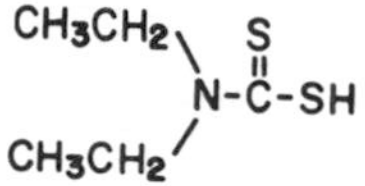 $(CH_3CH_2)_2N-C(=S)-SH$	Rabbit tissues, urine, and feces	Colorimetric assay	18
		Human urine	Crystallization studies	18
		Rabbit urine	Comparative column chromatography	17
	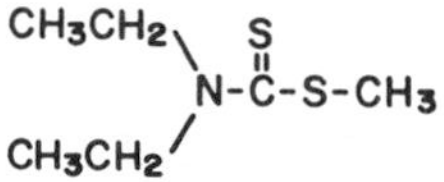 $(CH_3CH_2)_2N-C(=S)-S-CH_3$	Rat and mouse tissues in vitro	Comparative chromatography and isotope dilution	22
		Rat and mouse tissues in vivo	Comparative chromatography and isotope dilution	22

TABLE 1 (Cont.)

Parent compound and name	Metabolites			
	Proposed structure	Source	Evidence for structure	Reference
	Glucuronide (COOH, ring O) –S–C(=S)–N(CH_2CH_3)$_2$	Rat urine	Comparative chromatography and electrophoresis	12, 13
		Rat tissue	Comparative chromatography and electrophoresis	12
		Human urine	Comparative IR[d] and MP[e]	19
	SO_4^{2-}	Rat urine and tissue	Comparative chromatography and $BaSO_4$ precipitation	12, 13
	CS_2	Rat respiratory air	Properties of derivative	12, 13
		Human respiratory air	Colorimetric assay	21
		Rat liver in vitro	Colorimetric assay	20

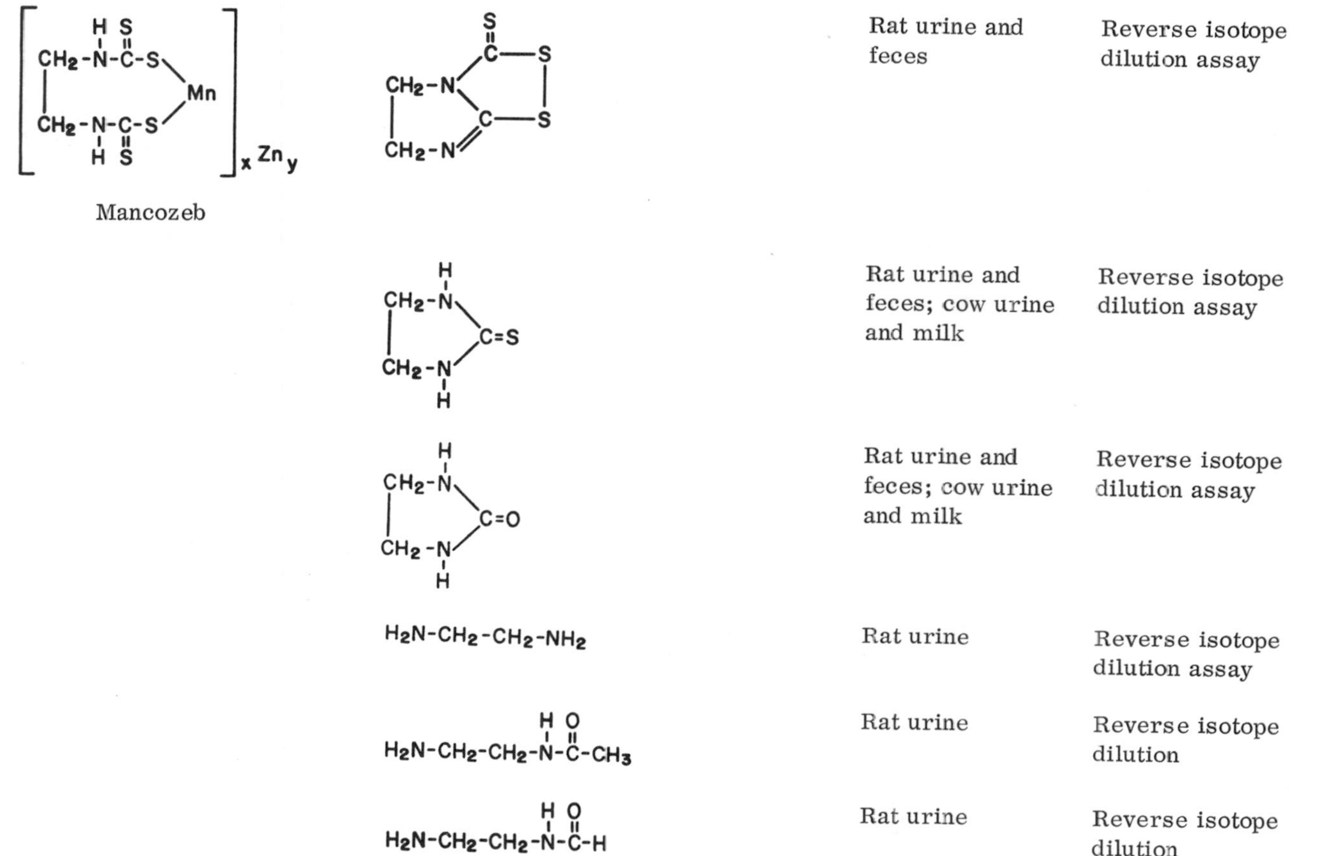

Compound	Metabolite	Sample	Method	Ref.
Mancozeb		Rat urine and feces	Reverse isotope dilution assay	25
		Rat urine and feces; cow urine and milk	Reverse isotope dilution assay	25
		Rat urine and feces; cow urine and milk	Reverse isotope dilution assay	25
	$H_2N\text{-}CH_2\text{-}CH_2\text{-}NH_2$	Rat urine	Reverse isotope dilution assay	25
	$H_2N\text{-}CH_2\text{-}CH_2\text{-}NH\text{-}C(=O)\text{-}CH_3$	Rat urine	Reverse isotope dilution	25
	$H_2N\text{-}CH_2\text{-}CH_2\text{-}NH\text{-}C(=O)\text{-}H$	Rat urine	Reverse isotope dilution	25

TABLE 1 (Cont.)

Parent compound and name	Metabolites: Proposed structure	Source	Evidence for structure	Reference
CH_2-N(H)-C(=S)-S–Mn–S-C(=S)-N(H)-CH_2 (ring) Maneb Manganous ethylene-1,2-bisdithiocarbamate	H_2N-CH_2-CH_2-NH_2	Rat urine and feces	Comparative chromatography	27
	CH_2-N(H)-C(=S)-N(H)-CH_2 (ring)	Rat urine and feces	Comparative chromatography	27
	CH_2-N(H)-C(=S)-S-C(=S)-N(H)-CH_2 (ring)	Rat urine and feces	Comparative chromatography	27
CH_2-N(H)-C(=S)-S–Zn–S-C(=S)-N(H)-CH_2 (ring) Zineb Zinc ethylene-1,2-bisdithiocarbamate	CH_2—CH_2, N=C, N, C=S, S—S (bicyclic)	Rat urine	Comparative spectrometry	28

	$\mathrm{CH_2{-}NH{-}C(=S){-}NH{-}CH_2}$ (ring)	Rat urine	Comparative spectrometry	28
$\mathrm{C_6H_4[NH{-}C(=S){-}NH{-}C(=O){-}OCH_2CH_3]_2}$ Thiophanate 1,2-Bis(3-ethoxycarbonyl-2-thioureido)benzene	$\mathrm{C_6H_4(NH{-}C(=S){-}NH{-}C(=O){-}OCH_2CH_3)(NH_2)}$	Mouse liver in vitro	Comparative TLC and color reaction	32
	$\mathrm{C_6H_4(NH{-}C(=S){-}NH{-}C(=O){-}OCH_2CH_3)(NH{-}C(=O){-}NH{-}C(=O){-}OCH_2CH_3)}$	Mouse liver in vitro	Comparative TLC	32
	$\mathrm{C_6H_4[NH{-}C(=O){-}NH{-}C(=O){-}OCH_2CH_3]_2}$	Mouse liver in vitro	Comparative TLC	32
	$\mathrm{C_6H_4(NH{-}C(=O){-}NH{-}C(=O){-}OCH_2CH_3)(NH_2)}$	Mouse liver in vitro	Comparative TLC and color reaction	32
	$\mathrm{C_6H_4(NH{-}C(=S){-}NH{-}C(=O){-}OCH_2CH_3)(NH{-}C(=O){-}NH_2)}$	Mouse liver in vitro	Comparative TLC	32

TABLE 1 (Cont.)

Parent compound and name	Metabolites: Proposed structure	Source	Evidence for structure	Reference
	$C_6H_4(NH\text{-}CO\text{-}NH\text{-}CO\text{-}OCH_2CH_3)(NH\text{-}CO\text{-}NH_2)$	Mouse liver in vitro	Comparative TLC	32
	2-aminobenzimidazole (NH_2)	Mouse and sheep liver in vitro; mouse urine and feces	Comparative TLC and color reaction	32
	HO-substituted 2-aminobenzimidazole (NH_2)	Mouse and sheep liver in vitro; mouse urine	Comparative TLC and color reaction	32
	[HO-substituted 2-aminobenzimidazole (NH_2)] Conj.	Mouse urine and feces	Comparative TLC and color reaction of hydrolysis product	32
	benzimidazole-2-yl $N(H)\text{-}C(=O)\text{-}OCH_2CH_3$	Mouse and sheep liver in vitro; mouse urine and feces	Comparative TLC	32

		Mouse feces; rat microsomes in vitro	Comparative TLC	30
	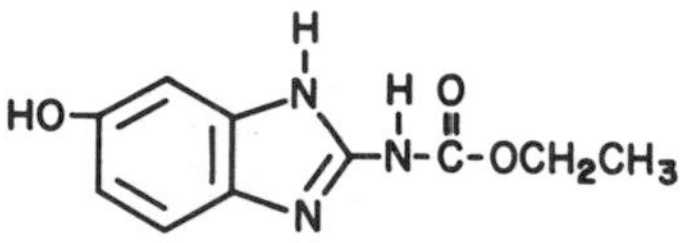	Mouse and sheep liver in vitro; mouse urine and feces	Comparative TLC and color reaction	32
	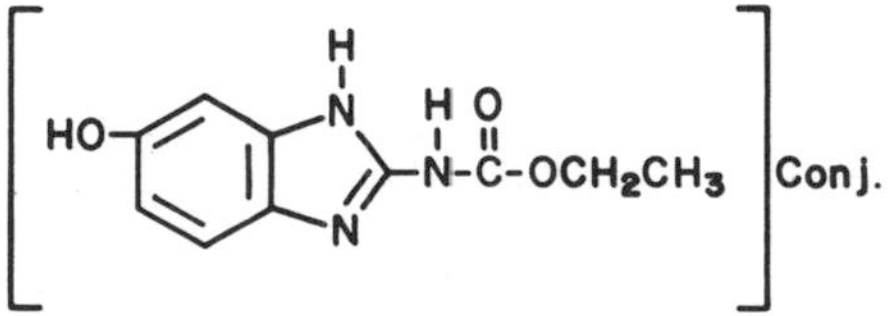 [ ] Conj.	Mouse urine and feces	Comparative TLC and color reaction of hydrolysis product	32
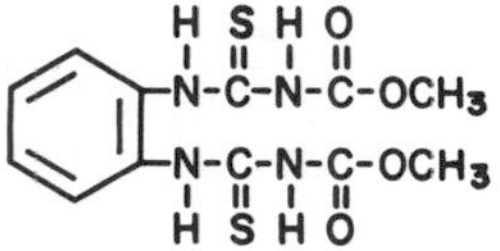 Thiophanate-methyl 1,2-Bis(3-methoxy carbonyl-2-thioureido)benzene	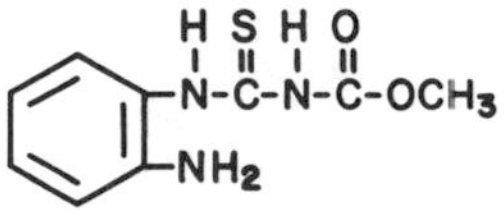	Mouse liver in vitro	Comparative TLC	32
	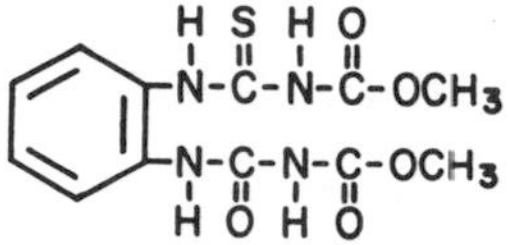	Mouse liver in vitro	Comparative TLC	32
	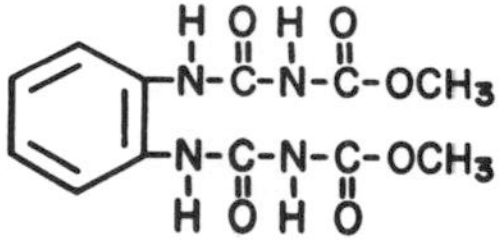	Mouse liver in vitro	Comparative TLC	32
		Mouse urine	Comparative TLC	31

TABLE 1 (Cont.)

Parent compound and name	Metabolites			
	Proposed structure	Source	Evidence for structure	Reference
	[benzene ring with -N(H)-C(=O)-N(H)-C(=O)-OCH$_3$ and -N(H)-C(=O)-N(H)-C(=O)-OCH$_3$] Conj.	Mouse urine	Comparative TLC after hydrolysis	31
	benzene ring with -N(H)-C(=O)-N(H)-C(=O)-OCH$_3$ and -NH$_2$	Rat microsomes in vitro	Comparative TLC	31
		Mouse liver in vitro	Comparative TLC and color reaction	32
	benzene ring with -N(H)-C(=S)-N(H)-C(=O)-OCH$_3$ and -N(H)-C(=S)-NH$_2$	Mouse liver in vitro	Comparative TLC	32
	benzene ring with -N(H)-C(=S)-N(H)-C(=O)-OCH$_3$ and -N(H)-C(=O)-NH$_2$	Mouse liver in vitro	Comparative TLC	32

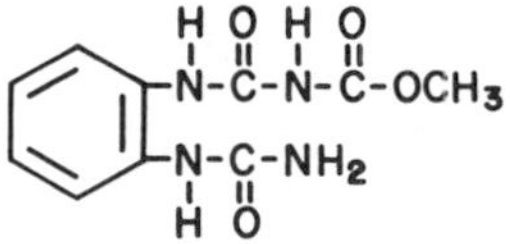	Mouse liver in vitro	Comparative TLC	32
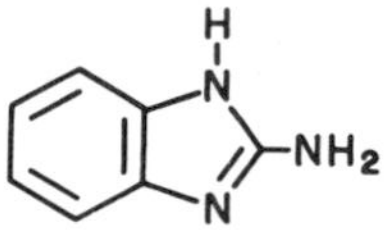	Mouse and sheep liver in vitro; mouse urine and feces; sheep urine and feces	Comparative TLC and color reaction	32
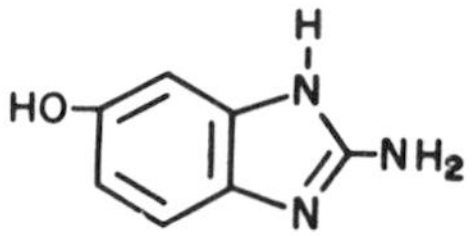	Mouse and sheep liver in vitro; mouse urine and feces; sheep urine and feces	Comparative TLC and color reaction	32
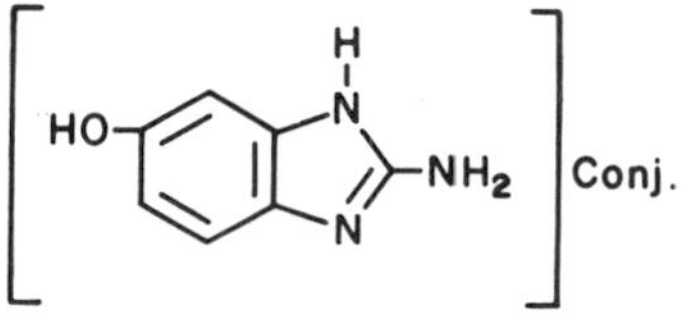 Conj.	Mouse urine and feces; sheep urine and feces	Comparative TLC and color reaction of hydrolysis product	32
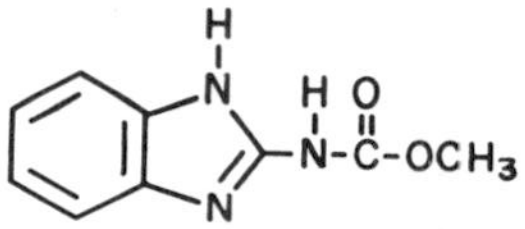	Rat microsomes in vitro; mouse urine	Comparative TLC	31
	Mouse and sheep liver in vitro; mouse urine and feces; sheep urine and feces	Comparative TLC	32

TABLE 1 (Cont.)

Parent compound and name	Metabolites: Proposed structure	Source	Evidence for structure	Reference
	[HO– (benzimidazole, N–H, N) –N(H)–C(=O)–OCH_3] Conj.	Mouse urine	Comparative TLC after hydrolysis	31
	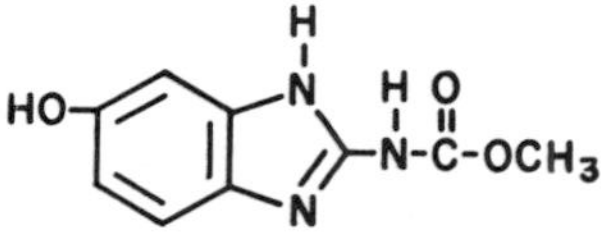 HO– (benzimidazole, N–H, N) –N(H)–C(=O)–OCH_3	Mouse and sheep liver in vitro; mouse urine and feces; sheep urine and feces	Comparative TLC and color reaction	32
		Mouse urine; rat microsomes in vitro	Comparative TLC	31
	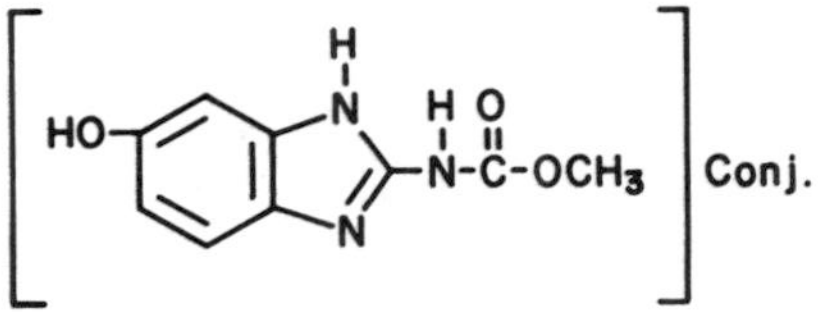 [HO– (benzimidazole, N–H, N) –N(H)–C(=O)–OCH_3] Conj.	Mouse urine and feces; sheep urine and feces	Comparative TLC and color reaction of hydrolysis product	32
		Mouse urine	Comparative TLC after hydrolysis	31

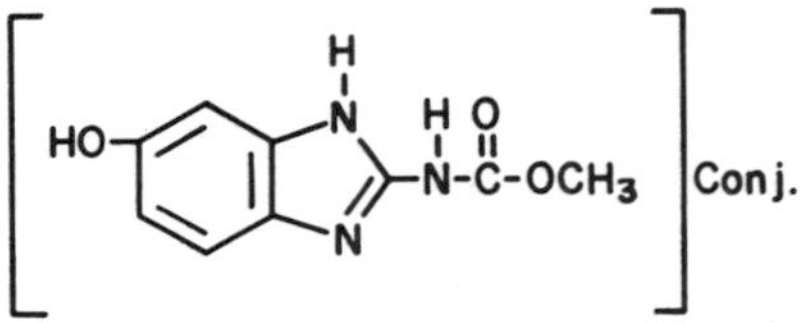

Compound	Metabolite	Source	Identification	Ref.
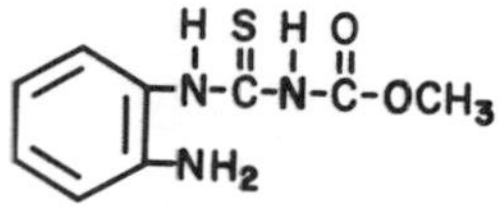 NF 48 2-(3-Methoxy carbonyl-2-thioureido)aniline	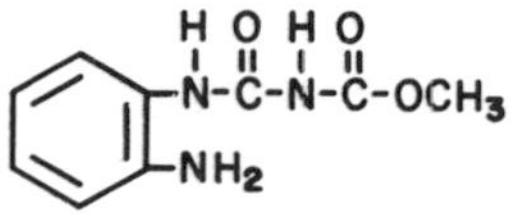	Mouse liver in vitro	Comparative TLC and color reaction	32
	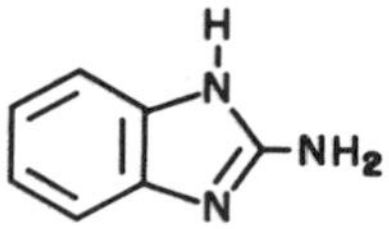	Mouse and sheep liver in vitro; sheep urine and feces	Comparative TLC and color reaction	31
	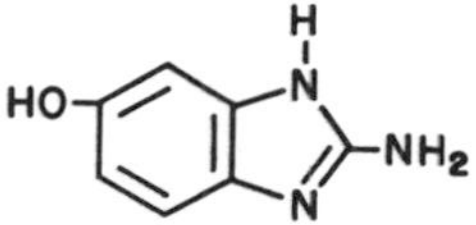	Mouse and sheep liver in vitro; sheep urine and feces	Comparative TLC and color reaction	31
	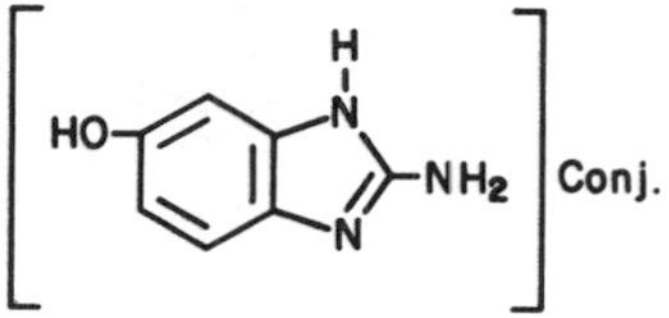 Conj.	Sheep urine and feces	Comparative TLC and color reaction of hydrolysis product	31
	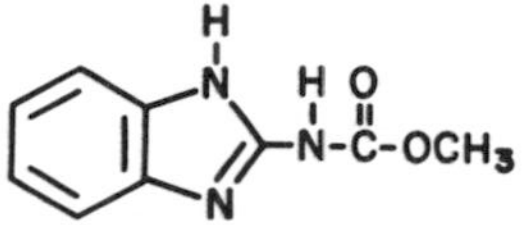	Mouse and sheep liver in vitro; sheep urine and feces	Comparative TLC	31

TABLE 1 (Cont.)

Parent compound and name	Proposed structure	Metabolites: Source	Evidence for structure	Reference
	HO, H, N, H O, N-C-OCH$_3$, N	Mouse and sheep liver in vitro; sheep urine and feces	Comparative TLC and color reaction	31
	[HO, H, N, H O, N-C-OCH$_3$, N] Conj.	Sheep urine and feces	Comparative TLC and color reaction of hydrolysis product	31

[a]TLC = thin-layer chromatography.
[b]GLC = gas-liquid chromatography.
[c]MS = mass spectrum.
[d]IR = infrared.
[e]MP = melting point.

feces [18], human urine [18], and rabbit tissues [18] after treatment with TTD. In a later study, the S-glucuronide of diethyldithiocarbamate was detected in the urine of humans given TTD [19]; also TTD was reported to be partly degraded to CS_2 [20, 21]. Strömme [12, 13] conducted detailed studies on the metabolism of [^{35}S]TTD in the rat. He reported that rats initially excreted primarily the [^{35}S]glucuronide of diethyldithiocarbamate in the urine after an intraperitoneal dose of [^{35}S]TTD; the total urinary excretion of this metabolite accounted for 57% of the ^{35}S given. Ethanol inhibited the formation of the S-glucuronide [12]. Detectable amounts of [^{35}S]diethyldithiocarbamate and [^{35}S]sulfate were also excreted in the urine, and approximately 2% of the ^{35}S was respired (presumably as $C^{35}S_2$) within 4 hr after dosing. One hr after the [^{35}S]TTD was given, all of the nonprotein bound ^{35}S in the plasma and liver was present as either $^{35}SO_4$ or the [^{35}S]glucuronide of diethyldithiocarbamate. However, much of the ^{35}S in the tissues was protein bound. Some findings indicated that the TTD reacted with serum proteins to form mixed disulfides [12]. An interesting study by Gessner and Jakubowski [22] showed that TTD was partly converted to the methyl ester of diethyldithiocarbamic acid (see Table 1) by rat and mouse tissues in vitro and in vivo. This metabolite was formed by an S-adenosylmethionine transferase enzyme in the microsomal fraction of both the liver and kidney. Some studies suggested that the sulfur in this compound was further metabolized to sulfate without intermediate cleavage to diethyldithiocarbamate.

Antonovich and Vekshtein [23] orally administered thiram to rats and observed the parent compound in blood and all major organs and glands of the body. Metabolites of thiram, if any, were not reported. The results of Robbins and Kastclic [24] suggested that thiram was degraded when incubated with rumen fluid in vitro and when orally administered to ruminants. However, all of their studies were conducted with unlabeled thiram, and the parent compound and metabolites were measured by degradation to H_2S or CS_2; determination of the latter components was complicated by interfering substances.

Rats orally dosed with [ethylene-^{14}C]mancozeb excreted 70.9% of the ^{14}C in the feces and 15.5% in the urine [25]. The parent compound accounted for most of the ^{14}C in the feces but was not detected in the urine. Ethylenebisisothiocyanate sulfide, ethyleneurea, and ethylenethiourea were present in both the urine and feces (see Table 1). In addition, the urine contained ethylenediamine and the acetyl and formyl derivatives of this compound (see Table 1). The nature of the radiolabeled residues in the rat tissues was not determined; but, interestingly, the concentration of ^{14}C in the thyroid was several times greater than that in all other tissues and organs analyzed. A cow dosed with [ethylene-^{14}C]mancozeb also excreted most of the ^{14}C in the feces and urine [25]. Ethyleneurea and ethylenethiourea were detected in both the urine and milk from this animal (see Table 1). Lyman [25] also reported that ^{14}C-labeled urea, oxalic acid, and glycine were present in the urine and that ^{14}C-labeled fat, protein, and lactose were present in the milk.

However, the characterization of these proposed metabolites was not described. The nature of the ^{14}C-labeled compounds in the tissues of the cow was not determined, but the thyroid had the highest specific activity. The concentration of ^{14}C in the thyroid of cows given [^{14}C]ethylenethiourea was several times higher than that in any of the other tissues analyzed [25]. Newsome [26] reported that ethylenethiourea concentrated in the thyroid of rats fed this compound.

Seidler et al. [27] dosed rats with [ethylene-^{14}C]maneb and reported that approximately 55% of the ^{14}C was excreted in the feces and urine within 3 days. The excreta contained ethylenediamine, ethylenebisthiuram monosulfide, ethylenethiourea (see Table 1), and unknown metabolites. One day after dosing, the rat tissues contained 1.2% of the ^{14}C given.

Rats exhaled CS_2 after being orally dosed with zineb and four metabolites were excreted in the urine [28]. The two major metabolites in the urine were identified as ethylenebisisothiocyanate sulfide and ethylenethiourea.

Information concerning the animal metabolism of thiocarbamate fungicides is not complete. Little concerning the animal metabolism of thiocarbamates such as thiram and metham has been reported. Perhaps the concern that some of these compounds may be degraded to metabolites that are carcinogenic [10, 25], goitrogenic, and teratogenic [26, 29] will stimulate additional metabolism studies.

The substituted thioureidobenzene fungicides that have been studied were at least partly absorbed from the gut and metabolized to a large number of metabolites that were excreted in the urine and feces. In vitro studies with preparations from both rodents and ruminants indicated that similar metabolites were formed.

Noguchi et al. [30] orally dosed mice with thiophanate and thiophanate-methyl that had been prepared with either a ^{14}C label in one of the thioureido groups or with a ^{35}S label. When [^{14}C]thiophanate was given, 49% of the ^{14}C was eliminated in the urine, 46% in the feces, and 6% in the expiratory air within 4 days. When the [^{35}S]thiophanate was given, 38% of the ^{35}S was eliminated in the urine, 50% in the feces, and 0.01% in the expiratory air. Of the radioactivity given as [^{14}C]thiophanate-methyl and as [^{35}S]thiophanate-methyl, 82.5 and 86.2%, respectively, was eliminated in the urine, and 19.6 and 19.2%, respectively, was eliminated in the feces. Time course studies showed that the ^{14}C and ^{35}S concentration in body tissues decreased with time but was still detectable 96 hr after dosing. The nature of the radiolabeled compounds in the tissues of mice dosed with thiophanate and thiophanate-methyl was not determined. One metabolite in the feces of mice dosed with thiophanate was identified as 2-ethoxycarbonylaminobenzimidazole (see Table 1); the same metabolite was formed when this fungicide was incubated with rat liver microsomes in vitro [30]. 5-Hydroxy-2-methoxycarbonylaminobenzimidazole was detected in the urine of mice treated with thiophanate-methyl and in a rat microsome preparation incubated with thiophanate-methyl (Table 1).

In a similar study, mice were orally dosed with thiophanate-methyl which had been prepared with a ^{35}S or ^{14}C label in either the benzene ring, the thioureido group, or a terminal methyl [31]. The elimination of radioactivity given as the four differently labeled compounds was similar. From 16.2 to 28.5% of the activity was eliminated in the feces and 66.9-88.9% in the urine, within 4 days after dosing. When the [$^{14}CH_3$]-labeled compound was given, 1.1% of the activity was eliminated in the respiratory air. Radioactivity was observed in the tissues of mice 96 hr after dosing with all of the radiolabeled thiophanate-methyl preparations, but the ^{14}C given as [phenyl-^{14}C]thiophanate-methyl was most rapidly eliminated from the tissues. The nature of the radiolabeled compounds in the tissues of these animals was not determined; however, 1,2-bis(3-methyl-carbonyl-2-ureido)benzene was present in mouse urine (both free and conjugated) and was formed by rat microsomes in vitro (see Table 1). Other metabolites identified in mouse urine and in rat liver microsome preparations included methylbenzimidazol-2-yl carbamate and methyl 5-hydroxybenzimidazol-2-yl carbamate. Numerous other radiolabeled metabolites were not identified.

More recently, Douch [32] reported extensive studies on the metabolism of thiophanate, thiophanate-methyl, and a related fungicide NF 48 in mice and sheep and in in vitro preparations. When the sheep and mice were orally dosed with these compounds, from 42 to 65% of the dose was accounted for as metabolites in the urine and feces collected for 72 hr after dosing. Both sheep and mice converted these fungicides into a variety of benzimidazole, ethyl benzimidazol-2-yl-carbamate, and methyl benzimidazol-2-yl-carbamate derivatives and their hydroxylation products (see Table 1). From 9 to 14% of the dose was eliminated in the urine and feces as conjugated metabolites. In vitro studies with sheep liver and mouse liver resulted in the production of many of the same metabolites (see Table 1). The in vitro metabolism to benzimidazole compounds required NADPH (nicotinamide adenine dinucleotide phosphate) and was inhibited by SKF 525-A and carbon monoxide.

III. ORGANOMETALLIC FUNGICIDES

Up to 95% of the ^{113}Sn given to sheep as [^{113}Sn]triphenyltin acetate was eliminated in the feces and 0.6% was eliminated in the urine [33, 34]. The amount of ^{113}Sn in the urine, blood, and milk decreased after dosing but was still detectable in these components 95 days after treatment and in other tissues 218 days after treatment. Some of the ^{113}Sn in the milk was present in inorganic form.

When plant material contaminated with triphenyltin acetate was fed to cows, most of the residues were excreted in the feces [35, 36]. Herok and Götte [33, 34] suggested, on the basis of graphic analysis of excretion curves, that the parent compound or the hydroxide was rapidly eliminated and that metabolites were more slowly eliminated. Studies with tri[phenyl-^{14}C]tin acetate would be of interest. When a related compound tri[phenyl-^{14}C]lead acetate was orally administered to rats, 25% of the ^{14}C was eliminated in the urine, 29% in the feces, and 20% in the expired air within 7 days [37].

A number of reports have been published concerning the toxicity of organotin compounds (particularly alkyltin), the binding of these compounds to mitochondria, their effect on oxidative phosphorylation, and their hemolytic activity [38-45]. However, published information concerning the animal metabolism of these compounds is very limited.

IV. HETEROCYCLIC NITROGEN FUNGICIDES

Sheep given an oral dose of [benzene-^{14}C]thiabendazole eliminated 73-77% of the ^{14}C in the urine and 13-16% in the feces within 48 hr [46]. The level of ^{14}C in the blood was highest about 5 hr after dosing and then declined. The level of radioactivity in tissues from sheep dosed with either [^{14}C]thiabendazole or [^{35}S]thiabendazole gradually declined from 6 hr until 16 days after dosing, and then residues were no longer detected. The urine from sheep dosed with both ^{14}C- and ^{35}S-labeled thiabendazole contained six radiolabeled components. The parent compound (present in only trace amounts) 5-hydroxy thiabendazole (5-OH-TBZ), the sulfate ester of 5-OH-TBZ, and the glucuronic acid conjugate of 5-OH-TBZ were detected in the urine (see Table 2). Two minor urinary metabolites were not identified and none of the ^{14}C in the feces and tissues was characterized. In a later study, goats, cattle, and pigs excreted approximately 65% of the radioactivity in the urine and approximately 20% in the feces within 48 hr after being orally dosed with either ^{14}C, ^{35}S, or ^{3}H-labeled thiabendazole [47]. When [^{14}C]-thiabendazole was given to humans, rats, and dogs [47] the average percentages of the ^{14}C excreted in the urine and feces, respectively, were as follows: man, 87 and 5% (48 hr); rat, 58 and 28% (48 hr); dog, 35 and 47% (72 hr). In man, the concentration of ^{14}C in the plasma was highest 1-2 hr after dosing and then declined. A large percentage of the ^{14}C in the plasma of humans was present as either the parent compound or as free and/or conjugated 5-OH-TBZ [48]. The ^{14}C concentration in dog and rat plasma was usually highest during the first 24 hr after dosing [48]. Results were similar when the plasma of goats, cattle, and pigs was monitored after they were dosed with radiolabeled thiabendazole [47]. Little or no thiabendazole was detected, but 5-OH-TBZ and conjugated forms of 5-OH-TBZ were found in the plasma of the calves and goats. Radioactivity was detected in tissues from calves, pigs, and goats and in the milk of goats and cows after oral dosing of these animals with radiolabeled thiabendazole; the amount of activity in these tissues generally decreased with time. The evidence reported suggested that some of the residues in the tissues were present as the parent compound, some as free and/or conjugated 5-OH-TBZ (see Table 2), and

TABLE 2

Heterocyclic Nitrogen Fungicide Metabolism in Animals

Parent compound and name	Metabolites			
	Proposed structure	Source	Evidence for structure[a]	Reference
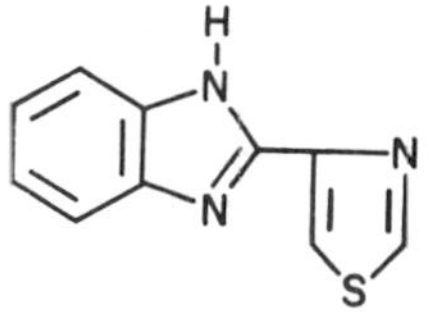 Thiabendazole 2-(4'-Thiazolyl)benzimadazole	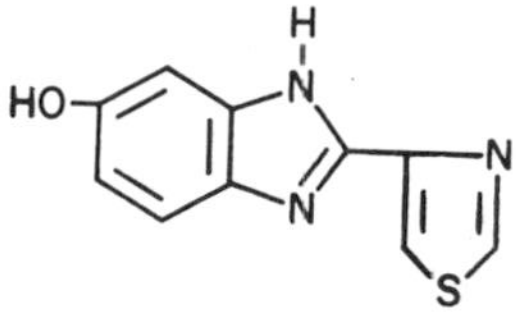	Sheep urine	Comparative IR spectroscopy and MP	46
		Human and dog plasma; rat, dog, and human urine	Fluorometric assay	48
		Calf and goat tissues; cow and goat milk; calf, pig, and goat urine	Fluorometric assay	47
	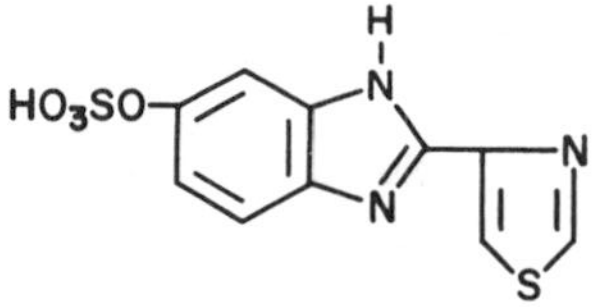	Sheep urine	Comparative electrophoresis and UV spectroscopy	46
		Human plasma and rat urine	Fluorometric assay after enzyme hydrolysis	48

TABLE 2 (Cont.)

Parent compound and name	Metabolites: Proposed structure	Source	Evidence for structure[a]	Reference
	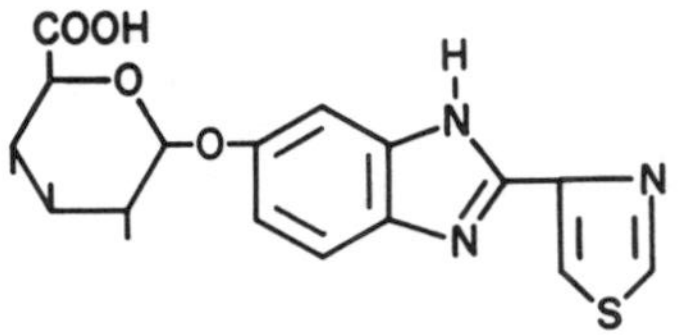	Sheep urine	Interpretation of UV, IR, NMR[b] spectra	46
		Rat urine	Fluorometric assay after enzyme hydrolysis	48
	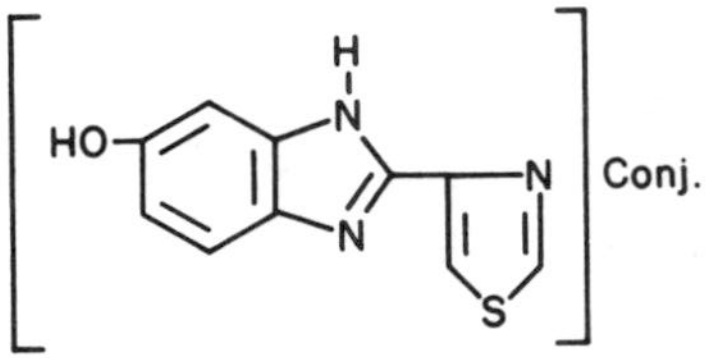 Conj.	Dog and bovine plasma; calf, pig, and goat urine; goat and bovine tissues	Fluorometric assay after hydrolysis	47
		Human rat, and dog urine	Fluorometric assay after hydrolysis	48
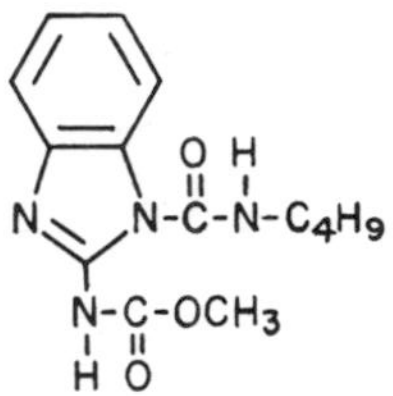 Benomyl Methyl 1-(butylcarbamoyl)-benzimidazol-2-yl carbamate	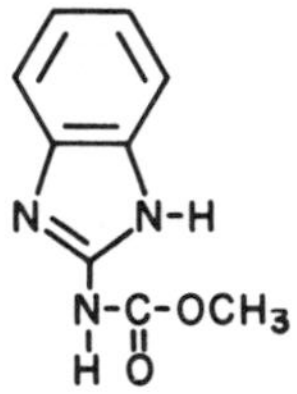	Dog feces; dog and rat tissue	Comparative chromatography and MS	50
		Rat, rabbit, and sheep tissues in vitro; rumen fluid in vitro; mouse, rabbit,	Comparative IR spectroscopy	53

Structure	Source	Identification	Ref.
	and sheep urine; mouse, rabbit, and sheep feces		
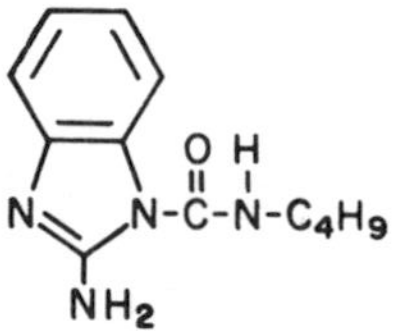	Rat, rabbit, and sheep tissue in vitro; rumen fluid in vitro	Comparative IR spectroscopy	53
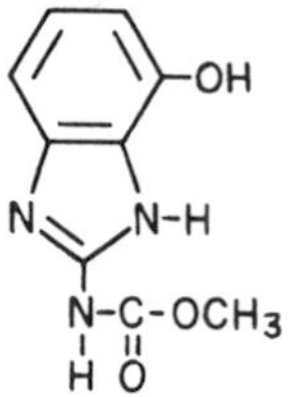	Cow urine	Comparative chromatography and MS	50
	Cow milk and feces; rat and dog tissues	Comparative chromatography	50
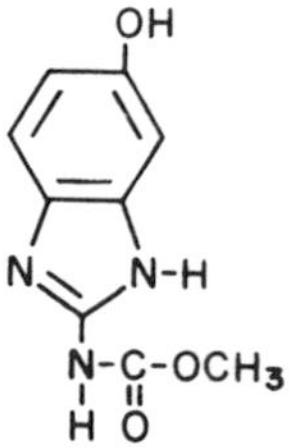	Rat urine	Elemental analysis, comparative MS and IR spectroscopy	51
	Dog and cow feces; chicken excreta; cow urine and milk; rat and dog tissues; chicken eggs	Comparative chromatography	50

TABLE 2 (Cont.)

Parent compound and name	Metabolites			
	Proposed structure	Source	Evidence for structure[a]	Reference
		Rat, dog, and sheep tissue in vitro; rumen fluid in vitro; mouse, rabbit, and sheep urine; mouse, rabbit, and sheep feces	Comparative IR	53
	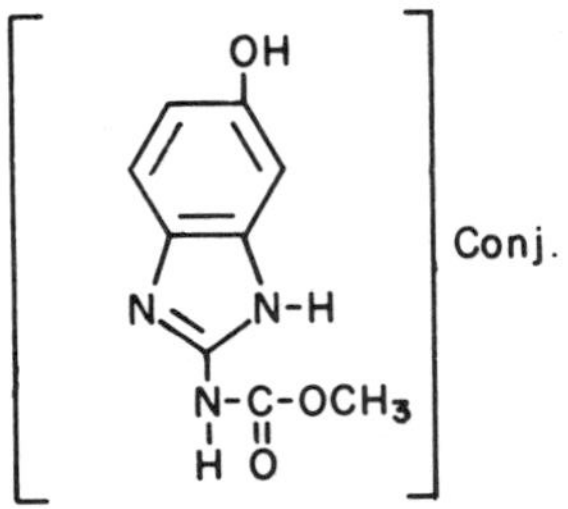 Conj.	Rat urine	Comparative TLC and MS of hydrolysis product	50
		Mouse, rabbit, and sheep urine; mouse, rabbit, and sheep feces	Comparative IR spectroscopy of hydrolysis product	53
	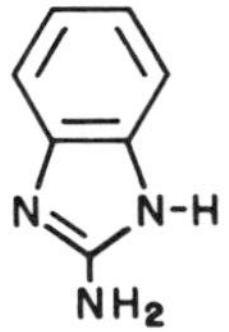	Rat, sheep, and rabbit tissues in vitro; rumen fluid in vitro; mouse, rabbit, and sheep urine; mouse, rabbit, and sheep feces	Comparative IR spectroscopy	53

	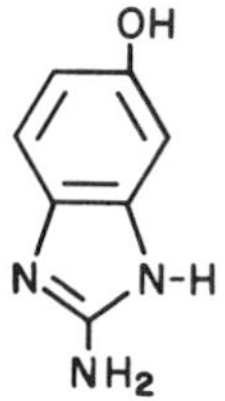 OH, N, N-H, NH_2	Rat, rabbit, and sheep tissues in vitro; rumen fluid in vitro; mouse, rabbit, and sheep urine; mouse, rabbit, and sheep feces	Comparative IR spectroscopy	53
	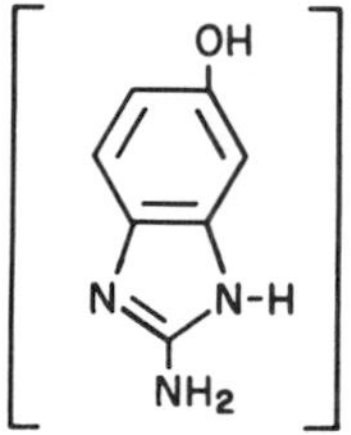 [OH, N, N-H, NH_2] Conj.	Mouse, rabbit, and sheep urine; mouse, rabbit, and sheep feces	Comparative IR spectroscopy of hydrolysis product	53
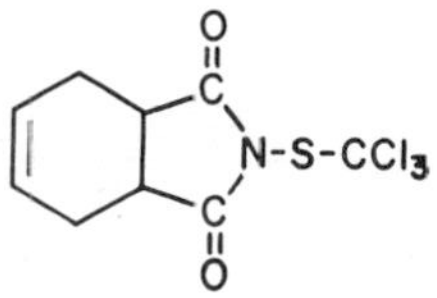 O=C, N-S-CCl_3, C=O Captan N-Trichloromethylthio-4-cyclohexene-1,2-dicarboximide	$HO_3S-CH_2-S-S-CH_2SO_3H$	Rat urine	Interpretation of IR, NMR, and X-ray-induced photo-electron spectrum	56
	$HO_3S-CH_2-S-S(=O)-CH_2SO_3H$	Rat urine	Interpretation of IR, NMR, MS and X-ray-induced photoelectron spectrum	56

TABLE 2 (Cont.)

Parent compound and name	Metabolites			
	Proposed structure	Source	Evidence for structure[a]	Reference
	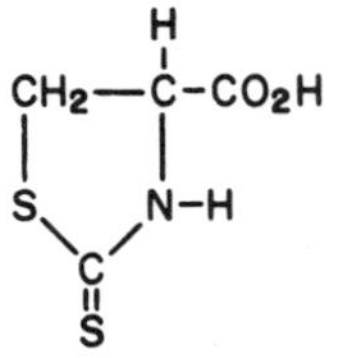	Rat urine	Comparative TLC and electrophoresis and interpretation of MS and UV spectrum	56
	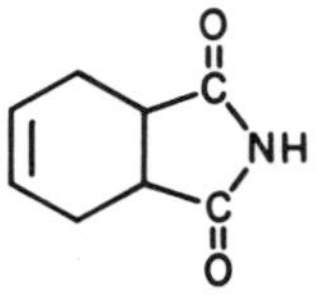	Rat blood	Comparative GLC	54
	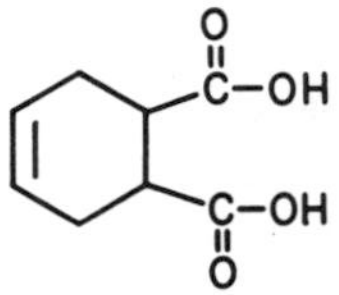	Rat blood	Comparative GLC	54

Compound	Metabolite	Source	Method of identification	Ref.
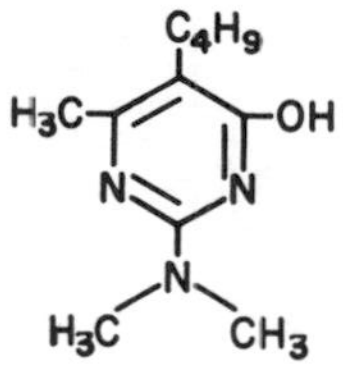 Dimethirimol 2-Dimethylamino-4-hydroxy-5-n-butyl-methylpyrimidine	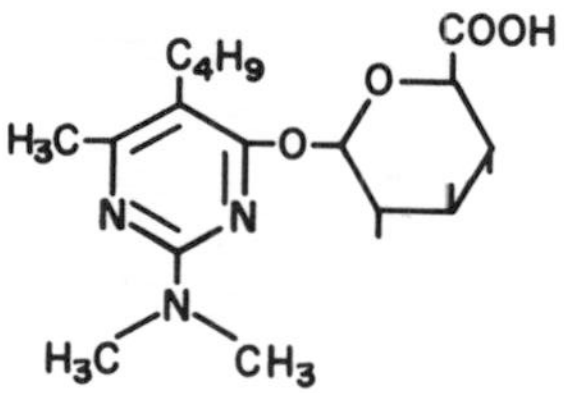	Rat bile	Characterization of hydrolysis products	57
		Rat bile and urine	Color tests, interpretation of IR and comparative IR spectroscopy of hydrolysis product	58
	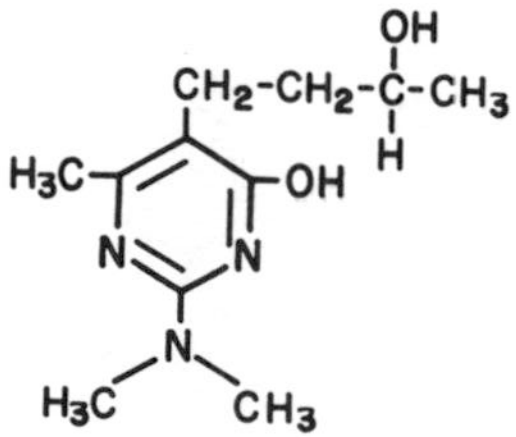	Rat and dog urine; rat and dog bile	Interpretation of MS and IR spectrum	58
	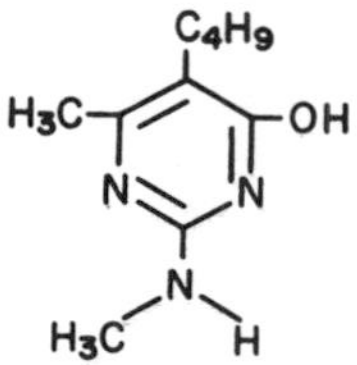	Rat and dog urine; rat and dog bile	Interpretation of MS	58
	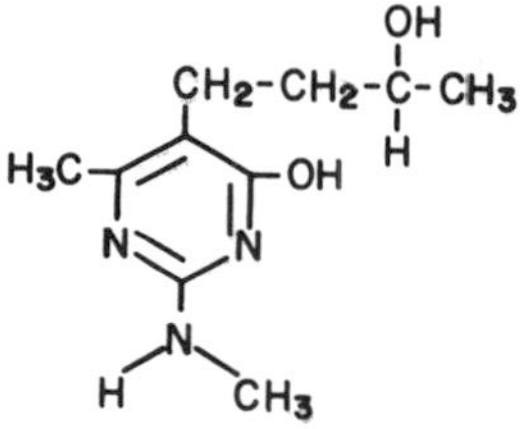	Rat and dog urine; dog bile	Interpretation of MS	58

TABLE 2 (Cont.)

Parent compound and name	Metabolites			
	Proposed structure	Source	Evidence for structure[a]	Reference
	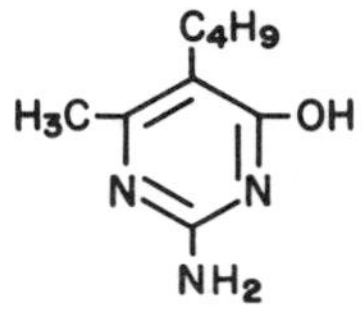 C_4H_9; H_3C; OH; N; N; NH_2	Rat and dog urine; rat and dog bile	Comparative MS and UV spectroscopy	58
	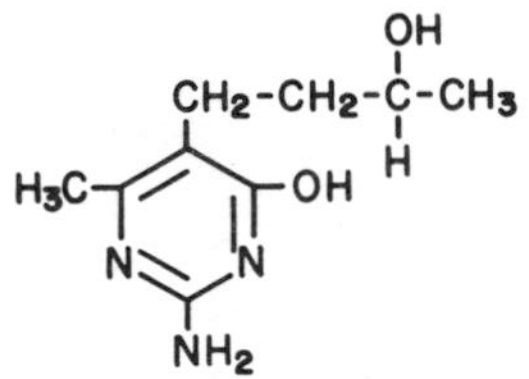 OH; CH_2-CH_2-C-CH_3; H; H_3C; OH; N; N; NH_2	Rat and dog urine; rat and dog bile	Interpretation of MS	58
	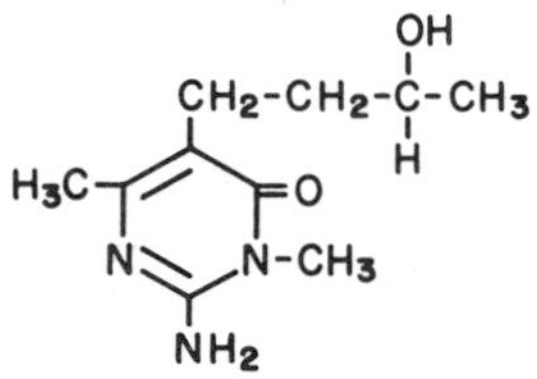 OH; CH_2-CH_2-C-CH_3; H; H_3C; O; N; N-CH_3; NH_2	Dog urine	Interpretation of MS, NMR, and UV spectra	58

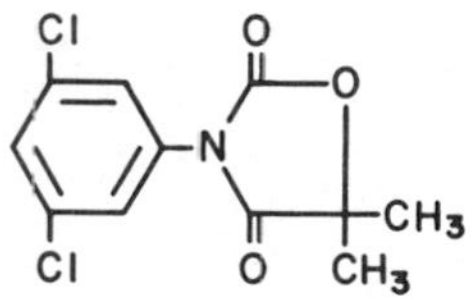 DDOD 3-(3', 5'-Dichorophenyl)- 5, 5-dimethyl oxazolidine-2, 4-dione	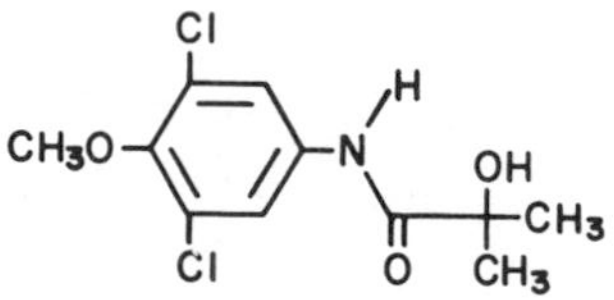	Rat urine	Comparative IR and NMR spectroscopy	60
	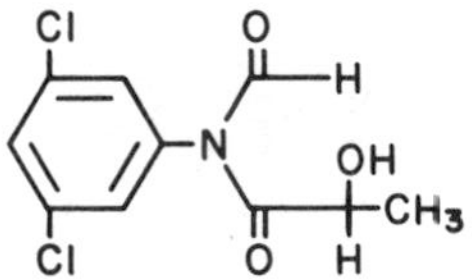	Rat urine	Comparative chromatography	60
	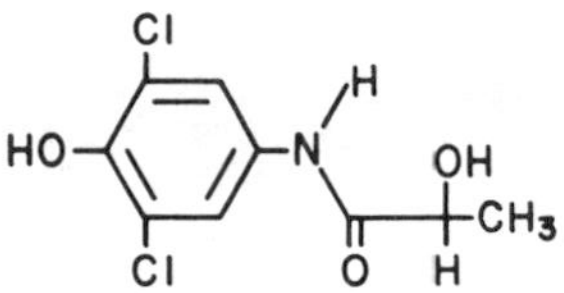	Rat urine	Comparative IR and NMR spectroscopy	60
	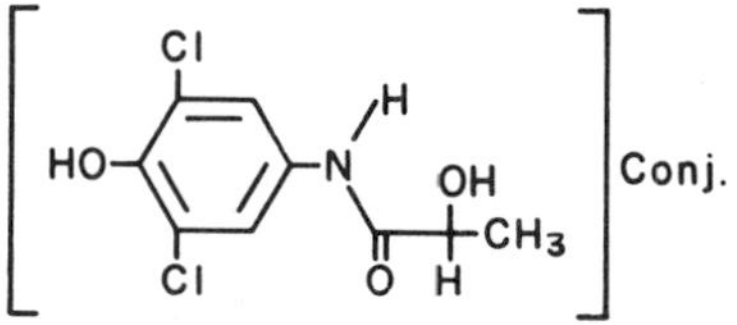 Conj.	Rat urine	Comparative IR and NMR of hydrolysis product	60

TABLE 2 (Cont.)

Parent compound and name	Metabolites			
	Proposed structure	Source	Evidence for structure[a]	Reference
(3,5-Cl-phenyl)-N-succinimide ring structure (Cl, Cl, O, O, N) DSI N-(3', 5'-Dichloro-phenyl)succinimide	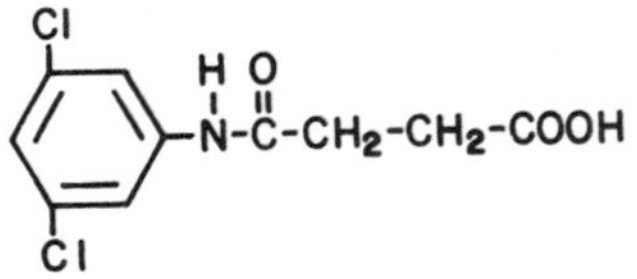 Cl, Cl-phenyl-N(H)-C(=O)-CH_2-CH_2-COOH	Rat and dog urine	Comparative TLC and IR of derivative	61
		Mouse, rat, rabbit, and dog liver homogenates in vitro	Comparative TLC	61
	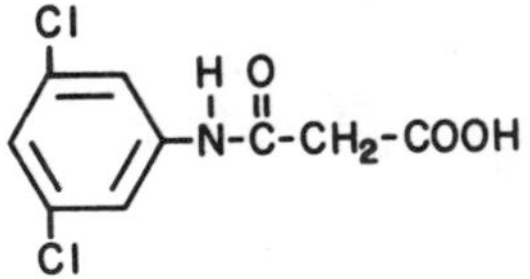 Cl, Cl-phenyl-N(H)-C(=O)-CH_2-COOH	Rat and dog urine	Comparative TLC and degradation studies	61
	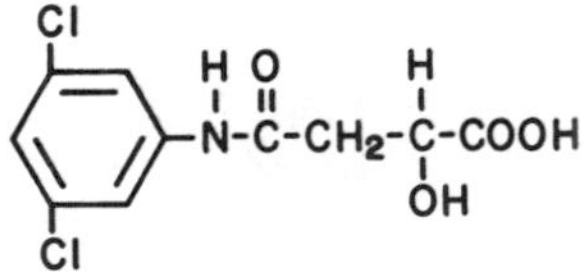 Cl, Cl-phenyl-N(H)-C(=O)-CH_2-C(H)(OH)-COOH	Rat and dog urine	Interpretation of MS and NMR of derivatives	61
	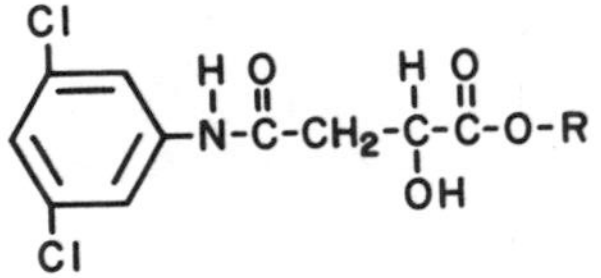 Cl, Cl-phenyl-N(H)-C(=O)-CH_2-C(H)(OH)-C(=O)-O-R	Rat and dog urine	Interpretation of MS and NMR of derivatives	61

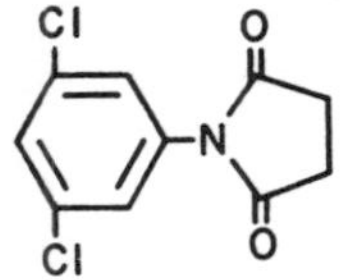

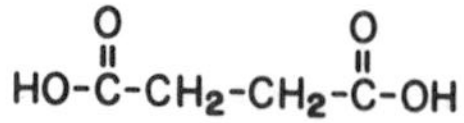

	$HO-\overset{O}{\overset{\|\|}{C}}-CH_2-CH_2-\overset{O}{\overset{\|\|}{C}}-OH$	Mouse, rat, rabbit, and dog liver homogenates in vitro	Comparative TLC	61
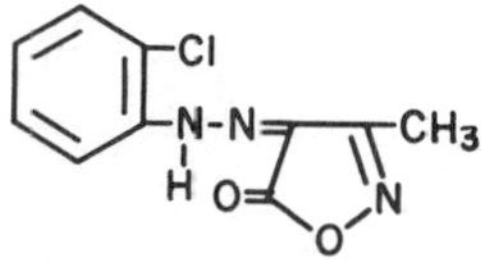 Drazoxolon 4-(2-Chlorophenylhydrazano)-3-methylisoxazol-5-one	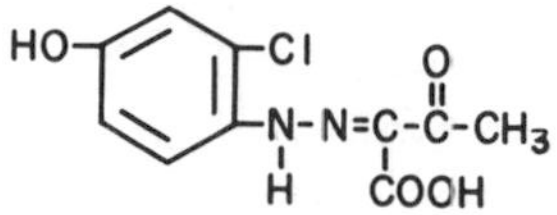	Dog urine	Comparative IR, UV, and NMR	62
	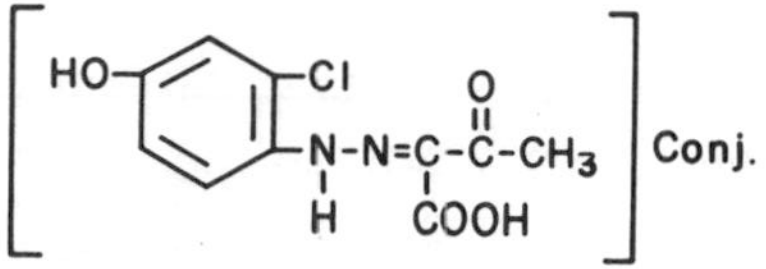 Conj.	Dog and rat urine; rat bile	Comparative IR, UV and NMR of hydrolysis product	62
	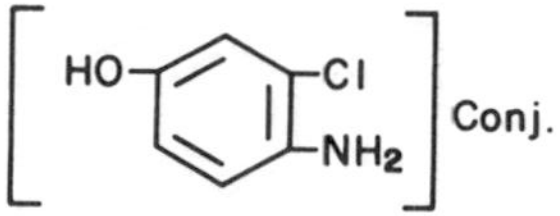 Conj.	Dog and rat urine	Comparative UV of hydrolysis product	62
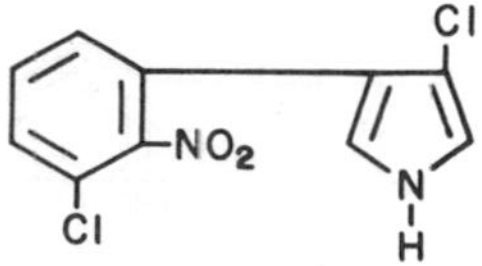 Pyrrolnitrin 3-Chloro-4-[2-nitro-3-chlorophenyl]pyrrole	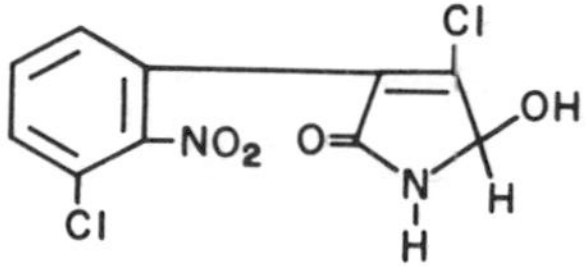	Rat liver in vitro	Comparative MS	63

TABLE 2 (Cont.)

Parent compound and name	Metabolites			
	Proposed structure	Source	Evidence for structure[a]	Reference
	Cl, NO_2, O, O, N, H, Cl	Rat liver in vitro	Comparative MS	63
	NO_2, O, O, N, H, Cl	Rat liver in vitro	Comparative MS	63

[a]For abbreviations, see Table 1.
[b]Nuclear magnetic resonance.

some as unidentified metabolites. Most of the ^{14}C in the milk (90% or more) was present as metabolites. Thiabendazole, free 5-OH-TBZ, and conjugated forms of 5-OH-TBZ were found in the urine of humans, rats, and dogs [48] and in that of calves, swine, and goats [47]. However, the percentage of the total ^{14}C in the urine accounted for by these compounds varied a great deal [48]. The parent compound and these metabolites accounted for 50% of the ^{14}C in human urine, 74% of the ^{14}C in rat urine, and only 23% of the ^{14}C in dog urine. The nature of the rest of the radiolabeled metabolites in the urine apparently has not been investigated further. Tocco et al. [48] have suggested that there may be ring cleavage or sulfoxidation of thiabendazole; some supporting evidence has been reported [49]. Nevertheless, further studies are needed to completely elucidate the animal metabolism of this compound.

Gardiner et al. [50] reported that a rat given a single oral dose of [^{14}C]benomyl eliminated 78.9% of the ^{14}C in the urine and 8.7% in the feces within 72 hr. In contrast, a dog given the same compound eliminated 16.2% of the ^{14}C in the urine and 83.4% in the feces. Little or no parent compound was detected in rat urine. Conjugated forms of methyl-5-hydroxybenzimidazol-2-yl carbamate (see Table 2) accounted for approximately 80% of the ^{14}C in the rat urine. This finding agreed with an earlier report [51]. Enzyme hydrolysis studies suggested these were probably glucuronic acid or sulfate ester conjugates. The nature of the rest of the ^{14}C in the rat urine and feces was not determined. The major ^{14}C-labeled metabolites in dog feces were identified as methyl benzimidazol-2-yl carbamate and methyl 5-hydroxybenzimidazol-2-yl carbamate (see Table 2). Minor metabolites in the dog feces and the ^{14}C-labeled compounds in the dog urine were not identified. Methyl 4-hydroxybenzimidazole and methyl 5-hydroxybenzimidazole were detected in the urine, feces, and milk from cows fed unlabeled benomyl [50]. Possibly these metabolites were eliminated as conjugates because the method used [52] involved a hydrolysis step. Gardiner et al. [50] also fed unlabeled benomyl to chickens and found methyl 5-hydroxybenzimidazole in eggs and excreta. Tissues of rats and dogs fed 2,500 ppm of unlabeled benomyl contained detectable levels of benomyl and/or methyl benzimidazol-2-yl carbamate, methyl 4-hydroxybenzimidazole, and methyl 5-hydroxybenzimidazole [50] as shown in Table 2. Results of studies by Douch [53] showed that liver and blood from mice, rabbits, and sheep, and rumen fluid from sheep all metabolized benomyl in vitro to a variety of metabolites including 2-aminobenzimidazole, 5-hydroxy-2-aminobenzimidazole, methyl benzimidazol-2-yl carbamate, 1-butylcarbamoyl 2-aminobenzimidazole, and methyl 5-hydroxybenzimidazol-2-yl carbamate (Table 2). SKF 525-A inhibited the formation of the hydroxylated metabolites in vitro.

Mice, rabbits, and sheep given unlabeled benomyl excreted 41-71% of the dose in the urine and 21-46% in the feces as metabolites within 96 hr [53]. No parent compound was detected in the excrement; however, the urine and feces of all three species contained the following metabolites: 2-aminobenzimidazole; free and conjugated 5-hydroxy-2-aminobenzimidazole; methyl benzimidazol-2-yl carbamate; and free and conjugated methyl 5-hydroxybenzimidazol-2-yl carbamate (see Table 2). In all 3 species about 20% of the dose was excreted as conjugates of hydroxylated metabolites. Thus, the animal metabolism of benomyl reported by Douch [53] was quite different from that reported by Gardiner et al. [50]. For instance, Douch [53] reported that methyl 5-hydroxybenzimidazol-2-yl carbamate was a minor metabolite in the urine of sheep, rabbit, and mouse; but Gardiner et al. [50] reported that this was the major urinary metabolite. Other qualitative and quantitative differences in the results of these two research groups may be explained, at least in part, by species differences and experimental procedures; however, further studies are needed to clarify these apparent inconsistencies.

Engst and Raab [54] gave rats ^{35}S-labeled captan and reported that essentially all of the ^{35}S was eliminated in the excreta within 3 days. The parent compound was not detected in the blood and organs of these rats; however, tetrahydrophthalamide and tetrahydrophthalic acid were detected in the blood (see Table 2). The nature of the ^{35}S-labeled metabolites in the urine was not determined. These authors noted a decrease in the SH level in red blood cells and a decrease in SH enzymes after administration of captan (5% of the LD-50), changes indicating that this compound interacted directly with sulfhydryl compounds.

Rats orally dosed with [^{35}S]captan eliminated 60.5% of the ^{35}S in the urine and 40% in the feces within 3 days; most of the activity was eliminated within 24 hr after dosing and primarily as metabolites. Less than 1% of the radioactivity remained in the animals 1 day after dosing [55].

DeBaun et al. [56] orally dosed rats with [methylthio-^{14}C]captan and found that 51.8% of the ^{14}C was eliminated in the urine, 22.8% in the expiratory air, and 15.9% in the feces within 4 days; the tissues contained 0.6% of the ^{14}C at the end of the experiment. When the same compound was given to rats as an intraperitoneal injection, the elimination of ^{14}C was much slower, suggesting that the gut was involved in the metabolism of this compound. The three major urinary metabolites of orally administered [methylthio-^{14}C]captan were identified as thiazolidine-2-thione-4-carboxylic acid, a salt of dithiobis(methanesulfonic acid), and the disulfide monoxide derivative of dithiobis(methanesulfonic acid) (see Table 2). Minor metabolites in the urine and residues in the tissues were not identified. Interestingly, radiolabeled thiazolidine-2-thione-4-carboyxlic acid was detected in the urine of rats simultaneously given unlabeled captan and either [^{14}C]cystine or [^{35}S]glutathione. [^{35}S]Dithiobis(methanesulfonic acid) was found in the urine of rats treated with [^{35}S]sulfite and unlabeled captan [56]. These and other supporting data led DeBaun et al. [56] to conclude that the trichloromethylthio moiety of captan was metabolized to thiophosgene which was metabolized further by either (1) oxidation and/or hydrolysis to CO_2; or (2) reaction with a cysteine moiety to yield thiazolidine-2-thione-4-carboxylic acid; or (3) reaction with sulfite to produce dithio(methanesulfonic acid).

There apparently is no information in the literature concerning the fate of much of the aromatic part of the molecule. Thus further animal metabolism studies with ring-labeled captan are needed.

Calderbank [57] reported that more than 80% of the ^{14}C fed to rats as [pyrimidine-^{14}C]dimethirimol was excreted in the urine within 48 hr. He indicated that five metabolites present in the rat urine resulted from either dealkylation of the dimethylamino group, oxidation of the N-butyl substituent, or both. Unfortunately, the evidence for these metabolites was not described. Another metabolite in the rat bile was identified as the O-glucuronide of dimethirimol.

Bratt et al. [58] orally dosed rats and dogs with [2-pyrimidine-^{14}C]dimethirimol and reported that the rats excreted 87.2% of the ^{14}C in the urine and 10.9% in the feces in 8 days; dogs excreted 66.5-88.6% of the ^{14}C in the urine and 9.0-13.8% in the feces in 4 days. Biliary cannulated rats eliminated in the bile, within 24 hr, 21.4-35.8% of the ^{14}C given orally as [^{14}C]dimethirimol. Radioactivity was also detected in the bile of dogs given [^{14}C]dimethirimol. No ^{14}C was detected in the fat of rats 24 hr after dosing (0.25 μCi given). At least 17 radiolabeled compounds were present in the dog urine, 10 in the dog bile, 16 in the rat urine, and 6 in the rat bile. The major metabolites in the urine (accounting for 55-75% of the dose) were characterized by spectral studies. These metabolites resulted from either sequential N-dealkylation, penultimate hydroxylation of the N-butyl group, or both (see Table 2). One metabolite was identified as the glucuronic acid conjugate of the parent compound.

One hour after an oral dose of UCC-974 (3,5-dimethyltetrahydro-1,3,5, 2H-thiadiazine-2-thione), 62-88% of the dose remained in the gut of rats as the parent compound [59]; traces of carbon disulfide were detected in the respiratory air after both oral and intravenous administration of this compound. Smyth et al. [59] have proposed that this compound is slowly hydrolyzed in the animal to carbon disulfide, formaldehyde, and methylamine, but evidence to support this proposal apparently has not been published.

Sumida et al. [60] orally dosed rats with either [^{14}C]DDOD or [^{3}H]DDOD at rates from 100 mg/kg to 3,000 mg/kg of body weight and then monitored the elimination of radioactivity in the respiratory air, urine, and feces. From 80 to 100% of the radioactivity was eliminated in the combined urine and feces during a 14-day collection period. When smaller doses were given, the amount of activity excreted in the urine and feces was approximately equal; when larger doses were given, most of the radioactivity was eliminated in the feces. Radioactive residues were highest in most tissues 12 hr after dosing, then slowly declined, but were still detectable after 48 hr. However, the amount of radioactivity in the adipose tissue increased gradually up to 48 hr after dosing. The nature of the radiolabeled compounds in the tissues was not determined. The feces contained the parent compound and one or more unidentified metabolites. At least eight radiolabeled metabolites were present in the urine. Of these, four resulted from hydroxylation at the 4' position of the benzene ring moiety and hydrolytic or oxidative modification of the oxazolidine ring moiety (see Table 2).

Within 48 hr after receiving an oral dose of [^{14}C]N-(3',5'-dichlorophenyl)-succinimide (DSI), rats eliminated 69.7-71.2% of the ^{14}C in the urine and 10.8-17.5% in the feces; within 6 days, they eliminated 5.9-10.8% in the respiratory air [61]. When [phenyl-^{3}H]DSI was given, 89.6% of the ^{3}H appeared in the urine and 12.9% in the feces. When [^{14}C]DSI was given orally to dogs, 32.4% of the ^{14}C appeared in the urine and 38.7% in the feces within 6 days. Whole-body autoradiograms showed that 4 hr after dosing, the ^{14}C given to rats as [carbonyl-^{14}C]DSI was widely distributed in the body; the concentration was highest in the kidney, lung, liver, and intestines. The nature of the ^{14}C-labeled compounds in the tissues was not determined. Rat urinary metabolites were identified as N-(3',5'-dichlorophenyl)succinamic acid (DSA), N-(3',5'-dichlorophenyl)-2-hydroxysuccinamic acid, N-(3',5'-dichlorophenyl)malonamic acid, and apparent derivatives of DSA. Approximately 25% of the urinary ^{14}C (present as two or more metabolites) was not identified. The metabolites identified in rat urine were also present in dog urine but in different proportions. When [^{14}C]DSI was incubated with rat and dog liver and kidney homogenates, two metabolites were formed; one was tentatively identified as DSA and the other as succinic acid by comparative chromatography. The microsomal fraction of the tissue homogenates was most active in metabolizing [^{14}C]DSI [61].

Daniel [62] reported that rats orally dosed with [4-isoxazol-^{14}C]drazoxalon eliminated about 75% of the ^{14}C in the urine, 13% in the feces, and 7% as [^{14}C]O_2 within 96 hr. When [phenyl-^{14}C]drazoxalon was given to rats, about 75% of the ^{14}C was eliminated in the urine and 15% in the feces within 96 hr. Another group of surgically modified rats given [^{14}C]drazoxalon orally excreted about 14% of the label in the bile in 48 hr. Dogs given a single oral dose of [4-isoxazol-^{14}C]drazoxalon excreted 35% of the label in the urine and 35% in the feces within 96 hr. 2-(2-Chloro-4-hydroxyphenylhydrazano)acetoacetic acid (I) was identified in the dog urine. The major metabolites in rat urine, dog urine, and rat bile were identified as conjugated forms, probably sulfate ester and glucuronic acid conjugates, of compound I (see Table 2). Conjugated forms of 4-amino-3-chlorophenol were also detected in the dog and rat urine. The nature of the radiolabeled compounds in the feces was not determined, and tissues were not analyzed for ^{14}C residues.

Murphy and Williams [63] reported that when rats were given [pyrrole-^{14}C]pyrrolnitrin intravenously, 29% of the ^{14}C was eliminated in the urine as the parent compound within 24 hr. When the same compound was given to rats with cannulated bile ducts, approximately 50% of the ^{14}C was excreted in the bile in 24 hr. No parent compound was detected in the urine, but at least six radioactive metabolites were detected by chromatographic analysis. Pyrrolnitrin was also metabolized by rat microsomes in vitro to a variety of products; three of these were identified, all containing an oxidized pyrrole ring (see Table 2). A number of major metabolites were not characterized.

V. AROMATIC AND RELATED FUNGICIDES

Eberts [64] gave a single dose of [^{14}C]dichloran to humans and rats and then collected urine and feces for ^{14}C analysis. Rats excreted 74.8% of the ^{14}C in the urine and 7.3% in the feces during the 14-day collection period. Humans excreted 75.2% of the ^{14}C in the urine and 31.6% in the feces within 7 days. One day after dosing, most of the ^{14}C still in the rat was present in the gastrointestinal tract and bladder, but detectable amounts were also present in the liver. Rat tissues analyzed 14 days after dosing reportedly contained no "measurable amount of radioactivity." However, the amount of ^{14}C given was not reported; therefore the sensitivity of the assays cannot be determined. Eberts [64] reported that the sulfate ester of 2,6-dichloro-4-hydroxyaniline accounted for approximately 85% of the ^{14}C in the urine. The rest of the activity was present as the parent compound and unidentified metabolites. In a later abstract, Eberts [65] reported that no parent compound was present in the urine of rats dosed with dichloran but that both free and conjugated 2,6-dichloro-4-hydroxyaniline (see Table 3) were present in the urine of these animals. Apparently, detailed descriptions of both of these studies by Eberts have not been published.

In similar studies, Maté et al. [66] gave rats [^{14}C]dichloran (oral and intraperitoneal) and found that 82-91% of the ^{14}C was excreted in the urine and 1-2% in the feces within 72 hr. Small amounts of ^{14}C were excreted in the bile after both oral and intraperitoneal dosing. After acid hydrolysis of the urine, 2,6-dichloro-4-hydroxyaniline accounted for 70.4% of the urinary ^{14}C and 4-amino-2,6-dichloroaniline accounted for 2.4%; the rest of the radiolabeled materials in the urine was not identified. The data suggested that 4-amino-2,6-dichloroaniline was not a precursor to 2,6-dichloro-4-hydroxyaniline. Maté et al. [66] have proposed a series of intermediates to account for the unusual replacement of the nitro group of 2,6-dichloro-4-nitroaniline with a hydroxyl group.

When rabbits were given 2 g of unlabeled pentachloronitrobenzene (PCNB) by stomach tube 46-62% of the dose was excreted unchanged in the feces, 4-14% in the urine as N-acetyl-S-pentachlorophenylcysteine, and 12-14% in the urine as free and conjugated pentachloroaniline (PCA) over a 72-hr collection period [67] (see Table 3). There was little or no evidence for the formation of pentachlorophenol (PCP). Tissues were not analyzed for the parent compound or metabolites. When a lactating cow was fed 5 ppm of unlabeled PCNB for 5 days, the parent compound was not detected (sensitivity of assay, 0.01 ppm) in the milk [68]. The urine collected from the time of the initial feeding until 2 days after the last feeding of PCNB contained PCA, accounting for 45% of the PCNB given. Kuchar et al. [69] fed

TABLE 3

The Metabolism of Aromatic and Related Fungicides in Animals

Parent compound and name	Metabolites: Proposed structure	Source	Evidence for structure[a]	Reference
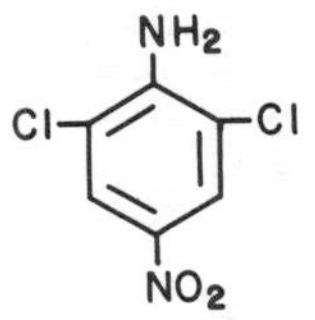 NH_2 Cl Cl NO_2 Dichloran, DCNA 2, 6-Dichloro-4-nitroaniline	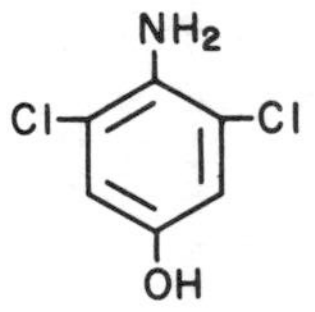 NH_2 Cl Cl OH	Rat urine	Comparative TLC, UV, IR, and NMR spectroscopy	65
		Mouse microsomes in vitro	Comparative TLC	66
		Rat urine	Reverse isotope dilution assay	66
	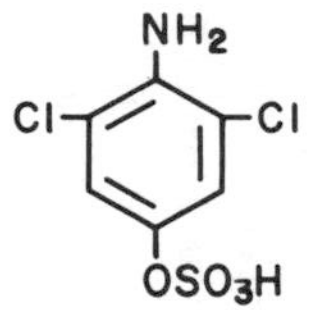 NH_2 Cl Cl OSO_3H	Rat urine	Comparative TLC, UV, IR, and NMR spectroscopy of hydrolysis product; synthesis in vivo	65
	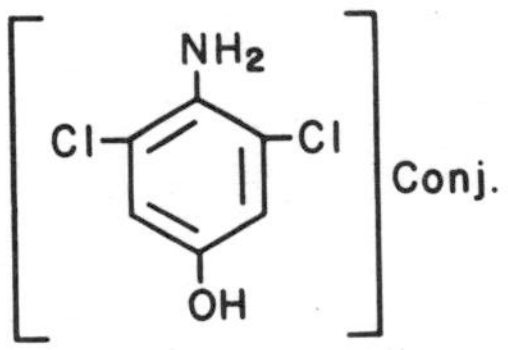 [NH_2 Cl Cl OH] Conj.	Rat urine	Comparative TLC, UV, and IR spectroscopy of hydrolysis product	65

	NH_2, Cl, Cl, NH_2	Mouse microsomes in vitro	Comparative TLC	66
		Rat urine	Reverse isotope dilution assay	66
NO_2, Cl, Cl, Cl, Cl, Cl PCNB Pentachloronitrobenzene	NH_2, Cl, Cl, Cl, Cl, Cl	Dog urine, feces, fat, and liver	Comparative GLC and IR spectroscopy	69
		Rabbit urine	Comparative MP	67
		Cow urine	Comparative GLC	68
		Rat, dog, and cow tissues, urine and feces	Comparative GLC	70
	CH_3, S, Cl, Cl, Cl, Cl, Cl	Dog urine, feces, liver, fat, kidney, and muscle	Comparative GLC and MS	69
		Rat, dog, and cow tissues, urine, and feces	Comparative GLC	70
	[NH_2, Cl, Cl, Cl, Cl, Cl] Conj.	Dog urine	Comparative GLC of hydrolysis product	69
		Rabbit urine	Comparative MP of hydrolysis product	67

TABLE 3 (Cont.)

Parent compound and name	Metabolites: Proposed structure	Source	Evidence for structure[a]	Reference
		Cow milk, kidney, and liver	Comparative GLC of hydrolysis product	70
	Cl, Cl, Cl, Cl, Cl; S-CH_2-C(COOH)(H)-N(H)-C(=O)-CH_3	Rabbit urine	Elemental analysis, MP, and degradation studies	58
NO_2; Cl, Cl, Cl, Cl	NH_2; Cl, Cl, Cl, Cl	Rabbit urine	Comparative MP	67
2,3,4,6-Tetrachloronitrobenzene	S-CH_2-C(COOH)-H; NH-C(=O)-CH_3; Cl, Cl, Cl, Cl	Rabbit urine	Elemental analysis and characterization of hydrolysis product	67

NO_2, Cl, Cl, Cl, Cl (structure) 2,3,4,5-Tetrachloronitrobenzene	NH_2, Cl, Cl, Cl, Cl (structure)	Rabbit feces and urine	Comparative MP	71
NO_2, Cl, Cl, Cl, Cl (structure) TCNB 2,3,5,6-Tetrachloronitrobenzene	NH_2, Cl, Cl, Cl, Cl (structure)	Rabbit urine	Comparative MP	71
	S-CH_2-C-H(COOH)-NH-C(=O)-CH_3, Cl, Cl, Cl, Cl (structure)	Rabbit urine	Elemental analysis and characterization of hydrolysis product	71
	NH_2, Cl, Cl, Cl, Cl, OH (structure)	Rabbit urine	Elemental analysis and MP	71

TABLE 3 (Cont.)

Parent compound and name	Metabolites			
	Proposed structure	Source	Evidence for structure[a]	Reference
	[NH$_2$, Cl, Cl, Cl, Cl, OH] Conj.	Rabbit urine	Elemental analysis and MP of hydrolysis product	71
Cl, Cl, Cl, Cl, Cl, CH_2OH Blastin 2,3,4,5,6-Pentachlorobenzyl alcohol	[Cl, Cl, Cl, Cl, Cl, CH_2OH] Conj.	Rat urine	Comparative IR spectroscopy, TLC, and GLC of hydrolysis product	73
	Cl, Cl, Cl, Cl, Cl, COOH	Rat urine	Comparative chromatography	73

PCP Pentachlorophenol	Conj.	Mouse urine	Comparative chromatography and isotope dilution of hydrolysis product	75
		Mouse urine	Comparative chromatography and isotope dilution	75
	Conj.	Mouse urine	Comparative chromatography and isotope dilution of hydrolysis product	75
Chloroneb 1,4-Dichloro-2,5-dimethoxybenzene		Beef liver in vitro	Comparative GLC	77
		Dog and rat urine	Comparative GLC and IR spectroscopy	78
		Cow urine and milk; dog feces and tissues	Comparative GLC	78

TABLE 3 (Cont.)

Parent compound and name	Metabolites: Proposed structure	Source	Evidence for structure[a]	Reference
	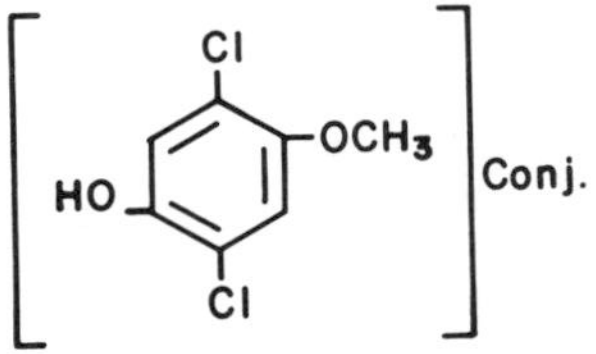	Cow urine	Comparative GLC of hydrolysis product	77
		Dog, rat, and cow urine	Comparative GLC and IR spectroscopy of hydrolysis product	78
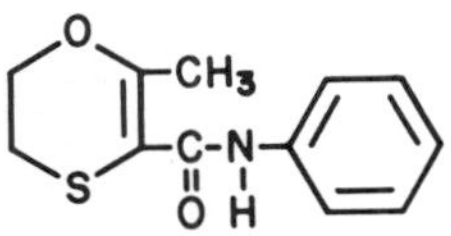 Carboxin 5, 6-Dihydro-2-methyl-1, 4-oxathiin-3-carboxanilide	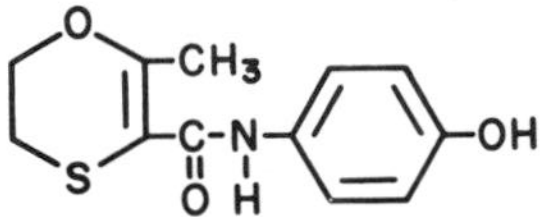	Rat and rabbit urine; rat bile	Comparative chromatography and MS	79
	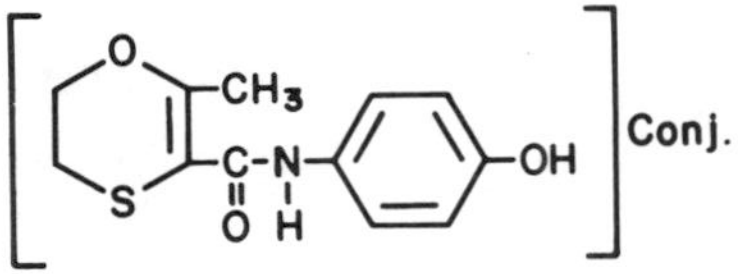	Rat bile; rat and rabbit urine	Comparative chromatography and MS of hydrolysis product	79
	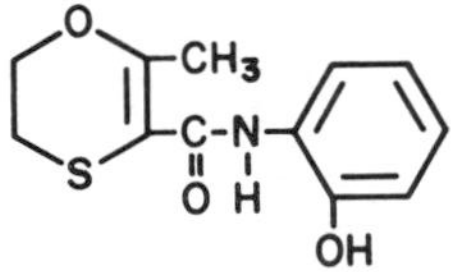	Rat and rabbit urine	Comparative chromatography and MS	79

	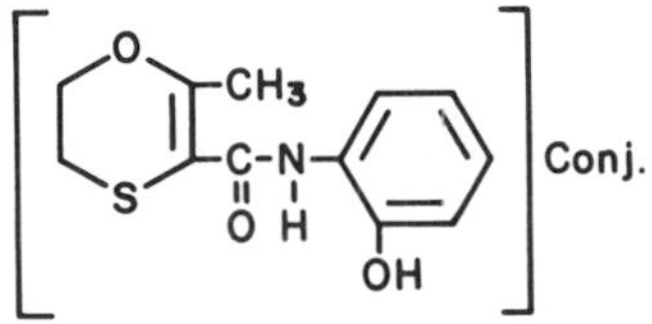	Rat and rabbit urine	Comparative chromatography and MS of hydrolysis product	79
	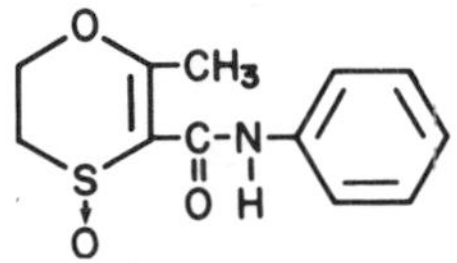	Rat urine	Comparative TLC	80
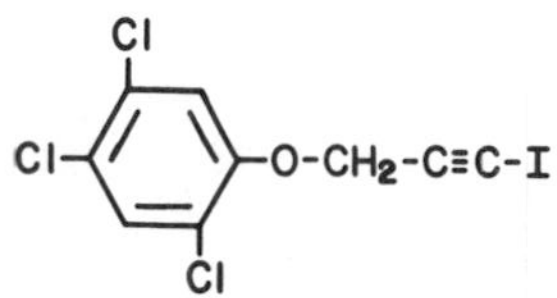 Haloprogin 2,4,5-Trichlorophenyl-γ-iodopropargyl ether	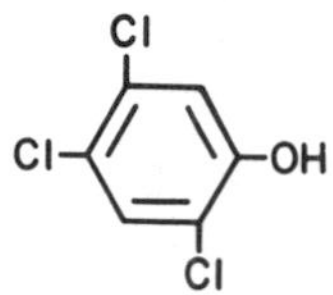	Rat urine	Comparative GLC and carrier crystallization	81
	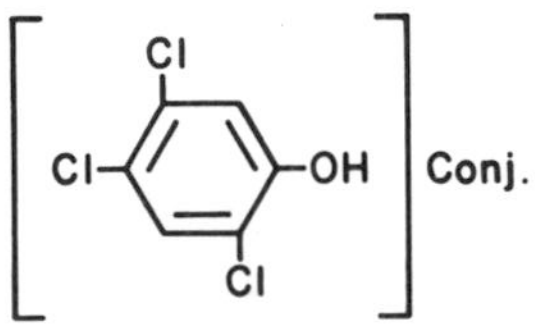	Rat urine	Comparative GLC of hydrolysis product	81

TABLE 3 (Cont.)

Parent compound and name	Metabolites			
	Proposed structure	Source	Evidence for structure[a]	Reference
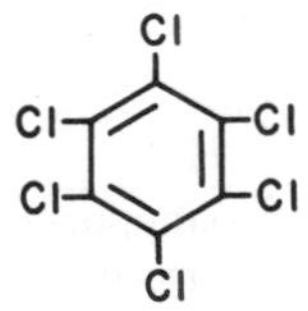 HCB Hexachlorobenzene	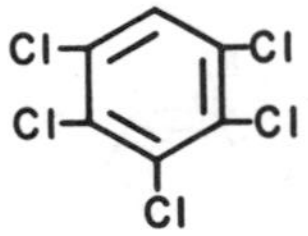	Rat urine	Isotope dilution; comparative TLC	92
	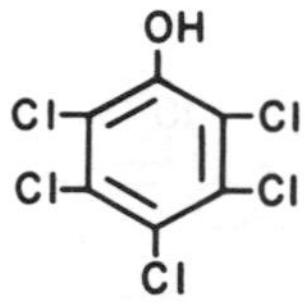	Rat urine	Isotope dilution, comparative TLC	92
		Rat liver microsomes in vitro	Comparative TLC	92
	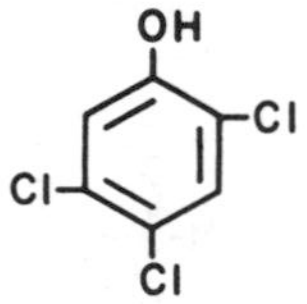	Rat urine	Comparative TLC	92

[a]For abbreviations, see Table 1.

unlabeled PCNB to dogs and rats for up to 2 years. Pentachloroaniline and methyl pentachlorophenylsulfide (MPCPS) were detected in a variety of tissues, as well as in the urine and feces from these animals. There was evidence for the presence of conjugated metabolites in dog urine. In addition, pentachlorobenzene (PCB) and hexachlorobenzene (HCB) were found in various tissues, but these probably resulted from contaminants in the PCNB fed to the animals [69]. In related studies, Borzelleca et al. [70] orally treated dogs, rats, and cows with technical grade PCNB; the tissues, milk, feces, and urine from these animals were analyzed for PCNB, PCA, and MPCPS and for contaminants in the technical grade of PCNB given. Pentachloronitrobenzene was present only at very low levels or was not detected in most tissues from these animals; however, trace amounts of PCNB were detected in the milk. It was also present in the feces and urine from the dog. Both PCA and MPCPS were detected in the urine, feces, and milk from these animals. Apparently no animal metabolism studies with radiolabeled PCNB have been reported and the studies with unlabeled PCNB have not accounted for all of the compound given. Thus other routes of metabolism of this compound cannot be ruled out. Also, the metabolite resulting from the replacement of the nitro group [69, 70] with a methylthio group is unique and warrants further investigation to determine the sources of the methylthio group.

Bray et al. [71] gave rabbits unlabeled 2,3,5,6-tetrachloronitrobenzene (TCNB) by stomach tube and reported that 59-78% of the dose was excreted unaltered in the feces. The urine contained 9-12% of the dose as 2,3,5,6-tetrachloroaniline, 2% as free 4-amino-2,3,5,6-tetrachlorophenol, 13% as conjugated forms of 4-amino-2,3,5,6-tetrachlorophenol, and 11% as N-acetyl-S-(2,3,5,6-tetrachlorophenyl)-L-cysteine (see Table 3). Tissues from these animals were not analyzed for residues.

When rabbits were orally treated with 2,3,4,5-tetrachloronitrobenzene, 27-36% of the dose was eliminated in the feces (partly as the parent compound and partly as 2,3,4,5-tetrachloroaniline). The urine also contained 2,3,4,5-tetrachloroaniline and other metabolites thought to be free and conjugated forms of 2-amino-3,4,5,6-tetrachlorophenol. The structure of the latter metabolites apparently has not been confirmed.

Betts et al. [67] reported that when 2,3,4,6-tetrachloronitrobenzene was given orally to rabbits, 31% of the dose was excreted in the urine as 2,3,4,6-tetrachloroaniline and 32% as N-acetyl-S-(2,3,4,6-tetrachlorophenyl)cysteine. Apparently, at least one additional unidentified metabolite was in the urine. Later, Betts et al. [72] reported on the animal metabolism of a number of related trichloronitrobenzenes. The metabolism of these compounds also involved mercapturic acid formation, reduction of nitro groups to give the corresponding aniline, and the formation of aminotrichlorophenols.

Rats excreted approximately 75% of a single oral dose of [^{14}C]blastin in the feces as the parent compound within 72 hr; three ^{14}C-labeled urinary metabolites accounted for approximately 20% of the dose [73]. The major urinary metabolite was identified as pentachlorobenzoic acid, and the other two urinary metabolites were apparently present as conjugates of the parent compound. The authors indicated that there was "no significant accumulation" of ^{14}C in the tissues of rats given 10 μCi of [^{14}C]blastin [73].

Bevenue and Beckman [74] have reviewed the literature dealing with PCP with particular emphasis on methodology, toxicology, and tissue residues. Many of the studies cited indicated that PCP can be absorbed by the skin and inhalation and through the gut. More recently, Jakobson and Yllner [75] gave mice [^{14}C]PCP by subcutaneous or intraperitoneal injection and found that 72-83% of the ^{14}C was excreted in the urine within 4 days. There was both gastric and biliary secretion of PCP, its metabolites or both and excretion in the feces (3.8-7.8%). A total of 11, 3, and less than 1% of the ^{14}C remained in the animals 4, 7, and 30 days, respectively, after injection of [^{14}C]PCP. Only trace amounts of ^{14}C were respired. Approximately 52% of the ^{14}C in urine collected from 0 to 24 hr after treatment was present as the parent compound, 14% as conjugated PCP, and 40% as free and/or conjugated tetrachlorohydroquinone (see Table 3). In a similar study, rats given an oral dose of [^{14}C]PCP respired little or no ^{14}C but excreted 68.3% of the ^{14}C in the urine and 9.2-13.2% in the feces (combined benzene and H_2O extracts) during a 10-day collection period [76]. The liver, kidney, and blood were the tissues with the highest ^{14}C concentration 40 hr after dosing. The nature of the ^{14}C-labeled compounds in the feces, urine, and tissues was not determined.

Gutenmann and Lisk [77] reported that 2,5-dichloro-4-methoxyphenol (DCMP) was formed when chloroneb was incubated with beef liver in vitro (see Table 3). This metabolite and the parent compound were not detected in the milk, urine, or feces when 5 ppm of chloroneb was fed to a cow. However, conjugated forms of DCMP, accounting for 44.3% of the dose, was excreted in the urine within 4 days. The fate of the remaining 55.7% of the dose was not determined. In a similar study, Rhodes and Pease [78] fed rats, dogs, and cows unlabeled chloroneb and found that free and conjugated DCMP were excreted in the urine. 2,5-Dichloro-4-methoxyphenol was detected in the milk of a cow and in the tissues of dogs fed chloroneb. Unfortunately, the quantitative importance of these metabolites was not reported; further studies with radiolabeled chloroneb would be of interest.

Waring [79] orally treated rats and rabbits with carboxin containing a ^{14}C label at either the 6-position of the heterocyclic ring or uniformly labeled in the aromatic moiety. Whole-body autoradiography studies showed that 2 hr after dosing, ^{14}C was present in the liver, intestinal tract, and salivary gland; 6 hr after dosing there was also activity in the kidney. Only trace amounts of ^{14}C were detected in the rats 48 hr after dosing (6.3 μCi). Of the ^{14}C given, 28-52% was excreted as the parent compound in the feces of the rats and 6-10% in the feces of the rabbits. Unaltered carboxin was also excreted in the urine of the rat (3-25%) and rabbit (0-6%). The urine of both rats and rabbits contained free and conjugated "*o*-hydroxycarboxin," free and conjugated "*p*-hydroxycarboxin" (see Table 3), and at least three minor unidentified metabolites. The rat bile contained trace amounts of free and conjugated forms of "p-hydroxycarboxin." Scission between the two ring systems by either the rat or rabbit was not evident. Chin et al. [80] gave rats a single oral dose of [phenyl-^{14}C]carboxin and reported that 60% of the ^{14}C was excreted in the urine and 35% in the feces within 96 hr; the rats contained 0.5% of the total ^{14}C administered 96 hr after dosing. They reported that most of the ^{14}C in the rat urine was present as the sulfoxide (see Table 3) on the basis of cochromatography studies. However, Waring [79] questioned that assignment of structure because he contended that the TLC system used would not have differentiated between the phenol and sulfoxide (Table 3).

Weikel and Bartek [81] reported that when animals were topically treated with [benzene-^{14}C]haloprogin, the cumulative 120-hr urinary excretion of ^{14}C was as follows: rats, 88.5%; rabbits, 92.7%; and pigs, 15.2%. The [^{14}C]haloprogin was poorly absorbed (19.7%) through the skin of pigs, but four hours after the topical treatment, the ^{14}C was widely distributed throughout the body; concentrations were highest in the spleen, lungs, liver, and kidneys. When [^{14}C]haloprogin was given to rats intravenously, 103.5% of the ^{14}C was eliminated in the urine within 24 hr. The relative ^{14}C concentrations in the various body parts of these rats changed with time after dosing but were still detectable 4 hr and, in some instances, 24 hr after dosing. The nature of the ^{14}C-labeled compounds in the tissues was not determined. Little or no parent compound was present in the urine, but at least four metabolites were present. Of these, two were identified as 2,4,5-trichlorophenol and a conjugated form of 2,4,5-trichlorophenol. Some evidence suggested that the conjugate was a sulfate ester.

A number of studies have shown that HCB accumulates in tissues of chickens [82, 83], birds [84, 85], sheep [86], humans [87], fish [88], and rats [89, 90]. The greatest concentration of HCB was in the fat [82, 83, 86]. Until the recent reports by Mehendale and Matthews [91] and Mehendale et al. [92], information concerning the metabolism of HCB in animals was very limited. When rats were given a single oral dose of [^{14}C]HCB, approximately 16% of the ^{14}C was eliminated in the feces and less than 1% in the urine in 7 days [92]. Approximately 70% of the ^{14}C was still in the tissues of the animals 7 days after dosing; most of these residues were in the fat. The ^{14}C residues in the tissues consisted primarily of the parent compound, but metabolites were also present in trace amounts. The ^{14}C in the feces was present as the parent compound and the ^{14}C in the urine was present as pentachlorophenol, tetrachlorobenzene (isomer not indicated), pentachlorobenzene, 2,4,5-trichlorophenol, and as several unidentified metabolites (see Table 3). Earlier studies by Stijve [93] suggested that chickens metabolized HCB to pentachlorophenol to a small extent. Some of the same metabolites were detected after HCB was incubated with tissue homogenates in vitro [92]. Dechlorination of HCB was not dependent on NADPH, but the formation of pentachlorophenol required this cofactor. Several microsomal enzymes were induced by HCB pretreatment [92, 94].

Wit and Van Genderen [95] orally dosed rats and dogs with [^{14}C]hexachlorophene and cows with unlabeled hexachlorophene to study the fate of this compound. The rabbits excreted approximately one-third of the ^{14}C in the urine and one-third in the feces as the parent compound, and one-third in the feces as an unidentified metabolite(s). Rats excreted a smaller percentage of the ^{14}C in the urine as the parent compound; the feces from rats contained both the parent compound and unidentified ^{14}C-labeled metabolites. The cows eliminated some of the hexachlorophene in the urine (0-2%) and feces (55-79%) as the parent compound within 4 to 5 days after dosing. The milk contained little (<0.8 ppm) or no hexachlorophene. The remainder of the compound administered was not accounted for. The tissues of the rats, rabbits, and cows used in these studies were not analyzed for the parent compound or the metabolites. When 5 ppm of hexachlorophene was fed to a lactating cow, no parent compound was detected in the milk (sensitivity of assay 0.02 ppm), but the feces and urine collected for 10 days contained 63.8 and 0.24% of the dose, respectively, as the parent compound [96]. The same workers reported that hexachlorophene was not metabolized by rumen fluid or bovine liver in vitro. Unfortunately the urine, feces, milk, and tissues were not analyzed for metabolites that may have been present. Mitin et al. [97] reported that relatively large amounts of ^{14}C were present in the tissues of guinea pigs and in milk of goats after administration of [^{14}C]hexachlorophene.

Gandolfi and Buhler [98] gave rats an intraperitoneal dose of [methylene-^{14}C]hexachlorophene and observed extensive enterohepatic circulation of ^{14}C. The parent compound and a metabolite, tentatively identified as hexachlorophene monoglucuronide, were present in the bile.

VI. ANTIBIOTIC FUNGICIDES

Barnes and Boothroyd [99] reported that rabbits and rats excreted within 48 hr 35.2 and 14.6%, respectively, of the ^{36}Cl given as a single oral dose of [^{36}Cl]griseofulvin. Less than 1% of the ^{36}Cl in the urine was present as the parent compound, but a major metabolite in both rabbit and rat urine was identified as 6-demethylgriseofulvin (see Table 4). The same metabolite was present in the urine of a man dosed with griseofulvin. Rats given an oral dose of [^{14}C]griseofulvin (^{14}C label in the 3, 4, 6, 8, 2', 4', and 6' positions) excreted 31% of the ^{14}C in the urine over a 96-hr period; when rats were given the same preparation intravenously, 55% of the ^{14}C was excreted in the urine [100]. Major ^{14}C-labeled metabolites in the urine of these animals were identified as free and conjugated 4-demethylgriseofulvin and free and conjugated 6-demethylgriseofulvin (see Table 4). After [^{14}C]griseofulvin was given intravenously, about 37% of the ^{14}C was detected in the intestine. This observation suggested that enterohepatic circulation of this compound and/or its metabolites was important. This observation was confirmed in a later study by Symchowicz et al. [101] who reported that

TABLE 4

The Metabolism of Griseofulvin in Animals

Parent compound and name	Metabolites: Proposed structure	Source	Evidence for structure[a]	Reference
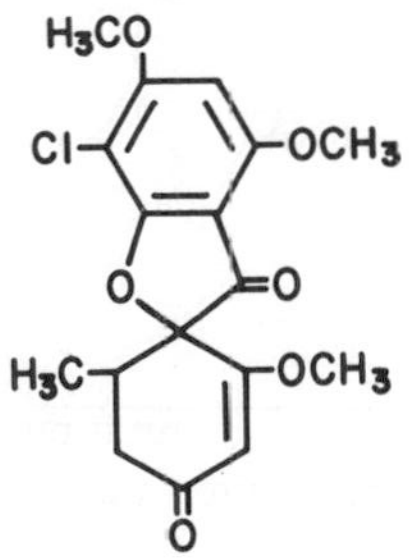 Griseofulvin 7-Chloro-4,6,2'-trimethoxy-6'-methylgris-2'-en-3,4'-dione	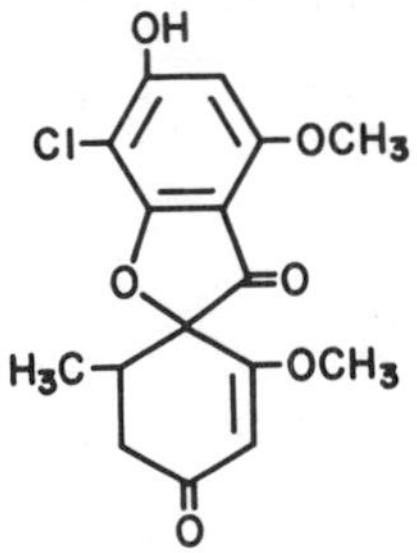	Rabbit, rat and human urine	Interpretation of UV and IR spectra	99
		Dog urine and bile	Comparative IR spectroscopy	102
		Rat liver in vitro	Comparative paper chromatography	104
		Rat urine	Comparative paper chromatography	100
		Rat urine and bile; dog bile	Comparative paper chromatography	101
	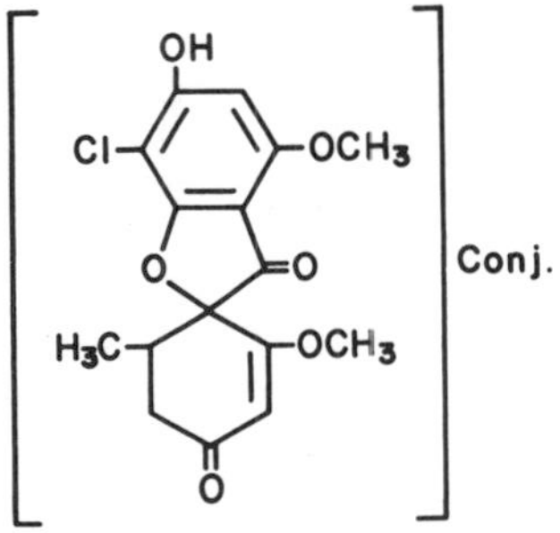 [ ] Conj.	Dog urine	Comparative IR spectroscopy of hydrolysis product	102
		Rat liver in vitro	Comparative paper chromatography of hydrolysis product	104
		Rat urine	Comparative paper chromatography of hydrolysis product	100

TABLE 4 (Cont.)

Parent compound and name	Metabolites: Proposed structure	Source	Evidence for structure[a]	Reference
		Rat and dog bile	Comparative paper chromatography of hydrolysis product	101
	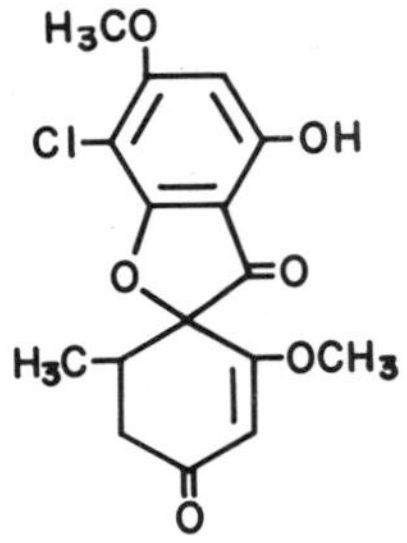	Rat liver in vitro	Comparative paper chromatography	104
		Rat urine	Comparative paper chromatography	100
		Rat urine and bile; dog bile	Comparative paper chromatography	101

Structure	Source	Method of identification	Reference
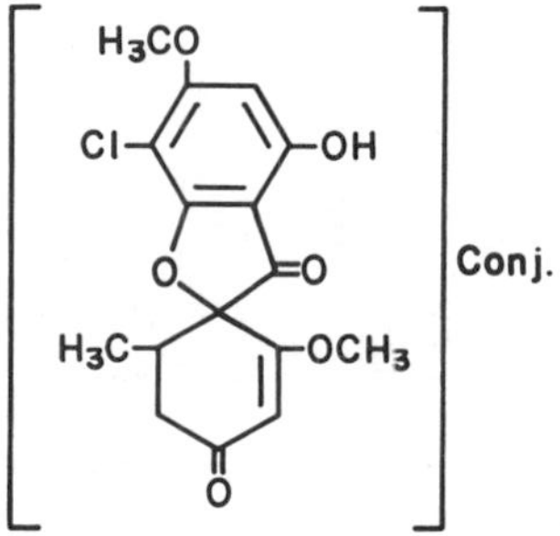 [H3CO, Cl, OH, O, O, H3C, OCH3, O] Conj.	Rat liver in vitro	Comparative paper chromatography of hydrolysis product	104
	Rat urine	Comparative paper chromatography of hydrolysis product	100
	Rat and dog bile	Comparative paper chromatography of hydrolysis product	101
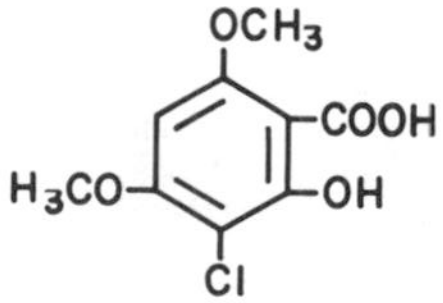 [OCH3, COOH, OH, Cl, H3CO]	Rabbit urine	Elemental analysis and interpretation of IR spectrum	103

[a]For abbreviations, see Table 1.

about 77% of the ^{14}C given intravenously to a biliary cannulated rat as [^{14}C]griseofulvin was secreted in the bile during a 24-hr period, and only 12% was excreted in the urine. The major metabolite in the rat bile was identified as 6-demethylgriseofulvin (see Table 4). In contrast, biliary excretion represented a minor pathway in the rabbit; only 11% of the ^{14}C was detected in the bile and 78% in the urine of the rabbit [101]. Two major metabolites in the urine of both intact and cannulated rats were identified as 4-demethylgriseofulvin and 6-demethylgriseofulvin. The major metabolite in the rabbit bile and the rabbit urine was identified as 6-demethylgriseofulvin.

Harris and Riegelman [102] reported that the disappearance of griseofulvin from the plasma of dogs after intravenous dosing was biexponential; the average half-life of the fast and slow exponential components was 4 and 47 min, respectively. The parent compound was not detected in the urine or bile of these animals, but approximately 40% of the dose was eliminated in the urine as 6-demethylgriseofulvin and 25% as a conjugate of the same metabolite. Less than 3% of the dose was eliminated in the bile of these dogs as 6-demethylgriseofulvin.

One additional metabolite in the urine of rabbits orally dosed with unlabeled griseofulvin was identified as 3-chloro-4,6-dimethoxysalicylic acid by Tomomatsu and Kitamura [103]. They did not report what percentage of the dose this metabolite accounted for, but most likely it was a minor component because it was not reported in the later studies by Barnes and Boothroyd [99].

In vitro studies by Symchowicz and Wong [104] showed that griseofulvin was metabolized by rat liver to free and conjugated 4-demethylgriseofulvin, free and conjugated 6-demethylgriseofulvin, as well as other unidentified metabolites. Surprisingly other tissues such as heart, kidney, lung, and skin did not metabolize griseofulvin.

VII. ORGANOPHOSPHATE FUNGICIDES

Uesugi et al. [105] and Tomizawa et al. [106] have reviewed the metabolism of edifenphos (Hinosan), Kitazin P, and Inezin by animals and a number of other organisms, although detailed reports about the animal metabolism of these compounds have not been published [107]. However, Uesugi et al. [105] reported that rats metabolized Hinosan mainly by cleavage of the P-S linkage to produce ethyl S-phenyl hydrogen phosphorothiolate. There is apparently some evidence for hydroxylation of Hinosan at the para position of the phenyl ring and hydrolysis to give diisopropyl hydrogen phosphate by animal systems. Uesugi et al. [105] also reported that rats cleaved the S-C linkage in kitazin to give O,O-diisopropyl hydrogen phosphorothiolate. It is hoped that the results of these studies will be published in detail so that the quantitative importance of these metabolites can be evaluated.

VIII. GENERAL DISCUSSION

With certain notable exceptions, most of the organic fungicides that have been studied in animals were readily absorbed from the gut. Some of these compounds were eliminated in the urine and feces, primarily as the parent compound, but most were at least partially metabolized by the animals. A large percentage of most of these fungicides, their metabolites, or both were eliminated in the urine and feces within 2 to 4 days after dosing. Some rather unique metabolic reactions have been reported in the animal metabolism of fungicides, such as the conversion of the nitro group in DCNA to a hydroxyl group, and the conversion of the nitro group in PCNB to a methylthio group. However, most of the animal-mediated transformations of fungicides reported (dealkylation, hydrolysis, oxidation, reduction, dehalogenation, hydroxylation, glucuronic acid conjugation, sulfate ester conjugation, and amino acid conjugation) are common to the animal metabolism of other types of xenobiotics [108-110].

Although the animal metabolism of a few fungicides has been studied in depth, this is not generally true for this group of pesticides. No published information was found concerning the animal metabolism of folpet, difolatan, dyrene, ethozal, diazoben, dodine, captafol, oxine-copper, and dichlone. Acknowledging that some reports were probably missed in this literature survey, nevertheless it is apparent that published information about the animal metabolism of many of the commonly used fungicides is limited. Some of the studies that have been reported were conducted with nonradiolabeled fungicides. In most of these studies only part of the administered material was accounted for by the chemical, spectrophotometric, or colorimetric techniques used. More complete information can usually be obtained when a radiolabeled (preferably ^{14}C-labeled) compound is used in animal metabolism because the radioactivity can be easily and accurately measured, whether it is present as the parent compound or as metabolites. Also, comparative metabolism studies with a compound synthesized with a radiolabel at strategically different locations in the molecule can give valuable information about the metabolism of the compound in question. For example, studies of this nature will often determine whether the linkage between two ring systems or functional groups are cleaved by the animal. Still another advantage of conducting animal metabolism studies with radiolabeled compounds is that it greatly simplifies the isolation, identification, and quantitation of metabolites. Future studies of the metabolism of fungicides by the animal would benefit by more extensive use of radiotracer techniques.

In many of the metabolism studies that have been reported, only a small percentage of the metabolites produced by the animal were identified. Often the quantitative importance of the identified metabolites was not reported. Inspection of Tables 1-4 shows that studies to determine the nature of the conjugating groups have usually been neglected. Usually, the polar conjugates were cleaved chemically or enzymatically and only the nonpolar hydrolysis product was characterized. Some workers have assigned structures to intact metabolites based on this information and enzyme hydrolysis studies. This approach can lead to false conclusions because of a number of possible problems, which include nonenzymatic hydrolysis, lack of enzyme specificity, failure of enzymes to hydrolyze some conjugates due to steric hindrance, and the fact that many of the preparations used often contain mixtures of two or more hydrolytic enzymes.

Tables 1-4 also show that many of the fungicide metabolites that have been reported were characterized by cochromatography only; such characterizations of structures can be considered as only tentative identifications. With instruments now available it is possible to obtain infrared, ultraviolet and mass spectral data on less than 10 μg of most compounds. Nuclear magnetic resonance spectra can be obtained on 100 μg or less of most pure compounds with spectrophotometers equipped with fast Fourier transform capabilities. Commercial analytical laboratories will perform these services for a relatively nominal fee. The need for definitive spectral data to verify structures justifies the time and expense of obtaining this type of information.

Because fungicides are applied to and metabolized by plants, other questions arise. For instance, what is the fate of plant fungicide metabolites when consumed by animals? Information to answer this question is extremely limited. Other areas of research that have received only limited emphasis include (1) studies on the comparative animal metabolism of fungicides; (2) studies to determine the nature of fungicide metabolites in the milk, eggs, and tissues from animals; (3) the animal metabolism of fungicides given by inhalation and dermal application; (4) the animal metabolism of fungicide photodecomposition products; and (5) the possible effect of other xenobiotics on the animal metabolism of fungicides. Thus, even though considerable valuable information concerning the animal metabolism of fungicides has been reported, it is obvious that much additional work is needed. Whether improved technology and continued concern for the safe usage of fungicides will result in more definitive and complete studies in this area of research remains to be seen.

REFERENCES

1. D. C. Torgeson (ed.), Fungicides, Vol. 2, Academic Press, New York, 1969.
2. W. T. Thomson, Agricultural Chemicals, Book IV, Fungicides, Thomson Publications, Davis, California, 1967.
3. J. G. Saha, Residue Rev., 42, 103 (1972).
4. T. W. Clarkson, Ann. Rev. Pharmacol., 12, 375 (1972).
5. L. Friberg and V. Jaroslav, Mercury in the Environment, Chemical Rubber Co. Press, Cleveland, Ohio, 1972.

6. H. M. Dekhuijzen, J. W. Vonk, and A. Kaars Sijpesteijn, International Symposium on Pesticide Terminal Residues, Tel-Aviv, Israel, 1971 (A. S. Tahori, ed.), Butterworths, London, 1971, pp. 233-241.
7. M. Sh. Vekshtein and I. I. Khitsenko, Gig. Sanit., 36, 23 (1971).
8. N. Izmirova and V. Marinov, Eksp. Med. Morfol., 11, 152 (1972); through Pestic. Abstr., 7, 74-0651 (1974).
9. N. Izmirova, Eksp. Med. Morfol., 11, 240 (1972); through Pestic. Abstr., 7, 74-0653 (1974).
10. G. Eisenbrand, O. Ungerer, and R. Preussmann, Food Cosmet. Toxicol., 12, 229 (1974).
11. J. R. Hodgson, T. R. Castles, E. Murrill, and C. C. Lee, Federation Proc., 33, 1836 (1974).
12. J. H. Strömme, Biochem. Pharmacol., 14, 393 (1965).
13. J. H. Strömme, Biochem. Pharmacol., 14, 381 (1965).
14. J. Hald, E. Jacobsen, and V. Larsen, Acta Pharmacol., 4, 285 (1948).
15. L. Eldjarn, Scand. J. Clin. Lab. Invest., 2, 198 (1950).
16. L. Eldjarn, Scand. J. Clin. Lab. Invest., 2, 202 (1950).
17. G. Domar, A. Fredga, and H. Linderholm, Acta Chem. Scand., 3, 1441 (1949).
18. H. Linderholm and K. Berg, Scand. J. Clin. Lab. Invest., 3, 96 (1951).
19. J. Kaslander, Biochim. Biophys. Acta, 71, 730 (1963).
20. C. S. Prickett and C. D. Johnston, Biochim. Biophys. Acta, 12, 542 (1953).
21. E. Merlevede and H. Casier, Arch. Intern. Pharmacodyn., 132, 427 (1961).
22. T. Gessner and M. Jakubowski, Biochem. Pharmacol., 21, 219 (1972).
23. E. A. Antonovich and M. Sh. Vekshtein, Gig. Primen. Toksikol. Pestits. Klin. Otravlenii, No. 8, 221 (1970); through Chem. Abstr., 77, 110262j (1972).
24. R. C. Robbins and J. Kastelic, J. Agr. Food Chem., 9, 256 (1961).
25. W. R. Lyman, International Symposium on Pesticide Terminal Residues, Tel-Aviv, Israel, 1971 (A. S. Tahori, ed.), Butterworths, London, 1971, pp. 243-256.
26. W. H. Newsome, Bull. Environ. Contam. Toxicol., 11, 174 (1974).
27. H. Seidler, M. Härtig, W. Schnaak, and R. Engst, Nahrung, 14, 363 (1970).
28. R. Truhaut, M. Fujita, N. P. Lich, and M. Chaigneau, C. R. Hebd. Séanc. Acad. Sci., Paris (D), 276, 229 (1973); through Chem. Abstr., 78, 1324642z (1973).
29. K. S. Khera, Teratology, 7, 243 (1973).
30. T. Noguchi, K. Ohkuma, and S. Kasaka, International Symposium on Pesticide Terminal Residues, Tel-Aviv, Israel, 1971 (A. S. Tahori, ed.), Butterworths, London, 1971, pp. 257-270.
31. T. Noguchi, Environmental Toxicology of Pesticides, Proceedings of a United States-Japanese Seminar, Oiso, Japan, October 1971

(F. Matsumura, G. M. Boush, and T. Misato, eds.), Academic Press, New York, 1972, pp. 607-625.
32. P. G. C. Douch, Xenobiotica, 4, 457 (1974).
33. J. Herok and H. Götte, Intern. J. Appl. Radiation Isotopes, 14, 461 (1963).
34. J. Herok and H. Götte, Zentr. Veterinaermed, 11, 20 (1964).
35. J. Brüggemann, K. Barth, and K. H. Niesar, Zentr. Veterinaermed, 11, 4 (1964).
36. J. Bruggemann, O. R. Klimmer, and K. H. Niesar, Zentr. Veterinaermed, 11, 40 (1964).
37. B. Williams, L. G. Dring, and R. T. Williams, Biochem. J., 127, 24p (1972).
38. M. S. Rose and W. N. Aldridge, Biochem. J., 106, 821 (1968).
39. M. S. Rose, Biochem. J., 111, 129 (1969).
40. W. N. Aldridge and M. S. Rose, FEBS Letters, 4, 61 (1969).
41. M. S. Rose and E. A. Lock, Biochem. J., 120, 151 (1970).
42. W. N. Aldridge and B. W. Street, Biochem. J., 118, 171 (1970).
43. W. N. Aldridge and B. W. Street, Biochem. J., 124, 221 (1971).
44. M. S. Rose, International Symposium on Pesticide Terminal Residues, Tel-Aviv, Israel, 1971 (A. S. Tahori, ed.), Butterworths, London, 1971, pp. 281-288.
45. K. H. Byington, R. Y. Yeh, and L. R. Forte, Toxicol. Appl. Pharmacol., 27, 230 (1974).
46. D. J. Tocco, R. P. Buhs, H. D. Brown, A. R. Matzuk, H. E. Mertel, R. E. Harman, and N. R. Trenner, J. Med. Chem., 7, 399 (1964).
47. D. J. Tocco, J. R. Egerton, W. Bowers, V. W. Christensen, and C. Rosenblum, J. Pharmacol. Exptl. Therap., 149, 263 (1965).
48. D. J. Tocco, C. Rosenblum, C. M. Martin, and H. J. Robinson, Toxicol. Appl. Pharmacol., 9, 31 (1966).
49. C. Rosenblum, D. J. Tocco, and E. E. Howe, Proceedings of 2nd Annual Oak Ridge Conference, Gatlinburg, Tenn., p. 58, April 1964; U.S. Atomic Energy Commission Report TID-7689.
50. J. A. Gardiner, J. J. Kirkland, H. L. Klopping, and H. Sherman, J. Agr. Food Chem., 22, 419 (1974).
51. J. A. Gardiner, R. K. Brantley, and H. Sherman, J. Agr. Food Chem., 16, 1050 (1968).
52. J. J. Kirkland, J. Agr. Food Chem., 21, 171 (1973).
53. P. G. C. Douch, Xenobiotica, 3, 367 (1973).
54. R. Engst and M. Raab, Nahrung, 17, 731 (1973).
55. H. Seidler, M. Härtig, W. Schnaak, and R. Engst, Nahrung, 15, 177 (1971).
56. J. R. DeBaun, J. B. Miaullis, J. Knarr, A. Milhailovski, and J. J. Menn, Xenobiotica, 4, 101 (1974).
57. A. Calderbank, Acta Phytopathologica Academiae Scientarum Hungaricae, 6, 355 (1971).

58. H. Bratt, J. W. Daniel, and I. H. Monks, Food Cosmetol. Toxicol., 10, 489 (1972).
59. H. F. Smyth, Jr., C. P. Carpenter, and C. S. Weil, Toxicol. Appl. Pharmacol., 9, 521 (1966).
60. S. Sumida, Y. Hisada, A. Kometani, and J. Miyamoto, Agr. Biol. Chem., 37, 2127 (1973).
61. H. Ohkawa, Y. Hisada, N. Fujiwara, and J. Miyamoto, Agr. Biol. Chem., 38, 1359 (1974).
62. J. W. Daniel, Biochem. J., 111, 695 (1969).
63. P. J. Murphy and T. L. Williams, J. Med. Chem., 15, 137 (1972).
64. F. S. Eberts, Proc. of Botran Symposium, Upjohn Company, Kalamazoo, Michigan, 1965, pp. 1-9.
65. F. S. Eberts, Pesticide Subdivision, 153rd Meeting, American Chemical Society, April 1967, Miami Beach, Florida.
66. C. Maté, A. J. Ryan, and S. E. Wright, Food Cosmet. Toxicol., 5, 657 (1967).
67. J. J. Betts, S. P. James, and W. V. Thorpe, Biochem. J., 61, 611 (1955).
68. L. E. St. John, J. W. Ammering, D. G. Wagner, R. G. Warner, and D. J. Lisk, J. Dairy Sci., 48, 502 (1965).
69. E. J. Kuchar, F. O. Geenty, W. P. Griffith, and R. J. Thomas, J. Agr. Food Chem., 17, 1237 (1969).
70. J. F. Borzelleca, P. S. Larson, E. M. Crawford, G. R. Hennigar, Jr., E. J. Kuchar, and H. H. Klein, Toxicol. Appl. Pharmacol., 18, 522 (1971).
71. H. G. Bray, Z. Hybs, S. P. James, and W. V. Thorpe, Biochem. J., 53, 266 (1953).
72. J. J. Betts, H. G. Bray, S. P. James, and W. V. Thorpe, Biochem. J., 66, 610 (1957).
73. M. Ishida, H. Sumi, and H. Oku, Residue Rev., 25, 139 (1969).
74. A. Bevenue and H. Beckman, Residue Rev., 19, 83 (1967).
75. I. Jakobson and S. Yllner, Acta Pharmacol. Toxicol., 29, 513 (1971).
76. R. V. Larsen, L. E. Kirsch, S. M. Shaw, J. E. Christian, and G. S. Born, J. Pharm. Sci., 61, 2004 (1972).
77. W. H. Gutenmann and D. J. Lisk, J. Agr. Food Chem., 17, 1008 (1969).
78. R. C. Rhodes and H. L. Pease, J. Agr. Food Chem., 19, 750 (1971).
79. R. H. Waring, Xenobiotica, 3, 65 (1973).
80. W. T. Chin, G. M. Stone, and A. E. Smith, International Symposium on Pesticide Terminal Residues, Tel-Aviv, Israel, 1971 (A. S. Tahori, ed.), Butterworths, London, 1971, pp. 271-279.
81. J. H. Weikel, Jr. and M. J. Bartek, Toxicol. Appl. Pharmacol., 22, 375 (1972).
82. M. Avrahami and R. T. Steele, N.Z. J. Agr. Res., 15, 476 (1972).
83. M. Avrahami and R. T. Steele, N.Z. J. Agr. Res., 15, 482 (1972).

84. J. G. Vos, H. L. van der Maas, A. Musch, and E. Ram, Toxicol. Appl. Pharmacol., 18, 944 (1971).
85. J. G. Vos, H. A. Breeman, and H. Benschop, Meded. Rijksfac. Landbouwwetensch. Gent, 33, 1263 (1968); through Chem. Abstr., 71, 59133q (1969).
86. M. Avrahami and R. T. Steele, N.Z. J. Agr. Res., 15, 489 (1972).
87. M. N. Brady and D. S. Siyali, Med. J. Australia, 1, 158 (1972).
88. J. L. Johnson, D. L. Stalling, and J. W. Hogan, Bull. Environ. Contam. Toxicol., 11, 393 (1974).
89. D. C. Villeneuve, W. E. J. Phillips, L. G. Panopio, C. E. Mendoza, G. V. Hatina, and D. L. Grant, Arch. Environ. Contam. Toxicol., 2, 243 (1974).
90. P. W. Albro and R. Thomas, Bull. Environ. Contam. Toxicol., 12, 289 (1974).
91. H. M. Mehendale and H. B. Matthews, Pesticide Division, 165th Meeting, American Chemical Society, Dallas, Texas, April 1973.
92. H. M. Mehendale, M. Fields, and H. B. Matthews, J. Agr. Food Chem., 23, 261 (1975).
93. T. Stijve, Mitt. Gebiete Lebensmittel Hyg., 62, 406 (1971).
94. M. D. Stonard and P. Z. Nenov, Biochem. Pharmacol., 23, 2175 (1974).
95. J. G. Wit and H. Van Genderen, Acta Phys. Pharmacol. Neerl., 11, 123 (1962).
96. L. E. St. John and D. J. Lisk, J. Agr. Food Chem., 20, 389 (1972).
97. V. Mitin, M. Delak, and M. Tranger, Vet. Arhiv., 39, 318 (1969); through Chem. Abstr., 73, 96226j (1970).
98. A. J. Gandolfi and D. R. Buhler, Xenobiotica, 4, 693 (1974).
99. M. J. Barnes and B. Boothroyd, Biochem. J., 78, 41 (1961).
100. S. Symchowicz and K. K. Wong, Biochem. Pharmacol., 15, 1595 (1966).
101. S. Symchowicz, M. S. Staub, and K. K. Wong, Biochem. Pharmacol., 16, 2405 (1967).
102. P. A. Harris and S. Riegelman, J. Pharm. Sci., 58, 93 (1969).
103. S. Tomomatsu and J. Kitamura, Chem. Pharm. Bull., 8, 755 (1960).
104. S. Symchowicz and K. K. Wong, Biochem. Pharmacol., 15, 1601 (1966).
105. Y. Uesugi, C. Tomizawa, and T. Murai, in Environmental Toxicology of Pesticides (F. Matsumura, G. M. Boush, and T. Misato, eds.), Academic Press, New York, 1972, p. 327.
106. C. Tomizawa, Y. Uesugi, and T. Murai, Radiotracer Studies of Chemical Residues in Food and Agriculture, International Atomic Energy Agency, Vienna, 1972, pp. 103-105.
107. Y. Uesugi, private communication, 1975.
108. R. T. Williams, Detoxication Mechanisms, 2nd ed., Chapman and Hall, Ltd., London, 1959.
109. C. O. Wilson, O. Gisvold and R. E. Doerge, Textbook of Organic, Medicinal and Pharmaceutical Chemistry, 6th ed., Lippincott, Philadelphia, 1971.
110. G. D. Paulson, Residue Rev., 58, 1 (1975).

Chapter 5

NONBIOLOGICAL CONVERSIONS OF FUNGICIDES

David Woodcock

Department of Agriculture and Horticulture
Long Ashton Research Station
University of Bristol
Bristol, England

I. INTRODUCTION

Although the title of this chapter implies a definite demarcation between biological and nonbiological transformations, this is very far from the truth. In environments such as soil or growing plants, the effects of air, light, water, and enzymes are often completely integrated, and only with the aid of complementary in vitro experiments can possibly separate contributions be evaluated with any degree of certainty. Before discussing the behavior of individual fungicides, it is perhaps appropriate to mention briefly the various types of reactions that may take place in environments predominantly aerobic and aqueous. These would include hydrolysis of alkyl or aryl esters, amides, phosphates, and carbamates, not only in a pesticidal formulation, particularly during storage under favorable conditions, but also on the plant surface and in the soil. Oxidation also takes place at carbon, sulfur, and nitrogen atoms and isomerizations involving pentavalent phosphorus are known. Epoxidation which occurs with certain insecticides is also possible, though there do not appear to be any reports as yet involving fungicides. Dimerization and polymerization are more common and while reduction is rare, there are several examples of photoreduction and photodehydrochlorination.

However, almost any of the known organic constituents of soil are potential reducing agents, especially under oxygen-deficient conditions, and indeed because of its nature, which provides reactive surfaces supplied with oxygen and water in varying amounts, soil affords an almost ideal medium for the degradation of pesticides. The recently discovered free radicals in soil humic acids [1] are potential reagents for a variety of reactions, and these same semiquinones also could serve as oxidizing agents.

The importance of light energy in the fate of pesticides has become increasingly recognized in the past decade, and all of the above reactions may be unobtrusively photopromoted, if not photoinitiated. The wavelength of ultraviolet (UV) light extends from 400-40 nm, but because of a thin layer of ozone in the earth's atmosphere, practically all the sun's emitted radiation below 290 nm is absorbed before it reaches the surface. In the light of the report of Gore et al. [2], who found that of 76 pesticides, 29 had no bands above 290 nm in their UV absorption spectra, and 25 had only a few, this could be critical in terms of terminal residue considerations. However, because of a not inconsiderable flux density of 10^{16} photons per cm^2 per month, a long-term effect at shorter wavelengths is possible, and indeed dieldrin with no absorption band as high as 290 nm does, in fact, break down in sunlight.

For a photoreaction to occur, an organic compound must absorb energy either directly, or through a photosensitizer (photochemically excited molecule), which may be deliberately introduced or which may already be present, e.g., on a plant surface. The important fact, however, is that the energy available in the UV region (see Table 1) is of the order needed for the fission of some typical covalent bonds in organic compounds (see Table 2).

TABLE 1

Solar Energy Available

Wavelength, nm	Energy, Kcal/mole
200	143
250	114
300	95
350	82
400	72

TABLE 2

Covalent-Bond Fission Energies

Bond	Energy, Kcal/mole
C-C, in ethane	84
C-H, in ethane	98
C-OH	92
$C-NH_2$	78
C-Cl	78
$O-CH_3$	77
O-H, in water	119
N-H	96

The environment of the pesticide molecule is also all important, for inter- as well as intramolecular reactions may occur. Thus adsorption on soil particles may change the maximum absorption wavelength and hence alter photodecomposition energy requirements. The effect of photosensitizers present on the leaf surface, in the soil, or even in the air may also be extremely important in relation to unexplained synergistic effects on which little, if any, work has been done.

Attempts to simulate the wavelength range and intensity of solar radiation in vitro have generally involved single-lamp sources and while photolysis products are generally identical for all light sources, a correlation factor between artificial and natural sunlight is necessary. The experimental approach has been discussed by Crosby [3]. Whereas low-pressure lamps produce 90% of their energy at 254 nm, medium-pressure lamps have this proportion of the available energy fairly evenly spread over the range of 290 to 366 nm. Despite arguments to the contrary, in vitro experiments are invaluable in that degradation products can be obtained in relative abundance, enabling their detection, isolation, and identification to be carried out more easily and more confidently than in experiments in vivo. Irradiation of solid deposits on glass or metal has some obvious merit but only the surface layer of a thick deposit may be affected, and therefore irradiation in solution is more frequently used.

Since pesticides in the atmosphere are exposed to more intensive UV radiation over a wider wavelength range, and to a variety of free radicals produced photochemically from other common pollutants such as sulfur dioxide and nitrogen tetroxide, investigations involving gaseous-phase irradiation seem highly desirable and many more investigations of this type surely will be instigated.

A brief note on photosensitization is perhaps pertinent here. This is the phenomenon whereby one photoexcited species acts as a donor and transfers its excitation (triplet state) energy to another (acceptor) species, which then undergoes a photochemical reaction as though it had absorbed the light energy directly. While many organic chemicals, including naturally occurring substances, induce or enhance such reactions, others act as quenching agents [4]. Examples of the former include chlorophyll, while the carotenoids have a quenching capacity in varying degree. Studies of interactions between pesticide and photosensitizer may reveal combinations of potential use in controlling the persistence of pesticide residues on treated crops and in the soil [5-7].

II. SOIL FUNGICIDES AND SEED STERILANTS

While this class of fungicide may be less affected by UV radiation than those applied to foliage, the soil in general can provide a near optimum environment for hydrolytic and oxidative decomposition.

A. Organomercurials

In early work on the fungicidal action of organomercurials, the formation of metallic mercury was often accepted as being due to chemical breakdown [8], but there is increasing evidence for at least concomitant microbial action [9, 10]. However, that 2-chloro-4-hydroxymercuriphenol (Semesan) is still fungicidally active after some months in steamed soil, although completely inactive after only 2-3 weeks in nonsterile soil, would seem to exclude the operation of purely chemical degradative processes. The success of methylmercury dicyandiamide as a soil fungicide results probably from its stability in soil, despite its water solubility and vapor action.

An in vitro study of the photodecomposition of organomercury pesticides by Shiina et al. [11], using light of wavelength >290 nm for a period equivalent to 7 summer days, demonstrated that those compounds (including phenylmercury acetate and chloride) which do not absorb below 290 nm were little affected, whereas two others with absorption bands >290 nm were degraded in varying degree. The extent of decomposition was measured by the hanging-drop culture method using spores of *Cochliobolus miyabeanus*.

B. Organosulfur Compounds

1. Sodium Methyldithiocarbamate (metham sodium)

Metham sodium (1) tends to be erratic in field performance and may be decomposed in a period as short as 2 weeks [12, 13]. Since soil sterilization has no effect on this oxidative degradation, it is now generally accepted that it is abiotic in character. Moreover, it seems likely that the variable results obtained in the field are linked with the critical role played by soil type, and it has been suggested that the increased aeration of metham sodium absorbed on soil particles speeds the decomposition process. A variety of breakdown products have been identified (see below), and it is assumed that the fungicidal action of metham sodium is due to methyl isothiocyanate, the production of which has been reported at both high and low pHs. Two other products should be noted. Turner and Corden [14] showed that more than 87% of metham sodium applied to a sandy loam soil (pH < 7) was accounted for as *N,N'*-dimethylthiourea (2), while Moje et al. [15] reported the detection of carbonyl sulfide in the atmosphere above soil of pH < 5 containing the fungicide.

$$\underset{(1)}{CH_3NHCSSNa} \xrightarrow{pH\ 9.5} CH_3NCS + S$$

$$\underset{(1)}{CH_3NHCSSNa} \xrightarrow{pH\ <7} CH_3NCS + CS_2 + H_2S + CH_3NH_2 + \underset{(2)}{CH_3NHCSNHCH_3}$$

$$\underset{(1)}{CH_3NHCSSNa} \xrightarrow{pH\ 3\text{-}5} COS + CH_3NCS$$

In a recent paper, Smelt and Leistra [16] demonstrated conclusively that the conversion of metham sodium to methyl isothiocyanate was dependent on temperature and soil type, and that a first-order rate equation described the decomposition of the latter compound. The half-life varied from a few days to a few weeks, again according to temperature and soil type. This substantiates earlier work by Ashley et al. [17].

2. Tetrahydro-3,5-dimethyl-2*H*-1,3,5-thiadiazine-2-thione (dazomet)

Dazomet (3), which may be regarded as a cyclic dithiocarbamate, also readily decomposes in the soil, yielding methylaminomethyl dithiocarbamate (4) which then degrades further to methyl isothiocyanate. This degradation is favored by increasing moisture content and high temperature, and according to Munnecke et al. [18, 19], there does not appear to be any evidence for microbial involvement.

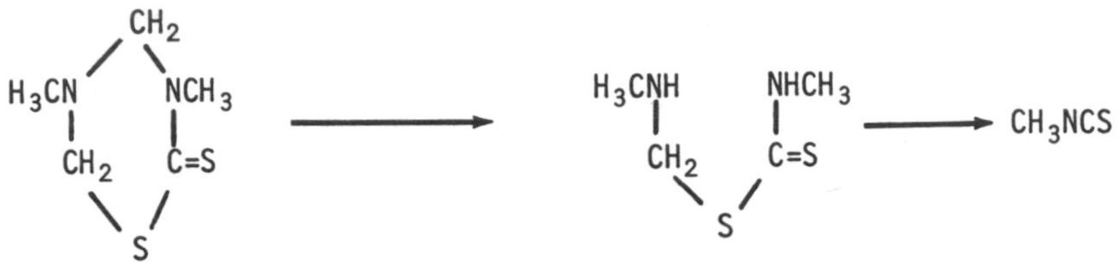

CH_2 H_3CN NCH_3 CH_2 C=S S → H_3CNH $NHCH_3$ CH_2 C=S S → CH_3NCS

(3) (4)

3. Disodium Ethylenebis(dithiocarbamate) (nabam)

Nabam (5) was one of the first dithiocarbamates to be tried as a soil fungicide. On aeration of its aqueous solution, a complex yellow precipitate is obtained; its main fungicidal components were elemental sulfur and ethylenethiuram monosulfide (etem) (6). This structure for etem, originally proposed by Thorn and Ludwig [20] and reaffirmed later by Thorn [21], continued to be used until quite recently [22, 23]. Renewed interest in the degradation product, however, prompted a reappraisal of the structure [24], and

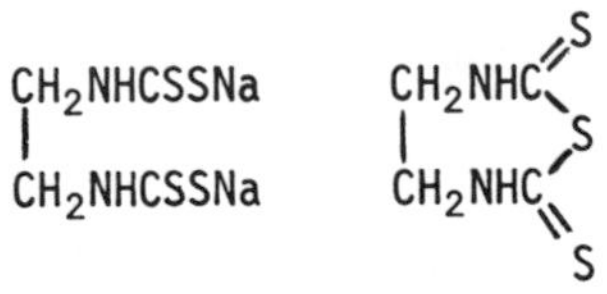

$CH_2NHCSSNa$ | $CH_2NHCSSNa$ $CH_2NHC(=S)$ | S $CH_2NHC(=S)$

(5) (6)

nuclear magnetic resonance (NMR) evidence ruled out the symmetrical structure. This together with mass spectral (MS) evidence (m/e 176) suggested that etem is 5,6-dihydro-3*H*-imidazo[2,1-c]-1,2,4-dithiazole-3-thione (7), an asymmetrical structure first considered by Thorn [21]. Such a structure is compatible with the facile formation of a complex with carbon disulfide, which has the resonance-stabilized structure now considered to be (8). Structure (7) for etem was later confirmed by Benson et al. [25] who employed MS, NMR, infrared (IR), and Raman spectral data, the latter significantly indicating the presence of an S-S bond.

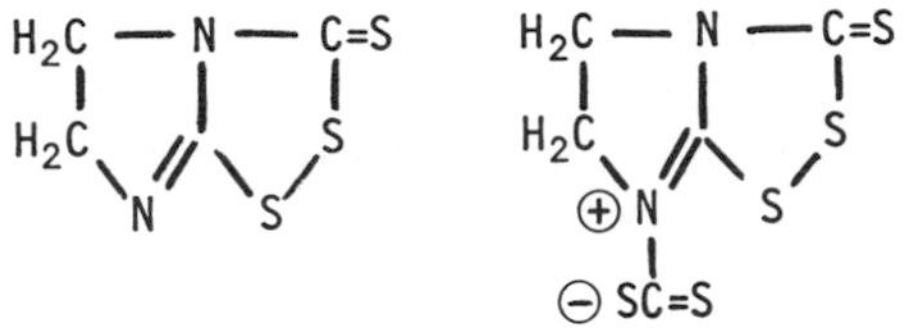

(7) (8)

Decomposition of nabam in the soil is also rapid and Munnecke [26] showed that a 1,000 ppm solution lost about one-half of its fungicidal activity in 2 days. This breakdown, which also leads to gaseous products, was shown to be independent of the presence of microorganisms, since the decay curves from sterile and nonsterile soils were very similar.

Etem is also decomposed by dilute acid, hot 1 N HCl giving a quantitative yield of ethylenediamine and carbon disulfide [27], while 0.25 N HCl results in a 90% yield of ethylenethiourea [28].

4. Ethylenethiourea

Ethylenethiourea (9) occurs in technical ethylenebisdithiocarbamates as an impurity and is also a breakdown product. In field experiments with manganese ethylenebis(dithiocarbamate) (maneb), Yip et al. [29] found that the ethylenethiourea initially present rapidly decreased to undetectable amounts. Vonk and Kaars Sijpesteijn [30] have also reported the ready uptake of ethylenethiourea by plants, and a minor conversion product was subsequently detected in cucumber and wheat seedlings. This has now been identified [31] as 2-imidazoline (10) which is assumed to be a terminal residue in these

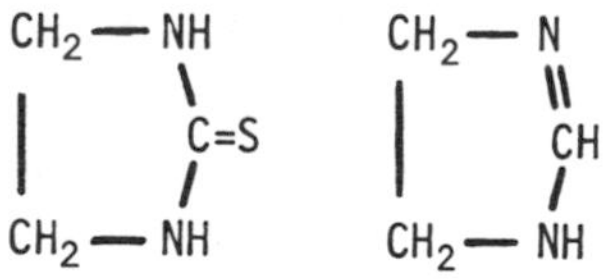

(9) (10)

plants. The mechanism of its formation seems at present rather obscure, but it appears to be nonenzymatic. 2-Imidazoline and ethylenethiourea have also been mentioned previously as intermediate degradation products of mancozeb applied to plants [32].

A recent report of tumorgenic activity in mice by Innes et al. [33] has prompted further investigations into the environmental fate of ethylenethiourea. The first by Cruickshank and Jarrow [34] demonstrated degradation on silica gel by radiation of wavelength >285 nm, mainly to ethyleneurea (11), though several other produts, including bisimidazolin-2-yl sulfide (12), were also formed. Ethyleneurea has been reported by Ingemells [35] to undergo

```
CH2 — NH          CH2 — N        N — CH2
|      \          |      \\     //     |
|       C=O       |       C-S-C        |
|      /          |      /      \      |
CH2 — NH          CH2 — NH       HN — CH2

   (11)                  (12)
```

photooxidation to hydantoin (13), but this latter compound was not detected among the products. Photodecomposition of ethylenethiourea was rapid, particularly in the presence of photosensitizers, and in aqueous solution a very slow photolysis was markedly accelerated by the presence of compounds such as 1- and 2-acetonaphthone, 1-naphthaldehyde, and methylene blue. Ross and Crosby [36] also found that ethylenethiourea in aqueous solution was virtually stable to sunlight, but the presence of dissolved oxygen and sensitizers such as acetone or riboflavin were found to cause rapid photolysis to ethyleneurea and glycine, possibly via hydantoin.

The persistence and degradation of [ethylene-^{14}C]ethylenethiourea in autoclaved and nonsterile soil of two types has been examined by Kaufman and Fletcher [37]. At rates as high as 200 ppm, essentially all of the compound was converted to ethyleneurea (11) in 8 days and a similar slower but steady conversion occurred in sterile soil; further degradation of the ethyleneurea with evolution of $^{14}CO_2$ occurred in nonsterile soil only. Four degradation products of ethylenethiourea were isolated and characterized, two being identified as hydantoin (13) and 1-(2'-imidazoline-2'-yl)imidazoline-2-thione (14).

```
                                  H2C — CH2
O=C — NH            H2C — N       |      |
|      \            |      \\     |      |
|       C=O         |       C — N       NH
|      /            |      /     \     /
H2C — NH            H2C — NH       C
                                   ||
                                   S

   (13)                  (14)
```

5. Dialkyldithiocarbamates

The decay curves for ferric dimethyldithiocarbamate (ferbam) in soil were found by Munnecke [26] to be comparable to those for nabam, though less steep, and the available evidence once again suggested that a physicochemical rather than a microbial breakdown process was operative. Whether this proceeds via tetramethylthiuram disulfide (thiram), which can be formed in phosphate buffer at pH 3-4, remains to be seen. However, Thorn and Richardson [38] have clearly demonstrated the importance of pH for the stability of dialkyldithiocarbamates in solution. They observed that the UV absorption spectrum of ferbam in phosphate buffer is that of the dimethyldithiocarbamate ion, but when the pH falls to 3, the curve is identical to that of thiram. Thus it is not surprising that when aqueous sodium dimethyldithiocarbamate is added to ferric sulfate at pH 3, thiram and not ferbam is obtained in quantitative yield. This instability certainly poses problems in mode-of-action studies.

Photochemical work involving dithiocarbamate fungicides is conspicuously scanty. Fission of the C-S bond using model compounds in benzene solution, was demonstrated by Okawara et al. [39] as long ago as 1960, but this work was in connection with photosensitive polymers. More recently Yamase et al. [40] showed the breakdown of sodium dimethyldithiocarbamate in aqueous solution at pH 8 under the influence of 254 nm radiation. The products were carbon disulfide and dimethylamine, and of the nine aromatic ketones examined as sensitizers of this photolysis, 2-acetonaphthone was the most effective. Other recent papers have described the irradiation of some benzoyldithiocarbamates, but perhaps the most useful work is that of Fitton et al. [41], who photoirradiated several 4-hydroxybenzyldithiocarbamates under various conditions. They suggested that the initial step is the homolytic fission of the C-S bond with the formation of dithiocarbamoyl and benzyl radicals, which then proceed to dimerize, the former giving a thiuram disulfide and the latter either dimerizing to a diphenylethane or dissociating with loss of a proton to a _p_-quinone methide. When irradiations were carried out in the presence of water, aromatic ketones were also produced probably arising via the _p_-quinone methide.

C. Other Aliphatic Compounds

Dithiocarbamates apart, there are few notable aliphatic fungicides, although formaldehyde, one of the oldest and a powerful bactericide and fungicide, had only limited use as a seed sterilant because of its phytotoxicity. It is still used, however, as a soil fumigant, the tendency of the usual 40% aqueous solution ("formalin") to polymerize being delayed by the addition of methanol.

More recently, 2,2-dibromo-2-cyanoacetamide (_15_), originally used as an effective seed sterilant [42], has been employed as an antimicrobial compound for industrial water treatment. As such it is eventually discharged into the environment and its fate is important. Exner et al. [43] have investigated rates and products of decomposition under a variety of conditions. Hydrolysis at pH 7.4 resulted in the ultimate release of ammonia, carbon dioxide, and Br^- via the degradation sequence shown in Figure 1.

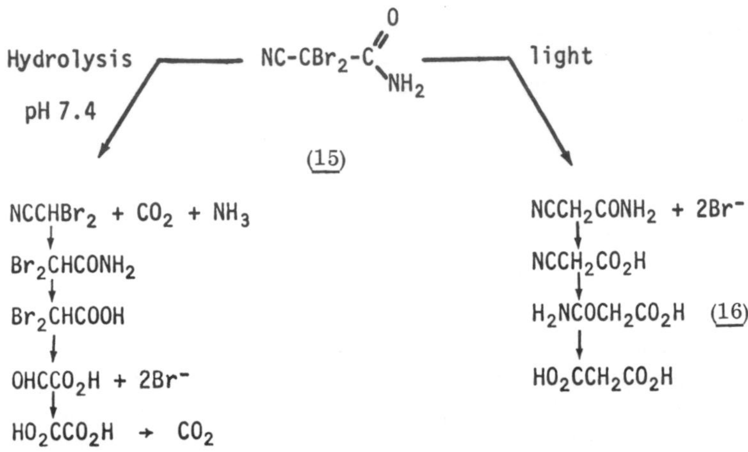

FIG. 1. Decomposition of 2,2-dibromo-2-cyanoacetamide.

Exposure of an aqueous solution containing 4,000 ppm to sunlight for 28 days resulted in almost complete decomposition with the formation of 92% of the theoretically possible Br^- and no isolatable organic matter. A 2,000 ppm solution exposed at a distance of 50 cm from a sunlamp, showed some 80% decomposition after 3 days, with concomitant formation of 90% of the theoretically possible Br^-. In this case, however, using IR and MS methods, residual organic material was identified as a mixture of cyanoacetic, malonic, and oxalic acids, together with probably the half-amide of malonic acid (16).

D. Aromatic Compounds

1. Chlorinated Benzenes

Photoreduction of chlorinated aromatic compounds has been reported relatively recently and the extent of this reaction as a major route of environmental loss of soil fungicides such as technazene (1,2,4,5-tetrachloro-3-nitrobenzene) and quintozene (pentachloronitrobenzene) is important. However, although a solution of quintozene in methanol or hexane on exposure to sunlight for 7 days became yellow, no decomposition products could be demonstrated using thin-layer (TLC) or gas-liquid (GLC) chromatography; irradiation of a thin film was similarly negative [44]. These results contrasted sharply with laboratory findings in which an examination of quintozene, which had been exposed in hexane solution to a low-pressure mercury arc lamp radiating at 254 nm, revealed the presence of four decomposition products in the rapidly discolored solution. These were in order of increasing

GLC retention time: 1,2,4,5-tetrachlorobenzene (17), pentachlorobenzene (18), and 2,3,4,6- and 2,3,4,5-tetrachloronitrobenzenes (19, 20); all four compounds were found to be present after only 5 min irradiation. After 56 hr they represented 20-23% of the initial quintozene, only 7% of which remained, and their relatively constant proportions indicated steady-state photolysis, which could be excluded when a borosilicate filter absorbing below 285 nm was interposed between the lamp and the solution.

(17) 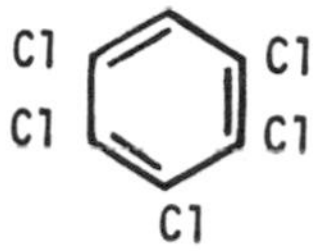 (18) 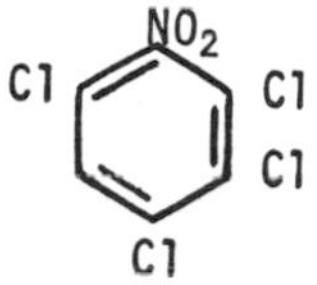 (19)

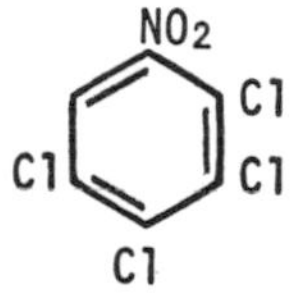

(20) 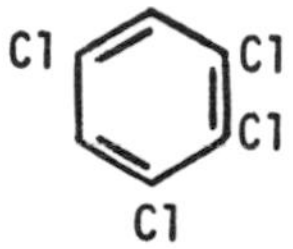 (21)

Although a solution of pentachlorobenzene in hexane or methanol did not discolor after 7 days exposure to sunlight, irradiation in hexane for 3 hr with a mercury lamp resulted in the formation of 1,2,4,5-tetrachlorobenzene (17) in 50% yield, together with some 13% of the isomeric 1,2,3,5-tetrachlorobenzene (21). This seems to confirm the irradiation fate of pentachlorobenzene, the slow disappearance of which corresponded to the appearance of the 1,2,4,5-tetrachloro compound.

Crosby and Hamadmad [44] have considered these results in terms of both free-radical and ionic-hydride transfer mechanisms. The effect of the electron-attracting substituent on the position of reductive dechlorination is consistent with the former, involving hydrogen abstraction from the solvent [45]. This is also supported by the fact that very similar proportions of identical products are obtained whether the irradiation was carried out in hexane, cyclohexane, acetone, or ethanol, and the detection of 1-, 2-, and 3-chlorohexanes in equal amounts when hexane was used as solvent is also in agreement with radical formation. A recent study of the photoreduction of some substituted nitrobenzenes in organic solvents under the influence of a high-pressure mercury lamp using rapid scan electron spin resonance (ESR) suggests that radicals of the type PhN($\dot{O}$)OR (where Ph = phenyl and R = solvent radical) are involved [46].

An ionic hydride transfer mechanism involving zwitterionic intermediates, and analogous to similar photosubstitution reactions involving the cyanide ion [47], can also be invoked to explain the products formed from quintozene, although the absence of products such as pentachloroaniline is contrary to the known affinity of the nitro group for hydride ions.

2. Fenaminosulf

Sodium-p-dimethylaminobenzenediazosulfonate (fenaminosulf) (22) is a promising seed dressing and soil fungicide for the control of species of Pythium, Aphanomyces, and Phytophthora [48], which is stable in alkaline solution. Exposure of its orange colored aqueous solution to sunlight, however, destroys its fungistatic activity with the progressive formation of p-dimethylaminophenol (23) and colored products. Hills and Leach [49]

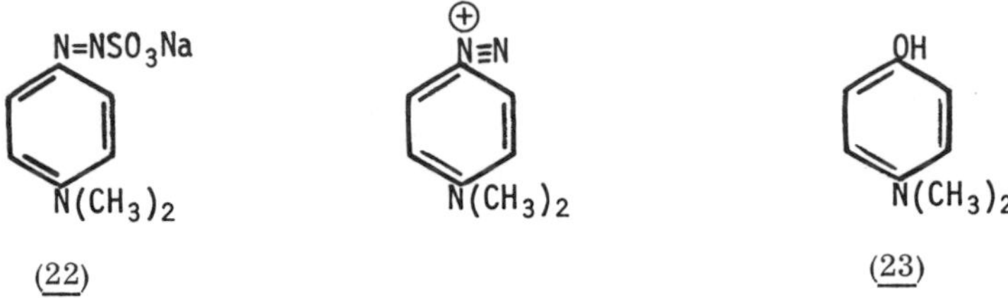

suggested a two-stage decomposition: photoreduction of the diazonium group with evolution of nitrogen and subsequent oxidation, followed ultimately by polymerization of the phenol formed. Later work by Crosby and Moilanen [50] confirmed that the rapid hydrolysis of fenaminosulf in aerated water was another example of a photonucleophilic reaction. p-Dimethylaminophenol was obtained in 99% yield, the remaining 1% being accounted for as dimethylaniline.

III. PROTECTANT FUNGICIDES

A. Organotin Compounds

The earliest reference to the breakdown of these compounds appears to be that of Kaars Sijpesteijn [51] who mentioned that under the influence of light and air trialkyl and triaryl tin compounds are gradually degraded to di- and monosubstituted derivatives, which are considerably less toxic, and ultimately to nontoxic inorganic tin. Kroller [52] has reported similar results, and Hardon et al. [53] and Bruggemann [54, 55] noted the disappearance of triphenyltin compounds a few days after spraying on crops.

A slightly conflicting report came from Tamura [56]. He found that whereas irradiation of tributyltin oxide, acetate, and chloride depressed their inhibitory activities against Pyricularia oryzae and a TLC examination of the irradiated materials showed the presence of dibutyltin compounds; the corresponding derivatives of triphenyltin appeared to be little affected. Chapman and Price [57], however, told rather a different story. They showed that irradiation of triphenyltin acetate (fentin acetate) on a glass plate at a distance of 20 cm from a medium-pressure UV lamp produced marked photodegradation, the three Sn-C bonds being readily cleaved (Table 3). It is clear that degradation proceeds faster at lower wavelengths and that the thickness of the irradiated layer within the range of 5-10 μm does not appreciably affect the nature of the products or the rate of breakdown.

These results have recently been substantiated by Barnes et al. [58] who clearly demonstrated the rapid photochemical degradation of fentin acetate to di- and monophenyltin compounds by irradiation with a high-pressure lamp emitting light of predominantly 365 nm wavelength for periods up to 1 hr. The experiments were carried out on Merck silica gel GF plates at a distance of 20 cm, followed by development using butanol:acetic acid (100:1 v/v). Irradiation for longer periods caused the complete breakdown of

TABLE 3

Degradation of Fentin Acetate by Ultraviolet Irradiation

Time, hr	Thickness of layer, μm	Percentage of Sn present as			
		Ph_3SnX	Ph_2SnX_2	$PhSnX_3$	Sn, inorg
		Wavelengths > 235 nm; 12 mW/cm^2			
5	10	64	19	9	12
5	5	33	27	24	18
25	10	21	17	15	27
25	5	18	26	22	36
60	5	11	31	14	44
		Wavelengths > 350 nm; 4 mW/cm^2			
85	10	63	25	7	6
100	10	39	20	19	19
100	10	37	13	13	24

fentin acetate into inorganic tin. Using the ^{14}C-labeled compound, they also showed rapid hydrolysis of triphenyltin acetate by distilled water, all of the radioactivity being transferred to the water in about 3 hr. Experiments using the radiolabeled compound in soil showed that although $^{14}CO_2$ evolution is brought about by microbial action, this may well be preceded by a preliminary abiotic Sn-C cleavage.

The degradation of triphenyltin chloride on the leaves of sugar beet has also been shown to proceed via di- and monophenyltin compounds to inorganic tin(IV), and this is almost certainly yet another example of photolytic decomposition [59].

B. Quinones

The photoreactivity of quinones is well known and the photodimer of thymoquinone has been known for over a century. The photoreaction of p-benzoquinone was reported by Ciamician and Silber [60] who found that in alcohol solution the main products were acetaldehyde, hydroquinone, and polymeric material, later shown by Hartley and Leonard [61] to approximate to a dimer [62]. More recently, the dimerization of dimethylbenzoquinones in sunlight involving the formation of cyclobutane rings has been described by Cookson et al. [63]. The trans-dimers formed are stable to further irradiation but the cis-dimers cyclize further through a second reaction of the same type; all the dimers dissociate to the original monomers on pyrolysis. Photocycloaddition has also been mentioned by Bryce-Smith and Gilbert [64] and several workers have observed the formation of hydroquinone and 2-hydroxy-p-benzoquinone in the photolysis of p-benzoquinone in aqueous solution [65-67].

1. Chloranil

Leighton and Dresia [68] observed that the photochemical stability of the quinones is improved by increased substitution and in the series p-benzoquinone and its mono-, di- and tetrachloro derivatives, the quantum efficiencies decreased in the order 0.505, 0.354, 0.256, and 0.095, respectively. The threshold wavelength region required for photodecomposition decreased with decreasing oxidation potential, which means that chloranil, 2,3,5,6-tetrachloro-p-benzoquinone, (0.73V) should be less stable than dichlone (0.42V). Measurements of the decomposition rates of the two compounds in dioxan solutions exposed to sunlight indicated that dichlone is indeed the more stable [69]. This is supported by field observations, for whereas chloranil, though a good seed protectant, does not persist on foliage, dichlone is an effective protectant both on seeds and leaves. More recently, Bredoux [70] has studied the photoreduction of chloranil. He found that irradiation of an alcoholic solution in sunlight resulted in the formation of tetrachlorohydroquinone (24) and 2,3,5-trichloro-6-hydroxy-p-benzoquinone

(25, R = H). Fisch and Hemmerlin [71] confirmed the photochemically induced replacement of chlorine by hydroxyl and reported yields of 68 and 28% for the tetrachloro and trichloro compounds, respectively. The latter yield is important since chloranil normally contains about 0.25% of 2,3,5-trichloro-6-hydroxy-p-benzoquinone as an impurity. They suggested a two-stage mechanism, involving the quinone ether (25, R = alkyl) which could be formed from a zwitterion such as (26) as a result of nucleophilic attack by the solvent used.

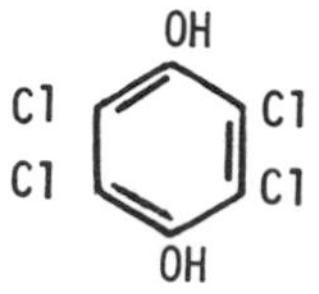

(24) 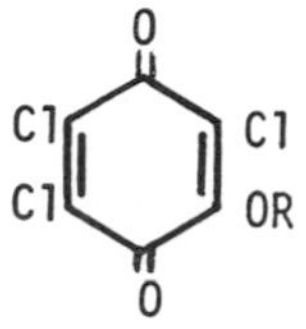 (25) 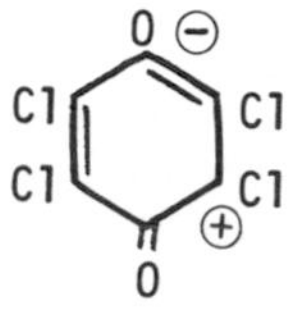 (26)

2. Dichlone

2,3-Dichloro-1,4-naphthoquinone (dichlone), a somewhat more effective seed fungicide than chloranil, is also subject to photodegradation. Using electron capture GLC assay, White et al. [72] demonstrated increases in the concentration of degradation and conversion products concomitant with decreased dichlone concentration in a benzene solution of the fungicide which had been kept for 12 months under normal laboratory conditions of light and temperature. The same effect could be obtained by exposing a freshly prepared benzene solution of dichlone to sunlight for 6 hr. Using IR, NMR, and MS techniques, the two photoproducts were identified as phthalic acid and 2-chloro-3-phenyl-1,4-naphthoquinone, the latter presumably being formed as a result of free-radical attack involving solvent.

Other substitution reactions involving either or both reactive chlorine atoms with functional groups such as -SH or $-NH_2$ present in amino acids, peptides, or proteins, have also been found to take place. Thus Owens [73] noted the inhibition of certain enzymes by SH groups, that is, simple thiols formed compounds of type (27), whereas with higher-molecular-weight thiols such as glutathione, steric hindrance precluded disubstitution, and the reaction which formed compounds of type (28) is irreversible. Owens

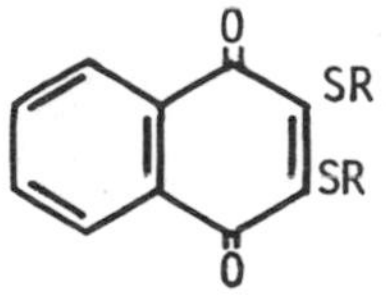

(27) 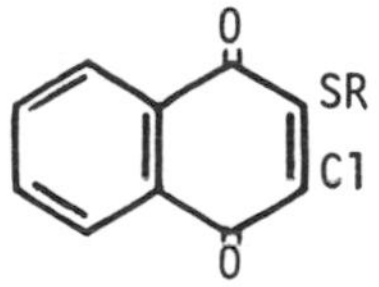 (28)

and Novotny [74] working with intact conidia of *Neurospora crassa* did indeed find strong evidence for the formation of compound (28, R = coenzyme A), though nothing appears to be known of its ultimate fate. Rapid reaction of dichlone with spore constituents has also been reported by Owens and Miller [75]. More recently, it has been suggested by Gause et al. [76] that the initial step in such reactions involves a semiquinone free radical.

C. Aromatic Compounds

1. Substituted Phenols

a. Pentachlorophenol. Irradiation of pentachlorophenol in hexane or methanol solution by sunlight caused a pale yellow color but neither TLC nor GLC revealed any decomposition products. Using a low-pressure lamp, Crosby and Hamadmad [44] showed that 60% of pentachlorophenol in hexane solution remained unchanged after 32 hr but some 30% of 2,3,5,6-tetrachlorophenol (29) had been formed as the major transformation product, together with about 10% of another phenolic material, assumed to be an isomer the presence of which was detected using GLC. By contrast, Haitt et al. [77] had earlier reported that the decomposition followed first-order kinetics, but there was no discussion as to the identity of the products formed.

Munakata and Kuwahara [78] confirmed that an aqueous solution of sodium pentachlorophenate was decomposed when subjected to solar radiation, the solution turning purple and the concentration being reduced to 50% in 10

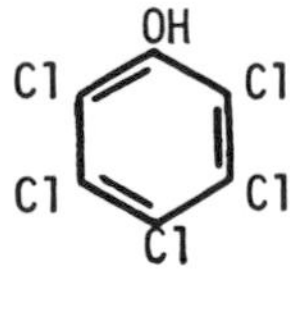

(29) 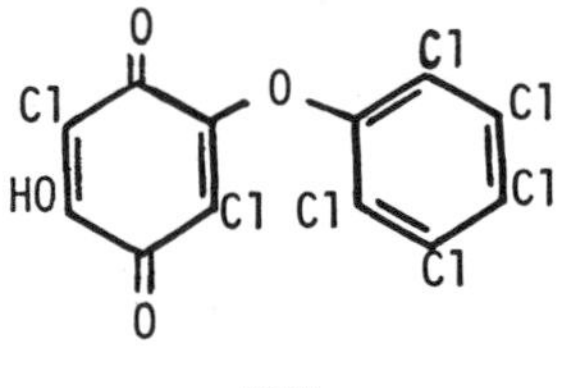 (30) 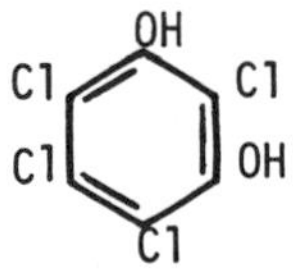 (31)

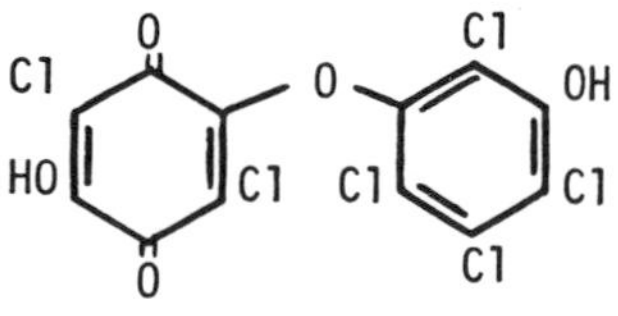

(32) 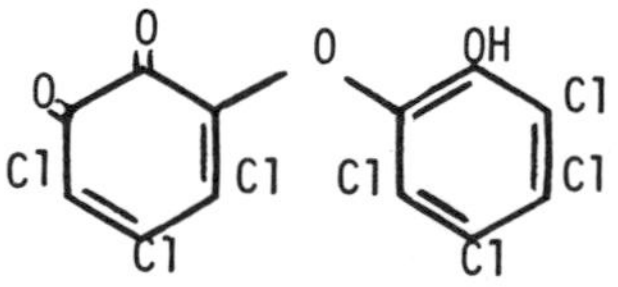 (33)

days. They carried out a column chromatographic investigation of the irradiated solution and showed the presence of six products (30-35), which did not include the previously reported 2,3,5,6-tetrachlorophenol (29). Four compounds (30, 32-34) were previously unknown, and in addition there was considerable resinification. It is interesting to note that the photodegradation products were in general more fungitoxic than sodium pentachlorophenate, but less phytotoxic and less harmful to fish.

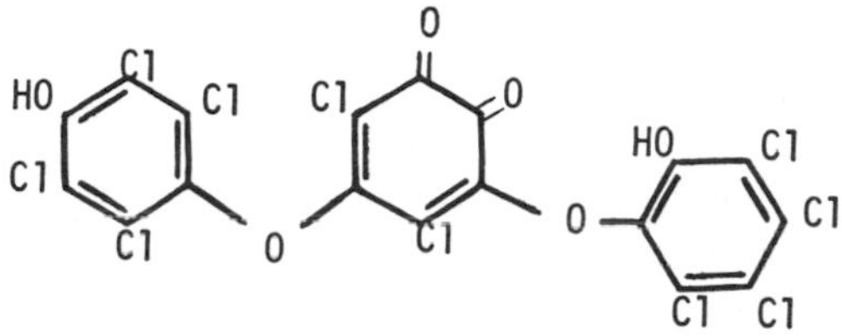

(34)

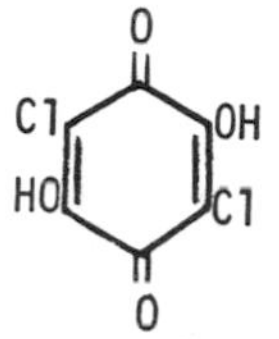

(35)

b. Nitrophenols. Casida and his group have studied the photodecomposition of dinoseb (36, R = H) and its acaricidal isopropylcarbonate, dinobuton (36, R = $COCH(CH_3)_2$) on snapbean leaves by sunlight [79, 80], using ^{14}C-(carbonyl)- and ^{14}C-(benzene ring)-labeled compounds. They examined the

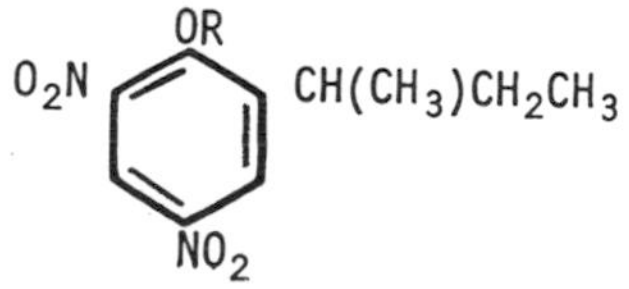

(36)

ether-soluble surface residues from leaves of plants which had been kept in a glasshouse for 1 day after application of the radio labeled dinobuton by two dimensional TLC. Such a fraction could be expected to contain any photodecomposition products, but in fact it consisted almost entirely of dinobuton with only trace levels (<0.2%) of dinoseb and at least eight other radioactive products. Seven days later, dinobuton was still the major product, but dinoseb had disappeared and other products had increased in amount.

The photosensitizing effect of rotenone and xanthen-9-one was most marked, both compounds increasing the rate at which the ether-soluble products were further decomposed. Both sensitizers resulted in the formation of 10 to 13 products present at levels above 0.2%, but there were noticeable differences in the products found in each case. Thus while 6-amino-2-sec-butyl-4-nitrophenol was detected only with rotenone sensitization and ester cleavage was more prominent leading to an accelerated loss of radioactivity from ^{14}C-(carbonyl)-labeled dinobuton, xanthen-9-one was more effective in promoting the decomposition of ^{14}C-(ring)-labeled dinobuton. There was also a suggestion that in the presence of xanthen-9-one, modification of the sec-butyl side chain took place, leading to the formation of one of the isomeric hydroxy-sec-butyl derivatives together with the corresponding 1-methylprop-1-enyl compound, but the radioactivity level was insufficient for cochromatography. Since this material was obtained from an acetone homogenate of bean leaves analyzed 4 hr after topical treatment with ^{14}C-(ring)-labeled dinobuton, it is not clear whether it is a true photoproduct or the result of plant metabolic activity.

The effect of UV radiation from a medium-pressure lamp on ethanolic solutions of cycloalkyldinitrophenols has been investigated by Watkins and Clark [81]. Using GLC-MS techniques, they found a variety of products as yet unidentified, including possibly 4-cyclohexenyl-2,6-dinitrophenol from the 4-cyclohexyl compound, and there was also evidence of isomerization.

2. Dichlofluanid

N'-Dichlorofluoromethylthio-N,N-dimethyl-N'-phenylsulfamide, dichlofluanid (37), a protective fungicide with a broad spectrum of activity is one of the fluoro compounds described by Kühle et al. [82]. It readily hydrolyzes under alkaline conditions in vitro to form N,N-dimethyl-N'-phenylsulfamide (38). This breakdown has also been demonstrated on strawberries and

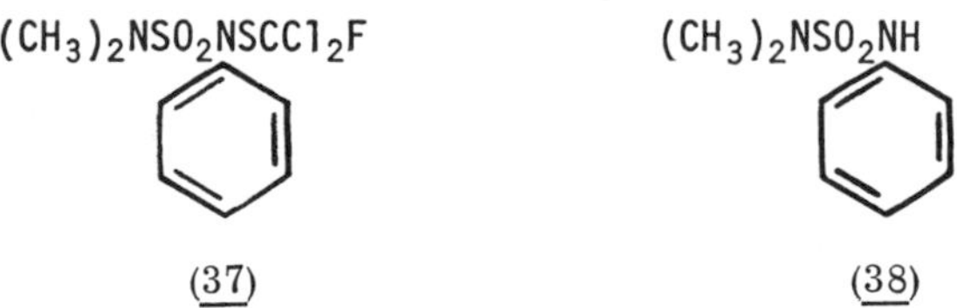

grapes under field conditions by Vogeler and Niessen [83]. The photolysis of dichlofluanid has been examined by Watkins and Clark [81] using a 100-W medium-pressure lamp, with irradiation under nitrogen. The nature of the solvent proved very significant. In methanol or benzene, conversion to a brown amorphous intractable solid is virtually complete in 22 hr; in acetone, the product was a brown oil. Identification of the former product by MS determination using direct probe has been hampered by its lack of volatility. The oily product, when subjected to GLC analysis, showed 12 major peaks. Using column chromatography, several fractions were obtained, the first of which proved to be phenyl isothiocyanate. A later fraction had a similar IR spectrum but with additional bands at 805 and 830 cm^{-1} and the mass spectrum showed a molecular ion m/e 267. The typical isotope pattern for two chlorine atoms together with the rest of the spectrum suggested the presence of a compound $C_6H_4(NCS)SCCl_2F$, but the exact orientation has not yet been assigned.

The formation of isothiocyanates can result from intramolecular rearrangement, but another possible mechanism would be cleavage of both substituents from the nitrogen atom leading to the formation of the $SCCl_2F$ radical which could then participate in nuclear substitution. Alternatively, the elimination of fluorine would presumably yield thiophosgene which could then react further to form other isothiocyanates. This production of isothiocyanates could be fundamental to the mode of action.

D. Heterocyclic Compounds

1. Substituted Phthalimides

These compounds are characterized by stability in aqueous solution except under alkaline conditions.

The deleterious effect of light on captan, N-(trichlormethylthio)-3a,4,7,7a-tetrahydrophthalimide (39) was recognized by Peynaud [84] soon after it was introduced as a protectant fungicide and its total destruction by UV light of wavelength 254 nm was commented on by Mitchell [85], though neither worker mentioned specific degradation products. Later work by Serra [86] showed that 50% of the fungicide was transformed into inactive material after 3 days of sunshine, or after 6 hr at a distance of 30 cm from a "lampe germicide." He also investigated the effect of light on the related compounds folpet, N-(trichloromethylthio)phthalimide (40) and captafol, N-(1,1,2,2-tetrachloroethylthio)-3a,4,7,7a-tetrahydrophthalimide) (41) and showed that after 8 hr irradiation, the extents of degradation were 33, 64, and 88% for folpet, captan, and captafol, respectively.

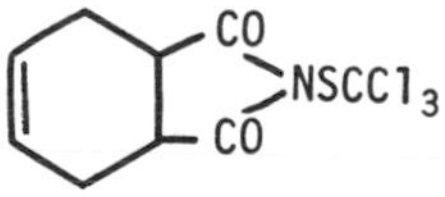

$NSCCl_3$ (39) $NSCCl_3$ (40) $NSCCl_2CHCl_2$ (41)

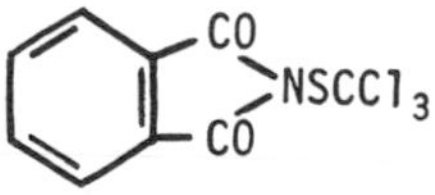

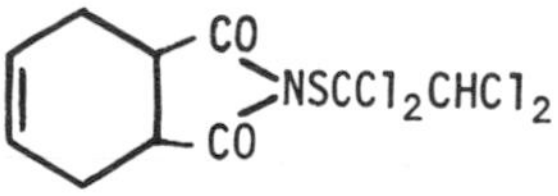

2. Substituted s-Triazines

Anilazine, 2,4-dichloro-6-(o-chloroanilino)-s-triazine) (42) perhaps better known by the trade name Dyrene, is a protective fungicide suitable for foliage application, particularly against dollar spot and Helminthosporium

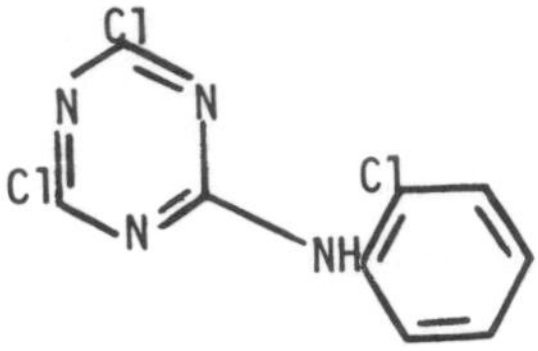

(42)

diseases of turf, but is ineffective as a seed protectant. The presence of reactive halogen atoms in the molecule, which are almost certainly responsible for its fungitoxicity, means that inactivation of the fungicide by nucleophilic substitution is also possible. The stability of anilazine in soil, monitored by a colorimetric method dependent upon the presence of reactive halogens in the molecule [87], was reported by Burchfield [88]. He found that the half-life, when applied to moist soil (17.5% water, pH 6.4) at a rate of 100 μg/g, was only 0.5 day at 26° C, and 2.5 days in air-dried soil. The presence of moisture thus accelerates the disappearance of the fungicide, but no evidence was offered as to whether this represented chemical hydrolysis, except insofar that no correlation with hydrolysis rates in aqueous buffer was observed. Reaction with organic compounds in soil seems another obvious possibility, and a glance at Table 4 shows the potential nucleophilicity of this medium in such a context.

TABLE 4

Important Nucleophilic Groups Found in Soil Organic Matter

Group	pK_a	Percent at pH 7.5
α-Carboxyl	3.0-3.2	99.99
Phosphoryl (primary)	2.0	99.99
Heterocyclic amine (guanine)	2.3	99.99
γ-Carboxyl (glutamyl)	4.4	99.99
Imidazolium	5.6-7.0	76-99
Phosphoryl (secondary)	6.0	96

IV. SYSTEMIC FUNGICIDES

A. Heterocyclic Compounds

1. Benzimidazoles

The breakdown of methyl 1-(butylcarbamoyl)benzimidazol-2-ylcarbamate, benomyl, Benlate (43), in aqueous solution to methyl benzimidazol-2-ylcarbamate, carbendazim, MBC (44), was first reported by Clemons and Sisler [89]. Using *Saccharomyces pastorianus*, to which benomyl is 30 times

(43) (44)

as toxic as carbendazim, they also demonstrated a 50% breakdown of benomyl in 1 hr on incubation at 30°C of the nutrient medium to which it had been added. This instability of benomyl was confirmed by other workers. Thus Sims et al. [90] were able to detect carbendazim, but not benomyl, in exctracts of cotton plants 4 weeks after application of the latter, while Peterson and Edgington [91] found that benomyl was transformed completely to carbendazim in bean plants in 5 days. Additional evidence has been furnished by several other research groups [92-95] and the conversion of benomyl under alkaline conditions to 3-butyl-*sec*-triazino[1,2a]benzimidazole-2,4(1*H*,3*H*)dione (46) and to 2-(3-butylureido)benzimidazole (47) has also

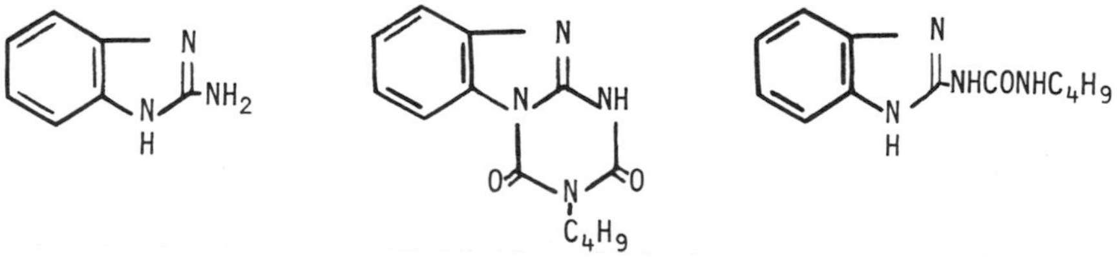

(45) (46) (47)

been described [95, 96]. More recently, however, Baude et al. [97], using the 2-^{14}C-labeled compound, have shown that benomyl does have some stability in typical aqueous suspensions used for foliar application. Since carbendazim will continuously form from benomyl during a normal assay procedure, they developed a method which enabled initial concentrations of both compounds to be estimated. This involved maceration of the crop, followed by treatment with boiling 1 N NaOH which converted benomyl to the stable urea derivative (47) and carbendazim to 2-aminobenzimidazole (45). After the usual cleanup procedure these derivatives could be separated by TLC and estimated by radioscanning or radioautography and scintillation counting. The results, given in Table 5, would seem to indicate that no significant change in the carbendazim content of the aqueous suspension of Benlate takes place over 49 hr at pH 7-8.

Baude et al. [97] also followed the breakdown of [^{14}C]Benlate on various plant seedlings exposed outdoors, and the results given in Table 6 provide evidence for a relatively slow breakdown of benomyl on foliar surfaces.

The influence of pH on the hydrolysis of benomyl in 50% aqueous methanol has been studied by Calmon and Sayag [98], who found evidence for nucleophilic attack on the butylcarbamoylcarbonyl group by a molecule of water at low pHs, and nucleophilic attack by OH^- on the same group at high pHs.

The stability of benomyl in organic solvents has been examined by Chiba and Doornbos [99]. They confirmed the observations made by Kilgore and White [93] that degradation to carbendazim took place in chloroform at room temperature, though rather surprisingly no mention was made of the exclusion of light and moisture. On the data given, breakdown in chloroform is less than in other solvents such as benzene, ether, ethanol, acetone, ethyl acetate, or methylene dichloride.

The thermal stability of benomyl has been the subject of comment by Chiba and Doornbos [99] and by Laski and Watts [100]. The latter workers obtained no detector response to 200 ng on a gas chromatograph with a column temperature of 180°C. Complete decomposition to carbendazim has been reported at 180°C and 70 eV during MS analysis, and only a trace of

TABLE 5

Stability of Benlate (0.12 kg/100 liter; 1 lb/100 gal) in Aqueous Suspension at 23°C

Hours	pH	Benomyl, %	Carbendazim, %
0	7.3	97	3
1	7.3	90	10
6	7.8	93	7
49	8.0	95	5

TABLE 6

Breakdown of Benomyl to Carbendazim on Foliar Surfaces

Plant	Days after spraying	Rainfall, inches	Total ^{14}C detected, μCi	Benomyl, %	Carbendazim, %
Apple	0	Trace	0.87	91	9
	3	0.2	0.35	91	9
	7	5.0	0.45	91	9
	13	5.6	0.34	81	19
	21	5.6		48	52
Cucumber	0	0	0.65	91	9
	3	0	0.59	60	40
	9	Trace	0.40	55	45
	17	Trace	0.14	71	29
	23	0.4	0.05	60	40
Banana	0	0	1.04	91	9
	3	0	0.60	86	14
	9	Trace	0.62	86	14
	17	Trace	0.28	75	25
	23	0.4	0.36	77	23
Orange	23	0.4	0.12	53	47
Grape	21	5.4		62	38

parent ion could be observed at 80°C and 20 eV. Carbendazim has also been reported to decompose on GLC columns [81].

The hydrolytic process also appears to be promoted by photolysis, for Kilgore and White [93] found that exposure to sunlight enhanced the conversion of benomyl to carbendazim; however, a recent report by Buchenauer [101] raises doubts about this effect. Even carbendazim is not absolutely photostable, for Watkins [102] has shown that the UV absorption spectrum of a sample of carbendazim exposed to sunlight, but protected from rain, showed a distinct change after several months. In laboratory irradiation experiments using a Hanovia 100 W medium-pressure lamp, some 80% decomposition of carbendazim in methanol took place in 4 days. Products so far identified include dimethyl oxalate in addition to mono- and dicarbomethoxyguanidines (48, 49) and there was also MS evidence for the presence of an acid salt of guanidine. Perhaps surprisingly, there was no evidence of any products of dehydrodimerization such as have been obtained in the

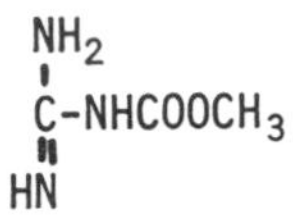

```
NH2
 |
 C-NHCOOCH3
 ||
HN

(48)

NHCOOCH3
 |
 C-NHCOOCH3
 ||
HN

(49)
```

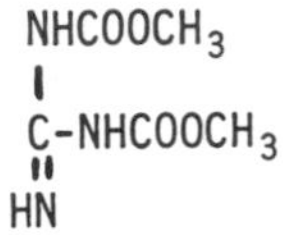

case of benzimidazole itself by Cole et al. [103]. They explained the formation of 2,4'- and 2,5'-bisbenzimidazoles by assuming the intermediacy of a benzimidazol-2-yl radical, either generated directly as a result of UV radiation or by rearrangement of a preformed benzimidazol-1-yl radical. Grossmann et al. [104] have also reported the instability of carbendazim under high-intensity radiation. In phosphate-citrate buffer (pH 5) more than 90% was transformed in 2 hr into another fungitoxic compound with a bimodal dosage response pattern of inhibition, but the metabolite was not stable and completely reverted to carbendazim in 7 days.

In a further report, Buchenauer [101] notes that the photoalteration of benomyl and carbendazim are promoted by stronger radiation intensity and high pH. Thus while carbendazim loses no fungitoxicity in solution at pH 5 or 7, it is inactivated at pH 9 with the formation of 2-aminobenzimidazole (45), though its formation probably represents photoassisted hydrolysis rather than true photodecomposition.

2. Carboxylic Acid Anilides

Among the structural features present in 2,3-dihydro-6-methyl-5-phenylcarbamoyl-1,4-oxathiin, carboxin (50, R = H) which makes it susceptible to degradation are the hetero sulfur atom, the amide linkage, and the aromatic ring. One of the first reports on its stability came from Chin et al. [105] who found that oxidation to the 4-oxide, oxycarboxin (51), took place not only in water and soil, but also in wheat, barley, and cotton plants. This was

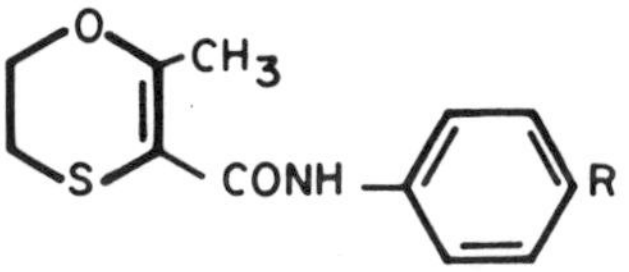

(50)

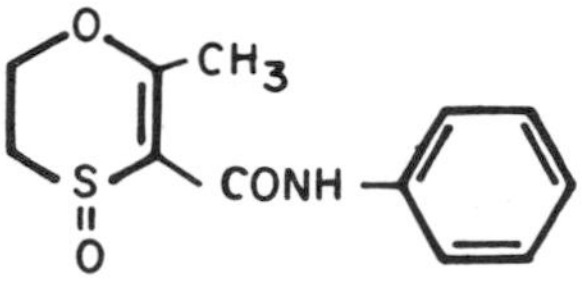

(51) 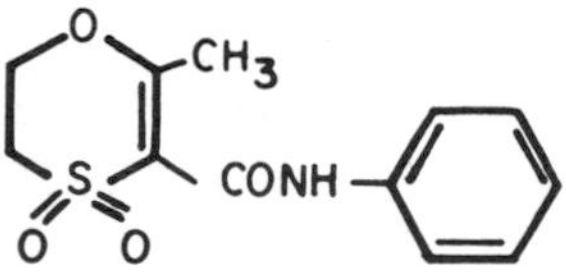 (52)

facilitated by low pH; the compound was relatively stable above pH 6. Further oxidation to the sulfone (52) was not detected and no hydrolysis appeared to take place. Observations using carboxin ^{14}C-labeled in either hetero or aromatic rings showed that the molecule was translocated intact in the xylem.

Later work has indicated that in bean and barley seedlings, carboxin was also metabolized to compounds which were readily bound to plant tissue. Thus Snel and Edgington [106], in confirming the oxidation to oxycarboxin (51) also noted that fission of the amide linkage took place in root tissues with the formation of aniline which was then involved in polymerization or condensation reactions. Lyr and Ritter [107], however, have commented on the resistance of the carboxamide link to proteases.

More recently, Briggs et al. [108] have shown that the major soluble metabolite of carboxin in barley seedlings and mature plants is 2,3-dihydro-6-methyl-5-(4'-hydroxyphenylcarbamoyl)-1,4-oxathiin (50, R = OH), and this phenolic metabolite complexes with lignin. They also showed that tissues with little lignin, especially the leaves, contained a compound believed to be a polymer of this phenol and although this polymerization may be enzyme catalyzed, it also occurs in vitro. Surprisingly they only found small amounts of the sulfoxide (51), often mentioned by earlier workers, and noted that its formation is unlikely to be enzymic in origin. The possibility of even more labile mono- and dihydroxy compounds accounting for both low R_f compounds observed and for some of the lignin-linked radioactivity, was also mentioned. Phenol production could result in increased fungitoxicity and thus explain earlier findings that carboxin inhibits respiration, photosynthesis [109], and the tricarboxylic acid cycle [110, 111].

The related 2,3-dihydro-6-methyl-5-phenylcarbamoyl-4H-pyran, pyracarbolid (53) which is also active against rust, smuts, and species of Rhizoctonia [112], has recently provided the first case of fission of the heterocyclic nucleus of this type of compound, albeit to a rather minute

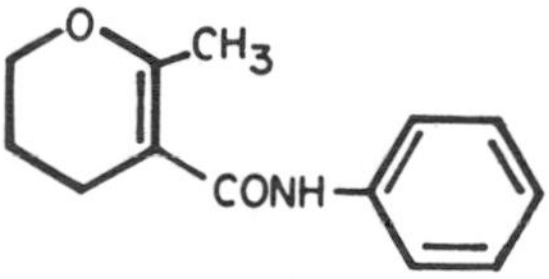

(53) 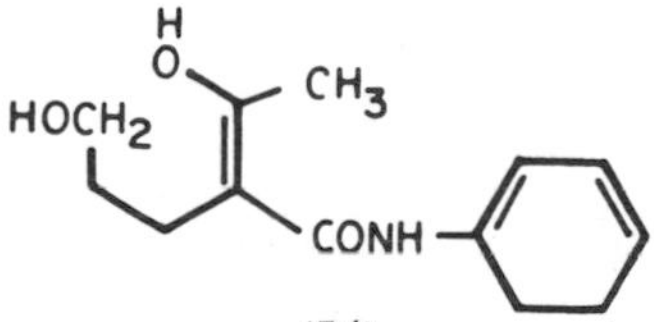 (54)

amount. Oeser et al. [113] using the ^{14}C-labeled compound (C-2 in pyran ring), showed that the 10% of the applied material which was translocated in wheat plants, was relatively stable, only 5% being metabolized. Of the three metabolites which could be detected by TLC after 17 days, one proved to be 2,6-dihydroxy-3-phenylcarbamoyl-hex-2-ene (54).

3. Other Heterocyclic Compounds

Two other systemic fungicides containing the benzimidazole nucleus, 2-(4-thiazoly)- and 2-(2-furyl)benzimidazole, thiabendazole (55) and fuberidazole (56), respectively, have very similar antifungal spectra but appear to have

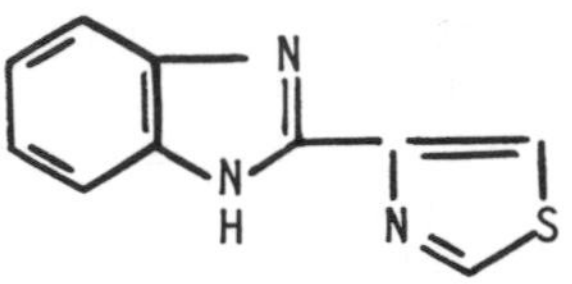

(55) 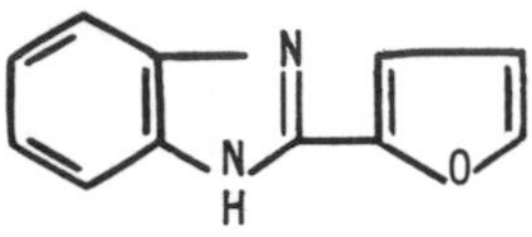 (56)

much greater stability than benomyl. Thus the former is stable at high and low pH and sublimes when heated to 310°C. So far, the only hint of degradation appears to be in preliminary experiments in an investigation into the influence of UV light [81]. Photodegradation in aqueous solution has been confirmed by Buchenauer [101], who showed that irradiation at wavelength 254 nm at a distance of 80 cm for 32 hr or with sunlight for 80 hr, caused some 40% decomposition; this reaction was independent of pH. There was evidence of sensitization by riboflavin and xanthene (at 20 ppm) but not by rhodamine B, and there was no evidence of metabolite formation. Solid thiabendazole was not inactivated by exposure to UV light for 32 hr, but after 80 hr sunlight, there was a 15% reduction in fungicidal activity. The disappearance of 75% of thiabendazole from the leaves of pepper plants reported by Ben-Aziz and Aharonson [114] is of possible significance, but is so far without explanation.

The systemic fungicide triforine, 1,4-di-(2,2,2-trichloro-1-formamidoethyl)piperazine (57) has been investigated by Fuchs and co-workers [115] and by von Bruchhausen and Stiasni [116]. Both used the ^{3}H-labeled compound and obtained essentially identical patterns of nonfungitoxic metabolic products on TLC. The suggestion of the former workers that piperazine (58) was a terminal residue on plants was confirmed by von Bruchhausen and Stiasni who showed that on day 8 after a soil-drench application of [^{3}H]triforine to barley plants, piperazine comprised 20% of the radioactivity of the four metabolites found in the shoots. More recently Fuchs and Ost [117] have used the ^{3}H- and ^{14}C-(CH)-labeled triforine in experiments to distinguish between nonenzymatic and enzymatic breakdown. They found that the compound decomposed rather rapidly in aqueous solution at room temperature and even in the dark more than 95% decomposition took place in 7 months. Isolation of several nonfungitoxic degradation products, followed by MS structure determinations, led to the scheme proposed for triforine breakdown shown in Figure 2.

In plants, Fuchs and co-workers found that decomposition of triforine was considerably slower than in aqueous solution, but it seemed likely that at least three of the four main radioactive products obtained using the ^{3}H- or ^{14}C-(CH)-labeled compound are identical with the nonenzymatic breakdown products formed in aqueous solution, the fourth one probably being a glycoside or other derivative of one of the nonenzymatic breakdown products.

There was also evidence of some change on irradiation of triforine in methanolic solution under nitrogen using a medium-pressure lamp [81]. This sensitivity to UV (254 nm) and sunlight has been recently investigated by Buchenauer [101]. He showed that in aqueous solution 50% inactivation took place in 3 and 30 hr, respectively, while 25% decomposition of solid triforine was caused by irradiation on glass for 80 hr. Irradiation in aqueous solution followed by bioautographic examination using *Cladosporium cucumerinum* showed that after exposures up to 4 hr, a photosensitive fungitoxic metabolite was present which was no longer detectable after 16 hr. Photodecomposition was accelerated by the presence of photosensitizers such as riboflavin and xanthene, and the shorter half-life on plant surfaces compared to glass could well be due to this, though the physical effect of moisture may also be implicated.

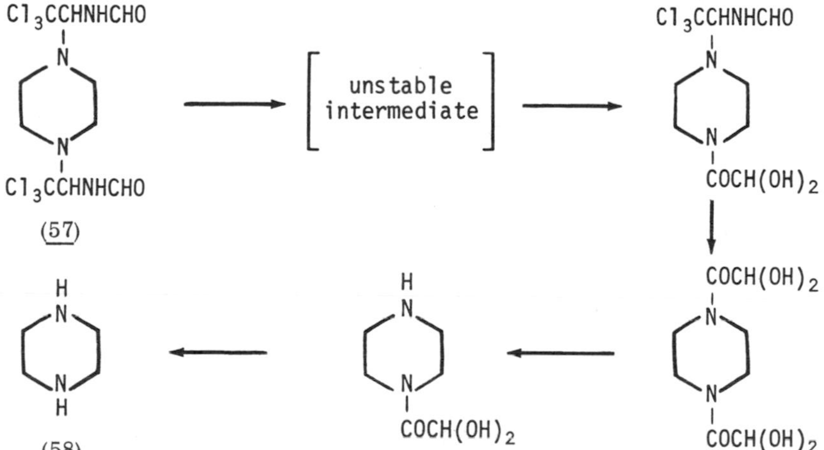

$Cl_3CCHNHCHO$ / N / N / $Cl_3CCHNHCHO$ (57) → [unstable intermediate] → $Cl_3CCHNHCHO$ / N / N / $COCH(OH)_2$ → $COCH(OH)_2$ / N / N / $COCH(OH)_2$ → H / N / N / $COCH(OH)_2$ → H / N / N / H (58)

FIG. 2. Decomposition of triforine.

3-(3′,5-Dichlorophenyl)-5,5-dimethyloxazolidine-2,4-dione (DDOD) (59), which is active against *Sclerotinia sclerotiorum* and *Botrytis cinerea* [118], has been the subject of uptake and distribution studies by Sumida et al. [119] who used the ^{14}C-(4-carbonyl)- or ^{3}H-(phenyl)-labeled compounds. They found that DDOD underwent considerable degradation in 14 days, not only in sterile or nonsterile nutrient solution, but also in distilled water at low concentrations (0.22-2.5 ppm). Opening of the heterocyclic ring (see Fig. 3) gave *N*-(3,5-dichlorophenyl)-*N*-α-hydroxyisobutyrylcarbamic acid (60) from which α-hydroxyisobutyryl-3,5-dichloroanilide (61) was obtained by decarboxylation. Comparable metabolism of DDOD in bean plants occurred to only a small extent if at all, but yet another metabolite, 3-(3′,5′-dichloro-4′-hydroxyphenyl)-5,5-dimethyloxazolidine-2,4-dione (62), was tentatively identified in the leaves of vines which had been injected with [^{14}C]DDOD. While degradation to the carbamic acid (60) and anilide (61) took place to a lesser extent in sterilized, as compared to nonsterilized soil, it could still be nonmicrobial in origin, for the former contained less water for the promotion of hydrolysis. Alteration of the physical nature of the soil particles was also mentioned as a possible cause of reduced breakdown. Soil treated with [^{3}H]DDOD was used to study the fate of the aromatic ring, and one minor degradation product was identified as 3,5-dichloroaniline (63).

Although there was some slight evidence of photodecomposition when thin or thick films of [^{14}C]DDOD were irradiated at 5,000 lux using a 500-W xenon lamp, it is not clear whether the loss of much of the radioactivity is due to sublimation of intact DDOD or to conversion to more volatile compounds.

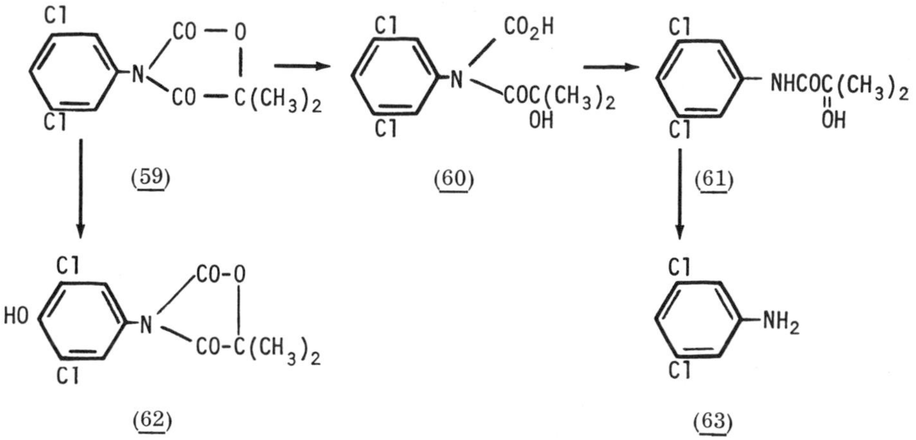

FIG. 3. Decomposition of 3-(3,5-dichlorophenyl)-5,5-dimethyl-1,3-oxazolidine-2,4-dione.

B. Aromatic Compounds

1. Substituted Thioureidobenzenes

Although these compounds are based on thiourea and contain a benzene ring, they are correctly categorized in this section and their relatively facile conversion to benzimidazoles is vital for their fungicidal activity. The transformation of 1,2-bis(3-ethoxycarbonyl-2-thioureido)benzene, thiophanate (64, R = C_2H_5) and the corresponding methoxy analogue, thiophanate-methyl (64, R = CH_3) to ethyl and methyl benzimidazol-2-ylcarbamates, respectively, takes place not only in refluxing 50% aqueous acetic acid in the presence of copper(II) acetate [120], but also more slowly in tap water, buffer, or sterile nutrient. This conversion has also been reported to take place in bean and cucumber plants [121] and more recently in the cotton plant, in which Buchenauer et al. [122] demonstrated almost complete conversion of thiophanate-methyl to carbendazim in 6 days.

Photochemical decomposition of the thiophanates is implied in the investigations of Soeda et al. [123] who used ^{14}C-(ring)-labeled thiophanate-methyl. Two products were obtained and identified by TLC cochromatography viz. carbendazim (44) and dimethyl *o*-phenylene-4,4'-bisallophanate (65). Half-lives of thiophanate-methyl on apple and vine leaves and on glass plates were 15.1, 12.2, and 2.8 days, respectively; proportionately more of the allophanate was formed on glass than on the leaves.

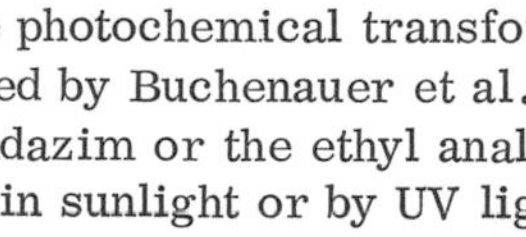

NHCSHCOOR
NHCSHCOOR

(64)

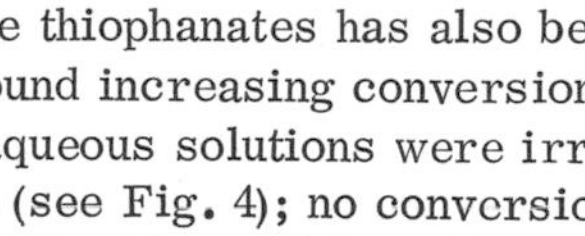

NHCONHCOOH$_3$
NHCONHCOOH$_3$

(65)

The photochemical transformation of the thiophanates has also been reported by Buchenauer et al. [124] who found increasing conversion to carbendazim or the ethyl analogue, when aqueous solutions were irradiated either in sunlight or by UV light of 254 nm (see Fig. 4); no conversion takes place as a result of incubation in the dark. Residues of these fungicides on the leaves of cotton plants after spray treatment were also shown to be transformed to the more fungitoxic benzimidazoles, and it was suggested that this may have been made possible by the effect of dew, since photolysis appears to be absent when the thiophanates are irradiated in the solid state. The greater rate of photochemical transformation which was found on glass as compared to leaves, confirmed the observation made by Soeda et al. [123], and it was thought to be due to the higher concentration of the fungicide suspension on the surface of the leaves.

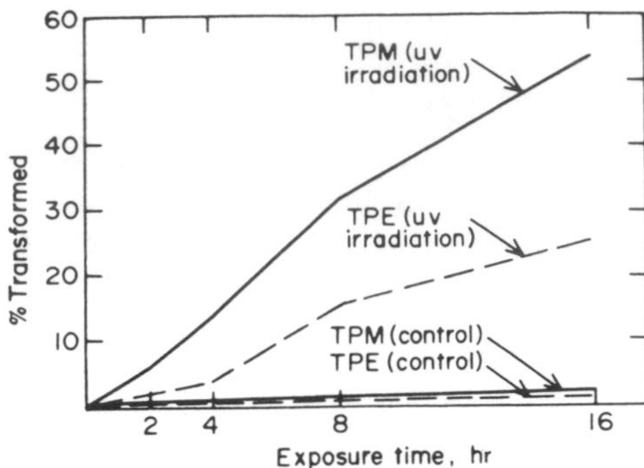

FIG. 4. Photolysis of thiophanate (TPE) and thiophanate-methyl (TMP) (initial concentration 25 ppm). From Buchenauer et al. [124], by courtesy of Pesticide Science.

2. Substituted Isothioureidobenzenes

This group of compounds, which includes N-(2-acetamidophenyl)-, N-(2-benzamidophenyl)-, and N-(2-methylureidophenyl)-N-(ethoxycarbonyl)-5-methylisothiourea (66, R = CH_3, phenyl, and $NHCH_3$, respectively), control the vascular diseases of cotton and tomato plants caused by *Verticillium albo-atrum* and *Fusarium oxysporum* f. sp. *lycopersici*, and may also depend for their activity on cyclization to form the benzimidazole nucleus. Thus Buchenauer [125] has reported the TLC identification of ethyl benzimidazol-2-ylcarbamate (67) in ethyl acetate extracts of shoots which had been treated 10 days previously by root drench with these compounds. This conversion also takes place by incubation of 25 ppm aqueous solutions for 12 hr.

SCH_3
$NHCNCOOC_2H_5$
NHCOR

(66)

N
N
H
$NHCOOC_2H_5$

(67)

3. Chloraniformethan

The photolysis of 1-(3,4-dichloroanilino)-1-formylamino-2,2,2-trichloroethane, chloraniformethan (68) has been studied by Watkins et al. [126]. Irradiation of a methanolic solution resulted in a reduction in intensity of the bands at 211 and 25 nm in the UV absorption spectrum, and a TLC examination showed very little unchanged chloraniformethan after 6 days. Removal of the solvent left a yellow oil from which ammonium chloride was obtained by acetone treatment. Of the eight peaks shown on GLC examination of the residual oil, one major peak was shown to be identical, on two columns, with that from 3,4-dichloroaniline.

More recently, however, Müller et al. [127] have published a detailed scheme for the photodegradation of chloraniformethan (see Fig. 5). They used a high-pressure mercury-vapor lamp ($\lambda > 230$ nm) to irradiate solutions of the compound in methanol, aqueous methanol, or acetone, and identified the products by IR, NMR, and MS. The key compound in the proposed scheme was the Schiff's base (69), a primary photolysis product which is believed to form the reactive ketenimine (70) as a result of photodehydrochlorination. Addition of water to this intermediate gave the anilide (71) which, in turn, by further photolysis of the C-C bond, generated the radical (72). This can be stabilized by hydrogen acceptance most readily from a solvent molecule, and its radical character was supported by the appearance of several typical derived products (76-78). The influence of the solvent was found to be important, and an enhanced hydrolysis of the Schiff's base (69) with the formation of 3,4-dichloroaniline, which had been reported earlier [126], and chloral during irradiation in methanol, was confirmed. 1,1-Bis(3,4-dichloroanilino)-2,2,2-trichloroethane (73) was formed from the Schiff's base and 3,4-dichloroaniline; products (74) and (75) which arose from reaction with methanol and formic acid, respectively, were formed only in methanol or aqueous methanol. The yields of the various products obtained by irradiation of a 1% solution of chloraniformethan are given in Table 7. Other products observed included aniline derivatives and diazopigments, together with large yields of polymeric, colored, and polar products (ca. 30%).

Although 3,4-dichloroaniline has been observed as an apparent terminal residue in these two studies, further photolysis of this compound to yield, inter alia, 3,3',4,4'-tetrachloroazobenzene has been reported in benzene solution with benzophenone as sensitizer [128], and in aqueous solution in the presence of riboflavin-5'-phosphate [129].

C. Antibiotics

Blasticidin S (79), produced by Streptomyces griseochromogenes [130], is now well established as a control agent, both protectant and chemotherapeutic, for Pyricularia oryzae which causes the disease rice blast [131]. Although a reasonably stable compound, it is subject to degradation or inactivation in the field not only by the rice plant and various microorganisms, but also by sunlight. ^{14}C-Labeled blasticidin has been used by

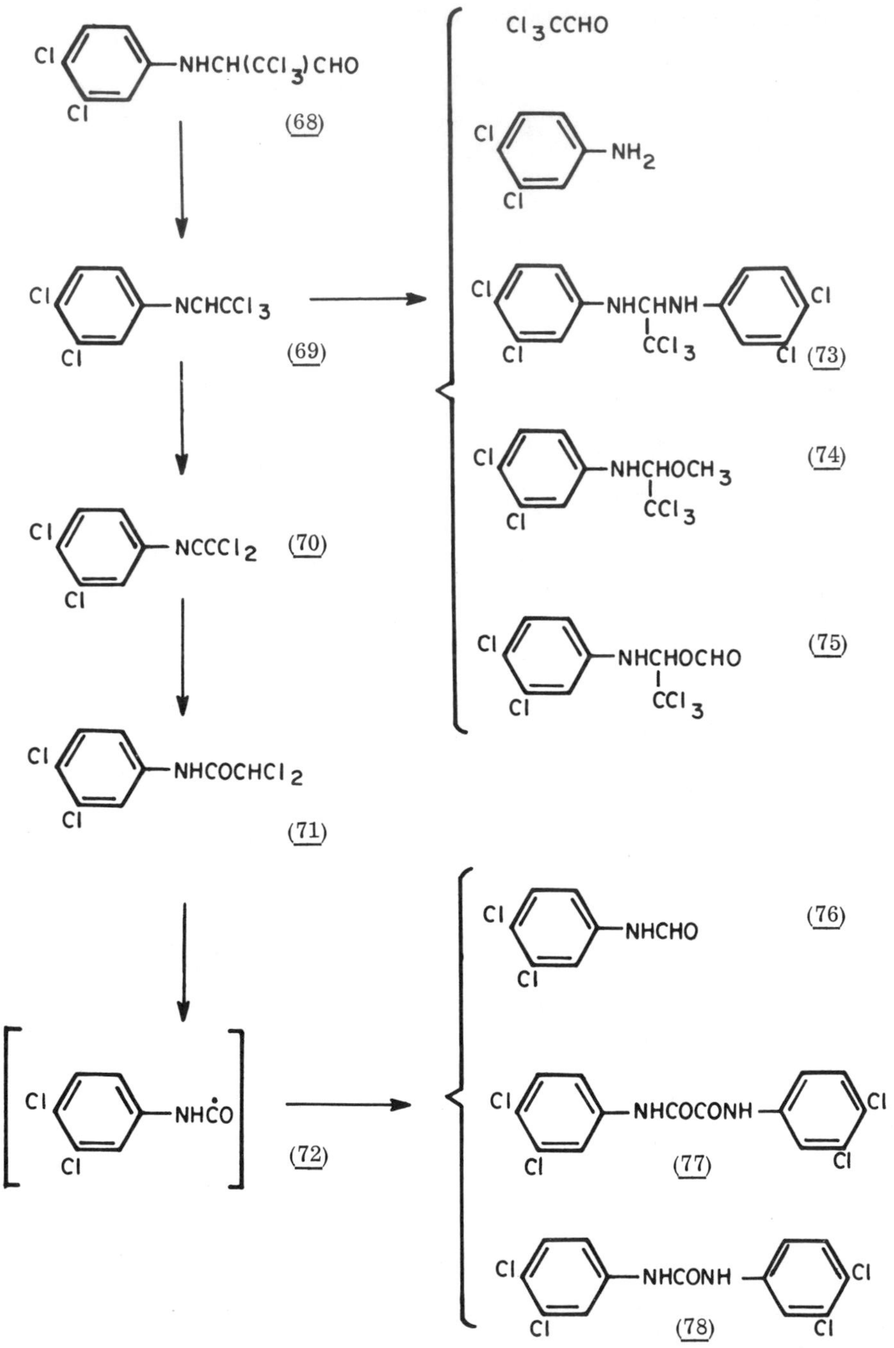

FIG. 5. Photodegradation of chloraniformethan.

TABLE 7

Irradiation Products of Chloraniformethan

Product	Yield, %		
	Methanol, 120 hr	Acetone 150 hr	Aqueous methanol 200 hr
(59)	2	3	2
(61)	<1	5	<1
(63)	<1	6	<1
(64)	6	12	5
(65)	9	2	8
Chloral	3	2	3
3,4-Dichloraniline	15	10	19
(66)	6	3	5

Misato et al. [132] to demonstrate degradation by sunlight both on glass plates and on rice plant leaves. The major metabolite was found to be cytosine (80) but cytomycin (81), together with various unknown substances, was also observed using TLC.

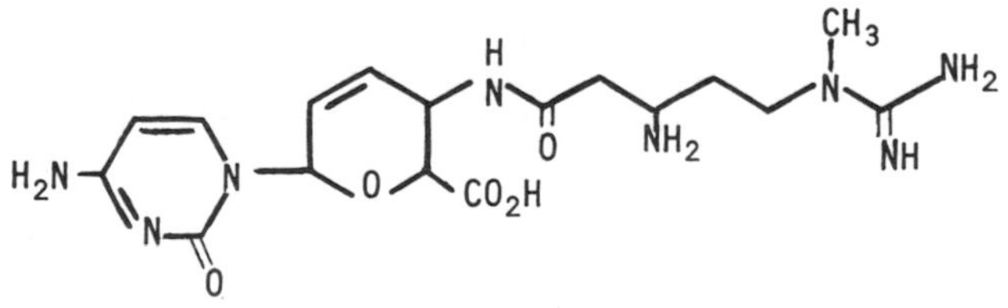

(79)

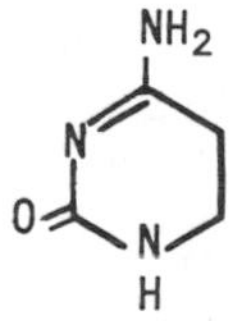

(80)

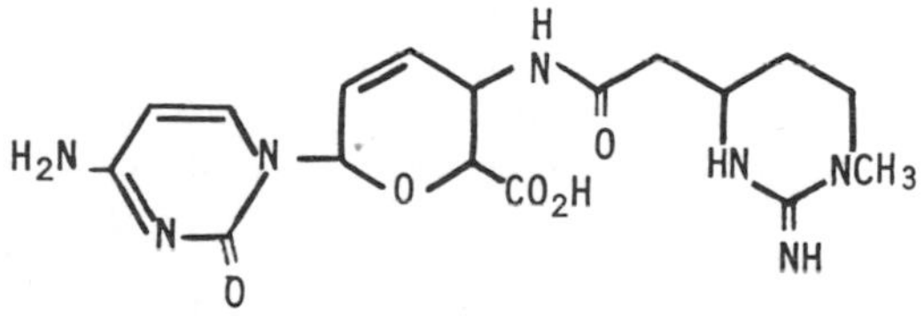

(81)

V. RECENT DEVELOPMENTS

The photostability of some of the carboxylic acid anilide fungicides has been examined by Buchenauer [133]. Irradiation of carboxin (50) [see Section IV.A.2], furcarbanil (82, R = phenyl), pyracarbolid (53), cyclafuramid (82, R = cyclohexyl), benodanil (83, R = I), mebenil (83, R = methyl), and

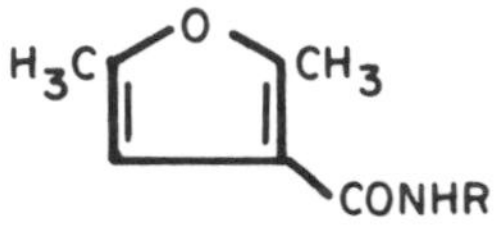

(82)

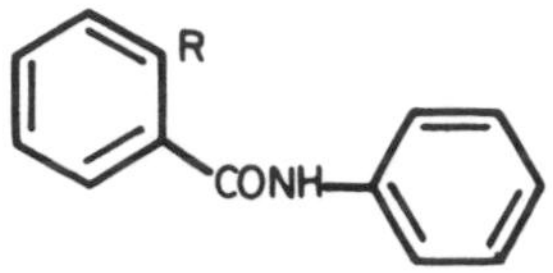

(83)

oxycarboxin (51) by UV light (254 nm) for 4 hr resulted in 97, 64, 58, 50, 20, 18, and 15% inactivation, respectively. Photoinactivation took place much more slowly in sunlight than in UV light, and stability increased in the following order: carboxin < furcarbanil < cyclafuramid < pyracarbolid < mebenil < benodanil < oxycarboxin. On the leaf surface of bean plants carboxin and furcarbanil were almost completely inactivated after 40 hr exposure to sunlight while cyclafuramid lost 85% of its activity; the activity of leaf deposits of pyracarbolid, mebenil, benodanil, and oxycarboxin similarly decreased by 83, 53, 50, and 41%, respectively.

There is now firm evidence of photometabolites from thiabendazole (55), and cleavage of the thiazole moiety has been reported by Jacobs et al. [134] and by Watkins [135]. Jacobs et al. found that while ^{14}C-(ring)-labeled thiabendazole was recovered unchanged from sugar beet plants grown in artificial light, only 78% of the radioactivity recovered from plants grown for 14 days in sunlight represented unchanged fungicide. Of the rest, benzimidazole (0.8%) was detected by combined GLC-MS and benzimidazole-2-carboxamide (1.2%) by reverse isotope dilution procedure, the remaining radioactivity being associated with polar and polymeric products. Degradation of thiabendazole was more extensive on glass plates exposed to sunlight or in aqueous methanolic solution subjected to high-intensity UV radiation, but no additional products were reported. Fission of the imidazole ring has been demonstrated conclusively by Watkins [135]. Irradiation (>290 nm) of a methanolic solution of thiabendazole resulted in the formation of thiazole-4-carboxamide (identified by comparison of IR and MS) and probably thiazol-4-ylamidine (MS), in addition to benzimidazole-2-carboxamide.

Photodimerization of the pyrimidine fungicides dimethirimol (84, R = NMe_2) and ethirimol (84, R = NHEt) has now been reported by Cavell et al.

[136]. When a nitrogen-flushed solution of dimethirimol in acetonitrile was photolyzed, using radiation from a medium-pressure lamp (300-320 nm), four products were separated using thick-layer chromatography. Mass spectral and other evidence suggested that dimerization had taken place across the 5,6-bond, probably by a triplet-state reaction, to give the *trans-anti*, *trans-syn*, *cis-anti*, and *cis-syn* isomers (85-88). Analogous results were obtained with ethirimol, but dimers were not detected in crops, soils, or water which had been treated with dimethirimol or ethirimol and exposed to sunlight.

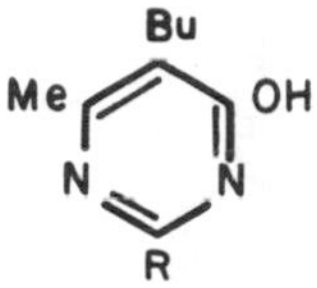

(84)

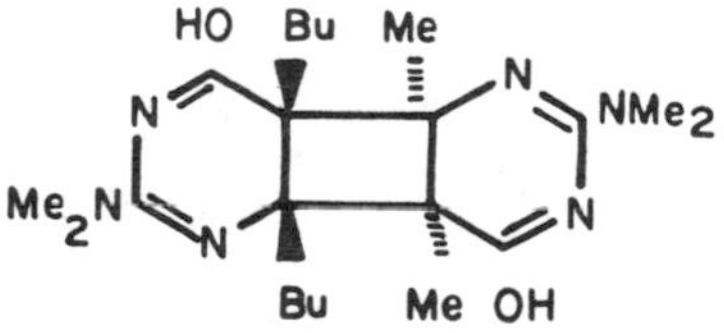

(85)

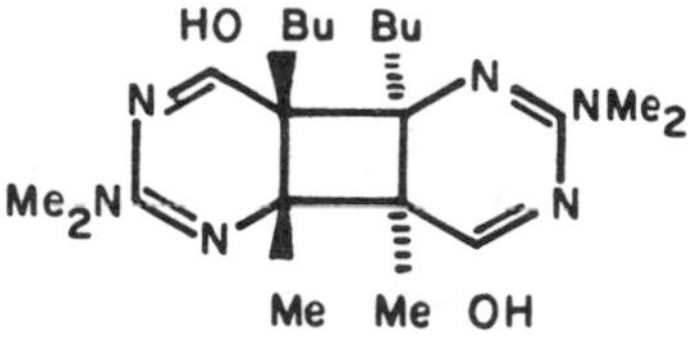

(86)

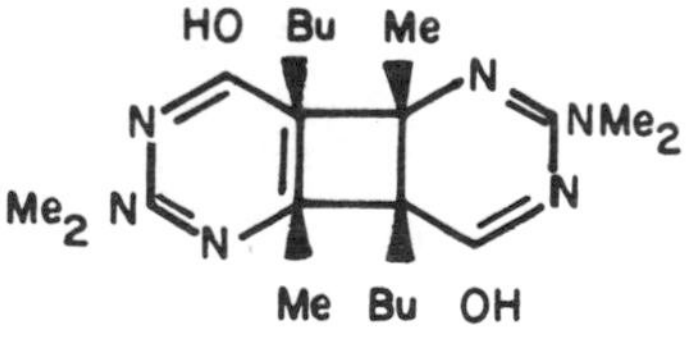

(87)

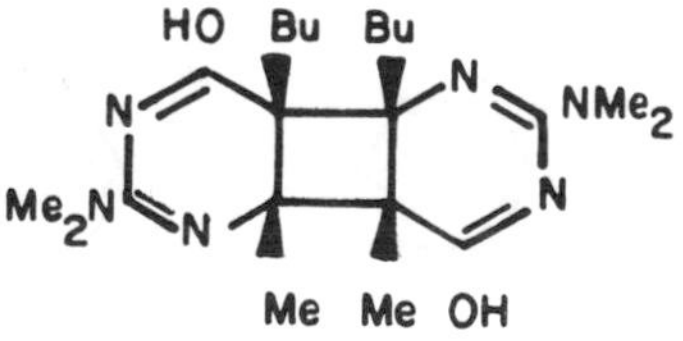

(88)

A comprehensive study of the degradation of *S-n*-butyl-*S'-p-tert*-butylbenzyl *N*-3-pyridyldithiocarbonimidate (S-1358, Denmert, Sumitomo Chemical Co.) in plants and soils and by light has been made by Ohkawa et al. [137] using the ^{14}C-(methylene)-labeled compound. Figure 6 shows the proposed photodegradation pathways, which are mostly also common to degrdation in plants and soils.

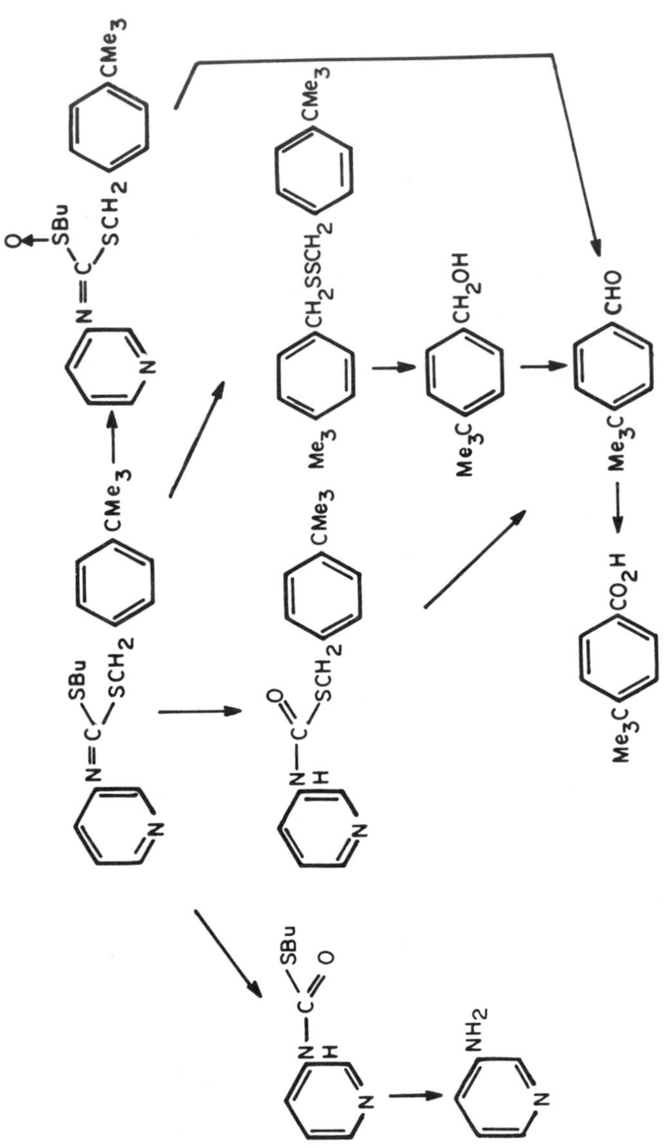

FIG. 6. Photodegradation pathways of Denmert.

REFERENCES

1. C. Steelink and G. Tollin, in Soil Biochemistry (A. D. McLaren and G. H. Peterson, eds.), Vol. 1, Marcel Dekker, New York, 1967, p. 147.
2. R. C. Gore, R. W. Hannah, S. C. Pattacini, and T. J. Porro, J. Assoc. Offic. Anal. Chemists, 54, 1040 (1971).
3. D. G. Crosby, Residue Rev., 25, 1 (1969).
4. C. S. Foote, Science, 162, 963 (1968).
5. G. W. Ivie and J. Casida, Science, 167, 1620 (1970).
6. G. W. Ivie and J. Casida, J. Agr. Food Chem., 19, 405 (1971).
7. L. Lykken, in Environmental Toxicology of Pesticides (F. Matsumura, G. M. Boush, and T. Misato, eds.), Academic Press, New York, 1972, p. 449.
8. R. H. Daines, Phytopathology, 26, 90 (1936).
9. Y. Kimura and V. L. Miller, J. Agr. Food Chem., 12, 253 (1964).
10. D. E. Munnecke and R. A. Solberg, Phytopathology, 48, 396 (1958).
11. H. Shiina, R. Nishiyama, M. Ichihashi, and K. Fujikawa, Nippon Nogei Kagaku Kaishi, 39, 481 (1964); Chem. Abstr., 63, 10594g (1965).
12. R. A. Gray, Phytopathology, 52, 734 (1962).
13. J. M. Walker, Rept. Glasshouse Crops Res. Inst. 1968, 1969, p. 80.
14. N. J. Turner and M. E. Corden, Phytopathology, 53, 1388 (1963).
15. W. Moje, D. E. Munnecke, and L. T. Richardson, Nature, 202, 831 (1964).
16. J. H. Smelt and M. Leistra, Pestic. Sci., 5, 401 (1974).
17. M. G. Ashley, B. L. Leigh, and L. S. Lloyd, J. Sci. Food Agr., 14, 153 (1963).
18. D. E. Munnecke, K. H. Domsch, and J. W. Eckert, Phytopathology, 52, 1298 (1962).
19. D. E. Munnecke and J. P. Martin, Phytopathology, 54, 941 (1964).
20. G. D. Thorn and R. A. Ludwig, Can. J. Chem., 32, 872 (1954).
21. G. D. Thorn, Can. J. Chem., 38, 2349 (1960).
22. R. Engst and W. Schnaak, Z. Anal. Chem., 248, 321 (1969).
23. J. W. Vonk and A. Kaars Sijpesteijn, Ann. Appl. Biol., 65, 489 (1970).
24. C. W. Pluijgers, J. W. Vonk, and G. D. Thorn, Tetrahedron Letters, No. 18, 1317 (1971).
25. N. R. Benson, R. D. Ross, J-Y. T. Chen, R. P. Barron, and D. Mastbrook, J. Assoc. Offic. Anal. Chemists, 55, 44 (1972).
26. D. E. Munnecke, Phytopathology, 48, 581 (1958).
27. D. G. Clarke, H. Baum, F. L. Stanley, and W. F. Hester, Anal. Chem., 23, 1842 (1951).
28. C. F. H. Allen, C. O. Edens, and J. Van Allan, Organic Synthesis (E. C. Horning, ed.), Coll. Vol. 3, Wiley, New York, 1955, p. 394.
29. G. Yip, J. H. Onley, and S. F. Howard, J. Assoc. Offic. Anal. Chem., 54, 1373 (1971).

30. J. W. Vonk and A. Kaars Sijpesteijn, Ann. Appl. Biol., 65, 489 (1970).
31. J. W. Vonk and A. Kaars Sijpesteijn, Pestic. Biochem. Physiol., 1, 163 (1971).
32. 1967 Evaluation of Some Pesticide Residues in Food, FAO/WHO Monograph, Rome, 1968, p. 178 (unpublished data submitted by Rohm & Haas).
33. J. R. M. Innes, B. M. Ulland, M. G. Valerio, L. Petrucelli, L. Fishbein, E. R. Hart, A. J. Pallotta, R. R. Bates, H. L. Falk, J. J. Gart, M. Klein, I. Mitchel, and J. Peters, J. Natl. Cancer Inst., 42, 1101 (1969).
34. P. A. Cruickshank and H. C. Jarrow, J. Agr. Food Chem., 21, 333 (1973).
35. W. Ingemells, J. Soc. Dyers Colourists, 79, 651 (1963).
36. R. D. Ross and D. G. Crosby, J. Agr. Food Chem., 21, 335 (1973).
37. D. D. Kaufman and C. L. Fletcher, 2nd Intern. Cong. Plant Pathology, Minneapolis, 1973, Abstr. 1018.
38. G. D. Thorn and L. T. Richardson, Meded. Landbouwhoogeschool Opzoekingsstat. Staat Gent, 27, 1175 (1962).
39. M. Okawara, H. Yamashina, K. Ishiyama, and E. Imoto, Kogyo Kagaku Zasshi, 66, 1383 (1963); Chem. Abstr., 60, 14628 (1964).
40. T. Yamase, H. Kokado, and E. Inone, Kogyo Kagaku Zasshi, 72, 162 (1969); Chem. Abstr., 70, 264 (1969).
41. A. O. Fitton, J. Hill, M. Qutob, and A. Thompson, J. Chem. Soc., 1972, p. 2658.
42. H. K. Nolan and I. Heckenbleikner, U.S. Pat. 2,419,888, American Cyanamid Co. (1947).
43. J. H. Exner, G. A. Burk, and D. Kyriacou, J. Agr. Food Chem., 21, 838 (1973).
44. D. G. Crosby and N. Hamadmad, J. Agr. Food Chem., 19, 1171 (1971).
45. W. Wolf and N. Karasch, J. Org. Chem., 30, 2493 (1965).
46. S. K. Wang and J. K. S. Wan, Can. J. Chem., 51, 753 (1973).
47. R. L. Letzinger and J. H. McCain, J. Amer. Chem. Soc., 88, 2884 (1966).
48. E. Ubschat, Angew. Chem., 72, 981 (1960).
49. F. J. Hills and L. D. Leach, Phytopathology, 52, 51 (1962).
50. D. G. Crosby and K. W. Moilanen, unpublished (1970), cited in Environmental Toxicology of Pesticides (F. Matsumura, G. M. Boush, and T. Misato, eds.), Academic Press, New York, 1972, p. 423.
51. A. Kaars Sijpesteijn, Meded. Landbouwhoogeschool Opzoekingsstat. Staat Gent, 24, 850 (1959).
52. E. Kroller, Deut. Lebensm. Rundschau, 7, 190 (1960).
53. J. H. Hardon, A. F. H. Besemer, and H. Brunik, Deut. Lebens. Rundschau, 38, 349 (1962).
54. J. Brüggemann, K. Barth, and K. H. Niesar, Zentr. Veterinarmed., 11, 4 (1964).

55. J. Brüggemann, O. R. Klimmer, and K. H. Niesar, Zentr. Veterinarmed., 11, 40 (1964).
56. H. Tamura, Nogyo Gijusta, No. 18, 1965, p. 135.
57. A. H. Chapman and J. W. Price, Intern. Pestic. Control, 14, 1 (1972).
58. R. D. Barnes, A. T. Bull, and R. C. Poller, Pestic. Sci., 4, 305 (1973).
59. K-D. Freitag and R. Bock, Pestic. Sci., 5, 731 (1974).
60. G. Ciamician and P. Silber, Ber., 34, 1530 (1901).
61. W. N. Hartley and A. G. G. Leonard, J. Chem. Soc., 95, 34 (1909).
62. P. A. Leighton and G. S. Forbes, J. Amer. Chem. Soc., 51, 3549 (1929).
63. R. C. Cookson, D. A. Cox, and J. Hudec, J. Chem. Soc., 1961, p. 4499.
64. D. Bryce-Smith and A. Gilbert, Proc. Chem. Soc., 1964, p. 3471.
65. F. Poupe, Collection Czech. Chem. Commun., 12, 225 (1947).
66. H. I. Joschek and S. I. Miller, J. Amer. Chem. Soc., 88, 3273 (1966).
67. K. C. Kurien and P. A. Robins, J. Chem. Soc. B, 1970, p. 855.
68. P. A. Leighton and W. F. Dresia, J. Amer. Chem. Soc., 52, 3556 (1930).
69. H. P. Burchfield and G. L. McNew, Contrib. Boyce Thompson Inst., 16, 131 (1950).
70. F. Bredoux, Bull. Soc. Chim., 1968, p. 4180.
71. M. H. Fisch and W. M. Hemmerlin, Tetrahedron Letters, No. 31, 1972, p. 3125.
72. E. R. White, W. W. Kilgore, and G. Mallett, J. Agr. Food Chem., 17, 585 (1969).
73. R. G. Owens, Contrib. Boyce Thompson Inst., 17, 221 (1953).
74. R. G. Owens and H. M. Novotny, Contrib. Boyce Thompson Inst., 19, 463 (1958).
75. R. G. Owens and L. P. Miller, Contrib. Boyce Thompson Inst., 19, 177 (1958).
76. E. M. Gause, D. A. Montalvo, and J. R. Rowlands, Biochim. Biophys. Acta, 141, 217 (1967).
77. C. W. Haitt, W. T. Haskins, and L. Oliver, Amer. J. Trop. Med., 9, 527 (1960).
78. K. Munakata and M. Kuwahara, Residue Rev., 25, 13 (1969).
79. S. K. Bandal and J. E. Casida, J. Agr. Food Chem., 20, 1235 (1972).
80. H. Matsuo and J. E. Casida, Bull. Environ. Contam. Toxicol., 5, 72 (1970).
81. D. A. M. Watkins and T. Clark, Ann. Rept. Long Ashton Res. Sta., 1972, 1973, p. 86.
82. E. Kühle, E. Klauke, and F. Grewe, Angew. Chem., 76, 807 (1964).
83. K. Vogeler and H. Niessen, Pflanzenschutz Nachrichten Bayer, 20, 534 (1967).
84. E. Peynaud, Phytiatr. Phytopharm., 3, 83 (1954).
85. L. C. Mitchell, J. Assoc. Offic. Agr. Chemists, 44, 643 (1961).

86. G. Serra, Phytiatr. Phytopharm., 13, 107 (1964).
87. H. P. Burchfield and E. E. Storrs, Contrib. Boyce Thompson Inst., 18, 319 (1956).
88. H. P. Burchfield, Contrib. Boyce Thompson Inst., 20, 205 (1959).
89. G. P. Clemons and H. D. Sisler, Phytopathology, 59, 705 (1969).
90. J. J. Sims, H. Mee, and D. C. Erwin, Phytopathology, 59, 1775 (1969).
91. C. A. Peterson and L. V. Edgington, Phytopathology, 59, 1044 (1969).
92. A. Fuchs, A. L. Homans, and F. W. de Vries, Phytopathol. Z., 69, 330 (1970).
93. W. W. Kilgore and E. R. White, Bull. Environ. Contam. Toxicol., 5, 67 (1970).
94. W. A. Maxwell and G. Brody, Appl. Microbiol., 21, 944 (1971).
95. J. M. Ogawa, E. A. Bose, B. T. Manji, E. R. White, and W. W. Kilgore, Phytopathology, 61, 905 (1971).
96. E. R. White, E. A. Bose, J. M. Ogawa, B. T. Manji, and W. W. Kilgore, J. Agr. Food Chem., 21, 616 (1973).
97. F. J. Baude, J. A. Gardiner, and J. C. Y. Han, J. Agr. Food Chem., 21, 1884 (1973).
98. J-P. Calmon and D. Sayag, Compt. Rend. Acad. Sci., Ser. C, 277, 875 (1973).
99. M. Chiba and K. Doornbos, Bull. Environ. Contam. Toxicol., 11, 273 (1974).
100. R. R. Laski and R. R. Watts, J. Assoc. Offic. Anal. Chem., 56, 328 (1973).
101. H. Buchenauer, unpublished results, 1975.
102. D. A. M. Watkins, Chemosphere, 3, 239 (1974).
103. E. R. Cole, G. Crank, and A-S. Sheikh, Tetrahedron Letters, No. 32, 1973, p. 2987.
104. F. Grossmann, H. Buchenauer, and L. V. Edgington, 2nd Intern. Congr. Plant Pathology, Minneapolis, 1973, Abstr. 1015.
105. W. T. Chin, G. M. Stone, A. E. Smith, and B. von Schmeling, Proc. 5th Brit. Insectic. Fungic. Conf., Brighton, 1969, p. 322.
106. M. Snel and L. V. Edgington, Phytopathology, 59, 1051 (1969).
107. H. Lyr and G. Ritter, 2nd Intern. Congr. Plant Pathology, Minneapolis, 1973, Abstr. 1019.
108. D. E. Briggs, R. H. Waring, and A. M. Hackett, Pestic. Sci., 5, 599 (1974).
109. L. W. Carlson, Can. J. Plant Sci., 50, 627 (1970).
110. D. E. Mathre, Phytopathology, 60, 671 (1970).
111. N. N. Ragsdale and H. D. Sisler, Phytopathology, 60, 1422 (1970).
112. H. Stingl, K. Härtel, and G. Heubach, 7th Intern. Congr. Plant Protection, Paris, 1970, p. 205.
113. H. Oeser, S. Gorbach, and E. Boerner, 3rd Intern. Congr. Pesticide Chemistry, Helsinki, 1974, Abstr. 262.

114. A. Ben-Aziz and N. Aharonson, Pestic. Biochem. Physiol., 4, 120 (1974).
115. A. Fuchs, M. Viets-Verweij, and F. W. de Vries, Phytopathol. Z., 75, 111 (1972).
116. V. von Bruchhausen and M. Stiasni, Pestic. Sci., 4, 767 (1973).
117. A. Fuchs and W. Ost, 3rd Intern. Congr. Pesticide Chemistry, Helsinki, 1974, Abstr. 260.
118. A. Fujinami, T. Ozaki, and S. Yamamoto, Agr. Biol. Chem., 35, 1707 (1971).
119. S. Sumida, R. Yoshihara, and J. Miyamoto, Agr. Biol. Chem., 37, 2781 (1973).
120. H. A. Selling, J. W. Vonk, and A. Kaars Sijpesteijn, Chem. Ind., 1970, p. 1625.
121. T. Noguchi, K. Ohkuma, and S. Kosaka, in Pesticide Terminal Residues (A. S. Tahori, ed.), Butterworths, London, 1971, p. 257.
122. H. Buchenauer, D. C. Erwin, and N. T. Keen, Phytopathology, 63, 1091 (1973).
123. Y. Soeda, S. Kosaka, and T. Noguchi, Agr. Biol. Chem., 36, 817, 931 (1972).
124. H. Buchenauer, L. V. Edgington, and F. Grossmann, Pestic. Sci., 4, 343 (1973).
125. H. Buchenauer, 2nd Intern. Congr. Plant Pathology, Minneapolis, 1973, Abstr. 0199.
126. D. A. M. Watkins, T. Clark, and E. Norton, Ann. Rept. Long Ashton Res. Sta. 1973, p. 93, 1974.
127. H. Müller, S. Gäb, and F. Korte, Chemosphere, 3, 157 (1974).
128. J. R. Plimmer and P. C. Kearney, 158th Meeting Amer. Chem. Soc., Division of Pesticide Chemistry, New York, Sept. 1969.
129. J. D. Rosen, M. Siewierski, and G. Winnett, J. Agr. Food Chem., 18, 494 (1970).
130. K. Fukunaga, T. Misato, and M. Awakawa, Bull. Agr. Chem. Soc. Japan, 19, 181 (1955).
131. T. Misato, Residue Rev., 25, 93 (1969).
132. T. Misato, I. Yamaguchi, and Y. Homma, in Environmental Toxicity of Pesticides (F. Matsumura, G. M. Boush, and T. Misato, eds.), Academic Press, New York, 1972, p. 587.
133. H. Buchenauer, Pestic. Sci., 6, 525 (1975).
134. T. A. Jacobs, J. R. Carlin, R. W. Walker, F. J. Wolf, and W. J. A. VandenHeuvel, J. Agr. Food Chem., 23, 704 (1975).
135. D. A. M. Watkins, Chemosphere, 5, 77 (1976).
136. B. D. Cavell, S. J. Pollard, and C. H. J. Wells, Chem. Ind., 1976, p. 566.
137. H. Ohkawa, R. Shibaike, Y. Okihara, M. Morikawa, and J. Miyamoto, Agr. Biol. Chem., 40, 943 (1976).

Chapter 6

PERMEATION AND MIGRATION OF FUNGICIDES IN FUNGAL CELLS

Donald V. Richmond

Department of Agriculture and Horticulture
Long Ashton Research Station
University of Bristol
Bristol, England

I. INTRODUCTION

Earlier chapters considered the obstacles encountered by fungicides before reaching the fungus, and succeeding chapters will deal with the sites affected by fungicides. This chapter describes only how fungicides become available to the fungus, enter fungal cells, and migrate within them.

Much research on the accumulation of fungicides by fungal spores was carried out at the Boyce Thompson Institute for Plant Research, Yonkers, New York, has been comprehensively reviewed by Miller [1], and will not be discussed in detail here. Miller's review was written before the introduction of the first commercially successful systemic fungicides, i.e., oxycarboxin, 2,3-dihydro-6-methyl-5-phenylcarbamoyl-1,4-oxathiin-4,4-dioxide [2] and benomyl, methyl 1-(butylcarbamoyl)benzimidazol-2-yl-carbamate [3]. Since then the emphasis in fungicide research has changed. Considerable attention is now being paid to the translocation of systemic fungicides within the host plant because advances in controlling plant diseases may depend on our ability to design compounds that are transported within the plant. Systemic fungicides also possess a high degree of selectivity between host and parasite. Their specific modes of action have encouraged biochemical studies to determine their primary effects on fungi. To protect the environment, there has been an increase in the study of the toxicology of fungicides and the transformations they undergo in plants, soil, and the atmosphere. These changes have recently resulted in less attention being paid to the mechanisms by which fungicides enter and accumulate within fungal cells. Nevertheless, it remains true that a toxicant must reach its site of action before an organism can be killed. This review aims to discuss the permeability of fungal cells to fungicides in the light of recent knowledge of the structure and functions of cells and cell components.

A. Structure of Fungal Cells

Although resembling other organisms in fine structure, fungi differ from them in nutrition, morphology, and cell wall composition. Many fungi form tubes, or hyphae, which grow only at the tips and are divided transversally by performated septa. As cytoplasm and nuclei can pass through these pores, materials can be transported along considerable lengths of hypha. Fungi propagate by spores which can be killed or prevented from germinating by protective fungicides. Fungal spores vary greatly in dimensions and external ornamentation which aid dispersal by air, water, or animals.

Fungi are eukaryotic having nuclei surrounded by a nuclear membrane. The cytoplasm contains mitochondria, vacuoles, endoplasmic reticulum, and ribosomes; other inclusions occur in specialized cells; plastids are absent. The functions of Golgi bodies which synthesize, transport, and secrete materials in plant and animal cells, are usually performed in fungi by the endoplasmic reticulum. The cytoplasm is surrounded by a semipermeable membrane, the plasmalemma, and usually a cell wall [4]. The walls of fungi differ widely in composition, but most consist mainly of chitin and glucan [5]. Walls assist in maintaining the integrity of the cell by providing mechanical support. They are usually considered completely permeable to solutes but may sometimes act as molecular sieves to prevent entry of toxicants [6]. The composition of spore walls may differ quantitatively from hyphal walls of the same organism [7] and may affect binding by ionic fungicides [8].

B. Function of Membranes

Membranes are important in almost all activities of cells. The plasmalemma prevents loss of soluble materials from the cell, excludes some solutes and permits the entry of others, and enables the cell to maintain an internal environment different from the external medium. Internal membranes separate the cell into compartments such as nuclei, mitochondria, and vacuoles, and are also the sites of important cell functions [9]. Cell membranes divide the cell into two regions, a cytoplasmic region including the contents of the nucleus and an "external" region which includes the lumen of the endoplasmic reticulum, the contents of vesicles, and the space between the nuclear membranes. This region is called external because it is potentially connected with the outside of the cell. Thus vesicles derived from the Golgi apparatus or the endoplasmic reticulum may secrete their contents outside the cell [10]. The nuclear membrane with its pores acts as a sieve restricting movement between nucleus and cytoplasm [11]. Some cell functions are not confined to a single membrane but involve a number of different membranes [12]. Membranes may interact with each other, fuse, and exchange materials to form a dynamic endomembrane system [13] in which the endoplasmic reticulum plays a major role [14]. Cellular organization is thus in a continuous state of flux, whereas an electron micrograph can represent the position of structures only at the time of fixation.

C. Structure of Membranes

Cell membranes are now considered to consist of a fluid mosaic of integral globular proteins embedded in a phospholipid bilayer [15, 16]. The globular proteins may be attached to oligosaccharides to form glycoproteins or to lipids to form lipoproteins. Some protein molecules protrude from both membrane faces, others are partially embedded and hence protrude from only one face. In addition, about 30% of the total protein consists of peripheral proteins loosely bound to the membrane surface. The structure is essentially dynamic and both lipid and protein molecules can move in the plane of the sheetlike bilayer. Membranes are asymmetric, and the two surfaces differ in structure and composition; most of the membrane protein occurs on the side of the membrane adjacent to the cytoplasm. When two membranes fuse, the cytoplasmic side of each membrane always remains on the cytoplasmic side and the external side is always external [10]. Although all membranes have a similar basic structure, their precise molecular organization is genetically determined and varies from tissue to tissue.

D. How Substances Enter Cells

Any theory of cell permeability has to explain how substances can enter a living cell at rates, and in amounts, dependent on their oil-water partition coefficients while the cell is also permeable to lipid-insoluble substances, such as sugars, amino acids, inorganic ions, and water. Plants living in water, for instance, are able to take up ions from very dilute solutions and accumulate them to many times their original concentration. Early theories of permeability were reviewed by Collander [17]; modern theories distinguish clearly between simple physical diffusion processes and active transport against a concentration gradient, for which the required energy is supplied by the metabolic activity of the cell. Active, unlike passive transport may be decreased in the absence of free oxygen or in the presence of certain enzyme inhibitors.

Essential nutrients are able to enter the cell by means of specific proteins embedded in the membrane. These proteins act either as passive filters or they may actively pump molecules across the membrane [10].

Early workers considered that water entered the cell through pores in the cell membrane but recent work suggests that entry occurs directly through the lipid phase of the membrane. Once within the cell, movement is not unimpeded, as many organelles are bounded by membranes surrounded by stationary layers of water which restrict water entry [18].

Ions enter cells actively or passively: passive transport results from either the diffusion potential or the Donnan potential across the membrane; active transport may involve a protein carrier, or pinocytosis. The protein carrier binds ions and ferries them across the membrane and pinocytosis occurs when ions bind to a small area of membrane which then becomes invaginated to form a vesicle which is transported within the cytoplasm [19], possibly by microtubules [20]. When these vesicles fuse with vacuoles, ions are transferred directly from an external solution into the vacuolar solution without coming into contact with other parts of the cell [21].

E. Entry of Fungicides into Fungi

Fungicides act on fungi in many ways, that is, they may enter the cell and be transported within in the same way as nutrients, or they may be adsorbed on the surface of spores or may enter the cytoplasm; uptake into the cell may be either passive or active. Active transport implies uptake against a concentration gradient by a mechanism deriving energy from cell metabolism. When the fungicide is strongly bound by cell constituents or reacts chemically with them, the fungicide is removed from the equilibrium system and the concentration gradient is illusory [22]. Reactive, nonspecific, toxicants may combine with cell components to cause general disorganization; whereas specific compounds will inhibit a single metabolic pathway. Even a resting fungal spore is metabolizing slowly, and changes produced in one cell function will ultimately affect many others. The fungal cell is a dynamic, living, system which responds to a toxicant by a complex series of biochemical and ultrastructural changes.

II. HOW FUNGICIDES REACH THE FUNGUS

Fungicides used in dormant sprays may be general cell poisons, very soluble in water. Protectant fungicides are poisons, usually only slightly soluble, able to penetrate fungal spores but not the leaves of green plants, whereas systemic fungicides have specific modes of action and can discriminate between host and parasite.

Dormant season sprays, applied as eradicants, need be neither selective nor persistent, and therefore highly toxic, soluble compounds such as sodium hydroxide have been used [23]. Such substances will not be further considered.

Apart from seed dressings and dormant season sprays, most agricultural fungicides are applied to leaves as an aqueous suspension or emulsion. Protective fungicides produce after drying a more or less continuous deposit of fungicide on the leaf surface. The fungicide is mobilized by rain or dew, in which it dissolves to form a toxic solution to kill any fungal spores nearby. Although the fungicide may be a highly reactive compound it is usually sparingly soluble in water and has a low vapor pressure and thus remains on the leaf for several weeks continuously supplying toxicant. The fungal spore is killed by the accumulation of fungicide while the plant remains unharmed because the compound is unable to penetrate the cuticle of the leaf; selective toxicity is due to selective permeability [24]. Some protective fungicides, especially those effective against powdery mildew, can penetrate spores directly in the vapor phase [25, 26].

Systemic fungicides may be applied as foliage sprays, seed dressings, soil drenches, or postharvest dips; woody plants may be treated by stem injection [27]. The systemic fungicides now in use are translocated in the apoplast of green plants. They move mainly in nonliving parts of cells, i.e., in the transpiration stream, to accumulate at the margins and tips of leaves [28]. Systemic fungicides such as oxycarboxin, effective both as protectants and eradicants, will control established infections [29], provided enough chemical reaches the infections. The systemic properties of fungicides vary in different species of plants [30], and some compounds may be only partially systemic in certain plants. In practice, a compound may have both systemic and protective properties.

A. Protective Fungicides

The fungicide on the leaf usually becomes available to the fungus by solution in water. Fungal spores may selectively absorb fungicide from dilute solutions [1]. Thus spores of *Monilinia fructicola* can accumulate copper to a concentration 4,000 times greater than that of the external solution [31]. By comparison, uptake by the leaf is very small. Negligible amounts of copper were absorbed by leaves from radioactive Bordeaux mixture [32], and apple leaves took up only 1 μg of captan per gram of leaf compared with 7,100 μg per gram of spores in the same time [33]. The spores of pathogens will, however, have to complete with nonpathogenic microorganisms for the available fungicide.

The fungicide in solution may be decomposed by hydrolysis, solar radiation, or reaction with leaf or spore exudates. Such exudates can dissolve copper from deposits of Bordeaux mixture [34, 35] to form soluble copper complexes less toxic than the free cupric ion [36]. Losses of fungicide from solution are replenished by the remaining insoluble deposit since fungicides are applied to leaves at concentrations far in excess of the toxic dose.

B. Systemic Fungicides

Systemic compounds may control diseases by killing the pathogen, preventing infection, altering host metabolism, or stimulating the resistance mechanisms of the host plant [37, 38]. The compound may itself be fungicidal or it may be inactive and be converted into an active compound within the plant. Consequently, in vitro tests may give no indication of fungicidal activity.

Good agreement exists between the effectiveness of benomyl in vitro and in vivo [39], whereas it is not possible to predict the in vivo activity of triforine, 1,4-di-(2,2,2-trichloro-1-formamidoethyl)piperazine, from in vitro tests. The effectiveness of benomyl against a fungus can often be predicted from the taxonomy of the fungus; whereas triforine is effective in vivo against many unrelated leaf-infecting fungi [40]. Differences in the behavior of the two fungicides are probably due to the way they move in the plant. Systemic compounds accumulate at certain sites according to the hydrophilic and lipophilic properties of the molecule [28]. Diseases will be controlled when these sites coincide with the sites of infection by the pathogen. Triforine is accumulated by the cuticle of the leaf and hence the fungicide is effective against leaf-infecting fungi [40].

III. DETECTION OF SITES OF ACTION AND ACCUMULATION

A. Morphological Effects

When the test tube dilution spore-germination technique [41] is used to determine the fungistatic activities of protective fungicides such as captan, _N_-(trichloromethylthio)-3a,4,7,7a-tetrahydrophthalimide, or dodine, dodecylguanidine acetate, individual fungal spores either germinate normally, produce shorter germ tubes, or show no germination. This suggests that these compounds are acting directly to inhibit spore germination. Some systemic fungicides have little effect on spore germination but are highly active in preventing hyphal growth or cell division. When spores are incubated in the presence of these compounds the spores germinate but the germ tubes are much distorted. Koch [42] suggested that such distortions may indicate systemic activity. Triarimol, α-(2,4-dichlorophenyl)-α-phenyl-5-pyrimidinemethanol, and triforine produce similar distortions in germinating conidia of _Cladosporium cucumerinum_ and _Aspergillus niger_ and have similar modes of action [43]. Compounds with different modes of action may, however, produce similar distortions [44]. These effects must therefore be interpreted with caution.

Studies of stained preparations have shown that benomyl and carbendazim, methyl benzimidazol-2-yl-carbamate, inhibit mitosis in fungi [45-47]; these results have been confirmed biochemically [45, 46].

B. Electron Microscopy

Studies with the electron microscope can give information about the effect of fungicides on the structural components of cells. Thus, spores of

Neurospora crassa showed changes in mitochondrial structure after treatment with captan, suggesting that the fungicide may have reacted with the SH groups that control the swelling of mitochondria [48]. Striking changes also occur in the endoplasmic reticulum of conidia of Botrytis fabae germinating in the presence of benomyl [49]. Lipoprotein inclusions in spores of Penicillium digitatum treated with 2-aminobutane are broken down suggesting that the fungicide stimulates the activity of hydrolytic enzymes [50].

Carboxin, 2,3-dihydro-6-methyl-5-phenylcarbamoyl-1,4-oxathiin, and oxycarboxin inhibit mitochondrial respiration [51, 52]. Mitochondria in the yeast Rhodotorula mucilaginosa became swollen after treatment with carboxin [53]. Similar effects are also produced in the mitochondria of rusts infecting plants, the roots of which were treated with oxycarboxin [54, 55].

Cycloheximide changes the pattern of wall synthesis in hyphae of Aspergillus nidulans; the hyphae do not elongate, but the walls become thicker. The antibiotic may interfere with the vesicles which transport wall precursors towards the hyphal tip [56].

C. Electron Microscope Autoradiography

The sites at which fungicides accumulate can be identified by autoradiography; thus, when yeast cells were grown in the presence of mercuric chloride labeled with ^{203}Hg large amounts of the metal were found in the cell wall [57].

D. X-Ray Microanalysis

The subcellular distribution of elements can be determined by the X-ray microanalyzer. By examining thin sections of biological tissue it is possible to combine a study of morphology with simultaneous chemical analysis of naturally occurring or introduced substances [58].

E. Microelectrophoresis

The ionic charges on surfaces of fungal spores can be studied by microelectrophoresis [59]. Surface groups can be identified by the shape of the pH/mobility curve, the pH at the isopotential point, and the effect of specific chemical and enzymic treatments on the pH/mobility curve [60]; surface lipid can be detected by the effect of sodium dodecyl sulfate on mobility [61]. Most fungal spores have a negatively charged surface which reacts with cationic fungicides such as dodine [62].

F. Chemical Studies

The uptake of fungicides from solution is usually measured either spectrophotometrically or radiochemically. A convenient device for the automatic recording of uptake is described by Haverkate, Brevoord, and Verloop [63]. However, loss of fungicide from solution does not necessarily measure uptake [22]. Spores of Glomerella cingulata, for example, can reduce thiram, tetramethylthiuram disulfide, to dimethyldithiocarbamate ions in the external solution [64]. Therefore the true fungicide uptake cannot be determined until cellular reactions have been clarified. Fungicides may be converted into other compounds within the cell. Thus captan is detoxified by intracellular spore thiols [65], cupric ions are reduced to cuprous [66], and mercuric to mercurous [67]. A nontoxic compound may also be converted into a toxic one; thus the nontoxic 5-fluorocytosine is taken up by yeast cells and converted within the cell by cytosine deaminase into the highly toxic 5-fluorouracil [68].

Kinetic studies using an automatic recorder can give useful information on mechanisms of uptake [63]. Active transport systems can be identified by studying the effects of temperature, metabolic inhibitors, salts, and oxygen supply on uptake, and by comparing the uptake of live with dead spores.

Treatment of spores with fungicides may also result in the efflux of substances into the solution. Thus when spores take up copper from an unbuffered solution of a copper salt, the solution becomes more acid as the sorbed copper is replaced by hydrogen ions [69]. Surface-active compounds may increase membrane permeability so that soluble cell constituents are lost. Thus polyene antibiotics rapidly release potassium from cells of Candida albicans [70]. When spores of Fusarium culmorum are treated with n-propyl-5-nitro-2-furylketone, the compound is slowly taken up, but a degradation product, n-propyl-5-amino-2-furylketone, is released into solution. The reduction of the original compound occurs within the spore [71].

The nature of the binding of fungicide can be studied by washing spores which have taken up fungicide with various solutions. A labeled fungicide taken up by a spore may be exchangeable with nonradioactive fungicide; metal ions may be removed by treatment with acid.

The distribution of fungicides within cells may be studied by macerating spores which have taken up labeled fungicide and measuring the radioactivity of the various fractions after differential centrifugation [1]. The method suffers from the disadvantage that redistribution of the labeled compound may occur after disintegration of the spores.

IV. SIGNIFICANCE OF UPTAKE

Dosage-response curves from fungistatic spore germination tests have been used to give information concerning the mode of action of fungicides. Fungicides having curves of the same slope were considered to have the same mode of action [72]. However, it has now been shown by workers at the Boyce Thompson Institute for Plant Research that no valid conclusions concerning mode of action can be drawn from such curves [1, 73, 74]. The fungicidal activity of compounds must be compared on a spore-weight basis since the concentration of fungicide in the external solution is not necessarily related to the concentration within the spore [75]. Extensive studies with radioisotopes have shown that fungal spores are able to accumulate large amounts of fungicides from solution and that fungicides are usually less toxic on a spore-weight basis than other biocides [76]. The toxicity of most protective fungicides is clearly due to their strongly selective accumulation by spores [22].

A. Relation Between Uptake and Fungitoxicity

There should be a direct relationship between uptake and toxicity for fungicides which are either highly specific or completely nonspecific poisons. The most effective fungicide in a related series of compounds should be the compound which is accumulated to the greatest extent. This is the case for a series of 5-nitrofuran derivatives [77] and for the n-alkyl thioethers of 2,5-dimercapto-1,3,4-thiadiazole [78]. However, few fungicides are either completely specific or completely nonspecific. A fungicide with a specific mode of action may also react with other cell components before reaching its active site or sites. The total amount of fungicide taken up by a fungal spore will therefore be the sum of the amount bound to active sites, the amount detoxified by reaction with cell components or bound to nonspecific sites, and the amount free within the cell. Since fungal spores of different species differ in size, structure, and composition, they are likely to take up different amounts of fungicide from solution. The relation between uptake and toxicity has been studied for captan which reacts with intracellular sulfhydryl compounds [65, 79]. The sulfhydryl content of spores of ten fungal species and of yeast cells was determined and compared with the uptake of captan by the fungi on an equivalent fresh-weight basis. The fungistatic activity of captan against eight of the species was also determined. There was no relation between captan uptake and toxicity or between captan toxicity and the sulfhydryl content of the cells. However, there was a general relation between captan uptake and sulfhydryl content. Species with the highest sulhydryl content had the highest captan uptake [80]. Most of the captan taken up by fungal spores is detoxified by reaction with soluble thiols [65]. When spores are pretreated with nontoxic concentrations of thiol reagents, the uptake of captan is reduced to a negligible amount and hence the toxicity of captan, on a spore-weight basis, is greatly increased. Unfortunately the fungistatic activity of captan is unaffected by the treatment [65], probably because the rate-determining step is the penetration of the cell membrane. Similar results have been obtained by Haverkate, Tempel, and

and Den Held [81] for 2,4,5-trichlorophenylsulfonylmethyl thiocyanate which also reacts with sulfhydryl groups.

B. Rate of Uptake

The rate of uptake is important to fungitoxicity. A fungicide such as captan is taken up slowly from solution and decomposed within the protoplast [82]. The rate of uptake is dependent upon the time required to penetrate the membrane and not upon the rate of reaction with cell constituents [65]. In contrast, dodine is rapidly adsorbed by spore walls by an ionic bonding mechanism [83]. Bartz and Mitchell showed the importance of uptake rate to the toxicity of dodine [84]. They found that the fungistatic ED_{50} of dodine to spores of _Colletotrichum orbiculare_ was 30 times greater than to _Venturia inaequalis_. However, when calculated on a spore-weight basis the toxicities were reversed. When the time required for each fungus to accumulate a toxic dose was calculated, it was found that _C. orbiculare_ absorbed the fungicide slowly, whereas _V. inaequalis_ rapidly absorbed a toxic dose and thus became the more sensitive organism.

The relation between time and toxicity was also studied by Turner [85] who found that the rates at which ten fungicides killed spores of _Monilinia fructicola_ varied greatly, although their fungistatic activities, as determined by slide germination tests, were similar.

C. Uptake and Selectivity

Fungi may not be susceptible to a particular fungicide because (a) their membranes are impermeable, (b) there are no binding sites, or (c) the fungicide is detoxified within the cell.

The sensitivity of oomycetes to streptomycin is probably due to the ability of the antibiotic to pass through their cell membranes since sensitive fungi take up nine times as much streptomycin as insensitive ones [86]. Sensitivity to griseofulvin is also directly related to uptake [87].

Yeasts vary in susceptibility to cycloheximide; the insensitive _Saccharomyces fragilis_ takes up less cycloheximide than the sensitive _S. pastorianus_, probably because of the absence of internal binding sites [88].

When insensitive fungi take up large amounts of fungicides, detoxification may take place within the cell. Fenaminosulf, β-dimethylaminobenzenediazo sodium sulfonate, is ten times as toxic to _Pythium_ as it is to _Rhizoctonia_ [89] because _Rhizoctonia_ is able to detoxify the fungicide by an enzyme system which is absent from _Pythium_ [90]. Macroconidia of _Fusarium solani_ can detoxify dodine and release the products into solution [91].

Insensitivity may also be due to a combination of low permeability and detoxication. Thus insensitive fungi take up less quintozene, pentachloronitrobenzene, from solution than sensitive fungi; in addition, they detoxify the fungicide and excrete the products into the medium [92]. Sensitive fungi take up four times as much ascochitine as insensitive fungi; highly insensitive fungi take up the same amount as moderately insensitive fungi but also have a detoxication mechanism [93]. Insensitive strains of fungi normally sensitive to fungicides have occasionally been isolated in the past [94-96], but such strains have become more frequent and important with the introduction of systemic fungicides [97, 98]. We know little about the mechanisms of insensitivity, but insensitivity to metals is not usually due to restricted uptake [99, 100]. The insensitivity of strains of Pyricularia oryzae to blasticidin-S [101] and strains of Cladosporium cucumerinum to 6-azauracil [102] may, however, be due to reduced uptake of the toxicants. Insensitivity to diphenyl is connected with changes in membrane permeability. The fungicide stimulates potassium uptake by conidia of Fusarium solani but a mutant, insensitive to diphenyl, has a potassium transport system which is not stimulated by the fungicide [103].

V. ROLE OF THE CELL WALL

Although the active sites of fungicides are within the cell, the cell wall can be important in modifying toxicity. A fungicide must pass through the wall to enter the cell and hence may become detoxified, by physical or chemical binding to wall components. Owens and Miller [104] found that only a small proportion of the copper ions taken up by conidia of Neurospora sitophila was bound to the cell wall, although Somers [69] found a greater proportion in conidial walls of Alternaria tenuis and Penicillium italicum. Mercury is taken up by yeast cells and strongly bound to the cell wall. When grown in the presence of mercuric chloride labeled with ^{203}Hg, yeast cells take up most of the mercury in the medium. As incubation continues, mercury moves from the wall to the cytoplasm. Isolated yeast cell walls are capable of binding their own weight of mercury. Mercury bound to the cell wall is not released by extraction with sodium hydroxide or ethylenediamine but a major fraction is extracted by treatment with the broad-spectrum proteolytic enzymes, Pronase and Helicase (snail enzyme). The mercury is thus probably bound to the structural protein of the wall. The presence of mercury in the cell wall was confirmed by autoradiography [57]. Metal cations may be absorbed on the surface of Neurospora ascospores and only enter the spores when germination begins [105, 106]. The interaction between metal cations and yeast cells also occurs in two stages, namely a nonspecific binding to anionic groups on the surface is followed by uptake via a specific transport system into a nonexchangeable pool within the cell [107, 108]. Dodine, a cationic surface-active fungicide, reacts in two stages with fungal spores. Uptake of the fungicide by spores is rapid, independent of temperature, and conforms to the Langmuir adsorption isotherm, suggesting an ionic bonding mechanism [83]. Dodine is detoxified by binding to spore walls which have first to become saturated with fungicide before the toxic reaction within the cell can occur [83, 109]. Electrophoretic studies show that the negative charge on conidia of *Neurospora crassa* is reduced to zero in the presence of dodine. Increasing dodine concentrations then give a positive charge to the spores. However, the spores are killed by concentrations of dodine lower than those causing a reduction in mobility. The toxic reaction does not, therefore, occur at the spore surface [62]. The sensitivity of *Candida albicans* to the polyene antibiotic, amphotericin methylester (AME) varies with the phase of growth; cells in the exponential phase are sensitive to 0.2 μg AME per ml, whereas cells in the stationary phase require 2.4 μg AME per ml for a similar response [110]. Since protoplasts of *C. albicans* derived from exponential- or stationary-phase cells do not show such changes of polyene sensitivity [111], the difference in the response of whole cells must be due to changes in the cell walls. Gale et al. [110] suggest that binding of AME to membrane sterols takes place by two routes, namely direct passage from external medium to the membrane without impedance by the wall, and an indirect route in which AME is first bound to nonspecific sites in the wall. The AME enters cells in the exponential phase mainly by the direct route. The highly cross-linked disulfides in the walls of stationary-phase cells impede direct access, thus AME reaches the membrane only after intermediate binding to nonspecific sites in the wall. Conidia of *Aspergillus fumigatus* also show changes in sensitivity to AME. Ungerminated conidia are insensitive to AME but rapidly become sensitive during the initial stages of germination [112].

Ferenczy et al. [113] examined the effect of six antifungal antibiotics including nystatin, cycloheximide, and griseofulvin, on the regeneration of protoplasts, germination of conidia, and growth of mycelia of *Syncephalastrum racemosum* and *Aspergillus japonicus*. There were no differences in sensitivity to the antibiotics between protoplasts and conidia or hyphae and hence the walls appear to play no role in modifying the toxicity of these antibiotics. Stutzenberger and Parle [114] found that thiabendazole and other related benzimidazole compounds were absorbed on the surface of conidia of *Pithomyces chartarum*. The absorption was rapid and independent of temperature and hence was considered to be an ion-exchange process. The authors suggest that the fungicide became available to the fungus when the spores germinated. The spore wall may also act as a mechanical barrier to the entry of fungicides. Gershon [6] has suggested that the toxicity of copper chelates of derivatives of 8-quinolinol could be explained on the assumption that the fungal spore wall acted as a barrier preventing certain chelates from entering the cell. The fungitoxicity of copper chelates to spores of five fungal species was determined; the compounds were either toxic at low concentrations or were inactive. Gershon [6] proposed that as the chelate molecules approached a spore they became oriented parallel to holes in the wall by electrostatic attraction; if the long axes of the chelates were shorter than the diameter of the holes, then the molecules could penetrate the walls; if the long axes were longer than the diameter of the holes, then the molecules would be excluded. He considered that the entry of the molecules into spores was dependent on the shape and charge distribution of the molecules.

VI. ROLE OF THE CELL MEMBRANE

The effects of the cell wall in modifying the action of ionic surface-active agents and polyene antibiotics was discussed in the preceding section. These compounds also destroy cell permeability and the effects of these and other fungicides on membrane functions are considered here. Fungicides may act on the cell membrane in two ways: metal cations may enter the cell by one of its normal transport systems to exchange with another ion within the cell, whereas other toxicants may act directly on the cell membrane to alter its permeability, thus allowing more toxicant to enter the cell or the cell to lose essential nutrients.

Fungal spores can lose up to 30% of their cell contents when suspended in distilled water without incurring any reduction in spore germination [115]. Spores can withstand further losses when treated with toxicants but, if the loss of solids is too great, germination is inhibited [1]. Miller and McCallan [76] studied the effects of a number of metal ions on the loss of phosphorus from conidia of *Aspergillus niger* labeled with ^{32}P. Silver and copper were the only metals that caused an appreciable loss of ^{32}P from the conidia. Tröger [116] found that silver and copper were also the most effective metal ions in releasing amino acids from conidia of *Fusarium decemcellulare*.

Phosphorus compounds are released from conidia of *Neurospora sitophila* treated with dyrene, 2,4-dichloro-6-(*o*-chloroanilino)-*s*-triazine, which reacts with amino and sulfhydryl groups [117]. Dichlone, 2,3-dichloro-1,4-naphthoquinone, a fungicide which also reacts readily with sulfhydryl compounds, induces a rapid loss of intracellular potassium from human erythrocytes. Sikka et al. [118] suggested that the fungicide altered membrane permeability by reacting with the sulfhydryl groups of the membrane. Captan reacts with sulfhydryl groups in fungal cells [119, 120] but does not seem to have any action on the cell membrane. However, ultrastructural studies on conidia of *Neurospora crassa* showed that changes occurred in the mitochondria after treatment with captan [48]. Mitochondria isolated from the livers of rats were severely swollen after treatment with captan and had no visible cristae [121]. Captan may react with the sulfhydryl groups in the mitochondrial membrane which control the swelling and contraction of mitochondria. Differences in the structure of mitochondrial and plasma membranes may enable captan to react preferentially with mitochondrial membranes.

Fungal conidia can accumulate large amounts of dodine before germination is affected [109, 122, 83]. Uptake of dodine at the ED_{50} was 1% of the cell weight for conidia of Neurospora crassa and 2.4% for Alternaria tenuis [83]. Live and dead conidia of N. crassa took up similar amounts of dodine, hence uptake is unrelated to the physical state of cell protein and is independent of cell metabolism [123]. Macroconidia of Fusarium solani behaved differently and were able to detoxify dodine and release a less toxic compound into the medium [91]. Cells of Saccharomyces pastorianus lost phosphorus compounds and amino acids when treated with dodine [122]. However, the release of cellular constituents from conidia of Monilinia fructicola and N. crassa only occurred in appreciable amounts at concentrations of dodine greater than the ED_{50} value [83, 91]. Somers [83] considered that dodine reacted with the cell membrane to alter its permeability, thus allowing more dodine to enter the cell and disrupt intracellular membranes.

Another cationic surface-active fungicide, glyodin (2-heptadecyl-2-imidazoline) is rapidly taken up by fungal spores. Spores can accumulate an ED_{50} dose in 30 sec and the final concentration within the spores can be 10,000 times greater than that in the external solution [124]. Most of the glyodin taken up by spores is associated with lipid and only small amounts are absorbed by the spore wall [125]. Extracellular secretions from spores of Aspergillus niger can retard uptake of glyodin by spores of Neurospora sitophila. Other surface-active compounds also retard glyodin uptake [124]. Glyodin alters the permeability of yeast cells. The cell membrane is normally impermeable to pyruvate ions, but pyruvate can enter yeast cells which have been treated with a toxic dose of glyodin [126]. The surface-active properties of glyodin may only play a secondary part in toxicity which is probably due to effects on nucleic acid or protein synthesis [127]. Riemersma [128] has studied the effects of the cationic surface-active compounds cetylpyridinium bromide and cetyltrimethylammonium bromide on yeast cells. If uptake exceeded a critical amount the membrane was completely destroyed and lysis occurred. Riemersma [129] suggested that the surface-active agents were bound to phospholipids. They altered membrane structure and thus produced an "all or none" response in individual cells.

Polyene antibiotics bind to membrane sterols and thus destroy the selective permeability of the membrane [130]. Membrane destruction leads to the loss of small ions such as potassium and permits the entry of protons to neutralize the charge so created. However, more protons enter cells of Candida albicans than are required to replace the lost potassium ions, hence internal acidification occurs and soluble material is precipitated in the cytoplasm [131]. The toxicity of the polyenes to cells of Acholeplasma laidlawii is due to the ability of the compounds to form hydrophobic complexes with membrane sterols such as cholesterol and ergosterol. A complex is formed by eight molecules of AME with eight cholesterol molecules and is oriented perpendicular to the membrane. The center of the complex forms a pore which penetrates halfway into the membrane; two such pores on either side of the membrane can then join to form a continuous conducting pore [132].

VII. INORGANIC FUNGICIDES

Miller [1] reviewed in detail the early work on the uptake and toxicity of metal ions; the effects of heavy metals on fungal cells have recently been summarized by Ross [96].

The first reactions of metals with fungal spores are likely to be with groups on the spore surface. If the primary fungistatic action of metals is dependent on nonspecific binding to spore walls, then there should be a general relationship between the toxicity of metal ions and their binding affinity as measured by the electronegativity value of the metal [133]. Somers [134] compared the toxicity of a large number of metal ions to fungal spores with the electronegativity of the metals. He found there was an approximately exponential relationship between elctronegativity and the logarithm of the metal ion concentration at the ED_{50}. A different view was taken by Owens and Miller [104] who considered that toxicity of metal ions was due to their accumulation within the cytoplasm. They disintegrated spores which had taken up sublethal amounts of labeled silver, cerium, zinc, cadmium, or mercury ions and examined the distribution of ions among the cell fractions. The metals accumulated mainly in the particulate fractions and little was found in the cell walls.

The two results show the importance of distinguishing between the fungistatic and fungicidal effects of toxicants. Some toxicants may be fungistatic at low concentrations and fungicidal at high. The two terms are then more applicable to the testing method than to the toxicant. Since most protective fungicides act by preventing spore germination, the fungistatic test may give a more reliable indication of field performance. The rate of uptake may be as important as the fungicidal activity. Thus the ED_{50} of cycloheximide against Saccharomyces pastorianus is 0.4 μg/g of cells [88], about 10,000 times less than that of copper [69], yet the fungistatic activity of the two toxicants is comparable. Copper is much more effective as a fungistat than as a fungicide. The ED_{50} in a fungistatic test was 1 μg/ml, whereas 22.4 μg/ml were required in a fungicidal test after 30 min incubation. The wide difference in toxicity suggests that the sites of action are different for the two processes [69]. The accumulation of copper by fungal spores appears to be mainly passive although a small amount of the uptake by Neurospora crassa may be linked to respiration [69].

The uranyl ion is reversibly bound to phosphoryl and carboxyl groups on the yeast cell membrane but does not enter the cell. The phosphoryl pathway in yeast involves simultaneous transport and phosphorylation of sugar.

When uranyl ions bind to phosphoryl groups transport of sugar is inhibited [135].

Nickel, cobalt, and zinc also bind reversibly to sites on the yeast membrane but these ions can, in addition, enter a nonexchangeable pool inside the cell by a system that transports magnesium and manganese. Uptake involves an active transport system which can obtain the required energy from fermentation reactions [136].

Mercury binds to the yeast-cell membrane but does not enter the cell; uptake of mercury is accompanied by loss of potassium from the cell [137]. There appears to be an "all or none" response of individual cells to mercury [138].

VIII. ALIPHATIC AMINES

Eckert et al. [139] found that 3-methylpyrrolidine and 2-aminobutane were the most active of 50 aliphatic amines in inhibiting germination of spores of _Penicillium digitatum_; most other amines were inactive. Salts of 2-aminobutane are used commercially to control rots of citrus fruits caused by species of _Penicillium_. The 2-aminobutane cation has a narrow spectrum of activity and is ineffective against bacteria. The levo is considerably more active than the dextro enantiomer suggesting that the cation binds to a specific site on, or in, the fungal spore. Hyphae of _P. digitatum_ accumulate tritium-labeled L-2-aminobutane to a concentration about 100 times greater than the external solution. The compound is not metabolized by the cell and uptake is inhibited by inhibitors of respiration and uncouplers of oxidative phosphorylation. This evidence suggests that L-2-aminobutane is one of the few fungicides that is actively transported into fungal cells [140, 141].

IX. FUNGICIDES REACTING WITH THIOLS

A number of protective fungicides, of which captan and folpet, _N_-(trichloromethylthio)phthalimide, are the most important, react strongly with sulfhydryl compounds (thiols), Richmond and Somers [33] determined the solubility, oleyl alcohol/water partition coefficient, fungistatic activity to _Alternaria tenuis_ and _Neurospora crassa_, and spore uptake by _N. crassa_ of captan and six analogues. The analogues varied widely in lipid solubility but this variation was not reflected in either fungitoxicity or spore uptake except for the two compounds with the lowest partition coefficients. The toxicity of the analogues to _Erysiphe graminis_ on oats followed a different pattern from their fungistatic activity to _N. crassa_. There appeared to be no correlation between lipid solubility and either protectant or eradicant action against _E. graminis_. The most effective eradicants had high water solubilities [142]. Lukens and Horsfall [143] examined the toxicity of captan and some related compounds to bean powdery mildew (caused by Erysiphe polygoni). They found a positive correlation between control of the powdery mildew and the oil/water partition coefficients of the compounds. There was also a positive and linear relationship between control and the rates of hydrolysis of the compounds but the relationship between control and the reactivity of the compounds to active groups was negative. The authors suggested that much of the fungicide entering fungal cells was detoxified by reaction with sulfhydryl compounds before reaching the sites of action and hence the less reactive compounds were more effective fungicides.

More recently Hansch and his coworkers [144, 145] have demonstrated the value of using physicochemical constants in regression analysis to separate the roles of hydrophobic, electronic, and steric effects of a series of compounds on biological response. The partition coefficient represents the hydrophobic interactions of molecules with cell constituents. Lien [146] has reexamined by this technique the earlier results on the toxicity of captan and related compounds. The toxicity of the imides and their $NSCCl_3$ compounds was related to the logarithms of the partition coefficients of the intact molecules. He suggested that the imido group of the compounds bound to the protein of a membrane and the $NSCCl_3$ group then took part in hydrophobic interactions with the lipid portion of the membrane to produce a conformational change resulting in increased membrane permeability. Unfortunately there is no evidence that captan has any effect on cell permeability. Captan neither releases purines and pyrimidines from conidial suspensions of Neurospora crassa [147] nor alters the fine structure of the conidial plasmalemma [48]. However, some of the $NSCCl_3$ compounds Lien [146] predicted would have fungicidal activity have been synthesized and are active [148]. The precise mode of action of captan and related compounds is still unknown although they react with both essential [149-151] and nonessential [65, 152, 153] sulfhydryl compounds in the cell and also with some proteins which lack a sulfhydryl group [154]. Controversy continues as to whether the reaction of captan and folpet with nonessential thiols is a part of the toxic mechanism [81, 153] or should be considered as a detoxication [65, 143]. Probably both processes are involved [151]. Chlorothalonil, 2,4,5,6-tetrachloroisophthalonitrile, is taken up rapidly for 30 min by cells of Saccharomyces pastorianus and then uptake slows down. The fungicide reacts first with soluble and later with insoluble thiols. Long exposure is required to kill the cells, hence fungitoxicity may be due to reactions with insoluble thiols [155]. Turner and Battershell [156, 157] applied the methods of regression analysis to a series of halogenated isophthalonitriles. A positive relationship was found between the fungitoxicity of the compounds and their reactivity with 4-nitrothiophenol but there was no correlation between fungitoxicity and partition coefficient. The results suggested that the fungitoxicity of the compounds was due to reaction with sulfhydryl groups. Turner later proposed that the rate of reaction of the fungicides with spore constituents was limited by their chemical reactivity rather than their ability to penetrate cell membranes [158]. The fungicides react first with reduced glutathione but toxicity is ultimately due to reaction with protein thiols [159]. Chlorothalonil is a noncompetitive inhibitor of glyceraldehyde-3-phosphate dehydrogenase which only reacts with the thiol groups of the enzyme. The mode of action of chlorothalonil closely resembles that of captan and folpet but pronounced differences in the chemical mechanism of reaction with cellular constituents are probably due to the chemical nature of the active groups [160].

Benzenediazocyanides also react nonspecifically with cell thiols. Uptake by conidia of _Fusarium culmorum_ was slow and was followed by the release of conversion products into the external solution. Solutions of the compounds in water are decomposed by light [161].

X. SYSTEMIC FUNGICIDES

Clemons and Sisler [162] have shown that carbendazim is taken up in very small amounts by conidia of _Neurospora crassa_ and sporidia of _Ustilago maydis_. Highly concentrated cell suspensions of both organisms took up 1.4 μg/g of cells in 30 min. The concentration of carbendazim in the cells was twice that in the external medium. Conidia of _Fusarium oxysporum_ also take up small amounts of benomyl [163]. Fungal spores take up much larger amounts of the related fungicide, thiabendazole; conidia of _Aspergillus oryzae_ accumulate up to 900 μg/g of spores. The thiabendazole taken up is distributed among many spore components but most is considered to occur free in the cytoplasm [164]. In contrast thiabendazole and related compounds are absorbed on the surface of conidia of _Pithomyces chartarum_ to the extent of 8,000 μg/g of spores [114]. This abnormal behavior may be due to the unusual surface of the conidium which is covered with crystals of the cyclic peptide sporidesmin [165].

Nachmias and Barash [172] found that cells of _Sporobolomyces roseus_ accumulated [^{14}C]carbendazim up to 24 μg/g of dry weight of cells. They considered that carbendazim entered the cells by an energy-dependent transport system specific for compounds containing a benzimidazole nucleus. Mutants of _S. roseus_ insensitive to benzimidazole fungicides took up much less [^{14}C]carbendazim than sensitive cells. They attributed insensitivity to impairment of the uptake system.

Davidse [173] has shown that the rate of conversion of carbendazim into the less toxic 5-hydroxybenzimidazol-2-yl-carbamate in mycelium of _Aspergillus nidulans_ is the same in sensitive and insensitive strains. Insensitivity to carbendazim is due neither to restricted uptake nor to increased detoxication. The toxicity of carbendazim to _A. nidulans_ is due to binding to fungal tubulin and insensitivity to a reduced affinity of the binding site for the fungicide [174].

The oxathiins are mainly effective against Basidiomycetes [166]. Mathre [167] determined the uptake of the oxathiin fungicide, oxycarboxin by sensitive and insensitive fungi. He found that the sensitive fungi, Rhizoctonia solani and Ustilago maydis took up nearly three times as much ^{14}C-labeled oxycarboxin as the insensitive fungi Fusarium oxysporum f. sp. lycopersici and Saccharomyces cerevisiae. He concluded that insensitive fungi were not attacked by oxycarboxin because they were unable to take up the fungicide. However, Lyr et al. [53] have shown that the selectivity of oxycarboxin to Basidiomycetes is not due to greater uptake by sensitive fungi; differences in uptake are due simply to differences in the lipid contents of the cells. It is now known that there is a biochemical reason for the specificity of oxycarboxin. Basidiomycetes have an alternative electron transport pathway and oxycarboxin forms a complex with the ion-sulfur protein which is the terminal oxidase of this pathway [168].

Chloroneb, 1,4-dichloro-2,5-dimethoxybenzene, is a soil fungicide that controls some soilborne, seedling diseases. Fungi insensitive to chloroneb, take up slightly more ^{14}C-labeled fungicide than sensitive fungi, hence insensitivity is not due to inability to take up the fungicide [169]. A chloroneb-insensitive mutant of Ustilago maydis was also insensitive to dicloran, 2,6-dichloro-4-nitroaniline, diphenyl, naphthalene, and pentachloronitrobenzene [170]. This suggests that the fungicide acts in a similar way to the aromatic hydrocarbons [171]. Chloroneb appears to act by preventing sporidia from dividing normally [170].

XI. CONCLUSIONS

Amount of uptake may only estimate fungicidal activity approximately because it comprises the sum of the amount of fungicide bound to the active site, the amount accumulated unaltered at nonspecific sites, and the amount detoxified; low uptake may be due to impermeability. However, measurements of uptake are essential for a complete understanding of the mode of action of a fungicide and should complement all investigations in which the sites of accumulation and detoxication as well as the biochemical effects of fungicides are studied.

Despite the reservations mentioned above, the newer systemic fungicides are far more toxic on a weight basis than older compounds. This increase in toxicity which is due to their greater specificity, has unfortunately, been accompanied by the emergence of insensitive strains of normally sensitive fungi (see Chapter 12). Since increasing development costs may restrict the introduction of new compounds, ways must be found of prolonging the life of existing compounds. The future therefore poses challenging problems to all concerned with fungicide research.

REFERENCES

1. L. P. Miller, in Fungicides (D. C. Torgeson, ed.), Vol. 2, Academic Press, New York, 1969, Chap. 1.
2. B. von Schmeling and M. Kulka, Science, 152, 659 (1966).
3. C. J. Delp and H. L. Klöpping, Plant Disease Reptr., 52, 95 (1968).
4. C. E. Bracker, Ann. Rev. Phytopathol., 5, 343 (1967).
5. S. Bartnicki-Garcia, Ann. Rev. Microbiol., 22, 87 (1968).
6. H. Gershon, Proc. 2nd Intern. IUPA Congr. Pesticide Chemistry, Tel Aviv, 1972, Vol. 5, p. 363.
7. P. R. Hahadevan and U. R. Mahadkar, Indian J. Exptl. Biol., 8, 207 (1970).
8. D. V. Richmond and E. Somers, Proc. 2nd Intern. IUPAC Congr. Pesticide Chemistry, 1972, Vol. 5, p. 309.
9. H. Tedeschi, Cell Physiology, Academic Press, New York, 1974, p. 4.
10. M. S. Bretscher, in Perspectives in Membrane Biology (S. Estrada-O and C. Gitler, eds.), Academic Press, New York, 1974, p. 3.
11. P. L. Paine, L. C. Moore, and S. B. Horowitz, Nature, 254, 109 (1975).
12. J. Meldolesi, Phil. Trans. Roy. Soc. Ser. B, 268, 39 (1974).
13. C. E. Bracker and S. N. Grove, Protoplasma, 73, 15 (1971).
14. D. H. Northcote, Phil. Trans. Roy. Soc. Ser. B, 268, 119 (1974).
15. S. J. Singer and G. L. Nicholson, Science, 175, 720 (1972).
16. M. S. Bretscher, Science, 181, 622 (1973).
17. R. Collander, in Plant Physiology (F. C. Steward, ed.), Vol. 2, Academic Press, New York, 1959, Chap. 1.
18. J. L. Oschman, B. J. Wall, and B. L. Gupta, Symp. Soc. Exptl. Biol., 28, 305 (1974).
19. J. F. Sutcliffe and D. A. Baker, Plants and Mineral Salts, Edward Arnold, London, 1974, pp. 29-32.
20. W. J. Cram, in Ion Transport in Plants (W. P. Anderson, ed.), Academic Press, London, 1973, p. 419.
21. R. A. Leigh, R. G. Wyn Jones, and F. A. Williamson, in Ion Transport in Plants (W. P. Anderson, ed.), Academic Press, London, 1973, p. 407.
22. E. Somers, SCI (Soc. Chem. Ind., London) Monogr., 29, 243 (1968).
23. H. Wormald, The Brown Rot Diseases of Fruit Trees, Tech. Bull. No. 3, H.M.S.O., London, 1954.
24. E. Somers, Sci. Progr., 50, 218 (1962).
25. K. J. Bent, Ann. Appl. Biol., 60, 251 (1967).
26. E. C. Hislop, Ann. Appl. Biol., 60, 165 (1967).
27. E. Evans, in Systemic Fungicides (R. W. Marsh, ed.), Wiley, New York, 1972, Chap. 9.

28. S. H. Crowdy, in Systemic Fungicides (R. W. Marsh, ed.), Wiley, New York, 1972, Chap. 5.
29. M. Snel and L. V. Edgington, Phytopathology, 60, 1708 (1970).
30. A. Verloop, Proc. 2nd Intern. IUPAC Congr. Pesticide Chemistry, Tel Aviv, 1972, Vol. 5, p. 389.
31. P. B. Marsh, Phytopathology, 35, 54 (1945).
32. E. Somers and D. V. Richmond, Nature, 180, 798 (1957).
33. D. V. Richmond and E. Somers, Ann. Appl. Biol., 50, 33 (1962).
34. R. L. Wain and E. H. Wilkinson, Ann. Appl. Biol., 30, 379 (1943).
35. J. T. Martin and E. Somers, Nature, 180, 797 (1957).
36. E. Somers, Nature, 206, 216 (1965).
37. A. E. Dimond, in Systemic Fungicides (R. W. Marsh, ed.), Wiley, New York, 1972, Chap. 6.
38. P. Langcake and S. G. A. Wickins, J. Gen. Microbiol., 88, 295 (1975).
39. G. J. Bollen and A. Fuchs, Neth. J. Plant Pathol., 76, 299 (1970).
40. C. A. Drandarevski and A. Fuchs, Meded. Fac. Landbwetenschap. Rijksuniv. Gent, 38, 1525 (1973).
41. American Phytopathological Society, Committee on Standardization of Fungicidal Tests, Phytopathology, 37, 354 (1947).
42. W. Koch, Pestic. Sci., 2, 207 (1971).
43. J. L. Sherald, N. N. Ragsdale, and H. D. Sisler, Pestic. Sci., 4, 719 (1973).
44. D. V. Richmond, Appl. Microbiol., 19, 289 (1975).
45. R. S. Hammerschlag and H. D. Sisler, Pestic. Biochem. Physiol., 3, 42 (1973).
46. L. C. Davidse, Pestic. Biochem. Physiol., 3, 317 (1973).
47. D. V. Richmond and A. Phillips, Pestic. Biochem. Physiol., 5, 367 (1975).
48. D. V. Richmond, E. Somers, and P. F. Millington, Ann. Appl. Biol., 59, 233 (1967).
49. D. V. Richmond and R. J. Pring, J. Gen. Microbiol., 66, 79 (1971).
50. A. I. Zaki, J. W. Eckert, and R. M. Endo, Pestic. Biochem. Physiol., 3, 7 (1973).
51. D. E. Mathre, Pestic. Biochem. Physiol., 1, 216 (1971).
52. G. A. White, Biochem. Biophys. Res. Commun., 44, 1212 (1971).
53. H. Lyr, G. Ritter, and G. Casperson, Z. Allg. Mikrobiol., 12, 271 (1972).
54. M. D. Simons, Phytopathology, 65, 388 (1975).
55. R. J. Pring and D. V. Richmond, Physiol. Plant Pathol., 8, 155 (1976).
56. E. Sternlicht, D. Katz, and R. F. Rosenberger, J. Bacteriol., 114, 819 (1973).
57. A. D. Murray and D. K. Kidby, J. Gen. Microbiol., 86, 66 (1975).
58. T. W. Davies and D. A. Erasmus, Science Tools, 20, 9 (1973).
59. D. V. Richmond and D. J. Fisher, Advan. Microbiol. Physiol., 9, 1 (1973).

60. D. J. Fisher and D. V. Richmond, J. Gen. Microbiol., 57, 51 (1969).
61. D. J. Fisher, P. J. Holloway, and D. V. Richmond, J. Gen. Microbiol., 72, 71 (1972).
62. E. Somers and D. J. Fisher, J. Gen. Microbiol., 48, 147 (1967).
63. F. Haverkate, J. W. Brevoord, and A. Verloop, Pesticide Sci., 3, 1 (1972).
64. L. T. Richardson and G. D. Thorn, Can. J. Bot., 39, 531 (1961).
65. D. V. Richmond and E. Somers, Ann. Appl. Biol., 57, 231 (1966).
66. E. Somers, in Symposium Reinhardsbrunn, Akademie-Verlag, Berlin, 1967, p. 325.
67. L. P. Miller and N. Russakow, Phytopathology, 55, 1068 (1965).
68. T. Arai, Postepy Hig. Med. Doswiadczalnej, 28, 649 (1974).
69. E. Somers, Ann. Appl. Biol., 51, 425 (1963).
70. S. M. Hammond, P. A. Lambert, and B. N. Kliger, J. Gen. Microbiol., 81, 325 (1974).
71. A. Verloop and F. Haverkate, Proc. 2nd Intern. IUPAC Congr. Pesticide Chemistry, Tel Aviv, 1972, Vol. 5, p. 347.
72. J. G. Horsfall, Principles of Fungicidal Action, Chronica Botanica, Waltham, Mass., 1956, p. 21.
73. S. E. A. McCallan, H. P. Burchfield, and L. P. Miller, Phytopathology, 49, 544 (1959).
74. R. G. Owens, Ann. Rev. Phytopathol., 1, 77 (1963).
75. S. E. A. McCallan and L. P. Miller, Advan. Pest Control Res., 2, 107 (1958).
76. L. P. Miller and S. E. A. McCallan, Proc. Intern. Conf. Peaceful Uses Atomic Energy, Geneva, 1956, Vol. 12, p. 170.
77. R. G. Owens and H. M. Novotny, Contr. Boyce Thompson Inst., 20, 151 (1959).
78. E. Somers, Can. J. Bot., 36, 997 (1958).
79. R. G. Owens and G. Blaak, Contrib. Boyce Thompson Inst., 20, 459 (1960).
80. D. V. Richmond and E. Somers, Ann. Appl. Biol., 52, 327 (1963).
81. F. Haverkate, A. Tempel, and A. J. Den Held, Neth. J. Pl. Path., 75, 308 (1969).
82. D. V. Richmond and E. Somers, Ann. Appl. Biol., 50, 45 (1962).
83. E. Somers and R. J. Pring, Ann. Appl. Biol., 58, 457 (1966).
84. J. A. Bartz and J. E. Mitchell, Phytopathology, 60, 345 (1970).
85. N. J. Turner, Contrib. Boyce Thompson Inst., 24, 227 (1970).
86. J. Vörös, Phytopathol. Z., 54, 249 (1965).
87. M. A. El-Nakeeb and J. O. Lampen, J. Gen. Microbiol., 39, 285 (1965).
88. E. W. Westcott and H. D. Sisler, Phytopathology, 54, 1261 (1964).
89. F. J. Hills and L. D. Leach, Phytopathology, 52, 51 (1962).
90. W. J. Tolmsoff, Phytopathology, 52, 755 (1962).
91. J. A. Bartz and J. E. Mitchell, Phytopathology, 60, 350 (1970).

92. T. Nakanishi and H. Oku, Phytopathology, 59, 1761 (1969).
93. T. Nakanishi and H. Oku, Phytopathology, 59, 1563 (1969).
94. J. Ashida, Ann. Rev. Phytopathology, 3, 153 (1965).
95. S. G. Georgopoulos and C. Zaracovitis, Ann. Rev. Phytopathology, 5, 109 (1967).
96. I. S. Ross, Trans. Brit. Mycol. Soc., 64, 175 (1975).
97. J. Dekker, in Systemic Fungicides (R. W. Marsh, ed.), Wiley, New York, 1972, Chap. 8.
98. W. Greenaway and F. R. Whatley, Current Advan. Plant Sci., 20, 335 (1975).
99. R. G. Turner, in Symposium No. 9 of the British Ecological Society, Sheffield, 1968, p. 399 (1969).
100. W. Greenaway, J. Gen. Microbiol., 73, 251 (1972).
101. K. T. Huang, T. Misato, and H. Asuyama, J. Antibiotics (Tokyo), 17, 71 (1964).
102. J. Dekker, Proc. 2nd Intern. IUPAC Congr. Pesticide Chemistry, Tel Aviv, 1972, Vol. 5, p. 305.
103. S. G. Georgopoulos, C. Zafiratos, and E. Georgiadis, Physiol. Plant, 20, 373 (1967).
104. R. G. Owens and L. P. Miller, Contrib. Boyce Thompson Inst., 19, 177 (1957).
105. R. J. Lowry, A. S. Sussman, and B. von Böventer, Mycologia, 49, 609 (1957).
106. A. S. Sussman, B. von Böventer-Heidenhain, and R. J. Lowry, Plant Physiol. (Lancaster), 32, 586 (1957).
107. A. Rothstein and A. D. Hayes, Arch. Biochem. Biophys., 63, 87 (1956).
108. A. Rothstein, A. D. Hayes, D. Jennings, and D. Hooper, J. Gen. Physiol., 41, 585 (1958).
109. L. P. Miller, Phytopathology, 50, 646 (1960).
110. E. F. Gale, A. M. Johnson, D. Kerridge, and T. Y. Koh, J. Gen. Microbiol., 87, 20 (1975).
111. S. M. Hammond and B. N. Kliger, Proc. Soc. Gen. Microbiol., 1, 45 (1974).
112. N. J. Russell, D. Kerridge, and E. F. Gale, J. Gen. Microbiol., 87, 331 (1975).
113. L. Ferenczy, F. Kevei, J. Zsolt, J. Teren, and I. Berek, Acta Facultas Med. Univ. Brno, 37, 95 (1970).
114. F. J. Stutzenberger and J. N. Parle, J. Gen. Microbiol., 73, 85 (1972).
115. L. P. Miller and S. E. A. McCallan, Science, 126, 1233 (1957).
116. R. Tröger, Arch. Mikrobiol., 33, 186 (1958).
117. H. P. Burchfield and E. E. Storrs, Contrib. Boyce Thompson Inst., 18, 429 (1957).
118. H. C. Sikka, E. H. Schwartzel, J. Saxena, and G. Zweig, Chem. Biol. Interactions, 9, 261 (1974).

119. M. R. Siegel, J. Agr. Food Chem., 18, 823 (1970).
120. D. V. Richmond and E. Somers, Ann. Appl. Biol., 62, 35 (1968).
121. B. D. Nelson, Biochem. Pharmacol., 20, 749 (1971).
122. I. F. Brown and H. D. Sisler, Phytopathology, 50, 830 (1960).
123. E. Somers, Mededel. Landbouwhogeschool Opzoekingssta. Staat. Gent, 28, 580 (1963).
124. L. P. Miller, S. E. A. McCallan, and R. M. Weed, Contrib. Boyce Thompson Inst., 17, 173 (1953).
125. L. P. Miller and S. E. A. McCallan, Proc. 4th Intern. Congr. Crop Protect., Hamburg, 1957, Vol. 2, p. 1379.
126. M. Kottke and H. D. Sisler, Phytopathology, 52, 959 (1962).
127. D. Kerridge, J. Gen. Microbiol., 19, 497 (1958).
128. J. C. Riemersma, J. Pharm. Pharmacol., 18, 602 (1966).
129. J. C. Riemersma, J. Pharm. Pharmacol., 18, 657 (1966).
130. J. O. Lampen, Symp. Soc. Gen. Microbiol., 16, 111 (1966).
131. S. M. Hammond, P. A. Lambert, and B. N. Kliger, J. Gen. Microbiol., 81, 331 (1974).
132. B. De Kruijff and R. A. Demel, Biochem. Biophys. Acta, 339, 57 (1974).
133. J. F. Danielli and J. T. Davies, Advan. Enzymol., 11, 35 (1951).
134. E. Somers, Ann. Appl. Biol., 49, 246 (1961).
135. A. Rothstein, in Effects of Metals on Cells, Subcellular Elements and Macromolecules (J. Maniloff, J. R. Coleman, and M. W. Miller, eds.), Charles C. Thomas, Springfield, Illinois, 1970, Chap. 18.
136. G.-F. Fuhrmann and A. Rothstein, Biochim. Biophys. Acta, 163, 325 (1968).
137. A. Rothstein, Federation Proc., 18, 1026 (1959).
138. H. Passow, A. Rothstein, and T. W. Clarkson, Pharmacol. Rev., 13, 185 (1961).
139. J. W. Eckert, M. L. Rahm, and M. J. Kolbezen, J. Agr. Food Chem., 20, 104 (1972).
140. J. W. Eckert, M. L. Rahm, and B. Bretschneider, Phytopathology, 60, 1290 (1970).
141. J. W. Eckert, Proc. 2nd Intern. IUPAC Congr. Pesticide Chemistry, Tel Aviv, 1972, Vol. 5, p. 241.
142. D. V. Richmond, E. Somers, and C. Zaracovitis, Nature, 204, 1329 (1964).
143. R. J. Lukens and J. G. Horsfall, Phytopathology, 57, 876 (1967).
144. C. Hansch, E. J. Lien, and F. Helmer, Arch. Biochem. Biophys., 128, 319 (1968).
145. C. Hansch and E. J. Lien, J. Med. Chem., 14, 653 (1971).
146. E. J. Lien, J. Agr. Food Chem., 17, 1265 (1969).
147. D. V. Richmond, M.Sc. Thesis, University of Bristol, 1961.
148. E. J. Lien, C-T. Kong, and R. J. Lukens, Pestic. Biochem. Physiol., 4, 289 (1974).

149. R. J. Lukens and H. D. Sisler, Phytopathology, 48, 235 (1958).
150. M. R. Siegel and H. D. Sisler, Phytopathology, 58, 1129 (1968).
151. M. R. Siegel, Pestic. Biochem. Physiol., 1, 225 (1971).
152. M. R. Siegel, J. Agr. Food Chem., 18, 819 (1970).
153. M. R. Siegel, J. Agr. Food Chem., 18, 823 (1970).
154. M. R. Siegel, Pestic. Biochem. Physiol., 1, 234 (1971).
155. P. G. Vincent and H. D. Sisler, Physiol. Plant, 21, 1249 (1968).
156. N. J. Turner and R. D. Battershell, Contrib. Boyce Thompson Inst., 24, 139 (1969).
157. N. J. Turner and R. Battershell, Contrib. Boyce Thompson Inst., 24, 203 (1970).
158. N. J. Turner, Contrib. Boyce Thompson Inst. Plant Res., 24, 357 (1971).
159. R. W. Tillman, M. R. Siegel, and J. W. Long, Pestic. Biochem. Physiol., 3, 160 (1973).
160. J. W. Long and M. R. Siegel, Chem.-Biol. Interactions, 10, 383 (1975).
161. F. Haverkate, A. Tempel, and C. W. Pluijgers, Rec. Trav. Chim. Pays-Bas, 90, 641 (1971).
162. G. P. Clemons and H. D. Sisler, Pestic. Biochem. Physiol., 1, 32 (1971).
163. J. R. Decallonne and J. A. Meyer, Phytochem., 11, 2155 (1972).
164. D. Gottlieb and K. Kumar, Phytopathology, 60, 1451 (1970).
165. W. S. Bertaud, I. M. Morice, D. W. Russell, and A. Taylor, J. Gen. Microbiol., 32, 385 (1963).
166. L. V. Edgington, G. S. Walton, and P. M. Miller, Science, 153, 307 (1966).
167. D. E. Mathre, Phytopathology, 58, 1464 (1968).
168. H. Lyr, T. Schewe, W. Müller, and D. Zanke, in Symposium Reinhardsbrunn, Akademie-Verlag, Berlin, 1975, p. 153.
169. W. K. Hock and H. D. Sisler, Phytopathology, 59, 627 (1969).
170. R. W. Tillman and H. D. Sisler, Phytopathology, 63, 219 (1973).
171. S. G. Georgopoulos and C. Zaracovitis, Ann. Rev. Phytopathol., 5, 109 (1967).
172. A. Nachmias and I. Barash, J. Gen. Microbiol., 94, 167 (1976).
173. L. C. Davidse, Pestic. Biochem. Physiol., 6, 538 (1976).
174. L. C. Davidse and W. Flach, J. Cell. Biol., 72, 174 (1977).

Chapter 7

INHIBITION OF FUNGAL CELL WALL SYNTHESIS AND CELL MEMBRANE FUNCTION

Tomomasa Misato and Kazuo Kakiki

Fungicide Laboratory
Institute of Physical and Chemical Research
Wako-Shi, Saitama, Japan

I. INHIBITORS OF FUNGAL CELL WALL SYNTHESIS

A. Introduction

Bacterial, yeast, fungal, and plant cells possess a wall which is very differentiated and is composed of characteristic macromolecules such as bacterial peptidoglycan, fungal chitin, and plant cellulose. Animal cells do not have such a structure. It has been confirmed that the primary site of action of penicillin [1-4] is cell-wall synthesis. Recently we have demonstrated that the polyoxin antifungal antibiotics inhibit fungal cell-wall synthesis [5-8]. Microbial cells with a defective wall as a result of the action of these antibiotics lose resistance to osmotic pressure, and bursting results in cell death.

B. Structure of Cell Wall

The gross morphology of fungi depends to a great degree upon the conformation of their cell wall. The wall provides stability to the shapes of vegetative hyphae and various reproductive structures. Electron microscopic observations [9-13] and biochemical analyses [14] have revealed that the cell wall structure of fungi consists of two systems: one is a microfibrillar system which is embedded in the other, referred to as an amorphous matrix. The fibrous texture of fungal walls was observed for the first time by Frey-Wyssling and Mühlethaler [15] in the sporangiophore of Phycomyces. Following the observations of the fibrillar structure in the Phycomyces wall, the presence of a fibrous structure was confirmed by Roelfsen [16] and Middlebrook and Preston [17]. The presence of these structures was recognized in other fungi as well as in yeast [18] and bacteria [19]. This fibrillar substance in the fungal cell wall was identified as chitin or cellulose corresponding to peptidoglycan in bacteria. The presence of chitin in yeast cell walls has been observed by many investigators but in yeast cells chitin appears to distribute specifically in the bud-scar regions [20].

Several electron microscopic observations [9-13] indicated that the cell wall of fungi is a multilayer structure as in Phycomyces. Agar and Douglas [11] also observed a multilaminate wall structure of Saccharomyces in which the laminations appeared to be prominent, particularly in the bud-scar region.

Some fungi, but not all, exhibit dimorphism such as yeast and mycelia-like forms in their growth process. The form is determined by nutritional and environmental conditions [21, 22]. Biochemical processes involved in the determination of morphological forms are little understood. The occurrence of an active protein disulfide reductase in the yeast form and its low activity in the mycelial form are consistent with the conclusion that the enzyme is necessary to provide sufficient plasticity for bud formation [23, 24]. The composition of the cell wall of this species may be dependent on the morphological form. Paracoccidioides brasitiensis shows thermal dimorphism. It is yeast-like and mycelial at 37° and 20°C, respectively [25]. Both forms contain chitin and glucan as the major polysaccharides. Only the mycelial form contains appreciable quantities of galactomannan, possibly a peptidogalactomannan [26]. The cell wall of Saccharomyces cerevisiae in the log phase can be removed by lytic enzymes from snail gut. In contrast, the yeast cell wall in the stationary growth phase cannot be effectively removed by the enzyme treatment [27]. Thus, the architecture of fungal and yeast cell walls change with the nutritional and cultural conditions.

C. Chemical Composition of Fungal Cell Wall

Isolation of fungal cell walls is usually accomplished by mechanical breakage of cells followed by several differential centrifugations and/or enzymic treatment. It is easy to obtain a relatively pure wall fraction from fungi, but it is difficult to separate the multilayer wall from the cytoplasmic membrane. Therefore protein and lipids contents of the cell wall fraction are often variable.

In the fungi, as in green plants, the cell wall is mainly composed of polysaccharides. Since the early investigation of Van Wisselingh [28], a number of investigators [29-34] have studied chemical constitution of fungal cell walls. Until quite recently, however, studies in the area have made little progress, due to a lack of analytical methods of extraction, purification, crystallization, and determination of higher-branching polysaccharides. A variety of polymers have been detected in fungal cell walls. The ubiquitous occurrence of chitin in fungal cell walls was revealed by the classical investigation of Van Wisselingh [28], who worked out cytochemical test procedures involving alkaline conversion of chitin to chitosan.

Although the cell wall of fungi is a dynamic complex, it is usually composed of 80-90% polysaccharides with most of the remainder consisting of protein [35-37] and lipids [38-41]. Wide deviations from these values are rare: however, the cell wall of the yeast *Saccharomyces guttalata* contains 40% protein [42]. Cell walls of fungi lack muramic acid containing peptidoglycan, teichoic acid, and lipopolysaccharide and are therefore unlike those of bacteria.

Polysaccharides of fungal cell walls are built from a variety of sugars. It has been reported that at least 11 monosaccharides occur in fungal walls, but only D-glucose, *N*-acetylglucosamine, and D-mannose, are consistently found in most fungi. Their relative proportions, however, vary remarkably from traces in certain organisms to principal components in others.

Physically, the fungal cell wall is a fabric of interwoven microfibrils embedded in or cemented by amorphous matrix substances. Chitin [43-45] and cellulose [46-48] are well known as the microfibrous or skeletal components of the wall of the vast majority of fungi, whereas, in most true yeast, the skeletal part is seemingly composed of "yeast" glucan [49-51] which differ from cellulose and chitin.

In recent investigations, evidence for the presence of unusual polysaccharides has come to light. Korn and Northcote [52] demonstrated the occurrence of glucosamine in forms distinct from chitin and chitosan in isolated cell walls of baker's yeast. Harold [53] has isolated a polymer of galactosamine from the walls of *Neurospora crassa*. This polymer is capable of binding polyphosphate. Brown and Linderberg [54] isolated a heteropolymer of mannose, galactose, and glucuronic acid from the cell walls of Pullularia pullulans. The most typical example of a fungal cell wall containing uronic acid is that of Mucor rouxii in which polymers of D-glucuronic acid account for 25% of the sporangiophore wall [55]. Skucas [56] found that Papillale and Allomyces sporangia were composed of pectic acid material including D-galacturonic acid.

Glucans composed of glucose residues but differing from cellulose are common constituents in fungal cell walls. "Yeast" glucan [49-51] has been investigated most intensively. The literature on this subject was reviewed by Nickerson et al. [57]. "Yeast" glucan can be crystallized by treatment with dilute acid. The acid conversion of glucan (hydroglucan) is accompanied by the formation of aggregates of microfibrils [58]. Nickerson and co-workers [57] extracted polysaccharide-protein complexes from isolated cell walls of Candida albicans. Mundkur [59] observed a thin layer of mannan which merged into a relatively thick layer of glucan in Saccharomyces cell walls. Eddy [60] demonstrated the presence of phosphomannan which constituted outer and inner strata of the yeast cell wall. His finding was supported by studies on the action of papain which brought about a release of mannan associated with nondialyzable phosphorous. Another glucan detected as a fungal cell wall constituent is callose, which is widely observed throughout the plant kingdom. Mangin [61] reported that this glucan was soluble in dilute alkali and had an affinity for basic dyes. The affinity of callose for basic dyes seems to be due to phosphate groups present in this polysaccharide but the details are not clear. Other polysaccharides have been detected in some fungal cell walls. The literature on these has been recently reviewed by Gander [62].

The presence of other types of cell wall polysaccharides has been suggested on the basis of the detection of monosaccharides in the hydrolysate of isolated cell walls. Galactose, fucose, xylose, and 6-deoxyhexose are found in all major classes of fungi. However, any polysaccharides containing these monosaccharides has not yet been characterized in detail.

Bacterial cell surface polysaccharides and peptidoglycans have been investigated extensively and the details have been reviewed [63-64]. In fungal cell walls, details of the mechanisms of biosynthesis and crosslinkage of polysaccharides are not obvious. Using crude enzyme systems, some studies indicated polymerizations of chitin from UDP-N-acetylglucosamine [65], mannan from guanidine diphosphomanose (GDP-mannose) [66], and β-1,3-glucan from uridine diphosphoglucose (UDP-glucose) [67].

D. Inhibition of Biosynthetic Enzymes by Fungicides

1. Polyoxins

Polyoxins [68-74] are nucleotide antibiotics consisting of at least 12 closely related active components. The antibiotics inhibit the growth of some filamentous fungi but are inactive against bacteria and yeast. Polyoxins have been widely used in Japan for the protection of plants from pathogenic fungi such as Alternaria kikuchiana, Pellicularia sasakii, and Cochliobolus miyabeanus.

Polyoxins do not affect the rate of spore germination, but they induce distortion of germ tubes and finally inhibit the growth of germ tubes of Neurospora crassa. Polyoxin D did not inhibit germination of C. miyabeanus spores, but caused distortion and rupture of the germ tubes. In the presence of 20% sucrose and polyoxin D, germ tubes were stabilized and growth continued but giant protoplast-like structures were formed. Diameters of "hyphae" induced by polyoxin D were 3 to 10 times those of normal mycelia. These induced structures were sensitive to osmotic shock and lysed in distilled water. When washed and transferred to liquid or solid media free of polyoxin, most of the protoplast-like structures converted to normal mycelial growth [75].

Using growing cells of C. miyabeanus, Sasaki et al. [5] found that polyoxin D did not inhibit the incorporation of radioactive precursors into protein and nucleic acids but it did inhibit the incorporation of $[^{14}C]$glucosamine into a cell-wall fraction, which was defined as soluble fraction after chitinase digestion and hydrolysis in 6 N HCl for 6 hr at 100°C. This finding was confirmed by investigations of Neurospora crassa [75] and Pyricularia oryzae [6, 76]. In contrast with above experiments, the uptake of $[^{14}C]$glucosamine into UDP-N-acetylglucosamine increased markedly in the presence of polyoxin D [6, 75-76] (see Fig. 1). When $[^{14}C]$glucose was used instead of $[^{14}C]$glucosamine, neither the accumulation of UDP-N-acetylglucosamine nor the inhibition of incorporation into the cell walls occurred in the presence of polyoxin D [75].

Polyoxin D did not inhibit UDP-N-acetylglucosamine pyrophosphorylase in a cell-free system isolated from N. crassa, but acted as an inhibitor of chitin synthetase competing with UDP-N-acetylglucosamine [75]. The Michaelis constant (K_m) of chitin synthetase from N. crassa for UDP-N-acetylglucosamine was calculated to be 1.43×10^{-3} M and the inhibitor constant (K_i) for polyoxin D was 1.40×10^{-6} M [75]. The effects of various N-aminoacyl derivatives of polyoxin C and other polyoxins on chitin synthetase from Pyricularia oryzae were investigated and the value of binding affinity ($-\Delta G_{bind}$) of the compounds for the enzyme were calculated from the K_i values [8]. In addition, the values of partial binding affinity ($-\Delta g$) for the several atoms, atomic groups, and moieties of the polyoxin J molecule were estimated as shown in Fig. 2 [8]. These results suggest that the carbamoyl polyoximid acid moiety of the antibiotic acts to stabilize the polyoxin-enzyme complex, and that the oxygen atom at C-1", the amino group at C-2", the hydroxyl groups at C-3" and C-4", the aliphatic carbon chain, and the terminal carbomyloxy group are involved in the stabilization [8]. The pK_i-pH plots for the ionized amino group at C-2" position indicate that this group may play a very important role in the binding of polyoxins to chitin synthetase [77]. The investigations on various nucleotides and nucleotide sugar suggest that this enzyme may have a specific binding site for the

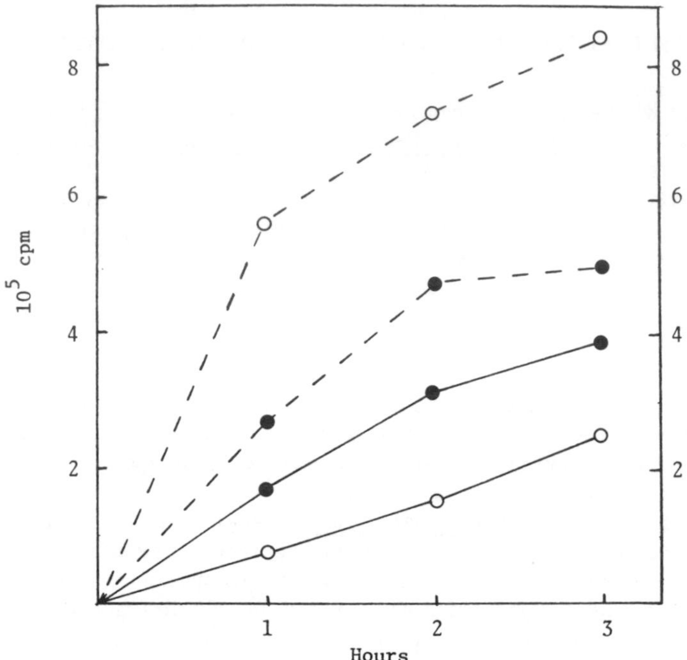

FIG. 1. Effects of polyoxin D on the incorporation of [^{14}C]glucosamine into chitin and the accumulation of UDP-*N*-acetyl-[^{14}C]glucosamine in *N. crassa*. ●, no polyoxin D; ○, 0.19 mM polyoxin D; - - - -, UDP-*N*-acetyl-[^{14}C]glucosamine; ———, chitin; cpm, counts per minute.

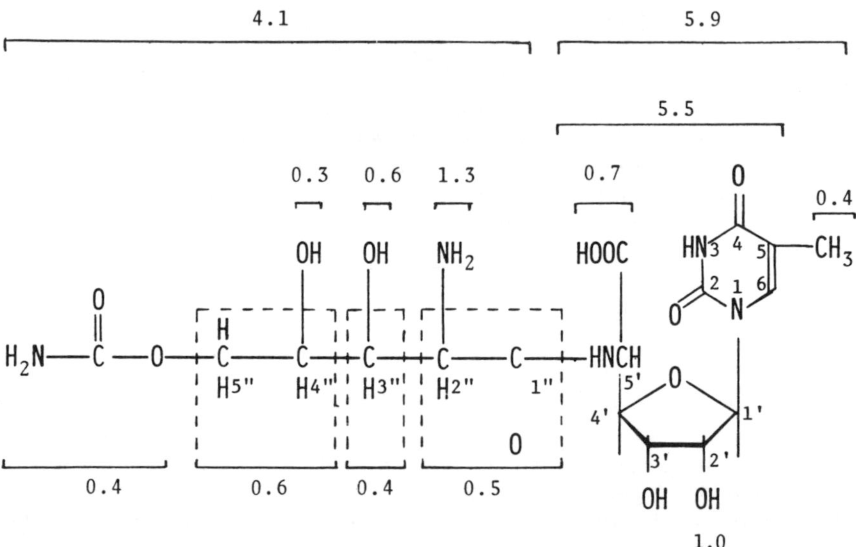

FIG. 2. Values of partial binding affinity, -Δg (in Kcal/mol at 25° C) of polyoxin J for chitin synthetase prepared from *P. oryzae*.

uridine moiety of UDP-*N*-acetylglucosamine, and for the pyrimidine nucleoside moiety of polyoxins [8, 77].

Polyoxins do not affect the growth of yeast or yeast protoplast. However, Keller and Cabib [78] found that chitin synthetase prepared from *S. cerevisiae* was inhibited by polyoxins. The same result was obtained with polyoxin-resistant strains of *Alternaria kikuchiana*. A number of strains of *A. kikuchiana* were isolated, and the effect of polyoxins on the chitin synthetic systems was investigated. Neither incorporation of [^{14}C]glucosamine into the cell wall, nor accumulation of UDP-*N*-acetyl-[^{14}C]glucosamine was inhibited by polyoxins. On the other hand, chitin synthetases prepared from the resistant strains were as sensitive to polyoxins as the enzymes from sensitive strains. Therefore, in yeast and resistant fungi polyoxins may be unable to penetrate into the cells [79].

Mitani and Inoue [80] found that glycyl-L-alanine, glycyl-D,L-valine and D,L-alanylglycin protected mycelial growth of *Pellicularia sasakii* from inhibition by polyoxins. Hori et al. [77] investigated the competitive action of dipeptides with polyoxins. Abnormal accumulation of UDP-*N*-acetylglucosamine and inhibition of chitin synthesis by polyoxins in intact cells of *P. oryzae* were both antagonized by the dipeptides. However, the peptides could not antagonize the inhibition of chitin synthetase by polyoxin B in a cell-free system. Therefore, the peptides may act as antagonist to the transport of polyoxins into the cell.

2. Kitazin-P and Other Organophosphorus Fungicides

Kitazin (*O*,*O*' diethyl-*S*-benzylphosphorothiolate) and related organophosphorus fungicides have been widely used as rice blast disease protectants in the field since 1965. The chemical structure of kitazin is very similar to insecticidal organophosphorus compounds which inhibit acetylcholinesterase in insects. Kitazin was the first organophosphorus compound used for rice blast control, but it has been replaced by kitazin P (see Fig. 3). Kitazin P is very soluble in water as compared with other organophosphorus fungicides and this proves to be advantageous for rice blast control; kitazin P can be absorbed through roots and leaf sheaths, and translocated to upper parts of the plant attacked by the rice blast fungus, *Pyricularia oryzae*.

Kitazin P and related organophosphorus fungicides inhibit the growth of filamentous fungi such as *Pyricularia oryzae*, *Cochliobolus miyabeanus*, and *Pellicularia sasakii* but are inactive against bacteria and yeasts.

Kakiki et al. [81], using resting cells of *P. oryzae*, found that kitazin did not inhibit the respiration and the incorporation of radioactive precursors

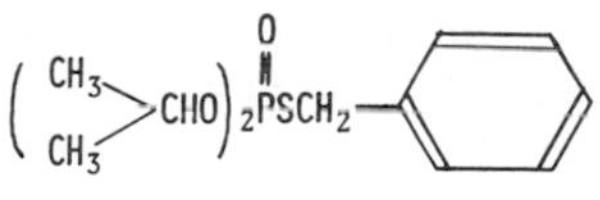

$$\left(\begin{matrix}CH_3\\CH_3\end{matrix}\!>\!CHO\right)_2\overset{O}{\overset{\|}{P}}SCH_2-C_6H_5$$

FIG. 3. Chemical structure of kitazin P.

into nucleic acids and protein, but it did inhibit incorporation of [^{14}C]glucosamine into the cell wall (see Fig. 4). Uchino et al. [82] found that kitazin caused swelling at the tips of growing mycelia in solid medium. Sudo and Tamari [83] investigated the composition of P. oryzae cells treated with kitazin and found that the carbohydrate content was remarkably lower than untreated mycelial cells. Kitazin induced slight leakage of small molecules from P. oryzae cytoplasm at higher concentrations [81].

All organophosphorus compounds put to practical use are phosphorothiolates. Phosphate-type compounds also have high inhibitory activities against mycelial growth, but thionate and dithionate type compounds do not show inhibitory activity. Therefore, fungal mycelia may not have the enzymes which catalyzed the conversion of thionates or dithionates to phosphates or thiolates, respectively [84].

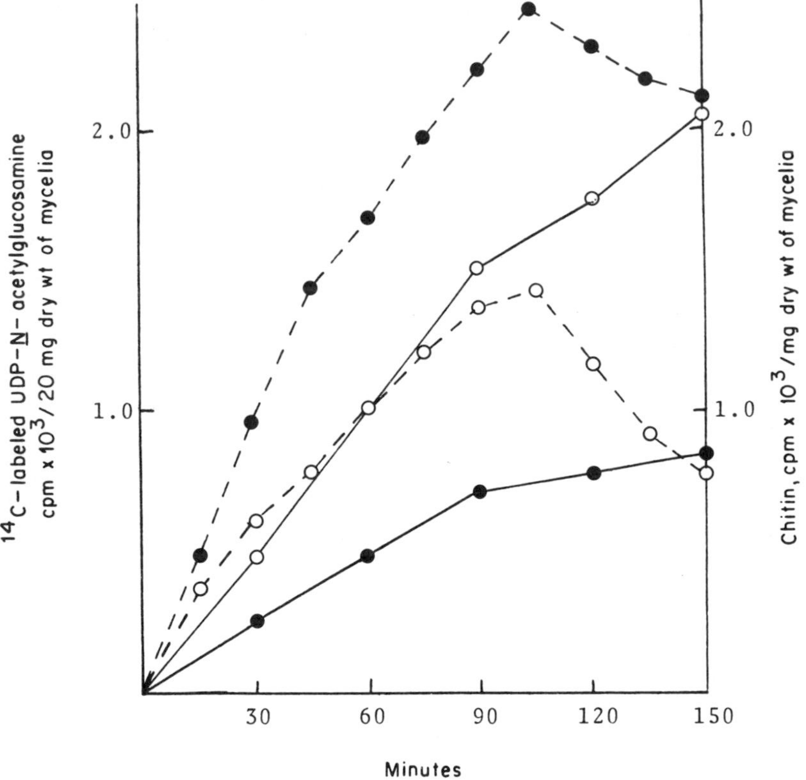

FIG. 4. Time course of the accumulation of UDP-N-acetyl[^{14}C]glucosamine and the incorporation of [^{14}C]glucosamine into chitin in P. oryzae. ●, kitazin P, 65 ppm; ○, control; - - - -, accumulation of UDP-N-acetyl-[^{14}C]glucosamine; ———, incorporation of [^{14}C]glucosamine into chitin; cpm, counts per minute.

Thiolate and phosphate analogues both inhibit incorporation of [^{14}C]glucosamine into the cell wall and cause marked abnormal accumulation of UDP-N-acetylglucosamine. The same results were obtained in mycelia treated with hinosan (O-ethyl-S,S'-diphenyl phosphorodithiolate) and inezin (O-ethyl-S-benzyl phosphorothiolate). On the other hand, ineffective analogues of thionates and dithionates do not cause such accumulation [84]. Kitazin P inhibits chitin synthetase prepared from P. oryzae, acting as noncompetitive inhibitor with UDP-N-acetylglucosamine [85].

It has been suggested that the antiacetylcholinestelase activity of organophosphorus insecticides is related to their Hamett constant. No correlation was found between chitin synthetase inhibitory activity and hydrolysis constant using inezin and its analogues [86]. Investigations using kitazin and its analogues supported results obtained with inezin. Inhibitory activities of kitazin analogues varied with length of alkoxy group or benzyl group [85] (Table 1). These results suggest that their inhibitory activities vary according to the full length of the molecule rather than the hydrolysis constant.

TABLE 1

Inhibition of Chitin Synthetase Prepared from P. Oryzae by Kitazin P and its Analogues

$$(RO)_2\overset{O}{\overset{\|}{P}}\cdot X(CH_2)_n C_6H_5$$

R	X	n	Concentration required for 50% inhibition, ppm
CH_3	S	1	645
C_2H_5	S	1	260
n-C_3H_7	S	1	47.5
n-C_4H_9	S	1	32.5
n-C_5H_{11}	S	1	52.5
n-C_6H_{13}	S	1	125
CH_3	O	1	1,850
C_2H_5	O	1	750
n-C_3H_7	O	1	105
n-C_4H_9	O	1	32.5

TABLE 1 (Cont.)

R	X	n	Concentration required for 50% inhibition, ppm
n-C_5H_{11}	O	1	52.5
n-C_6H_{13}	O	1	75.0
CH_3	S	2	360
CH_3	S	3	92.5
CH_3	S	4	62.5
CH_3	S	5	42.5
CH_3	O	2	1,100
CH_3	O	3	550
CH_3	O	4	300
C_2H_5	S	2	95.0
C_2H_5	S	3	62.5
C_2H_5	S	4	30.0
C_2H_5	S	5	30.0
C_2H_5	O	2	400
C_2H_5	O	3	135
C_2H_5	O	4	65.0
i-C_3H_7	S	1	75.0

II. INHIBITORS OF CELL MEMBRANE FUNCTION

A. Cell Membrane

Cells are enclosed with a number of surface layers. The number and appearances of these layers differ from one species to another. The outermost layer is recognized as a cell wall which is composed of polysaccharides and glycoproteins. The innermost layer has a characteristic triple-layer appearance which is referred to as a cytoplasmic or plasma membrane. Isolated membranes are composed of about 40% lipids and 60% protein. There is always difficulty in preparing membranes free of contamination with nonmembrane materials. Methods for the preparation of bacterial membranes have been reviewed by Salton [87] and Steck [88]. There are no well-established techniques for preparation of fungal cell membranes. The best preparations so far obtained are those from mycoplasmas and certain gram-positive bacteria. Cell walls and membranes of gram-positive bacteria can be dissociated readily, and contamination with wall material is minimized in the membrane preparation. Mycoplasmas are surrounded by only a single membranous layer. They are devoid of a rigid cell wall or envelope which protects the cytoplasmic membrane from osmotic pressure. These organisms also lack intracellular membranes such as mesosomes. Mycoplasma membrane fractions can be prepared relatively free from cytoplasmic contaminations after gentle cell disruption by osmotic shock, but it is difficult to obtain very high purity (more than 95%).

The cell envelopes of gram-negative bacteria consist of three layers: the cytoplasmic or inner layer membrane, the peptidoglycan (cell wall), and the outer membrane. With gram-negative bacteria, it is difficult to separate the inner from the outer membrane.

Salton [87] has summarized literature on the composition of 13 membrane preparations from mycoplasma and gram-positive bacteria. The preparations contain 6-37% lipids, 40-85% protein, 0.2-19% hexose, and 0.8-15% RNA on the basis of dry weight. Common lipids constituents of membranes are phospholipids, glycolipids, spinogolipids, and sterols in eukaryotic cells. These compounds have both polar and nonpolar regions. Such structures are well suited for the formation of interfaces between a polar solvent and the hydrophobic region in the lipid molecules.

There is no general characteristic of the lipid composition of bacteria. Wide variations in the nature of phospholipids and fatty acids are found in different species. In some species the phospholipid fraction consists of one or two components (*Azotobacter* sp. contains phosphatidylethanolamine and *Staphylococcus aureus* contains only phosphatidylglycerol and diphosphatidylglycerol) [87]. On the other hand, *Pseudomonas* sp. may contain four major phospholipids: phosphatidylglycerol, phosphatidylethanolamine, phosphatidylserine, and phosphatidylcholine. Glycolipids contain mono-, di-, and trisaccharides as their carbohydrate moiety [89]. Both gram-positive and gram-negative bacteria contain cyclopropane acids from negligible amounts to 20% of the total lipid [90]. The fatty acids in some gram-negative species consist mainly of even-numbered saturated acids (C-16, 18, and 20), while gram-positive species contain a high proportion of branched C-15 and C-17 chain acids [91].

Our knowledge of membrane protein is as yet rudimentary; clearly, the difficulties are attributed to cytoplasmic contamination. Nevertheless, membrane proteins are currently under intensive and fruitful investigation and the serious methodological problems have now been overcome. Up to 50 distinct proteins associated with the envelope of *E. coli* have been demonstrated by sodium dodecylsulfate-polyacrylamide gel electrophoresis (SDS-PAGE) [92]. A number of enzymes, particularly those participating in the electron-transport system, are located in membrane preparations from gram-positive bacteria.

Machtinger and Fox [93] summarized membrane-associated enzymes and membrane structural proteins of gram-positive or gram-negative bacteria. Concerning the localization of biologically active membrane proteins in inner and outer membranes, a definitive bias to one of these fractions is found. The inner membrane is apparently the site of almost all enzymes involved in oxidative phosphorylation [94] and of "carrier" proteins [95] involved in membrane transport systems.

Bacterial membranes appear to be free steroids, whereas membranes from eukaryotic cells in general contain steroids as essential structural components [96].

Mycoplasmas have proved to be useful for studies of the influence of cholesterol and fatty acid compositions on the physical properties and functions of membranes [97-98]. The effect of fatty acid composition and cholesterol content on the permeability of glycerol through cell and derived liposomes has been studied with Mycoplasmia laidlawii B. The rate of glycerol permeation at any temperature between 0° and 25°C increased as the fatty acid supplement varied from elaidic to oleic or linoleic acids. On the other hand, addition of cholesterol decreased the rate of permeation at any given temperature in the range of 10 to 40°C [99]. The decrease in permeability resulting from the addition of cholesterol depends upon monolayer of extracted phospholipids [100].

It should be emphasized that microbial cell membranes like endoplasmic systems are not a stable structure but undergo a dynamic modification particularly during growth and division of the cell.

B. Transport Across Membrane

The biological membrane serves as osmotic barrier between the intracellular and external media. The barrier is permeable to water but other metabolic substances and drugs pass through the membrane by specific mechanisms. The nature of the mechanisms for transport of specific substance may vary according to cell type.

Some drugs act on the cell surface to alter its physical and chemical states and thereby affect membrane permeability, but most drugs act upon enzyme activities in the cell interior. Therefore, many drugs must pass through the cell membrane in order to exert their activity in the cell. Investigations of membrane transport have been mainly achieved in bacterial and mammalian cells. Transport processes of bacteria are usually divided into four classes:

1. Passive (Simple) Diffusion

A substance diffuses through the membrane according to concentration or electrical gradients. Lipid-soluble drugs may diffuse freely through the cell membrane by their concentration gradients. Examples of passive diffusion through microbial membrane appear to be rare.

2. Facilitated Diffusion

Facilitated diffusion, suggested by Osterhout [101], is the simplest and commonest type of carrier process. A substance is insoluble in the hydrophobic barrier but reacts with a carrier within the membrane to form a carrier-substance complex which can pass through the membrane and release the free substance at the hydrophobic outer layer. The energy for translocation is provided by the solute concentration gradients and no metabolic energy is required. The carrier affects only the rate of movements and not the final equilibrium. Such a concept would allow higher efficiency and greater possibility for regulation than simple diffusion. Kaback [102] suggested that the accumulation of certain sugars by bacteria represented a form of facilitated diffusion.

3. Active Transport

If metabolic energy is coupled with the facilitated diffusion and there is an accumulation of the transported substance, the transport system is active. Inhibition of energy generation or coupling results in immediate and direct inhibition of transport. Transport of amino acids [103] and lactose [104] in E. coli are examples of this process.

4. Group Translocation

In contrast to both facilitated diffusion and active transport, group translocation involves solute modification mediated by a carrier during the transport process. This mechanism suggests that the carrier molecules behave like enzymes catalyzing group-transfer reactions. In contrast to the active transport for lactose in E. coli [104], many sugars enter bacterial cells by group transfer. In many bacteria, this process has been shown to involve the phosphorylation of sugars as they pass through the membrane. The phosphoenolpyruvate-glycose phosphotransferase system (PTS) mediates this uptake process [105].

C. Components of the Transport System

Theories of facilitated and active transports evoke the existence of carriers without reference to the nature or complexity of these substances. The general study on carrier transport in bacterial systems has shifted from an indirect kinetic examination to a genetic and biochemical dissection. In the past few years, this approach has resulted in considerable progress toward an understanding of molecular components involved in transport systems. The characterization of the phosphoenolpyruvate-glycose phosphotransferase system, and the isolation and characterization of the M protein and many binding proteins are major contributions in this field.

The first experimental advance was made in 1957 by Cohen and Monad [106] during the course of studies of cryptic mutants of E. coli unable to utilize lactose as a carbon source. These mutants possessed a normal component of β-galactosidase but lactose was unable to be incorporated into the cell since a specific component of the transport system was missing. This component is formed in wild-type cells when protein synthesis is induced in the presence of a suitable galactoside. The presence of a specific and inducible protein which forms a part of the transport system has been established and this component is called "galactoside permease or M protein" [107].

In recent years, a number of attempts have been made to demonstrate directly the existence of such proteins. Fox and Kennedy [108] suggested the presence of a transport protein, M protein, which is a product of the lactose y gene. They found that the "galactoside permease or M protein" in E. coli is inhibited by N-ethylmaleimide and can be protected from this inhibition by thiodigalactoside. This experiment suggests that the sulfhydryl inhibitor reacts at some substrate-binding site for which thiodigalactoside has a higher affinity. Fox and Kennedy proposed the use of this substrate-protection phenomenon to label specifically the permease protein. In principle, cells were treated with N-ethylmaleimide in the presence of thiogalactoside, which would react with all SH groups other than those of the M protein and with N-ethylmaleimide in the absence of the protective agent. The technique of labeling with radioactive N-ethylmaleimide has been extended by Johnes and Kennedy [109] for use in the isolation and characterization of the M protein. By using this technique the following finding were obtained [108]: The M protein is located in the membrane fraction of induced cell. It is only induced by the lactose operon and is always present in lactose constitutive strains.

The M protein isolated is a single polypeptide, which has a molecular weight of about 35,000 and contains the active sulfhydryl group with which N-ethylmaleimide reacts. This M protein is the first isolated permease [107].

Application of osmotic shock to gram-negative bacteria results in impairment of a number of transport activities and in release of proteins into shock medium [110]. Some of these proteins have specific binding affinities for transport substrates; arabinose-, arginine-, galactose-, histidine-, leucine-, phosphate-, and sulfate-binding proteins were found in osmotic shock medium of E. coli [111]. In each case, substrate binding is highly specific. The binding proteins for sulfate [112] and leucine [113] have been crystallized. These binding proteins are apparently the permease for each of the substrates.

The phosphoenolpyruvate-glycose phosphotransferase system (PTS) was first described by Kunding et al. [105] as an enzyme system from E. coli.

The system can phosphorylate many sugars according to the general reaction shown as follows:

$$\text{Sugar} + \text{phosphoenolpyruvate} \xrightarrow[\text{PTS}]{\text{Mg}^{2+}} \text{sugar-PO}_3^{2-} + \text{pyruvate}$$

This reaction is specific for phosphoenolpyruvate, although a variety of sugars can serve as acceptors. The enzyme system has been demonstrated in many bacterial species including E. coli [105], S. aureus [114], and Salmonella typhimurium [115]. Enzyme I, histidine protein (HPr), factor III, and enzyme II are the four protein components of this enzyme system that have been identified [116]. It has been possible to show that the reaction is the sum of at least two separate reactions as follows:

$$\text{PEP} + \text{HPr} \underset{\text{Mg}^{2+}}{\overset{\text{I}}{\rightleftharpoons}} \text{phospho-HPr} + \text{pyruvate}$$

$$\text{phospho-HPr} + \text{sugar} \xrightarrow[\text{III}]{\text{II}} \text{sugar-P} + \text{HPr}$$

The bulk of enzyme I and HPr are found in the soluble fraction and have been purified to a nearly homogeneous state. The HPr has a molecular weight of about 9,100. Enzyme I has not been extensively characterized. The HPrs from E. coli and S. typhimurium are similar as judged by their amino acid composition and electrophoretic behavior and show good cross-reactivity in enzyme assay. The reaction P-HPr ⟶ sugar-P seems to vary according to organisms and sugars. Enzyme II for β-glucosides in E. coli is inducible and has been partially resolved from the intact membrane by extraction with a detergent. This isolated protein has a molecular weight of about 200,000. The lactose-enzyme II system in S. aureus requires the addition of a soluble protein (lactose factor III). Hanson and Anderson [117] report that factor III of the fructose system in Aerobactor aerogenes seemed to be a protein, but the factor has not been extensively characterized. The details of PTS system were summarized by Simoni [111] in 1972.

D. Inhibition of Cell Membrane Function by Fungicides

1. Surface-Active Agents

Surface-active agents promote biological activity of fungicides and are always ingredients of fungicide preparations. However, some surface-active agents themselves have fungicidal activity. Surface-active agents are classified into three types according to the electric charge in solution: cationic, anionic, and neutral. Fungicidal substances belong mainly to cationic or anionic types. The former have been investigated more than the latter.

Cationic surface-active agents possess a hydrophobic group, such as a hydrocarbon chain or alkylbenzene ring, together with a positive charged hydrophilic group (quaternary ammonium, sulfonium, and others). Cetyl trimethylammonium bromide (CTAB), cetyl pyridinium chloride (CPC), dodine (dodecyl guanidine), glyodin (2-heptadecyl imidazoline) and some polypeptide antibiotics are grouped in this class. Using a homologous series, it has been shown that bacteriocidal activity is related to chain length. The optimal length varies with test organisms or test substances. Baker et al. [118] suggested that detergent action disorganized cell membranes and denatured essential proteins. Knox et al. [119] found that lactic dehydrogenase in a cell-free system from E. coli was inhibited by detergents at a bacteriocidal concentration for the intact cell. Therefore, they suggested that the inhibition of essential enzymes could account for the inhibition of cell metabolism and change in membrane permeability by detergents. Brown and Sisler [120] demonstrated that dodine strongly inhibited the oxidation of glucose by S. pastorianus and caused the leakage of cellular constituents. Salton [121] demonstrated that small molecular metabolites were released from both gram-positive and gram-negative bacteria by CTAB and observed that cell death produced by the detergents occurred simultaneously with leakage. He suggested that the primary cause of cell death by detergents was attributable to the effect on sensitive enzymes. Many investigators [122, 123] studied detergent action on bacterial protoplasts or intact cell, and found that detergent did not cause a rapid lysis of intact bacterial cells, but treatment of protoplast membrane with CTAB, SDS, and other cationic detergents resulted in the formation of relatively small lipoprotein. Salton [124] summarized the literature concerning detergent action on cells.

Kuhn and Dann [125] studied effects of a series of sulfonium iodides (general formula, $RS^+(CH_3)_2I$) on bacterial cells and found that a compound ($R = C_{16}$) was most effective against E. coli and another compound ($R = C_{12}$) was most active in killing of cells of S. aureus. In investigations of the effects of alkyl pyridinium bromide on S. aureus, Domodarian and Sivaraman [126] found that linolenyl and oleyl derivatives were, respectively, about 9 and 2.5 times more active than stearyl derivative. Gilby and Few [122] investigated the action of a series of C_{12}-n-alkyl ionic detergents upon M. lysodeikticus and its protoplast, and found the following relationship between bacteriocidal lytic activity and ionic group:

$$-NH_3^+ > -N^+(CH_3)_3 > -SO_4^- > -SO_3^-$$

2. Phenol

Phenol inhibits enzyme activities and releases small molecular metabolites from inside the cell. Especially, certain nitro and halogenated phenols are well known to act as uncoupling agents of the energy transfer system. In this section, we will discuss the relation of structures of amylphenol and alkylphenols to their biological activity.

Early works concerning the effect of chemical substitution on antibacterial activity have been summarized by Suter [127], Hugo [128], and Hamilton [129]. Among a series of homologous phenols or cresols, the optimal bacteriocidal activities against gram-negative and gram-positive bacteria were obtained with n-amylphenol and n-hexyl or n-heptyl phenol. The position of the alkyl residue on the benzene ring has little effect. A compound containing a straight chain is more effective than an equivalent branched-chain analogue. Uesugi and Suzuki [130] investigated the fungicidal activities of a series of alkyl- or acylphenols and found that the difference in activity between O and P isomers was marked with acylphenol but was very slight with alkylphenol. The optimal chain length of the O-acylphenols was shorter than that of the P isomers.

Gale and Taylor [131] and Haydon [132] found that phenol released small molecular metabolites from bacterial cells. Uesugi and Fukunaga [133] also found that butylphenol liberated a large quantity of ninhydrin-positive substances from fungal cells at very low concentrations which had little effect on the energy transfer system. The phenol finally caused cell death. Thus it is evident that phenol alters permeabilities of microbial membranes.

3. Polyenes

Polyenes are a group of macrolide antibiotics which selectively damage the permeability of membranes of sensitive cells. Thus the polyenes are active against yeasts, a wide variety of fungi, and other eukaryotic cells. Their cell membranes are found to contain sterols. However, the antibiotics do not act on bacteria and blue-green algae membranes which contain no significant amount of sterol. Polyenes are impractical as agricultural fungicides because they are readily decomposed by light.

All the polyene antibiotics possess a large lactone ring containing a hydroxylated portion and a system of conjugated bonds, for example filipin (see Fig. 5). The extent of damage produced by the antibiotics is dependent upon the size of ring. In general, polyene antibiotics containing a small ring structure produce greater membrane damage, although those with larger rings show greater antifungal activity. The polyenes do not, in general, burst protoplasts of sensitive cells. Like other substances affecting membrane permeability, polyenes produce leakage of small molecular materials. The degree of damage can be determined by the amounts and the variety of molecules released. Lampen et al. [134] investigated the effect of nystatin and N-acetylcandicidin on leakage of intracellular constituents from sensitive yeast cells and their protoplasts. Sensitive yeast cells bound the antibiotic at 30° but not at 0°C. However, protoplasts could bind at either 0° and 30°C and protoplast membrane fragments took up the antibiotic very rapidly at 0°C.

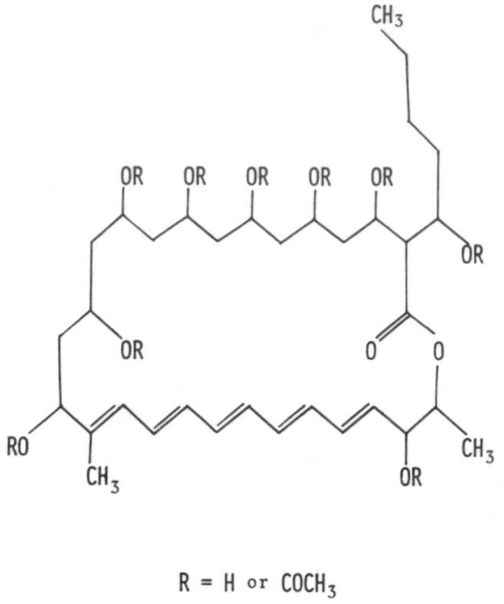

R = H or $COCH_3$

FIG. 5. Chemical structure of filipin.

The inhibition of fungal growth by polyene antibiotics can be prevented by the addition of sterol [135-137]. This protection appears to be due to the direct interaction between sterols and polyene antibiotics. This would indicate that the interaction occurred between polyenes and membrane sterol. This view is supported by the fact that microorganisms containing sterols are sensitive to the polyenes. Mycoplasmia gallisepticum cells, which require sterols for growth, were unaffected by filipin in a sterol-free medium, whereas their growth in the presence of sterols was rapidly inhibited by the antibiotic. More direct evidence for a binding reaction between polyene and sterols was obtained by using ^{14}C-labeled polyenes [134]. Yeast protoplasts and intact cells were disrupted by various procedures, and amounts of [^{14}C]nystatin bound to various cell fractions were determined. The amounts of the radioactive polyene were proportional to sterol content in all fractions. Further treatment of protoplast membranes with digition released to bound polyene. Kinsky [138] reported that polyene-binding ability of N. crassa membranes decreased upon extraction of the membrane with organic solvents and could be restored again by the addition of sterols. The interaction between polyenes and monolayer membranes with various composition was investigated by Demel et al. [139] and Kinsky [140]. Filipin and nystatin did not penetrate into a monolayer of lipids extracted from insensitive bacteria which mainly consisted of phospholipids and did not contain sterols. On the other hand, the polyenes rapidly penetrated into a monolayer of lipids extracted from beef erythrocytes which contained cholesterol. The same results were obtained in investigations using other polyenes. These results indicate that polyenes react specifically with sterols and produce reorientation of sterols within the membrane. This reorientation would bring about disorganization of the membrane structure.

4. Other Inhibitors of Cell Membrane Function

Metal ions and maneb, manganese ethylenebis(dithiocarbamate), are widely used crop-protectant fungicides. These fungicides also release small-molecular-weight metabolites from the inside of the cell. Miller and McCallan [141] and Tröger [142] investigated the interaction between metal ions and several fungi. Silver was taken up rapidly by fungal spores and completely inhibited spore germination, but copper, zinc, and cadminium reduced germination appreciably only after some hours of contact with the spores. Silver and copper released a large quantity of phosphorus and ninhidrin positive compounds into the ambient solution. Wedding and Kendrick [143] and Morehart and Crossan [144] reported that sodium _N_-methyldithiocarbamate and maneb caused the loss of soluble cell constituents from treated mycelia and related the effect of the dithiocarbamates on permeability to inhibition of growth of the mycelium.

REFERENCES

1. J. R. Duguid, _Edinburgh Med. J._, _53_, 401 (1946).
2. J. T. Park and M. Johnson, _J. Biol. Chem._, _170_, 585 (1949).
3. A. D. Gardner, _Nature_, _146_, 837 (1940).
4. J. T. Park, _J. Biol. Chem._, _194_, 194 (1952).
5. S. Sasaki, N. Ohta, I. Yamaguchi, S. Kuroda, and T. Misato, _J. Agr. Chem. Soc. Japan_, _42_, 633 (1952).
6. N. Ohta, K. Kakiki, and T. Misato, _Agr. Biol. Chem._, _34_, 1224 (1970).
7. A. Endo and T. Misato, _Biochem. Biophys. Res. Commun._, _37_, 718 (1969).
8. M. Hori, K. Kakiki, and T. Misato, _Agr. Biol. Chem._, _38_, 691 (1974).
9. J. M. Aronson and R. D. Preston, _Proc. Roy. Soc. London_, _B152_, 346 (1960).
10. J. M. Aronson and R. D. Preston, _J. Biophys. Biochem. Cytol._, _8_, 247 (1960).
11. H. D. Agar and H. C. Douglas, _J. Bacteriol._, _70_, 427 (1955).
12. B. Mundkur, _Exptl. Cell Res._, _20_, 28 (1960).
13. E. Vitols, R. J. North, and A. W. Linnane, _J. Biophys. Biochem. Cytol._, _9_, 689 (1961).
14. K. Horikoshi, _J. Agr. Chem. Soc. Japan_, _41_, 1221 (1967).
15. A. Frey-Wyssling and K. Mühlethaler, _Vieteljahresschr. Naturforsch._ Ges. Zuerich, _95_, 45 (1950).
16. P. A. Roelfsen, _Biochim. Biophys. Acta_, _6_, 357 (1951).
17. M. J. Middlebrook and R. D. Preston, _Biochim. Biophys. Acta_, _9_, 32 (1952).

18. A. L. Houwink and D. R. Kreger, J. Microbiol. Serol., 19, 1 (1953).
19. E. Kellenberger and A. Ryter, J. Biophys. Biochem. Cytol., 4, 323 (1958).
20. E. Cabib, R. Ulane, and R. Bowers, J. Biol. Chem., 248, 1451 (1973).
21. G. L. Gilardi and N. C. Laffer, J. Bacteriol., 83, 219 (1962).
22. S. Bartnicki-Garcia and W. J. Nickerson, J. Bacteriol., 84, 829 (1962).
23. W. J. Nickerson and C. W. Chung, Amer. J. Bot., 41, 114 (1954).
24. W. J. Nickerson, J. Gen. Physiol., 37, 483 (1954).
25. F. Kanetsuna and L. M. Carbonell, J. Bacteriol., 101, 675 (1970).
26. F. Kanetsuna and L. M. Carbonell, J. Bacteriol., 106, 946 (1971).
27. L. M. Carbonell, J. Bacteriol., 100, 1076 (1969).
28. C. Van Wisselingh, Jahrb. Wiss. Botan., 31, 619 (1898).
29. F. W. Chattaway, M. R. Holmes, and A. J. E. Barlow, J. Gen. Microbiol., 51, 367 (1968).
30. C. M. Crook and I. R. Johnston, Biochemical J., 83, 325 (1962).
31. S. Bartnicki-Garcia and W. J. Nickerson, Biochim. Biophys. Acta, 58, 102 (1962).
32. K. Horikoshi and S. Iida, Biochim. Biophys. Acta, 83, 197 (1964).
33. J. M. Aronson and L. Machilis, Amer. J. Bot., 46, 292 (1959).
34. S. Bartnicki-Garcia, J. Gen. Microbiol., 42, 57 (1966).
35. S. Bartnicki-Garcia and E. Reyes, Arch. Biochem. Biophys., 108, 125 (1964).
36. R. R. Mahadevan and E. L. Tatum, J. Bacteriol., 90, 1073 (1965).
37. M. S. Manocha and J. R. Colvin, J. Bacteriol., 94, 202 (1967).
38. K. G. H. Dyke, Biochim. Biophys. Acta, 82, 374 (1964).
39. H. Hurst, J. Exptl. Biol., 29, 30 (1952).
40. G. Kessler and W. J. Nickerson, J. Biol. Chem., 234, 2281 (1959).
41. M. Novaes-Ledieu, A. Jimenez-Martinez, and J. R. Villaneuva, J. Gen. Microbiol., 47, 237 (1967).
42. M. Shifrine and H. J. Phaff, J. Microbiol. Serol., 24, 274 (1958).
43. R. Frey, Ber. Schweiz Botan. Ges., 60, 199 (1950).
44. G. Van Iterson, K. H. Meyer, and W. Lotman, Rec. Trav. Chim., 55, 61 (1936).
45. J. M. Aronson and L. Machilis, Amer. J. Bot., 46, 292 (1959).
46. K. Mühlethaler, Amer. J. Bot., 43, 673 (1956).
47. K. Gezelius, Exptl. Cell Res., 18, 425 (1959).
48. M. Toama and K. B. Ranby, Exptl. Cell Res., 12, 256 (1957).
49. R. E. Moneno, F. Kanetsuna, and L. M. Carbonell, Arch. Biochem. Biophys., 130, 212 (1969).
50. D. J. Manners and J. C. Patterson, Biochem. J., 98, 19c (1966).
51. D. R. Kreger, Biochim. Biophys. Acta, 13, 1 (1954).
52. E. D. Korn and D. H. Northcote, Biochem. J., 75, 12 (1960).
53. F. M. Harold, Biochim. Biophys. Acta, 57, 59 (1962).

54. R. G. Brown and L. Linderberg, Acta Chim. Scand., 21, 2383 (1967).
55. S. Bartnicki-Garcia, Ann. Rev. Microbiol., 22, 87 (1968).
56. G. P. Skucas, Amer. J. Bot., 53, 1006 (1966).
57. W. J. Nickerson, G. Falcone, and G. Kessler, in Macromolecular Complexes (M. V. Perterson, ed.), Ronald Press, New York, 1961, p. 205.
58. A. L. Houwink and D. R. Kreger, J. Microbiol. Serol., 19, 1 (1953).
59. B. Mundkur, Exptl. Cell Res., 20, 28 (1960).
60. A. A. Eddy, Proc. Roy. Soc. London, B149, 425 (1958).
61. L. Mangin, Compt. Rend., 110, 644 (1890).
62. J. E. Gander, Ann. Rev. Microbiol., 28, 103 (1974).
63. M. J. Osban, Ann. Rev. Biochem., 38, 501 (1969).
64. L. Glaser, Ann. Rev. Biochem., 42, 91 (1973).
65. E. P. Camargo, C. P. Dietrich, D. Sonneborn, and J. L. Strominger, J. Biol. Chem., 242, 3121 (1967).
66. P. Babczinski and W. Tanner, Biochem. Biophys. Res. Commun., 54, 1119 (1973).
67. M. C. Wang and S. Bartnicki-Garcia, Biochem. Biophys. Res. Commun., 24, 832 (1966).
68. K. Isono, S. Suzuki, M. Tanaka, T. Nanbara, and K. Shibuya, Tetrahedron Letters, 6, 425 (1970).
69. K. Isono, K. Kobinata, and S. Suzuki, Agr. Biol. Chem., 32, 792 (1968).
70. K. Isono, J. Nagastu, Y. Kawashima, and S. Suzuki, Agr. Biol. Chem., 29, 848 (1965).
71. K. Isono, J. Nagastu, K. Kobinata, K. Sasaki, and S. Suzuki, Agr. Biol. Chem., 31, 190 (1967).
72. K. Isono and S. Suzuki, Agr. Biol. Chem., 32, 1193 (1968).
73. S. Suzuki, K. Isono, J. Nagastu, Y. Kawashima, K. Yamagata, K. Sasaki, and K. Hashimoto, Agr. Biol. Chem., 30, 817 (1966).
74. S. Suzuki, K. Isono, J. Nagastu, J. Mizutani, K. Kawashima, and T. Mizuno, J. Antibiotics (Tokyo), Ser. A, 18, 131 (1965).
75. A. Endo, K. Kakiki, and T. Misato, J. Bacteriol., 104, 189 (1970).
76. M. Hori, K. Kakiki, S. Suzuki, and T. Misato, Agr. Biol. Chem., 35, 1280 (1971).
77. M. Hori, K. Kakiki, and T. Misato, Agr. Biol. Chem., 38, 699 (1974).
78. F. A. Keller and E. Cabib, J. Biol. Chem., 246, 160 (1971).
79. M. Hori, M. Eguchi, K. Kakiki, and T. Misato, J. Antibiotics (Tokyo), Ser. A., 27, 260 (1974).
80. M. Mitani and Y. Inoue, J. Antibiotics (Tokyo), Ser. A., 21, 492 (1968).
81. K. Kakiki, T. Maeda, H. Abe, and T. Misato, J. Agr. Biol. Chem. Soc. Japan, 43, 37 (1969).
82. I. Uchino and T. Yamakawa, Rept. Zenkoren Tech. Center (Japan), 6, 1 (1964).

83. T. Sudo and K. Tamari, Proc. Meet. Agr. Chem. Soc. Japan (Tokyo), 1969, p. 14.
84. T. Maeda, H. Abe, K. Kakiki, and T. Misato, Agr. Biol. Chem., 34, 700 (1970).
85. N. Muramastu, K. Kakiki, and T. Misato, unpublished data (1972).
86. K. Kakiki and T. Misato, unpublished data (1972).
87. M. R. J. Salton, Ann. Rev. Microbiol., 21, 417 (1967).
88. L. Steck, in Membrane Molecular Biology (C. F. Fox and A. D. Keith, eds.), Sinauer Associates, Inc., Stamford, Connecticut, 1972, p. 76.
89. D. E. Brundish, N. Shaw, and J. Baddiley, Biochem. J., 99, 546 (1966).
90. M. Blumer, T. Chase, and S. W. Watson, J. Bacteriol., 99, 366 (1969).
91. J. P. Brown and B. J. Cosenza, Nature, 204, 802 (1964).
92. I. B. Holland and S. Tuckett, J. Supramol. Struct., 1, 77 (1972).
93. N. A. Machtinger and C. F. Fox, Ann. Rev. Biochem., 42, 575 (1973).
94. T. Miura and S. Mizushima, Biochim. Biophys. Acta, 193, 268 (1969).
95. C. F. Fox, J. H. Law, N. Stukagoshi, and G. Wilson, Proc. Natl. Acad. Sci. USA, 67, 598 (1971).
96. J. H. Law and W. R. Snyder, in Membrane Molecular Biology (C. F. Fox and A. D. Keith, eds.), Sinauer Associates, Inc., Stamford, Connecticut, 1972, p. 8.
97. D. M. Engelman, J. Mol. Biol., 58, 153 (1971).
98. A. W. Rodwell and J. E. Peterson, J. Gen. Microbiol., 68, 173 (1971).
99. R. N. McElhaney, J. De Gier, and L. M. Van Deenen, Biochim. Biophys. Acta, 219, 245 (1970).
100. R. A. Demel, S. C. Kinsky, C. B. Kinsky, and L. M. Van Deenen, Biochim. Biophys. Acta, 150, 655 (1968).
101. W. J. V. Osterhout, Proc. Natl. Acad. Sci. USA, 21, 125 (1935).
102. H. R. Kaback, J. Biol. Chem., 243, 3711 (1968).
103. H. R. Kaback, Ann. Rev. Biochem., 39, 561 (1970).
104. T. H. Wilson, M. Kusch, and E. R. Kashket, Biochem. Biophys. Res. Commun., 40, 1409 (1970).
105. W. Kunding, S. Ghosh, and S. Roseman, Proc. Natl. Acad. Sci. USA, 52, 1067 (1964).
106. G. N. Cohen and J. Monad, Bacteriol. Rev., 21, 169 (1957).
107. Y. Anraku, J. Biol. Chem., 243, 3116 (1968).
108. C. F. Fox and E. P. Kennedy, Proc. Natl. Acad. Sci. USA, 54, 891 (1965).
109. T. D. H. Johnes and E. P. Kennedy, J. Biol. Chem., 244, 5981 (1969).

110. L. A. Heppel, J. Gen. Physiol., 54, 95s (1969).
111. R. D. Simoni, in Membrane Molecular Biology (C. F. Fox and A. D. Keith, eds.), Sinauer Associates, Inc., Stamford, Connecticut, 1972, p. 311.
112. A. B. Pardee, Science, 156, 1627 (1967).
113. J. R. Piperno and L. Oxender, J. Biol. Chem., 241, 5732 (1966).
114. J. B. Egan and M. L. Morse, Biochim. Biophys. Acta, 112, 63 (1966).
115. R. D. Simoni, M. Levinthal, F. D. Kundig, W. Kundig, B. Anderson, P. E. Hartman, and S. Roseman, Proc. Natl. Acad. Sci. USA, 58, 1963 (1967).
116. M. H. Saier, R. D. Simoni, and S. Roseman, J. Biol. Chem., 245, 5870 (1970).
117. T. E. Hanson and R. L. Anderson, Proc. Natl. Acad. Sci. USA, 61, 269 (1968).
118. Z. Baker, R. W. Harrison, and B. F. Miller, J. Exptl. Med., 74, 249 (1941).
119. W. E. Knox, V. H. Auerbach, K. Zarudnaya, and M. Spirtes, J. Bacteriol., 58, 443 (1949).
120. I. F. Brown and H. D. Sisler, Phytopathology, 50, 830 (1960).
121. M. R. J. Salton, J. Gen. Microbiol., 5, 391 (1951).
122. A. R. Gilby and A. V. Few, Nature, 179, 442 (1957).
123. M. R. J. Salton and A. Netschey, Biochim. Biophys. Acta, 107, 531 (1965).
124. M. R. J. Salton, J. Gen. Physiol., 52, 227s (1968).
125. R. Kuhn and O. Dann, Ber. Deut. Chem. Ges., 73, 1092 (1940).
126. M. Domodarian and C. Sivaraman, J. Bacteriol., 65, 89 (1953).
127. C. M. Suter, Chem. Rev., 28, 269 (1941).
128. W. B. Hugo, J. Pharm. Pharmacol., 9, 145 (1957).
129. W. A. Hamilton, Biochem. J., 118, 46p (1970).
130. Y. Uesugi and T. Suzuki, Bull. Natl. Inst. Agr. Sci. (Tokyo), Ser. C, 17, 193 (1964).
131. E. F. Gale and E. S. Taylor, J. Gen. Microbiol., 1, 77 (1947).
132. D. A. Haydon, Proc. Roy. Soc. London, B145, 383 (1956).
133. Y. Uesugi and K. Fukunaga, Ann. Phytopathol. Soc. Japan, 33, 168 (1967).
134. J. O. Lampen, P. M. Arnow, Z. Borowska, and A. I. Laskin, J. Bacteriol., 84, 1152 (1962).
135. D. Gottlieb, H. E. Carter, and J. H. Soneker, Phytopathology, 50, 594 (1960).
136. J. O. Lampen, P. M. Arnow, and R. S. Safferman, J. Bacteriol., 80, 200 (1960).
137. J. O. Lampen, J. W. Gill, P. M. Arnow, and I. Magana-Plaza, J. Bacteriol., 86, 945 (1963).
138. S. C. Kinsky, Arch. Biocheim. Biophys., 102, 180 (1963).

139. R. A. Demel, F. J. L. Crombag, L. L. M. Van Deenen, and S. C. Kinsky, Biochim. Biophys. Acta, 150, 1 (1968).
140. S. C. Kinsky, J. Haxby, C. B. Kinsky, R. A. Demel, and L. L. M. Van Deenen, Biochim. Biophys. Acta, 152, 174 (1968).
141. L. P. Miller and S. E. A. McCallan, Agr. Food Chem., 5, 116 (1957).
142. R. Tröger, Arch. Microbiol., 29, 430 (1958).
143. R. T. Wedding and J. B. Kendrick, Jr., Phytopathology, 49, 557 (1959).
144. A. L. Morehart and D. F. Crossan, Toxicol. Appl. Pharmacol., 4, 720 (1962).

Chapter 8

EFFECT OF FUNGICIDES ON ENERGY PRODUCTION AND INTERMEDIARY METABOLISM

H. Lyr

Institute of Plant Protection Research
Kleinmachnow, Academy of Agricultural Sciences
of the German Democratic Republic

I. INTRODUCTION

Energy supply is one of the key factors for living systems. The famous plant physiologist Pfeffer compared a living cell with a leaking boat which can be prevented from sinking only by permanently pumping water under energy consumption. By analogy, any attack on the energy function of an organism results in a disturbance of the whole metabolism and lead finally to death when energy is below the level for maintenance of the essential cell structures. Shortage of energy supply results at first in a stoppage of growth without killing the organism. This would be defined as a fungistatic effect. It can be reversed or converted into a fungicidal effect if the relationship between synthesis and breakdown reactions becomes negative for a critical period of time, which depends on temperature. Fungicides become more toxic with increased temperature. In a fungal mycelium, apical or spore-producing cells can remain alive or even grow when the total mycelial dry weight is decreasing, indicating a preponderance of catabolic processes. This happens to the debit of energy and material resources of dying or aging cells. When energy production by fermentation or respiration is affected by fungicides, nearly all other metabolic activities, the synthesis of such as protein, RNA, and DNA, enzyme induction, and lipid and cell-wall syntheses are impaired.

Under certain circumstances, a lowering of adenosine triphosphate (ATP) level by uncoupling fungicides or substances attacking the mitochondrial functions can induce the synthesis of special enzymes or enhance cell wall synthesis. In several fungi the synthesis of the enzymes tyrosinase or laccase can be induced by pentachlorophenol, a potent uncoupler of oxidative phosphorylation [1, 54]. In some <u>Mucorales</u> cell-wall formation is enhanced when energy production by respiration is inhibited [2, 3]. At low ATP levels some syntheses are apparently inhibited, while others are stimulated and this surely has ecological significance.

Energy can be produced by fermentation and respiration, but the latter is much more effective and yields 38 moles of ATP for one mole of glucose, whereas the former produces only 2 moles of ATP. Aerobic dissimilation of carbohydrates is predominant in fungi, although many, but not all, possess the ability for fermentation.

Inhibition of fermentation or respiration by fungicides has often been described. Many chemical compounds are active in this respect (copper, mercurials, arsenicals, quinones, phenols, sulfur, captan, dithiocarbamates, tin-organics and others). For details, see articles by Martin, Owens, Kaars-Sijpesteijn et al., Lukens, Rich, Corden, and Byrde in the valuable review by Torgeson [4], or consult Webb [5]. The inhibitory effects, although different for various compounds, can mainly be attributed to disturbance of permeability or chemical reaction with SH, OH, COOH, phosphate or amino groups of sensitive enzymes in the fermentation or respiration chain. It is evident that these effects are rather unspecific, although sensitivity of enzymes to such inhibitors may differ considerably, depending on the presence or absence of sensitive groups. The effect of reactions with essential SH groups on enzyme activity may differ from enzyme to enzyme and even for the same enzyme in different species. Therefore, a certain specificity may result even on the basis of an unspecific reaction. This is best established for the sulfhydryl activated enzymes. Mercurials such as <u>p</u>-chloromercuribenzoate are potent inhibitors, reacting quantitatively with SH groups. For example, amylase of higher plant origin, which is SH activated, is sensitive, whereas amylase from fungal sources is nearly insensitive to SH blockers [12]. The reactivity of SH groups of enzymes with various sulfhydryl reagents also differs. The reason is not yet clear, but the steric position of the SH group within the enzyme molecule or its state of oxidation can influence reactivity.

A reaction with sulfhydryl groups has been described for several fungicides including copper, mercurials, quinones [6], sulfur, ferbam [6], captan [7, 8], thiram [6], folpet [9], and tetrachloroisophthalonitrile [33, 58]. Compounds such as quinones and folpet apparently also attack amino groups [7, 10].

It is difficult, therefore, to determine the primary mechanism of action of such compounds within the cell, because several metabolic pathways may be affected simultaneously.

This form of multisite action which is characteristic for protective fungicides has probably prevented the development of genetic resistance to these compounds. Only a low degree of tolerance to these compounds can be achieved by fungi through defense mechanisms such as the formation of surplus thiols. It is not surprising that fermentation and respiration are impaired by these fungicides, because these pathways possess several essential SH-activated enzymes.

Such an unspecific mechanism of action would also lead to phytotoxicity, if these compounds entered the plant cell. Therefore, the classic fungicides are mainly protective, hindering spore germination on the leaf surface without contacting the host cells which are protected by a cuticular barrier.

It is interesting that some of the newer, more specific, and often systemic fungicides exhibit a specific attack in the respiratory system which shall be described in more detail in Section III. This is surprising, because according to classical biochemical opinion the respiratory system is quite similar in most organisms. Therefore little or no selectivity in the action of these fungicides should be expected. For systemic fungicides, which act intratherapeutically, the sensitivity of a fungal plant pathogen to a fungicide needs to be much higher than that of the host plant. Modern fungicides even exhibit selectivity within the fungi. From a practical point of view, this is a big disadvantage, but it opens the possibility for a better understanding of the biochemical differences between fungal species and families. At present, we are just beginning to investigate the reasons for selectivity of systemic fungicides, which can theoretically be based on several features. Within the scope of this chapter only the mechanism of action, that is, the mode of interaction of a fungicide with its target (receptor) within the cell, can be discussed.

II. NONSPECIFIC EFFECTS ON ENERGY PRODUCTION

A. Inhibition of Fermentation by Fungicides

Degradation of carbohydrates is for numerous fungal species the main energy source. This is true for many phytopathogenic and wood destroying fungi and soil-inhabiting species. The cleavage of monosaccharides involves two well-known metabolic pathways, the glycolytic Embden-Meyerhof-Parnass (EMP) pathway and the hexose monophosphate shunt (HMP). Both are present in fungi, although to a different extent, and their relative activity can change in the course of development. According to Ritter, Kluge, and Lyr [11], a measurable glycolytic degradation in absence of oxygen occurs in numerous fungal species, but not in all. In addition to the EMP pathway, all fungi use the HMP pathway which serves mainly for the production of reducing equivalents and materials for several syntheses. A number of fungi are obligate aerobes with a low glycolytic potency and a low activity of the enzyme phosphofructokinase (PFK), the key enzyme in the initial portion of the EMP pathway. Fermentation or acid production cannot be demonstrated under anaerobic conditions in these species. But surprisingly, they exhibit a significant sensitivity to the systemic fungicide carboxin (DCMO) (Table 1). Comparative investigations revealed that there is no causal connection, but only a convergence in the evolution of these features, since several exceptions could be found. It is remarkable that these properties are specific for some taxa. For instance, Helotiales, Sordariales, Xylariales, Chaetomiales, and the Basidiomycetes prefer the HMP pathway as obligate aerobic organisms, whereas Peronosporales, Eurotiales and the true Endomycetes possess a well-functioning EMP pathway. Some families such as the Hypocreales are not homogeneous in this respect; a phylogenetic background seems probable.

Theoretically a fungistatic action could be achieved by inhibition of enzymes responsible for the breakdown of polymeric carbohydrates which are the precursors of monosaccharides. Such enzymes, secreted by many fungi, include cellulase, xylanase, mannanase, amylase, pectinase, invertase, etc. Comparative investigation demonstrated that these enzymes, which are probably glycoproteins, are rather insensitive to fungicides and other inhibitors [12, 13]. Pectinase is affected by phenols [12] and more specifically by quinones such as rufianic acid [14]. This compound was proposed for phytotherapeutical purposes, but its effect is not strong enough for practical application. A further disadvantage of such a strategy of fungicidal action is that enzyme synthesis is not directly impaired and therefore new ectoenzymes can be secreted by the mycelium, which can overcome external inhibitors. Moreover, because natural substrates consist of a complex of carbohydrates, several different enzymes must be inhibited to give a fungistatic effect. Oxidized polyphenols which occur in the heartwood of many trees nonspecifically inhibit enzymes such as pectinase and cellulase [67]. Such inhibition alone does not provide an adequate basis for fungitoxicity, but it may support the defense mechanisms of other heartwood toxins.

A specific inhibition of enzymes of the EMP or HMP pathway by fungicides has not yet been demonstrated. At present no specific inhibitors for these metabolic pathways are known. Therefore the significance of an inhibition of one of these pathways in fungi possessing both is obscure. An

TABLE 1

Estimation of the Proportion of EMP and HMP Pathways, Phosphofructokinase (PFK) Activity, Fermentation, Acid Production, and Carboxin Sensitivity in Various Fungal Species[a]

Organism	Degradation of glucose, %		PFK, IU/g protein	Fermentation	Acid production	Carboxin sensitivity
	EMP	HMP				
Botrytis cinerea	20	80	3.5	-	-	+
Verticillium albo-atrum	26	74	2.3	-	-	+
Pleurotus ostreatus	12	88	4.3	-	-	+
Rhizoctonia solani	30	70	2.7	-	-	+
Ustilago maydis	19	81	0.8	-	-	+
Aspergillus niger	49	51	26.4	++	+	-
Fusarium oxysporum	36	64	26.6	+	+	-
Cochliobolus carbonum	41	59	75.8	+	-	-
Saccharomyces cerevisiae	87	13	37.5	++	+	-

[a]Taken from Ref. 11.

Key: + = moderate, ++ = strong; - = no activity.

unspecific inhibition of CO_2 production often occurs under the influence of fungicides, but the significance of this for the total fungicidal effect is difficult to estimate because other metabolic or synthetic processes can be impaired at the same time.

An indirect inhibition of glycolysis can be caused by sublethal damage to the cytoplasmic membrane, which results in a pathological leakage of potassium ions from the cell. This can be caused by membrane-active substances such as Ag^+, Hg^+, Cu^{2+}, Cu^+, and uranyl ions, bis(tributyltin) oxide, and other organic tin compounds or polyenic antibiotics (nystatin, candicidin, amphotericin, fungimycin, etc.). They apparently impair permeability or the K/H-ion pump mechanism by reaction with sterols or essential SH and phosphate groups of the membrane, so that the normal, higher K^+-ion con centration within the cell is lost. The key enzyme of the glycolytic pathway, phosphofructokinase, is activated by K ions. Their loss, therefore, results in a decrease of fermentation potency. Addition of higher concentrations of K or NH_4 ions to the medium reverses this effect. Because most of the heavy metal ions and compounds mentioned above react unspecifically, this special effect is best demonstrated with larger polyenic antibiotics such as nystatin which selectively interact with sterols of the membrane without membrane disruption [15, 103] or by δ-hexachlorocyclohexane (δ-HCH), a rather inert substance [16]. Membrane damage, of course also affects other processes within the cell, but fermentation is quite sensitive.

It is a remarkable fact that copper ions inhibit glycolysis more strongly under anaerobic conditions than under aerobic conditions [17]. According to Schlegel [18], addition of ascorbic acid has an effect similar to anaerobic conditions. It can be demonstrated (Lyr, unpublished data), that Cu^+ ions at low concentrations damage membrane integrity much more than Cu^{2+} ions. This is also true for the 1:1 copper complex with thujaplicin [19], which is a potent inhibitor of fermentation and respiration. The greater potency of Cu^+ ions is due to a more effective formation of complexes with sulfhydryl groups. The physiological effect of Cu^+ ions is as strong as that of Ag^+ or Hg^+ ions.

A direct, although unspecific inhibition of fermentation can be achieved by reaction of fungicides with SH groups of enzymes in the glycolytic pathway. As already mentioned, the sensitivity of single enzymes to SH inhibitors differs for the various types of blocking agents. This depends on the reactivity of the sulfhydryl groups, their position within the macromolecule, and their significance for the enzyme activity. Table 2 gives some examples of enzymes with essential SH groups and their inhibition by fungicides in vitro. Inhibition of fermentation has been frequently described for copper [17], mercurials and captan [32], quinones [30], arsenite [31], and β-thujaplicin and Cu-β-thujaplicin [19].

The physiological effect of an inhibition of glycolysis under normal growth conditions is difficult to judge, because an inhibition of respiration can often be observed at the same time. Quinones have a multiple effect on

TABLE 2

Some Important Fungal Enzymes in Fermentative and Respiratory Pathways[a]

Enzymes	Fungicides[b]						References
	JB	Cap.	Hg	Cu	S	Qu	
Fermentation							
Pyruvate-kinase			++				20,28
Hexokinase		0	++	0	+		21,26,64
Pyruvate-decarboxylase	+	+	++			++	7,20,22,25
Respiration							
Cytochrome c oxidase	0	0	0	++	++		21,23
Succinic dehydrogenase		0	+	++	+		20,21,43
iso-Citric dehydrogenase		+		+	+		21
Fumarase	++	+	++	++	+		21,24
Malate dehydrogenase			+				20,27

[a]Enzyme activity depends on free sulfhydryl groups. Inhibition by unspecific fungicides (+ = moderate, ++ = strong inhibition).
JB = o-iodosobenzoate, Cap. = captan, Hg = mercurials, Cu = copper, S = lime sulfur, Qu = quinones.

proteins including tanning and reaction with groups other than SH. In captan and folpet the $SCCl_3$ group is the toxic principle by virtue of its high reactivity with cell thiols and the production of thiophosgene, a decomposition product [29], which can react with SH, NH_2, and OH groups. Residues of copper or captan on grapes may be a handicap for wine fermentation because of their inhibitory effect on glycolytic enzymes of yeasts.

Fluoride ions are also potent inhibitors of glycolysis. They act by forming stable complexes with magnesium and phosphate ($FMgPO_4$ complex) in certain enzymes. The inhibition of 2-phosphoroglycerate dehydratase (enolase) by fluoride has been thoroughly investigated [20].

B. Inhibition of Respiration by Fungicides

Under aerobic conditions the final products of the HMP or EMP pathway are metabolized via the citric acid cycle to CO_2 and H_2O with a high yield of energy in form of ATP. This process proceeds within the mitochondria, which have much in common with bacteria, although a strong interdependence with cell syntheses has developed during evolution, perhaps in a symbiontic state. Fungal mitochondria vary greatly in size and may often have complicated shapes [1, 2, 11]. Their structure is quite similar to those found in plant and animal cells (see Fig. 1), and they possess an outer and an inner membrane. The latter is folded and forms the typical cristae, where most enzymes of the respiratory chain are situated. Phospholipids are an important component of the inner membrane and are possibly responsible for the binding of the lipophilic enzymes of the respiratory chain. Oomycetes, in contrast to the higher fungi, have a divergent structure of mitochondria. Instead of cristae they have tubuli as invaginations of the inner membrane (see Fig. 2). The significance of this difference is still unknown.

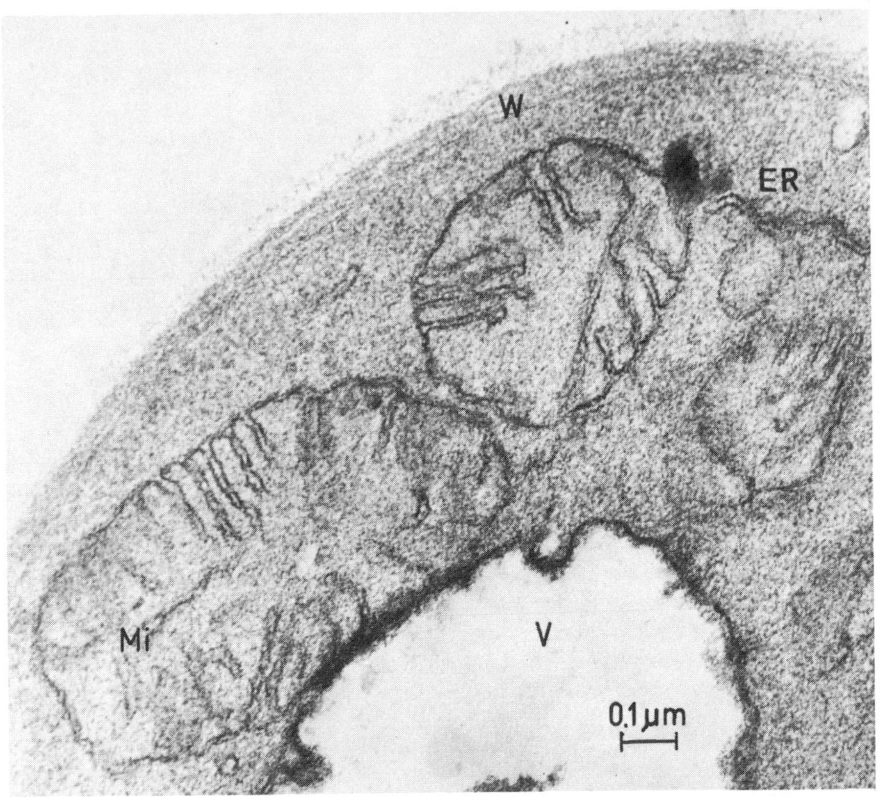

FIG. 1. Mitochondria (Mi) of Rhodotorula mucilaginosa with enveloping outer membrane and folded inner membrane forming cristae (cristae mitochondriales). V = vacuole, ER = endoplasmic reticulum, W = cell wall with lamellated structure. $KMnO_4$ fixation; magnification 90,000. (Taken from Ref. 73.)

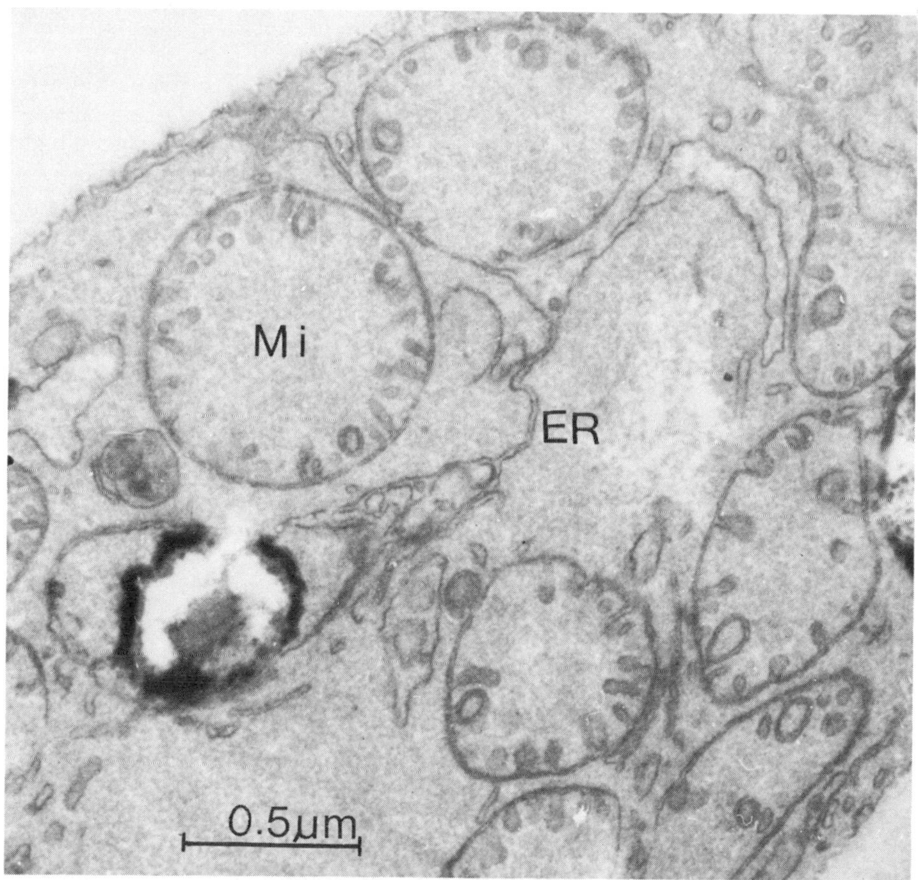

FIG. 2. Typical mitochondrial structure of an Oomycete (Phytophthora erythroseptica). The inner membrane forms typical tubular invaginations. $KMnO_4$ fixation; magnification 51,600. (Courtesy Dr. H. Fehrmann.) (Abbreviations as in Fig. 1.)

Protein synthesis by mitochondria is inhibited selectively by the antibacterial antibiotic chloramphenicol and is insensitive to cycloheximide which inhibits cytoplasmic protein synthesis in fungal cells. At higher concentrations, chloramphenicol inhibits NADH oxidation in yeast mitochondria possibly by inhibiting NADH dehydrogenase [59]. High concentrations in the growth medium of Rhodotorula or Pythium result in an inhibition of growth and a progressive disorganization of mitochondria [61]. The synthesis of cytochromes a, a_3, b, and c_1 and of succinic dehydrogenase is decreased under the influence of chloramphenicol [61]. This demonstrates that mitochondria have their own special system of protein synthesis, which has much in common with that of bacteria.

Respiration is measured by oxygen consumption. The oxygen receptor of the respiratory chain is cytochrome oxidase, a copper-containing enzyme which is poisoned by cyanide, azide, CO, and H_2S. This enzyme has a high affinity for oxygen and therefore respiration can be performed at low oxygen pressures. Very common in fungi is a cyanide insensitive oxygen consumption, which was first observed in higher plants. Here a so-called "alternative terminal oxidase" is operating. While normal respiration produces ATP as the energy compound, the "alternative pathway" is ineffective in this respect and respiration yields only heat. Details will be discussed in connection with the description of the mechanism of action of the carboxins.

Most of the unspecifically acting fungicides described above also inhibit respiration. The respiratory system contains several sulfhydryl enzymes which are sensitive to substances reacting with SH groups (Table 2). The nonheme-iron-sulfur proteins of site I and II possess SH groups in Fe complexes. The thiol-containing coenzyme A is very sensitive to several fungicides. Captan [34], thujaplicin and its 1:1 copper complexes [19], as well as other copper complexes, can react with this component of the respiratory system and thus strongly inhibit acetate oxidation. Ziram [35], nabam or ethylene diisothiocyanate [36] and other thiocarbamates [37], sorbic acid [62], Ag, Cu, and Hg ions and malachite green [38] also inhibit oxygen consumption with glucose or acetate as substrates. Many compounds exhibit a multisite action at appropriate concentrations; therefore it is difficult to indicate the primary point of attack. Whereas coenzyme A seems to be very sensitive to SH reagents, other systems such as cytochrome oxidase are less affected.

Although sulfur is one of the oldest fungicides, its mechanism of action is not very clear because of the many reactions of sulfur. Most likely, it acts in elemental or reduced form rather than in the form of SO_2 or other oxidized compounds such as pentathionic acid [106]. In its elemental form it can compete, according to Tweedy and Turner [107], with oxygen in the respiratory chain. It probably accepts electrons from the cytochrome b region (oxidation-reduction potential of cytochrome b = -0.040, of elemental sulfur = +0.140), and is reduced quantitatively to hydrogen sulfide. By this mechanism two phosphorylating sites within the respiratory chain are partly inactive. This means that the ATP level is lowered and many syntheses are impaired. The degradation of endogenous substrates is stimulated, but with low efficiency for ATP formation. The situation is similar in this respect to the action of uncoupling agents, but their mechanism of action is quite different. Surely, the H_2S produced will undergo many other reactions within the cell, which could include chelation of metals, such as copper or iron, and interference with -S-S- bridges of proteins. Therefore, a multisite and rather unspecific effect of lime sulfur can be expected. Selectivity of elemental sulfur among the fungi, which is well known in practice, may be due to differences in penetration, to structural insensitivity, or other tolerance mechanisms.

Membrane-active compounds may affect respiration indirectly by causing leakage of metabolites. This is known for polyenes [15, 42], triterpeneglycosides [39], and steroidglycosides [40]. However, isolated mitochondria of yeast are not sensitive to polyenes [41], probably due to the relatively low sterol content of mitochondrial membranes.

III. SPECIFIC EFFECTS ON ENERGY PRODUCTION

A. Uncoupling of Oxidative Phosphorylation

Inhibition of respiration can be achieved by blocking the electron transport chain of mitochondria, which is measurable by a reduced oxygen consumption. Another very effective mode of disturbance of energy production is the uncoupling of oxidative phosphorylation. In this case, electron flux is actually enhanced, resulting in an increase in respiration. Uncoupling means that the energy of the redox change in the respiration chain is not conserved in the form of ATP, but is liberated as heat. Normally, three molecules of ATP can be formed from ADP + P_i by the oxidation of two hydrogen atoms in the form of NADH. This is the most effective method of energy production. As Figure 4 shows, there exist three coupling sites responsible for energy transfer from potential differences between members of the respiratory chain into the high-energy phosphate bonds of ATP. The mechanism of this transfer has not yet been elucidated, although many models have been proposed. Very often uncoupling results in a stimulation of a mitochondrial ATP-ase which seems to be linked with the mitochondrial membrane and perhaps with the coupling process itself. Therefore the ATP content of the cell rapidly diminishes when oxidative phosphorylation is uncoupled. As a consequence, only glycolysis can contribute to energy production and the energy level within the cell is low. In aerobic organisms, uncoupling results in a complete breakdown of energy production. All ATP-dependent syntheses and metabolic processes are stopped.

Many compounds have been described as effective uncouplers of oxidative phosphorylation. Their specificity is very different. Whereas some must be regarded as quite unspecific because they affect still other processes within the cell, some are highly specific. Specificity depends to a high degree on the concentration of these compound. Uncoupling is very often achieved at low concentrations. Table 3 gives a survey on the diversity of substances which have been described as uncouplers. Their mechanism of action is apparently different, but all hinder ATP formation within the mitochondria at appropriate concentrations. This phenomenon is not restricted to fungicides; some of the compounds are used as medicinals or herbicides.

TABLE 3

Examples of Compounds Which Uncouple Oxidative Phosphorylation

Class	Example	References
Antibiotics	Showdomycin, gramicidin	48,51,60
Nitrophenols	2,4-Dinitrophenol	52,53
Halophenols	Pentachlorophenol	54
Benzimidazoles	4,5,6,7-Tetrachloro-2-trifluoromethyl-benzimidazole	57
Salicylanilides	2,5-Dichloronitrosalicylanilide	55
Quinones	2-Hydroxy-3-alkyl-1,4-naphthoquinone	45
Arsenite	Sodium arsenite	44
Malonitriles	Benzalmalononitrile, drazoxolon	46,101,102
Copper salts	$CuSO_4$, Cu-thujaplicin complex	43
Fatty acids	Oleic acid	56

It is generally agreed that lipophilic and weak acid properties are common features of many uncouplers. They exhibit their activity in the region of the inner mitochondrial membrane where the respiratory enzymes are located. Unsaturated fatty acids are much stronger uncouplers than saturated fatty acids. They can be liberated enzymatically within the mitochondria from phospholipids or neutral lipids and can be found in aged mitochondria. They could play a natural role in the regulation of mitochondrial activity, especially in the lysis of aged mitochondria with low energy production. Because copper ions, arsenite, and the antibiotic showdomycin have uncoupling properties and are effective SH-group reagents, one can assume that thiol groups are involved in the "coupling" system as proposed by Hadler et al. [48]. The uncoupling effect of several phenols and polyphenols is well known [49]. The compound 2,4-dinitrophenol is one of the most specific and widely used agents in biochemical research. Chlorinated phenols such as pentachlorophenol are very potent uncouplers, but have side effects on membranes which lead to a disintegration of the cell. Weinbach and Garbus [50] point out that many uncouplers can bind to proteins. Whether this property contributes to their mechanism of action is not clear.

The naturally occuring heartwood toxin of <u>Thuja plicata</u>, thujaplicin, and related compounds such as nootkatin, see Figure 3 (3, 4), proved to be uncouplers of oxidative phosphorylation in wood-destroying fungi [63, 64] at very low concentrations (10^{-6}–10^{-7} M). Pentachlorophenol as well as thujaplicins induce the enzyme laccase in some fungi. Probably still other functions can be effected by thujaplicin. Raa and Goksoyr [18, 65] assume that formation of acetyl coenzyme A from acetate is blocked by this substance. Thujaplicins are the most potent inhibitors of the enzyme tyrosinase known at present [66]. This action is surely not correlated with their main mechanism of action, but it may protect the fungicides from detoxication by fungal tyrosinases.

Uncouplers of oxidative phosphorylation (chlorinated phenols) are used in practice as fungicides mainly for wood protection because of their great stability and broad toxicity toward fungi and insects.

In the case of such unspecific compounds as copper and arsenite, it is difficult to say whether uncoupling contributes to a high degree to their overall toxicity because the intracellular distribution and concentrations are not known. In general, fungal mycelia are quite sensitive to uncouplers.

B. Interference with the Function of the Electron Transport System in Mitochondria

1. Carboxin (DCMO, Vitavax)

Schmeling and Kulka [68] introduced the oxathiins, a new class of compounds [Fig. 3 (1, 2)] with systemic mobility within the water transport system of the plant and high specificity toward Basidiomycetes. Although

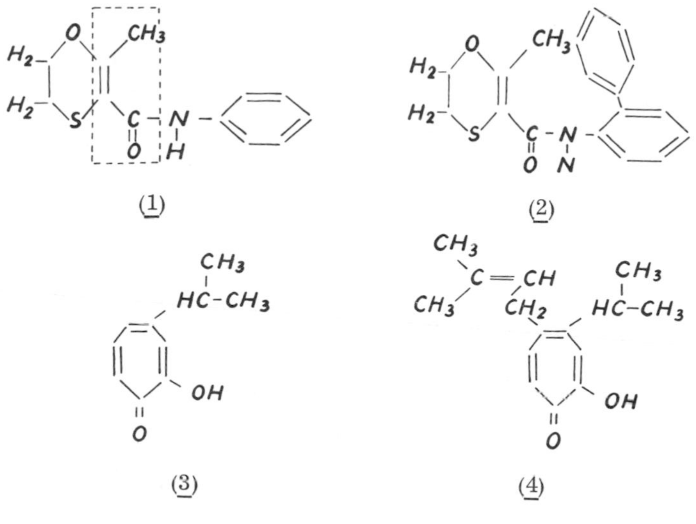

FIG. 3. Chemical structures of carboxin (DCMO) (1), F 427 (2), thujaplicin (3), and nootkatin (4).

sensitivity can also be demonstrated for some orders of Ascomycetes (Helotiales, Sordariales, Chaetomiales, Dothiorales) (Table 1), selectivity of action among the fungi is outstanding. Use of carboxin as a seed dressing for barley and wheat to control Ustilago sp. is now practical on a large scale. Several investigations have dealt with the mechanism of action of the main compound carboxin (DCMO), which is at present the systemic fungicide most thoroughly investigated in this respect. Several authors demonstrated that respiration is strongly inhibited with acetate as the carbon source, and significantly but less strongly inhibited with glucose [69-72]. Succinate accumulates within the cell [69]. Differential uptake of the fungicide by sensitive and insensitive fungal species was mainly the consequence of a different lipid content of the cells and proved not to be the reason for selectivity of action [73]. White [111,112] concluded in investigations with isolated mitochondria that succinodehydrogenase (SDH) or a closely related reaction is primarily inhibited. However, Lyr et al. [73] and Mathre [74] observed that isolated SDH is not inhibited by DCMO. This was confirmed by Schewe et al. [75]. Further investigations revealed that the site of action of DCMO is the nonheme-iron-sulfur protein ($FeSP_P$) of the succinodehydrogenase complex (see Fig. 4). Carboxin attacks the respiratory chain at the substrate side of cytochrome b and ubiquinone and at the oxygen side of the point of action of the respiratory inhibitor, 2-thenoyltrifluoracetone (TTFA). NADH oxidase is quite insensitive to DCMO which, in contrast to the chemically related salicylic acid anilide, does not uncouple oxidative phosphorylation [75]. Under the influence of DCMO [73], mitochondria show strong swelling and structural disintegration after some hours (see Fig. 5).

A comparison was made of inhibition of SDH activity by several derivatives of carboxin in fungal mitochondria and ETP (electron transporting particles) of a sensitive species (Trametes versicolor, a Basidiomycete) and of an insensitive species (Trichoderma viride, an Ascomycete) and in ETP from beef heart preparations. These studies lead to the conclusion that it is not penetration through the mitochondrial membrane, but structural differences of the receptor (the $FeSP_P$ region) that are probably responsible for the selective action toward some groups of fungi [76]. The lipid partition coefficient is not directly related to activity, but other features of the molecule are important. From structure-activity analyses a model of the receptor region for DCMO was proposed [76] (see Fig. 6). A more detailed model is under investigation. It explains most of the necessary properties of an effective carboxin derivative.

Generally a variation in the oxathiin ring is of minor importance in selectivity as compared to a variation in the aniline ring system. A remarkable change of selectivity occurs upon introduction of a phenyl ring at the o-position of the aniline ring. This compound, F 427, see Figure 3 (2), has been described by Edgington and Barron [77]. It has a broader activity against Ascomycetes and Phycomycetes than carboxin, although a significant selectivity is still retained. This could be confirmed by our own investigations

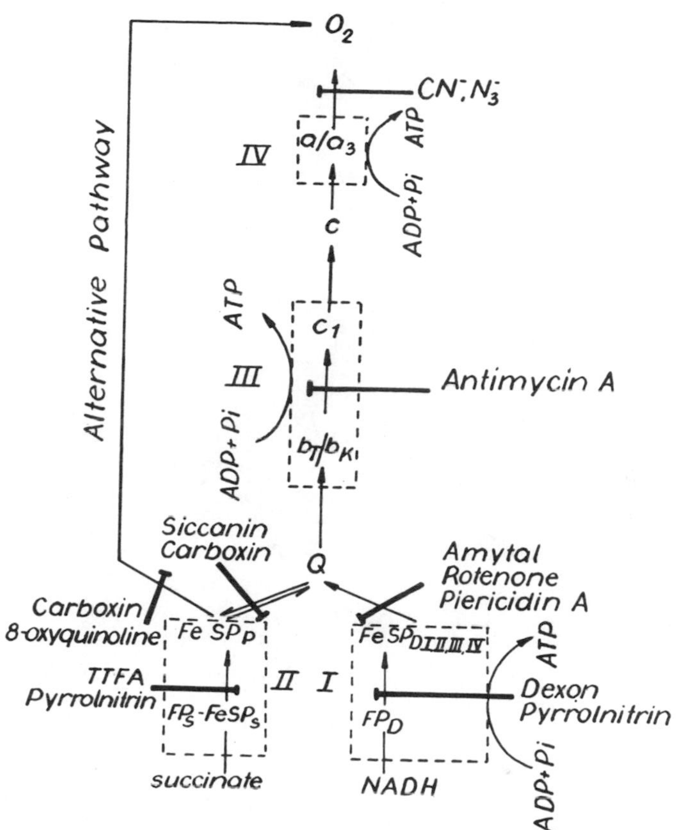

FIG. 4. Scheme of the respiratory chain showing the alternative pathway and the sites of energy conservation. The points of attack of some fungicides are indicated. FeSP = nonheme-iron-sulfur proteins, FP = flavoproteins, Q = ubiquinone; b_1, c_1, c, a = corresponding cytochromes, Pi = inorganic phospate.

with isolated mitochondria [76,114]. Within groups of fungi, e.g., the Basidiomycetes or Ascomycetes, and in plants and higher animals, small structural differences within the receptor region seem to occur which are responsible for the relative, but not absolute selectivity of carboxin. The Basidiomycetes are inhibited at concentrations of about 10^{-6} M, whereas inhibition of insensitive fungi or higher plants requires 10-100 times higher concentrations [78]. The receptors may possess a different amino acid sequence, according to the phylogenetic background which is responsible for the binding affinity and orientation. The active center is apparently the same in all species. In this case, we have for the first time detailed information on the basis of selectivity of a systemic fungicide.

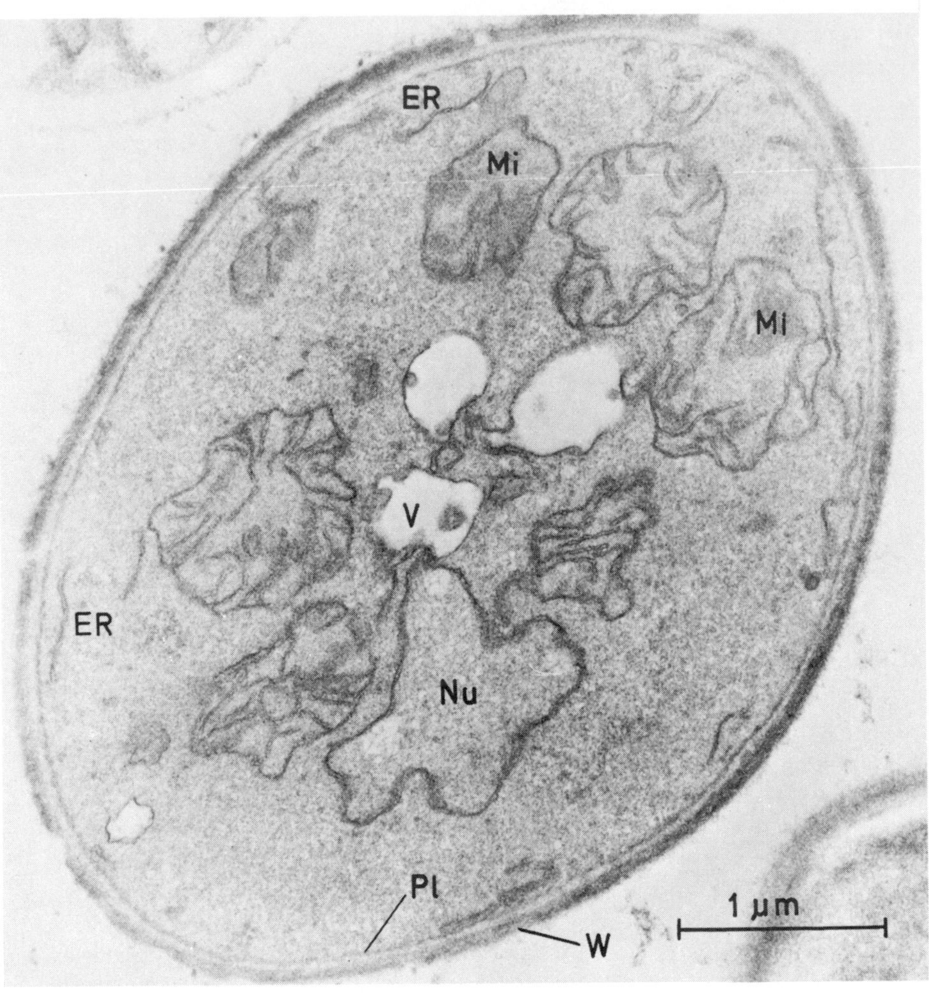

FIG. 5. Mitochondria (Mi) of Rhodotorula mucilaginosa 2 hr after application of 1×10^{-5} M carboxin. The mitochondria and cristae are irregularly shaped (compare with Fig. 1). ER = endoplasmic reticulum, V = vacuole, Nu = nucleus, Pl = plasmalemma, W = cell wall. $KMnO_4$ fixation; magnification 36,000. (Taken from Ref. 73.)

Carboxins are highly active in obligate aerobic organisms. As already mentioned, there is no absolute correlation, but only a phylogenetic convergence between preponderance of the HMP pathway and a carboxin-sensitive receptor structure in the $FeSP_P$ region of mitochondria.

Another remarkable fact is the inhibition of the so-called "alternative pathway" of respiration by carboxin which was clearly demonstrated by Sherald and Sisler [79]. This effect was investigated in detail by Lyr and

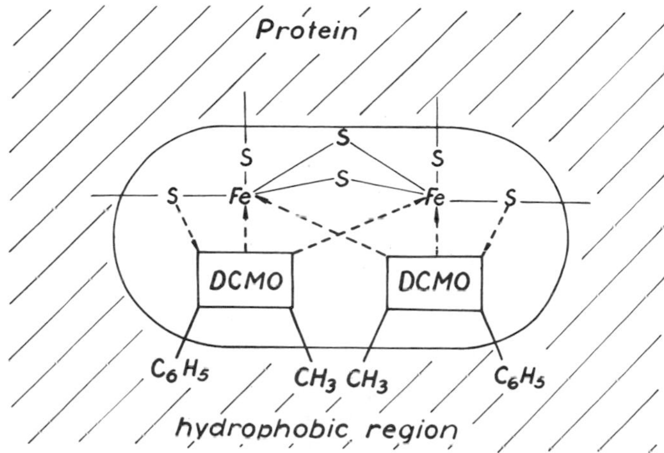

FIG. 6. Scheme of the proposed structure of the receptor for carboxin (DCMO) within complex II of the respiratory chain.

Schewe [80] and led to a new concept of the "alternative oxidase" in fungi and plants. From our experimental results and data available in the literature the following hypothesis was proposed:

The nonheme-iron-sulfur protein ($FeSP_P$) which is associated with the succinate-ubiquinone-reductase complex (complex II) and which is the locus of attack of carboxin, is the "alternative terminal oxidase." This is possible because of its capacity for autoxidation. The prerequisite and limiting step for autoxidation is the formation of an oxygenated complex with the reduced form of $FeSP_P$. Oxygen should react directly with free coordination points of the central Fe^{2+} atoms, which have a coordination number of 4 in iron-sulfur proteins, although Fe complexes can reach a maximal coordination number of 6. This coordinative binding of oxygen is inhibited by certain chelating compounds (e.g., benzohydroxamic acid, 8-oxyquinoline, TTFA, or carboxin). The mechanisms of action of these compounds differ in detail; therefore only carboxin and TTFA inhibit the normal pathway of respiration as well, whereas benzohydroxamic acid and 8-oxyquinoline at appropriate concentrations are without effect on this pathway (see Fig. 7).

The "alternative pathway" yields mainly heat and only a small amount of ATP. It seems to be an ancient part of the respiration system, whereas the more effective but more complicated cytochrome chain is a newer development of the mitochondrium. The wide distribution of the alternative pathway in the plant and fungal kingdoms [80] speaks for an ecological importance, although no proof exists for this. Such a function could be heat production where energy consumption for growth is limited by low temperature, or the production of metabolites by keeping the citric acid cycle working through removal of reducing equivalents. Some fungi and higher plants are able to grow even at temperatures below 0°C.

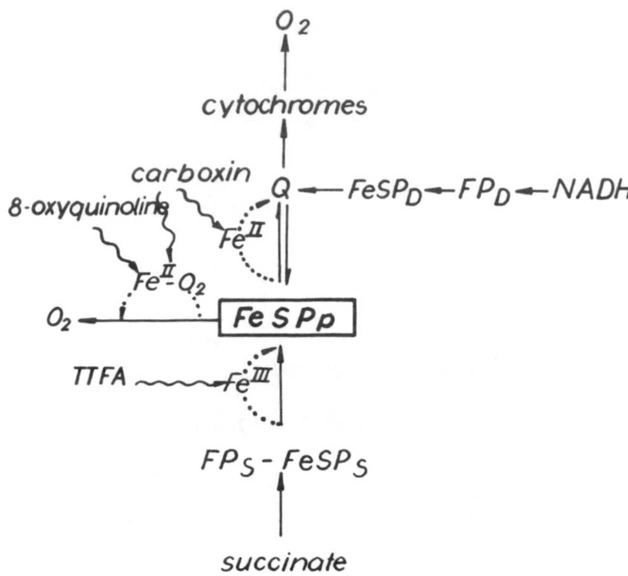

FIG. 7. Illustration of the points of attack of carboxin and 8-oxyquinoline within complex II showing their differential action on the normal respiratory chain and the alternative pathway. $FeSP_P$ = nonheme-iron-sulfur protein of complex II (succinodehydrogenase complex).

The occurrence of the "alternative pathway of respiration" is probably determined by the capacity of the $FeSP_P$ part of SDH for autoxidation. This seems to vary under certain cultural conditions or age in fungi [108], and in higher animals the system has apparently been completely lost. Small structural variations in the active center can account for this property. The affinity of the "alternative terminal oxidase" for oxygen is much lower than that of cytochrome oxidase [79]. Georgopoulos et al. [81] were successful in selecting mutants of Ustilago maydis resistant to carboxin. A single-gene mutation affects the behavior of the succinodehydrogenase system of mitochondria. The authors conclude that "some component of the system itself has been modified," which seems to be the FeSP part as described above.

2. Dexon

Another selective fungicide, mainly active in vivo against Oomycetes is dexon, p-dimethylaminobenzodiazosulfonic acid, see Figure 8, (5). It moves within the xylem [82] and gives good protection against Aphanomyces euteiches in pea plants [83] but is readily inactivated, even within the plant, by irradiation [84]. Detailed investigations by Tolmsoff [85] revealed that

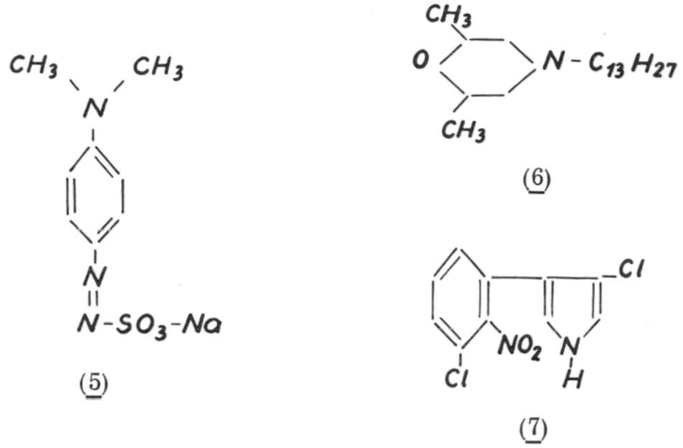

FIG. 8. Chemical structures of Dexon (5), Tridemorph (6), and Pyrrolnitrin (7).

the selective toxicity to Oomycetes, as compared to other fungi (*Pythium* is about 10-30 times more sensitive than *Rhizoctonia* or *Fusarium* and other Ascomycetes), is mainly due to penetration barriers. Disintegration of the cells by detergents or other means abolishes selectivity at the subcellular level. Tolmsoff [85] concluded that dexon strongly inhibits NADH-cytochrome c oxidase and that the site of action is similar to that of rotenone. Investigations with ETP from beef heart mitochondrial preparations by Halangk and Schewe [86] revealed that inhibition in the NADH oxidase region is dependent on electron flux in the respiratory chain. Since NADH-ubiquinone reductase, NADH ferricyanide, and NADH-menadione reductase (which do not react via the FeSP part of complex I) are also inhibited, the site of action must be nearer the substrate side than that of rotenone. The point of attack seems to be the FMN part of NADH dehydrogenase (see Fig. 4). It is possible that an intermediary flavosemiquinone reacts irreversibly by a radical mechanism with the active form of the dexon molecule.

Several other flavin enzymes were investigated regarding their inhibition by dexon (see Table 4). The data show that some flavin enzyme systems are sensitive to dexon, whereas others are not. Succinodehydrogenase, for example, is insensitive. Dexon seems to be a group reagent for some NADH-dehydrogenases with flavin structure, but not for all flavin-containing enzyme systems. The structural and functional prerequisites for its inhibitory action are not yet known, nor is there a clear correlation of the nature of the prostetic group (FMN or FAD) or the stereospecificity of dehydrogenation and inhibitory action. It is only certain that the presence of a metal component is not important for the inhibition by dexon.

In vivo, inhibition of the respiratory chain at complex I seems to be most important. The significance of the inhibition of other systems has not

TABLE 4

Sensitivity of Some Flavin Enzyme Systems to Dexon

Sensitive systems	Insensitive systems
NADH dehydrogenase of the mitochondrial respiratory chain (soluble and structurally bound)	Succinodehydrogenase
NADH dehydrogenase from the outer surface of the inner membrane of plant mitochondria	NADH-cytochrome b_5 reductase from microsomes (rat liver)
DT diaphorase from rat liver cytosol (NADH as substrate)	Soluble NADH oxidase from *Escherichia coli*
	Dicoumarol insensitive NADH dehydrogenase from rat liver cytosol
	DT diaphorase from rat liver cytosol (NADPH as substrate)

yet been investigated, but the results indicate that a multisite inhibition is highly probable. This would diminish the danger of rapid acquisition of resistance to this compound.

In the dark, the compound is rather stable, which also holds for soil conditions; therefore a practical application as a soil fungicide or for seed dressing is possible. Some bacteria such as *Pseudomonas fragi* carry out a reductive degradation which yields *N*,*N*-dimethyl-*p*-phenylenediamine as the first degradation product [87].

3. Tridemorph

Tridemorph, see Figure 8 (*6*), is a valuable systemic fungicide for cereal mildews, especially barley mildew caused by *Erysiphe graminis* [88]. It also acts on several other fungi with differing intensity. The ED_{50} varies from 0.02 ppm for *Erysiphe graminis* to 300 ppm for *Fusarium oxysporum*. Very little is known regarding the mechanism of action and the cause for this selectivity. Some facts point to the possibility that energy production is primarily affected. At a concentration of 10^{-4} M, oxygen consumption in vivo is inhibited about 30%. Combination with 8-oxyquinoline results in an additive effect in a certain range. A strong synergistic effect is achieved with antimycin A at concentrations where oxygen consumption by the latter is not yet impaired. The combination completely inhibits respiration [89]. In ETP of isolated beef heart mitochondria, tridemorph exhibits a strong inhibition of the NADH oxidase system (ED_{50}, 7×10^{-6} M) and of the succinate-cytochrome c oxidoreductase system (ED_{50}, 2.3×10^{-5} M). Using spectroscopy to study various parts of the electron transport system, three points of attack could be demonstrated [90]. One point of attack seems to be analogous to that of rotenone and another to that of antimycin A. Furthermore, cytochrome oxidase is inhibited. This means that tridemorph in vitro preferably inhibits the sites of energy conservation. Therefore, ATP synthesis could be impaired and this might account for growth inhibition and inhibition of other processes [89]. Whether this mechanism is operative in vivo as the primary mechanism of action still remains to be proven.

4. Pyrrolnitrin

This compound, see Figure 8 (7), has been used experimentally for the therapy of animal mycoses, but it is less effective than amphotericin B or 5-fluorocytosine [91]. According to Tripathi and Gottlieb [92], the incorporation of [^{14}C]glucose into nucleic acids and proteins of intact cells of Saccharomyces cerevisiae is inhibited as well as the incorporation of [^{14}C]-uracil, [^{3}H]thymidine, and [^{14}C]amino acids into RNA, DNA, and protein, respectively. However, the in vitro protein synthesis of Rhizoctonia solani and Escherichia coli is insensitive to this fungicide. A damage of the cell membrane could be excluded. Further investigations revealed that the primary site of attack is the respiratory chain. At growth-inhibiting concentrations, pyrrolnitrin inhibits the activity of succinodehydrogenase, NADH oxidase, succinate-cytochrome c reductase, NADH-cytochrome c reductase, and succinate-coenzyme Q reductase. The antimycin A-insensitive reduction of dichlorophenol-indophenol and tetrazolium dyes is inhibited as well. The site of inhibition probably lies before cytochrome c and between succinodehydrogenase and ubiquinone (see Fig. 4). A direct reaction with the succinoflavoprotein is improbable. Respiration of Saccharomyces cerevisiae is totally inhibited at 50 μg/ml of the fungicide. Wong et al. [93] concluded that the primary site of action is a block of electron transfer between the flavoprotein of NADH dehydrogenase and cytochrome b in Microsporum gypseum, at the same locus where rotenone blocks electron flux. This could be the nonheme-iron-sulfur region (FeSP) of complex I. Other systems are attacked at higher concentrations. In contrast to these results, Lambowitz and Slayman [94] emphasize that pyrrolnitrin completely uncouples oxidative phosphorylation in Neurospora crassa at lower concentrations (160 μM) and inhibits electron transfer in the flavin region of NADH- or succinodehydrogenase at higher concentrations (200-400 μM). With more than 500 μM the function of cytochrome oxidase is impaired. The action of this compound is rather unselective. Animal respiration is inhibited as well. Similar inhibition of respiration was found in Candida utilis, which was explained as an effect on the mitochondrial membrane. Further investigations are needed to elucidate the exact mechanism of attack in the mitochondrial system. Nose and Arima [95] state that pyrrolnitrin forms complexes with phospholipids of the bacterial cell membrane and inhibits growth by this mechanism.

5. Thiabendazole

Many results point to the fact that thiabendazole, see Figure 9 (8), acts primarily as an inhibitor of nuclear division, as is well established for the benzimidazole fungicide, benomyl, and related compounds. But Allan and Gottlieb [96] described an inhibition of respiration by thiabendazole in *Aspergillus*, which could be a side effect of this fungicide. More detailed investigations, however, are needed to clarify whether this is also an important mechanism responsible for a fungicidal effect.

6. Siccanin

Siccanin is an antibiotic isolated from cultures of *Helminthosporium siccans*. The chemical structure was established by Hirai et al. [97]. It inhibits the growth of various fungi especially *Trichophyton*, *Epidermophyton*, and *Microsporum*. Nose and Endo [98] demonstrated that respiration is primarily affected and that succinodehydrogenase is the most sensitive site of action (see Fig. 4). A concentration of 0.03 μg/ml caused 50% inhibition of SDH activity, but had no effect on the exchange reaction between inorganic phosphate and ATP or ADP and ATP. Phosphorylation was inhibited to the same extent as respiration; therefore the phosphorous/oxygen ratio did not change. No effect was observed on the adenosine triphosphorylase of *Trichophyton mentagrophytes*. Several other cell functions were also impaired, but this is explained as secondary effects resulting from respiratory

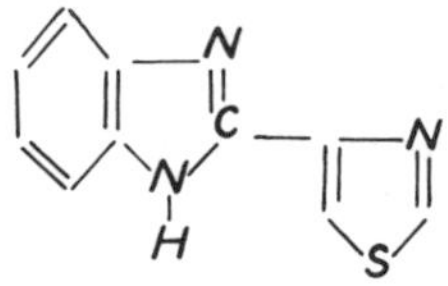

(8)

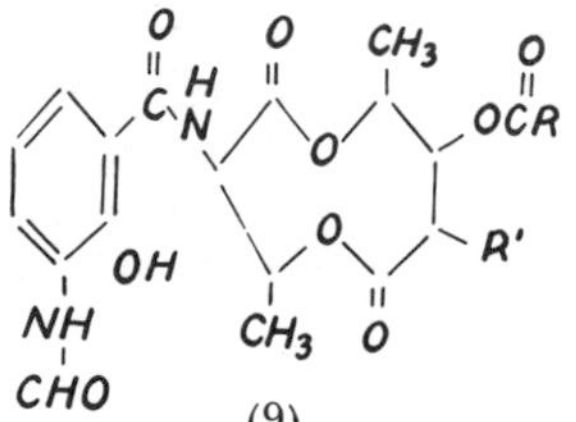

(9)

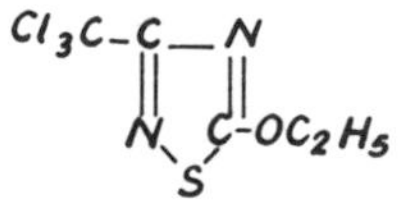

(10)

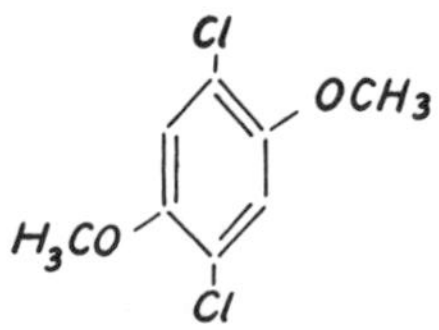

(11)

FIG. 9. Chemical structures of thiabendazole (8), antimycin A (9), terrazole (10), and chloroneb (11).

inhibition. The exact point of attack in the succinodehydrogenase system (complex II) has not yet been characterized.

7. Piericidin A and Antimycin A

Piericidin A, a highly specific inhibitor of the electron flux through complex I of the respiratory chain of fungi, acts at the same site as rotenone or amytal. It seems to combine with the FeSP region of this complex. The inhibition probably involves multiple binding of high specificity [99]. The energy conservation site of complex I and the piericidin-sensitive site are not identical as has been shown in *Torulopsis utilis* [100]. Piericidin A as well as antimycin A, see Figure 9 (9), which blocks electron flux between cytochrome b and c (see Fig. 4), are very toxic antibiotics for the respiratory system of several groups of organisms. Although some differences in inhibition in vivo have been described [109], the antibiotics are not used as fungicides in practice but merely as biochemical tools of high specificity.

C. Other Effects

1. Terrazole

Terrazole, see Figure 9 (10), is a systemic fungicide of rather high specificity toward Oomycetes. It inhibits spore germination and hyphal growth of sensitive organisms at concentrations of 1-10 ppm in the medium. The compound produces vacuolization and a deformation of hyphal tips. Addition of amino acids, nucleic acids, fatty acids, vitamins, cholesterol, ergosterol, or lecithin do not reverse the inhibitory effect of terrazole on the growth of *Mucor mucedo* [104]. Production of CO_2 and oxygen consumption are only slightly impaired at lower concentrations, and membrane permeability shows no severe disturbance. Electron microscope observations reveal a significant lysis of mitochondria, which is not absolutely uniform within the mitochondrial population (see Figs. 10-12). Especially the inner membrane is attacked, as indicated by a swelling of the cristae and an increasing vacuolization resulting in complete lysis. The outer membrane has a tendency for enlargement. The plasmalemma of the cell membrane forms invaginations and has an abnormal pattern. The cell wall is often thickened up to 700%. Comparative investigations indicate that the primary effect is on the mitochondria [105]. The phosphorus/oxygen ratio decreases and in some experiments a surplus of inorganic phosphate can be detected, although oxygen consumption is inhibited only moderately. The mechanism of action is not yet fully elucidated, but there are indications that phospholipases are directly or indirectly activated, and that these enzymes may be responsible for the attack on the mitochondrium. The reason of the relatively high specificity toward Oomycetes needs further investigations [113].

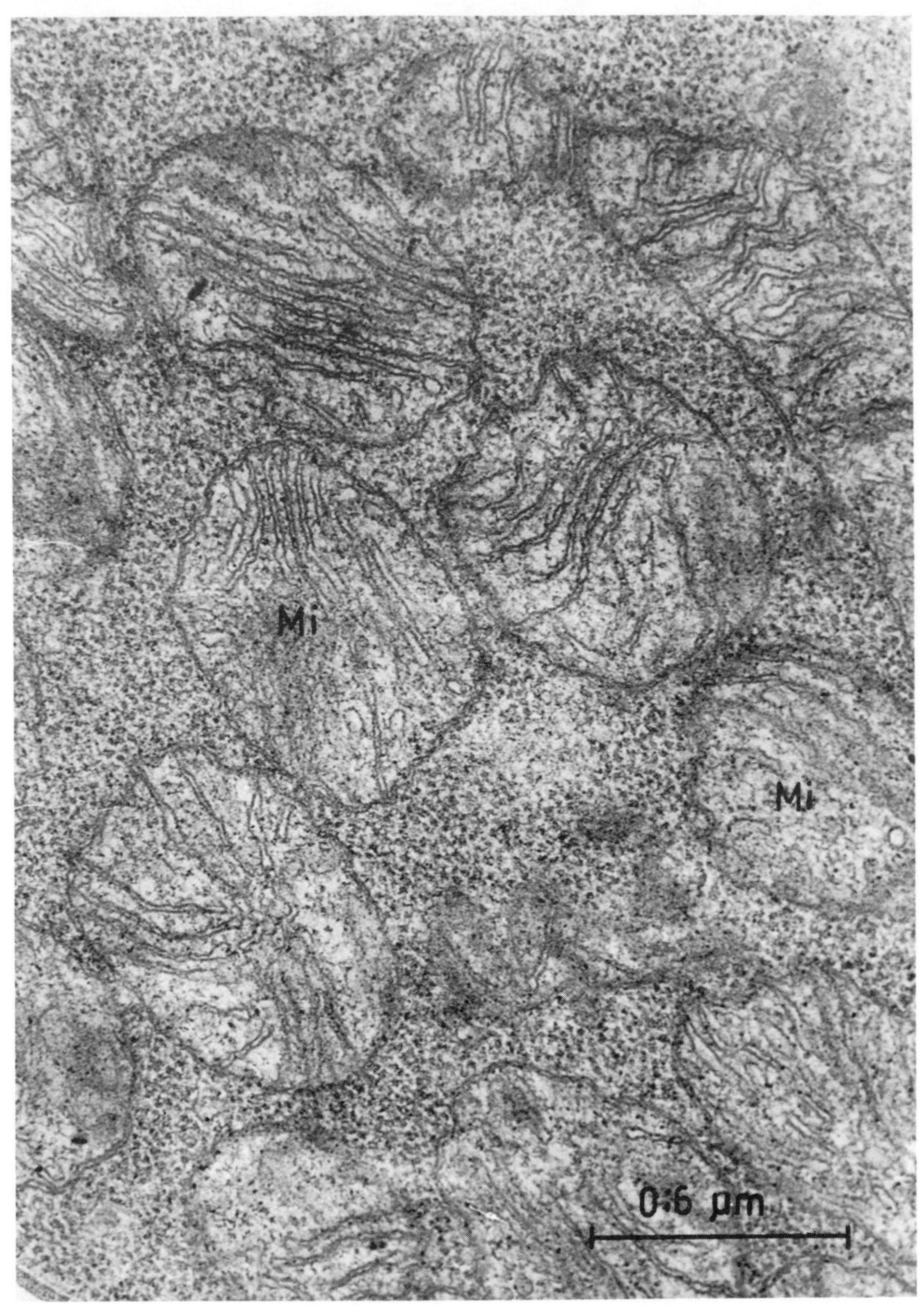

FIG. 10. Normal mitochondrial structure in *Mucor mucedo*. The cytoplasm is filled with densely packed ribosomes which are also present within the mitochondria. OsO_4 fixation; magnification 60,000. (Photo by G. Casperson)

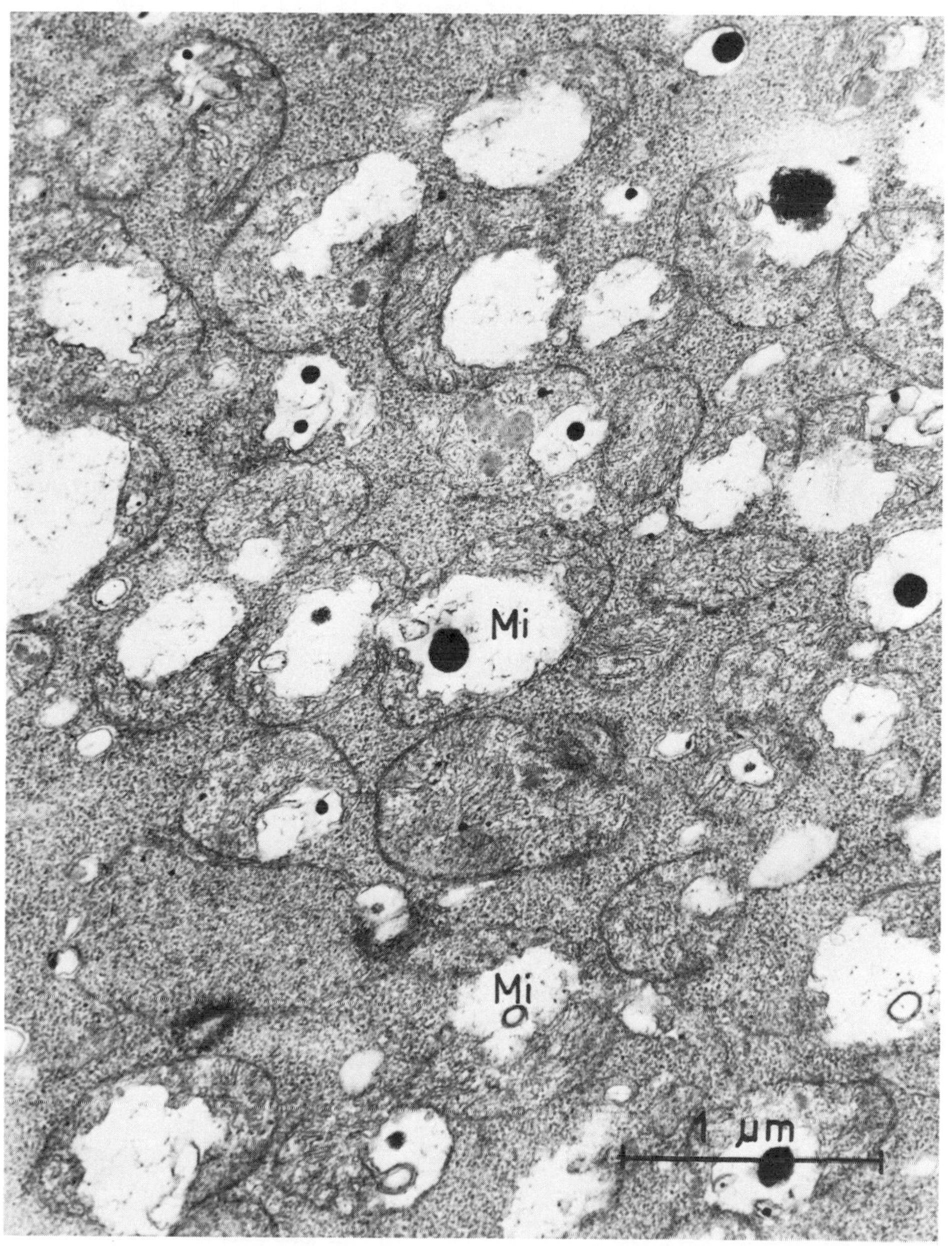

FIG. 11. Structure of mitochondria in Mucor mucedo grown in a glucose asparagine medium containing 10 ppm terrazole. The mitochondria (Mi) show varying stages of lysis. OsO_4 fixation; magnification 36,000. (Photo by G. Casperson)

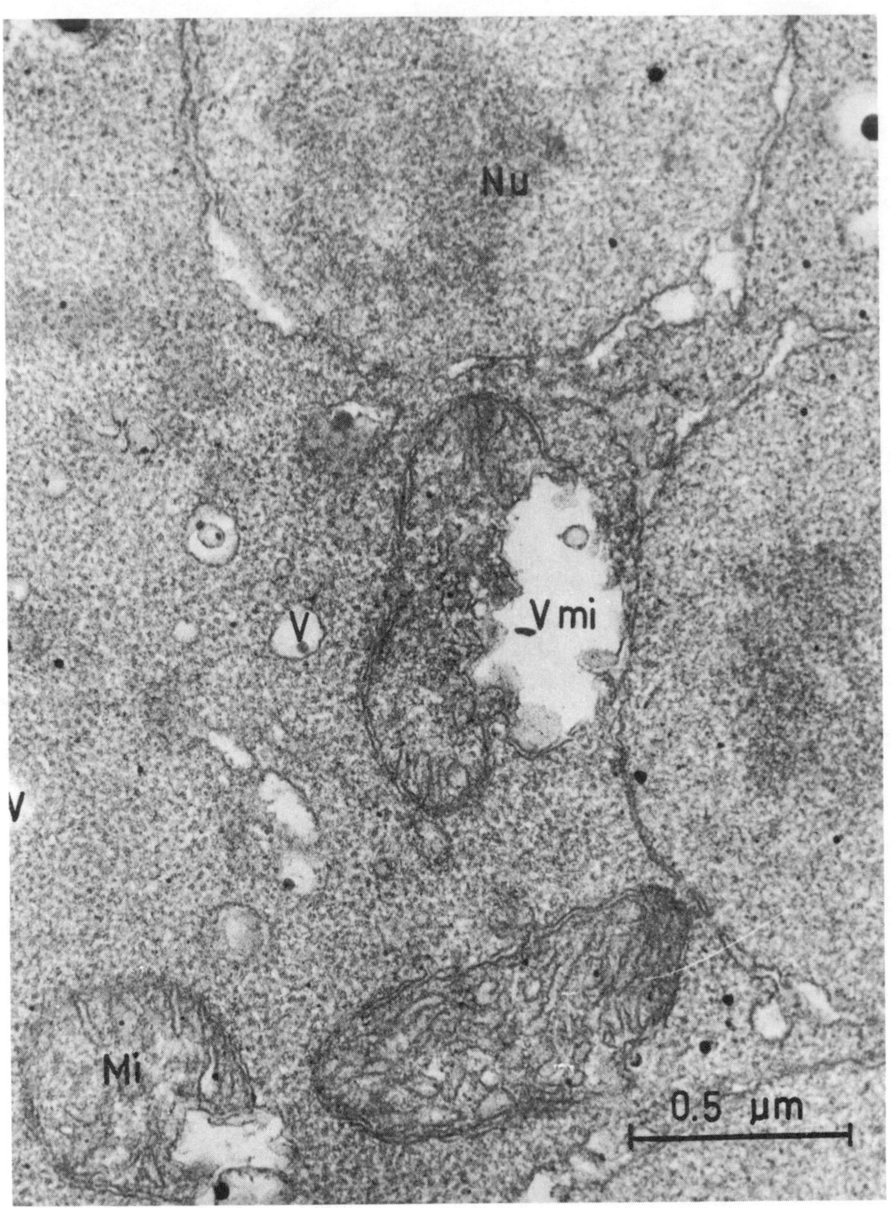

FIG. 12. Detailed illustration of the effect of 10 ppm of terrazole on the mitochondrial structure of Mucor mucedo. The membranes in particular are attacked by the fungicide. The inner mitochondrial membrane forming the cristae is most sensitive. The membrane begins to swell and later disintegrates. Lysis begins at certain regions of the mitochondrium. The more resistant outer membrane enlarges and forms large vacuoles (Vmi). The nuclear envelope membranes separate from each other. Nu = nucleus, V = cell vacuole. OsO_4 fixation; magnification 60,000. (Photo by G. Casperson)

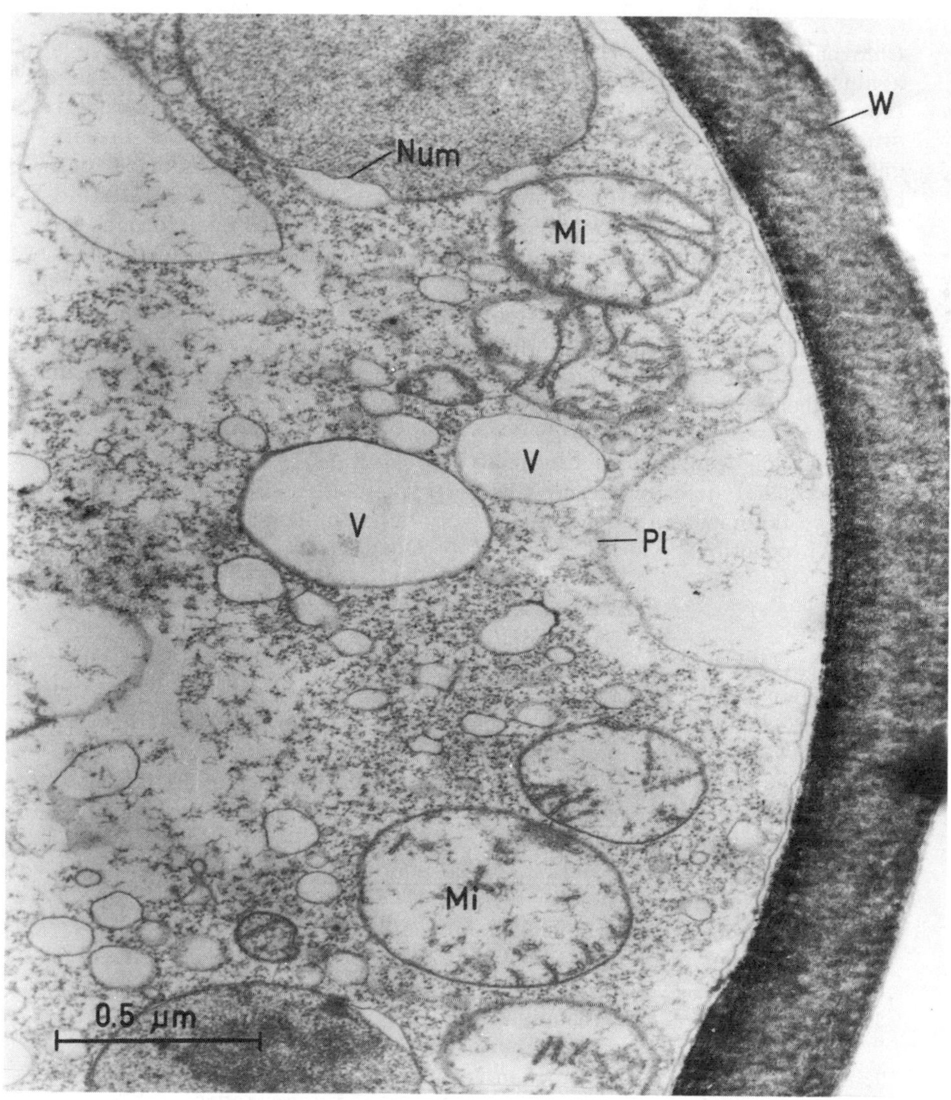

FIG. 13. Mitochondria of <u>Mucor mucedo</u> after treatment with 30 ppm chloroneb for 1 hr. The fungicide induces a lysis of the inner membrane system of the mitochondria (Mi), vacuolization (V) of the cytoplasm, a swelling of the nuclear envelope (Num), and an invagination of the plasmalemma (Pl). W = cell wall of the germinating spore. OsO_4 fixation; magnification 60,000. (Photo by G. Casperson)

2. Chloroneb

Chloroneb, see Figure 9 (11), is a systemic fungicide which is effective against Oomycetes. Although its mechanism of action is not yet elucidated, there are indications that it has lytic effects on the mitochondrial structure similar to those of terrazole described above. The compound was investigated by Tillman and Sisler [110], who came to the conclusions that it inhibits cell division in a still unexplained manner. The ultrastructure of the very sensitive fungus Mucor mucedo is markedly changed by chloroneb (see Fig. 13). Most remarkable is the vacuolization of the mitochondria and a pathological thickening of the cell wall. The effect resembles that of terrazole (see Fig. 12). The pathological apposition growth of the cell wall seems to be a secondary effect which may result from an inhibited ATP formation. However, Tillman and Sisler found no inhibition of oxygen consumption in their organisms, therefore the mode of action of chloroneb needs further investigations.

The action described for terrazole and chloroneb is an example of an unconventional attack on the respiratory system. Although not absolutely specific, it demonstrates one of a variety of possibilities of inhibition of energy production.

REFERENCES

1. H. Lyr and W. Luthardt, Nature, 207, 753 (1965).
2. B. E. Schultz, G. Kraepelin, and W. Hinkelmann, J. Gen. Microbiol., 82, 1 (1974).
3. G. Casperson and H. Lyr, Z. Allgem. Mikrobiol., 15, 481 (1975).
4. D. C. Torgeson, Fungicides, Vol. 1, Academic Press, New York, 1969.
5. J. L. Webb, Enzyme and Metabolic Inhibitors, Vol. 3, Academic Press, New York, 1966.
6. R. G. Owens, Contr. Boyce Thompson Inst., 17, 221 (1953).
7. R. G. Owens and H. M. Novotny, Contr. Boyce Thompson Inst., 20, 171 (1959).
8. R. J. Lukens, Ph.D. Thesis, University of Maryland, College Park, 1958.
9. M. R. Siegel, Pestic. Biochem. Physiol., 1, 225 (1971).
10. M. R. Siegel, Pestic. Biochem. Physiol., 1, 234 (1971).
11. G. Ritter, E. Kluge, and H. Lyr, Z. Allgem. Mikrobiol., 13, 243 (1973).
12. H. Lyr, Enzymologia, 23, 231 (1961).
13. J. A. Gascoigne and M. M. Gascoigne, Biological Degradation of Cellulose, Butterworths, London, 1960.
14. F. Grossmann, Naturwissenschaften, 49, 138 (1962).

15. J. O. Lampen, in Biochemical Studies of Antimicrobial Drugs, 16th Symposium of the Society of General Microbiology (B. A. Newton and P. E. Reynolds, eds.), Cambridge University Press, Cambridge, Mass., 1966, p. 111.
16. H. Lyr, Z. Allgem. Mikrobiol., 7, 373 (1967).
17. R. Tröger, Arch. Mikrobiol., 48, 282 (1964).
18. H. G. Schlegel, Flora (Jena), 141, 1 (1954).
19. J. Raa and J. Goksoyr, Physiol. Plantarum, 18, 159 (1965).
20. O. Hoffmann-Ostenhof, Enzymologie, Springer-Verlag, Vienna, 1954.
21. R. J. W. Byrde, J. T. Martin, and D. J. D. Nicholas, Nature, 179, 638 (1956).
22. M. W. Foote, J. E. Little, and T. J. Sproston, J. Biol. Chem., 181, 481 (1949).
23. C. J. Sih, P. B. Hamilton, and S. G. Knight, J. Bacteriol., 75, 623 (1958).
24. G. Favelukes and A. O. M. Stoppani, Biochim. Biophys. Acta, 28, 654 (1958).
25. A. O. M. Stoppani, A. S. Actis, J. O. Defarrari, and E. L. Gonzalez, Biochem. J., 54, 378 (1953).
26. A. Medina and D. J. D. Nicholas, Biochem. J., 66, 573 (1957).
27. M. A. Johnson and D. S. Frear, Phytochemistry, 2, 75 (1963).
28. S. Washio and Y. Mano, J. Biochem. (Tokyo), 48, 874 (1960).
29. R. J. Lukens and H. D. Sisler, Phytopathology, 48, 235 (1958).
30. O. Hoffmann-Ostenhof and E. Kriz, Monatsh. Chem., 80, 720 (1949).
31. K. Dresel, Biochem. Z., 192, 351 (1928).
32. T. C. Montie and H. D. Sisler, Phytopathology, 52, 94 (1962).
33. N. J. Turner and R. Battershell, Contr. Boyce Thompson Inst., 24, 203 (1970).
34. R. G. Owens and G. Blaak, Contr. Boyce Thompson Inst., 20, 459 (1960).
35. H. D. Sisler and N. L. Marshall, J. Wash. Acad. Sci., 47, 321 (1957).
36. J. J. Habrekke and J. Goksoyr, Physiol. Plantarum, 23, 517 (1970).
37. H. L. Klöpping, Ph.D. Thesis, University of Utrecht, 1951.
38. S. E. A. McCallan and L. P. Miller, Contr. Boyce Thompson Inst., 18, 483 (1957).
39. R. A. Olsen, Physiol. Plantarum, 24, 534 (1971).
40. L. Ferenczy and F. Kevei, in Mechanisms of Action of Fungicides and Antibiotics, 1st Symposium Reinhardsbrunn, Akademie-Verlag, Berlin, 1967, p. 59.
41. D. Gottlieb, H. E. Carter, J. H. Sloneker, L. Wu, and E. Gaudy, Phytopathology, 51, 321 (1961).
42. E. Borowski and B. Cybulska, in Mechanisms of Action of Fungicides and Antibiotics, 1st Symposium Reinhardsbrunn, Akademie-Verlag, Berlin, 1967, p. 39.

43. H. Lyr, in Mechanisms of Action of Fungicides and Antibiotics, 1st Symposium Reinhardsbrunn, Akademie-Verlag, Berlin, 1967, p. 27.
44. A. Fluharty and D. R. Sanadi, J. Biol. Chem., 236, 2772 (1961).
45. J. L. Howland, Biochim. Biophys. Acta, 77, 659 (1963).
46. C. R. Bovell, L. Packer, and G. R. Schonbaum, Arch. Biochem. Biophys., 104, 458 (1964).
47. W. N. Aldridge, Biochem. J., 69, 367 (1958).
48. H. I. Hadler, B. E. Claybourn, and T. P. Tschang, J. Antibiotics (Tokyo), 23, 276 (1970).
49. M. Lieberman and J. B. Biale, Plant Physiol., 31, 420 (1956).
50. E. C. Weinbach and J. Garbus, Nature, 221, 1016 (1969).
51. R. D. Hotchkiss, Advan. Enzymol., 4, 153 (1944).
52. W. F. Loomis and F. Lipmann, J. Biol. Chem., 173, 807 (1948).
53. E. C. Slater, in Proc. 5th Intern. Congr. Biochemistry, Vienna, Pergamon Press, Oxford, 1963, p. 325.
54. H. Lyr and H. Ziegler, Phytopathol. Z., 36, 146 (1959).
55. R. Gönnert, J. Johannis, E. Schraufstätter and R. Strute, Med. Chem. Abhandl., Med. Chem. Forschungsstätten Farbenfabriken Bayer, 7, 540 (1963).
56. B. C. Pressman and H. A. Lardy, Biochim. Biophys. Acta, 21, 458 (1956).
57. R. B. Beechey, Biochem. J., 98, 284 (1966).
58. P. G. Vincent and H. D. Sisler, Physiol. Plantarum, 21, 1249 (1968).
59. B. Mackler and B. Haynes, Biochim. Biophys. Acta, 147, 317 (1970).
60. D. G. Smith and R. Marchant, Arch. Mikrobiol., 60, 262 (1968).
61. G. D. Clark-Walker and A. W. Linnane, J. Cell. Biol., 34, 1 (1967).
62. K. Harada, R. Higuchi, and I. Utsumi, Agr. Biol. Chem. (Tokyo), 32, 940 (1968).
63. H. Lyr, Flora (Jena), 150, 227 (1961).
64. H. Lyr, Flora (Jena), Abt. A, 157, 305 (1966).
65. J. Raa and J. Goksoyr, Physiol. Plantarum, 19, 840 (1966).
66. W. Luthardt, Acta Biol. Med. Ger., 19, 199 (1967).
67. H. Lyr, Phytopathol. Z., 52, 229 (1965).
68. B. Schmeling and M. Kulka, Science, 152, 659 (1966).
69. D. E. Mathre, Phytopathology, 58, 1464 (1968).
70. D. E. Mathre, Phytopathology, 60, 671 (1970).
71. N. N. Ragsdale and H. D. Sisler, Phytopathology, 60, 1422 (1970).
72. H. Lyr, W. Luthardt, and G. Ritter, Z. Allgem. Mikrobiol., 11, 373 (1971).
73. H. Lyr, G. Ritter, and G. Casperson, Z. Allgem. Mikrobiol., 12, 275 (1972).
74. D. E. Mathre, J. Agr. Food Chem., 19, 872 (1971).
75. T. Schewe, S. Rapoport, G. Böhme, and W. Kunz, Acta Biol. Med. Ger., 31, 73 (1973).

76. H. Lyr, T. Schewe, D. Zanke, and W. Müller, in Mechanisms of Action of Systemic Fungicides, 4th Symposium Reinhardsbrunn, Akademie-Verlag, Berlin, 1975, p. 153.
77. L. V. Edgington and G. L. Barron, Phytopathology, 57, 1156 (1967).
78. D. E. Mathre, Bull. Environ. Contam. Toxicol., 8, 311 (1972).
79. J. L. Sherald and H. D. Sisler, Plant Cell Physiol. (Tokyo), 13, 1039 (1972).
80. H. Lyr and T. Schewe, Acta Biol. Med. Ger., 34, 1631 (1975).
81. S. G. Georgopoulos, E. Alexandri, and M. Chrysayi, J. Bacteriol., 110, 809 (1972).
82. F. J. Hills, Phytopathology, 52, 389 (1962).
83. J. E. Mitchell and D. J. Hagedorn, Phytopathology, 61, 978 (1971).
84. F. J. Hills and L. D. Leach, Phytopathology, 52, 51 (1962).
85. W. J. Tolmsoff, Abstract of Dissertation in Phytopathology, 52, 755 (1962).
86. W. Halangk and T. Schewe, in Mechanisms of Action of Systemic Fungicides, Symposium Reinhardsbrunn, Akademie-Verlag, Berlin, 1975, p. 177.
87. N. G. K. Karanth, S. G. Bhat, C. S. Valdyanathan, and V. N. Vasantharajan, Appl. Microbiol., 27, 43 (1974).
88. E. H. Pommer, S. Otto, and J. Kradel, Proc. 5th Brit. Insectic. Fungic. Conf., Vol. 2, 1970, p. 347.
89. H. Bergmann, H. Lyr, E. Kluge, and G. Ritter, in Mechanism of Action of Systemic Fungicides, 4th Symposium Reinhardsbrunn, Akademie-Verlag, Berlin, 1975, p. 183.
90. W. Müller and T. Schewe, in Mechanisms of Action of Systemic Fungicides, 4th Symposium Reinhardsbrunn, Akademie-Verlag, Berlin, 1975, p. 189.
91. R. S. Gordee and T. R. Matthews, Appl. Microbiol., 17, 690 (1969).
92. R. K. Tripathi and D. Gottlieb, J. Bacteriol., 100, 310 (1969).
93. D. T. Wong, J. Horng, and R. S. Gorde, J. Bacteriol., 106, 168 (1971).
94. A. M. Lambowitz and C. W. Slayman, J. Bacteriol., 112, 1020 (1972).
95. M. Nose and K. Arima, J. Antibiotics (Tokyo), 22, 135 (1969).
96. P. M. Allam and D. Gottlieb, Phytopathology, 60, 1282 (1970).
97. K. Hirai, S. Nosoe, K. Tsuda, Y. Iitaka, K. Ishibashi, and M. Shirasaka, Tetrahedron Letters, No. 23, 1967, p. 2177.
98. K. Nose and A. Endo, J. Bacteriol., 105, 176 (1971).
99. M. Gutman, T. P. Singer, and J. E. Casida, Biochem. Biophys. Res. Commun., 37, 615 (1969).
100. R. A. Clegg and P. B. Garland, Biochem. J., 124, 135 (1971).
101. V. H. Parker and L. A. Summers, Biochem. Pharmacol., 19, 315 (1970).

102. S. Muraoka and H. Teryada, Biochim. Biophys. Acta, 275, 271 (1972).
103. P. Liras and J. O. Lampen, Biochim. Biophys. Acta, 372, 141 (1974).
104. H. Lyr, B. Laussmann, and G. Casperson, Z. Allgem. Mikrobiol., 15, 345 (1975).
105. G. Casperson and H. Lyr, Z. Allgem. Mikrobiol., 15, 381 (1975).
106. B. G. Tweedy, in Fungicides, Vol. 2 (D. C. Torgeson, ed.), Academic Press, New York, 1969, p. 119.
107. B. G. Tweedy and N. Turner, Contr. Boyce Thompson Inst., 23, 255 (1966).
108. M. F. Henry, M. C. Hamaide-Deplus, and E. J. Nyns, J. Microbiol. Serol., 40, 79 (1974).
109. K. Singh and S. Rakhit, J. Antibiotics, 24, 704 (1971).
110. R. W. Tillman and H. D. Sisler, Phytopathology, 63, 219 (1973).
111. G. A. White, Biochem. Biophys. Res. Commün., 44, 1212 (1971).
112. G. A. White and G. D. Thorn, Pestic. Biochem. Physiol., 5, 380 (1975).
113. B. Radzuhn, Ph.D. Thesis, Humboldt Universität, Berlin (1977).
114. W. Müller, T. Schewe, H. Lyr, and D. Zanke, Z. Allgem. Mikrobiol., 17, 321 (1977).

Chapter 9

INHIBITORS OF LIPID SYNTHESIS

Nancy N. Ragsdale

Department of Botany
University of Maryland
College Park, Maryland

I. INTRODUCTION

Inhibitors of lipid synthesis offer interesting new possibilities for the control of plant pathogens. Knowledge of the action of metabolic inhibitors on lipid metabolism may be quite useful in future development of fungicides. This chapter discusses existing fungicides, as well as other compounds which have primary sites of action in the areas of lipid metabolism, involving sterols, gibberellins, carotenoids, sex hormones, fatty acids, and acetogenins.

Acetate is the basic unit in the biosynthesis of these metabolites (see Fig. 1). Systems biosynthesizing fatty acids and acetogenins form an acetoacetyl unit by condensation of malonyl and acetyl moieties. Subsequent chain elongation is carried out by further reactions with malonyl moieties. The basic difference between these two pathways is that fatty acid synthesis is reductive, whereas acetogenin synthesis is nonreductive. The initial stage of sterol, gibberellin, and carotenoid biosynthesis is frequently called the "isoprenoid pathway." Isoprenoid refers to the basic 5-carbon unit that is formed from mevalonic acid after the loss of one carbon. Subsequent reactions involve condensations of 5-carbon units. Larger moieties may condense, as in the reaction of two farnesyl units to form squalene, or may react with another 5-carbon unit as in the formation of geranylgeranyl pyrophosphate.

The functions of a number of the compounds produced by these various pathways are not fully understood; nor, for that matter, are the relationships between the pathways themselves. One of the useful aspects of metabolic

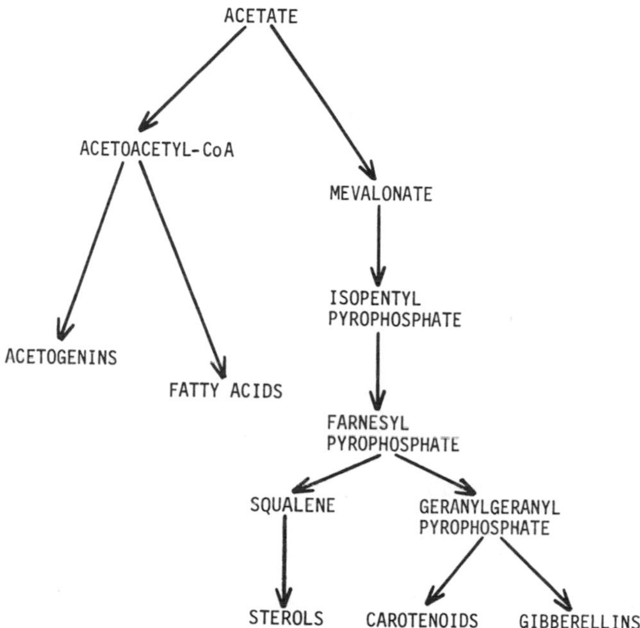

FIG. 1. Scheme of lipid biosynthesis.

studies with fungitoxic compounds is the elucidation of biochemical pathways and the establishment of relationships as well as the roles of the products.

II. STEROLS

A. Biosynthesis

The biosynthesis and role of sterols in fungi and other organisms have been the subjects of many investigations in recent years. Although a complete understanding of the functions of sterols in the cell has not been achieved, their importance in membrane structure and function [1] and their role in sexual reproduction [2] have been partially elucidated. A great deal of progress has also been made in determining biosynthetic pathways [3-5]. The first cyclic intermediate in fungal sterol synthesis is considered to be lanosterol [3]. Between lanosterol and ergosterol, which is the major sterol in most fungi, the following reactions take place (see Fig. 2):

1. Introduction of a methyl group at C-24 accompanied by a double bond shift from C-24(25) to C-24(28).

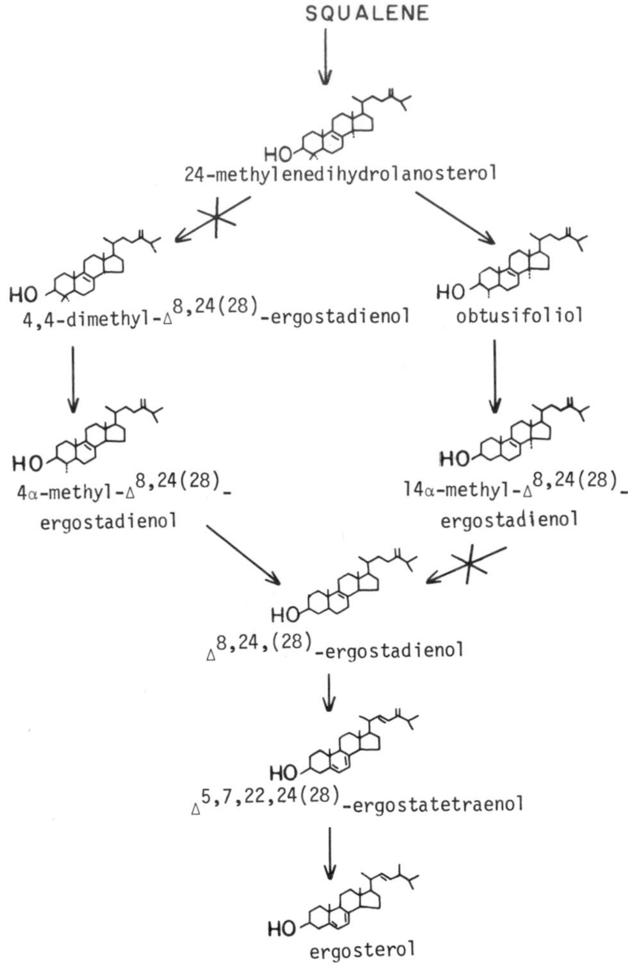

FIG. 2. Scheme of sterol biosynthesis indicating major points on inhibition by the fungicides triarimol, triforine, and S-1358.

2. Removal of three methyl groups on the steroid nucleus (two at C-4 and one at C-14).
3. Introduction of a double bond at C-22.
4. Reduction of the C-24(28) double bond.
5. Double bond shift from C-8(9) to C-7.
6. Introduction of a double bond at C-5(6).

The sequence of these reactions has not been completely elucidated. A variety of pathways may exist within a given organism, but it appears that there are preferential sequences [6, 7]. For example, in yeasts the C-24 methylation usually does not occur until certain methyl groups on the steroid nucleus are removed [6]. However, in a number of fungi, such as *Aspergillus fumigatus* [8, 9] and *Ustilago maydis* [10], C-24 methylation occurs prior to the removal of these methyl groups. Indeed, it was suggested in the case of *U. maydis* that C-24 methylation might occur before the cyclization process [10].

B. Triarimol, Triforine, and S-1358

1. General Effects

Within the past few years primary sites of action within the sterol biosynthetic pathway have been demonstrated for the newly developed fungicides triarimol (1) triforine (2), and S-1358 (3), see Table 1 and Figure 3 [11-13].

TABLE 1

Inhibitors of Sterol Synthesis

Name used in text	Chemical name	Other names
Triarimol (1)	α-(2,4-Dichlorophenyl)-α-phenyl-5-pyrimidinemethanol	EL-273
Triforine (2)	*N*,*N*'-Bis-(1-formamido-2,2,2-trichloroethyl)piperazine	CELA-W524
S-1358 (3)	*N*-3-Pyridyl-*S*-*n*-butyl-*S*-*p*-*t*-butylbenzyl imidodithiocarbonate	
Fenarimol (4)	α-(2-Chlorophenyl)-α-(4-chlorophenyl)-5-pyrimidinemethanol	EL-222
AY-9944	*trans*-1,4-Bis(2-chlorobenzylaminomethyl)cyclohexane dihydrochloride	
SKF-525-A	β-Diethylaminoethyldiphenylpropyl acetate hydrochloride	
SKF-3301-A	2,2-Diphenyl-1-(β-dimethylaminoethoxy) pentane hydrochloride	
Triparanol (6)	2-(*p*-Chlorophenyl)-1-[*p*-(2-(diethylaminoethoxy)phenyl]-1-(*p*-tolyl)ethanol	MER-29

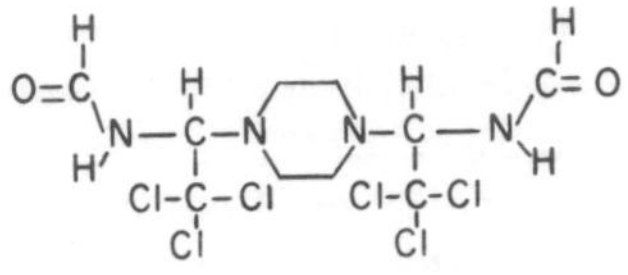

(1) triforine

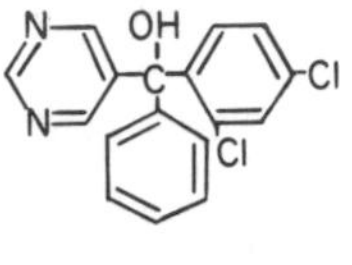

(2) triarimol

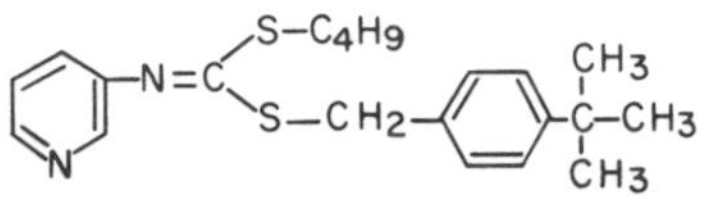

(3) S-1358

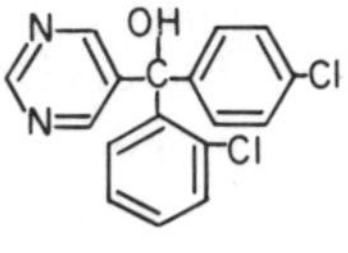

(4) fenarimol

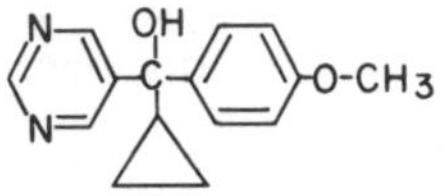

(5) ancymidol

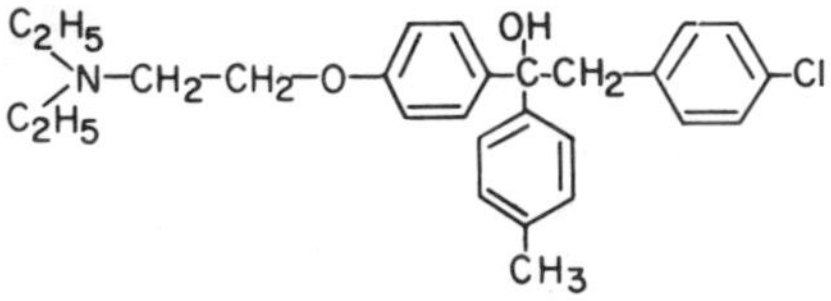

(6) triparanol

FIG. 3. Chemical structure of several compounds affecting sterol biosynthesis.

Although these compounds are quite different structurally, their antifungal spectra as well as their patterns of growth inhibition are similar. They are particularly effective against powdery mildew fungi [14-16]. All three fungicides characteristically fail to inhibit spore germination or initial cell growth and subsequently produce abnormal growth patterns. Hyphae are frequently swollen and/or excessively branched [12, 13]. Sporidia of *U. maydis*, which normally multiply by budding, become abnormally large, branched, and multicellular [17]. Each cell of the multicelled sporidia contains a nucleus indicating that mitosis continues even though sporidial multiplication ceases [17].

Triarimol, triforine, and S-1358 rapidly interfere with advanced stages of sterol metabolism. However, several other vital processes are only mildly inhibited for a considerable time after ergosterol synthesis has been strongly curtailed. None of the fungicides appreciably affect early stages of dry-weight increase, respiratory metabolism, or protein and nucleic acid syntheses [8, 12, 13, 17-21]. The compound S-1358 does not inhibit cell wall synthesis in *Monilinia fructigena* [13]. There is little or no effect during early stages of growth on various classes of lipids other than sterols [8, 13, 17]. However, in cultures treated with triarimol or triforine for longer periods there is a marked accumulation of free fatty acids [8, 10, 17, 18]. Preliminary data indicate that this accumulation results from reduced incorporation of fatty acids into triglycerides and phospholipids and possibly from degradation of existing phospholipids as well [22]. The fact that S-1358 does not produce this effect [23] may indicate a difference in precise mode of action from triarimol or triforine.

2. Effects on Sterol Synthesis

In sensitive organisms treated with triforine, triarimol, or S-1358, there is a rapid decline in C-4 desmethyl sterols (primarily ergosterol) with a concomitant increase in methyl and dimethyl sterols [8, 13, 17]. Incorporation of [^{14}C]acetate into the sterol fraction is consistent with these observations; labeling of C-4 desmethyl sterols is strongly inhibited, whereas radioactivity of the methyl and dimethyl sterols increases dramatically [8, 10, 13, 17, 24].

All three fungicides cause accumulation of 24-methylenedihydrolanosterol and obtusifoliol (see Fig. 2) [8, 10, 24]. An additional sterol, 14α-methyl-$\Delta^{8,24(28)}$-ergostadienol (see Fig. 2) accumulates after several hours in fungi treated with triarimol or triforine [8, 10]. No sterols are biosynthesized in the pathway beyond those with a C-14 methyl group; therefore, one may conclude that a primary site of action of these fungicides is the C-14 demethylation.

Further studies based on interconversion of residual sterols and labeled intermediates in the pathway between 14-methyl sterols and ergosterol indicate that triarimol also inhibits two reactions in the side chain, namely, introduction of the C-22 double bond and reduction of the C-24(28) double bond [8, 10]. Perhaps these reactions in the side chain would take place if the C-14 methyl were removed. This is supported by the fact that no naturally occurring 14-methyl sterols with a C-22 unsaturation have been isolated. Although numerous statements have been made concerning the independence of side-chain from nuclear reactions, there is undoubtedly a preferred synthetic sequence in a given organism. There is probably some specificity in the sequence of side-chain reactions as well. Failure to reduce the C-24(28) double bond might result from a failure to introduce the C-22 double bond. In *Saccharomyces cerevisiae*, for example, introduction of the C-22 double bond is specifically favored by the presence of the C-24(28) double bond [25]. The C-24(28) double bond is reduced after this reaction [25].

3. Site of Action

a. Mixed-function Oxidases. Interference with C-14 demethylation indicates that mixed-function oxidase activity is affected. However, since other mixed-function oxidase reactions such as fatty acid desaturation [10] and C-4 sterol demethylation [8, 10, 24] do not appear to be strongly curtailed, there is evidently no general interference with this type of activity. Inhibition more directly involves the C-14 methyl sterol demethylase. The methyl sterol demethylase consists of a number of enzymes and associated electron carriers, and it is quite conceivable that the C-4 and C-14 demethylases differ in some respects. In the case of C-4 demethylation, the C-4 hydroxyl group acts as an electron receptor. It has been proposed that during C-14 demethylation a hydroxyl group inserted at C-15 serves as an electron receptor and is subsequently removed to form a transient intermediate with C-14(15) double bond [26]. Possibly the enzymes peculiar to the C-14 type demethylation may be sensitive to the action of these toxicants.

Recently it has been shown that alkylimidazoles inhibit cholesterol synthesis [27]. An earlier study indicated that arylimidazoles bind to cytochrome P-450 through an interaction of the imidazole with the heme moiety of the cytochrome and interactions of the aryl groups with adjacent substrate sites [28]. The toxic effects of triarimol, triforine, and S-1358 may well result from a combination of this nature with cytochrome P-450.

Further evidence that inhibition of ergosterol synthesis is due to interference with mixed-function oxidases is provided by studies of petite mutants of S. cerevisiae [29-32]. A block in porphyrin synthesis leads to lowered activity of hematin enzymes such as those associated with mixed-function oxidation, resulting in a failure to synthesize ergosterol and the accumulation of methylated sterol intermediates. A similar phenomenon occurs in Saccharomyces carlsbergensis treated with thiamine [33]. The effect is overcome by the addition of pyridoxine, which suggests that thiamine interferes with heme synthesis [33].

b. Studies with Mutants and Antagonists. The suggestion that all three fungicides have a common site of action is supported by experimental data with Cladosporium cucumerinum mutants [8, 12]. The two mutants selected for resistance to triarimol are also tolerant to triforine [12] and to S-1358 [8]. Triarimol-resistant mutants of A. fumigatus are also resistant to triforine, but the wild type is not sensitive to S-1358 and the mutants could not be tested for cross resistance to this compound. Actually, the data from the studies with the A. fumigatus mutants support the proposal that triarimol has more than one site of inhibition. Although the mycelial growth of the mutants is resistant to triarimol, sporulation remains sensitive. Triforine, on the other hand, has no effect on sporulation. Thus it is possible that a metabolite of ergosterol involved in the sporulation process is not synthesized in the presence of triarimol. This theory is strengthened by the observation that addition of ergosterol does not overcome toxic effects of triarimol in cultures of C. cucumerinum [12] or U. maydis [11]. On the other hand, toxicity of triforine to C. cucumerinum is alleviated when ergosterol is added [12].

In the case of the C. cucumerinum and A. fumigatus mutants, selection was made for tolerance to triarimol [8, 12]. If triarimol inhibits several reactions including those inhibited by triforine and S-1358, it is logical to expect that mutants resistant to triarimol would also be resistant to the other two compounds. On the other hand, mutants selected for tolerance to triforine or S-1358 might not be tolerant to triarimol.

Additional evidence suggests that triarimol not only has more than one site of action but also that different organisms may be affected at different sites. Toxicity of triarimol to C. cucumerinum and U. maydis is reduced by two different groups of compounds [12]. Progesterone, testosterone, farnesol, vitamin A, and β-carotene are effective antagonists in C. cucumerinum but not in U. maydis. A number of fatty acids, including palmitic, oleic, and linoleic, are moderately effective in reducing triarimol toxicity to U. maydis but not to C. cucumerinum. While the nature of action of these antagonists is not fully understood, the fact that different compounds are effective in the two organisms suggests that two different sites of action are involved.

c. Sterol Carrier Protein. Preliminary data indicate that triarimol affects sterol synthesis somewhat differently in Chlorella sorokiniana than in fungi [34]. In this alga, triarimol strongly inhibits ergosterol synthesis, but none of the methyl intermediates accumulate. Although ergosterol is normally the major sterol, growth of the organism is not impaired. A later stage of sterol synthesis appears to be affected. Other studies in algae comparing inhibitors of sterol synthesis suggest there is a lack of specificity from one alga to another with regard to sites of inhibition by a particular compound [35].

Numerous investigations have led to the proposal that the water-insoluble intermediates in sterol synthesis are bound to a sterol-carrier protein (SCP) while undergoing enzymatic conversions and then are transported to their functional sites before they are released from SCP [4, 36]. One group of investigators has suggested that there are several species of SCP [37]. Triarimol, triforine, and compounds of this type may bind to an SCP and interfere with the SCP-sterol-enzyme complex [8, 10]. The toxicants could attach to a particular region of an SCP and prevent sterol attachment or specific enzymatic reactions such as the C-14 demethylation and, in the case of triarimol, certain reactions in the side chain. Slight differences in SCP from individual organisms could account for resistance or variations in the particular stage of sterol synthesis that is blocked.

4. Conclusions

The accumulation of sterol intermediates in cultures treated with these fungicides indicates that only specific aspects of sterol synthesis are strongly inhibited [8, 10, 24]. In A. fumigatus treated with triforine there is actually more total sterol present than in control cultures [8]. Whether all intermediates that accumulate are part of the primary pathway is debatable. In most fungi the C-14 methyl is the first of the methyl groups to be removed [15]. However, since this demethylation is blocked in treated cultures while C-4 demethylation proceeds, sterols are produced which may not be intermediates in the main pathway. Why lanosterol does not accumulate in treated cultures is somewhat of a puzzle. Perhaps the transmethylation reaction at C-24 heavily favors formation of 24-methylenedihydrolanosterol and thus very little lanosterol is ever present.

The large quantity of sterol intermediates that accumulates in the presence of these toxicants indicates a failure in the control mechanisms governing sterol biosynthesis. The pattern of accumulation suggests that in normal cells an end product, perhaps ergosterol or some derivative thereof, exerts a negative feedback effect. When this end product is not present, synthesis continues and an excessive accumulation of intermediates results. It would be interesting to examine the effects of adding ergosterol or various steroids on the sterol patterns in treated cultures.

C. Other Sterol Inhibitors

Studies of the above three fungicides represent the most detailed examinations to date of the nature and ramifications of sterol inhibition in fungi. There are several analogues of triarimol including a promising new fungicide, fenarimol (4), known also as EL-222 [38] (see Table 1 and Fig. 3), which undoubtedly have the same basis of toxicity as triarimol. Aside from studies with triarimol, triforine, and S-1358, investigations of sterol inhibition in fungi have primarily involved two areas:

1. Examinations of compounds which have hypocholesteremic activity in mammals.
2. Use of antimetabolites to examine control mechanisms of sterol biosynthesis.

Examples from each of these categories will be discussed briefly.

1. Hypocholesteremic Compounds

Investigations with hypocholesteremic compounds have been primarily surveys of gross effects on various aspects of fungal growth and on the reversal of inhibitory effects. Antifungal activity has been demonstrated for a number of compounds including triparanol [see Fig. 3 (6)], AY-9944, SKF-525-A, and SKF-3301-A (see Table 1) [39-42]. Addition of oleic acid to cultures of Sordaria fimicola inhibited by AY-9944 or SKF-3301-A restores vegetative growth, but sexual reproduction remains inhibited [39]. Inhibition of sexual reproduction in Cochliobolus carbonum by SKF-3301-A can be overcome by addition of squalene or sterols [41]. Triparanol, a well-known inhibitor of cholesterol biosynthesis in mammals [43], also inhibits multiplication of S. cerevisiae cells [42]. This latter effect can be overcome by the addition of lecithin, lauric or oleic acids, ergosterol, or squalene [42]. Triparanol and triarimol are structurally similar and affect lipid synthesis in U. maydis in much the same manner [22]. However, the mutants mentioned previously as tolerant to triarimol are not tolerant to triparanol [12]. Possibly triparanol affects the same sites as triarimol in addition to other sites as well. In this case one would expect mutants tolerant to triparanol to be tolerant to triarimol.

2. Antimetabolites

Several studies have been made of the control aspects of sterol synthesis in fungi using compounds which are known or expected to block particular reactions. A mevalonic acid analogue, 3-hydroxy-3-methyl-5-phenyl-4-pentenoic acid, causes 50% inhibition of ergosterol synthesis in S. cerevisiae [44]. Undoubtedly this compound competes with mevalonic acid in the sterol biosynthetic pathway. In another study, homocysteine was used to block methylation of the sterol side chain in an effort to determine which sterol intermediate in S. cerevisiae reacts with S-adenosylmethionine [45]. The compound which accumulates, $\Delta^{7,24}$-cholestadienol, is suggested as the natural acceptor of the methyl group in ergosterol biosynthesis in yeast [45].

The principle of feedback inhibition of sterol synthesis was utilized in a third investigation. The effects of a number of bile acids, biosynthesized from cholesterol, on the incorporation of [^{14}C]acetate into the nonsaponifiable fraction in cell-free extracts of S. cerevisiae were examined [46]. It had been previously demonstrated that cholesterol itself is ineffective in regulating cholesterol synthesis in vitro [47]. Several bile acids meeting precise structural requirements blocked acetate conversion to sterols by inhibiting the enzyme, β-hydroxy-β-methylglutaryl-CoA reductase. Although bile acids have not been shown to occur in fungi, this study promotes the concept that in fungi some sterol derivative, rather than the primary sterol itself, may be responsible for control of sterol biosynthesis.

D. Inhibitors Interacting with Membranes

In addition to the compounds discussed thus far, the inhibitory action of two classes of chemicals, the polyenes and saponins, is associated with sterols. However, these compounds do not directly block sterol synthesis; toxicity results from destruction of membrane integrity [48-51]. It is debatable whether toxicity results strictly from binding to membrane sterols as originally proposed [49, 50, 52] or from interaction with membranes possessing specific structural characteristics conferred by certain sterols [53, 54]. Since the subject of membranes is discussed in another chapter, these inhibitors will be mentioned here only briefly.

1. Polyenes

The overall toxic effects of polyenes differ considerably from those of triforine, triarimol, or S-1358. Fungicidal concentrations of polyenes completely inhibit initial growth; even low concentrations can cause a loss of weight [55]. A number of studies in yeasts indicate resistance can be primarily attributed to two factors associated with membrane structure, that is, alteration of sterol quality [30, 32, 56-59] or growth conditions [60, 61]. In a number of cases, mutants with the greatest degree of resistance contain predominantly Δ^8-sterols as opposed to $\Delta^{5,7}$- and Δ^7-sterols in more susceptible strains [58, 59]. The resistance of petite mutants is attributed to an altered sterol pattern which is similar to those of fungal cells treated with triarimol, triforine, or S-1358 [30, 32]. However, these petite mutants have slow growth rates which could well be responsible for their tolerance. The addition of ergosterol improves growth, but the effects of the polyene are difficult to measure in this situation due to complex formation between ergosterol and the antibiotic [30]. Undoubtedly a better understanding of polyene action will develop with advances in membrane technology.

2. Saponins

Saponins are of considerable interest not only because of their polyene-like activity [49, 50], but also because many are naturally occurring disease-resistance factors [51]. One of the better known compounds in this group is avenacin, the resistance factor to *Ophiobolus graminis* var. *graminis* in oat roots [62, 63]. Aesin, an antifungal agent from horse chestnut [64], and the saponins from lucerne (alfalfa) [65] have been the subjects of a number of investigations. The antifungal activity of all these compounds is related to the types and quantities of sterols present in the fungi which, in turn, affect membrane properties [53, 66-69].

E. Conclusions

The membrane-sterol relationship is considered critical for the toxicity of polyenes and many saponins. The ultimate toxicity of compounds such as triarimol, triforine, and S-1358 undoubtedly results from the lack of suitable sterols for membrane structure and function. These inhibitors which block sterol synthesis are lipophilic in nature. Although their structures vary considerably, their seemingly similar modes of action probably result from their movement within the cell to the endoplasmic reticulum where they inhibit sterol synthesis and possibly other syntheses occurring there.

III. GIBBERELLINS

A. General Information

Most research on gibberellins has been concerned with the study of their growth-regulatory activities in higher plants. However, one must remember that the original work on these compound involved their isolation from the fungus Gibberella fujikuroi, which was associated with a disease of rice seedlings [70]. Fungi are frequently used in basic studies of gibberellin biosynthesis. Effects of higher plant growth retardants on gibberellin biosynthesis are often observed in fungi in conjunction with the effects on elongation in higher plants. Although gibberellins appear to have no role in normal fungal growth [71, 72], they may be involved in pathogenicity [73]. For example, the application of individual growth regulators, including gibberellin, increases susceptibility of cucumber to scab disease [74]. Another study involving a number of Fusarium species indicates that gibberellin production is associated with pathogenicity [75].

B. Biosynthesis

The initial stages of gibberellin biosynthesis involve a segment of the isoprenoid pathway shared with sterol and carotenoid biosynthesis (Fig. 1). The pathway diverges after the formation of farnesyl pyrophosphate. The gibberellin-carotenoid branch forms geranylgeranyl pyrophosphate after which the pathway of these two syntheses diverges (Fig. 4). The latter stages of gibberellin biosynthesis involve a series of cyclizations and oxidations [76, 77] many of which are analogous to those in sterol biosynthesis [5].

C. AMO-1618, CCC, and Phosfon D

Research concerning the action of growth retardants on gibberellin biosynthesis in fungi has primarily involved the compounds AMO-1618, CCC, Phosfon D (see Table 2) and their analogues tested in G. fujikuroi (Fusarium moniliforme). Several studies indicate that concentrations of retardants which strongly suppress gibberellin biosynthesis have no effect on increases in mycelial dry weight [71, 72, 78, 79]. Investigations with CCC and AMO-1618 indicate that these compounds do not instigate breakdown of existing gibberellins [72, 78] but rather that they interfere with new synthesis at a stage preceding formation of the gibberellin precursor, kaurene (see Fig. 4) [79]. More recently, the point of inhibition has been traced to an area of the biosynthetic pathway between geranylgeranyl pyrophosphate and and kaurene [80]. This observation is supported by the accumulation of

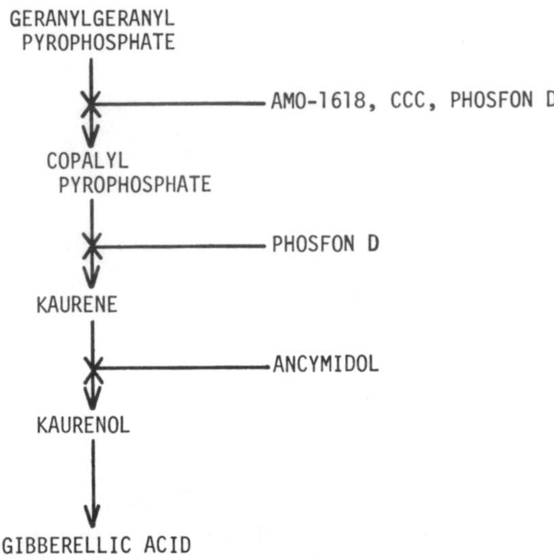

FIG. 4. Scheme of gibberellin biosynthesis indicating major points of inhibition by AMO-1618, CCC, Phosfon D and ancymidol.

TABLE 2

Inhibitors of Gibberellin, Carotenoid, Fatty Acid, and Acetogenin Synthesis

Name used in text	Chemical name	Other names
AMO-1618	2'-Isopropyl-4'-(trimethylammonium chloride)-5'-methylphenylpiperidine carboxylate	
CCC	(2-Chloroethyl)trimethylammonium chloride	Cycocel
Phosfon D	Tributyl-2,4-dichlorobenzylphosphonium chloride	
Ancymidol (5)[a]	α-Cyclopropyl-α-(4-methoxyphenyl)-5-pyrimidinemethanol	EL-531
DPA	Diphenylamine	
CPTA	2-(4-Chlorophenylthio)triethylamine hydrochloride	
Cerulenin	(2S)(3R)2,3-Epoxy-4-oxo-7,10-dodecadienoylamide	
Dichlorvos	2,2-Dichlorovinyl dimethyl phosphate	Vapona

[a]See Figure 3.

geranylgeranyl pyrophosphate and the incorporation of radiolabeled kaurene into gibberellin in the presence of the retardants [80].

Early work indicates that Phosfon D has no effect on gibberellin biosynthesis in F. moniliforme [71]. However, a later investigation shows that the fungus rapidly degrades the inhibitor [78]. An examination of the conversion of copalyl pyrophosphate (an intermediate between geranylgeranyl pyrophosphate and kaurene) to kaurene in a cell-free system from G. fujikuroi shows this step is blocked by Phosfon D (see Fig. 4) [81]. Comparative studies in the same system indicate that CCC and AMO-1618 do not strongly inhibit this particular conversion [81]. Since all three compounds are known to inhibit the conversion of geranylgeranyl pyrophosphate to kaurene in sensitive whole systems [81], it is suggested that Phosfon D inhibits the two cyclizations involved in the conversion of geranylgeranyl pyrophosphate to kaurene, whereas AMO-1618 and CCC inhibit only one, that is, the conversion to copalyl pyrophosphate [81].

D. Ancymidol

As mentioned earlier, the initial phase of the gibberellin biosynthetic pathway is identical with that of sterols while the latter segment involves a number of reactions resembling those in sterol biosynthesis. Any inhibitor of that portion of the pathway preceding farnesyl pyrophosphate would therefore inhibit gibberellin and sterol as well as carotenoid biosynthesis. One might also expect that inhibitors of the latter stages of gibberellin synthesis have an effect on sterol synthesis and vice versa. A good example of such an inhibitor may be the growth retardant ancymidol [see Fig. 3 (5), and Table 2]. This compound is an analogue of triarimol, an inhibitor of ergosterol synthesis in fungi [8, 11]. Although ancymidol is known primarily for its growth regulatory effects in higher plants [82-85], it is weakly fungitoxic [12]. The two compounds most likely have a similar mode of fungitoxic action because mutants of C. cucumerinum resistant to triarimol are also resistant to ancymidol [12]. It has also been shown that triarimol, like ancymidol, produces growth regulatory effects in higher plants [86, 87] and markedly reduces the amount of extractable gibberellin [84].

One possible explanation of antifungal and antigibberellin activity of both triarimol and ancymidol is an effect on mixed-function oxidase activity. Ancymidol blocks conversion of kaurene to kaurenol in higher plants, whereas a much higher concentration inhibits to a lesser extent in the pathway between mevalonic acid and kaurene [85]. The conversion of kaurene to kaurenol is similar to the first step of a sterol demethylation reaction [26, 76]. Actually the methyl group involved in the kaurene conversion may be most nearly analogous to the sterol C-14 methyl group since in either case there is no adjacent hydroxyl group to serve as an electron receptor. Thus it is possible that these two compounds affect a similar type of mixed-function oxidase activity in sterol and gibberellin biosynthesis.

E. Mode of Action of Growth Retardants

Actually the relationship between gibberellins and sterols is not completely understood insofar as growth regulation in higher plants is concerned, a fact which has led to some controversy concerning the mechanism of action of growth retardants. Some studies indicate that retardants affect sterol synthesis [88, 89] and that growth retardation can be overcome by the addition of sterols as well as gibberellin [88]. It is suggested that these inhibitors produce their effects through a more general inhibitory action on isoprenoid biosynthesis [88]. However, there are several factors, such as the type or part of organism and its phase of growth as well as transport properties of the retardant that could influence action of growth regulators. Inhibition of either gibberellin or sterol synthesis could involve the other through membrane function. Sterols are a vital part of the membrane structure and gibberellins affect membrane properties [90], possibly by forming a complex with phospholipids [91].

IV. CAROTENOIDS

A. Biosynthesis

Carotenoids are, for the most part, tetraterpenoids with varying ring structures and degrees of unsaturation. The initial phase of biosynthesis, as mentioned previously, is identical to that of sterols and gibberellins. Two geranylgeranyl pyrophosphates condense to yield the 40-carbon polyene, phytoene, which is referred to as a colorless carotenoid [92]. The colored carotenoids, such as lycopene and β-carotene, are biosynthesized from phytoene through a series of desaturation and cyclization reactions (see Fig. 5).

B. Functions

Although not all the roles of carotenoids in fungi are known, a number have been investigated. One of the best documented functions is photoprotection of cells from degradative effects of visible light [93]. Photoprotection is a property related to the number of conjugated double bonds in the carotenoid molecule [94]. Carotenoids also protect against damage from ultraviolet (UV) irradiation as has been demonstrated in a comparison of albino and wild-type strains of *Neurospora crassa* [95]. The albino strain and the wild type in which carotenoid synthesis is inhibited are more sensitive than normal to UV irradiation. Carotenoids may also play a role in sporulation. This function may be that of a light-sensitive compound involved in a regulatory system as observed in sporulation of *Leptosphaeria michotii* [96] or

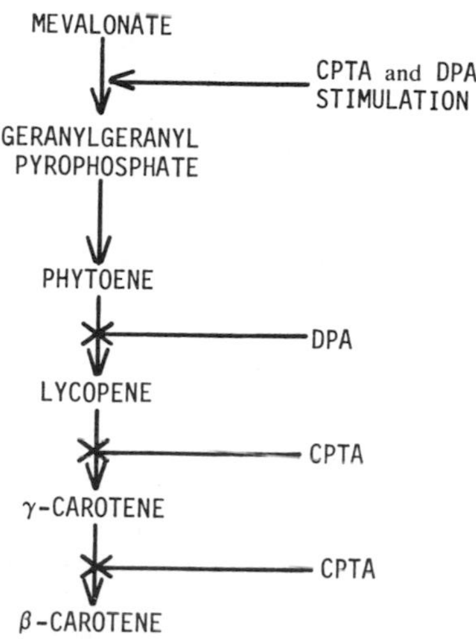

FIG. 5. Scheme of carotenoid biosynthesis indicating major points of inhibition by DPA and CPTA.

that of a hormone precursor affecting sexual reproduction as noted in the Mucorales [97]. Recently it has been suggested that carotenoids may also be involved in stabilizing membranes structurally [98].

C. Diphenylamine (DPA)

A number of compounds interfere with the production of the highly unsaturated or cyclized carotenoids. The best known among these is DPA, which was reported some years ago to inhibit chromagenesis in a number of fungi and bacteria [99]. Later studies indicate that DPA inhibits synthesis of highly conjugated carotenoids, and this results in an accumulation of several colorless carotenoids, primarily phytoene (see Fig. 5) [100-102].

Further investigations show that DPA specifically inhibits carotenoid biosynthesis [102-105]. Concentrations which block the synthesis of highly unsaturated carotenoids such as β-carotene have no appreciable effects on growth [103] or lipid synthesis [104, 105]. Diphenylamine actually appears to stimulate formation of phytoene and related compounds [102, 104, 105] resulting in a greater quantity of total carotenoids in treated than in control cultures [104, 105]. The chemical nature of carotenoids which accumulate leads to the suggestion that DPA inhibits desaturation reactions [102]. From comparisons of the chemical structure of phytoene with DPA and benzophenone, a similar-type inhibitor [106], it is deduced that the dihydro forms of the inhibitors resemble a portion of the phytoene molecule and compete with the natural substrate for the dehydrogenase [107].

D. 2-(4-Chlorophenylthio)triethylamine Hydrochloride (CPTA)

Another compound which noticeably affects carotenogenesis is CPTA (see Table 2). Early experiments with Blakeslea trispora and Phycomyces blakesleeanus show that this compound causes an accumulation of lycopene, an aliphatic carotenoid (see Fig. 5) [108]. Further examinations reveal that although synthesis of β-carotene is somewhat reduced in treated cultures, total polyene synthesis increased 2.5-fold with a 16-fold increase in γ-carotene, which contains one ring in its structure [109]. When DPA was added in addition to CPTA, there was a 14-fold stimulation of polyene production and accumulation of phytoene, lycopene, and γ-carotene [109]. Apparently the abundance of polyenes overcomes to some extent the inhibitory effects of DPA. The proposal is made that CPTA may act as a derepressor of a regulatory gene affecting biosynthesis in the carotenoid pathway [109].

It is suggested that CPTA interferes with cyclization at several levels (see Fig. 5) [110]. The buildup of γ-carotene [109] would seem to indicate that the first cyclization may not be as severely limited as the second, which is necessary to form β-carotene. The lack of substantial inhibition of cyclizations in general may result from the large accumulation of precursors, which probably antagonize the toxicity to some extent. An in vitro type study should clarify this point.

E. Conclusions

An interesting sidelight is a recent study in which benzophenone, a DPA-type inhibitor of carotenoid synthesis, was substituted at the 4-position with a β-diethylaminoethoxy group [111]. With this substitution it stimulates rather than inhibits desaturation and interferes instead with cyclization [111]. This establishes an interesting structure-activity relationship for further investigation.

Although there are no fungicides with modes of action based on inhibition of carotenoid synthesis, there are a number of herbicides which act by this mechanism. The inhibitory effects of these compounds on carotenoid synthesis lead to bleaching of chlorophyll and eventual death of the plant [112, 113]. It would be interesting to know whether the fungal inoculum potential is lowered in areas where such herbicies have been used.

V. SEX HORMONES

Although sex hormones undoubtedly function in most fungi, relatively few have been characterized. As a result, knowledge is rather limited concerning biosynthetic pathways and metabolic inhibitors. For the interested reader a recent review summarizes the current information [114]. The discussion here is devoted to a number of the better-known fungal sex hormones present in lipid extracts.

A. Steroid Hormones

Antheridiol [see Fig. 6 (7)] is the best known steroid hormone outside the animal kingdom. This compound, which is produced by female cells of Achlya [115], promotes formation of antheridia and subsequently the synthesis of hormone B, in all probability a steroid [114], by male cells [115-117]. Hormone B promotes formation of oogonia by female cells [115-117]. The only compound known to interfere specifically with the action of antheridiol is an isomer [118] which inhibits formation of antheridia when added to cultures in quantities far exceeding the natural levels of antheridiol [114].

Steroid hormones are probably also involved in sexual reproduction in Phytophthora and Pythium. It appears likely that the sterols, which are required for this process [119, 120], are converted to specific steroid hormones [121, 122]. The sterol-induced sexual reproduction in Pythium periplocum can be annulled by the addition of estradiol [123] or polyene antibiotics [124]. The polyenes most likely complex with the sterols and prevent further metabolism. Estradiol may act as an analogue and compete with the natural hormone for a reaction site. It would be interesting to examine the effects of compounds such as triarimol on reproduction in these fungi. A study of this type would be helpful in determining the extent to which supplemental sterols are metabolized. There is also the possibility that compounds like triarimol act as hormone analogues. Some of the general effects of these compounds bear a striking resemblance to those produced by a peptide factor [125, 126] which acts as a hormone in yeast.

B. Trisporic Acid

Trisporic acid [see Fig. 6 (8)], a derivative of β-carotene [97], is a sex hormone found in mated cultures of the Mucorales [114]. The stimulatory effects of this compound on carotenogenesis indicate that it is most likely an enzyme derepressor [127] and thus self-amplifying [114]. The conversion of β-carotene to trisporic acid probably involves mixed-function oxidases [128]. Mated cultures of B. trispora treated with barbiturates, which induce an increase in the level of mixed-function oxidases, possess substantially higher levels of carotenoids and trisporic acid than normally present. Unmated cultures treated with barbiturate show no increase in carotenogenesis. Therefore it was concluded that the increased mixed-function oxidase activity in mated cultures stimulates synthesis of trisporic acid, which, in turn, stimulates carotenogenesis [128]. As might be expected, compounds such as DPA which inhibit synthesis of β-carotene also inhibit

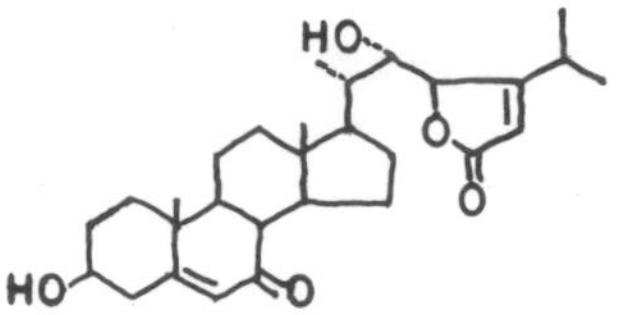

(7) antheridiol

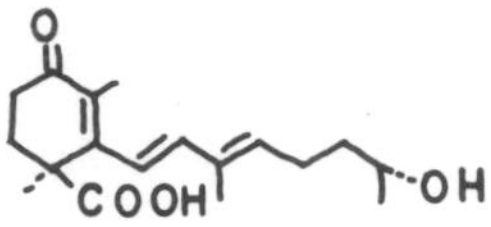

(8) trisporic acid

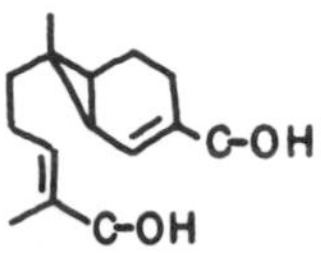

(9) sirenin

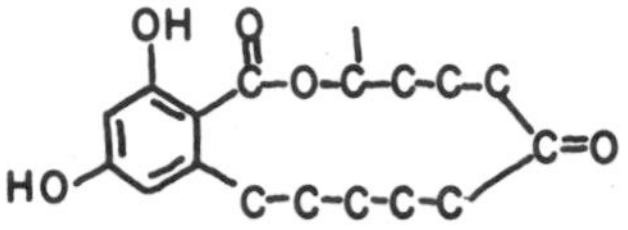

(10) zearalenone

FIG. 6. Chemical structures of several fungal sex hormones.

sexual reproduction in these fungi [114]. This phenomenon is the basis of a useful assay system for selecting inhibitors of carotenogenesis.

C. Sirenin

Sirenin [see Fig. 6 (9)], a hormone produced by female gametes of Allomyces [129], is a 15-carbon compound probably biosynthesized through a series of reactions from farnesyl pyrophosphate [114]. Although a number of isomers have been synthesized, none appear to antagonize the action of sirenin [130]. Inhibitors of the mevalonate-farnesyl segment of the isoprenoid pathway would undoubtedly inhibit sirenin synthesis, but they would inhibit syntheses of other vital metabolites as well. A better understanding of the conversion of farnesyl pyrophosphate to sirenin might facilitate the development of specific inhibitors of sirenin biosynthesis.

D. Zearalenone

Zearalenone [see Fig. 6 (10)], a fungal acetogenin (discussed below) toxic to animals, is reported to be essential for perithecial formation in Fusarium roseum [131, 132]. This process is inhibited by the insecticide dichlorvos at concentrations which have no perceptible effect on vegetative growth [133]. Since inhibition can be reversed by the addition of zearalenone to the culture medium, it is suggested that dichlorvos inhibits perithecial formation by interfering with zearalenone biosynthesis, the site of hormonal action, or both [133].

VI. FATTY ACIDS

Fatty acids are important intermediates with a variety of functions ranging from an energy source to vital components of membranes. In the cell, turnover of fatty acids is considerable and accumulation of free fatty acids is usually small. Although investigations of inhibitors of fatty acid synthesis in fungi are limited, thorough knowledge of biosynthesis has made possible a full investigation of the antifungal antibiotic, cerulenin.

Cerulenin (see Table 2), an inhibitor isolated from the culture filtrate of Cephalosporium caerulens [134], is essentially an amide analogue of a 12-carbon fatty acid with two double bonds [135]. Although fungal growth is ultimately inhibited severely by cerulenin, the compound has little effect initially on protein, nucleic acid, and cell-wall syntheses; nor does it affect to any extent the oxidation of glucose, citrate, or succinate in a nongrowth medium [136]. However, incorporation of [^{14}C]acetate into both the fatty acid and nonsaponifiable lipid fractions is strongly inhibited almost immediately [136]. An analysis of the sterol fraction shows that the quantity of ergosterol relative to dry weight declines steadily over a period of time [136]. It has also been noted that inhibitory effects on growth can be reversed by the addition of ergosterol and various fatty acids [137].

Further studies using a cell-free system from S. cerevisiae reveal that although cerulenin strongly inhibits [^{14}C]acetate incorporation into fatty acids and nonsaponifiable lipids, it does not inhibit incorporation of [^{14}C]-mevalonate into the latter [138]. Therefore, the decline in sterol synthesis must result from a block in the pathways prior to mevalonic acid [138]. Further studies compared the effects of cerulenin with avidin, a known inhibitor of the acetyl-CoA carboxylase reaction in yeast. Avidin, in contrast to cerulenin, has no effect on [^{14}C]acetate incorporation into the nonsaponifiable fraction; whereas, cerulenin does not interfere with the synthesis of malonyl-CoA in fatty acid synthesis [138]. These investigations indicate that in yeast cerulenin prevents the condensation of acetyl-CoA and malonyl-CoA to form acetoacetyl-CoA (see Fig. 7). Chain elongation of supplemented

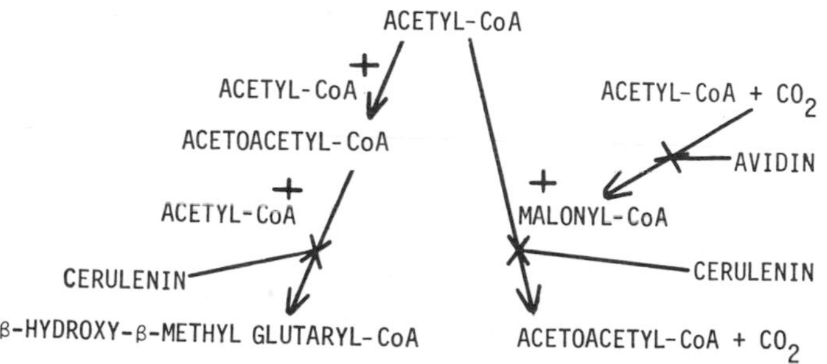

FIG. 7. Scheme of initial steps in fatty acid and sterol biosynthesis indicating points of inhibition by cerulenin and avidin.

fatty acids does not appear inhibited in yeast [137]. However, it appears that in bacteria formation of acetoacetyl-CoA and chain elongation are inhibited [139, 140].

The failure to synthesize sterols in yeast treated with cerulenin may be accounted for by the inhibition of β-hydroxy-β-methylglutaryl-CoA synthetase [141]. There is no inhibition of the acetoacetyl-CoA thiolase activity in sterol synthesis [141]. It can be concluded therefore that cerulenin has more than one site of action in the area of lipid synthesis (see Fig. 7).

VIII. ACETOGENINS

A. General Information

Acetogenins, also called polyketides, have received recognition primarily through one group of compounds in this class, the aflatoxins. These metabolites are quite toxic to animals, and because of their presence in foodstuffs a great deal of research on acetogenins has been devoted to controlling aflatoxin production or eliminating it when present [142, 143]. Other acetogenins which are toxic to animals include emodin, citrinen, ochratoxin, and zearalenone. There are also a number of toxins in this class associated with fungal pathogenesis [144]. These include skyrin and diaporthin (isolated from the chestnut blight organism), pyriculol (associated with rice blast disease), and alternariol (produces secondary symptoms in early blight). Many fungal pigments, such as anthroquinones, belong to this class as well.

B. Biosynthesis

Acetogenins, as the name implies, are synthesized from acetate units [145, 146] via the polyacetate (polyketide) pathway. The early stages of biosynthesis are similar to those of fatty acid synthesis with malonyl and acetyl units involved (see Fig. 1) [147, 148]. The pivotal point between the two pathways appears to be at the acetoacetyl intermediate [149]. From this stage the most obvious distinction is that fatty acid synthesis is reductive, whereas acetogenin biosynthesis is nonreductive and results in the formation of linear polyketides [149]. The polyketide chains, which vary in length, are cyclized and subsequently modified in various ways [149]. The polyacetate pathway functions primarily during that portion of growth called "idiophase," which occurs after the logarithmic stage of development [149]. The mechanisms of activation of this pathway are not known, nor for the most part, are the roles acetogenins play in the cells that produce them.

C. Control of Aflatoxins

1. Methods in Practice

As mentioned previously, much research on acetogenins has concerned aflatoxins. Efforts have been made to eliminate aflatoxins in foodstuffs by destroying them with various degradative agents [150-152] or by preventing fungal growth [153, 154]. However, none of these methods have been entirely satisfactory, and research has continued in an effort to find inhibitors of aflatoxin biosynthesis.

2. Influence of Metals

Metal balance plays an important role in aflatoxin biosynthesis [155], and this factor offers one possible means of controlling production. Several investigations indicate that zinc is required for aflatoxin production [156-159] and that omission of this element sharply reduces synthesis [157]. There is conflicting evidence concerning the effect of cadmium. One study indicates that cadmium stimulates aflatoxin synthesis and lowers the concentration of zinc required [158], whereas another study reports that cadmium is an inhibitor [159]. Actually, conflicting data could arise from variations in factors such as culture conditions or growth phase. Other metals which diminish aflatoxin synthesis in varying degrees include copper, iron, manganese, mercury, and silver [159]. Barium is effective in delaying synthesis for a few days [158]. It is interesting to note that concentrations of metals sufficient to increase or diminish aflatoxin biosynthesis do not have parallel effects on growth as measured by mycelial weight [157-159].

3. Dichlorvos

At present, the most effective chemical for the control of aflatoxin biosynthesis is dichlorvos (see Table 2), an organophosphorus insecticide used to protect harvested grain from insect infestation. When grain is treated with dichlorvos prior to fungal inoculation, aflatoxin production is prevented [160]. However, if chemical concentrations are lowered or application is made after fungal inoculation, dichlorvos is not effective [160]. A somewhat analogous situation exists when dichlorvos is added to a fungal culture in defined medium. The pesticide inhibits aflatoxin production 90% without affecting fungal growth when it is added prior to the initiation of toxin production, but it has negligible effects when added after idiophase has begun, indicating a possible interference with the induction of synthesis [161]. Since several other organophosphorus insecticides are much inferior to dichlorvos in controlling aflatoxin production, it is concluded that the mechanism of inhibition in this case is unlike that by which these compounds, as insecticides, inhibit acetylcholinesterase [161]. Aflatoxins are not the only acetogenins affected by dichlorvos. It also interferes with the biosynthesis of zearalenone [133] and ochratoxin [162].

4. Other Chemical Inhibitors

Other compounds which, at rather high concentrations, inhibit aflatoxin production include p-aminobenzoic acid, sulfanilimide, anthranilic acid, potassium sulfite, and dimethyl sulfoxide [163, 164]. These compounds have little or no effect on fungal growth at concentrations which inhibit aflatoxin synthesis [163, 164].

D. Pigments

Lack of pigmentation has been repeatedly observed to accompany inhibition of aflatoxin biosynthesis [157, 158, 165]. For example, conidia are white rather than the usual green when *Aspergillus flavus* is treated with dimethyl sulfoxide [165], and mycelium is white rather than yellow when nutritional requirements for aflatoxin production are lacking [158]. These pigments, like aflatoxin, are probably acetogenins and their absence may indicate inhibition of toxin production by the polyacetate pathway.

E. Functions

The function of acetogenins are not fully understood. Some may have hormonal activity. As mentioned in the section on sex hormones, zearalenone appears to be involved in perithecial formation [131, 132]. Pigments may protect from UV irradiation [165] or function in photosensitive regulatory systems. Other roles such as safety-valve shunts to utilize accumulating metabolites and metal binding have been proposed. Several of the toxins are involved in fungal pathogenesis. Present knowledge makes it rather difficult to predict the long-term results of inhibiting acetogenin biosynthesis.

ACKNOWLEDGMENTS

The author would like to acknowledge support during the writing of this chapter from National Institutes of Health Grant AI 00225. I am grateful to Drs. H. D. Sisler, L. R. Krusberg, and G. W. Patterson for their helpful criticism in preparing this manuscript.

VIII. RECENT DEVELOPMENTS

Several recently developed fungicides evidently act by a mechanism similar or identical to that of triarimol, triforine, and S-1358. Triadimefon [MEB 6447, 1-(4-chlorophenoxy)-3,3-dimethyl-(1,2,4-triazol-1-yl)-2-butanone] and fluotrimazole [BUE 0620, bis-phenyl-(3-trifluoromethylphenyl)-1-(1,2,4-triazolyl)-methane] strongly inhibit ergosterol synthesis in Ustilago avenae with an accompanying accumulation of C-4 methyl and dimethyl sterol intermediates [166, 167]. Triadimefon produces similar effects in U. maydis [168]. Strains of C. cucumerinum tolerant to triadimefon are also tolerant to triarimol and triforine [166]. Triadimefon, like triarimol, exhibits growth regulatory effects in higher plants [169]. Growth retardation is accompanied by an increase in chlorophyll content. Both effects are overcome by the addition of gibberellic acid.

Imazalil, 1-[2-(2,4-dichlorophenyl)-2-(2-propenyloxy)ethyl]-1H-imidazole nitrate, also interferes with sterol biosynthesis [170]. Treated cultures of Aspergillus nidulans have increased concentrations of C-4 methyl and dimethyl sterols and a sharp reduction in synthesis of C-4 desmethyl sterols. It appears likely that the imidazole derivatives, clotrimazole [bis-phenyl-(2-chlorophenyl-1-imidazolyl)-methane] and miconazole [1-{2-(2,4-dichlorophenyl)-2-[(2,4-dichlorophenyl)methoxy]}ethyl -1H-imidazole mononitrate] have modes of action similar to that imazilil.

Further evidence has accumulated, indicating that these compounds, which interfere with sterol biosynthesis, affect enzymes that are peculiar to C-14 demethylation. In cell-free homogenates of S. cerevisiae, an organism in which C-14 demethylation precedes those at C-4[6, 7], C-4 demethylation does not occur in the presence of S-1358 or triarimol, presumably because these compounds block C-14 demethylation [171]. There is very little inhibition of C-4 demethylation when a C-14 desmethyl sterol substrate is used.

Demethylation at C-14 is also blocked in cholesterol biosynthesis by rat-liver subcellular fractions [172]. Although it is suggested that triarimol may be an inhibitor of cytochrome P-450 [172], evidence suggests that only a specific type of P-450 such as that involved in sterol C-14 demethylation is affected.

REFERENCES

1. W. R. Nes, Lipids, 9, 596 (1974).
2. J. W. Hendrix, Ann. Rev. Phytopath., 8, 111 (1970).
3. J. D. Weete, Fungal Lipid Biochemistry, Plenum Press, New York, 1974.
4. M. E. Dempsey, Ann. Rev. Biochem., 43, 967 (1974).
5. L. J. Goad and T. W. Goodwin, Progr. Phytochem., 3, 113 (1972).
6. M. Fryberg, A. C. Oehlschlager, and A. M. Unrau, J. Amer. Chem. Soc., 95, 5747 (1973).
7. D. H. R. Barton, J. E. T. Corrie, P. J. Marshall, and D. A. Widdowson, Bioorg. Chem., 2, 363 (1973).

8. J. L. Sherald and H. D. Sisler, Pestic. Biochem. Physiol., 5, 477 (1975).
9. G. Goulston, L. J. Goad, and T. W. Goodwin, Biochem. J., 102, 15c (1967).
10. N. N. Ragsdale, Biochim. Biophys. Acta, 380, 81 (1975).
11. N. N. Ragsdale and H. D. Sisler, Biochem. Biophys. Res. Commun., 46, 2048 (1972).
12. J. L. Sherald, N. N. Ragsdale, and H. D. Sisler, Pestic. Sci., 4, 719 (1973).
13. T. Kato, S. Tanaka, M. Ueda, and Y. Kawase, Agr. Biol. Chem., 38, 2377 (1974).
14. I. F. Brown, Jr., Proc. 7th Intern. Congr. Plant Protection, Paris, 1970, p. 206.
15. P. Schicke and K. H. Veen, Proc. 5th Brit. Insectic. Fungic. Conf., Brighton, England, 1969, Vol. 2, p. 569.
16. T. Kato, S. Tanaka, S. Yamamoto, Y. Kawase, and M. Ueda, Ann. Phytopathol. Soc. Japan, 41, 1 (1975).
17. N. N. Ragsdale and H. D. Sisler, Pestic. Biochem. Physiol., 3, 20 (1973).
18. H. D. Sisler and N. N. Ragsdale, in Systemic Fungicides (H. Lyr and C. Polter, eds.), Akademie-Verlag, Berlin, 1975, pp. 101-108.
19. L. D. Houseworth, E. W. Brunton, and B. G. Tweedy, Phytopathology, 61, 896 (1971).
20. A. Kaars Sijpesteijn, in Systemic Fungicides (R. W. Marsh, ed.), Wiley, New York, 1972, pp. 132-155.
21. E. W. Brunton, Ph.D. Thesis, University of Missouri, Columbia, Mo., 1972.
22. N. N. Ragsdale and H. D. Sisler, Ann. Proc. Amer. Phytopathol. Soc., 2, 55 (1975).
23. T. Kato, unpublished work, 1975.
24. T. Kato, S. Tanaka, M. Ueda, and Y. Kawase, Agr. Biol. Chem., 39, 169 (1975).
25. T. A. Jarman, A. A. L. Gunatilaka, and D. A. Widdowson, Bioorg. Chem., 4, 202 (1975).
26. G. J. Schroepfer, Jr., B. N. Lutsky, J. A. Martin, S. Huntoon, B. Fourcans, W. H. Lee, and J. Vermilion, Proc. Roy. Soc. London, 180, 125 (1972).
27. K. H. Baggaley, S. D. Atkin, P. D. English, R. M. Hindley, B. Morgan, and J. Green, Biochem. Pharmacol., 24, 1902 (1975).
28. C. F. Wilkinson, K. Hetnarski, and L. J. Hicks, Pestic. Biochem. Physiol., 4, 299 (1974).
29. M. Bard, R. A. Woods, and J. M. Haslam, Biochem. Biophys. Res. Commun., 56, 324 (1974).
30. E. G. Gollub, P. Trocha, P. K. Liu, and D. B. Sprinson, Biochem. Biophys. Res. Commun., 56, 471 (1974).

31. F. Karst and F. Lacroute, Biochem. Biophys. Res. Commun., 59, 370 (1974).
32. P. J. Trocha, S. J. Jasne, and D. B. Sprinson, Biochem. Biophys. Res. Commun., 59, 666 (1974).
33. J. Nagai and H. Katsuki, Biochem. Biophys. Res. Commun., 60, 555 (1974).
34. C. Frasinel and G. W. Patterson, unpublished work, 1975.
35. G. W. Patterson, P. J. Doyle, L. G. Dickson, and J. T. Chan, Lipids, 9, 567 (1974).
36. T. J. Scallen, M. V. Srikantaiah, B. Seetharam, E. Hansbury, and K. I. Gavey, Fed. Proc., 33, 1733 (1974).
37. T. J. Scallen, B. Seetharam, M. V. Srikantaiah, and E. Hansbury, J. Amer. Oil Chem. Soc., 51, 516A (1974).
38. I. F. Brown, Jr., H. M. Taylor, and H. R. Hall, Ann. Proc. Amer. Phytopathol. Soc., 2, 31 (1975).
39. C. G. Elliott, J. Gen. Microbiol., 56, 331 (1969).
40. G. Matolcsy, M. Hamrán, and B. Bordás, Pestic. Sci., 4, 267 (1973).
41. R. R. Nelson, D. Huisingh, and R. K. Webster, Phytopathology, 57, 1081 (1967).
42. S. Aaronson, Proc. Soc. Exptl. Biol. Med., 136, 61 (1971).
43. J. Avigan, C. Steinberg, H. E. Vroman, M. J. Thompson, and E. J. Mosettig, J. Biol. Chem., 235, 3123 (1960).
44. J. M. Stewart and D. W. Woolley, Biochemistry, 3, 1998 (1964).
45. H. Hatanaka, N. Ariga, J. Nagai, and H. Katsuki, Biochem. Biophys. Res. Commun., 60, 787 (1974).
46. H. Hatanaka, A. Kawaguchi, S. Hayakawa, and H. Katsuki, Biochim. Biophys. Acta, 270, 397 (1972).
47. M. D. Siperstein, Amer. J. Clin. Nutr., 8, 645 (1960).
48. S. C. Kinsky, J. Bacteriol., 92, 889 (1961).
49. R. R. Dourmashkin, R. M. Daugherty, and R. J. C. Harris, Nature, 194, 1116 (1962).
50. A. M. Glauert, J. T. Dingle, and J. A. Lucy, Nature, 196, 953 (1962).
51. R. A. Olsen, Physiol. Plantarum, 24, 534 (1971).
52. J. O. Lampen, in Biochemical Studies of Antimicrobial Drugs (B. A. Heston and P. E. Reynolds, eds.), 16th Symposium of the Society for General Microbiology, Cambridge University Press, Cambridge, Mass., 1966, pp. 111-130.
53. Y. Assa, B. Gestetner, I. Chet, and Y. Henis, Life Sci., 11, 637 (1972).
54. C. C. HsuChen and D. S. Feingold, Biochem. Biophys. Res. Commun., 51, 972 (1973).
55. S. C. Kinsky, in Antibiotics (D. Gottlieb and P. D. Shaw, eds.), Vol. 1, Springer, New York, 1967, pp. 122-141.
56. R. A. Woods, J. Bacteriol., 108, 69 (1971).
57. L. W. Parks, F. T. Bond, E. D. Thompson, and P. R. Starr, J. Lipid Res., 13, 311 (1972).

58. M. Fryberg, A. C. Oehlschlager, and A. M. Unrau, Arch. Biochem. Biophys., 160, 83 (1974).
59. S. J. Kim, K. J. Kwon-Chung, G. W. A. Milne, W. B. Hill, and G. W. Patterson, Antimicrob. Agents Chemother., 7, 99 (1975).
60. C. C. HsuChen and D. S. Feingold, Nature, 251, 656 (1974).
61. P. Venables and A. D. Russell, Antimicrob. Agents Chemother., 7, 121 (1975).
62. E. M. C. Turner, J. Exptl. Bot., 7, 80 (1956).
63. J. V. Maizel, H. J. Burkhardt, and H. K. Mitchell, Biochemistry, 3, 424 (1964).
64. B. Wolters, Naturwissenschaften, 53, 253 (1966).
65. B. Gestetner, Y. Assa, Y. Henis, Y. Birk, and A. Bondi, J. Sci. Food Agr., 22, 168 (1971).
66. B. Gestetner, Y. Assa, Y. Henis, Y. Tencer, M. Rotman, Y. Birk, and A. Bondi, Biochim. Biophys. Acta, 270, 181 (1972).
67. R. A. Olsen, Physiol. Plant., 25, 204 (1971).
68. R. A. Olsen, Physiol. Plant., 28, 507 (1973).
69. R. A. Olsen, Physiol. Plant., 29, 145 (1973).
70. B. B. Stowe, F. H. Stodola, T. Hayashi, and P. W. Brian, in Plant Growth Regulation (R. M. Klein, ed.), Iowa State Press, Ames, Iowa, 1961, pp. 465-471.
71. H. Kende, H. Ninnemann, and A. Lang, Naturwissenschaften, 50, 599 (1963).
72. H. Ninnemann, J. A. D. Zeevaart, H. Kende, and A. Lang, Planta, 61, 229 (1964).
73. N. Gogala, Biol. Vestn., 20, 55 (1972).
74. O. M. Van Andel, Neth. J. Plant Pathol., 74, Suppl. 1, 113 (1968).
75. I. Hirol, J. Fac. Agr., Tottori Univ., 5, 1 (1969).
76. C. A. West and R. R. Fall, in Plant Growth Substances 1970 (D. J. Carr, ed.), Springer-Verlag, Berlin, 1972, pp. 133-142.
77. T. A. Geissman, A. J. Verbiscar, B. D. Phinney, and G. Cragg, Phytochemistry, 5, 933 (1966).
78. H. Harada and A. Lang, Plant Physiol., 40, 176 (1965).
79. B. E. Cross and P. L. Myers, Phytochemistry, 8, 79 (1969).
80. M. F. Barnes, E. N. Light, and A. Lang, Planta, 88, 172 (1969).
81. I. Shechter and C. A. West, J. Biol. Chem., 244, 3200 (1969).
82. E. E. Tschabold, H. M. Taylor, J. D. Davenport, R. E. Hackler, E. V. Krumkalns, and W. C. Meredith, Plant Physiol., 46, S19 (1970).
83. A. C. Leopold, Plant Physiol., 48, 537 (1971).
84. J. B. Shive, Jr. and H. D. Sisler, Plant Physiol., 57, 640 (1976).
85. R. C. Coolbaugh and R. Hamilton, Plant Physiol., 57, 245 (1976).
86. R. C. Seem, H. Cole, Jr., and N. L. Lacasse, Plant Disease Reptr., 56, 386 (1972).
87. W. J. Manning and P. M. Vardaro, Phytopathology, 63, 1415 (1973).
88. T. J. Douglas and L. G. Paleg, Plant Physiol., 54, 238 (1974).

89. J. R. Sabine, L. G. Paleg, and T. J. Douglas, Australian J. Biol. Sci., 26, 113 (1973).
90. A. Wood and L. G. Paleg, Australian J. Plant Physiol., 1, 31 (1974).
91. A. Wood, L. G. Paleg, and T. M. Spotswood, Australian J. Plant Physiol., 1, 167 (1974).
92. T. W. Goodwin, in Carotenoids (O. Isler, ed.), Birkhäuser Verlag, Basel, Switzerland, 1971, pp. 577-636.
93. N. I. Krinsky, in Photophysiology (A. C. Giese, ed.), Vol. 3, Academic Press, New York, 1968, p. 123.
94. R. Y. Stanier, Brookhaven Symp. Biol., 11, 43 (1959).
95. S. A. C. Morris and R. E. Subden, Mutation Res., 22, 105 (1974).
96. S. Jerebzoff-Quintin and S. Jerebzoff, Compt. Rend., Ser. D, 278, 1239 (1974).
97. D. J. Austin, J. D. Bu'Lock, and G. W. Gooday, Nature, 223, 1178 (1969).
98. L. Huang and A. Haug, Biochim. Biophys. Acta, 352, 361 (1974).
99. M. S. Kharash, E. A. Conway, and W. Bloom, J. Bacteriol., 32, 533 (1936).
100. G. Turian, Helv. Chim. Acta, 33, 1988 (1950).
101. T. W. Goodwin, Biochem. J., 50, 550 (1952).
102. T. W. Goodwin, in Advances in Enzymology (F. F. Nord, ed.), Vol. 21, Interscience Publishers, New York, 1959, pp. 295-368.
103. T. W. Goodwin, M. Jamikorn, and J. S. Willmer, Biochem. J., 53, 531 (1953).
104. J. A. Olson and H. Knizley, Jr., Arch. Biochem. Biophys., 97, 138 (1962).
105. F. H. Foppen and O. Gribanovski-Sassu, Biochim. Biophys. Acta, 176, 357 (1969).
106. R. Herber, B. Maudinas, and J. Villoutreix, Phytochemistry, 11, 3461 (1972).
107. H. C. Rilling, Arch. Biochem. Biophys., 110, 39 (1965).
108. C. W. Coggins, Jr., G. L. Henning, and H. Yokoyama, Science, 168, 1589 (1970).
109. W. J. Hsu, H. Yokoyama, and C. W. Coggins, Jr., Phytochemistry, 11, 2985 (1972).
110. M. Elahi, T. H. Lee, K. L. Simpson, and C. O. Chichester, Phytochemistry, 12, 1633 (1973).
111. W. Hsu, S. M. Poling, and H. Yokoyama, Phytochemistry, 13, 415 (1974).
112. E. R. Burns, G. A. Buchanan, and M. C. Carter, Plant Physiol., 47, 144 (1971).
113. A. Ben-Aziz and E. Koren, Plant Physiol., 54, 916 (1974).
114. G. W. Gooday, Ann. Rev. Biochem., 43, 35 (1974).
115. G. P. Arsenault, K. Biemann, A. W. Barksdale, and T. C. McMorris, J. Amer. Chem. Soc., 90, 5635 (1968).

116. J. R. Raper and A. J. Haagen-Smit, J. Biol. Chem., 143, 311 (1942).
117. A. W. Barksdale and L. L. Lasure, Appl. Microbiol., 28, 544 (1974).
118. T. C. McMorris, J. Org. Chem., 35, 458 (1970).
119. C. G. Elliott, M. R. Hendrie, B. A. Knights, and W. Parker, Nature, 203, 427 (1964).
120. J. W. Hendrix, Science, 144, 1028 (1964).
121. C. G. Elliott and B. A. Knights, Biochim. Biophys. Acta, 360, 78 (1974).
122. J. W. Hendrix, Can. J. Microbiol., 21, 735 (1975).
123. J. W. Hendrix and S. M. Guttman, Science, 161, 1252 (1968).
124. J. W. Hendrix and D. K. Lauder, J. Gen. Microbiol., 44, 115 (1966).
125. W. Duntze, V. MacKay, and T. R. Manney, Science, 168, 1472 (1970).
126. E. Throm and W. Duntze, J. Bacteriol., 104, 1388 (1970).
127. A. M. Thomas, R. C. Harris, J. T. O. Kirk, and T. W. Goodwin, Phytochemistry, 6, 361 (1967).
128. J. D. Bu'Lock and D. J. Winstanley, J. Gen. Microbiol., 69, 391 (1971).
129. L. Machlis, Nature, 181, 1790 (1958).
130. L. Machlis, Plant Physiol., 52, 527 (1973).
131. C. P. Eugenio, Phytopathology, 60, 1055 (1970).
132. J. C. Wolf and C. J. Mirocha, Can. J. Microbiol., 19, 725 (1973).
133. J. C. Wolf, J. R. Lieberman, and C. J. Mirocha, Phytopathology, 62, 937 (1972).
134. A. Matsumae, S. Nomura, and T. Hata, J. Antibiotics, Ser. A, 17, 1 (1964).
135. Y. Sano, S. Nomura, Y. Kamio, S. Ōmura, and T. Hata, J. Antibiotics (Tokyo), Ser. A, 20, 344 (1967).
136. S. Nomura, T. Horiuchi, S. Ōmura, and T. Hata, J. Biochem., 71, 783 (1972).
137. J. Awaya, T. Ohno, H. Ohno, and S. Ōmura, Biochim. Biophys. Acta, 409, 267 (1975).
138. S. Nomura, T. Horiuchi, T. Hata, and S. Ōmura, J. Antibiotics (Tokyo), Ser. A, 25, 365 (1972).
139. G. D'Agnolo, I. S. Rosenfeld, J. Awaya, S. Ōmura, and P. R. Vagelos, Biochim. Biophys. Acta, 326, 155 (1973).
140. D. Vance, I. Goldberg, O. Mitsuhashi, and K. Block, Biochem. Biophys. Res. Commun., 48, 649 (1972).
141. T. Ohno, T. Kesado, J. Awaya, and S. Ōmura, Biochem. Biophys. Res. Commun., 57, 1119 (1974).
142. L. A. Goldblatt, Aflatoxin: Scientific Background, Control, and Implications, Academic Press, New York, 1969.
143. C. J. Mirocha and C. M. Christensen, Ann. Rev. Phytopathol., 12, 303 (1974).
144. L. D. Owens, Science, 165, 18 (1969).
145. M. Biollaz, G. Büchi, and G. Milne, J. Amer. Chem. Soc., 92, 1035 (1970).

146. D. P. H. Hsieh and R. I. Mateles, Biochim. Biophys. Acta, 208, 482 (1970).
147. M. T. Lin, D. P. H. Hsieh, R. C. Yao, and J. A. Donkersloot, Biochemistry, 12, 5167 (1973).
148. S. R. Gupta, H. R. Prasanna, L. Viswanathan, and T. A. Venkitasubramanian, J. Gen. Microbiol., 88, 317 (1975).
149. R. W. Detroy, E. B. Lillehoj, and A. Ciegler, in Microbial Toxins (A. Ciegler, S. Kadis, and S. J. Ajl, eds.), Vol. 6, Academic Press, New York, 1971, pp. 68-90.
150. T. Shantha and V. S. Murthy, J. Food Sci. Technol., 12, 20 (1975).
151. L. B. Bullerman, H. M. Barnhart, and T. E. Hartung, J. Food Sci., 38, 1238 (1974).
152. A. Ciegler, E. B. Lillehoj, R. E. Peterson, and H. H. Hall, Appl. Microbiol., 14, 934 (1966).
153. M. Elbereir, G. Krug, K. S. Grunert, and C. Franzke, Nahrung, 19, K3 (1975).
154. A. Ciegler, Lloydia, 38, 21 (1975).
155. N. D. Davis, U. L. Diener, and V. P. Agnihotri, Mycopathol. Mycol. Appl., 31, 251 (1967).
156. B. F. Nesbitt, J. O'Kelly, K. Sargeant, and A. Sheridan, Nature, 195, 1062 (1962).
157. R. I. Mateles and J. C. Adye, Appl. Microbiol., 13, 208 (1965).
158. E. G. H. Lee, P. M. Townsley, and C. C. Walden, J. Food Sci., 31, 432 (1966).
159. P. B. Marsh, M. E. Simpson, and M. W. Trucksess, Appl. Microbiol., 30, 52 (1975).
160. H. R. Gundu-Rao and P. K. Harein, J. Econ. Entomol., 65, 988 (1972).
161. D. P. H. Hsieh, J. Agr. Food Chem., 21, 468 (1973).
162. M. T. Wu and J. C. Ayres, J. Agr. Food Chem., 22, 536 (1974).
163. N. D. Davis and U. L. Diener, Appl. Microbiol., 15, 1517 (1967).
164. G. A. Bean, W. L. Klarman, G. W. Rambo, and J. B. Sanford, Phytopathology, 61, 380 (1971).
165. G. A. Bean, G. W. Rambo, and W. L. Klarman, Life Sci., 8, 1185 (1969).
166. H. Buchenauer, Pestic. Biochem. Physiol., in press.
167. H. Buchenauer, Z. Pflanzenkrankh. Pflanzenschutz., 83, 363 (1976).
168. P. Leroux and M. Gredt, Phytopathol. Z., 86, 276 (1976).
169. H. Buchenauer and F. Grossmann, Neth. J. Plant Pathol., in press.
170. M. R. Siegel, unpublished work, 1976.
171. T. Kato, Abstracts, Symposium on Internal Therapy of Plants, Wageningen, The Netherlands (1976).
172. K. A. Mitropoulos, G. F. Gibbons, C. M. Connell, and R. A. Woods, Biochem. Biophys. Res. Commun., 71, 892 (1976).

Chapter 10

EFFECT OF FUNGICIDES ON NUCLEIC ACID SYNTHESIS AND NUCLEAR FUNCTION

J. Dekker

Laboratory of Phytopathology
Agricultural University
Wageningen, The Netherlands

I. INTRODUCTION

This chapter discusses fungicides which interfere primarily with nucleic acid synthesis or nuclear function or presumably do so. Compounds inhibiting these processes indirectly and multisite inhibitors as a rule are not included.

Among the antifungal compounds which interfere with nucleic acid syntheses or nuclear function are natural products, such as antibiotics, and chemicals synthesized for commercial purposes. These compounds may interfere with the synthesis or interconversion of nucleotides (Section II), the replication or transcription of deoxyribonucleic acid, DNA (Section III), or with mitosis (Section IV). The biochemical effect often occurs at one specific site. It will be discussed how far a relationship exists between this specific action and the degree of biological selectivity. In addition, the phenomenon of fungicide resistance will be considered, particularly in relation to the mechanism of action of the inhibitors.

Several fungicides mentioned in this chapter do not show promise for practical use against pathogenic fungi due to lack of selective action, low activity against fungi, or for other reasons. These compounds are discussed because they are of interest for theoretical considerations. In addition, some compounds have been included, of which it is not yet sure or even doubtful whether they act primarily on nucleic acid synthesis or nuclear function.

II. SYNTHESIS OF NUCLEOTIDES

A. General

Nucleotides are the building blocks of nucleic acids; each of them is composed of a purine or pyrimidine base, a pentose sugar, and phosphoric acid. Deoxyribonucleic acid contains the purine bases adenine and guanine, the pyrimidine bases cytosine and thymine, the pentose sugar deoxyribose, and phosphoric acid. Ribonucleic acid (RNA) contains the same constituents, except that the pyrimidine uracil is substituted for thymine and the pentose ribose for deoxyribose.

Inhibitors may interfere with nucleotide synthesis at different sites. They may inhibit de novo purine or pyrimidine biosynthesis, often by interaction with enzymes operating in these pathways and in competition with natural substrates. In addition, they may interfere with conversion of purine and pyrimidine bases into nucleotides or with interconversion of nucleotides. Finally, analogues may be converted into fraudulent nucleotides impairing normal nucleotide function, and, eventually, even be incorporated into nucleic acids or into coenzymes in place of the normal nucleotides.

Many compounds which are known to interfere with nucleotide synthesis have been studied, mainly in connection with cancer chemotherapy. Few of these, however, appear to possess a pronounced selective action against fungi.

B. Antimetabolites of Purines

8-Azaguanine (1), 8-azaadenine, 8-azamethylpurine, and 2,6-diaminopurine were active against powdery mildew on cucumber leaf discs floating on a solution of the compounds to be tested. Systemic action, however, was limited and no disease control was obtained by application of these compounds

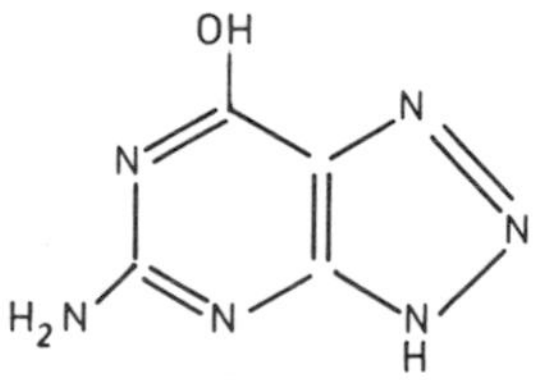

(1) 8-azauguanine

to the roots of intact plants [1]. 8-Azaguanine inhibits growth also of *Aspergillus nidulans* [2]; mutants resistant to this compound and to purine, 2-thioxanthine, and 2-thiouric acid appeared defective in at least one step of purine uptake or breakdown [3]. Tolerance of mutants of *Neurospora crassa* to 8-azauguanine appeared to depend upon mutation in one gene at each of three different loci [4]. The mechanism of action of 8-azaguanine against fungi has not yet been studied. Information on the activity of this compound on other systems is discussed by Roy-Burman [5]. Incorporation of 8-azaguanine into fraudulent RNA has been reported to occur in tobacco mosaic virus, *Bacillus cereus*, and certain neoplasms. The conclusion that this is the primary site of action is supported by the fact that neoplasms resistant to azaguanine appeared to be unable to synthesize guanylic and 8-azaguanylic acid.

Nothing is known about the mechanism of antifungal activity of 8-azaadenine, 8-azamethylpurine, and 2,6-diaminopurine, compounds which have shown activity against neoplasms, bacteria, and multiplication of various viruses [6]. Upgrading of 2,6-diaminopurine may be essential for activity, since resistance in bacteria to this compound appeared accompanied by loss of adenylic phosphorylase. 8-Azaadenine and 8-azamethylpurine may be upgraded even to the di- and triphosphate, thus impairing normal nucleotide interconversion and function.

C. Antimetabolites of Pyrimidines

6-Azauracil (2), 6-azacytosine, 6-azathiouracil, dithiouracil, and 2,4-dithiopyrimidine have shown systemic activity against cucumber powdery mildew on leaf disks floating on an aqueous solution of these compounds [1].

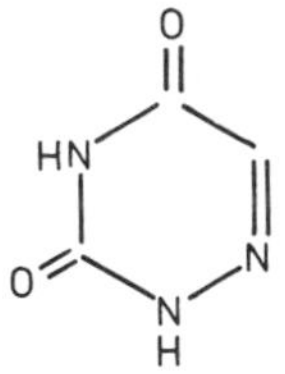

(2) 6-azauracil

Only 6-azauracil (AzU) provided systemic control of cucumber powdery mildew after root application. It appeared systemically active also against cucumber scab, caused by *Cladosporium cucumerinum*, and bean rust, caused by *Uromyces appendiculatus* [7]. Activity of AzU against *Colletotrichum lindemuthianum* was found by Matolcsy and Doma [8].

From studies with C. cucumerinum it appeared that not AzU, but 6-azauridine monophosphate (AzUMP), which is synthesized by the fungus via 6-azauridine (AzUR), is the fungitoxic principle. AzUMP exerts its fungitoxic action by inhibition of the enzyme orotidine-5'-decarboxylase, thus preventing pyrimidine biosynthesis. Accordingly, UV-induced AzU-resistant strains were obtained, which were unable to convert AzU into AzUR due to lack or reduced activity of the enzyme uridine phosphorylase; these strains were, as expected, sensitive to AzUR and AzUMP [9].

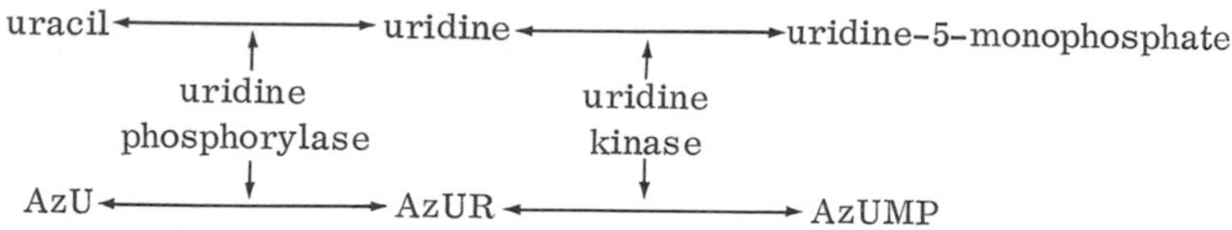

A second group of UV-induced AzU-resistant strains appeared to be tolerant also to AzUR and AzUMP. Evidence was obtained that both reduced conversion of AzUR into AzUMP and decreased permeability, especially for AzUMP, contributed to this type of resistance [10]. Some AzU may be incorporated into RNA, although this seems to be of minor importance. Similar results were obtained in studies with C. lindemuthianum [11]. Conversion of AzU into its di- and triphosphates and into RNA has been reported in other living systems, leading to other possibilities for metabolic activity [12, 13].

Selective action with respect to the combination of host and parasite cannot be explained by a difference in the formation of the toxic principle, since it was shown that also cucumber plants readily convert AzU into AzUR and AzUMP. Two factors are supposed to operate in host-parasite specificity, namely an extremely high orotidine-5'-decarboxylase activity in cells of cucumber leaves which renders the plant less vulnerable than the fungus, and a more rapid rate of RNA synthesis in the fungus during infection than in full-grown plant parts. Accordingly, it was observed that, at high doses of AzU, the youngest still-expanding leaves of cucumber were injured most [14]. When AzU was applied at one-week intervals in an apple orchard, excellent control of powdery mildew was obtained, but at the end of the season malformed young shoots were observed [15]. AzU, which has moreover been proved to be toxic to various other types of living organisms, therefore cannot be considered to be sufficiently selective in its action for control of plant diseases.

Results from work with bacteria show that 6-azacytosine and 6-azacytidine are deaminated to the corresponding uracil analogues; accordingly, cross resistance in bacteria to 6-azauracil and 6-azacytidine was found [16].

D. Dimethirimol and Related Compounds

1. Development and Structure

Dimethirimol, 5-n-butyl-2-dimethylamino-4-hydroxy-6-methylpyrimidine (3), was first reported by Elias et al. in 1968 [17] to provide systemic control of powdery mildew in cucurbits, caused by Sphaerotheca fuliginea. Shortly thereafter, the closely related compound ethirimol, 5-n-butyl-2-ethylamino-4-hydroxy-6-methylpyrimidine (4), was introduced for control of barley powdery mildew, caused by *Erysiphe graminis* [18]. A third compound, bupirimate, *n*-5-butyl-2-ethylamino-6-methylpyrimidine-4-yl dimethylsulfamate (5), has been introduced more recently [19].

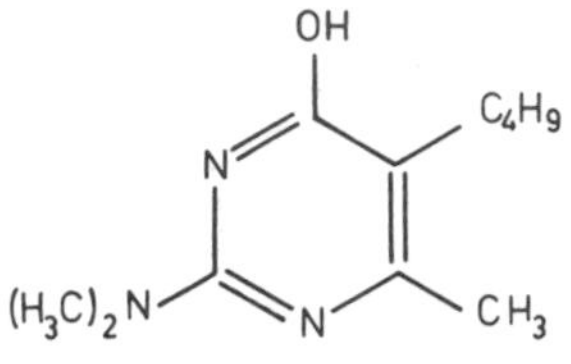

(3) dimethirimol 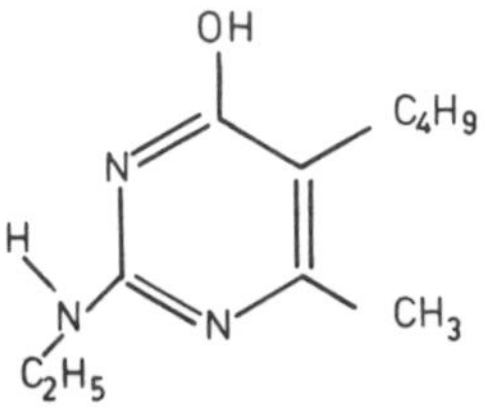 (4) ethirimol 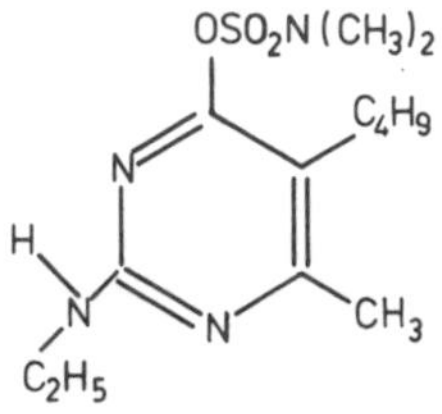 (5) bupirimate

2. Biological Activity

Dimethirimol and ethirimol are very selective with respect to control of plant diseases, providing only control of powdery mildew diseases. Cucumber powdery mildew appeared more sensitive to dimethirimol than to ethirimol; the reverse was true for barley powdery mildew. Mildew of vine, caused by *Uncinula necator*, and of apple, caused by *Podosphaera leucotricha*, appeared less sensitive to these compounds than cucumber and barley powdery mildew. Bupirimate is active against various powdery mildews and has proved very effective in field trials against apple powdery mildew. The three compounds are active only against the powdery mildew fungi. In vitro, spore germination of *E. graminis* was inhibited by dimethirimol and ethirimol [20, 21] and elongation of germ tubes of *Sphaerotheca fuliginea* was reduced 50% at concentrations as low as 0.5×10^{-8} M ethirimol and 10^{-7} M dimethirimol [22].

3. Mechanism of Action

Bent [21] observed that the effect of dimethirimol and ethirimol in vitro and in the plant was antagonized by folic acid and purines, and to a lesser extent by thymine. Calderbank [23] assumed that these compounds act as noncompetitive enzyme inhibitors and interfere with tetrahydrofolate-directed C-1 metabolism, which plays a role in the synthesis of purines, thymidine, and amino acids. Possibly they might act as antagonists of the coenzyme pyridoxal, which is involved in reactions in which C-1 units are transferred to tetrahydrofolic acid. No pertinent data, however, are available about the mechanism of action of these compounds.

4. Uptake, Transport, and Fate in Plants

Dimethirimol and ethirimol are taken up by the roots of various plants and transported upwards in the xylem. No phloem transport was observed.

Consequently only leaves present at the time of application are protected and no downward translocation can be demonstrated [23]. In nonwoody plants the fungicides rapidly accumulate in the leaves, and concentrate in the leaf margin. In woody plants little movement of the fungicides from the region of the veins occurred, and, accordingly, little protection of apple and vine powdery mildew was obtained [24].

Ethirimol is rapidly degraded when administered to barley plants via the roots, with a half-life of 3 to 4 days. Ethirimol glucoside, desethylethirimol, and hydroxylated butylethirimol and a few unidentified products were recovered. Demethirimol is also rapidly transformed in cucumber plants into a series of related products, e.g., the desmethyl derivative, several glucosides, and an amino compound [24].

5. Disease Control and Development of Resistance

Dimethirimol was introduced in 1968 for use against powdery mildew on cucumber in greenhouses in England and Holland, and was widely used for this purpose in 1969. A suspension of this fungicide, applied to the soil around the base of the plant, provided continuous protection of the whole plant for several weeks. In the autumn of 1969, and more so in the spring of 1970, the results in some greenhouses were disappointing. This appeared due to the development of dimethirimol-resistant strains in *Sphaerotheca fuliginea*. As a consequence, the use of this chemical for control of cucumber powdery mildew in greenhouses was abandoned [25]. Recently it was found that treatment of cucumber with a combination of dimethirimol and benomyl did not give rise to development of resistance in powdery mildew [26].

Ethirimol is active against barley powdery mildew not only by foliar sprays, but also by seed dressing, the latter application being even more successful than the former [27]. Apparently the chemical is taken up during germination of the seedling and transported to the aerial parts of the plant. By this treatment and by application of ethirimol as granules in the furrow, a long-lasting protection of susceptible varieties against powdery mildew was obtained [28, 29]. It was observed that ethirimol-tolerant forms of *Erysiphe graminis* were present in the barley crop, and most frequently in crops treated with this fungicide. In order to reduce the frequency of tolerant forms in the pathogen population at the beginning of the spring season, it is recommended not to treat winter barley with ethirimol [30].

Ethirimol is now used as a seed dressing for control of barley powdery mildew.

E. Aristeromycin

Aristeromycin (6), isolated from a culture filtrate of *Streptomyces citricolor* [31], appeared to be an adenine analogue [32]. It was found to be active against some pathogenic fungi and bacteria, but has little activity against

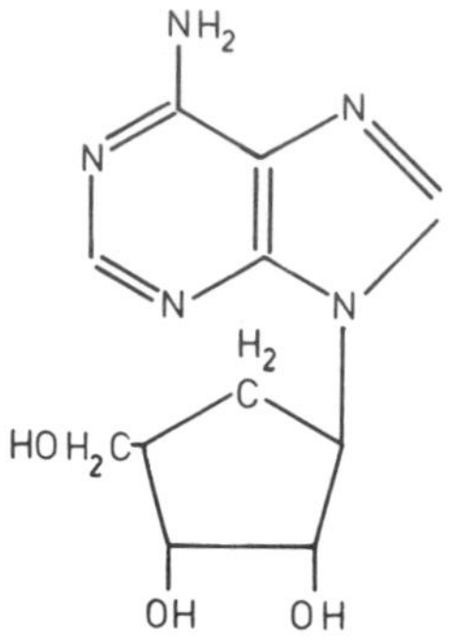

(6) aristeromycin

most other microorganisms. Particularly sensitive are *Piricularia oryzae*, *Pellicularia filamentosa* f. sp. *solani*, *Ophiostoma fimbriata*, *Helminthosporium sigmoideum* var. *irregulare*, *Fusicladium levieri*, *Elsinoe fawcetti*, *Cephalosporium gramineum*, and *Alternaria kikuchiana*, and among the bacteria, *Xanthomonas oryzae* and *X. pruni*. In the greenhouse a curative effect was found against rice blast caused by *P. oryzae*. Sporulation of this fungus on leaves of rice appeared inhibited. Aristeromycin was also active against bacterial leaf blight of rice caused by *Xanthomonas oryzae* [31]. Its mechanism of action was studied in *X. oryzae*. Growth inhibition of this bacterium by aristeromycin was antagonized partially by adenosine. deoxyadenosine, adenine, and inosine, but antagonism by adenine was most pronounced. It has been suggested that the antibiotic interferes with formation of adenylic acid from adenine and adenosine [33].

F. Azaserine and 6-Diazo-5-oxo-L-norleucine (DON)

Azaserine (7) was discovered in culture filtrates of *Streptomyces fragilis* in 1956, and DON (8) in culture filtrates of an unidentified *Streptomyces* sp.

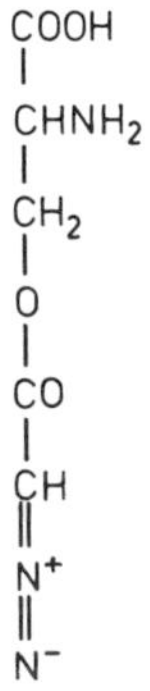

(7) azaserine 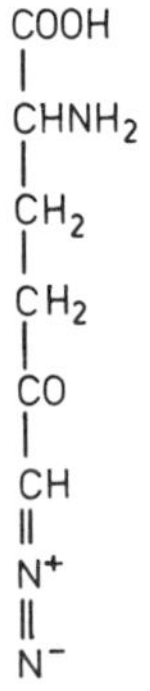 (8) DON

shortly thereafter. Both antibiotics show activity against some fungi and bacteria, and also against tumors. They are inhibitors of de novo purine biosynthesis. Structurally related to glutamine, they inhibit the conversion of formylglycinamide ribonucleotide into formylglycinamidine, a reaction which requires glutamine as a substrate. It is reported that azaserine combines irreversibly with a sulfhydryl group on the enzyme which catalyzes this conversion. Also other glutamine-requiring reactions may be inhibited, but these do not seem to be primary sites [34].

G. Glyodin

The fungicidal properties of glyodin, 2-heptadecyl-2-imidazoline acetate (9) were first reported in 1946 [35]. It shows a broad spectrum of antifungal activity. Recently it was shown to be active as a foliar protectant against

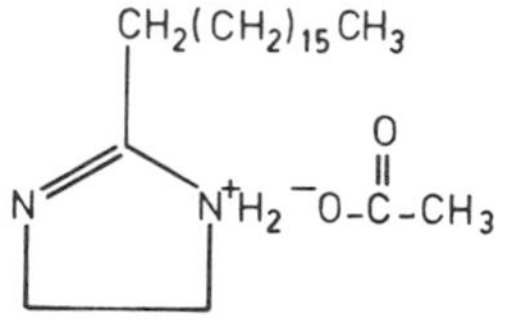

$CH_2(CH_2)_{15}CH_3$

N $\quad N^{+}H_2 \; {}^{-}O-\overset{O}{\overset{\|}{C}}-CH_3$

(9) glyodin

tobacco mosaic virus [36]. The literature concerning its mechanism of action has been reviewed by Sisler [37]. At low doses its activity is antagonized by guanine and xanthine, which suggests that purine biosynthesis may be inhibited. It causes, however, also membrane damage, which might even be of primary importance for fungicidal action.

Glyodin has been in practical use as a protective fungicide against foliar plant diseases, but is nowadays not much used for this purpose.

III. REPLICATION AND TRANSCRIPTION OF DNA

A. General

Deoxyribonucleic acid consists of two helical polynucleotide chains coiled around a common axis to form a double helix with the pentose sugar and phosphate as backbones. The chains are held together by hydrogen bonds between pairs of the organic bases adenine and thymine and guanosine and cytosine.

Replication of DNA, which is catalyzed by the enzyme DNA polymerase, is essential for genetic continuity. During this process the double-stranded parent DNA unwinds and the separated strands serve as templates for the alignment of new nucleotides to form new strands complimentary to the parental strands, so that the original helix is reproduced.

The second biological function of DNA is to serve as a template for the production of RNA, which is catalyzed by the enzyme RNA polymerase. Three types of RNA are formed: (a) messenger RNA, which codes for the production of proteins on the ribisome; (b) transfer RNA, a smaller molecule involved in selecting and transporting amino acids; and (c) ribosomal RNA.

Compounds may interfere in several ways with the replication and transcription of DNA. They may form a complex with DNA by an irreversible covalent binding or by a reversible noncovalent binding. In the latter case binding occurs usually by intercalation of the compound between two nucleotides of the DNA helix. Certain chemicals even may break one or both DNA strands. Also inhibition of the enzymes DNA and RNA polymerase occur. One or more of these events may prevent replication or transcription. Most compounds which act at one or more of these sites in various biological systems appear to be antibiotics, but only few of these show fungicidal activity.

B. Bleomycins

Bleomycins are a group of basic glycopeptide antibiotics, produced by *Streptomyces verticillus*, with activity against some bacteria, viruses, and tumors [38]. More than 13 members of this group have now been discovered. The chemical structure of constituents of bleomycin A_2, mol wt 1,400, has been reported [39]. From a fermentation broth of Streptomyces E 327, active against rice-sheath blight (*Pellicularia sasakii*), five substances were isolated and identified as bleomycin A_5, A_6, B_2, B_4, and B_6, respectively [40]. Bleomycin B_2, which was the most active compound, showed high activity in vitro against *P. sasakii*, but was somewhat less active against *Bacillus subtilis*. When applied as a 20 μg/ml aqueous spray on rice plants in pots inoculated with *P. sasakii* this compound provided a 50-86% control of the disease; an equal concentration of commercially available bleomycin provided 91% control. No signs of phytotoxicity were observed.

The mechanism of action has been studied with *E. coli*, Hela cells, and with purified DNA [41]. Bleomycins show binding to DNA, twice as much to single-strand as to double-strand DNA. In the presence of a sulfhydryl compound, such as 2-mercaptoethanol, it causes scissions in the sugar-phosphate backbone of single-strand and to a lesser extent also in double-strand DNA. High concentrations of bleomycin even may cause degradation of DNA to the level of nucleotides [42].

In studies with *E. coli*, a stimulation of DNAse activity by bleomycin was observed which might be related to the scission of DNA strands. Inhibition of polynucleotide ligase (which presumably plays a role in the replication process), prepared from T_4 phage-infected *E. coli* was also observed [43].

Inhibition of DNA transcription by bleomycins was studied in E. coli and in phage T_7 [44]. Antimitotic action, probably a secondary effect, was reported in Hela cells [45].

The action of bleomycin is antagonized by the structurally related antibiotic phleomycin, which indicates that they may occupy the same binding site on DNA [41].

C. Phleomycin

Phleomycin is a copper-containing polypeptide antibiotic, first isolated from Streptomyces verticillus [46]. A culture filtrate of S. cinnamomeus f. azacoluta also appeared to contain this antibiotic [47]. In vitro it shows activity against some bacteria, viruses, and tumors, but not against fungi. In spite of this, it gave systemic control of certain fungal diseases, namely in experiments with bean rust [48, 49] and wheat stripe rust [50].

Tanaka et al. [51] showed that phleomycin inhibits DNA synthesis in E. coli. This appeared to be due to a binding of phleomycin to DNA, which in turn inhibits replication by DNA polymerase. Binding occurs at the adenine-thymine rich regions, probably to the thymine bases [52]. In E. coli, coumarin [53] and purines [54] amplify the antibiotic activity of phleomycin. It is suggested that these compounds, by binding preferentially to single-stranded DNA, may enhance local denaturation around phleomycin-thymine complexes; this might result in nuclease-induced DNA breakage and degradation. Phleomycin causes also single-strand breaks in purified adenovirus and phage DNA [55].

Recently Sleigh and Grigg [56] obtained information on the effect of phleomycin on fungal DNA. Incubation of phleomycin with DNA and single-strand-specific endonuclease from Neurospora crassa resulted in degradation of DNA. Addition of caffeine caused a greater breakdown than that expected from a combined effect. It is suggested that caffeine may increase denaturation of single-stranded DNA by binding at phleomycin-thymine complexes on the DNA. Strand breakage induced by phleomycin was also observed in animal cell lines [57].

It is not clear in what way phleomycin acts against rust diseases, since it did not show a direct effect on the pathogens concerned. It might act on the plant, increasing its resistance to the pathogen. That phleomycin may influence plants, was reported by Kihlman et al. [58], who found effects on chromosome structure and nucleic acid synthesis in Vicia faba.

D. Lomofungin

Lomofungin, 1-carboxymethoxy-5-formyl-4,6,8-trihydroxyphenazine (10), is an antibiotic produced by Streptomyces lomodensis [59]. It is active against bacteria, yeasts, and filamentous fungi. Growth of Saccharomyces

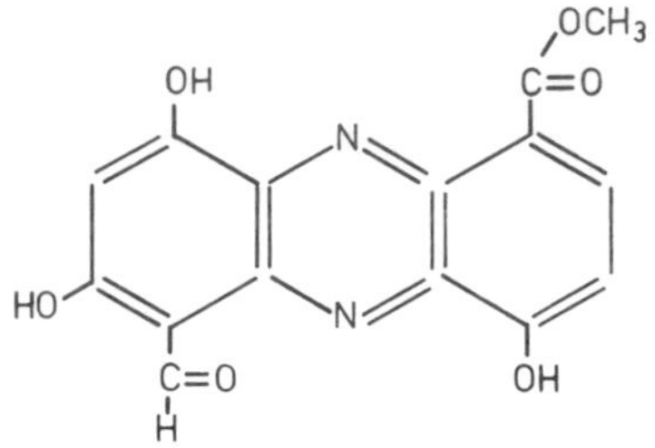

(10) lomofungin

cerevisiae, *S. fragilis*, and *S. pastorianus* was inhibited completely by 10 μg/ml lomofungin, growth of *Penicillium oxalicum* almost completely. The synthesis of RNA and DNA was inhibited by concentrations of 4 μg/ml of lomofungin, and protein synthesis was reduced in intact cells but not in cell-free systems of *Rhizoctonia solani* and *S. cerevisiae* [60]. In studies with spheroblasts of *S. cerevisiae* it was shown that nucleic acid synthesis is inhibited almost immediately upon addition of the antibiotic, but that protein synthesis is allowed to proceed at a normal rate for about 20 min before it becomes inhibited. These data suggest that in *S. cerevisiae* lomofungin acts primarily as an inhibitor of nucleic acid synthesis and that inhibition of protein synthesis is a secondary effect. It was shown that the antibiotic induces a complete breakdown of polyribosomes and a sharp increase of 80-S monomers, apparently by causing ribosomes to "run off" from messenger RNA [61]. A rapid and selective inhibition of the synthesis of high-molecular-weight RNA by lomofungin was demonstrated in *Schizosaccharomyces pombe* [62] and inhibition of three DNA-dependent RNA polymerases in extracts from *Saccharomyces cerevisiae* [63]. Inhibition of growth and RNA synthesis by lomofungin in yeast cells was prevented by Zn and Cu ions, which complex with the antibiotic and prevent its uptake by the cell. It is suggested that lomofungin acts by chelating with the firmly bound Zn of RNA polymerase, thus preventing formation of the polymerase-DNA initiation complex [64].

E. Edeine

Edeine is a polypeptide antibiotic, produced by a spore-forming strain of *Bacillus brevis*, and discovered in 1959. It is a broad-spectrum antibiotic, inhibiting gram-positive and gram-negative bacteria, some fungi, including yeasts, mammalian neoplastic cells, mycoplasmas, and the reproduction of certain coliphages. In experiments with *E. coli B* it was shown that edeine inhibits the synthesis of DNA, whereas the synthesis of proteins is not affected significantly and the synthesis of RNA is even slightly stimulated. Since inhibition of DNA synthesis prevents cell division and RNA and proteins accumulate, the bacteria become long and filamentous. In a study with DNA and DNA polymerase it was found that edeine does not bind to the DNA template, but interacts presumably with the DNA polymerizing enzymes [65].

In studies with *Neurospora crassa*, edeine-resistant strains were obtained; resistance appeared recessive and unstable [66]. In the alga *Euglena gracilis*, a hereditary loss of plastids was caused by edeine, a phenomenon which was also produced by some other inhibitors of DNA synthesis [67].

F. Thiolutin

Thiolutin (11) is a pyrrol antibiotic, produced by *Streptomyces albus*, and discovered in 1952. It shows activity against bacteria and fungi. A concentration of 1-5 μg/ml inhibits growth of the following fungi, pathogenic to humans: *Histoplasma capsulatum*, *Homodendrum pedrosolii*, *Plastomyces dermatitidis*, *Plastomyces brasiliensis*, *Cryptococcus neoformans*, *Epidermophyton floccosum*, *Microsporum audonini*, *Candida albicans*, *Coccidioides immitis*, and *Trichophyton rubrum* [68]. It is also active against various fungi and bacteria pathogenic to plants, but has been disappointing in experiments against plant diseases, and possibilities for practical use have not been found [69]. At a concentration of 2 μg/ml thiolutin inhibited growth of *S. cerevisiae*. In vivo RNA and protein synthesis were inhibited, but protein synthesis continued for a short time after RNA synthesis was completely stopped. In cell-free systems from yeast thiolutin did not affect polypeptide formation, but appeared to be a potent inhibitor of DNA-dependent RNA polymerase [70].

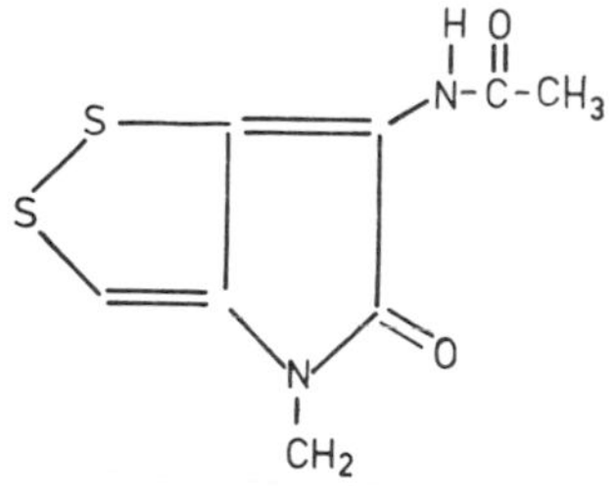

(11) thiolutin

G. Phytoactin

Phytoactin is a polypeptide antibiotic, produced by an unidentified *Streptomyces* sp. [71]. It is active against gram-positive bacteria and fungi, and its use for control of plant diseases has been explored. Activity of phytoactin against white pine blister rust, caused by *Cronartium ribicola* and various other diseases, has been reported [72]. Translocation of phytoactin in white pine seedlings was demonstrated [73] which opened the possibility for curative action. Reports on its effectiveness against this disease were, however, conflicting. It was concluded that neither aerial application nor basal stem application were satisfactory [74].

From studies with Saccharomyces pastorianus it was learned that RNA synthesis is most strongly affected [75]. According to these authors the mode of action resembles that of actinomycin, which binds to DNA and inhibits DNA-dependent RNA polymerase.

H. Actinomycins

Antibiotics belonging to the group of actinomycins, which are produced by various species of Streptomyces, were first discovered in 1940. Members of this group have shown cytostatic, antibacterial, and antifungal activity. Growth of Penicillium luteum purpurogenum, Aspergillus clavatus, Candida albicans, Trichophyton mentagrophytes, and Cryptococcus neoformans was inhibited by 0.2-2 μg/ml of actinomycin [76]; growth of Neurospora crassa was also reduced [77]. The actinomycins are highly toxic polypeptides. The structure and the mechanism of action of actinomycin D (12), which has been studied extensively as an anticancer agent and used as a tool is studies

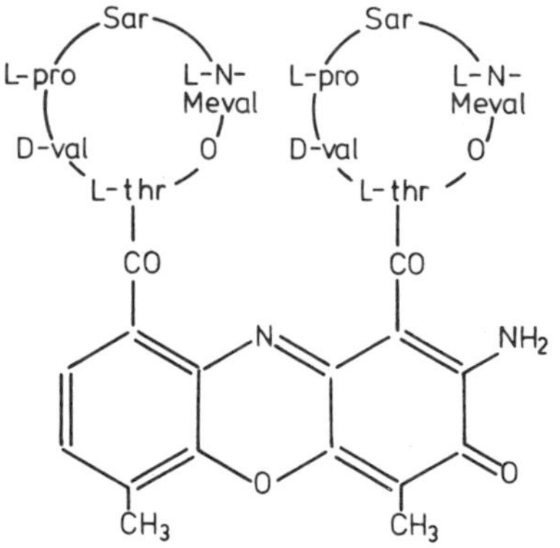

(12) actinomycin D

L-thr = L-threonine, D-val = D-valine, L-pro = L-proline, Sar = sarcosine, L-N-meval = L-N-methylvaline

on nucleic acid metabolism, have been elucidated. It consists of a phenoxazone ring system linked to two cyclic pentapeptides, each composed of L-methylvaline, sarcosine, L-proline, D-valine, and L-threonine. Actinomycin

D binds to the DNA helix by intercalation of its phenoxazone ring system between guanine-cytosine sequences, the peptide portion of the actinomycin molecule, which lies in the narrow groove of the DNA helix, being hydrogen bonded to deoxyguanosine on opposite DNA chains. In this way actinomycin blocks the transcription of DNA into RNA [78].

I. Daunomycin and Related Antibiotics

The antibiotics belonging to this group contain a tetrahydrotetracinquinone chromophore linked to a sugar. Antifungal activity has been reported for three antibiotics from this group, namely daunomycin (13), a metabolite of *Streptomyces peucetius*, and cinerubin and aklavin, both formed by species of *Streptomyces*. These antibiotics are also active against gram-positive bacteria, certain viruses, and experimental tumors. Their mechanism of

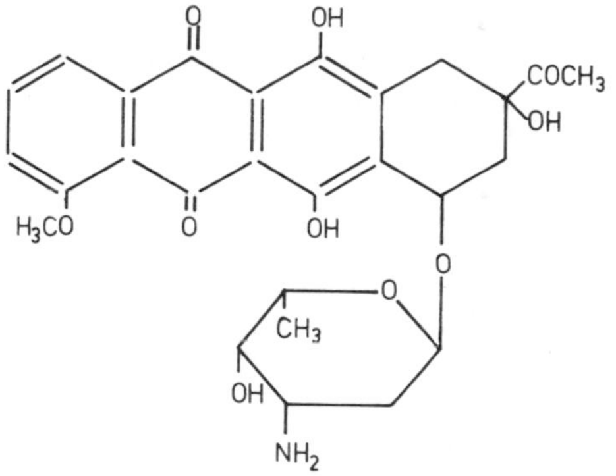

O
OH
$COCH_3$
OH
H_3CO
O
OH
O
O
CH_3
OH
NH_2

(13) daunomycin

action is discussed by Di Marco [79]. It was demonstrated that daunomycin and cinerubin react with DNA to form complexes. Binding is indicated by increase of the melting point and increase in viscosity, and decrease of the sedimentation coefficient of the DNA-daunomycin complex. These properties and the finding that daunomycin causes uncoiling of the supercoiled structure of closed circular duplex DNA by 12° [80] indicate that it intercalates between the DNA base pairs, in this way inhibiting duplication and transcription. Mitosis is also inhibited, presumably because daunomycin-bound DNA is unable to function properly in the mitotic process [81, 82].

Daunomycin induced cytological aberrations also in plants, namely aberrand mitosis and nonspecific chromatic breaks in root meristems of *Vicia faba* [83]. Although the action of this type of antibiotics is rather specific, it apparently is not very selective.

J. Cycloheximide

Cycloheximide was discovered in 1946 in a culture broth of Streptomyces griseus. It appeared toxic to a wide range of organisms, including fungi, algae, plants, protozoa, and animals, but inactive against bacteria; it also shows slight activity against some animal tumors. Cycloheximide has shown a curative action against a number of fungal plant diseases, but because of phytotoxicity its use for this purpose has been very limited. Promising results were only obtained when extremely low concentrations were sufficient for disease control, or when the treated plant parts were relatively resistant to the antibiotic, for example, trunks of white pine carrying blister rust cankers. The attempts to use cycloheximide for control of plant diseases have been reviewed [69, 84].

Cycloheximide has long been considered to act primarily on protein synthesis [85]. Recently, however, it has been indicated that the antibiotic may also act directly on RNA synthesis. It was shown that cycloheximide in Achlya bisexualis, an aquatic Oomycete, inhibits RNA polymerase I [86]. After introducing extremely low concentrations of this antibiotic in the medium, the capacity of this organism to synthesize RNA decreased rapidly, with a half-life of 1.4 min [87]. It appeared that immediate inhibition of protein and RNA synthesis occurred only when cycloheximide or its derivative cycloheximide oxime were added. A number of other analogues, however, did inhibit only protein synthesis [88]. Strong evidence has been obtained that inhibition of RNA synthesis by cycloheximide does not occur as a consequence of interrupted protein synthesis, but by some other mechanism. For a full discussion of the action of this fungicide see Chapter 11.

IV. NUCLEAR FUNCTION

A. General

In eukaryotic organisms most of the cell's hereditary material, i.e., DNA, is arranged on chromosomes located in the nucleus. The mechanism by which cells reproduce involves nuclear division (mitosis) and cytoplasm division (cytokinesis). In mitosis, four main phases can be discerned: prophase, metaphase, anaphase, and telophase. During the prophase the chromosomes, which lie extended in the nucleus, shorten and thicken by coiling. Toward the end of the prophase the chromosomes migrate to the equatorial plane of the nucleus. At the same time a spindle is formed, which, in cells of plants and fungi, occurs between the poles of the dividing mucleus. The spindle consists of a number of microtubules, each assembled from numerous protein subunits. At metaphase the spindle is fully developed, and the nuclear membrane and nucleoli have disappeared. At this stage the chromosome occurs in duplicate, that is, two chromatides held together at one point, the centromere, which is attached to the spindle. During anaphase the chromatides of each chromosome start to move to opposite sides along the spindle by a sliding filament mechanism [89]. Once the chromatides have reached the opposite poles of the spindle, telophase begins; the spindle disappears and the chromosomes lengthen by uncoiling. Nuclear envelopes appear around both groups of chromosomes, so that two new nuclei are formed. In plant cells, division is completed by the growth of a wall. This formation of daughter cells is followed by a period of growth during which both the nucleus and the cytoplasm enlarge. During this phase, which is designated G-1, chromosomes are fully extended, and protein and RNA synthesis proceed. Subsequently a period of DNA synthesis follows, the S-phase, which results in the doubling of the chromosomes. This again is followed by a G-2 phase, of which the metabolic significance is not yet understood. The cycle is completed by mitosis.

In fungi mitosis takes place inside the intact nuclear envelope, which becomes constricted around the two daughter nuclei at telophase. This and the small size of the fungal nuclei hamper observations with the optical microscope. In the plant pathogen *Fusarium oxysporum* an extra nuclear spindle-pole body was observed at each pole and between these bodies the intranuclear spindle [90]. Further details of mitosis in fungi remain to be elucidated.

Various fungicides appear to interfere with the process of mitosis. Not included in this section are those fungicides which inhibit mitosis as a result of action at other sites, such as the antibiotics daunomycin and bleomycin (see Section III).

B. Griseofulvin

1. Discovery and Structure

Griseofulvin (14) was first isolated as a metabolic product from a culture of *Penicillium griseofulvum* in 1939 [91]. Later Brian et al. [92] discovered that the culture filtrate of *Penicillium janczewskii*, a fungus isolated from soil, caused severe stunting and malformation of *Botrytis allii* germ tubes, and at lower dilutions a helical waving of the hyphal tip. The active compound appeared to be identical with the previously discovered griseofulvin,

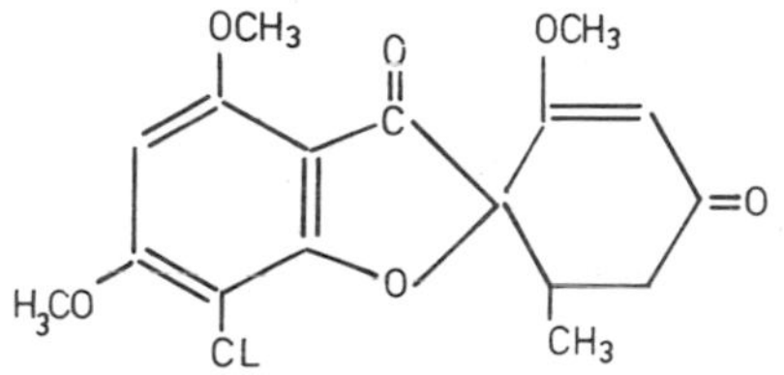

(14) griseofulvin

and was later found to be produced by various other species of Penicillium. The structure of griseofulvin was determined by Grove et al. [93]. Dechlorogriseofulvin and bromogriseofulvin have also been reported to be produced by species of Penicillium [94, 95]. A large number of griseofulvin analogues have been prepared, of which the 2'-n-propoxy derivative appeared most promising in vitro and for plant disease control [96]. Griseofulvin is fairly stable in aqueous solutions but it loses antifungal activity in daylight in the presence of p-aminobenzoic acid [97].

2. Biological Activity

Griseofulvin shows a wide spectrum of antifungal activity, causing twisting, stunting, excessive branching, abnormal swelling, and distortion of hyphae at low concentrations. It does, however, not inhibit growth of fungi belonging to the Oomycetes, and certain yeasts and bacteria. Some fungi, e.g., Botrytis alii, are extremely sensitive to griseofulvin, reacting with helical waving of the germ tubes at concentrations as low as 0.1-0.2 μg/ml. At higher concentrations of 5-10 μg/ml stunting of the germ tubes results, but even in saturated solutions the actual percentage of germination is not reduced, which shows that griseofulvin is more a fungistatic than a fungicidal agent [98]. Powdery mildew fungi are also very sensitive to griseofulvin; application of griseofulvin at low concentration to the roots of wheat seedling inoculated with conidia of Erysiphe graminis resulted in malformation or even lack of formation of haustoria and as a consequence complete growth inhibition of the fungus [99]. Dermatophytic fungi appeared to be very sensitive to griseofulvin [100]. It is remarkably nontoxic to plants, mammals, and humans.

3. Mechanism of Action

For many years it was assumed that griseofulvin might interfere with chitin synthesis in the cell wall, a theory which has now been abandoned. Paget and Walpole [101] observed an accumulation of cells at metaphase in the root meristem of broad bean exposed to this antibiotic. Nuclei in the hyphal tips of dermatophytes treated with griseofulvin were abnormally shaped and very large, suggesting that the daughter nuclei did not separate during cell division [102]. A study of the effect of griseofulvin on a diploid strain of Aspergillus nidulans revealed that spindles were damaged and distorted within 30 sec to 1 min after treatment [103]. Low concentrations of griseofulvin increased the frequency of somatic segregation in this fungus due to chromosome nondisjunction, which presents genetic evidence in support of the assumption that it interferes with nuclear function. Inhibition of mitosis by griseofulvin was also reported for the fungus Basidiobolus ranarum. Abnormal metaphase configurations were formed; daughter nuclei appeared to move only a very short distance apart after metaphase and then reverted to the early interphase state by reenclosure in the nuclear envelope [104].

Unlike colchicine, griseofulvin does not bind to tubulin, as was shown in studies in vitro with brain extracts. It is suggested that griseofulvin may affect some aspect of microtubule function, possibly a process vital to the sliding of microtubules which is assumed to be necessary for the separation of chromosomes [105]. The action of griseofulvin has been compared with that of methylbenzimidazol-2-yl carbamate (MBC), a compound which also interferes with mitosis. The ben-1 mutant of *A. nidulans*, described by Hastie and Georgopoulos [106], which is highly resistant to MBC, appeared as sensitive to griseofulvin as the wild strain. It is concluded [107] that, although MBC and griseofulvin interfere with chromosome segregation, their sites of action with the mitotic apparatus are probably different.

Abnormalities in nuclear division due to exposure to griseofulvin have also been reported for mammalian cell cultures [108]. The mechanism of selective action of griseofulvin among eukaryotic organism still needs to be elucidated. Development of high-level, stable tolerance in fungi to this selective antibiotic has not yet been observed.

4. Uptake and Fate in Plants

Griseofulvin is readily absorbed via roots or aerial parts and translocated in plants [109]. Possibilities for systemic disease control with this antibiotic have been widely explored and found for many diseases. Indications are that control was obtained by a direct action of the unchanged molecule upon the parasite inside the plant; all morphological abnormalities of the hyphae which occur in vitro were also observed in the plant. Furthermore, pure griseofulvin was recovered from the tips of broad bean plants which had been treated by root application. When taken up by the roots, it even seems more active against various powdery mildews than when applied as a spray. Griseofulvin is only slowly degraded in plants [109].

5. Disease Control

In 1957, griseofulvin was released by the British government for use as an agricultural fungicide. It had shown systemic activity against a large number of fungal plant diseases [84, 69], and expectations were especially high for control of Botrytis and powdery mildew diseases. Due to high cost and moderate performance when applied as a spray or dust, griseofulvin has not developed into a widely used control agent of fungal plant diseases. A limited use for control of apple blossom blight, caused by *Sclerotinia mali*, and Fusarium wilt of melon is reported from Japan, but costs are too high for large-scale application [110].

It is interesting to note that griseofulvin, after its development for agricultural purposes, was found to be useful in the medical field. In 1958, the effectiveness of griseofulvin against mycoses in guinea pigs was reported [111], which led to its development as a medicament against fungal diseases of skin, hair, and nails of humans. After oral administration, growth of the fungi in the skin cells is inhibited. Actual killing of the fungus is not necessary, since the keratinized cells are eventually thrown off. Significant protection was obtained against the development of inflammatory lesions by Trichophyton mentagrophytes [112]. A review of the treatment of mycoses with griseofulvin is given by Hildrick-Smith [113].

C. Benzimidazoles and Thiophanates

1. Development and Structure

Thiabendazole, 2-(4-thiazolyl)-1H-benzimidazole (15), originally known as an anthelminthic for veterinary purposes, was first introduced as an agent for systemic control of plant diseases by Staron and Allard [114]. A few years later, Schumann [115] found fuberidazole, 2-(2-furyl)benzimidazole (16), active as a seed disinfectant against Fusarium nivale on wheat.

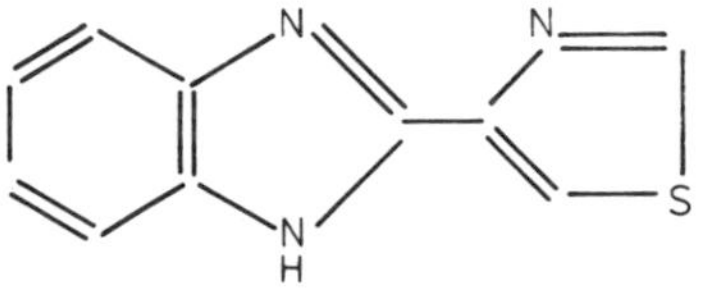

(15) thiabendazole

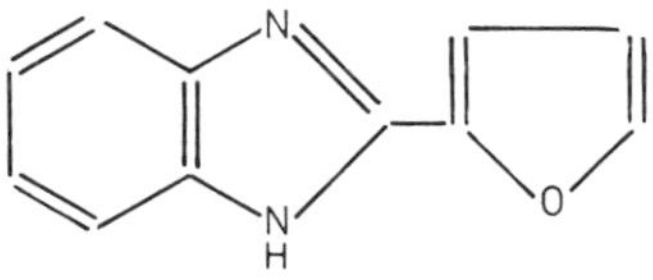

(16) fuberidazole

Again one year later, Delp and Klopping [116] introduced the systemic fungicide benomyl, methyl 1-(butylcarbamoyl)-2-benzimidazolecarbamate (17), with essentially the same spectrum of activity as thiabendazole and fuberidazole, but far more active against most plant pathogens. Clemons and Sisler [117] found that in dilute aqueous solutions of benomyl the compound methylbenzimidazol-2-yl carbamate, MBC or carbendazim (18) is formed, which is now considered to be the main fungitoxic principle of benomyl.

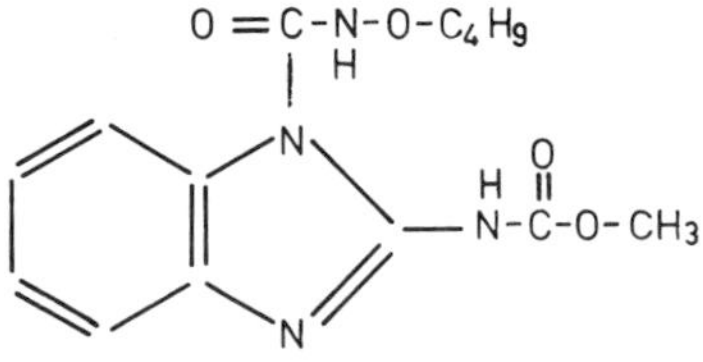

(17) benomyl

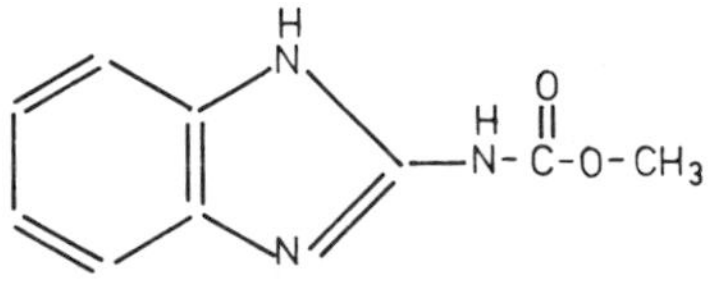

(18) MBC

A few years after the introduction of benomyl, a new class of systemic fungicides, the thiophanates, was introduced [118]: thiophanate,

1,2-bis(3-ethoxycarbonyl-2-thioureido)benzene (19); thiophanate-methyl, 1,2-bis(3-methoxycarbonyl-2-thioureido)benzene (20); and NF 48, 2-(3-methoxycarbonyl-2-thioureido)aniline (21).

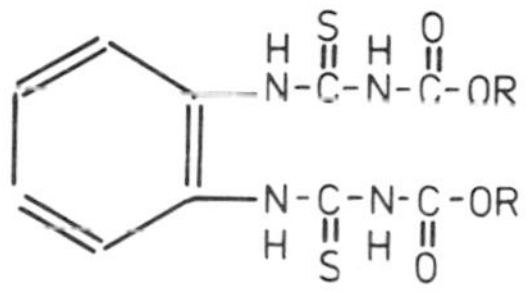

(19) R = C_2H_5, thiophanate
(20) R = CH_3, thiophanate methyl

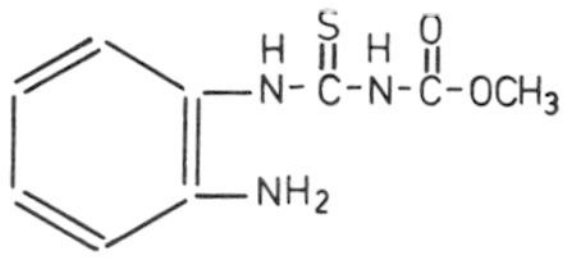

(21) NF 48

It soon was found that thiophanate-methyl and NF 48 could be converted into MBC, and thiophanate into ethylbenzimidazol-2-yl carbamate, EBC [119, 120]. A correlation was found between the formation of MBC from thiophanate-methyl and the fungitoxicity of the latter, from which it was concluded that not thiophanate-methyl itself, but its derivative MBC is responsible for fungicidal activity.

2. Biological Activity

The spectrum of antifungal activity of benomyl has been investigated by various research groups [121, 122]. Most Ascomycetes and, to a lesser extent, the Basidiomycetes are sensitive to benomyl. Within the form class Deuteromycetes, a correlation was found between sensitivity to benomyl and the morphogenesis of conidia. Fungi belonging to the Sphaeropsidales, the Melanconiales, and to those species among the Moniliales which form blasto-, phialo-, and aleuriospores, were highly sensitive to benomyl, whereas fungi which form porospores and annellospores were tolerant. Also Oomycetes and Zygomycetes were insensitive to benomyl. Although benomyl is generally not phytotoxic and thus remarkably selective in its action, various other living systems seem to be influenced by it, for example, earth worms [123], mammalian tissue culture [124], aphids [125], nematodes [126], algae, daphnia, and fish [127]. In some cases, adverse effects on plants have also been found [128]. Growth-regulating activity has been reported on celery [129] and in soybean-tissue culture [130]. Bacteria are insensitive.

The spectrum of antifungal activity of thiophanates is similar to that of benomyl [131], which is not surprising since the activity of benomyl and thiophanate-methyl is due to conversion of these compounds to the same fungitoxic principle, MBC. Thiophanate is converted to EBC, which shows the same spectrum of activity as MBC. Thiabendazole and fuberidazole, although in vitro less active than MBC, have a similar antifungal spectrum.

3. Mechanism of Action

a. Benomyl. As has been mentioned earlier, MBC is responsible for most of the fungitoxic action of benomyl. The mechanism of action of MBC will be discussed below. A difference in toxicity, however, has been found toward *Saccharomyces pastorianus* by Hammerschlag and Sisler [132]. In studies with *Ustilago maydis* and *S. cerevisiae* they [133] found that the toxicity of benomyl may be attributed to two breakdown products, namely MBC and butylisocyanate (BIC). The latter inhibits respiration of these fungi in a manner similar to benomyl. They concluded that differential effects of benomyl preparations and MBC on fungi should be ascribed to BIC.

b. MBC. This compound was not lethal to conidia of *Neurospora crassa* or sporidia of *Ustilago maydis* in a buffer solution, but it killed cells of both organisms in a medium supporting growth [134]. Although, at 1 μg/ml, germ tube development of conidia of *N. crassa* was prevented, the DNA content doubled and a 12-fold weight increase still occurred. Optical density of sporidial suspensions of *U. maydis*, treated with 8 μg/ml of MBC, increased threefold before growth stopped after 6 hr, but there was no increase in cell number. Cells of *S. pastorianus* treated with 10 μg/ml MBC did not separate; DNA, RNA, and protein synthesis were inhibited only after a lag period. It was shown that MBC inhibits cytokinesis or mitosis [133].

The mechanism of action of MBC was further investigated by Davidse [135], who used *Aspergillus nidulans* as a test organism. When mycelium of this fungus was exposed to a 4 μM aqueous solution of MBC, inhibition of dry-weight increase became evident only after 4 hr. Inhibition of DNA and RNA became apparent after 2 hr and was progressive with time so that after 8 hr synthesis of these compounds was inhibited almost completely. Increase of the number of nuclei per germ tube, however, stopped immediately after addition of MBC, while the DNA content of the treated samples was allowed to almost double. Staining of hyphae and microscopic observation showed no normal phases of mitosis, as were found in control hyphae. From these results it was concluded that the antifungal action of MBC can be ascribed to inhibition of mitosis. The approximate doubling of DNA after mitotic arrest in *A. nidulans* [135] and *Neurospora crassa* [134] can be explained by assuming that in these fungi the G-1 phase is long and the G-2 phase short, so that in unsynchronized cultures most nuclei are in the G-1 phase. When MBC is added, these nuclei can go through the S-phase and thus synthesize DNA before the division cycle is blocked in mitosis.

In several ways the antimitotic activity of MBC in A. nidulans showed a striking resemblance to the antimitotic activity of colchicine in human cell cultures, which prevents assembly of the spindle by binding to protein subunits of the spindle microtubules. In order to study whether MBC might act in a similar way, use was made of a number of UV-induced mutants of A. nidulans with a changed sensitivity to MBC [136]. Cell-free extracts of MBC-resistant (R), MBC-sensitive (003, wild type), and MBC-extra sensitive (strain 186) strains of A. nidulans were incubated with 2-[^{14}C]MBC and subjected to gel filtration. The amount of radioactivity present in the fractions containing macromolecules was high in strain 186, moderately high in strain 003, and zero in strain R. This illustrates a close correlation between the binding of MBC to some macromolecule and sensitivity of the strain to MBC. On the basis of retention properties of the macromolecule-MBC complex on DEAE ion exchangers and the molecular weight, which was estimated at 100,000 daltons, it seems likely that the MBC-binding macromolecule is identical with microtubular protein [137]. It is assumed that this binding may prevent the assembly of microtubular subunits into functional spindle fibers resulting in inhibition of mitosis. Resistance to MBC in A. nidulans might then be due to mutationally induced lower affinity of the microtubular protein to MBC. That natural resistance may also be due to low affinity is indicated by the observation that no binding of MBC occurred in extracts of Alternaria brassicola, which is insensitive to benomyl and MBC. The latter may be a valuable tool for studies of the mitotic process in fungi, as compared with that in other living systems.

c. Thiabendazole and Fuberidazole. Benomyl-resistant strains of Penicillium brevicompactum and P. corymbiferum are cross resistant to thiabendazole and fuberidazole [138]. Acquired resistance of A. nidulans to thiabendazole is determined by mutation in the same gene as resistance to benomyl, which is located on chromosome VIII, 5 units from orn-B 7, marking ornithine deficiency, and 37 units from ts-D 15, marking a specific temperature sensitivity. Although cross resistance between benomyl and thiabendazole appears to be the rule in A. nidulans, exceptions to this rule appeared to occur. A small percentage of UV-induced mutants appeared resistant to only one of these fungicides; thiabendazole-resistant strains were found which were even more sensitive to benomyl than the wild strain [136]. It is presumed that mutation in the above mentioned gene usually leads to a minor structural change in the microtubular protein which influences affinity to MBC and thiabendazole to the same extent, but in rare cases the mutation may result in a change in these proteins which influences affinity to these compounds differently [137,139]. These results are consistent with the hypothesis that the mechanism of action of thiabendazole is the same as that of benomyl. This probably also holds for fuberidazole.

d. Thiophanates. It has been shown that thiophanate-methyl and thiophanate are converted into MBC and EBC, respectively, which are responsible for the antifungal activity [140]. The observations that thiophanate-methyl and thiophanate possess lower antifungal activities than their respective metabolites MBC and EBC is explained by the slow rate of conversion of the former compounds into the latter [120]; EBC is less toxic than MBC, presumably because of a lesser binding to the spindle protein subunits.

4. Uptake, Transport, and Metabolism in Plants

Benzimidazole and thiophanate fungicides may be taken up by various plant parts, such as the leaves after spray or dust application, the roots after a soil drench, or by the stem or trunk after injection. Translocation of these compounds occurs predominantly in the apoplast, although some uptake and transport of MBC in the symplast has been reported [141], and usually little or no downward transport occurs. In several cases studied no movement from the leaves into the fruit was observed [142, 143].

Benomyl and thiophanate-methyl both are converted into MBC, which is considered to be the primary fungitoxic principle. Differences in performance observed between these compounds might be due to the rate of conversion [144, 145], transcuticular and translaminar movement [146], retention in the roots and subsequent delivery to the aerial parts [147], and formation of fungitoxic products other than MBC [133]. The rate of uptake and transport in plants and the persistence of these compounds or their fungitoxic conversion products are essential for curative or systemic control of fungal plant diseases.

The uptake, translocation, and distribution of the benzimidazole and thiophanate fungicides is more extensively discussed in Chapter 2, and their biological conversion in Chapter 3 of this volume.

5. Disease Control

The introduction of benomyl (Benlate) meant a breakthrough in the field of systemic control of fungal plant diseases. In contrast to other systemic fungicides, it shows a broad spectrum of antifungal activity in vitro as well as against plant diseases. It is easily taken up by plants, but its action is selective in such a way, that little or no pronounced phytotoxic effects are observed. An extremely large number of plant diseases respond to benomyl treatment, among which are seed- and soilborne diseases (Vol. 1, Chapter 6), vascular wilts (Vol. 1, Chapter 7), foliar and fruit diseases (Vol. 1, Chapter 8), and postharvest diseases (Vol. 1, Chapter 9). As control of fungal diseases by benomyl closely parallels its in vitro activity, diseases caused by pathogens belonging to the Oomycetes (such as species of *Pythium* and *Phytophthora*) and by those fungi which form porospores and annellospores (such as species of *Scopulariopsis*, *Alternaria*, and *Helminthosporium*) are not controlled. The same holds for carbendazim (MBC), the thiophanates, and thiabendazole. A list of diseases against which thiophanates are active is given by Kikkawa [148].

The selective biological activity of these fungicides (see Section IV.C.3) may also have adverse effects. They may interfere with microbiological balances in the soil in which pathogens as well as nonpathogens play a role by antibiosis and compete for food and space. Pathogens which are relatively tolerant to these fungicides may become more active after the fungicide has eliminated sensitive antagonists. Evidence for such phenomena has been obtained by various authors. After treatment of rye with benomyl against root and foot diseases, an increase of sharp eyespot, caused by *Rhizoctonia solani*, was observed [149]. An increase in gangrene, caused by benomyl-insensitive *Phoma* sp., was observed in potatoes treated with this compound at the time of planting [150]. Incidence of wilting of Chinese asters planted in soil contaminated with *Phytophthora cryptogea* was initially higher in benomyl-treated than in untreated soil [151]. These data indicate that in the presence of a selective fungicide a shift in the soil microflora may take place with consequences for disease incidence.

Another phenomenon which may endanger the use of fungicides with a selective and specific action is the development of fungicide resistance. Since the introduction of benomyl and related compounds for control of plant diseases, emergence of resistance to these fungicides has been reported frequently. Already in 1969, a decreasing effect of benomyl on cucumber powdery mildew was noticed [152]. Shortly after the application of benomyl had become a common practice for control of *Botrytis cinerea* in Dutch cyclamen greenhouses, highly resistant isolates appeared [153], so that the use of this compound had to be abandoned. The use of benomyl against *Cercospora betae* in Greece was stopped for the same reason [154]. Numerous other cases of acquired resistance have been reported in recent years [155].

The readiness with which a fungal population may become resistant to these fungicides may depend on the type of the fungus, the type of disease, the frequencies and mode of application of the fungicide, and the competitive ability of the newly emerged resistant strains [156]. It does not seem likely that the phenomenon of acquired resistance will preclude the use of benomyl and related compounds for control of plant diseases. The different aspects of acquired resistance to fungicides are discussed in this volume, Chapter 13.

D. Chloroneb

Chloroneb, 1,4-dichloro-2,5-dimethoxybenzene (**22**), was introduced in 1967 as a systemic fungicide. It has a limited spectrum of antifungal activity, inhibiting growth of *Rhizoctonia solani*, *Sclerotium rolfsii*, certain species of *Pythium*, and only a few other fungi [157]. When applied to the seed or as a soil drench, it is taken up by the roots and accumulated in the roots, lower stem, and cotyledons of cucumber, bean, and cotton plants, thus at places where need for protection against soil fungi is greatest [158].

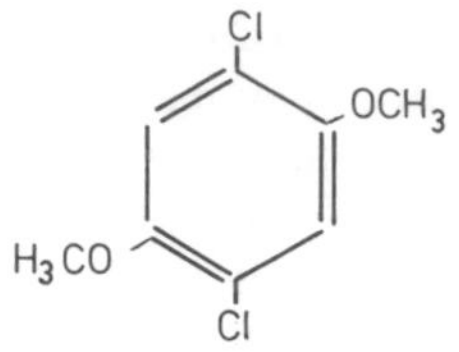

Cl, OCH_3, H_3CO, Cl

(22) chloroneb

After uptake of ^{14}C-labeled chloroneb by the roots of cotton seedlings, the unaltered fungicide moves into the parts of the plants above ground; after application of the fungicide to single cotyledons, translocation to the growing point and lateral transport to the nontreated cotyledons was observed [159]. Chloroneb moved into soybean seedling grown from treated seed [160]. In beans the fungicide was metabolized into 2, 5-dichloro-4-methoxyphenol and its β-D-glucoside [161].

Chloroneb may be used as a systemic fungicide for the treatment of seed and for an in-furrow soil treatment at planting time against various diseases, such as those caused by *Rhizoctonia solani* in cotton [162], soybeans, and beans and *R. bataticola* in *Phaseolus mungo* [163]. Diseases caused by some species of *Pythium* may also be controlled by chloroneb, such as *P. ultimum* in peas [164] and *P. aphanidermatum* in various crops [165-167]. Chloroneb is furthermore active against root rot of sugar beet, caused by *Sclerotium rolfsii* [168], and against *Typhula* blight of turf grass [169]. Attempts to control diseases, caused by *Fusarium*, *Phytophthora*, and other fungi have generally been less successful.

Studies with *Rhizoctonia solani* [170] suggested that chloroneb might act directly or indirectly on DNA synthesis. Additional investigations with *Ustilago maydis* [171] revealed that multiplication of sporidia of this fungus was prevented by chloroneb. Protein and nucleic acid synthesis in cultures treated with the fungicide proceeded at the normal rate until failure of sporidial division; thereafter, increase of DNA, RNA, and protein was strongly inhibited, apparently as a consequence from prevention of cell division. Georgopoulos [172] found that chloroneb, like benomyl and griseofulvin, increases somatic segregation of heterozygous diploid colonies of *Aspergillus nidulans*, which is considered to be an indication that the toxicity of the compound results from interference with mitosis at least in *A. nidulans*. Development of resistance to this fungicide has been reported. A moderately resistant mutant of *U. maydis* was isolated, which grew normally in medium containing 16 μg of chloroneb per ml, whereas growth of the wild type is prevented at 8 μg/ml. Tolerance was not due to differences in uptake or metabolism of the toxicant. It followed the pattern for single-gene resistance [171].

V. DISCUSSION

Most of the antifungal compounds, which are now known to act primarily and specifically on nucleic acid synthesis, are antibiotics and analogues of purine and pyrimidine. These chemicals may interfere with the synthesis of nucleotides (6-azauracil, 8-azaguanine, azaserine, aristeromycin) or may prevent replication or transcription of DNA by inhibition of DNA- and RNA-polymerizing enzymes (lomofungin, edeine, thiolutin, cycloheximide), or may bind to the DNA template (bleomycin, phleomycin, phytoactin, daunomycin, actinomycin D). The last two compounds act in a very specific way by intercalation between two nucleotides of the DNA helix.

Although the inhibitors of nucleic acid synthesis exert a rather specific action, often at one site in the metabolism of the organism, none of them show a high degree of biological selective action. As the processes involved in nucleic acid synthesis are essentially similar in living organisms, there seems to be little opportunity for discrimination between different organisms. Mutational changes resulting in resistance to this type of fungicide do not seem to occur readily, possibly because these mutations may be lethal. A certain degree of selective action, however, may possibly be due to differences in permeability of the protoplast membrane for the inhibitor, or to differences in the capacity of cells to detoxify or upgrade the chemical. In addition, the rate of metabolic activity may be of influence.

The compounds which interfere with nuclear function show another picture; presumably they all interfere with mitosis. They probably do so by binding to tubulin, a protein which constitutes the microtubules of the spindle. According to this hypothesis, the effect on nucleic acid synthesis is an indirect one. These compounds, in contrast to the direct inhibitors of nucleic acid synthesis, show a high degree of biological selective action. It still has to be shown whether this selective action is due to intrinsic differences at the site of action between sensitive and less sensitive or insensitive organisms, or to other factors. Ultraviolet-induced fungal mutants resistant to MBC were readily obtained. It was indicated that in these cases changes at the site of action were responsible for decreased sensitivity to this compound. It is not yet clear why resistance to MBC, but not to griseofulvin, can be readily induced by UV or other mutagenic agents.

ACKNOWLEDGMENTS

The author is indebted to Ir. T. Hijwegen for help with the literature and to Ir. L. C. Davidse and Dr. J. Eckert for a critical reading of the manuscript.

REFERENCES

1. J. Dekker, Mededel. Landbouwhogeschool Opzoekingsstat. Staat Gent, 27, 124 (1962).
2. A. T. Bull and B. M. Faulkner, Nature, 203, 506 (1964).
3. A. J. Darlington and C. Scazzocchio, J. Bacteriol., 93, 937 (1967).
4. G. R. Hoffman and H. V. Malling, J. Gen. Microbiol., 83, 319 (1974).
5. P. Roy-Burman, Analogues of Nucleic Acid Components, Springer-Verlag, Berlin-Heidelberg, New York, 1970.
6. M. Earl Balis, Antagonists and Nucleic Acids, North Holland Publishing Company, Amsterdam, 1968.
7. J. Dekker and A. J. P. Oort, Phytopathology, 54, 815 (1964).
8. G. Matolcsy and S. Doma, Acta Phytopathol. Acad. Sci. Hungar., 2, 361 (1967).
9. J. Dekker, in Wirkungsmechanismen von Fungiziden and Antibiotika (M. Girbardt, ed.), Akademie-Verlag, Berlin, 1967, pp. 333-339.
10. H. M. Dekhuijzen and J. Dekker, Pestic. Biochem. Physiol., 1, 11 (1971).
11. G. Matolcsy and S. Doma, Acta Phytopathol. Acad. Sci. Hungar., 4, 353 (1969).
12. R. W. Brockman and E. P. Anderson, in Metabolic Inhibitors (R. M. Hochster and J. H. Quastel, eds.), Academic Press, New York and London, 1963, pp. 239-285.
13. G. Matolcsy, World Rev. Pest Control, 10, 50 (1971).
14. J. Dekker, Neth. J. Plant Pathol., 74, Suppl. 1, 127 (1968).
15. J. Dekker and G. S. Roosje, Neth. J. Plant Pathol., 74, 219 (1968).
16. E. Bresnick, S. Singer, and G. H. Hitchings, Biochim. Biophys. Acta, 37, 251 (1960).
17. R. S. Elias, M. C. Shephard, B. K. Snell, and J. Stubbs, Nature, 219, 1160 (1968).
18. R. M. Bebbington, D. H. Brooks, M. J. Geoghegan, and B. K. Snell, Chem. Ind. (London), 1969, p. 1512.
19. K. J. Bent, private communication, 1975.
20. K. J. Bent, Endeavour, 28, 129 (1969).
21. K. J. Bent, Ann. Appl. Biol., 66, 103 (1970).
22. B. G. van't Land and J. Dekker, Neth. J. Plant Pathol., 78, 242 (1972).
23. A. Calderbank, Acta Phytopathol. Acad. Sci. Hungar., 6, 355 (1971).
24. B. D. Cavell, R. J. Hemmingway, and G. Teal, Proc. 6th Brit. Insect. Fungic. Conf. Brighton, 1971, Vol. 2, p. 431.
25. K. J. Bent, A. M. Cole, A. J. W. Turner, and M. Woolner, Proc. 6th Brit. Insect. Fungic. Conf. Brighton, 1971, Vol. 1, p. 274.
26. M. H. Ebben and D. M. Spencer, Proc. 7th Brit. Insect. Fungic. Conf., Brighton, 1973, Vol. 1, p. 211.
27. D. H. Brooks, Outlook on Agriculture, 6, 122 (1970).

28. L. V. Edgington, E. Reinbergs, and M. C. Shephard, Can. J. Plant Science, 52, 693 (1972).
29. H. W. Johnston, Can. Plant Disease Survey, 52, 82 (1972).
30. M. S. Wolfe and A. Dinoor, Proc. 7th Brit. Insect. Fungic. Conf., Brighton, 1973, Vol. 1, p. 11.
31. T. Kusaka, H. Yamamoto, M. Muroi, T. Kishi, and K. Mizuno, J. Antibiotics (Tokyo), 21, 255 (1968).
32. T. Kishi, M. Muroi, T. Kusaka, M. Nishikawa, K. Kamiya, and K. Mizuno, Chem. Pharm. Bull., 20, 940 (1972).
33. T. Kusaka, J. Antibiotics (Tokyo), 24, 756 (1971).
34. R. F. Pittillo and D. E. Hunt, in Antibiotics (D. Gottlieb and P. D. Shaw, eds.), Vol. 1, Mechanism of Action, Springer-Verlag, Berlin-Heidelberg-New York, 1967, pp. 481-493.
35. R. H. Wellman and S. E. A. McCallan, Contrib. Boyce Thompson Instit., 14, 151 (1946).
36. H. S. Chow and E. M. Rodgers, Phytopathology, 63, 1428 (1973).
37. H. D. Sisler, Ann. Rev. Phytopathol., 7, 311 (1969).
38. H. Umezawa, K. Maeda, T. Takeuchi, and Y. Okami, J. Antibiotics (Tokyo), 19, 200 (1966).
39. T. Takita, Y. Muraoka, K. Maeda, and H. Umezawa, J. Antibiotics (Tokyo), 21, 79 (1968).
40. M. Shimura, T. Watanabe, T. Ohashi, T. Shomura, T. Niida, and Y. Sekizawa, J. Antibiotics (Tokyo), 23, 166 (1970).
41. H. Suzuki, K. Nagai, E. Akutsu, J. Hiroshi, N. Tanaka, and H. Umezawa, J. Antibiotics (Tokyo), 23, 473 (1970).
42. C. W. Haidle, Mol. Pharmacol. J., 7, 645 (1971).
43. H. Yamaki, H. Suzuki, K. Nagai, N. Tanaka, and H. Umezawa, J. Antibiotics (Tokyo), 24, 178 (1971).
44. K. Shishido, J. Antibiotics (Tokyo), 26, 501 (1973).
45. D. N. Wheatly, G. C. Mueller, and K. Kajiwara, Brit. J. Cancer, 29, 117 (1974).
46. K. Maeda, H. Kosaka, K. Yagishita, and H. Umezawa, J. Antibiotics (Tokyo), 9, 82 (1956).
47. T. G. Pridham, L. A. Lindenfelser, O. L. Shotwell, F. H. Stodola, R. G. Benedict, C. Foley, R. W. Jackson, W. J. Zaumeyer, W. H. Preston, and J. W. Mitchell, Phytopathology, 46, 568 (1956).
48. J. W. Mitchell, B. C. Smale, E. J. Daly, W. H. Preston, T. G. Pridham, and E. S. Sharpe, Plant Disease Reptr., 43, 431 (1959).
49. B. C. Smale, M. D. Montgillon, and T. G. Pridham, Plant Disease Reptr., 45, 244 (1961).
50. S. H. Purdy, Plant Disease Reptr., 48, 159 (1964).
51. H. Tanaka, H. Yamaguchi, and H. Umezawa, Biochem. Biophys. Res. Commun., 10, 171 (1963).
52. A. Falaschi and A. Kornberg, Feder. Proc., 23, 940 (1964).
53. G. W. Grigg, M. J. Edwards, and D. J. Brown, J. Bacteriol., 107, 599 (1971).

54. A. M. Angyal, G. W. Grigg, R. J. Badger, D. J. Brown, and J. H. Lister, J. Gen. Microbiol., 85, 163 (1974).
55. R. Stern, J. A. Rose, and R. H. Friedman, Biochemistry, 13, 307 (1974).
56. M. J. Sleigh and G. W. Grigg, FEBS Letters, 39, 35 (1974).
57. R. M. Friedman, R. Stern, J. A. Rose, J. Natl. Cancer Instit., 52, 693 (1974).
58. B. A. Kihlman, G. Odmark, and B. Hartley, Mutation Res., 4, 783 (1967).
59. L. E. Johnson and A. Dietz, Appl. Microbiol., 17, 755 (1969).
60. D. Gottlieb and G. Nicolas, Appl. Microbiol., 18, 35 (1969).
61. M. Cannon, J. E. Davis, and A. Jimenez, FEBS Letters, 32, 277 (1973).
62. F. S. Fraser, J. Creanor, and J. M. Mitchison, Nature, 244, 222 (1973).
63. F. R. Cano, S. C. Kuo, and J. O. Lampen, Antimicrob. Agents Chemotherapy, 3, 723 (1973).
64. K. Pavletich, S. C. Kuo, and J. O. Lampen, Biochem. Biophys. Res. Commun., 60, 942 (1974).
65. Z. Kurylo-Borowska, in Antibiotics (D. Gottllieb and P. D. Shaw, eds.), Vol. 2, Biosynthesis, Springer-Verlag, Berlin-Heidelberg-New York, 1967, pp. 342-350.
66. B. Beetz and W. Koingmueller, Naturwissenschaften, 60, 301 (1973).
67. L. Ebinger, J. Gen. Microbiol., 71, 35 (1972).
68. H. Seneca, J. M. Kane, and J. Rockenbach, J. Antibiot. Chemotherapy, 2, 357 (1952).
69. J. Dekker, in Fungicides, Vol. II (D. C. Torgeson, ed.), Academic Press Inc., New York, 1969, pp. 579-635.
70. A. Jimenez, D. J. Tipper, and J. Davies, Antimicrob. Agents Chemotherapy, 3, 729 (1973).
71. J. S. Ziffer, S. J. Ishihara, T. J. Cairney, and A. W. Chow, Phytopathology, 47, 539 (1957).
72. V. D. Moss, Forest Sci., 7, 380 (1961).
73. A. E. Harvey, S. O. Graham, L. D. Becker, and D. H. Brown, Phytopathology, 56, 148 (1966).
74. C. D. Leaphart and E. F. Wicker, Plant Disease Reptr., 52, 6 (1968).
75. J. P. Lynch and H. D. Sisler, Phytopathology, 57, 367 (1967).
76. H. C. Reilly, A. Schatz, and S. A. Waksman, J. Bacteriol., 49, 585 (1945).
77. W. Kersten, H. Kersten, and H. M. Rauen, Nature, 187, 60 (1960).
78. H. M. Sobell, S. C. Jain, T. D. Sakore, and C. E. Nordman, Nature, 231, 200 (1971).
79. A. Di Marco, in Antibiotics (D. Gottlieb and P. D. Shaw, eds.), Vol. 1, Mechanism of Action, Springer-Verlag, Berlin-Heidelberg-New York, 1967, pp. 190-210.

80. W. I. Pigram, W. Fuller, and M. E. Davies, J. Mol. Biol., 80, 361 (1973).
81. J. H. Goldberg and P. A. Friedman, Ann. Rev. Biochem., 40, 775 (1971).
82. F. Ziewitz, S. Wekke, W. A. Linden, H. Baisch, and J. Straatman, Strahlentherapie, 147, 514 (1974).
83. M. A. Bempong, Amer. J. Genet. Cytol., 15, 587 (1973).
84. J. Dekker, Ann. Rev. Microbiol., 17, 243 (1963).
85. M. R. Siegel and H. D. Sisler, Biochim. Biophys. Acta, 103, 558 (1965).
86. W. E. Timberlake, L. McDowell, and D. H. Griffin, Biochem. Biophys. Res. Commun., 46, 942 (1972).
87. W. E. Timberlake and D. H. Griffin, Biochim. Biophys. Acta, 349, 39 (1974).
88. W. E. Timberlake and D. H. Griffin, Biochim. Biophys. Acta, 353, 248 (1974).
89. J. R. McIntosh, P. K. Hepler, and D. G. Van Wie, Nature, 224, 659 (1969).
90. J. R. Aist and P. H. Williams, J. Cell Biology, 55, 368 (1972).
91. A. E. Oxford, H. Raistrick, and P. Simonart, Bioch. J., 33, 240 (1939).
92. P. W. Brian, P. J. Curtis, and H. G. Hemming. Brit. Mycol. Soc. Trans., 29, 173 (1946).
93. J. F. Grove, D. Ismay, J. MacMillan, T. P. C. Mulholland, and M. A. Rogers, Chem. Ind. (London), 1951, p. 219.
94. J. MacMillan, J. Chem. Soc., 1954, p. 2585.
95. J. MacMillan, Chem. Ind. (London), 1951, p. 179.
96. R. Crosse, R. McWilliams, and A. Rhodes, J. Gen. Microbiol., 34, 51 (1964).
97. J. Dekker and I. Tulleners, Meded. Landbouwhogeschool Opzoekingsstat. Staat Gent, 27, 124 (1962).
98. P. W. Brian, Ann. Botany (London), 13, 59 (1949).
99. J. Dekker and R. G. van der Hoek-Scheuer, Neth. J. Plant Pathol., 70, 142 (1964).
100. H. Ziegler, Z. Allgem. Mikrobiol., 3, 211 (1963).
101. G. E. Paget and A. L. Walpole, Nature, 182, 1320 (1958).
102. T. R. Thyagarajan, O. P. Srivastava, and V. V. Vora, Naturwissenschaften, 50, 524 (1963).
103. S. H. B. Crackower, Can. J. Microb., 18, 683 (1972).
104. K. Gull and A. P. J. Trinci, Nature, 244, 292 (1973).
105. L. M. Grisham, L. Wilson, and K. G. Bensch, Nature, 244, 294 (1973).
106. A. C. Hastie and S. G. Georgopoulos, J. Gen. Microbiol., 67, 371 (1971).
107. A. Kappas and S. G. Georgopoulos, J. Bacteriol., 119, 334 (1974).

108. L. Larizza, G. Simoni, F. Tredici, and L. DeCarli, Mutation Res., 25, 123 (1974).
109. S. H. Crowdy, J. F. Grove, H. G. Hemming, and K. C. Robinson, J. Exptl. Bot., 7, 42 (1956).
110. T. Misato, Japan Pestic. Inform., 1, 15 (1969).
111. J. C. Gentles, Nature, 182, 476 (1958).
112. A. M. Allen, J. H. Reinhardt, W. A. Akers, and D. Gunnison, Arch. Dermatol., 108, 233 (1973).
113. G. Hildick-Smith, in Drill's Pharmacology in Medicine (J. R. DiPalma, ed.), 4th ed., McGraw-Hill, New York, 1971, pp. 1755-1769; Chem. Abstr., 77, 13467u (1972).
114. T. Staron and C. Allard, Phytiatr.-Phytopharmacie, 13, 163 (1964).
115. G. Schumann, Z. Pflanzenkrankh. Pflanzenschutz, 74, 155 (1967).
116. D. J. Delp and H. L. Klöpping, Plant Disease Reptr., 52, 95 (1968).
117. G. P. Clemons and H. D. Sisler, Phytopathology, 59, 705 (1969).
118. E. Aelbers, Meded. Landbouwhogeschool Opzoekingsstat, Staat Gent, 36, 126 (1971).
119. H. A. Selling, J. W. Vonk, and A. Kaars Sijpesteijn, Chem. Ind. (London), 1970, p. 1625.
120. J. W. Vonk and A. Kaars Sijpesteijn, Pestic. Sci., 2, 160 (1971).
121. G. J. Bollen and A. Fuchs, Neth. J. Plant Pathol., 76, 299 (1970).
122. L. V. Edgington, K. L. Kew, and G. L. Baron, Phytopathology, 61, 42 (1971).
123. A. Stringer and H. C. Lyons, Pestic. Sci., 5, 189 (1974).
124. J. A. Styles and R. Garner, Mutation Res., 26, 177 (1974).
125. B. Hinz and F. Daebeler, Arch. Phytopathol. Pflanzenschutz, 9, 337 (1973).
126. R. Cook and P. A. York, Plant Disease Reptr., 56, 261 (1972).
127. E. M. den Tonkelaar and J. H. Canton, private communication.
128. L. R. Schreiber and W. K. Hock, Phytopathology, 62, 499 (1972).
129. T. H. Thomas, Ann. Appl. Biol., 74, 233 (1973).
130. K. G. M. Skene, J. Hort. Sci., 47, 179 (1972).
131. G. J. Bollen, Neth. J. Plant Pathol., 78, 55 (1972).
132. R. S. Hammerschlag and H. D. Sisler, Pestic. Biochem. Physiol., 2, 123 (1972).
133. R. S. Hammerschlag and H. D. Sisler, Pestic. Biochem. Physiol., 3, 42 (1973).
134. G. P. Clemons and H. D. Sisler, Pestic. Biochem. Physiol., 1, 32 (1971).
135. L. C. Davidse, Pestic. Biochem. Physiol., 3, 317 (1973).
136. J. M. van Tuyl, L. C. Davidse, and J. Dekker, Neth. J. Plant Pathol., 80, 165 (1974).
137. L. C. Davidse, in Systemic Fungicides (H. Lyr and C. Polter, eds.), Akademie-Verlag, Berlin, 1975, pp. 137-143.

138. G. J. Bollen, Meded. Landbouwhogeschool Opzoekingsstat. Staat Gent, 36, 1188 (1971).
139. L. C. Davidse, J. Cell Biol., 72, 177 (1977).
140. Y. Yasuda, S. Hashimoto, Y. Soeda, and T. Noguchi, Ann. Phytopathol. Soc. Japan, 39, 49 (1973).
141. Z. Solel, J. M. Schooley, and L. V. Edgington, Pestic. Sci., 4, 713 (1973).
142. R. Lafon and J. C. Boniface, Phytiatr. Phytopharm., 20, 45 (1971).
143. G. E. Brown and L. G. Albrigo, Phytopathology, 62, 1434 (1973).
144. P. M. Upham and C. J. Delp, Phytopathology, 63, 814 (1973).
145. H. Young and K. R. W. Hammett, New Zealand J. Sci., 16, 535 (1973).
146. Z. Solel and L. V. Edgington, Phytopathology, 63, 505 (1973).
147. A. Fuchs, G. A. van den Berg, and L. C. Davidse, Pestic. Biochem. Physiol., 2, 191 (1972).
148. K. Kikkawa, Japan Pestic. Inform., 10, 80 (1972).
149. E. P. van der Hoeven and G. J. Bollen, Acta Botan. Neerl., 21, 107 (1972).
150. C. Logan, Ann. Appl. Biol., 28, 261 (1974).
151. G. A. van den Berg and G. J. Bollen, Acta Botan. Neerl., 20, 256 (1971).
152. W. T. Schroeder and R. Provvidenti, Plant Disease Reptr., 53, 271 (1969).
153. G. J. Bollen and G. Scholten, Neth. J. Plant Pathol., 77, 83 (1971).
154. S. G. Georgopoulos and C. Dovas, Plant Disease Reptr., 57, 321 (1973).
155. J. Dekker, in Systemic Fungicides (R. W. Marsh, ed.), Wiley, New York, 1972, pp. 156-174.
156. J. Dekker and L. C. Davidse, Environ. Quality Safety, Suppl. Vol. III, 410 (1975).
157. M. J. Fielding and R. C. Rhodes, Cotton Disease Council Prov., 27, 56 (1967).
158. R. C. Rhodes, H. L. Pease, and R. K. Brentley, J. Agr. Food Chem., 19, 745 (1971).
159. B. T. Kirk, J. B. Sinclair, and E. N. Lambremont, Phytopathology, 59, 1473 (1969).
160. P. N. Thapliyal and J. B. Sinclair, Phytopathology, 61, 1301 (1971).
161. G. D. Thorn, Pestic. Biochem. Physiol., 3, 137 (1973).
162. I. E. M. Darrag and J. B. Sinclair, Phytopathology, 59, 1102 (1969).
163. N. K. Jain and M. N. Khare, Mysore J. Agr. Sci., 6, 461 (1972).
164. L. T. Richardson, Plant Disease Reptr., 57, 3 (1973).
165. R. H. Littrell, J. D. Gay, and H. D. Wells, Plant Disease Reptr., 53, 913 (1969).
166. P. Parvatha Reddy, Mysore J. Agr. Sci., 6, 193 (1972).

167. R. S. Cox, Plant Disease Reptr., 53, 912 (1969).
168. A. N. Mukhopadhyay and R. P. Thakur, Plant Disease Reptr., 55, 630 (1971).
169. J. M. Vargas and J. B. Beard, Plant Disease Reptr., 54, 1075 (1970).
170. W. K. Hock and H. D. Sisler, Phytopathology, 59, 627 (1969).
171. R. W. Tillman and H. D. Sisler, Phytopathology, 63, 219 (1973).
172. S. G. Georgopoulos, A. Kappas, and A. C. Hastie, Phytopathology, 66, 217 (1976).

Chapter 11

EFFECT OF FUNGICIDES ON PROTEIN SYNTHESIS

Malcolm R. Siegel

Department of Plant Pathology
University of Kentucky
Lexington, Kentucky

I. INTRODUCTION

Inhibitors of specific metabolic functions have been successfully used to define biosynthetic pathways in cells. Perhaps the best examples are those used in whole cells and cell-free systems to elucidate the various mechanisms of protein biosynthesis.

Pestka [1, 2] has shown that inhibitors of protein synthesis can be classified on the basis of the ribosomal subunit they affect together with the source of ribosomes (i.e., from prokaryotic or eukaryotic cells). A large majority of the inhibitors which have been studied act only on ribosomes from prokaryotes. Indeed, even some of the inhibitors which are reported to affect ribosomal function in eukaryotes and prokaryotes or in eukaryotes alone, do so only in cell-free systems.

It is interesting to note that almost all of the inhibitors of protein synthesis are naturally occurring rather than synthetic compounds. Because of the narrow biological selectivity that is inherent in antibiotics, these compounds may not inhibit protein synthesis in fungi. Consequently, only a small number of the compounds studied can be considered to be antifungal agents, when they are defined as inhibitors of both growth and protein synthesis in cells.

The literature on the effect of inhibitors on protein synthesis is quite extensive [1-9]. Because there are fewer antifungal inhibitors, this review will deal primarily with the mechanism of action of the inhibitors and not with the mechanisms of protein synthesis.

A summary of the various processes by which proteins are synthesized will be followed by a detailed discussion of antifungal inhibitors of cytoplasmic protein synthesis. Compounds which inhibit protein synthesis only in fungal extracts or in fungal mitochondria will be mentioned briefly.

II. MECHANISMS OF PROTEIN SYNTHESIS

There are two protein synthesizing systems found in fungi. Synthesis in the cytoplasm is associated with the 80-S ribosomes, while synthesis in the mitochondria is associated with 55-S to 80-S ribosomes. Protein synthesis in the cytoplasm and mitochondria is regulated by DNA in the nucleus and mitochondria, respectively. The synthesis of proteins in mitochondria closely resembles that found in prokaryotic cells (bacteria and blue-green algae), as indicated by its sensitivity to antibacterial antibiotics [10]. The effect of various inhibitors on the biosynthesis of mitochondrial proteins will be discussed in Section IV.

The biosynthesis of proteins has received considerable attention and there are numerous reviews of this subject [3, 11-19]. The following discussion is a summary of the mechanisms of protein synthesis. Individual references to the voluminous literature on the subject will not be cited here but can be found in the various reviews.

The main features of the biosynthetic process can be described briefly. The ribosomes, particles that contain proteins and RNA (ribosomal RNA = rRNA), are the sites of protein synthesis. Ribosomes (cytoplasmic) from eukaryotes and prokaryotes exhibit sedimentation coefficients of 80 S and 70 S, respectively. The particles from eukaryotes are made up of 40-S and 60-S subunits; those from prokaryotes of 30-S and 50-S subunits. The 60-S subunit contains two species of rRNA and the 40-S subunit one species of rRNA. The rRNAs and various protein units not only function as structural components of the ribosome, but also have a functional role in the process of protein synthesis.

The genetic information is transferred to the ribosomes by a single-stranded RNA (messenger RNA = mRNA). The sequence of ribonucleotides in mRNA is specified during the process of transcription by the base sequence of the DNA. Therefore, mRNA is complimentary to one of the strands of the DNA. The nucleotide sequence in mRNA determines the amino acid sequence in the polypeptide chain. This reaction occurs in the presence of a third form of RNA (transfer RNA = tRNA) which transfers amino acids to the ribosomal site. There is at least one specific tRNA for each of the 20 amino acids used in protein synthesis. Each tRNA contains three nucleotides (triplet anticodon) that are complimentary to the triplet codon of the mRNA. The triplet codon of the mRNA specifies the next amino acid, as it is attached to its tRNA, to be incorporated into the elongating polypeptide chain. While peptide bond formation occurs on the ribosomes, the specificity of the system (amino acid sequence) is determined by codon-anticodon interactions.

In the synthesis of proteins, four distinct steps have been recognized: aminoacylation of tRNA, initiation, elongation, and termination of the polypeptide chains. The general mechanisms by which proteins are synthesized appear to be similar in both prokaryotes and eukaryotes. There are some minor differences, particularly in chain initiation and termination, and in the designation and chemical characteristics of the various soluble factors.

A. Formation of Aminoacyl-tRNA Complexes

While the nucleotide base sequences of the individual tRNAs vary, all have a somewhat general size and shape; each tRNA must bind to the same locus on the ribosome and be oriented so that the aminoacyl group is in the appropriate position in regard to its addition to the elongating polypeptide chain. Transfer RNAs are generally considered to be arranged in a cloverleaf pattern. Certain regions are base paired by internal hydrogen bonding, others are not paired, but are looped or nonlooped. Two regions of particular importance are the anticodon loop, which, as previously mentioned, contains the triplet of nucleotides complimentary to those in mRNA, and the acceptor end, to which the amino acid is attached. The acceptor end of all tRNAs contains a nonlooped sequence of three nucleotides, pCpCpA.

Amino acids are esterified by specific aminoacyl-tRNA synthetases through their carboxyl groups to the 2'- or 3'-hydroxyl group of the ribose moieties of the terminal adenylic acid residue of the tRNAs. The splitting of the pyrophosphate bond of the terminal ATP provides the energy for the synthetic process. The reaction for each amino acid (aa) can be summarized as:

$$\text{aa} + \text{ATP} + \text{tRNA} \rightleftharpoons \text{aminoacyl-tRNA} + \text{AMP} + \text{PPi} \quad (1)$$

The individual synthetase must obviously select the correct amino acid and tRNA to insure the fidelity of translation of the genetic information in the mRNA.

The summary reaction can be further subdivided into the "activation step" (2), in which the synthetase first recognizes the individual amino acid, and the "transfer step" (3), which consists of aminoacylation with the cognate tRNA (aa-tRNA).

$$\text{aa} + \text{ATP} + \text{enzyme} \rightleftharpoons \text{aa-AMP-enzyme} + \text{PPi} \quad (2)$$

$$\text{aa-AMP-enzyme} + \text{tRNA} \rightleftharpoons \text{aa-tRNA} + \text{AMP} + \text{enzyme} \quad (3)$$

B. Peptide Chain Initiation

The initiation process occurs in the presence of the ribosomal subunits, mRNA, initiator aminoacyl-tRNA, GTP, and several soluble protein initiation factors (M_1, M_2, and M_3). The 40-S subunit, derived by dissociation of the 80-S ribosome following chain termination, binds the soluble protein factor M_3. The factor serves a dual function in that it binds the mRNA and prevents the association of the 60-S subunit (4a).

$$40\text{S} + M_3 + \text{mRNA} \longrightarrow 40\text{S-}M_3\text{-mRNA} \quad (4a)$$

There is an initiator site on the 40-S subunit which binds the initiator aminoacyl-tRNA. In eukaryotes the initiator tRNA is methionyl-tRNA which recognizes the codons AUG or GUG of the mRNA. There are two species of methionyl-tRNA found in the cytoplasm of eukaryotic cells; however, it is only one of these species (met-tRNA_f) which serves as the initiator tRNA. The binding of met-tRNA_f to the 40-S initiator site is facilitated by the soluble protein factor M_1 or EIF1 (eukaryotic initiation factor) and M_2 (4b, 4c). The function of the M_2 protein factor is less clear, but may involve conformational changes of the ribosome necessary for binding of the met-tRNA_f.

$$40\text{S-}M_3\text{-mRNA} + (M_1, M_2) \longrightarrow M_{1,2}\text{-40S-}M_3\text{-mRNA} \quad (4b)$$

$$M_{1,2}\text{-40S-}M_3\text{-mRNA} + \text{met-tRNA}_f \longrightarrow \text{40S-}M_3\text{-mRNA-}M_{1,2}\text{-met-tRNA}_f \quad (4c)$$

Binding of met-$tRNA_f$, mRNA, and the protein factors to form the 40-S initiation complex is rapidly followed by the addition of the 60-S subunit forming the 80-S ribosome (4d).

$$\text{40S-}M_3\text{-mRNA-}M_{1,2}\text{-met-tRNA}_f + \text{60S} \longrightarrow \text{80S-mRNA-met-tRNA}_f + M_1 + M_2 + M_3 \quad (4d)$$

The met-$tRNA_f$ now occupies the peptidyl or donor site (P-site) on the ribosome. Formation of the 80-S ribosome has resulted in a new site, the aminoacyl or acceptor site (A-site), for the incoming aminoacyl-tRNA. After formation of the 80-S ribosome, the various protein initiation factors are released (4d).

In prokaryotes the initiator aminoacyl-tRNA is formylated and is designated as fmet-$tRNA_f$. The initiation factors are IF_1, IF_2, and IF_3. Similar to M_3, IF_3 binds mRNA to the 30-S subunit; IF_2, similar to M_1, promotes binding of fmet-$tRNA_f$ to the 30-S subunit; and IF_1, similar to M_2, promotes catalytic use of IF_2. All three factors bind to the 30-S initiation complex but are released upon formation of the 70-S ribosome. Release of IF_2 is dependent on the hydrolysis of GTP.

C. Peptide Chain Elongation

The process of polypeptide chain elongation involves the following components: 80-S ribosome, aminoacyl-tRNA, mRNA, peptidyltransferase, and various elongation factors. The entire process may be divided into three consecutive steps.

1. Binding of Aminoacyl-tRNA

Aminoacyl-tRNA binds to the ribosomal A-site as a complex of GTP and a soluble elongation factor, EF_1 (5a).

$$\text{aa-tRNA} + \text{GTP} + EF_1 \longrightarrow \text{aa-tRNA-GTP-}EF_1 \quad (5a)$$

Binding of the complex to the A-site of the mRNA ($\underset{A}{\longleftarrow\!\!\!-\!\!\!-\!\!\!-}$) proceeds in conjunction with codon-anticodon interactions (...). The P-site of the mRNA ($\underset{P}{\longleftarrow\!\!\!-\!\!\!-\!\!\!-}$) is occupied with peptidyl-tRNA (pept-tRNA) and its codon, +++ (5b).

$$\text{aa-tRNA-GTP-EF}_1 + \frac{\overset{\text{pept-tRNA}}{+++} \qquad \cdots}{\text{P} \qquad\qquad \text{A}} \longrightarrow$$

$$\frac{\overset{\text{pept-tRNA}}{+++} \qquad \overset{\text{aa-tRNA}}{\cdots}}{\text{P} \qquad\qquad \text{A}} + \text{GDP} + \text{EF}_1 + \text{Pi} \qquad (5b)$$

The hydrolysis of GTP supplies the energy for the binding processes.

2. Peptide Bond Formation

Peptide bond formation, as catalyzed by peptidyltransferase (PTf), a 60-S subunit protein, occurs between the carboxyl group of the terminal amino acid of the peptidyl-tRNA and the amino group of the aminoacyl-tRNA (6). The peptide chain has thus been transferred from the tRNA at the P-site to the tRNA at the A-site, which now contains one additional amino acid.

$$\frac{\overset{\text{pept-tRNA}}{+++} \qquad \text{PTf} \qquad \overset{\text{aa-tRNA}}{\cdots}}{\text{P} \qquad\qquad\qquad \text{A}} \longrightarrow \frac{\overset{\text{tRNA}}{+++} \qquad \overset{\text{pept-aa-tRNA}}{\cdots}}{\text{P} \qquad\qquad \text{A}} \qquad (6)$$

3. Translocation

The process of translocation involves removal of the deacylated tRNA from the P-site, movement of the peptidyl-tRNA from the A- to P-site, and movement of the mRNA by one codon (xxx) to the A-site. A soluble elongation factor (EF_2) and hydrolysis of GTP, which supplies energy, mediate the translocation processes. The GTP binds with EF_2 and then reacts with the 80-S ribosome to form a stable complex (7a).

$$\text{EF}_2 + \text{GTP} \longrightarrow \text{EF}_2\text{-GTP} + \text{80S} \longrightarrow \text{EF}_2\text{-GTP-80S} + \text{Pi} \qquad (7a)$$

The remaining steps in translocation now occur with the A-site becoming available for receiving another aminoacyl-tRNA-GTP-EF_1 complex (7b).

$$\frac{\overset{\text{tRNA}}{+++} \qquad\qquad \overset{\text{pept-aa-tRNA}}{\cdots}}{\text{P} \quad \text{EF}_2\text{-GTP} \qquad \text{A}} \longrightarrow \frac{\overset{\text{pept-aa-tRNA}}{\cdots} \qquad \text{xxx}}{\text{P} \qquad\qquad \text{A}}$$

$$+ \text{tRNA} + \text{EF}_2 + \text{GDP} \qquad (7b)$$

In prokaryotes the elongation factors EF-Tu and EF-Ts are comparable to EF_1, and EFG is comparable to EF_2. EF-Tu forms a complex with GTP and aminoacyl-tRNA. EF-Ts participates in the recycling of EF-Tu by promoting the formation of EF-Tu-GTP and the release of GDP from EF-Tu-GDP. EFG is like its counterpart EF_2, except no stable EFG-GTP complex is formed. However, during translocation EFG binds to the 70-S ribosome and GTP hydrolysis is required for EFG activity.

D. Peptide Chain Termination

Peptide chain termination occurs when mRNA containing the nonsense codons UAA, UGA, or UAG moves to the ribosomal A-site. This process prevents binding of an aminoacyl-tRNA-GTP-EF_1 complex to the A-site. In the presence of a soluble release factor (Rf), GTP and peptidyltransferase hydrolysis of the peptide chain from the tRNA at the P-site occurs. The Rf proteins must first bind to the ribosome in the presence of GTP and recognize the release codons (8a). This complex then activates peptidyltransferase which causes peptide chain hydrolysis (8b).

$$\underset{\text{P}\qquad\quad\text{A}}{\overset{\text{pept-tRNA}}{\frac{\cdots\qquad\text{xxx}}{\qquad\quad\text{UAA}}}} + \text{Rf} + \text{GTP} \longrightarrow \underset{\text{P}\qquad\qquad\qquad\text{A}}{\overset{\text{pept-tRNA}}{\frac{\cdots\qquad\text{PTf}\quad\text{xxx}}{\text{Rf-GTP}\quad\text{UAA}}}} \qquad \text{(8a)}$$

$$\underset{\text{P}\qquad\qquad\qquad\text{A}}{\overset{\text{pept-tRNA}}{\frac{\cdots\qquad\text{PTf}\quad\text{xxx}}{\text{Rf-GTP}\quad\text{UAA}}}} \longrightarrow \underset{\text{P}\qquad\qquad\qquad\text{A}}{\overset{\text{t-RNA}}{\frac{\cdots\qquad\qquad\quad\text{xxx}}{\text{Rf-GTP}\quad\text{UAA}}}} + \text{peptide} \qquad \text{(8b)}$$

The function of the GTP, with its hydrolysis to GDP, is to assist in the binding and dissociation of the Rf factor (8c).

$$\underset{\text{P}\qquad\qquad\qquad\text{A}}{\overset{\text{tRNA}}{\frac{\cdots\qquad\qquad\quad\text{xxx}}{\text{Rf-GTP}\quad\text{UAA}}}} \longrightarrow \underset{\text{P}\qquad\quad\text{A}}{\overset{\text{tRNA}}{\frac{\cdots\qquad\text{xxx}}{\qquad\quad\text{UAA}}}} + \text{Rf} + \text{GDP} + \text{Pi} \qquad \text{(8c)}$$

After peptidyl-tRNA hydrolysis and release of Rf, the ribosomal complex containing the deacylated tRNA and mRNA dissociates into its individual components (9).

$$\overset{\text{tRNA}}{\frac{\cdots\quad\text{xxx}}{\text{P}\qquad\text{A}}} \longrightarrow 40\text{S} + 60\text{S} + \text{mRNA} + \text{tRNA} \qquad \text{(9)}$$

In prokaryotes, two soluble release factors differ in codon specificity. Rf_1 recognizes UAA or UAG, and Rf_2 recognizes UAA or UGA. The interaction of Rf_1 and Rf_2 with bacterial ribosomes is stimulated by another soluble release factor, Rf_3. This factor interacts with GDP and GTP to cause dissociation of Rf_1 or Rf_2 from the ribosome.

III. ANTIFUNGAL INHIBITORS

A. Glutarimide Antibiotics

1. Chemistry and Biology

This group of compounds is composed of some naturally occurring antibiotics isolated from various streptomycetes, and synthetic derivatives which have a B-(2-hydroxyethyl)glutarimide moiety attached to a cyclic or acyclic ketone. The biology and chemistry of the glutarimide group have been reviewed elsewhere [19, 20].

The glutarimides can be grouped according to their degree of biological activity. They can be considered as being highly toxic, slightly toxic, or generally nontoxic. In the latter two groups belong the naturally occurring compounds inactone, actiphenol, synthetic and naturally occurring stereoisomers of cycloheximide, and various synthetic derivatives of cycloheximide.

The biologically active glutarimides include cycloheximide, acetoxycycloheximide, streptovitacin A (1), and streptimidone (2). In addition, the incompletely characterized glutarimide antibiotics protomycin, fermicidin, and niromycin have biological activity.

The glutarimides have a diverse range of biological activity, almost exclusively among eukaryotic organisms. Selectivity of these antibiotics among species of fungi, algae, higher plants and animals has been reported [19, 20]. These compounds also have antiviral and antitumor activity.

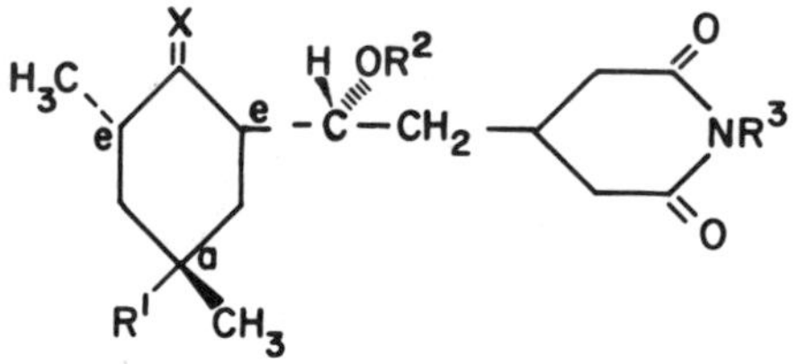

X, H_3C, e, e, H, OR^2, $-C-CH_2$, O, NR^3, O, a, R^1, CH_3

(1)

	X	R^1	R^2	R^3
Cycloheximide	O	H	H	H
Oxime	NOH	H	H	H
Semicarbazone	$N-NHCONH_2$	H	H	H
Acetate	O	H	$COCH_3$	H
Acetoxycycloheximide	O	$OCOCH_3$	H	H
Streptovitacin A	O	OH	H	H

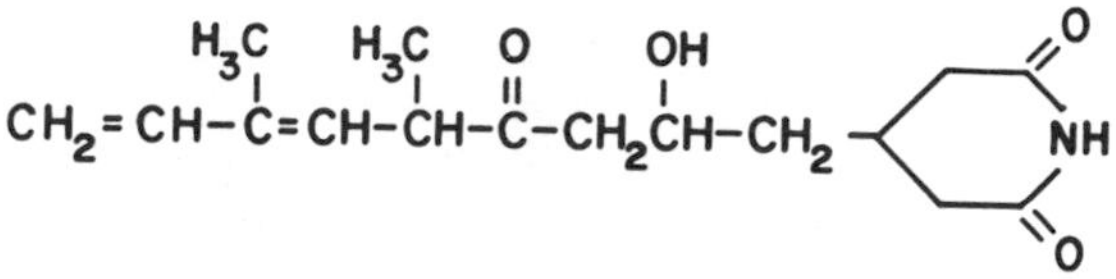

(2) streptimidone

Cycloheximide, B-[2,(3,5,dimethyl-2-oxocyclohexyl)-2-hydroxyethyl]-glutarimide, is the best known member of this group and its chemistry and mode of action have been studied extensively. The compound has limited uses in controlling plant diseases caused by fungi because of its phytotoxicity [21]. The general toxicity of the glutarimides also precludes their use in medicine.

Studies on the toxicity of the stereoisomers of cycloheximide and related glutarimides indicate that there is an optimum structural requirement for biological activity [22, 23]. There are four asymmetric centers in the structure of the cycloheximide (3). The substituents on the cyclohexanone ring

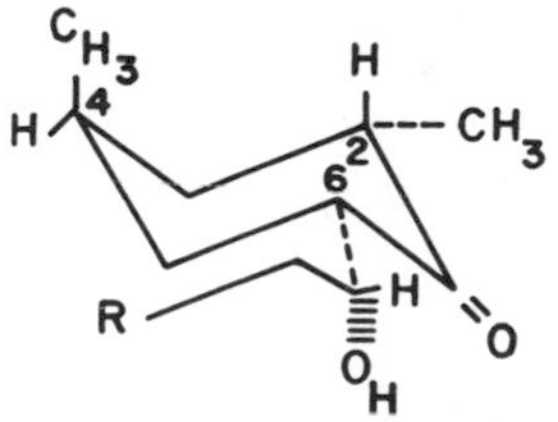

(R = glutarimide)

(3) l-cycloheximide (2e, 4a, 6e)

In this stereochemical representation the numbering of the cyclohexanone ring begins at the carbonyl carbon atom. In the chemical designation of cycloheximide the numbering of the ring begins at the carbon atom attached to the hydroxyethylglutarimide moiety.

are oriented 2-equatorial (e), 4 axil (a), and 6e [24]. Other stereoisomers studied include: d-isocycloheximide (2e, 4e, 6e); d-naramycin B (2a, 4e, 6e); l-α-epi-isocycloheximide (2e, 4e, 6a); and d-cycloheximide (the enantiomer of l-cycloheximide). The data indicate that with compounds containing cyclic ketones there is an almost absolute configurational requirement for toxicity. In the fungal assay system only l-cycloheximide exhibited significant in vivo (10^{-8} M) and in vitro (10^{-7} M) activity. However, a cyclohexanone ring is not absolutely necessary for activity. Streptimidone, which contains an aliphatic side chain, also exhibited toxicity.

Substitution of chemical groups on the ketone carbonyl (as OH), hydroxyl (as acetate, oxime, and semicarbazone), or imide nitrogen (as methyl) of the 1-cycloheximide molecule eliminates toxicity (1) [22, 23]. The biological activity reported for the oxime, semicarbazone, and acetate derivatives is based on enzymatic or H^+ catalyzed hydrolysis to cycloheximide [25].

Not all substitutions eliminate toxicity. Substitution of the cyclohexanone ring at the 4 position with acetate (acetoxycycloheximide) or hydroxyl (streptovitacin-A) actually enhances the inhibition of in vitro protein synthesis by these compounds [22, 23].

Structure-activity studies have led Siegel et al. [22] to postulate that proper configuration of the ketone carbonyl in relation to the hydroxyethylglutarimide portion of the molecule (3) is required for a three-point attachment at the site of action. This attachment to the substrate, which has been identified as the 60-S ribosomal subunit [26, 27], may involve weak linkages such as hydrogen bonding and Van der Waals' forces because toxicity is readily reversible in whole cells [28-30] or cell-free systems [30]. Some modification of this hypothesis may be necessary as streptovitacin A is reported not to diffuse out of treated reticulocytes [29].

It is apparent that substitutions at the 4 position of the cyclohexanone ring must also affect permeation of the compounds in different organisms [22, 23]. Acetoxycycloheximide and streptovitacin A are not as effective as cycloheximide in inhibiting growth of cells of *Saccharomyces pastorianus*. However, both compounds inhibit protein synthesis in intact mammalian cells at concentrations of 10^{-7} to 10^{-8} M.

2. Inhibition of Protein Synthesis

Kerridge [31] was the first to report that cycloheximide is a potent inhibitor of protein synthesis in whole cells of *S. carlsbergensis*. Inhibition of protein synthesis in cell-free systems from *S. pastorianus* [32] and rat liver [30] was reported in 1963 and 1964, respectively. Since 1964, cycloheximide and other glutarimide antibiotics have been used in numerous in vivo and in vitro studies for a variety of purposes which involve the basic premise that they are inhibitors of only protein synthesis. The criteria for this assumption are based on the following observations: (a) concentrations of the antibiotic which inhibit growth also inhibit protein synthesis in whole cells [19]; (b) the genetic basis for resistance in certain organisms is associated with the ribosomes and their protein-synthesizing function [5, 7, 19]; and (c) compounds (actinomycins) which inhibit DNA-dependent RNA synthesis (transcription) also inhibit protein synthesis (translation) in intact cells as effectively as cycloheximide [5, 19]. However, since actinomycins act directly on the transcription process they must be added to cells prior to the time when cycloheximide would be added to produce a comparable inhibitory effect on protein synthesis [5].

Apparently the basic premise that cycloheximide inhibits protein synthesis in most organisms under most conditions is valid. However, glutarimide antibiotics at concentrations of 10^3 to 10^{-5} M have been reported to inhibit other metabolic processes. Cycloheximide has been reported to inhibit the mechanism of energy transfer (specifically at site 1) of the electron transport chain in animal mitochondria [33], membrane transport in Euglena gracilis [34], and RNA synthesis (specifically RNA polymerase 1) in the Phycomycetes [35, 36] (see Chapter 10 for a further discussion of the effects of cycloheximide on nucleic acid synthesis).

The researcher should be aware of the potential effects of the glutarimide antibiotics. The use of these compounds at high concentrations as biochemical tools to inhibit only protein synthesis, especially in whole cell studies, may be inappropriate.

The glutarimide antibiotics were first shown to inhibit protein synthesis by affecting the transfer of amino acids from aminoacyl-tRNAs into protein [30, 37-39]. The compound did not interfere with either amino acid activation or transfer to tRNA. Inhibition of protein synthesis occurs in the cytoplasm but not in the mitochondria [10].

The ribosomes were shown to contain the binding or reactive site for the antibiotics [40-42]; the site was later shown to be the 60-S subunit [26, 27, 43]. The reactive subunit can be determined in two different ways: (a) ribosomal subunits (40-S and 60-S) prepared from either the same species or different species of resistant and susceptible yeast are mixed in vitro as "hybrid" reconstituted ribosomes and then incubated in the presence or absence of antibiotic [26, 27]; and (b) ribosomes from a susceptible species of yeast are dissociated into subunits, incubated separately in the presence or absence of antibiotic and then reconstituted [43]. In all cases, inhibition of protein synthesis occurs only in those systems which contain "hybrid" reconstituted ribosomes with either resistant or fungicide treated 60-S subunits. It has been suggested that resistance involves specific alteration of a single protein component of the 60-S subunit [27].

The genetics of resistance to cycloheximide has been reviewed [5, 19] and will be discussed in Chapter 12.

The glutarimide antibiotics either prevent or slow the movement of the ribosomes along the mRNA (chain elongation). Polysome breakdown (GTP dependent) to monomers does not occur in the presence of the antibiotics [30, 44-50] which do not cause premature release of peptides from polysomes [30, 47, 49]. This is in contrast to puromycin which accelerates the release of monomers from polysomes and causes the premature detachment of peptides during protein synthesis [1, 6, 7].

Cycloheximide [46, 47] and streptovitacin A [49] protect reticulocyte polysomes from NaF degradation if the antibiotics are added prior to the addition of the fluoride. Similarly, cycloheximide inhibits the breakdown of mouse liver polysomes following administration of actinomycin D or ethionine [50]. These effects on polysomes suggest that the process of peptide chain elongation is inhibited by the glutarimide antibiotics. However, some investigators have indicated that cycloheximide also inhibits chain initiation [47, 50-54].

Cycloheximide inhibits the binding of deacylated phe-tRNA (phe - phenylalanine) to a mixture of ribosomes and polyuridylic acid (poly-U) [53]. Polysome reaggregation from monomers was reported to be inhibited at lower concentrations of cycloheximide than peptide chain elongation [54]. Formation of the initiation complex (consisting of poly-U, phe-tRNA and 80-S reticulocyte ribosomes) is more sensitive to cycloheximide than the chain elongation process [53]. Lastly, cycloheximide inhibits polysome reassembly of NaF-induced monomers [46, 47, 50].

Cycloheximide does not appear to inhibit the conventional steps in peptide chain initiation. Some of the effects reported for chain initiation may be due to the inhibition of peptide chain elongation.

Inhibition of the initiation process has only been reported for poly-U directed polyphenylalanine synthesis and not for synthesis of other polypeptides. Lodish et al. [56] demonstrated that cycloheximide does not inhibit the initiation of fmet-$tRNA_f$ (E. coli) into globin chains on reticulocyte ribosomes over a wide range of concentrations, but does inhibit peptide chain elongation. Inhibition of binding of deacylated phe-tRNA has only been demonstrated with 80-S and not with the 40-S subunit [43]. The reactive site for the antibiotics has been shown to be the 60-S subunit, which is not involved in the binding of initiator tRNA. Cycloheximide at 4×10^{-6} M prevents NaF-induced polysome breakdown. Reassembly of the NaF-induced monomers into polysomes occurs in the presence of 4×10^{-6} M cycloheximide, but is inhibited at 1.5×10^{-2} M concentrations of the antibiotic. The rate of initial polysome reformation, from heat or chemically induced monomers, is considered to be critical in determining what processes in protein synthesis are inhibited [55]. In the presence of cycloheximide the rate of initial polysome reformation is slow. This type of an effect would be expected of inhibitors of chain elongation. Inhibitors of chain initiation would completely inhibit the initial rate of polysome reformation while inhibitors of chain termination would have no effect.

Both translocation [52-54, 57] and transpeptidation [58, 59], two steps in the peptide chain elongation process, have been reported to be inhibited by cycloheximide and other glutarimide antibiotics. These compounds have been shown to block Ef_2 GTP-dependent translocation of peptidyl-tRNA from the A (aminoacyl) site to the P (peptidyl) site on the ribosome [52-54, 57] and to prevent the release of deacylated tRNA from the ribosome [53]. Cycloheximide has no effect on ribosome-dependent GTPase associated with Ef_2 factor [52], and the inhibitory effect on translocation can be reversed by the addition of excess GTP and Ef_2 [53] and sulfhydryls [54].

Translocation was measured by the reaction of puromycin with ribosome bound peptidyl-tRNA. Formation and release of peptidyl puromycin only occurs when peptidyl-tRNA is in the P-site on the ribosome. If the peptidyl-tRNA is in the A-site it must be translocated before it can react with puromycin. The glutarimide antibiotics inhibit the reaction of puromycin only when peptidyl-tRNA is at the A-site, but not at the P-site. The puromycin reaction is also an indicator of peptidyltransferase activity associated with peptide bond formation. The glutarimide antibiotics do not inhibit the reaction of puromycin with ribosome bound peptidyl-tRNA [52-54, 57], acetyl-aminoacyl-tRNA [43, 55], and aminoacyl-oligonucleotide fragments of tRNA [43, 60]. Cycloheximide also does not inhibit peptidyltransferase activity associated with chain termination as measured by hydrolysis of fmet-$tRNA_f$ (E. coli) bound to eukaryotic ribosomes [55, 61].

On the other hand, Pestka et al. [58] and Schneider and Maxwell [59] conclude that cycloheximide does inhibit transpeptidation (peptide bond formation) when the process is measured as the reaction of nascent polypeptides on purified ribosomes with puromycin to form peptidyl puromycin in the absence of Ef_2 factor. Cycloheximide was shown under these conditions to be a mixed inhibitor (competitive and noncompetitive) of the puromycin reaction. Known inhibitors of translocation (fusidic acid and emetine) preincubated with polysomes did not prevent cycloheximide inhibition of peptidyl puromycin synthesis [58]. Cundliffe et al. [62] have also shown that cycloheximide inhibits the reactivity of puromycin with ribosome-bound nascent polypeptide.

The discrepancies in the reports on the effects of the glutarimide antibiotics on the puromycin reaction have not been resolved. Pestka et al. [58] suggested that the action of cycloheximide as a mixed inhibitor reflects the heterogenicity of peptidyl-tRNA on the ribosomes. If peptidyl-tRNA can reside in one of two ribosomal sites, in each of which it can react with puromycin, then the two phase inhibition reported for the glutarimide antibiotics may be valid. However, this explanation is inconsistent with the current donor-acceptor model of protein synthesis (Section II) which states that when peptidyl-tRNA is in the P-site it must be able to react with puromycin, and if it is in the A-site it cannot react with puromycin. Pestka [2] has offered an alternative model for protein synthesis that would reflect the heterogenicity of peptidyl-tRNA on the ribosome and may explain the action of cycloheximide and other antibiotics which are mixed inhibitors of transpeptidation.

B. Anisomycin

1. Chemistry and Biology

Anisomycin, an antibiotic isolated from cultures of various streptomycetes, is toxic to a wide range of eukaryotic organisms, but not to species of bacteria [63]. Certain species of yeast are particularly sensitive to anisomycin [63, 64]. Its structure (4) has been established as 2-p-methoxyphenylmethyl-3-acetoxy-4-hydroxypyrrolidine [65]. Structural specificity studies of isomers and analogues of anisomycin indicate that removal of the acetoxy group (position 3), acetylation or quaternization of the imide nitrogen, and bromination of the p-methoxyphenyl moiety greatly reduce or completely destroy the in vivo and in vitro activity of this antibiotic [63, 66].

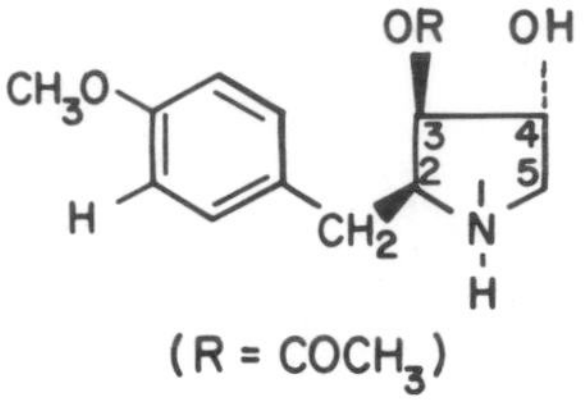

$$CH_3O$$ … (R = $COCH_3$)

(4) anisomycin

Mutants of S. cerevisiae resistant to anisomycin have been reported [26]. Every resistant strain apparently failed to take up the antibiotic since they were all sensitive to anisomycin when tested in cell-free systems.

2. Inhibition of Protein Synthesis

Anisomycin is a potent inhibitor of protein synthesis on 80-S ribosomes from yeast at concentrations of 10^{-6} M [43, 55, 60, 61, 67] and from mammalian cells at 10^{-6} to 10^{-8} M [59, 60, 63, 66]. DNA synthesis is partially inhibited and RNA synthesis unaffected in Hela cells [63]. Anisomycin was first reported by Grollman [63] to inhibit protein synthesis subsequent to the formation of aminoacyl-tRNA. It also prevents polysome disassembly to monomers and detachment of nascent peptides [62, 63]. The inhibition of protein synthesis is reversible in Hela cells [63].

The localization of the anisomycin reactive ribosomal site has been shown to be the 60-S subunit [43]. Inhibition of polyphenylalanine synthesis on ribosomes from yeast occurs only in systems containing "hybrid" reconstituted ribosomes composed of 60-S subunits that are pretreated with anisomycin [43].

Using [^{3}H]anisomycin, Barbacid and Vazquez [66] demonstrated that the antibiotic specifically binds to the 60-S subunit of ribosomes isolated from yeast and human tonsil. A maximum of one molecule of anisomycin was bound per ribosome. Deacetylanisomycin, which lacks the acetate group at the 3' position of the pyrrolidine ring, appears to have the same mode of action as anisomycin, but has a 350-fold lesser affinity for the ribosome. Certain inhibitors of peptide bond formation, such as the sesquiterpene antibiotics of the trichodermin group, completely prevent [^{3}H]anisomycin binding to the ribosome.

The antibiotic has no effect on chain initiation, as it neither binds to nor prevents the binding of phe-tRNA and acetyl phe-tRNA to the 40-S subunit [43].

Anisomycin is a potent inhibitor of peptidyltransferase activity [43, 55, 58-62, 67-70]. Peptidyltransferase mediated peptide bond formation is monitored in the following systems: (a) anisomycin inhibits the binding of aminoacyl oligonucleotide fragments of tRNA, CACCA-leu or CACCA-leu-acetyl (leu = leucine) in the absence of GTP and Ef_2, with the P and A ribosomal sites, respectively, of the peptidyltransferase center [43, 70]; and (b) anisomycin inhibits puromycin reactivity with eukaryotic ribosome bound peptidyl-tRNA [58, 59], aminoacyl-tRNA (fmet-$tRNA_f$, leu-tRNA, acetyl leu-tRNA) [43, 60, 66], and aminoacyl oligonucleotide fragments of tRNA (CACCA-leu or CACCA-leu-acetyl) [43, 70].

Anisomycin also inhibits peptidyltransferase activity associated with peptide chain termination as determined by the hydrolysis of fmet-$tRNA_f$ (*E. coli*) bound to eukaryotic ribosomes. Depending on the components in the assay system, product formation in the presence of anisomycin was partially inhibited, completely inhibited, or stimulated (see Table 1). However, there is some evidence to support the conclusion that the antibiotic functions primarily as an inhibitor of peptidyltransferase activity associated with peptide bond formation. Wei et al. [55] have demonstrated that anisomycin is a much more potent inhibitor of chain elongation, as measured by poly-U-directed polyphenylalanine synthesis, than of chain termination as measured by the hydrolysis reaction. The hydrolysis reaction which occurs in the presence of Rf, GTP, and release codon is a system which contains the components found in cells. Partial inhibition of chain termination in this system was also reported by Beaudet and Caskey [61]. An explanation of the complete inhibition of the hydrolysis reaction in the presence of ethanol and the stimulation of the reaction in the presence of acetone has been offered [68, 69].

It has been suggested that since anisomycin binds to the 60-S subunit, changes can occur in the conformation of the peptidyltransferase center which block entry of a larger nucleophile (ethanol) into the peptidyltransferase site but which facilitate entry of a small nucleophile (OH group of acetone). Under these conditions anisomycin inhibits fmet-ethyl ester formation in the presence of ethanol and stimulates fmet release in the presence of acetone.

TABLE 1

Effects of Anisomycin on Peptidyltransferase Catalyzed Reactions in Peptide Chain Termination

Substrate	Additions	Product	Effect on hydrolysis	Reference
fmet-$tRNA_f$	GTP, Rf, UAA[a]	fmet	Partial inhibition	67, 71
fmet-$tRNA_f$	Ethanol, tRNA, or CAA[b]	fmet-ethyl ester	Complete inhibition	65, 66
fmet-$tRNA_f$	Acetone, tRNA, or CAA	fmet	Stimulation	65, 66

[a]Release codon.
[b]Trinucleotide which corresponds to the 3' nucleotide sequence of tRNA.

C. Trichothecene Antibiotics

1. Chemistry and Biology

The trichothecenes comprise a group of closely related sesquiterpenoid mycotoxins produced by various species (primarily Trichothecium, Myrothecium, and Fusarium species) of the imperfect fungi which cause postharvest decay on grains and fruits. The chemical and biological properties of these compounds have been extensively reviewed by Bamburg and Strong [71].

Members of the trichothecenes all share a 4-B-hydroxy (acyloxy)-12,13-epoxy-Δ^9-trichothecene structure. They can be divided into three groups, according to the absence or presence of various ring substitutions. Structures (5-7) represent only those which will be mentioned in this text. Structure (7) illustrates only verrucarin A. There are other verrucarins as well as roridins which differ in the R group.

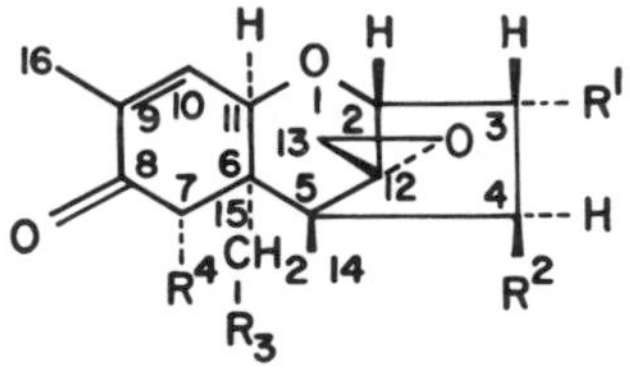

(5)

	R^1	R^2	R^3	R^4
Trichothecolone	H	OH	H	H
Trichothecin	H	$OOCCH{=}CHCH_3$	H	H
Nivalenol	OH	OH	OH	OH
Fusarenone	OH	OAC	OH	OH
Crotocin (epoxide at C-8)	H	$OOCCH{=}CHCH_3$	H	H

Except for trichothecin, none of the trichothecenes exhibit antibacterial activity. The trichothecenes exhibit a broad biological activity against eukaryotic cells. Toxicity to fungi, protozoa, insects, plants, and animals has been reported [71]. Several of the trichothecenes are highly cytotoxic and are implicated in mycotoxicoses of animals [72]. Most of the trichothecenes exhibit fungistatic activity and many are fungitoxic.

Generally, structure activity studies indicate that hydroxyl substituents at the R_2 group reduce fungitoxic, phytotoxic, and dermatitic activity of the

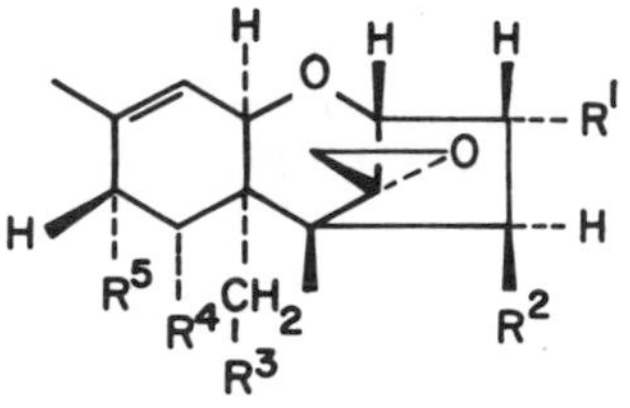

(6)

	R^1	R^2	R^3	R^4	R^5
Trichodermin	H	OAC	H	H	H
Trichodermol	H	OH	H	H	H
Verrucarol	H	OH	OH	H	H
Scirpentriol	OH	OH	OH	H	H
Calonectrin	OAC	H	OAC	H	H

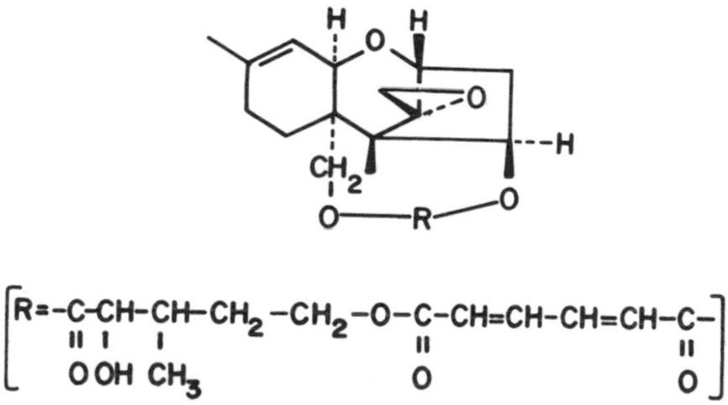

(7) verrucarin A

trichothecenes (i.e., alcohols of trichodermin, trichothecin, and crotocin). However, acetylation of the hydroxyl group restores activity. Esterification of some of the hydroxyl substituents is correlated with high toxicity. When the parent alcohols were injected into animals they were as active as the esterified compounds. This suggests that the alcohols cannot penetrate the waxy surfaces of plants and the oily surfaces of animal skin.

Hydrogenation of the olefinic bond at the C-9 and C-10 position, removal of the epoxide ring, or rearrangement of the trichothecene nucleus also results in a complete loss of biological activity.

Recently Wei and McLaughlin [73] investigated a number of trichothecenes as inhibitors of protein synthesis in rabbit reticulocytes. Their data agree generally with the previously discussed structural requirements for toxicity. The alcohols (trichodermol, trichothecalone, scirpentriol) are inhibitors of in vivo protein synthesis. Conversion of the hydroxyl group at R_2 to a carbonyl group destroys toxicity. Configuration of the hydroxyl group is important, as epitrichodermol is without activity. Esterification of trichodermin at the R_2 results in a substantial increase in activity. Reduction of the hydrogen bond at C-9 and C-10 or introduction of an epoxy ring between C-7 and C-8 results in a loss of activity.

2. Inhibition of Protein Synthesis

Nivalenol was the first of the trichothecenes to be reported to inhibit poly-U-directed polyphenylalanine synthesis in reticulocytes [74]. In mammalian culture cells nivalenol inhibits first protein synthesis and then DNA synthesis; whereas RNA synthesis is not severely affected [75, 76]. Similar results on protein and DNA synthesis have been reported for fusarenone in protozoa [77] and mouse fibroblasts (L-cell) [78]. All the trichothecenes do not have the same site of action as inhibitors of the processes of protein synthesis. Table 2 lists the trichothecenes as either inhibitors of

TABLE 2

Site of Action of the Trichothecenes on the Processes of Protein Synthesis

Compound	Chain initiation, Reference	Chain elongation or termination,[a] Reference
Trichodermin		70, 79, 80, 81 55, 62, 73
T-2 toxin	62	81
Nivalenol	62	70
Fusarenone	78	70
Trichodermol		70, 73
Verrucarin A	62, 73	70
Trichothecolone		73
Crotocin		73
Trichothecin		73
Scirpentriol	73	
Calonectrin	73	

[a]Chain elongation and termination catalyzed by peptidyltransferase activity.

peptidyltransferase activity associated with chain elongation and chain termination or as inhibitors of peptide chain initiation. Unfortunately, some of the compounds listed are reported to have more than one site of action. These discrepancies have not been reconciled.

Trichodermin has been investigated more than any of the other trichothecenes and there is general agreement that this compound is an inhibitor or peptidyltransferase catalyzed peptide bond formation (chain elongation) and peptidyl hydrolysis (chain termination), see Table 2. Trichodermin inhibits protein synthesis in yeast [79] and mammalian cells [55, 62, 73] without affecting polysomes in treated cells. Trichodermin also prevents puromycin-induced polysome dissociation and release of nascent peptides from ribosomes [62]. [Acetyl-^{14}C]trichodermin binds to eukaryotic ribosomes, but not to ribosomes from E. coli [80]. A single ribosomal site interaction was observed. Anisomycin completely prevents the binding of trichodermin and, therefore, apparently shares the same binding site [66]. Other members of the trichothecenes (trichodermol, trichothecin, and fusarenone) also completely inhibit the binding of labeled trichodermin to the ribosomes.

Inhibition of peptidyltransferase activity associated with chain elongation was demonstrated in a number of assay systems. Trichodermin inhibits the reaction of puromycin with ribosome bound fmet-tRNA$_f$ (E. coli) [55, 70, 73, 81] and aminoacyl oligonucleotide fragments of tRNA [70]. Trichodermin partially inhibits substrate binding of either aminoacyl or acetyl aminoacyl oligonucleotide fragments of tRNA to the A- and P-sites, respectively, of the peptidyltransferase center. Trichodermin does not affect either codon-anticodin recognition EF_1 catalyzed GTP binding of aminoacyl-tRNA to ribosomes [55, 70]. EF_2 dependent GTPase activity and binding of EF_2 to the ribosome [55], or EF_2-GTP dependent translocation of ribosome bound acetyl phe-tRNA from the A- to P-site, as determined by the puromycin reaction [70].

Peptidyltransferase catalyzed peptide chain termination, as measured by hydrolysis of ribosome bound fmet-tRNA$_f$ (E. coli), is inhibited by trichodermin [55, 73, 81]. Release factor (Rf) dependent hydrolysis of formylmethionine from ribosome complexes is prevented where GTP and codon recognition is not required (see Table 1) [81]. Trichodermin affects neither formation of an Rf-UAA ribosome complex nor ribosome dependent GTPase activity of Rf [81].

Trichodermin also inhibits chain termination as determined by reformation of polysomes from heat or chemically induced monomers [55]. In this assay system the antibiotic has no effect on the initial rate of reformation. This contrasts with cycloheximide which partially inhibits the rate of polysome reformation.

Nivalenol, T-2 toxin [62], verrucarin A [62, 73], and fusarenone [78] were reported to cause rapid and almost quantitative breakdown of polysomes in yeast spheroplasts and mammalian cells. Polysome breakdown could be prevented by cycloheximide, anisomycin, and trichodermin. This strongly suggests that these trichothecenes are inhibitors of chain initiation. However, using the chain elongation and termination assays (the reaction of puromycin with ribosome bound aminoacyl-tRNA and hydrolysis of fmet-tRNA) T-2 toxin [81], fusarenone, nivalenol, and verrucarin A [70] were reported to be inhibitors of peptidyltransferase activity.

Wei and McLaughlin [73] have measured the activity of various trichothecenes in a series of assay systems to determine whether the compounds are inhibitors of chain initiation or elongation and termination. Trichodermin, trichodermol, crotocin (R_2 substitutions), and trichothecolone and trichothecin (R_2 substitutions and C-8 modification) are all inhibitors of chain elongation and termination. According to the percentage of polysomes still intact at 90% concentration which give inhibition of protein synthesis, these simple esters of trichodermol may be considered inhibitors of termination [55]. Introduction of substitutions at the R_1, R_2, and R_3 position (scirpentriol, $R_1R_1R_3$ = OH; calonectrin, R_1 = OAC; R_2 = H; R_3 = OAC; verrucarins, R_1 = H; R_2 and R_3 = dicarboxylic acid ester) change the mechanism of action of these antibiotics to inhibitors of the initiation step in protein synthesis. Calonectrin was shown to inhibit initiation, while 15-desacetylcalonectrin inhibited elongation-termination. Therefore acetylation of the R_3 hydroxyl group must change the mode of action.

There appear to be two sites on the trichothecene molecule which determine whether a compound is an inhibitor of chain initiation. In this category are compounds with the diester groups between C-15 and C-4 and compounds with substituents on the C-3.

The trichothecenes have similar structures and a single ribosomal binding site [80], yet inhibit two distinct steps in protein synthesis (chain initiation and peptidyltransferase activity). Wei and McLaughlin [73] have suggested two models to explain these differences in inhibition: (a) all trichothecenes bind to the same ribosomal site and inhibition of initiation or elongation-termination is dependent upon how the various substituent groups interfere with the reactions of the appropriate initiation or elongation-termination factors; and (b) the trichothecenes interfere specifically with termination or initiation because they are able to bind to the ribosome only at that particular point in the ribosome-polysome cycle which involves the particular process.

D. Blasticidin S

1. Biology and Chemistry

Blasticidin S (8) is a member of the 2-aminohexose pyrimidine nucleosides. Also included in this group are gougerotin (9), amicetin, and bamicetin. Of all these antibiotics, only blasticidin S has antifungal activity. All are effective antibacterial agents.

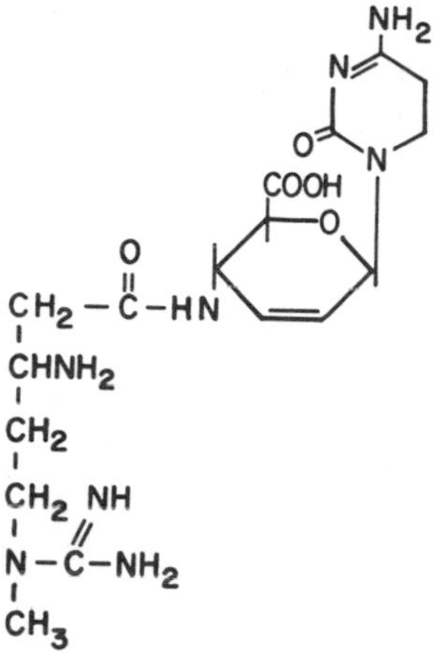

(8) blasticidin S

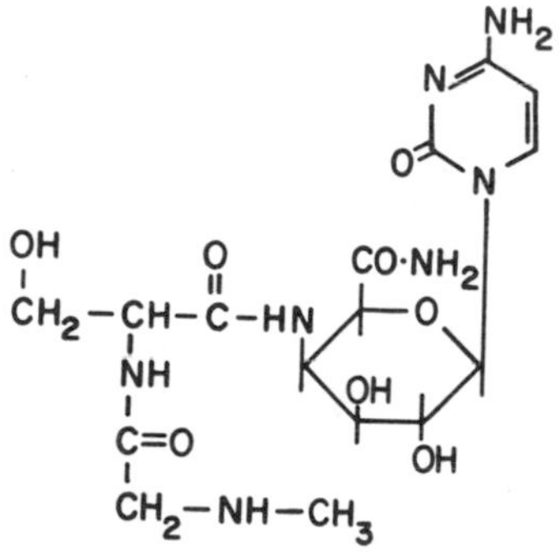

(9) gougerotin

Various aspects of the biology and chemistry of blasticidin S have been reviewed [20, 21, 82]. Blasticidin S was isolated from a soil actinomycete (Streptomyces griseochromogenes). Its chemical structure consists of a novel nucleoside (cytosinine) and the amino acid blastic acid. Total synthesis of the cytosinine portion of the antibiotic has been reported [83].

Blasticidin S inhibits many gram-negative and gram-positive bacteria. Its effect on fungi are variable and much less pronounced, except for its activity against Pyricularia oryzae which is highly sensitive. The compound can be phytotoxic on rice and other plants when repeatedly sprayed at concentrations of 5-20 μg/ml. Mammalian toxicity and antitumor activity have been reported.

Blasticidin S is a commercial systemic fungicide used in the control of rice blast disease caused by P. oryzae [21]. Phytotoxicity has been reduced in commercial preparations by the use of the monobenzyl aminobenzenesulfonate derivative of blasticidin S.

Two mechanisms of resistance to blasticidin S in fungi have been reported [82]. In P. sasokii, a naturally resistant species, the protein synthesizing system in whole cells and cell-free extracts is tolerant to the antibiotic. In this example, resistance is probably based on a genetic change at the reactive site of action (ribosome). On the other hand, a mutant of P. oryzae exhibits tolerance based on impermeability to the antibiotic. Protein synthesis in whole cells is unaffected while in a cell-free system it is highly sensitive.

Structure activity studies using derivatives and closely related compounds of blasticidin S indicate that removal of the aminoacyl portion from the cytosine moiety, substituting uracil for cytosine, and changing the aminoacyl group by substituting other amino acids, greatly reduces or destroys the in vivo and in vitro activity of the antibiotic [84, 86].

Based on similar modes of action and structure, Lichtenthaler and Trummlitz [85] suggested the term "aminoacyl-4-aminohexosylcytosine" be used to designate blasticidin S, gougerotin, amicetin, and bamicetin. Using Dreiding stereomodels, four major characteristic structural features, common to each, were found to be required for inhibition of protein synthesis by the antibiotics within this group: (a) The spatial arrangement of one oxygen and three nitrogen atoms is the same in the respective nucleobase portions; (b) an amide bond is found in the same distance from the first structural feature and within an identical spatial arrangement, representing the connecting link between the respective aminoacyl residues and the 4-aminohexuronic or 4-aminobenzoic acid moieties; (c) the carbonyl portion of the peptide link is part of a further structural and steric conformity within the antibiotics. The four antibiotics are capable of developing identical steric arrangement of a hydroxyl or an amino acid group in relation to the rest of the molecule; and (d) a terminal N-methylamino moiety on the aminoacyl portion of the molecule is present in the structures of these antibiotics.

While there are obvious similarities in structural characteristics, these antibiotics do differ in biological specificity and in effectiveness as inhibitors of various protein synthesizing systems. For example, ribosomes from eukaryotic cells appear to be more sensitive to blasticidin S than bacterial ribosomes, whereas bacterial ribosomes are more sensitive to gougerotin and amicetin [1].

It should also be noted that the 4-aminohexose pyrimidine nucleosides have some structural similarity to puromycin. Puromycin is also a substituted aminonucleoside. This antibiotic acts as a structural analogue of the aminoacyl-adenosine end of tRNA and takes part in the peptidyltransferase reactions by acting as an acceptor substrate [1, 3, 4, 7]. While the 4-aminohexose pyrimidine nucleosides have been shown to act on the peptidyltransferase center, their specific site of action differs from that of puromycin.

2. Inhibition of Protein Synthesis

Blasticidin S was reported to inhibit incorporation of amino acids into protein in extracts from P. oryzae [87] and E. coli [88]. The site of action was shown to occur subsequent to formation of aminoacyl-tRNA. Increasing concentrations of aminoacyl-tRNA do not reverse blasticidin S inhibition [89]. Inhibition of poly-U-directed polyphenylalanine synthesis also occurs in extracts from E. coli [89, 90] and yeast [43].

[^{14}C]Blasticidin S reacts with a single binding site on the 50-S ribosomal subunit [90]. Binding of blasticidin S is inhibited by gougerotin. Battaner and Vazquez [43], using "hybrid" reconstituted ribosomes, demonstrated that blasticidin S binds to the yeast 60-S ribosomal subunit. The binding of one molecule of [^{3}H]gougerotin, at saturation levels of the antibiotic, to yeast ribosomes is completely inhibited by blasticidin S [91]. It is apparent from these results that blasticidin S and gougerotin share and compete for the same single binding site on the larger ribosomal subunit.

Blasticidin S inhibits peptidyltransferase activity associated with chain elongation. This type of inhibition has been demonstrated in a number of assay systems. Blasticidin S inhibits the reaction of puromycin with ribosomal bound peptidyl-tRNA in extracts from E. coli and mammalian cells [68, 84, 92]. Both gougerotin and blasticidin S are mixed inhibitors of the puromycin reaction [58, 92]. At low concentrations they are competitive while at higher concentrations they are noncompetitive. This two-phase (heterogenicity) reactivity with the ribosome is similar to cycloheximide inhibition of peptidyl puromycin synthesis (Section B.2).

Blasticidin S inhibits the reaction of puromycin on ribosomes preincubated with fusadic acid (to inhibit translocation) [58]. Blasticidin S inhibits the reaction of puromycin with acetyl-phe-tRNA and CACCA-leu-acetyl bound on the P-site of yeast ribosomes [43]. The antibiotic does not inhibit binding of CACCA-leu-acetyl to the donor (P) site of the peptidyltransferase center [43]. Partial inhibition (30-48%) of binding of aminoacyl oligonucleotides to the acceptor (A) site was observed [43, 93]. Both gougerotin and blasticidin S inhibit the enhanced binding of substrates to the P-site which is induced by sparsomycin ("sparsomycin reaction").

If it is assumed that blasticidin S and gougerotin have similar mechanisms of action, then data for the latter antibiotic become pertinent to the discussion on the mode of action of blasticidin S. Gougerotin inhibits peptide chain elongation in bacterial and mammalian extracts without promoting polysome breakdown [94]. The puromycin reaction with ribosomal bound peptidyl-tRNA and acetyl aminoacyl oligonucleotide fragments of tRNA is inhibited. Binding of this substrate to the donor (P) site is enhanced by gougerotin [95-98]. On the other hand, the antibiotic inhibits the binding of aminoacyl oligonucleotide fragments to the acceptor (A) site [89, 93, 96]. Most of these reactions are quite similar to and supplement the data for the action of blasticidin S.

Based on similarities in structure, binding sites, and mode of action, the 4-aminohexose pyrimidine nucleosides are postulated to react with the peptidyltransferase center preventing the attachment of 3'-terminal CCA of aminoacyl-tRNA to the ribosomal acceptor (A) site.

Fewer data are available concerning the action of these antibiotics on peptidyltransferase activity associated with peptide chain termination. Gougerotin strongly inhibits GTP, Rf, and codon-directed hydrolysis of E. coli ribosome bound fmet-$tRNA_f$ [61, 68]. Inhibition of the hydrolysis reaction on reticulocyte ribosomes is less pronounced.

E. Kasugamycin

1. Biology and Chemistry

Kasugamycin is a water-soluble base isolated from Streptomyces kasugaensis [99]. The chemical structure (10) has been determined [100, 101] and consists of D-inositol linked to kasugamine. The antibiotic can be classified as an aminoglycoside (streptomycin group); however, the molecule lacks an inosamine moiety. Partial [102] and total [103] synthesis of kasugamycin has been reported.

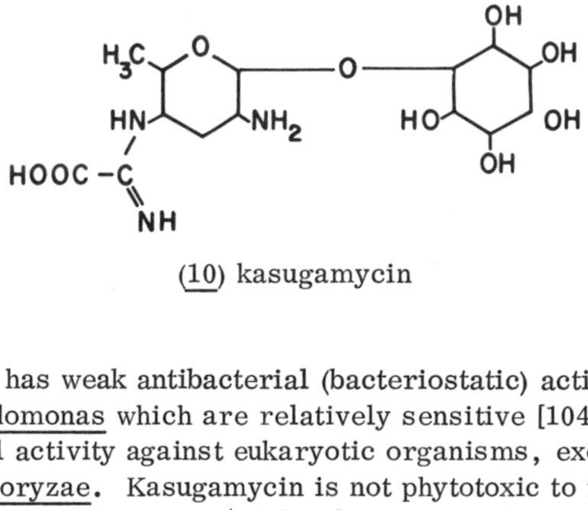

(10) kasugamycin

Kasugamycin has weak antibacterial (bacteriostatic) activity except for species of Pseudomonas which are relatively sensitive [104]. The antibiotic has no biological activity against eukaryotic organisms, except for the fungal pathogen P. oryzae. Kasugamycin is not phytotoxic to rice plants at concentrations as high as 300 μg/ml [105]. The compound exhibits no overt toxicity to animals [21].

Kasugamycin is a commercial fungicide exhibiting systemic activity and is used to control the rice blast pathogen [21]. A large margin exists between curative and phytotoxic dosage. For this reason the compound has an advantage, when used as a fungicide, over blasticidin S which is phytotoxic at lower concentrations [102]. An interesting aspect of the antifungal activity of the antibiotic is that its toxicity is expressed only in the plant or in the presence of juice from the plant [104, 105]. Spore germination, appressorium formation, and penetration are not inhibited on treated leaf surfaces.

In agar plate tests antifungal activity is expressed at pH values below 6 in the presence of plant juice. The activity was not observed in the presence of plant extract at pH 7.0 [104, 105]. Inhibition of protein synthesis does occur in a cell-free system from mycelium of P. oryzae grown in the absence of plant juice [106].

The rice plant extracts may exert an influence on the organism by changing or inducing changes in cell-membrane permeability. Induced changes in potential cellular detoxification processes or the structure of kasugamycin, which would affect uptake in P. oryzae, cannot be ruled out.

Mutants of P. oryzae resistant to kasugamycin have been readily produced in the laboratory [107, 108] and have been reported to occur in the field [108].

2. Inhibition of Protein Synthesis

The mode of action of kasugamycin has been studied primarily in bacterial cell-free systems. There is only one report of its action in fungal cells.

The antibiotic, at a concentration of 1.5×10^{-3} M, inhibits 63% of the poly-U-directed polyphenylalanine synthesis on ribosomes from P. oryzae [106]. Inhibition (43%) of phe-tRNA binding to poly-U and ribosomes was also demonstrated. Kasugamycin did not cause miscoding in this system as did other aminoglycoside antibiotics (e.g., streptomycin). The antibiotic was postulated to inhibit binding of aminoacyl-tRNA to ribosomes.

In bacterial cell-free systems, kasugamycin inhibits protein synthesis by reacting with the 30-S ribosomal subunit [109-113]. The antibiotic does not cause miscoding in bacterial cells [111, 114]. Localization of the site of action was determined with "hybrid" reconstituted ribosomes, constructed by exchanging 30-S and 50-S subunits, from kasugamycin resistant and sensitive strains of E. coli [112]. Only 70-S ribosomes containing 30-S subunits from a resistant strain were resistant to the action of the antibiotic. All the mutant strains were sensitive to streptomycin.

The 30-S ribosomal subunits from a resistant strain were found to be deficient in methylation of the 16-S RNA [110]. Methylation of the 16-S RNA from resistant strains followed by reconstitution into 30-S subunits converted the subunit into a kasugamycin-sensitive particle [115].

Binding of fmet-$tRNA_f$ to 30-S ribosomal subunits directed by poly-AUG or exogenous mRNA is inhibited by 2×10^{-5} M kasugamycin, but not by other aminoglycoside antibiotics [109]. The antibiotic also prevents 70-S complex formation as measured by the inhibition of binding of fmet-$tRNA_f$ to ribosomes in the presence of poly-AUG and exogenous mRNA [109].

Inhibition of poly-U-directed binding of phe-tRNA to 70-S ribosomes also occurs, but only at concentrations higher than that required for the inhibition of fmet-$tRNA_f$ binding to 30-S ribosomal subunits [109, 111]. The use of higher concentrations to inhibit poly-U-phe-tRNA activity on ribosomes produces results similar to those reported for the fungal cell-free system.

Peptide chain initiation is more sensitive than chain elongation, as indicated by the fact that protein synthesis reactions involving polysomes and endogenous mRNA are relatively unaffected by a broad range of concentrations (60-1,000 μM) of kasugamycin. The exogenous mRNA and 30-S subunit or ribosomal system is much more sensitive to the antibiotic [113]. Kasugamycin does not inhibit chain elongation (peptide bond formation or translocation) as measured by the reaction of puromycin on 70-S and 50-S subunits with peptidyl-tRNA [92, 116].

F. Miscellaneous Inhibitors

The inhibitors to be discussed in this section can be divided into three categories: (a) dicloran which exhibits antifungal activity and is postulated to be an inhibitor of protein synthesis in whole cells; (b) canavanine, tetracycline, streptomycin, and chloramphenicol which exhibit antifungal activity against selective fungal species and are known inhibitors of either cytoplasmic and mitochondrial or only mitochondrial protein synthesis; and (c) toxicants which exhibit no antifungal activity but are inhibitors of protein synthesis in fungal cell-free systems.

1. Dicloran

Dicloran (2,6-dichloro-4-nitroaniline) is a synthetic organic fungicide used to control diseases caused by species of *Botrytis*, *Monilinia*, *Rhizopus*, *Sclerotinia*, and *Sclerotium* [117]. It has been reported to inhibit protein synthesis in *Rhizopus arrhizus* [118] and to be toxic to cells of *Nicotiana tabacum* grown in tissue culture [119]. Inhibition of protein synthesis in cell-free systems has not been reported for this compound.

Dicloran at 2 μg/ml inhibits incorporation of [^{14}C]leucine into protein by 36%, but has no effect on the synthesis of DNA and RNA in *R. arrhizus*. In tobacco cells grown in tissue culture medium containing 10 μg/ml dicloran, only slight decreases (11.8%) in protein content were reported.

As was pointed out by Sisler [5] in an earlier review, the data do not support a definite assignment of the toxic action of dicloran to a site in the pathway of protein synthesis. A 36% reduction in the incorporation of leucine into protein would seem to be too low to account for the toxicity of the fungicide.

Experiments with *Aspergillus nidulans* suggest that dicloran has a nuclear site of action, affecting hereditary processes by inducing somatic segregation [174].

2. Canavanine

Canavanine, a naturally occurring amino acid structurally similar to arginine, is distributed primarily in the Lotoreae, a subfamily of the Leguminosae [120, 121]. Many leguminous seeds contain large quantities (5% dry weight) of canavanine [122].

Canavanine has potent antimetabolic properties against a wide range of prokaryotes and eukaryotes. Inhibitory effects with species of *Neurospora* [123], *Aspergillus* [124], and yeast [125-127] have been demonstrated. For example, growth of *Neurospora* was inhibited by 10^{-6} M canavanine. Inhibition was reversed by arginine. Similar results have been reported with bacteria [128].

While a variety of effects by canavanine on the metabolism of DNA, RNA, and protein can occur in numerous organisms, canavanine toxicity apparently involves protein synthesis and protein metabolism [129, 175]. Loftfield [8] pointed out that homologues of amino acids that catalyze aminoacyl adenylate formation either turn out to be competitive inhibitors of the corresponding tRNA esterification or successfully "fool" the aminoacyl synthetase by becoming attached to the tRNA and being incorporated into protein. Canavanine, as a homologue for arginine, is a competitive inhibitor of arginyl-tRNA synthetase isolated from bacteria [130,131] and Jack bean [132]. Canavanine also attaches to Arg-tRNA and is incorporated into protein in bacteria [130, 131, 133, 134].

The substitution of canavanine for arginine to form abnormal protein in bacteria has been demonstrated for alkaline phosphatase [134]. Only 25% of the enzyme, isolated from canavanine-treated bacteria, was in the active dimer.

Bacterial mutants, which contain altered arginyl-tRNA synthetase and are resistant to the inhibitory effects of canavanine, have been isolated [130, 133]. The altered synthetase has a diminished ability to activate and attach canavanine to the arg-tRNA. This results in a decreased incorporation of canavanine into protein and hence less functionally altered protein is synthesized.

While the mode of action of canavanine has been determined primarily in bacteria, it appears that the toxic mechanisms may be the same in fungi. Wilkie [127] has demonstrated that canavanine inhibits the synthesis of proteins in both cytoplasm and mitochondria of yeast. Mitochondrial protein synthesis was 100 times more sensitive to canavanine than that occurring in the cytoplasm. The arginyl-tRNA synthetases were not isolated in this system.

3. Tetracycline

The tetracyclines are classified as inhibitors of protein synthesis in both prokaryotes and eukaryotes [2]. Antifungal activity of the tetracycline group has been reported for species of Phytophthora and Pythium and other miscellaneous fungi [135]. In fungi, tetracycline is an inhibitor of cytoplasmic [43] and mitochondrial protein synthesis [5, 10, 136].

Tetracycline has been studied most intensively in bacterial cell-free systems. While the antibiotic binds preferentially to the 30-S ribosomal subunits (one molecule per subunit particle), some binding to the 50-S subunit also occurs [137, 138].

Tetracycline inhibits nonenzymatic binding of aminoacyl-tRNA to the ribosomal acceptor (A) site [139-141]. The antibiotic inhibits peptide chain termination as measured by release (hydrolysis) of ribosomal bound fmet [142-144]. Inhibition occurs by preventing codon-dependent binding of the release factors to the ribosomal A-site.

Tetracycline has two effects in whole bacterial cells. At a concentration of 2×10^{-4} M polysome breakdown occurs without substantial inhibition of binding of aminoacyl-tRNA, while at 8×10^{-4} M polysomes are preserved and binding of aminoacyl-tRNA is inhibited [145, 146].

High concentrations of tetracycline have been reported to inhibit fungal growth [135]. However, Hendrix [147] has observed that concentrations as low as 10^{-5}-10^{-6} M completely inhibit growth of species of Phytophthora and Pythium.

Poly-U-directed polyphenylalanine synthesis on ribosomes from S. cerevisiae was inhibited 50% by 10^{-4}-10^{-5} M tetracycline [43]. In this study, tetracycline was shown to inhibit binding of ac-phe-tRNA and phe-tRNA on 40-S ribosomal subunits to the P- and A-sites, respectively. Inhibition of binding was greatest at the ribosomal A-site. Tetracycline effectively inhibits the reaction of puromycin with ribosomal bound ac-phe-tRNA. The inhibition of transpeptidation (peptidyltransferase activity) also occurs on bacterial ribosomes [2]. However, it appears that inhibition of transpeptidation occurs only at higher tetracycline concentrations ($>10^{-4}$ M) and depends on the conditions of the assay system. The most sensitive ribosomal site in both prokaryotes and eukaryotes appears to be the aminoacyl acceptor (A) site.

4. Streptomycin

Streptomycin has been considered to be an inhibitor of protein synthesis only in prokaryotes [1]. However, antifungal activity of this antibiotic has been reported for species of Oomycetes [6, 135] and yeasts [135]. Sensitivity to streptomycin may be related to increased uptake. Pythium and Phytophthora species absorb 2 to 9 times as much antibiotic as insensitive Mucorales species and antibiotic tolerant strains of Aphanomyces [148].

There is no indication that streptomycin can inhibit cytoplasmic protein synthesis in fungi or other eukaryotes. Inhibition of protein synthesis in mitochondria is probably the site of action of this antibiotic [10, 149].

Streptomycin is one of the most intensively studied antibiotics in bacterial cell-free systems. One molecule of [^{3}H]dihydrostreptomycin binds per 30-S subunit [150]. No binding to the 50-S subunit occurs. Ribisomal 30-S subunits from resistant bacteria do not bind streptomycin. A single ribosomal 30-S subunit protein has been identified (S-12) as its binding site [151, 152]. Ribosomes from resistant bacteria contain altered S-12 protein. This protein does not bind streptomycin by itself, but requires the completed 30-S subunit.

It has been demonstrated, by numerous researchers, that streptomycin exhibits pleiotropic effects on protein biosynthesis. It inhibits peptide chain elongation by interfering with the binding of aminoacyl-tRNA to the ribosomal A-site [153]. Streptomycin apparently halts peptidyl-tRNA in the P-site, causing a distortion in both the P- and A-sites which in turn affects efficient binding of the aminoacyl-tRNA [154, 155]. Streptomycin causes the slow release of peptidyl-tRNA from the ribosomes with formation of monosomes [154]. These streptomycin-treated monosomes are irreversibly inactivated and cannot form new initiation complexes in whole cells or cell-free extracts [155, 156]. As would be expected with an inhibitor which acts on the ribosomal A-site, streptomycin blocks the release of peptides or fmet during chain termination [142].

Streptomycin also causes misreading of the genetic message which results in miscoded amino acids being incorporated into protein. This phenomenon has been reported in both whole cells and cell-free systems and apparently occurs at the 70-S ribosomal A-site which is involved in codon-anticodon interactions [1, 3].

While both tetracycline and streptomycin bind to the acceptor site, the ribosomal proteins involved in binding appear to be different since streptomycin-resistant mutants are still sensitive to tetracycline [158].

5. Chloramphenicol

Chloramphenicol is a potent inhibitor of protein synthesis in prokaryotes [1]. In eukaryotes inhibition of cytoplasmic protein synthesis does not occur. Instead, protein synthesis in mitochondria has been found to be sensitive to the antibiotic [10, 136, 149].

The spectrum of antifungal activity of chloramphenicol is very similar to that of tetracycline and streptomycin [135]. However, species of *Phytophthora* and *Pythium* are generally less sensitive to chloramphenicol than to tetracycline [147].

Chloramphenicol is an inhibitor of peptide chain elongation and transpeptidation (peptidyltransferase activity). The antibiotic inhibits the reaction of puromycin with a variety of ribosomal bound tRNA substrates [90, 159]. It has also been reported that the antibiotic inhibits the nonenzymatic attachment of aminoacyl oligonucleotides to the ribosomal A-site, but not of acetyl aminoacyl oligonucleotides to the P-site [95, 160].

As was previously discussed for the antibiotic cycloheximide, a two-phase inhibition (competitive and noncompetitive) of chloramphenicol activity by puromycin exists [92]. This suggests that peptidyl-tRNA exists in two states on the ribosome with respect to its reaction with chloramphenicol and its reaction with puromycin. Two ribosomal binding sites for chloramphenicol have been reported [160]. This may not be at variance with the report of only a single 50-S ribosomal subunit binding site [161], since a different range of concentrations of the antibiotic was used in each study.

Coupled with chloramphenicol inhibition of protein synthesis is the decreased rate of polysome breakdown and subunit exchange [162, 163]. Ribosomes can move slowly along the mRNA even though peptide bond formation is inhibited by the antibiotic.

Resistance to chloramphenicol in bacteria is based on the formation of an enzyme which causes activation by catalyzing acetylation of the antibiotic [164]. Acetylation of chloramphenicol may also be an effective detoxification mechanism in eukaryotes.

6. Inhibitors of Protein Synthesis in Cell-Free Systems

While there are only a limited number of antifungal compounds which are inhibitors of both growth and protein synthesis in whole cells, there are a number of toxicants which do not generally exhibit antifungal activity but are nevertheless effective inhibitors of protein synthesis in cell-free systems [43, 165].

The primary reason for the lack of antifungal activity is probably based on the impermeability of cells to these compounds. For example, mutants of S. cerevisiae have been isolated which have become sensitive to emetine and fusidic acid because of increased permeability [26]. However, detoxification mechanism in whole cells may also be a basis for a lack of in vivo activity of some of the inhibitors.

TABLE 3

Effect of Inhibitors on the Processes of Protein Synthesis in Cell-Free Systems from Fungi

Process	Inhibitor	Component affected
Initiation	Aurintricarboxylic acid	40-S Subunit
	Pactamycin	40-S Subunit
Elongation		
aa-tRNA binding	Edeine	40-S Subunit
Translocation	Fusidic acid	Ef_2, 60-S Subunit
	Tylophora alkaloids[a]	60-S Subunit
	Emetine	80-S, mRNA
Transpeptidation (peptidyltransferase)	Sparsomycin	60-S Subunit
Termination	Puromycin[b]	60-S Subunit
(peptidyltransferase)	Sparsomycin	60-S Subunit

[a]Cryptopleurine, tylophorine, tylocrebine.

[b]Premature peptide release and polysome breakdown.

The inhibitors listed in Table 3 are those which have been reported to significantly inhibit (over 50%) cytoplasmic protein synthesis in fungal cell-free systems at concentrations of 10^{-4}-10^{-7} M. Many of these compounds have been classified as inhibitors of protein synthesis in eukaryotes or in both prokaryotes and eukaryotes. Thus, most of these inhibitors permeate to eukaryotic cells other than fungi.

For additional information on the specific site of action of the inhibitors discussed in this section see Refs. 1-4, 7, 9, 43, and 165.

IV. INHIBITORS OF MITOCHONDRIAL PROTEIN SYNTHESIS

The biogenesis of mitochondria in eukaryotes involves a complex series of biosynthetic processes, including the synthesis of proteins, lipids, nucleic acids, and small molecules. It is beyond the scope of this review to discuss in detail the biogenesis of mitochondria or the effects of inhibitors on mitochondrial protein synthesis. Recent reviews on these subjects are available [10, 136, 149, 166-168].

It is clear that mitochondria are formed and synthesize specific proteins in close cooperation with nuclear DNA and the cytoplasmic protein-synthesizing system. Yet, the cytoplasmic processes are quite distinct from those in the mitochondria. The two systems are physically separated, their DNA usually differ in base composition, and the mRNA translated and protein synthesized in one system are not duplicated in the other.

Because the mitochondrial DNA is relatively small, the bulk of the proteins, including ribosomal proteins, are coded and synthesized in the cytoplasm and transported into the mitochondria. Using cytoplasmic "petite" mutants of *S. cerevisiae* a number of authors [10, 168] have concluded that respiration deficient mitochondria lack cytochromes aa_3, b, and c_1, as well as an energy transfer system. The mutants have lost portions or all of their mitochondrial DNA. These observations imply that the ribosomes in the cytoplasm produce most of the mitochondrial proteins. In addition, there are data which suggest that certain of the cytochrome oxidases and peroxidases are synthesized in the cytoplasm but are not present in "petite" mutants because the corresponding protein components were not synthesized in the mitochondria. Close cooperation between the cytoplasm and mitochondria is therefore evident.

Inhibitors are one of the major tools used to delineate mitochondrial protein synthesis. Species of yeast and *Neurospora* have been used extensively in studies involving inhibitors of cytoplasmic protein synthesis (cycloheximide and anisomycin) and mitochondrial protein synthesis (chloramphenicol, erythromycin, lincomycin, and other antibacterial antibiotics). The results of these studies are similar to those involving cytoplasmic "petite" mutants and also indicate that the mechanisms of protein synthesis in mitochondria approximates that found in prokaryotes [10, 136, 149, 166-168].

The existence of a phylogenetic difference in sensitivity to various antibiotics of the mitochondrial protein-synthesizing system in eukaryotes has been demonstrated [169, 170]. Ribosomes isolated from mammalian, plant, and fungal (species of Neurospora and yeast) mitochondria have sedimentation coefficients of 55 S, 66 S, and 73-80 S, respectively [168-171]. It has been suggested that the differences in antibiotic sensitivity are due to changes in the mitochondrial ribosome which occurred during evolution. This evolutionary change in the 55-S ribosome involves the loss of reactivity with certain antibiotics which effectively inhibit protein synthesis on yeast mitochondrial ribosomes [170]. On the other hand, the lack of antibiotic sensitivity has also been attributed [169] to changes in the permeability barrier in the mammalian mitochondria membrane (see Ref. 168 for a resume of the comparison of the characteristics of mitochondria and bacteria).

It is apparent that species of yeast, Neurospora, and other fungi have mitochondrial protein-synthesizing systems sensitive to a large number of inhibitors whose modes of action are similar to those found in prokaryotes. However, the inhibition of mitochondrial protein synthesis is not generally associated with fungitoxicity. There is one group of fungi (Oomycetes) whose growth and synthesis of mitochondrial proteins are inhibited by chloramphenicol and other antibacterial antibiotics. The reasons why toxicity to certain antibiotics occurs in this group of fungi are not known, although several hypotheses have been suggested.

Since certain species of yeast and Neurospora can grow as facultative anaerobes (using the glycolytic pathway for a source of energy), it might be expected that inhibition of mitochondrial protein synthesis would not result in toxicity to these organisms [5, 10, 168]. Conversely, species of Pythium and Phytophthora are not generally considered to be facultative anaerobes and the antibiotics should not only inhibit mitochondrial protein synthesis but also growth. While this hypothesis for differential toxicity of antibacterial inhibitors in fungi is attractive, it is not the only explanation.

Yu and Steward [172] have recently reported that species of Candida, which are "petite" negative and are obligate aerobes, can survive the toxic action of chloramphenicol and erythromycin. Inhibition of mitochondrial protein synthesis occurs only in logarithmic-phase cells and not in stationary-phase or slow-growing cells. This phasic response of cells is due neither to changes in permeability to the inhibitor nor to changes in the sensitivity of the mitochondrial protein synthesizing system. The phasic response to the inhibitors is apparently related to the loss of regulatory coupling between mitochondrial and cytoplasmic protein-synthesizing systems involved in the synthesis of mitochondrial cytochromes.

Lastly, Hendrix [173] has suggested that the Pythiaceae could contain intracellular, symbiotic mycoplasma or L-phase bacteria which are sensitive to antibacterial inhibitors. This would imply that while the inhibitors affect mitochondrial protein synthesis they are toxic because they also kill the symbiants.

A number of hypotheses have been suggested to explain the differential sensitivity of certain species of fungi to antibacterial inhibitors of mitochondrial protein synthesis. Further research should help resolve this phenomenon.

V. SUMMARY AND CONCLUSION

There are only a limited number of compounds which are inhibitors of both growth and protein synthesis in fungal cells. Inhibition of the cytoplasmic protein synthesizing system in fungi results in cellular death. Certain inhibitors of protein synthesis in prokaryotes are effective against the mitochondrial protein synthesizing system found in fungi and other eukaryotes. However, inhibition of mitochondrial protein synthesis does not necessarily result in inhibition of growth. The ability of a fungal organism to exist as a facultative anaerobe may be of considerable importance in surviving the action of inhibitors of mitochondrial protein synthesis.

The antifungal compounds discussed in this chapter were found to effectively inhibit one or more of the four major steps in protein synthesis (Table 4).

TABLE 4

Effect of Antifungal Inhibitors on the Processes of Protein Synthesis

Process	Inhibitor	Component affected
Formation of aa-tRNA	Canavanine	Arginyl-tRNA synthetase
Initiation	Kasugamycin	40-S subunit
	Trichothecenes[a]	40-S subunit
Elongation		
aa-tRNA binding	Tetracycline	40-S subunit and mt ribosome[b]
	Streptomycin	mt ribosome[b]
Transpeptidation (peptidyltransferase)	Glutarimides	60-S subunit
	Anisomycin	60-S subunit
	Trichothecenes[c]	60-S subunit
	Blasticidin S	60-S subunit
	Chloramphenicol	mt ribosome[d]
Translocation	Glutarimides	60-S subunit
Termination	Streptomycin	(see elongation)
	Tetracycline	(see elongation)
(peptidyltransferase)	Blasticidin S	60-S subunit
	Anisomycin	60-S subunit
	Trichothecenes[c]	60-S subunit

[a]Verrucarin A, scirpentriol, calonectrin.
[b]Inhibitor reacts with smaller ribosomal subunit from mitochondria (mt).
[c]Trichodermin, trichodermol, trichothecalone, crotocin, trichothecin.
[d]Inhibitor reacts with larger ribosomal subunit from mitochondria.

Because of their selective action, resistance to these inhibitors is, for the most part, based on genetic changes at the ribosomal sites of action. However, even with the existence of resistant mutants and data from structure-activity relationship studies, the molecular basis for the action of many of the inhibitors remains somewhat obscure.

It is possible that studies further elucidating the structure and function of the ribosome may eventually lead to a better understanding of the action of selective inhibitors of protein synthesis in fungi and other organisms.

REFERENCES

1. S. Pestka, Ann. Rev. Microbiol., 25, 487 (1971).
2. S. Pestka, in Methods in Enzymology (K. Moldave and L. Grossman, eds.), Vol. 30, Academic Press, New York, 1974, pp. 261-289.
3. A. Kaji, in Progress in Molecular and Subcellular Biology (F. E. Hahn, ed.), Vol. 3, Springer-Verlag, New York, 1973, pp. 85-143.
4. C. T. Caskey, in Metabolic Inhibitors (R. M. Hochster, M. Kates, and J. H. Quastel, eds.), Vol. IV, Academic Press, New York, 1973, pp. 131-177.
5. H. D. Sisler, Ann. Rev. Phytopathol., 7, 311 (1969).
6. D. Gottlieb and P. D. Shaw, Ann. Rev. Phytopathol., 8, 371 (1970).
7. E. F. Gale, E. Cundliffe, P. E. Reynolds, M. H. Richmond, and M. J. Waung, The Molecular Basis of Antibiotic Action, Wiley, New York, 1972, p. 278.
8. R. B. Loftfield, in Metabolic Inhibitors (R. M. Hochster, M. Kates, and J. H. Quastel, eds.), Vol. IV, Academic Press, New York, 1973, pp. 107-130.
9. D. Vazquez, Basic Life Sci., 1, 339 (1974).
10. G. Schatz and T. L. Mason, Ann. Rev. Biochem., 43, 51 (1974).
11. M. Nomura, A. Tissieres, and P. Lengyel (eds.), Ribosomes, Cold Spring Harbor Monograph Series, Cold Spring Harbor, New York, 1974, p. 930.
12. S. Ochoa and R. Mazurnder, in The Enzymes (P. D. Boyer, ed.), Vol. 10, Academic Press, New York, 1974, pp. 1-52.
13. J. Lucas-Lenard and L. Beres, in The Enzymes (P. D. Boyer, ed.), Vol. 10, Academic Press, New York, 1974, pp. 53-86.
14. W. P. Tate and C. T. Caskey, in The Enzymes (P. D. Boyer, ed.), Vol. 10, Academic Press, New York, 1974, pp. 86-118.
15. J. Lucas-Lenard and F. Lipmann, Ann. Rev. Biochem., 40, 409 (1971).
16. C. G. Kurland, Ann. Rev. Biochem., 41, 377 (1972).
17. R. Haselkorn and L. B. Rothman-Denes, Ann. Rev. Biochem., 42, 397 (1973).
18. D. Soll and P. R. Schimmell, in The Enzymes (P. D. Boyer, ed.), Vol. 10, Academic Press, New York, 1974, pp. 489-538.

19. H. D. Sisler and M. R. Siegel, in Antibiotics (D. Gottlieb and P. D. Shaw, eds.), Vol. I, Springer-Verlag, New York, 1967, pp. 283-307.
20. T. Korzybski, Z. Zowszyk-Gindiffer, and W. Zurylowicz, Antibiotics, Origin, Nature, and Properties, Vol. I, Pergamon Press, New York, 1967.
21. J. Dekker, World Rev. Pest Control, 10, 9 (1971).
22. M. R. Siegel, H. D. Sisler, and F. Johnson, Biochem. Pharmacol., 15, 1213 (1966).
23. H. L. Ennis, Biochem. Pharmacol., 17, 1197 (1968).
24. F. Johnson, N. A. Starkovsky, and W. D. Gurowitz, J. Amer. Chem. Soc., 87, 3492 (1965).
25. H. D. Sisler, M. R. Siegel, and N. N. Ragsdale, Phytopathology, 57, 1191 (1967).
26. A. Jimenez, B. Littlewood, and J. Davies, in Molecular Mechanisms of Antibiotic Action on Protein Biosynthesis and Membranes (E. Munoz, F. Garcia-Ferrandiz, and D. Vazquez, eds.), Elsevier Scientific Pub. Co., New York, 1972, pp. 292-306.
27. S. Roa and A. P. Grollman, Biochem. Biophys. Res. Commun., 29, 696 (1967).
28. B. W. Coursen and H. D. Sisler, Amer. J. Bot., 47, 541 (1960).
29. B. Colombo, L. Felicetti, and C. Baglioni, Biochim. Biophys. Acta, 119, 109 (1966).
30. H. L. Ennis and M. Lubin, Science, 146, 1474 (1964).
31. D. Kerridge, J. Gen. Microbiol., 19, 497 (1958).
32. M. R. Siegel and H. D. Sisler, Nature, 200, 675 (1963).
33. A. J. Garber, M. Jomain-Baum, L. Salganicoff, E. Farber, and R. W. Hanson, J. Biol. Chem., 248, 1530 (1973).
34. W. R. Evans, J. Biol. Chem., 246, 6144 (1971).
35. W. E. Timberlake and D. H. Griffin, Biochim. Biophys. Acta, 349, 39 (1974).
36. P. A. Horgen and D. H. Griffin, Proc. Natl. Acad. Sci., U.S., 68, 338 (1971).
37. M. R. Siegel and H. D. Sisler, Biochim. Biophys. Acta, 87, 83 (1964).
38. I. Widuczynski and A. O. M. Stoppani, Biochim. Biophys. Acta, 104, 413 (1965).
39. L. L. Bennett, Jr., V. L. Ward, and R. W. Brockman, Biochim. Biophys. Acta, 103, 478 (1965).
40. M. Pongratz and W. Klingmuller, Mol. Gen. Genet., 124, 359 (1973).
41. M. R. Siegel and H. D. Sisler, Biochim. Biophys. Acta, 103, 558 (1965).
42. D. Cooper, D. V. Banthorpe, and D. Wilkie, J. Mol. Biol., 26, 347 (1967).
43. E. Battaner and D. Vazquez, Biochim. Biophys. Acta, 254, 316 (1971).
44. F. O. Wettstein, H. Noll, and S. Penman, Biochem. Biophys. Acta, 87, 525 (1964).

45. A. R. Williamson and R. Schweet, J. Mol. Biol., 11, 358 (1965).
46. W. Godchaux, S. D. Adamson, and E. Herbert, J. Mol. Biol., 27, 57 (1967).
47. B. Colombo, L. Felicetti, and C. Baglioni, Biochem. Biophys. Res. Commun., 18, 389 (1965).
48. C. P. Stanners, Biochem. Biophys. Res. Commun., 24, 758 (1966).
49. L. Felicetti, B. Colombo, and C. Baglioni, Biochim. Biophys. Acta, 119, 120 (1966).
50. A. C. Trakatellis, M. Montjar, and A. E. Axelrod, Biochemistry, 4, 2065 (1965).
51. S. Y. Lin, R. D. Mosteller, and B. Hardesty, J. Mol. Biol., 21, 51 (1966).
52. W. McKeehan and B. Hardesty, Biochem. Biophys. Res. Commun., 36, 625 (1969).
53. T. G. Obrig, W. Culp, W. McKeehan, and B. Hardesty, J. Biol. Chem., 246, 174 (1971).
54. B. S. Baliga, A. W. Pronczuk, and H. N. Munro, J. Biol. Chem., 244, 4480 (1969).
55. C. M. Wei, B. S. Hansen, M. H. Vaughan, Jr., and C. S. McLaughlin, Proc. Natl. Acad. Sci. U.S., 71, 713 (1974).
56. H. F. Lodish, D. Houseman, and M. Jacobson, Biochemistry, 10, 2348 (1971).
57. B. S. Baliga, S. A. Cohen, H. N. Munro, FEBS Lett., 8, 249 (1970).
58. S. Pestka, H. Rosenfeld, R. Harris, and H. Hintikka, J. Biol. Chem., 247, 6895 (1972).
59. J. A. Schneider and E. Maxwell, Biochemistry, 12, 475 (1973).
60. R. Neth, R. E. Monro, G. Heller, and E. Vazquez, FEBS Lett., 6, 198 (1970).
61. A. L. Beaudet and C. T. Caskey, Proc. Natl. Acad. Sci. U.S., 68, 619 (1971).
62. E. Cundliffe, M. Cannon, and J. Davis, Proc. Natl. Acad. Sci., U.S., 71, 30 (1974).
63. A. P. Grollman, J. Biol. Chem., 242, 3226 (1967).
64. J. E. Lynch, A. R. English, H. Bauch, and H. Deligianis, Antibiot. Chemotherap., 4, 844 (1954).
65. J. J. Beereboom, K. Butler, F. C. Pennington, and I. A. Solomons, J. Org. Chem., 30, 2334 (1964).
66. M. Barbacid and D. Vazquez, J. Mol. Biol., 84, 603 (1974).
67. D. Vazquez, E. Battaner, R. Neth, G. Heller, and R. E. Munro, Cold Spring Harbor Symp. Quant. Biol., 34, 369 (1969).
68. C. T. Caskey, A. L. Beaudet, E. M. Scolnick, and M. Rosman, Proc. Natl. Acad. Sci. U.S., 68, 3163 (1971).
69. V. T. Innanen and D. M. Nicholls, Biochim. Biophys. Acta, 361, 221 (1974).
70. L. Carrasco, M. Barbacid, and M. Vazquez, Biochim. Biophys. Acta, 312, 368 (1973).

71. J. R. Bamburg and F. M. Strong, in Microbial Toxins (S. Kadis, A. Ciegler, and S. J. Ajl, eds.), Vol. III, Academic Press, New York, 1971, pp. 207-292.
72. S. G. Yates, in Microbial Toxins (S. Kadis, A. Ciegler, and S. J. Ajl, eds.), Vol. VII, Academic Press, New York, 1971, pp. 191-206.
73. C. M. Wei and C. S. McLaughlin, Biochem. Biophys. Res. Commun., 57, 838 (1974).
74. Y. Ueno, M. Hosoya, Y. Morita, I. Ueno, and T. Tatsuno, J. Biochem. (Tokyo), 64, 479 (1968).
75. Y. Ueno and K. Fukushina, Experimentia, 24, 1032 (1968).
76. K. Ohtsubo, M. Yamada, and M. Saito, Japan. J. Med. Sci. Biol., 21, 185 (1968).
77. N. Nakano, Japan J. Med. Sci. Biol., 21, 351 (1968).
78. K. Ohtsubo, P. Kaden, and C. Mittermayer, Biochem. Biophys. Acta, 287, 520 (1972).
79. M. E. Stafford and C. S. McLaughlin, J. Cell Phys., 82, 121 (1973).
80. M. Barbacid and D. Vazquez, Eur. J. Biochem., 44, 437 (1974).
81. W. P. Tate and C. T. Caskey, J. Biol. Chem., 248, 7970 (1973).
82. T. Misato, in Antibiotics (D. Gottlieb and P. D. Shaw, eds.), Vol. I, Springer-Verlag, New York, 1967, pp. 434-439.
83. T. Kondo, H. Nakai, and T. Goto, Tetrahedron, 29, 1801 (1973).
84. H. Yamaguchi and N. Tanaka, J. Biochem. (Tokyo), 60, 632 (1966).
85. F. W. Lichtenthaler and G. Trummlitz, FEBS Lett., 38, 237 (1974).
86. M. Kawana, D. G. Streeter, R. J. Rousseau, and R. K. Robins, J. Med. Chem., 15, 561 (1972).
87. T. K. Huang, T. Misato, and H. Asuyama, J. Antibiot., 17A, 65 (1964).
88. H. Yamaguchi, C. Yamamoto, and N. Tanaka, J. Biochem. (Tokyo), 57, 667 (1965).
89. C. Coutsogeorgopoulos, Biochemistry, 6, 1704 (1967).
90. T. Kinoshita, N. Tanaka, and H. Umezawa, J. Antibiot., 23A, 288 (1970).
91. M. Barbacid and D. Vazquez, Eur. J. Biochem., 44, 445 (1974).
92. S. Pestka, J. Biol. Chem., 247, 4669 (1972).
93. R. Harris and S. Pestka, J. Biol. Chem., 248, 1168 (1973).
94. S. R. Casjens and A. J. Morris, Biochim. Biophys. Acta, 108, 677 (1965).
95. M. L. Celma, R. E. Monro, and D. Vazquez, FEBS Lett., 6, 273 (1970).
96. J. Cerna, F. W. Lichtenthaler, and J. Rychlik, FEBS Lett., 14, 45 (1971).
97. A. Jimenez, R. E. Monro, and D. Vazquez, FEBS Lett., 7, 103 (1970).
98. A. E. Herner, I. H. Goldberg, and L. B. Cohen, Biochemistry, 8, 1335 (1969).

99. H. Umezawa, Y. Okami, T. Hashimoto, Y. Suhara, M. Hamada, and T. Takeuchi, J. Antibiot., 18A, 101 (1965).
100. Y. Sukara, K. Maeda, and H. Umezawa, Tetrahedron Lett., 12, 1239 (1966).
101. T. Idekawa, H. Umezawa, and Y. Iitaka, J. Antibiot., 19A, 49 (1966).
102. S. Yasuda, T. Ogasawara, S. Kawabata, I. Iwataki, and T. Matsumoto, Tetrahedron, 29, 3141 (1973).
103. Y. Suhara, F. Sasaki, G. Koyama, and K. Maeda, J. Amer. Chem. Soc., 94, 6501 (1972).
104. T. Ishiyama, I. Hara, M. Matsuoka, K. Sato, S. Shimada, R. Izawa, T. Hashimoto, M. Hamada, Y. Okami, T. Takeuchi, and H. Umezawa, J. Antibiot., 18A, 115 (1965).
105. M. Hamada, T. Hashimoto, T. Takahashi, S. Yokoyama, A. Miyake, T. Takeuchi, Y. Okami, and H. Umezawa, J. Antibiot., 18A, 104 (1965).
106. H. Masukawa, N. Tanaka, and H. Umezawa, J. Antibiot., 21A, 73 (1968).
107. K. Ohmori, J. Antibiot., 20A, 104 (1967).
108. T. Yamaguchi, Japan Pest. Inform. Bull. No. 18, 1974, p. 5.
109. A. Okuyama, N. Machiyama, T. Konishita, and N. Tanaka, Biochem. Biophys. Res. Commun., 43, 196 (1971).
110. T. L. Helser, J. E. Davies, and J. E. Dahlberg, Nature (New Biol.), 233, 12 (1971).
111. N. Tanaka, H. Yamaguchi, and H. Umezawa, J. Biochem. (Tokyo), 60, 429 (1966).
112. P. F. Sparling, Science, 167, 56 (1970).
113. P. C. Tai, B. J. Wallace, and B. D. Davis, Biochemistry, 12, 616 (1973).
114. N. Tanaka, Y. Yoshida, K. Sashita, H. Yamaguchi, and H. Umezawa, J. Antibiot., 19A, 65 (1966).
115. T. L. Helser, J. E. Davies, and J. E. Dahlberg, Nature (New Biol.), 235, 6 (1972).
116. N. Tanaka, in Molecular Mechanisms of Antibiotic Action on Protein Biosynthesis and Membranes (E. Munoz, F. Garcia-Ferrandiz, and D. Vazquez, eds.), Elsevier Publishing Co., New York, 1972, pp. 279-291.
117. W. T. Thomson, Agricultural Chemicals, Vol. 4, Thomson Publishing Co., Indianapolis, 1973, pp. 120-122.
118. D. J. Weber and J. M. Ogawa, Phytopathology, 55, 159 (1965).
119. J. Lewis, D. J. Weber, and S. Venketeswaran, Phytopathology, 59, 93 (1969).
120. E. A. Bell, in Phytochemical Ecology (J. B. Harborne, ed.), Academic Press, New York, 1972, pp. 163-177.
121. G. A. Rosenthal, J. Exptl. Bot., 25, 609 (1974).

122. G. A. Rosenthal, Plant Physiol., 46, 273 (1970).
123. N. H. Horowitz and A. M. Srb, J. Biol. Chem., 174, 371 (1948).
124. E. A. Childs, J. C. Ayres, and P. E. Koehler, Mycologia, 63, 181 (1971).
125. J. B. Walker, J. Biol. Chem., 212, 207 (1950).
126. E. J. Miller and S. J. Harrison, Nature, 166, 1035 (1950).
127. D. Wilkie, J. Mol. Biol., 47, 107 (1970).
128. B. E. Volcani and E. E. Snell, J. Biol. Chem., 174, 893 (1948).
129. T. E. Weaks, Jr. and G. E. Hunt, Physiol. Plant., 29, 421 (1973).
130. I. H. Hirshfield and H. P. J. Bloemers, J. Biol. Chem., 244, 2911 (1969).
131. S. K. Mitra and A. H. Mehler, J. Biol. Chem., 242, 5490 (1967).
132. C. C. Allende and J. E. Allende, J. Biol. Chem., 239, 1102 (1963).
133. L. S. Williams, J. Bacteriol., 113, 1419 (1973).
134. J. Attias, M. J. Schlesinger, and S. Schlesinger, J. Biol. Chem., 244, 3810 (1969).
135. P. H. Tsao, Ann. Rev. Phytopathology, 8, 157 (1970).
136. A. W. Linnane, J. M. Haslam, H. B. Lukins, and P. Nagley, Ann. Rev. Microbiol., 26, 163 (1972).
137. R. E. Connamacher and H. G. Mandel, Biochim. Biophys. Acta, 166, 475 (1968).
138. I. H. Maxwell, Mol. Pharmacol., 4, 25 (1968).
139. G. Suarez and D. Nathans, Biochem. Biophys. Res. Commun., 18, 743 (1965).
140. S. Sakkar and R. E. Thach, Proc. Natl. Acad. Sci., U.S., 60, 1479 (1968).
141. K. Igarashi and A. Kiji, Eur. J. Biochem., 14, 41 (1970).
142. E. Scolnick, R. Tomphins, C. T. Caskey, and M. Nirenberg, Proc. Natl. Acad. Sci. U.S., 61, 768 (1968).
143. Z. Vogel, A. Zamir, and D. Elson, Biochemistry, 8, 5161 (1969).
144. J. Goldstein and C. T. Caskey, Proc. Natl. Acad. Sci. U.S., 67, 537 (1970).
145. E. Cundliffe, Biochem. Biophys. Res. Commun., 33, 247 (1968).
146. C. Gurgo, D. Apirion, and D. Schlessinger, J. Mol. Biol., 45, 205 (1969).
147. J. W. Hendrix, private communication, 1974.
148. J. Voros, Phytopathol. Z., 54, 249 (1965).
149. R. Sanger, Cytoplasmic Genes and Organelles, Academic Press, New York, 1972.
150. A. Kaji and Y. Tanaka, J. Mol. Biol., 32, 221 (1968).
151. M. Ozaki, S. Misushima, and N. Nomura, Nature, 222, 333 (1969).
152. F. N. Chang and J. G. Flaks, Antimicrob. Agr. Chemotherap., 2, 294 (1972).
153. H. Kaji, I. Suzuka, and A. Kaji, J. Biol. Chem., 241, 1251 (1966).
154. J. Modolell and B. D. Davis, Nature, 224, 345 (1969).

155. J. Modolell and B. D. Davis, Proc. Natl. Acad. Sci. U.S., 67, 1148 (1970).
156. L. Luzzato, D. Apirion, and D. Schlessinger, J. Mol. Biol., 42, 315 (1969).
157. J. C. LeLong, M. A. Cousin, D. Gros, M. Greenberg, L. Manago, and F. Gros, Biochem. Biophys. Res. Commun., 42, 530 (1971).
158. N. W. Seeds, J. A. Retsma, and T. W. Conway, J. Mol. Biol., 27, 421 (1967).
159. S. Pestka, Arch. Biochem. Biophys., 136, 80 (1970).
160. J. L. Lessard and S. Pestka, J. Biol. Chem., 247, 6909 (1972).
161. R. Fernandez-Munoz, R. E. Monro, R. Torres-Pinedo, and D. Vazquez, Eur. J. Biochem., 23, 185 (1971).
162. R. Kaempfer, Proc. Natl. Acad. Sci., U.S., 61, 106 (1968).
163. C. Gurgo, D. Apirion, and D. Schlessinger, FEBS Lett., 3, 34 (1969).
164. Y. Suzuki and S. Okamoto, J. Biol. Chem., 242, 4722 (1967).
165. A. P. Grollman and M. T. Huang, Federation Proc., 32, 1673 (1973).
166. S. G. Van Den Bergh, P. Borst, L. M. Van Deenen, J. C. Riemersma, E. C. Slater, and J. M. Tager (eds.), Mitochondria/Biomembranes, North Holland, Amsterdam, 1972.
167. P. Borst, Ann. Rev. Biochem., 41, 333 (1972).
168. D. Lloyd, The Mitochondria of Microorganisms, Academic Press, New York, 1974.
169. N. G. Ibrahim, J. P. Burke, and D. S. Beattie, J. Biol. Chem., 249, 6806 (1974).
170. N. R. Towers, Life Sci., 14, 2037 (1974).
171. R. Datema, A. Agsteribbe, and A. M. Kroon, Biochim. Biophys. Acta, 335, 386 (1974).
172. R. S. T. Yu and P. R. Stewart, Cytobios, 9, 175 (1974).
173. J. W. Hendrix, Proc. Amer. Phytopathol. Soc., 1, 207 (1975).
174. S. G. Georgopoulos, A. Kappas, and A. C. Hastie, Phytopathology, 66, 217 (1976).
175. G. Rosenthal, Quart. Rev. Biol., in press.

Chapter 12

DEVELOPMENT OF FUNGAL RESISTANCE TO FUNGICIDES

S. G. Georgopoulos*

Department of Biology
Nuclear Research Center "Demokritos"
Athens, Greece

---

*Present address: Department of Plant Pathology, Agricultural College of Athens, Votanicos, Athens, Greece

I. INTRODUCTION

Severe limitations may be imposed on the chemical control of pathogenic or otherwise undesirable organisms by the appearance of less sensitive forms of the target species. But apart from its practical importance, adaptation to toxic chemicals has raised some very basic biological questions. It is, therefore, not surprising that a great deal of work has been devoted to the subject. Much can be learned particularly from the related field of adaptation to antibacterial agents. Helpful reviews include those of Moyed [1], Falkow et al. [2], Kiser et al. [3], Shaw [4], Franklin and Snow [5], Reynard [6], and Benveniste and Davies [7].

The literature on the development of resistance to antifungal compounds by fungi offers different viewpoints. For reasons which will be discussed later, the subject became of considerable practical importance only rather recently. A limited number of papers has been published on resistance to antifungal compounds used in agriculture or medicine, and information, on practical aspects in particular, is scanty. On the other hand, a few fungi have had a considerable share in studies of the genetics and biochemistry of modified sensitivity to antimetabolites and other inhibitors of general interest to cell biology but of little or no practical importance as fungicides. This chapter is mainly a summary of recent information on acquired resistance to commercial fungicides and an outline of the most interesting aspects of the relevant studies with inhibitors of no immediate practical importance. Work adequately considered in previous reviews [8-14] will not be discussed in detail here.

II. DEFINITIONS

Sevag [15] defined development of "tolerance or resistance to a drug" as "a stable and inheritable adjustment by a cell to external or endogenous toxic or inhibitory agents." This definition could be accepted to include acquired resistance to fungicides by fungi. Although the terms "resistance" and "tolerance" may not be synonyms in other instances, e.g., in describing response of plant varieties to fungal infections [16, 17], no clear distinction between the two has been made in the literature on antimicrobial action. The definition does not include unstable resistance resulting from temporary noninheritable changes following exposure to a toxic compound. Such phenotypic adaptation was the major subject of the early reports on resistance, and some examples were reviewed by Georgopoulos and Zaracovitis [9].

Continuous exposure during growth under practically optimal conditions in the laboratory to sublethal concentrations of a toxicant may cause adaptive changes which somewhat increase tolerance. It is doubtful, however, that the requirements for such adaptive changes can be met in nature. Furthermore, because of its instability, nongenetic adaptation is unlikely to create any serious problems. Therefore, in this chapter, fungicide resistance will be treated as an inheritable characteristic and fungicide resistant strains will be called "mutants."

Resistant mutants are described by their degree or level of resistance which can be defined as the magnitude of the mutational change in sensitivity. This, of course, requires an accurate measurement of sensitivity of both mutant and wild type by the same appropriate method [18]. Some workers compare the maximal noninhibitory concentrations. In this case, the degree of resistance is often overestimated because complete inhibition of the mutant can never be achieved due to solubility limitations and the sensitivity to toxicant concentrations below saturation is not measured [9]. Even in the case of a highly soluble toxicant, however, more precise and important information can be obtained if the dosage response curves for mutant and wild type are derived from appropriate experiments and are compared as to position and slope.

"Cross-resistance" means resistance to two or more toxicants mediated by the same genetic factor. This term should not be misused. A strain resistant to several compounds may carry several mutations. In such a case a positively correlated cross-resistance cannot be assumed. Recombination analysis should show whether resistance to all compounds is controlled by the same locus. Usually a mutant selected on one toxicant is resistant only to toxicants which are chemically related or have a very similar mechanism of action, but in rare cases a pleiotropic mutation, affecting a more general cell property may give resistance to unrelated toxicants. On the other hand, a mutation may increase resistance to one toxicant and sensitivity to another. To describe this situation, the term "collateral sensitivity" was introduced by Szybalski and Bryson [19] and is used mainly in the medical literature [20]. Entomologists speak of "negatively correlated cross-resistance" in this case [21]. It is important to note here that the two terms have the same meaning because both have been used in studies with fungi [22, 23].

III. GENETICS OF FUNGICIDE RESISTANCE

In many of the reported cases of acquired resistance to a fungicide the phenomenon has not been analyzed genetically. In others the genetic nature of sensitivity differences may have been established but the study has been restricted to one or a very few resistant mutants, thus the picture of the genetic control is far from complete. However, the few more detailed studies available indicate that the genetics of fungicide resistance is complex.

A. Extrachromosomal Control

In bacteria, drug resistance often results from the acquisition of extrachromosomal circular DNA molecules (plasmids) carrying resistant genes which are transmissible between bacterial cells of many different genera [24]. The spread of resistance among bacteria by this means is considered to be a more serious public health problem than in the case of chromosomally controlled resistance [24]. By contrast, plasmid-mediated resistance to chemicals has not been demonstrated in fungi. On the other hand, fungi possess mitochondria with DNA that does play a hereditary role. In *Neurospora*, for example, the mitochondria exhibit genetic continuity after interhyphal transplantation [25]. In each cell, the two genetic systems, that of the nucleus and that of the mitochondrion, appear to be distinct [26] but the cooperation of both is required for the synthesis of a functional mitochondrion. It is, therefore, reasonable to state that resistance to compounds which are toxic because they interact with mitochondrial systems, such as those of mitochondrial protein synthesis, electron transport, and phosphorylation, may be controlled by nuclear or mitochondrial genes or nucleocytoplasmic interactions.

Of the agriculturally important fungicides, the trisubstituted tin derivatives probably act by selective inhibition of oxidative phosphorylation in mitochondria [27]. Lancashire and Griffiths [28] isolated a number of mutants of *Saccharomyces cerevisiae* resistant to triethyltin and cross-resistant to other trialkyltin biocides, and to triphenyltin. In all of these mutants resistance was inherited in a non-Mendelian fashion indicating extrachromosomal control. By contrast, no extrachromosomal mutants for resistance to carboxin and other agriculturally important carboxamide fungicides are known. These fungicides act by inhibition of mitochondrial electron transport in the succinic dehydrogenase region [29-35] but no mitochondrial genes seem to be involved in the control of sensitivity differences. All carboxin-resistant mutants of both *Ustilago maydis* [33, 35] and *Aspergillus nidulans* [36, 37] that have been studied are chromosomal. Thiram is another agricultural fungicide to which resistance in *Cochliobolus carbonum* has been suspected, but not proved, to be under cytoplasmic or nuclear-cytoplasmic control [38].

The genetics of resistance to more widely known inhibitors of mitochondrial function has been most extensively studied in *S. cerevisiae* [22, 39-46] and has greatly contributed to the present knowledge of mitochondrial genetics. Linnane et al. [47] and Avner and Griffiths [48] have listed the conditions which must be fulfilled before antibiotic resistance determinants are assigned to yeast mitochondrial DNA. In most cases, each such extrachromosomal gene is specific for resistance to one inhibitor or to a few inhibitors of similar action only. It is interesting to note that more than one loci in the mitochondrial genome may be involved in the control of resistance to the same antibiotic. Thus, Kleese et al. [45] recognized two cytoplasmically inherited loci conditioning resistance to chloramphenicol, a mitochondrial protein synthesis inhibitor. Similarly, Avner and Griffiths [48] found two allelism groups for extrachromosomal resistance to oligomycin, an inhibitor of phosphorylation. Chromosomal mutations, on the other hand, affecting sensitivity to inhibitors of mitochondrial function in yeast are usually pleiotropic. Thus, oligomycin, in addition to the cytoplasmic mutants mentioned above, selects mutants under nuclear or nucleocytoplasmic control. These mutants are resistant not only to phosphorylation inhibitors, such as triethyltin, but also to uncouplers and inhibitors of mitochondrial protein synthesis [48, 49]. Rank and Bech-Hansen [22] recently described a chromosomal mutation in S. cerevisiae, conferring resistance to 12 inhibitors of mitochondrial function (oligomycin, chloramphenicol, and antimycin A which inhibits electron transport) and collateral sensitivity to five others (neomycin, ethidium bromide, and acriflavine). An alteration of the structure of the mitochondrial inner membrane affecting interaction with many unrelated inhibitors is believed to be responsible for this pleiotropy [22]. Nucleocytoplasmic interaction is implicated [47] in the case of a mutant of yeast, originally [42] isolated as resistant to inhibitors of mitochondrial protein synthesis and later found [43] to be cross-resistant to oligomycin.

Little has been done on the genetics of resistance to compounds affecting mitochondrial function in other fungi. In A. nidulans, oligomycin [50, 51] or chloramphenicol [215] resistance results from chromosomal and only very rarely from extrachromosomal mutations. In the same species, all of 37 independently isolated mutants for resistance to pyrrolnitrin, an inhibitor of electron transport, were chromosomal [37]. Some work on the inheritance of resistance to antibiotics inhibiting protein synthesis in mitochondria has recently been done with species of Phytophthora. In P. dreschleri chloramphenicol resistance segregated in a Mendelian fashion indicating chromosomal control [52, 53]. In P. capsici Mendelian ratios for streptomycin resistance were not obtained from mutant × wild-type crosses but extrachromosomal control could not be proved conclusively [54]. For details of the nuclear and extranuclear inheritance of resistance to toxicants acting by interference with the biogenesis and function of mitochondria in fungi, the reader is referred to the recent, comprehensive account by Lloyd [55].

B. Chromosomal Control

Cytoplasmic DNA cannot be involved in the control of resistance to fungicides which do not act on mitochondrial systems. Therefore, most of the fungicide resistances which have been investigated have been shown to result from mutation of chromosomal genes. In most cases where a considerable number of independently isolated mutants were studied and mutant × mutant crosses were performed, more than one loci for resistance to each fungicide were identified (see Table 1).

TABLE 1

Chromosomal Multigenic Systems for Fungicide Resistance

Fungicide	Organism	Number of chromosomal resistance loci	References
Aromatic hydrocarbons	Nectria haematococca	5	56, 57
	Aspergillus nidulans	2	58
Dodine	Nectria haematococca	4	59
	Venturia inaequalis	2	60, 61
Benzimidazoles	Aspergillus nidulans	2	62
Carboxamides	Ustilago maydis	3	33, 35, 63
	Aspergillus nidulans	3	36, 37, 64
Cycloheximide	Saccharomyces cerevisiae	8	65
	Neurospora crassa	3	66, 67

Acriflavine	*Aspergillus nidulans*	3	68
Oligomycin	*Saccharomyces cerevisiae*	2	69
Nystatin	*Saccharomyces cerevisiae*	3	70, 71
	Neurospora crassa	3	72
Edeine	*Neurospora crassa*	2	73
p-Fluorophenylalanine	*Aspergillus nidulans*	10	74, 75
8-Azaguanine	*Neurospora crassa*	3	76, 77
5-Fluorouracil	*Saccharomyces cerevisiae*	4	78
Sorbose	*Aspergillus nidulans*	2	79
	Neurospora crassa	6	80
Chloramphenicol	*Aspergillus nidulans*	3	215
Imazalil	*Aspergillus nidulans*	8	310

Resistance genes in such multigenic systems may be numerous, since their number appears to be limited only by the number of mutants investigated. In the case of aromatic hydrocarbons, for example, three loci for resistance (cnb$_{1-3}$) were recognized [56] by the analysis of 12 mutants of Nectria haematococca (syn. Hypomyces solani) but the number of loci was increased to five when an additional 100 mutants were examined [57]. Since the frequency of mutants at any one locus in sectors from fungicide treated colonies varied from 42% for cnb$_1$ to 1% for cnb$_5$, Georgopoulos and Panopoulos [57] pointed out that additional loci for resistance might have been recognized had it been possible to study larger numbers of mutants. Similarly, analysis of seven cycloheximide-resistant mutants of Neurospora crassa by Hsu [66] recognized two resistance loci, but when Vomvoyanni [67] studied 27 such mutants, 10 of these could not be assigned to either allelism group which shows that at least one more locus is involved.

In species on which much mutation work has been done, an estimate of how many loci may mutate to give resistance to a given toxicant can be obtained indirectly. This method, originally used for other purposes [81, 82], compares the frequency of resistant mutants (res. mut.) with that of auxotrophic mutants (aux. mut.) of all types obtained in the same induction experiment in the following way:

$$\frac{\text{frequency of res. mut.}}{\text{frequency of aux. mut.}} \times \text{number of aux. loci} = \text{number of res. loci}$$

With this method Srivastava and Sinha [75] have obtained an approximate number of 28 loci for p-fluorophenylalanine resistance in A. nidulans in which the number of all auxotrophic loci is close to 70. In the same fungus, Hastie [83] estimates 3-4 loci for resistance to benomyl. This method cannot be applied to species not extensively studied from the point of view of auxotrophic mutations. Furthermore, it is open to several possible sources of error. For example it assumes equal mutability of loci and equal probability of recovery of all kinds of mutants. However, it is a method which can provide some approximation.

In at least three cases the phenotypes of mutants at the same locus controlling resistance to a fungicide were not identical, indicating multiple alleles. In cycloheximide resistance of N. crassa [67] and carboxin resistance of U. maydis [311] such evidence is derived from sensitivity differences in the absence of modifying genes. In A. nidulans multiple alleles for resistance to benzimidazole-type fungicides must be assumed on the basis of differences in the cross-resistance patterns [84], as will be further discussed in Section IV. It should be remembered in this regard, however, that allelism is usually tested for by analysis of a limited number of progeny from each mutant × mutant cross which might not be sufficient to exclude the possibility of closely linked genes. In the case of resistance to benzimidazoles in A. nidulans it has been conclusively shown that a mutant highly resistant to benomyl and sensitive to thiabendazole, is allelic to another mutant which is highly sensitive to benomyl and resistant to thiabendazole [312]. In fact, the two mutations were shown to be 16 nucleotides apart.

It is, of course, to be expected that resistance to a toxicant may arise by mutation of any one of several genes. Firstly, not only modification of a sensitive site but also a number of other changes can lead to resistance, as will be discussed in Section IV. As an example, the various mutational alterations leading to resistance to 5-fluoropyrimidines in S. cerevisiae [78] can be mentioned. Secondly, even if only modification of the target site within the cell were involved, it cannot be excluded that more than one gene might contribute to a modified site resulting in resistance. From the field of agricultural fungicides, resistance to carboxin and to carboxin analogs may be cited as an example [33, 35, 63]. Of two types of mutations for carboxin resistance in U. maydis, both affect respiratory electron transport, which is the process inhibited by the fungicides [29, 30, 33], but only the one eliminates the antimycin A- and cyanide-insensitive pathway which is operative in the wild type.

With this reasoning one would expect that different mutations should give different levels of resistance. This is actually the case with most fungicide resistances in which multigenic systems have been studied, including resistance to cycloheximide in S. cerevisiae [65] and N. crassa [85], to dodine in N. haematococca [59], to benomyl in A. nidulans [62], to carboxin in U. maydis [63], A. nidulans [37], and to acriflavine [68], and to p-fluorophenylalanine [75] in A. nidulans. Resistance to the aromatic hydrocarbon-type fungicides in N. haematococca [57, 86] appears to be an exception because strains carrying resistance genes at different loci do not seem to differ in their fungicide tolerance levels. However, since the degree of resistance to aromatic hydrocarbon compounds is generally small, it may be a matter of the "resolving power" of the method employed.

Resistance loci vary in their mutability as indicated by examples mentioned earlier in this section [57, 67]. Most workers record induced mutation frequencies. Thus, from a sample of 8 million survivors of an ultraviolet (UV) treatment of conidia of A. nidulans, Hastie and Georgopoulos [62] obtained nine strains resistant to benzimidazole fungicides. Of these, five were mutants at the ben-1 and four at the ben-2 locus. In a study with 10 different fungi, van Tuyl [310] found the UV-induced mutation frequency for benomyl resistance to vary from 1.2 to 100 per 10 million survivors. In Fusarium oxysporum, one in 4.6×10^5 survivors of a UV treatment which killed 95% of the conidia was resistant to the benzimidazole fungicides [87]. In such cases, recording the mutagen and the dose is important.

To speculate on the probability of resistance development in nature, the spontaneous mutation frequencies would be more useful. In Venturia inaequalis almost one in 10^6 conidia tested seemed to carry a mutation for resistance to dodine [88]. Nystatin-resistant cells of Candida albicans appeared at a frequency of approximately 10^{-7} in the absence of mutagenic treatment [89].

The spontaneous mutation rate for p-fluorophenylalanine resistance in A. nidulans has been calculated as 3.8×10^{-3} [90]. Of course, this should be taken as the sum of the mutation frequencies of the many loci which control this resistance [75]. For oligomycin resistance in the same organism the estimated spontaneous mutation rate was 2×10^{-8} [50]. In P. capsici [54] and P. cactorum [91] stable streptomycin resistance occurs spontaneously at a high frequency (ca. 10^{-4}). Chromosomal and extrachromosomal erythromycin resistance appeared at equally high frequency in S. cerevisiae [39]. Although the rates for many of the agricultural fungicides may be lower, if the potential is present in a given species, resistance development in the field should not be difficult considering the very high numbers of spores produced by fungi. The question of whether the fungicide may act as the mutagen will be treated later in this section.

C. Gene Interaction

1. Interallelic Interaction

If more than one loci may mutate to give resistance to a fungicide, then starting with a sensitive strain, a single mutational event will leave sensitive genes at all but one of these loci even in the case of a uninucleate haploid cell. The degree of resistance which will be obtained will thus be the result of the interallelic interaction between resistant and sensitive genes. Addition of a second resistant gene at another locus will disturb this interaction and may lead to a different tolerance level. Positive interaction between resistance genes at different loci has been found in cycloheximide resistance of S. cerevisiae [65] and N. crassa [66], and in dodine resistance of N. haematococca [59] where double-resistant recombinants from dihybrid crosses are more resistant than either parent. It should be noted that in Neurospora, the double-resistant recombinant studied by Hsu [66] was capable of very limited growth in the absence of cycloheximide and this finding may prove very important for an understanding of the role of ribosomal components modified by these mutations [85]. In homocaryotic haploids of A. nidulans resistant to benzimidazole derivatives [62] and of N. haematococca resistant to hydrocarbon fungicides [56] no positive interaction has been observed, while in other instances of fungicide resistance the problem has not been studied. In the absence of more data on biochemical genetics of fungicide resistances it is premature to suggest that the absence of gene interaction indicates a similarity in the action of the genes and its presence a dissimilarity.

2. Intraallelic Interaction

The patterns of gene interaction become more complicated in heterocaryotic and in heterozygous diploid cells. The intraallelic interaction has been studied in several cases with respect to fungicide resistance. Whether resistance will be dominant over sensitivity in a heterozygous diploid may depend on the fungicide and also on the interacting alleles. This is quite reasonable because the interacting gene products may be different in each case. A few examples are given below.

In diploid strains of A. nidulans, heterozygous for either of two unlinked genes for benzimidazole resistance, this resistance appeared to be recessive [62]. In the same fungus PCNB resistance also behaved as a recessive character [58], while for p-fluorophenylalanine resistance mutant genes at five of ten known loci were dominant over their wild-type alleles [75]. Three loci for carboxin resistance have been studied in Aspergillus in this regard and in all cases resistance is reported to be semidominant [37]. In S. cerevisiae a mutation for nystatin resistance was found dominant [89], but of six genes for cycloheximide resistance three were semidominant, two recessive, and one dominant [65]. In Ustilago hordei dominance of both benomyl and carboxin resistance has been reported [92], but in U. maydis carboxin resistance was found semidominant [35]. Undoubtedly an accurate determination of the degree of resistance of the heterozygous strain in comparison with that of its component haploids is required before dominance can be decided upon. Few studies on gene interaction have been conducted with cell-free systems. A diploid strain of S. cerevisiae, heterozygous for resistance to cycloheximide (which acts by inhibition of protein synthesis), was phenotypically sensitive but its ribosomes exhibited an intermediate degree of resistance in vitro [93]. It was suggested that the heterozygous cells have a mixed complement of sensitive and resistant ribosomes. In vivo, mixed polysomes would be nonfunctional because movement along the messenger RNA would be blocked by the sensitive ribosomes. In vitro, however, where excess polyuridylic acid (poly-U) was used, polysome formation would be minimized and resistant ribosomes would accomplish polypeptide synthesis [93]. In a diploid strain of U. maydis both the resistant and the sensitive allele for carboxin resistance were expressed in vivo and in vitro where the sensitivity of the succinic dehydrogenase complex was studied [35]. In this case the differential temperature sensitivity of mutant and wild-type succinic dehydrogenase activities was utilized and the mixed nature of the mitochondrial preparations from the heterozygous cells was proved [35].

In the heterocaryotic condition the association of two types of nuclei in a common cytoplasm makes dominant-recessive relations between genes possible but the situation is different from that of the heterozygous diploid because the ratio of the two genes is not necessarily 1:1. That the fungicide itself may affect the nuclear ratio is shown by the work of Warr and Roper [94] who used a semidominant allele for cycloheximide resistance and showed adaptive changes of the nuclear ratio in A. nidulans heterocaryons grown with various concentrations of the antibiotic. In this case the proportion of the resistant nuclei increased with increasing cycloheximide concentration until a plateau was reached which apparently represented the nutritional limits of the combination of biochemical markers present in the heterocaryon. Resistance to members of the aromatic hydrocarbon group in *Rhizopus stolonifer* [95] and *Botrytis cinerea* [96] and probably to organomercurials in *Pyrenophora avenae* [97] has been suggested to be due to heterocaryotic effects. In *Botrytis*, for example, while sensitive strains produce only sensitive conidia, resistant strains growing in the absence of the fungicide produced about 50% resistant and 50% sensitive spores which clearly shows their heterocaryotic nature. By contrast, strains of other fungi (e.g., *N. haematococca* [98]), resistant to the same aromatic hydrocarbon fungicides, are definitely homocaryotic. To come back to *Botrytis*, resistant strains produced only resistant spores when grown in the presence of the toxicant [96] which again proves the adaptability of the nuclear ratio in the heterocaryon. An interesting but not fully understood case of resistance resulting from heterocaryotic association of nuclei, neither of which seemed to carry resistance genes in the homocaryon, has been reported in *Thanatephorus cucumeris* [99].

Often heterocaryotic colonies produce homocaryotic sectors better adapted to the presence of the fungicide. Dissociation may, of course, be prevented if mutants carrying complementary biochemical requirements are used for the synthesis of the heterocaryon and the heterocaryotic condition is imposed by the use of nonsupplemented medium. This method, however, does not prevent even drastic changes in nuclear ratio. Pittenger and Atwood [100], for example, showed that a requirement of a *N. crassa* heterocaryon may be satisfied if the wild-type allele is present even in a small proportion of the nuclei. In the light of such knowledge, Hsu's [66] conclusion that both of the cycloheximide-resistant genes studied by him are dominant in *Neurospora* heterocaryons, growing on minimal medium may be subject to criticism. In the case of benomyl, *Neurospora* heterocaryons tolerated the same level of fungicide as the resistant homocaryons but the growth rates on benomyl medium were not equal [101]. Since only one biochemical requirement was present in each of the components of the heterocaryon, safe conclusions cannot be reached in this case either.

A special case is the dicaryon of many basidiomycetes such as the smuts and rusts. If a fungicide-resistant gene is present in only one of the two nuclei, the 1:1 gene ratio is maintained in every cell. If the dicaryotic phase is the pathogenic one, it is very important to determine whether resistance is dominant. In corn seedlings infected with a mixture of sporidia of two strains of *U. maydis*, of which one was sensitive and one was resistant to carboxin, it appeared that in the dicaryon resistance was semidominant [35]. In the dicaryon of *Coprinus lagopus*, on the other hand, genes for ethionine resistance were recessive to their wild-type alleles [102].

3. Modification of Gene Action

The existence of modifiers, i.e., genes which do not affect fungicide sensitivity when present in the wild type but which modify the effect of specific resistance genes in a positive or negative way, has been recognized in a few cases. Cycloheximide resistance of S. cerevisiae [65] and N. crassa [66] and dodine resistance of N. haematococca [59] are good examples. If a strain which carries a modifier in addition to an unlinked gene for resistance is crossed to a wild-type strain, 50% of the progeny is sensitive. The resistant progeny is of two types: one with resistance of the same level as in the resistant parent, and the other with higher or lower resistance, depending on the modifier.

D. The Mutagenicity of Fungicides and the Possibility of Fungicide-Induced Resistance

In addition to well-known mutagens which may possess antifungal properties, genetic activity has been reported for a number of commercial fungicides. In many cases, chromosome damage and mitotic abnormalities have been observed in a variety of test systems. The effects of acenaphthene [103], dichloran [104], ferbam [105], Panogen 15 [106], ziram [107], griseofulvin [108], and benomyl [109, 110] on higher plant or animal cells can be cited as examples. Of these compounds some, if not all, affect fungal nuclei similarly, as has been shown by cytological [111-115] as well as genetic [116-119] investigations. In fact, their fungitoxicity seems to be due to their action on the genetic apparatus as discussed in detail in Chapter 10.

Recently, a number of reports have appeared discussing the induction of point (gene) mutations by fungicides. Compounds reported to be mutagenic are: captan in Escherichia coli [120, 121], Salmonella typhimurium [120], and N. crassa [122]; pentachloronitrobenzene (PCNB) in E. coli [121]; ferbam in Aspergillus niger [105]; benomyl in E. coli and Fusarium oxysporum [123], but not A. nidulans [116]; and methyl benzimidazole-2-ylcarbamate (MBC, the breakdown product of benomyl and of thiophanate-methyl) in S. typhimurium [124]. Folpet and metiram (a mixture containing zineb) have been claimed to induce mitotic gene conversion in S. cerevisiae [125].

The ability of fungicides to cause genetic changes has its impact on the environment, but what is important from our point of view is that toxicants which possess genetic activity may induce resistance. Obviously, induction of resistance by a fungicide cannot be assumed only because the change in sensitivity is recognized after some exposure of a population of cells to the chemical [9]. Suitable methods such as the "fluctuation test" of Luria and Delbrück [126] are available for the study of this problem. However, there is no reason why a fungicide which can induce other types of gene mutations in a given fungus should not similarly affect resistance loci if such loci are present. In a few instances specific induction of resistance (directed mutation) has been reported. In yeast, for example, cupric ions have been claimed to specifically induce two types of mutation for copper resistance [127].

Compounds which affect chromosomes and the mitotic process but are incapable of inducing point mutations should not induce resistance in haploid cells. In heterozygous diploid cells, however, if resistance is recessive or semidominant, reduced sensitivity will result if the fungicide can induce haploidization, mitotic crossing over, or hemizygosis. Thus, phenotypically sensitive diploid colonies of A. nidulans, heterozygous for either of two genes for resistance to benzimidazole fungicides, produce abundant resistant mitotic segregants soon after exposure to one of these compounds [62].

IV. BIOCHEMICAL MECHANISMS

A number of authors [1, 5, 10, 128] have attempted to list the possible biochemical changes which may underlie mutational resistance to a toxicant. These changes may lead to: (a) inactivation of the toxicant; (b) decreased conversion of a nontoxic into a toxic compound; (c) decreased membrane permeability to the toxic molecule; (d) modification of the cellular component with which the toxicant interacts (sensitive site); (e) increased levels of the inhibited enzyme; (f) decreased requirement for the product of the inhibited reaction; (g) increased concentration of an antagonist to the toxicant; and (h) operation of an alternate pathway bypassing the inhibited reaction. It should be understood that for the same toxicant different mechanisms may give rise to an increment of resistance in different organisms or in the same organism, depending on the mutant gene. For example, resistance to the antimetabolite 2-deoxy-D-glucose in a mutant of S. cerevisiae results from an increase in the activities of phosphatases with fairly high specificity for 2-deoxy-D-glucose-6-phosphate, a detoxification mechanism [129]. In mutants of C. lagopus, on the other hand, the resistant phenotype results from a block to the transport of the inhibitory analogue, an impermeability mechanism [130]. In different mutants of S. cerevisiae, resistance to 5-fluorocytosine may arise from loss of cytosine-specific permease, cytosine deaminase or uridine-5-phosphate pyrophosphorylase as well as from increased de novo synthesis of pyrimidines [78]. Of course, not all of the biochemical mechanisms of resistance are of great importance. In bacterial resistance to most of the important antibiotics, for example, only mechanisms (a), (c), and (d) are significant [7]. Mechanisms (f) and (h), in particular, do not seem to be operative in antibiotic resistance [1]. In fungi a mutational difference in sensitivity to antimycin A is related to the operation of an alternate pathway, but in this case it is the wild type which is resistant [63]. Mutational creation of an alternate pathway in a species which is truly negative would be difficult, but if the ability is present and suppressed, mutation at the suppressor locus would lead to one-step resistance. In this section the most important biochemical mechanisms will be discussed with emphasis primarily on antifungal compounds.

A. Detoxification

Destruction or inactivation of a toxic compound may take place outside the cell. Resistance development may thus be the result of the acquisition of an ability to form and excrete an enzyme such as β-lactamase (penicillinase) which hydrolyzes the β-lactam bond in the molecule of penicillins or cephalosporins, converting them to antibiotically inactive forms [5]. In the field of antifungal compounds, noncharacterized, inducible detoxifying systems have been found responsible for the unstable, and apparently nongenetic, resistance of dermatophytes to nystatin, 2-chloro-4-nitrophenol, and 5,7-dichloro-8-oxyquinoline [131, 132]. Detoxification can take place not only by chemical transformation but also by binding to compounds of no vital importance for cell economy before the toxicant can exert its effect. From the field of agricultural fungicides, one example of resistance development by the latter mechanism will be given below.

Organomercury compounds were generally successful in the seed treatment against *Pyrenophora avenae* until about 1960. In more recent years, failures have been reported [133-135] from various countries and it is now accepted that these are due to distinct resistance to organomercurials by many isolates of the fungus [97]. The work of Greenaway [136, 137] has shown that these isolates owe their tolerance to a preexisting pool of metabolically inessential compounds which bind and inactivate the toxic ions. Red anthraquinone pigments seem to be responsible for such inactivation, since they are capable of removal of toxicant from aqueous solution, presumably by chelation [136]. This resistance has not been subjected to genetic analysis but it is possible that single-gene differences account for differences in both sensitivity to mercurials and size of the intracellular pool of pigments [97]. Although a resistant strain has not been derived from a sensitive one experimentally, apparently this has happened in nature. It appears, therefore, that mutations for an elevated production of the toxicant-binding pigments are responsible for resistance to organomercurial fungicides. Isolates resistant to mercuric chloride have also been recognized in *P. avenae* [138]. This resistance does not appear to result from increased levels of pigments. A detoxification mechanism, based on a pool of low-molecular-weight thiols, has been suggested although no difference between resistant and sensitive isolates in the amount of extractable nonprotein thiol could be established [139]. In bacterial strains with plasmidborne genes for resistance to mercuric chloride, a detoxification mechanism exists which is based on the conversion of mercury to volatile forms [140, 141].

Fungicide resistance, resulting from an ability for chemical change of a fungicidal compound into nontoxic derivatives, is often responsible for naturally occurring insensitivity of wild-type strains of fungal species. The high insensitivity of several species to the antibiotic ascochitin appears to be a result, at least in part, of biological reduction into the less fungitoxic dihydro derivative [142]. Similarly, a positive correlation between pentachloronitrobenzene insensitivity and the ability for conversion into pentachloroaniline and pentachlorothioanisole has been reported [143]. That such a conversion mechanism may be responsible for sensitivity differences between strains of the same species is shown by the example of two differentially sensitive strains of *Botrytis cinerea* [144]. Resistance to dodine may also be due to a detoxification mechanism, suggested to be responsible for the relative insensitivity of conidia of wild-type *Fusarium solani* f. sp. *phaseoli* [145]. Such conidia took up ^{14}C-labeled dodine from solutions that inhibited germination and released a ^{14}C-labeled derivative that was markedly less toxic. In this case the amount of bound radioactivity increased to a maximum and then dropped [145]. In preliminary experiments with the related fungus *N. haematococca*, this drop appeared sharper in conidia of a dodine-resistant mutant [59] which might indicate a more effective detoxification.

B. Decreased Conversion

The analogues of purine and pyrimidine bases provide classical examples of compounds which are not toxic unless they are upgraded (converted to nucleotides). The conversions are catalyzed by the same enzymes that convert the normal bases to nucleotides. These enzymes normally allow cells to scavenge exogenous supplies of the natural purines and pyrimidines for the synthesis of nucleic acids. However, in the cell nucleotides can be produced by other pathways in which the bases are not intermediates. Therefore, the enzymes which upgrade purines and pyrimidines are optional and can be lost by mutation without seriously impairing growth. Such a mutation, however, results in the failure to convert one or more base analogues to nucleotides, and thereby confers resistance to the analogues. This has long been known both in microorganisms and in animal cell lines [146].

A very good example of such a mechanism of resistance to a base analogue in a plant pathogenic fungus is provided by the studies of Dekker [147, 148] with 6-azauracil (AzU). Although this compound is not used commercially, it has proved experimentally effective in the systemic control of scab and powdery mildew of cucumber [149]. In the scab fungus *Cladosporium cucumerinum* inhibition of spore germination and growth was found to result not from direct action of AzU, but from 6-azauridine-5'-phosphate (AzUMP) which is synthesized by the organism via 6-azauridine (AzUR). Metabolism of orotic acid, an intermediate in the synthesis of uridine-5'-phosphate (UMP), was shown to be inhibited by AzUMP, presumably by action on the enzyme orotidine-5'-phosphate (OMP) decarboxylase. The decisive importance of metabolic conversion of AzU for toxicity was demonstrated by the use of a UV-induced strain (strain III) which was resistant to AzU but not to AzUR or AzUMP. The mechanism of resistance in this case was shown to be the inability of the resistant cells to convert AzU to AzUR. Apparently loss of activity of the enzyme uridine phosphorylase in strain III also led to inability to convert the normal metabolite uracil into uridine but this was of no importance to the organism [147].

Unlike strain III, another resistant strain (R) or C. cucumerinum does not lack uridine phosphorylase activity but is resistant to AzUR and AzUMP as well as AzU [150]. In this strain the presence of an altered uridine kinase apparently leads to a decreased conversion of AzUR to AzUMP with the result of a poorer inhibition of OMP-decarboxylase. This mechanism is sufficient to explain resistance to AzU and AzUR, but for resistance to AzUMP, a mechanism of decreased permeability or detoxification was suggested [150]. One might suspect that strain R carries more than one mutation. Pleiotropy is, of course, another possibility.

Another base analogue, 5-fluorocytosine, is very valuable for the chemotherapy of fungal infections in man [151, 152] but its use has been severely limited by the development of resistance. The various mechanisms by which resistance to this antimetabolite may develop were mentioned at the beginning of this section. Grenson [153] and Jund and Lacroute [78] have studied biochemical mechanisms of resistance to other base analogues as well in S. cerevisiae.

Of the fungicides which do not act as antimetabolites, pyrazophos is toxic only if it is converted by the fungus into 2-hydroxy-5-methyl-6-ethoxycarbonylpyrazole-(1,5-α)-pyrimidine (PP). Fungi such as Pythium debaryanum and S. cerevisiae which cannot convert pyrazophos into PP are insensitive [154]. In view of this fact, one would expect that mutational alteration of the ability for metabolic conversion would result in resistant strains of sensitive fungi. However, such mutants have not yet been reported.

C. Decreased Permeability

Fungal cells are surrounded by a semipermeable membrane and unless a fungicide acts on the surface of this membrane it must penetrate it before it can reach other sites of action. This penetration can take place in a number of ways [155], but if permeability of the membrane to a particular type of toxic molecule is subject to genetic control, then a mutation reducing permeability will confer resistance. In such a case, the toxicant sensitivity of cell-free preparations from mutant cells should not differ substantially from that of preparations from wild-type cells. Thus the protein-synthesizing system of a blasticidin-S resistant strain of Pyricularia oryzae was sensitive to the antibiotic and it was concluded that resistance in vivo resulted from decreased penetration of the compound through the cytoplasmic membrane of the resistant cells [156]. Similarly, the in vitro protein synthesis of N. crassa was strongly inhibited by edeine but this inhibition was not affected by a mutation to edeine resistance [157]. The in vivo resistance of this mutant was also tentatively attributed to a block in edeine uptake. Of course, the possibility of removal of some toxicant-inactivating factor in the preparation of the cell-free systems from mutant cells should not be overlooked in such cases.

It should be stressed that a difference between mutant and wild-type cells in the amount of fungicide removed from ambient solution does not constitute proof of a decreased permeability mechanism. Such a difference may result from a decrease in the number, affinity, or activity of the intracellular receptors of the compound. In other instances, impaired uptake is indeed due to alterations in membrane activity, the loss of stereospecific concentrating systems or permeases. The examples of 5-fluorocytosine resistance in S. cerevisiae [78] and of 2-deoxy-D-glucose resistance in C. lagopus [130] resulting from loss of permease activity were mentioned at the beginning of this section. In the latter case, the transport system proved to be of rather wide specificity and was involved in the uptake of at least four hexoses including fructose. One class of p-fluorophenylalanine-resistant mutants of A. nidulans seem to have an impaired capacity for the transport of several amino acids and their analogues [158]. Similarly, a pair of allelic genes in S. cerevisiae control the activity of an amino acid permease system responsible for the accumulation of all the amino acids and their analogues. Strains carrying the mutant gene are resistant to a number of amino acid analogues as a result of reduced activity of the permease [159].

Of the agricultural fungicides, carbendazim (methyl benzimidazole-2-yl-carbamate) has been reported to enter cells of Sporobolomyces roseus by the operation of an active transport system which is dependent on energy, pH, temperature, and fungicide concentration [313]. Impairment of this uptake system has been suggested as the basis of carbendazim resistance in a strain of this species. Another example is provided by polyoxin resistance of Alternaria kikuchiana [223, 314]. In this case, sensitive and resistant strains do not seem to differ in the nonspecific adsorption of the antibiotic on cell walls. However, great differences were found in the amounts which were accumulated in the cytoplasm by an apparently active transport system with a very pronounced pH dependence.

Decreased permeability also of mitochondrial and nuclear membranes may confer resistance if the target site of the fungicide is inside these structures. The need for distinguishing between reduced permeability and reduced binding at the site must be stressed also in this case. In a mutant of Torulopsis utilis, for example, antimycin resistance was not a transport phenomenon but appeared to result from reduced binding of the inhibitor to the mitochondrial respiratory particle [160]. Bunn et al. [42], on the other hand, have described a number of S. cerevisieae mutants resistant to mitochondrial protein synthesis inhibitors as probable mitochondrial membrane mutants. Evidence for such a barrier to permeability at the level of the mitochondrial membrane had been obtained earlier by Thomas and Wilkie [39] who showed that in some chromosomal mutants resistance was dependent upon existence of intact mitochondrial membrane.

D. Site Modification

Mutational alteration of the cellular site at which a toxicant exerts its effect so that the site has a reduced affinity for, or is otherwise less affected by the toxic molecule, is undoubtedly the most interesting mechanism for resistance development. In fact, some of the most fascinating studies in biochemical genetics have used such mutations for the understanding of cellular mechanisms. If the site with which the toxicant interacts has been resolved to a single cellular component, i.e., one protein, the gene action may be elucidated rather easily. The problem is a more difficult one to solve in case of toxicants which act on complex structures such as the ribosomes or the complexes of the mitochondrial electron transport chain. In this case, the use of resistant mutants may help to determine which component of a complex is the important one for toxicity. Such studies with bacterial ribosomes have revealed not only which ribosomal subunit is involved but often whether RNA or protein and even which protein interacts with the antibiotic [7]. In the following discussion, site modification as a mechanism of fungicide resistance will be treated according to the process affected.

1. Nucleic Acid Synthesis and Nuclear Function

Today no commercial fungicide is known to act selectively by inhibition of DNA and RNA synthesis. Of the fungicides discussed by Sisler [161] as possible inhibitors of nucleic acid synthesis in 1969, none has proved in the meantime to have such a mechanism of action, excluding, of course, the analogues of purines and pyrimidines. In fact, of these compounds, the benzimidazole fungicides have been shown to prevent mitosis, their effect on DNA and RNA synthesis being indirect [114, 115]. There is evidence that the action of chloroneb also is not directly related to the synthesis of nucleic acids [162]. Resistance to base analogues, on the other hand, is usually the result of decreased permeability or decreased ability to upgrade the analogue or elevated production of the normal base, as discussed earlier.

Inhibitors acting on DNA have been surveyed rather recently and have been grouped according to their preferential inhibition of DNA-dependent synthesis of DNA or RNA [163]. Some of these inhibitors are toxic to fungal cells although they are not used as fungicides. In A. nidulans, for example, a 70% inhibition of colony growth was obtained by a concentration of 50 μM actinomycin D in the agar medium [119]. Phytoactin, another polypeptide antibiotic, probably acting like actinomycin in binding to DNA and preventing DNA-dependent RNA polymerase, is also toxic to several fungi and inhibits RNA synthesis in Saccharomyces pastorianus [164]. Fungal mutants resistant to these antibiotics have apparently not been reported. Fungal resistance to acriflavin, an acridine derivative, has long been known [68], but references to studies of its biochemical basis do not seem to be available. Resistance to ethidium bromide, an intercalator of double-stranded DNA, has recently been studied in S. cerevisiae [165]. In this organism, the compound inhibits mitochondrial function by interacting with mitochondrial DNA and inhibiting its replication and transcription. However, resistance does not result from modification of these systems but rather from a change in the mitochondrial membrane [165].

In contrast to nucleic acid synthesis, interference with chromosome separation at mitosis acquired major importance as a mechanism of toxicity of agricultural fungicides with the elucidation of the action of benzimidazole derivatives. Of these, the mechanism of action of methyl benzimidazol-2-yl carbamate (MBC, carbendazim) is the most thoroughly studied [114, 115, 166], since it is the important fungitoxic product of benomyl and thiophanate-methyl [167, 168]. Details on the mechanism of fungitoxicity of the benzimidazole fungicides are given in Chapter 10. It is sufficient to state here that the toxicity of MBC at least to S. cerevisiae, Ustilago maydis, and A. nidulans is due to its antimitotic activity. The mechanism of action of other benzimidazole fungicides appears generally similar to that of MBC, since the ability to induce mitotic segregation in diploid A. nidulans is a property common to all [116, 117].

Davidse [169, 170] made a very thorough study of the biochemical genetics of MBC resistance in A. nidulans. In this fungus two loci conditioning sensitivity to benzimidazole derivatives are known and mutation at each results in a different tolerance level [62]. Van Tuyl et al. [84] studied a large number of mutants at the high resistance locus ben-1 and made the interesting observation that positively correlated cross-resistance between benomyl, MBC, and thiabendazole (TBZ) is not a rule without exception. Rare mutations within this locus may give resistance to benomyl but not to TBZ, or increased resistance to TBZ with increased sensitivity to benomyl (negative correlation). This provided Davidse [169] with two allelic mutants, one of high resistance and one of high sensitivity to MBC, to compare to the wild type. On the basis of the similarity between the antimitotic activity of MBC in fungi and that of colchicine in animal and plant cells, Davidse speculated that MBC binds to spindle microtubule subunits and prevents assembly into normal fibers. This hypothesis seems to be correct, since gene-controlled sensitivity differences in A. nidulans correlate very well with the amount of MBC bound to a macromolecule present in cell-free extracts [169]. The magnitude of these differences was approximately the same, independently of whether mycelium was incubated with MBC in vivo and then extracted or cell-free extracts of mycelium were incubated with MBC in vitro. Although the macromolecule involved has been only partially purified and characterized, there is good evidence that it is identical with the spindle microtubular protein (tubulin) in A. nidulans. For example, its electrophoretic properties are similar to those of brain tubulin. Incorporation of labeled fungal protein into microtubules in a heterologous assembly system suggests copolymerization of fungal and brain tubulin [169]. Furthermore, MBC binding to this fungal protein was competitively inhibited by known antitubulins such as oncodazole and colchicine. The electrophoretic analysis of partially purified preparations of the MBC-protein complex showed similarities to mammalian tubulin monomers [170].

Since mutational modification of MBC sensitivity does not affect growth in fungicide-free media, normal assembly and functioning of microtubules is apparently not affected. Probably mutations to resistance and increased sensitivity, within the ben-1 locus, change the amino acid composition of the MBC-binding site on the tubulin molecule and this leads to decreased or increased affinity for the toxicant. The effect of mutation at the ben-2 locus which gives a lower level of tolerance to benzimidazoles has not been studied.

While the work with A. nidulans has conclusively shown involvement of tubulin modification in resistance to benzimidazole, work with two other species suggests two other mechanisms of resistance. The suggested impairment of an active transport system for MBC uptake was mentioned earlier in this chapter [313]. Recent work with Verticillium malthousei suggests that benomyl tolerance is related to natural acid production in this species [315]. The unidentified acid was believed to lower the pH of the medium and reduce benomyl effectiveness. It cannot be excluded that different resistance mechanisms operate in different organisms.

The toxicity of another group of fungicides, namely the aromatic hydrocarbons appears to be, at least partly, the result of interference with the hereditary processes [9]. Hexachlorobenzene, pentachloronitrobenzene (PCNB), 2,6-dichloro-4-nitroaniline (dicloran), diphenyl, sodium o-phenylphenate, and chloroneb are some of the commercial fungicides of this group. Until recently the only reason to group these compounds together was the cross-resistance of mutants of various fungi selected on media containing any member of the group [162, 174, 175]. The aromatic hydrocarbon nucleus is a common characteristic but not sufficient to include a fungicide in this group; daconil (tetrachloroisophthalonitrile), for example, does have such a nucleus but apparently a different biological activity [119, 162]. Recent work [119] has shown that probably all aromatic hydrocarbon fungicides induce sectoring in A. nidulans diploids. This property provides a second characteristic for the group. Action on chromosomes and mitosis in other organisms had earlier been suggested for some of the aromatic hydrocarbon fungicides [18, 104], but such action had not been accepted as the cause of their fungitoxicity. Effects on membrane permeability [176], protein synthesis [177], and wall composition [178-180] have been reported. The aromatic hydrocarbons probably act as structurally nonspecific, Ferguson-type toxicants which accumulate at interfaces. They may, however, induce somatic segregation at concentrations comparable to those at which specific inhibitors of nuclear processes have similar effects. Such concentrations are often only slightly inhibitory to growth [119] which indicates that nuclear function is one of the most sensitive processes, if not the most sensitive of all to the aromatic hydrocarbons.

Sensitive fungi develop resistance to aromatic hydrocarbon fungicides very easily. However, little is known about the resistance mechanism. Tillman and Sisler [162], working with a chloroneb resistant mutant of *U. maydis*, found that the mutation did not affect uptake or metabolism of chloroneb. Since all aromatic hydrocarbon fungicides induce somatic segregation, modification of a nuclear component would be sufficient to explain a reduced sensitivity of nuclear function, if not of growth, to all members of the group. It should be stated that the mechanism by which sectoring is induced may differ for the benzimidazoles and the hydrocarbons. Mutants resistant to benzimidazoles are not cross-resistant to hydrocarbons [181]. Furthermore, while the benzimidazole fungicides seem to induce sectoring in diploid *A. nidulans* by the mechanism of nondisjunction of chromosomes, members of the aromatic hydrocarbon group definitely cause chromosome breaks [182] and may also have other effects. This is in agreement with the nonspecific nature of the latter compounds, but in this case resistance resulting from mutation of a single gene which does not control permeability or detoxification is difficult to understand. Tillman and Sisler [162] suggest that partition in cellular regions is the important factor and that a mutation may make partitioning less favorable for all members of this group of fungicides.

The antifungal antibiotic griseofulvin has also long been known to inhibit mitosis [183] but it was not until rather recently that its antimitotic activity was conclusively demonstrated in fungi [112, 113]. Griseofulvin also induces somatic segregation. Growth inhibition of *A. nidulans* increases in proportion with the number of induced sectors [118]. Most likely, antimitotic action is the cause of fungitoxicity. Although some "training" of fungi to tolerate serially higher griseofulvin concentrations seems possible [183] no mutationally well-defined resistance seems to have been reported. A *ben*-1 mutant of *A. nidulans* resistant to MBC was not cross-resistant to griseofulvin [118], although it apparently had a modified spindle protein [169]. Probably there are differences between the two compounds in their mechanism of interference with nuclear function. In fact, MBC seems to have a colchicine-type mechanism [114] while griseofulvin does not seem to prevent the assembly of microtubules [184] its effect being rather similar to that of isopropyl-*N*-phenyl carbamate [113]. Undoubtedly much interesting work remains to be done on the action of fungicides affecting fungal nuclei and on the possible mutational modifications which influence their effects.

2. Protein Synthesis

Several amino acid analogues may, in a sense, be regarded as inhibitors of protein synthesis. Mutational resistance to such analogues, if not related to permease activity [159, 185] may result from modification of an intracellular site, namely the corresponding tRNA (transfer RNA) or activating enzyme. A good example is ethionine resistance in a mutant (E_2) of Coprinus lagopus studied by Lewis [102]. In this mutant the gene for resistance to ethionine is the structural gene for the methionine- and ethionine-activating enzyme. In the mutated form, the sites of the enzyme have been altered so that they effectively fit methionine but not ethionine. By contrast, a temperature-sensitive mutant of N. crassa is resistant to ethionine not because of an altered activating enzyme or tRNA but because it has lost repression control over its methionine production [186, 187]. The normal metabolite is thus synthesized in excessive quantities compared to the amount of the competitive inhibitor provided by the medium.

Before discussing inhibition of protein synthesis by selective inhibitors in eucaryots, it is necessary to distinguish between cytoplasmic (80-S ribosomes) and mitochondrial (70-S ribosomes) protein synthesis. The latter system resembles the one found in bacteria in a number of ways, including sensitivity to inhibitors. For the most part, a compound inhibiting the functions of 80-S ribosomes has no effect on 70-S ribosomes and vice versa, but there are certainly exceptions to this rule. As explained in Chapter 11, inhibition of cytoplasmic protein synthesis in fungi is lethal, but inhibition of mitochondrial protein synthesis does not necessarily cause inhibition of growth. To select mutants resistant to the inhibitors of the latter category it is often necessary to prevent anaerobiosis. In the following discussion resistance to cytoplasmic protein synthesis inhibitors will be treated first.

Cycloheximide (Actidione), now the most popular inhibitor of cytoplasmic protein synthesis, inhibits a stage after the formation of a peptide bond on the ribosomes [188-191]. The mechanism of resistance to this fungicide has been studied in S. cerevisiae [93] and N. crassa [85, 192], although mutational decrease of sensitivity is also known in A. nidulans [94]. In S. cerevisiae, out of eight genes which may mutate to cycloheximide resistance, at least one, of a low resistance level, is expressed in vitro [93]. This resistance is associated with the ribosomes, which is in agreement with the conclusion previously reached by Siegel and Sisler [190] on the basis of species specific differences in cycloheximide sensitivity in yeasts. Other resistance genes in S. cerevisiae are not expressed in vitro, their action probably being associated with some impermeability or detoxification mechanism [93].

The system in N. crassa appears more suitable for the study of modification of ribosomal sites associated with cycloheximide resistance. All of three genes studied give high resistance in vitro and all control ribosomal properties [85]. In the same fungus, the recovery of resistant mutants following UV-irradiation can be greatly increased if the irradiated conidia are incubated for several hours prior to the exposure to the fungicide [193]. This effect of preincubation was attributed [192] to the time required for the synthesis of new, insensitive ribosomes. However, in the case of toxicants inhibiting protein synthesis, preincubation would help recovery of resistant mutants with any modified component for which new protein synthesis is required. In wild-type S. fragilis, resistance is associated with the 60-S subunit [194] and this appears to be the case with resistant mutants of S. cerevisiae [195]. Components of this subunit, therefore, must be controlled by the genes for cycloheximide resistance. According to the current view, these include one species of RNA and a considerable number of different proteins [196]. It is believed that all components act in a cooperative way in supporting ribosomal functions. Alterations of individual components may affect conformation of ribosomes and may have pleiotropic effects. It is important to note that in single-gene mutants of N. crassa, protein synthesis proceeds normally in the absence of cycloheximide in spite of ribosome modification. In fact, the only fungal 80-S ribosomal mutations isolated so far which do not seem to impair ribosome function [67] are those selected for resistance to inhibitors. However, cycloheximide-resistance mutations interact in double resistant homocaryotic recombinants [66] and since all genes involved modify ribosomes, a different component is probably coded for by each gene. Furthermore, the fact that such recombinants may grow poorly in the absence of cycloheximide suggests that alteration of more than one component may become an adverse factor for ribosome function [85].

The mechanism of resistance to cryptopleurine, a phenanthrene alkaloid from plants, and to trichodermin, a fungal toxin, has recently been studied thoroughly in S. cerevisiae [197-199]. The first compound inhibits primarily protein synthesis in cell-free preparations, apparently acting during the peptide elongation phase. Resistant mutants of yeast are recessive, map at a locus linked to mating type, and are cross-resistant to the phenanthrene alkaloids tylophorine and tylocrebrine. Resistance is expressed in vitro and is associated with the ribosomes. Grant et al. [198] used various combinations of subunits to reconstitute ribosomes and showed that only those combinations containing 40-S subunits from mutant cells were resistant to the inhibitor. The resistance locus, therefore, specifies a component of the 40-S subunit which seems to be important for stability, since 40-S subunits from mutant cells lose activity in vitro much faster than those from wild-type cells [197, 198]. The locus for cryptopleurine resistance provides the first genetic marker for the 40-S ribosomal subunit in any eucaryot. Trichodermin specifically inhibits the protein synthesis of eucaryots, apparently by interference with the peptidyl transferase reaction on the ribosomes. Only one mutant of S. cerevisiae resistant to trichodermin has been studied and this has a modified 60-S subunit [199]. As might be expected, cryptopleurine-resistant mutants were not cross-resistant to cycloheximide or trichodermin [198]. It is yet unclear whether any of these mutations affect ribosomal proteins, ribosomal RNA, or modifying enzymes.

Blasticidin S is agriculturally more important than the inhibitors of protein synthesis mentioned above, for it is used commercially to control rice blast caused by Pyricularia oryzae [200]. In cell-free extracts of this fungus, blasticidin S inhibits the same stage of protein synthesis as cycloheximide [200]. Unlike cycloheximide, however, it is toxic to bacteria. It must be assumed that in sensitive fungi, blasticidin S inhibits both cytoplasmic and mitochondrial protein synthesis. Only one fungal strain resistant to blasticidin S seems to have been studied from the point of view of resistance mechanism and this appeared to be a permeability mutant [156]. However, resistance by site modification should be possible since the protein-synthesizing system of the resistant species, Pellicularia sasakii, was found [201] to be resistant to the antibiotic. Cross-resistance experiments with mutants resistant to other protein synthesis inhibitors might be helpful in this respect. Kasugamycin, which is also used for the control of P. oryzae in Japan, is another antibiotic acting by inhibition of protein synthesis. Resistant strains have been produced in the laboratory [202] and were recently found in the field [203]. There do not seem to be any reports on the mechanism of kasugamycin resistance in fungi available. In Escherichia coli, resistance results from mutations that inactivate an RNA methylase which dimethylates adjacent adenines of the 16-S ribosomal RNA [204]. Recently, however, a new kasugamycin-resistance mutation was reported which is associated with alternation of the ribosomal protein S2 [316].

Anisomycin has an effect on the cell-free protein-synthesizing system of yeast similar to that of cycloheximide [205]. Resistant strains have been reported [195, 206]. In yeast, all anisomycin-resistant strains tested seemed to be impermeable to the antibiotic, since they all gave sensitive in vitro protein-synthesizing systems [195]. Vomvoyanni [67] found that the cycloheximide-sensitive (cytoplasmic) system of N. crassa is strongly inhibited by anisomycin but cycloheximide-resistant mutant systems were not cross-resistant to this antibiotic. The same systems were found quite sensitive to fusidic acid [67], an antibacterial antibiotic inhibiting the same stage of protein synthesis but by action on one of the supernatant enzymes rather than the ribosomes [207]. Again no positive correlation between sensitivity to fusidic acid and cycloheximide was observed [67]. Finally, edeine, an oligopeptide antibiotic, seems to inhibit both bacterial and fungal protein synthesizing systems but, in this case also, the only resistant fungal strain that has been studied appears to be a permeability mutant [73, 157].

A number of antibacterial antibiotics, including chloramphenicol, tetracycline, and erythromycin, specifically inhibit the mitochondrial protein-synthesizing system in fungi. Resistance has been extensively studied in S. cerevisiae by assessing the ability of the organism to grow on a nonfermentable substrate in the presence of each of the antibiotics. Chromosomally inherited whole-cell resistance to chloramphenicol and tetracycline was found by Linnane et al. [40] to be associated with a specific alteration in cell permeability to both drugs. Rank [46], on the other hand, considers that the pleiotropic chromosomal genes for resistance to these and other inhibitors of mitochondrial function code for a fundamental component of mitochondrial inner membrane. Cytoplasmically determined chloramphenicol and micamycin resistance is lost upon isolation of mitochondria. Bunn et al. [42] believe that such mutants possess mitochondrial inner membranes which are impermeable to the antibiotics in vivo, but which are made leaky by the process of isolation. It is agreed that mitochondrially inherited resistance to erythromycin is the result of alterations in the mitochondrial ribosome [40, 44]. In erythromycin-resistant mutants of bacteria, the modified component is a protein of the large ribosomal subunit [208]. It does not seem to be known whether the same is true for the mitochondrial ribosomes of yeast [44]. In A. nidulans most, but not all, mutants mapping at one of the three known chromosomal loci for chloramphenicol resistance are highly sensitive to cycloheximide [215]. Such negatively correlated cross-resistance between a cytoplasmic and a mitochondrial protein-synthesis inhibitor does not seem to be known for other fungi and is worth further investigation.

Convincing evidence for selective action on the protein-synthesizing system is not available for any of the synthetic fungicides. In experiments with whole cells, dicloran inhibited protein synthesis in Rhizopus arrhizus at concentrations having no effect on the synthesis of DNA and RNA [177]. Mutants resistant to this compound and to other members of the aromatic hydrocarbon group are frequent [9, 11], but the mechanism is not yet known. In the experiments with R. arrhizus, the inhibition of incorporation of amino acids into protein definitely seems too moderate to account for fungitoxicity. Experiments with cell-free systems have not been carried out. In addition, dicloran has a definite effect on nuclear function as was discussed earlier in this section, and even if resistance developed through modification of an intracellular site, it would not necessarily have to be a site related to protein synthesis.

3. Energy Production

Under this heading we shall discuss resistance to toxicants which act by inhibition of (a) glycolysis, (b) the tricarboxylic acid (TCA) cycle, (c) respiratory electron transport, and (d) oxidative phosphorylation, all together representing a very large and diverse group of inhibitors. Kaars Sijpesteijn [27] classifies most of the nonsystemic agricultural fungicides as inhibitors of energy production. Many of these, including captan, folpet, nabam, maneb, zineb, and other fungicides acting as thiol reagents, act by multisite inhibition. Development of resistance to these, at least by mutational site alteration, is unlikely. There are, however, certain nonsystemics which may inhibit energy production by acting on a specific site. Trisubstituted tin fungicides, for example, selectively inhibit oxidative phosphorylation in mammals, binding to a specific site in mitochondria [209]. The sensitivity of S. cerevisiae to low levels of triethyltin (TET) sulfate is consistent with its action on mitochondrial energy conservation, although at higher concentrations the toxicant may affect ion transport [28]. In this organism a number of cytoplasmic TET-resistant mutants have been isolated and have been found cross-resistant to other trisubstituted (but not disubstituted) tin fungicides. Cross resistance tests with other inhibitors of mitochondrial function permitted grouping of these mutants into three classes. Of these, only one seems to involve a change in the actual binding site for TET. Mutations of the other two classes may cause a general conformational change in structural protein or lipid component of the inner mitochondrial membrane leading to multiple resistance to inhibitors [28].

Kaars Sijpesteijn [27] classified all the systemic agricultural fungicides as inhibitors of biosynthesis. Since then it has been shown that this is not true in the case of carboxin and other systemic carboxamide fungicides [29-35]. These fungicides inhibit energy production by preventing the transport of electrons specifically from succinate to coenzyme Q (CoQ) in the mitocondrial respiratory chain (complex II). The mechanism of carboxin resistance has been studied in U. maydis (33, 35, 311], A. nidulans [37], and Ustilago hordei [317]. In the first two organisms, chromosomal genes for resistance to carboxamides have been identified, while in U. hordei this resistance is reported to be polygenic [318].

Although a much higher carboxin concentration is needed for in vivo inhibition of wild-type A. nidulans than for U. maydis, mitochondrial preparations of both organisms are equally sensitive to very low concentrations of the fungicide. While this difference may be due to permeability or to the presence of a detoxification mechanism in whole cells of A. nidulans [37], sensitivity differences between mutant and wild-type strains of both organisms are the same in vivo and in vitro. Preparations from mutants (oxr-1A and oxr-1B) of U. maydis are not only resistant to carboxin but are also easily distinguished from wild-type preparations in the absence of the fungicide because they lose their succinic dehydrogenase activity at a much faster rate [33, 35, 311]. This difference is restricted to the carboxin-sensitive segment of the electron transport system. Aging does not differentially affect other enzyme activities of mutant and wild-type preparations. The same is true also for the carboxin-insensitive portion (approximately 50%) of complex II activity, assayed with 2,6-dichlorophenol-indophenol in the presence of phenazine methosulfate [35]. Therefore the resistance mutation modifies the site of action of the fungicide. Probably the oxr-1 locus codes for a component of complex II which is responsible for carboxin sensitivity and is at the same time important for stability, but the chemical nature of this component is not known. Recent studies with mammalian preparations indicate that carboxamides interact with an Fe-S center of succinic dehydrogenase [319] and it is not unlikely that such a center is modified by the carboxin-resistance mutations in fungi. This idea is supported by the observed [311] cross-resistance between carboxin and thenoyltrifluoroacetone (TTFA), a known iron chelator the action of which seems to be similar to that of carboxamides [320]. It is worth noting that in the heterozygous diploid, and probably also in the dicaryotic cell, both the wild-type and the mutant component are synthesized and appear to have an equal opportunity to participate in enzyme arrays of complex II [35].

In A. nidulans, accelerated loss of succinic dehydrogenase activity in mitochondrial preparations from carboxin-resistant mutants has not been reported [37]. Of three mutations at different loci, none appears to cause decreased permeability of the cytoplasmic or the outer mitochondrial membrane to the fungicide. It has been suggested that mutation at one locus (cbx-B) alters the site of action of carboxin because there is a small concentration dependence of resistance in this mutant. For the remaining two mutations (cbx-A and cbx-C) some effect on the mitochondrial inner membrane, not necessarily at the exact binding site of carboxin, was considered possible [37]. It is very interesting that the isolation of double-resistant recombinants was attempted in A. nidulans but apparently without success. In all three mutant × mutant crosses, no progeny class of higher resistance was obtained, which might be due to the absence of interallelic positive interaction. But in at least two of these crosses the ratio of wild type to resistant progeny in a random sample was approximately 1:1 instead of 1:3. If this is true, then probably the double recombinant class is not viable [37]. This is reminiscent of cycloheximide resistance in N. crassa where the double recombinant grows poorly [66], as discussed earlier in this section, and it probably means that more than one mutational modifications of complex II in a haploid cell impair its function. In the plant parasitic dicaryon of U. maydis the presence of two nonallelic genes for carboxin resistance, one in each of the two nuclei, does not seem to affect normal behavior [35].

In U. maydis, but not in A. nidulans, another class of mutants (ants), of low carboxin resistance, are devoid of the antimycin-insensitive alternate electron transport pathway present in wild-type cells. One of these mutants has been studied and has been found chromosomal in nature [63]. The ability of carboxin to select for such mutants indicates that its site of action is somehow related to the operation of the alternate pathway. Perhaps this site is at or near the branching point between the normal and the alternate electron transport. In U. maydis this point is definitely before cytochrome b because the alternate pathway does not readily reoxidize reduced cytochrome b [63]. In Euglena gracilis it has been found [210] that a similar alternate electron transport pathway is actually a pathway for succinate oxidation which only indirectly can oxidize NADH. In such a case the branching must be located within complex II. In fact, the terminal oxidase in this pathway may be a component of complex II [210]. Lyr and Schewe [321] suggest that this oxidase is the Fe-S protein of complex II. Involvement of Fe-S proteins in the alternate respiratory chain is gaining much support in recent years [322]. In such a case the ants mutation may affect an Fe-S protein of complex II responsible for both carboxin sensitivity and alternate respiration. This hypothesis is supported by the observation that the complex II activity from ants mutant cells appeared to be less stable than the one from wild-type cells [33].

In U. maydis and A. nidulans, mutations for in vivo resistance to carboxamides are also responsible for a similar resistance of the succinic dehydrogenase system in vitro but the results with U. hordei appear to be in disagreement. In the latter organism, carboxin resistance is claimed [317] to result only partially from carboxin-tolerant succinic dehydrogenase. Increased activity of the glyoxylate pathway and a tolerant system of phosphate uptake were thought to be mainly responsible for in vivo resistance. If carboxin-resistant strains of U. hordei really carry more than one mutant gene [318], it is conceivable that they may differ from the wild type in more than one cellular component.

Other inhibitors of respiratory electron transport not of immediate importance as agricultural fungicides include rotenone and antimycin A. Rotenone inhibits transport of electrons between NADH and coenzyme Q (complex I) while the action of antimycin A has been placed in the bc_1 region of the cytochrome chain (complex III). Most fungi are resistant to rotenone but in S. cerevisiae chromosomally controlled sensitive mutants have been obtained [211]. At least one of the two mutations studied seems to control the NADH dehydrogenase complex of the inner membrane of mitochondria, probably by affecting a mitochondrial protein capable of rotenone inactivation in the wild type [211]. Unlike rotenone, antimycin A is highly toxic to fungi which do not possess an insensitive electron transport pathway. The first described fungal mutant resistant to antimycin A was one of Venturia inaequalis which grew atypically in the absence of the antibiotic and had also lost pathogenicity [212]. No information on the effect of this mutation on respiratory activity in vivo and in vitro has been given. More recently, two antimycin-resistant mutants of the yeast Candida utilis have been described [160, 213]. In both cases chromosomal genes were involved and in vitro resistance to antimycin A and hydroxyquinoline N-oxide, another inhibitor of complex III, was demonstrated. Respiratory particles prepared from mutant cells could bind less antimycin than similar particles obtained from wild-type cells. Grimmelikhuijzen and Slater [213], however, do not consider the difference in antimycin binding sufficient to explain quantitatively the differential sensitivity of respiration to the antibiotic, for which no satisfactory explanation seems to be available. In any case, the alteration of the binding properties suggests a change in the structure of complex III, but what component of the complex is altered by the mutation is not known. It is worth noting that in one of these mutants the activity of succinate oxidase was lower, suggesting a close relationship between complex II and complex III [213].

The mechanism of fungal resistance to oxidative phosphorylation inhibitors has been studied in a number of cases. Oligomycin is the best known inhibitor of the adenosine triphosphatase complex, of which a number of components are known [214]. Three studies on the biochemical mechanism of oligomycin resistance will be mentioned here.

In one class of mutants of S. cerevisiae studied by Shannon et al. [216], mitochondrial DNA was found to control a component of the membrane fraction of the ATPase complex. This assumption was based on the oligomycin sensitivity of hybrid reconstituted ATPase complexes. When the membrane fraction was taken from resistant cells the complex was resistant to the antibiotic, independently of the source of the enzyme and the "oligomycin sensitivity-conferring protein." These cytoplasmic mutations were apparently of two general classes, namely those which confer resistance to amphotericin B and those which do not. In the same study, two other types of oligomycin-resistance mutations were identified, a nuclear and an extranuclear, which confer resistance in vivo but have no effect on the sensitivity of isolated mitochondria and which probably control some impermeability or detoxification mechanism [216].

In A. nidulans, oligomycin resistance results from chromosomal and only very rarely extrachromosomal mutations [50, 51]. In none of the mutants studied was ATPase resistance in vitro sufficient to account for the high resistance in vivo. While in nuclear mutants an impermeability mechanism may be operative, for the single mitochondrial mutant studied, an altered cytochrome spectrum was recognized. It was characterized by increased amounts of cytochrome c but not of cytochromes b and a. Thus the mutation is believed to specify a property of the mitochondrial inner membrane [51]. It must be noted that this mutant shows impaired growth ability in the absence of oligomycin [51].

Goffeau et al. [217] have described a very interesting oligomycin-resistant mutant of the "petite-negative" yeast Schizosaccharomyces pombe. This mutant is pleiotropic and, in addition to lack of oligomycin sensitivity, it shows weakened binding of ATPase to the mitochondrial membrane, loss of cytochrome $a+a_3$ absorption spectra, and decreased cytochrome oxidase and succinate/cytochrome c oxidoreductase activities. The deficiencies in this mutant could be closely simulated by inhibition of mitochondrial protein synthesis in the wild-type strain, but lack of such synthesis in the mutant could not be proved by in vitro experiments. The action of the mutant gene has not, therefore, been clarified.

For additional information on the mechanism of resistance to antifungal compounds acting by inhibition of oxidative electron transport or phosphorylation, Lloyd's [55] book can be consulted.

4. Cell Walls and Membranes

Since some of the components of fungal cell walls are absent in higher plants, selective inhibition of the synthesis of such components was early recognized as an attractive possibility for the control of plant pathogenic fungi [218]. However, conclusive evidence of fungitoxicity resulting from inhibition of fungal wall synthesis is available today only for the polyoxins. These antibiotics act as competitive inhibitors for uridine diphosphate-N-acetylglucosamine in the chitin synthetase reaction, the last step in the synthesis of chitin [219-221]. A mixture of polyoxins is widely used as an agricultural fungicide in Japan, and field development of resistant strains of *Alternaria kikuchiana* has been described [222].

The genetics of this resistance is not known. Recent studies [223, 314] indicate that with intact cells there is good correlation between polyoxin sensitivity and amount of inhibition of chitin synthesis or uridine diphosphate-N-acetylglucosamine accumulation. However, particulate chitin synthetase preparations from various strains of *A. kikuchiana* were all strongly inhibited by polyoxin. Resistance, therefore, did not seem to result from modification of the sensitive enzyme but rather from decreased permeability [223, 314] as was mentioned earlier in this section. Undoubtedly a study of the biochemical genetics of polyoxin resistance in a more suitable fungus would be very interesting.

Of the fungicides recently introduced as selective against the Oomycetes which possess nonchitinous cell walls, prothiocarb is suspected [224] of acting by inhibiting synthesis of a component of such walls, composed of cellulose and glucans. Such a mechanism of action remains to be proved experimentally but, in this case also, a study of possible resistance by mutational modification of wall-synthesizing systems would be of interest.

Selective inhibition of the biosynthesis of constituents of fungal membranes is also an interesting possibility for fungitoxicity. A number of recently studied systemic fungicides seem to act in this way. These include triarimol [225, 227] and its analogue fenarimol, triforine [226, 228], Denmert or S-1358 [323, 324], and triadimefon [325]. Imazalil could tentatively be included in the group, for imazalil-resistant mutants of *A. nidulans* were without exception resistant to fenarimol [310]. Although these compounds may not be structurally related, they all inhibit the synthesis of ergosterol in fungi. Triarimol, for example, appears to have three sites of action in ergosterol biosynthesis [227], not all shared with triforine [228]. The toxicity of these fungicides may result from the lack of proper sterols for the structure of membranes, since it is doubtful whether the sterols synthesized in the presence of these inhibitors can effectively substitute for ergosterol whose synthesis is inhibited [227]. In *Cladosporium cucumerinum* [226] and *Aspergillus fumigatus* [228] cross-resistance between triarimol, triforine, and Denmert has been demonstrated. In the mycelium of a resistant strain of the latter fungus, triarimol caused much less inhibition of ergosterol synthesis and less accumulation of ergosterol precursors but whether this reflects an alteration of one or more of the target sites is not clear [228]. In vitro studies on the resistance mechanism would be very interesting.

Membranes are involved in fungitoxicity of compounds which interact with sterols in the membrane structure. Such toxicants are the polyene antibiotics nystatin, amphotericin B, candicidin, and filipin, which bind to sterol components of membranes altering membrane permeability so that essential constituents leak out [229]. The detoxification of polyenes reported [131, 132] to be responsible for unstable resistance in dermatophytes was mentioned earlier in this section. Polyene resistance has also been reported in Candida albicans [89, 230, 231], S. cerevisiae [70, 71, 89], and N. crassa [72, 232]. In these cases resistance seems to be associated with an altered sterol content. Thus, Woods [71] suggests that the mutations for nystatin resistance in S. cerevisiae may represent blocks in the pathway of ergosterol biosynthesis. At least one of these mutations seems to eliminate ergosterol completely. Still, the growth yields of the mutants were not significantly lower than those of the wild type [71]. If ergosterol is really optional to yeast it would be very interesting to know whether fungicides acting on ergosterol biosynthesis are toxic to this species and what the sensitivity of the polyene resistant mutants would be. In Neurospora, polyenes were used successfully to obtain similar sterol mutants [72, 232]. In this case, mutants with considerably reduced or no ergosterol in their mycelial extracts grew poorly compared to the wild-type strain, particularly at high temperatures. Wild-type N. crassa is sensitive to triforine [226]. It would be interesting to know the triforine and triarimol sensitivity of the polyene resistant sterol mutants of this fungus.

The reader is referred to the recent review by Hamilton-Miller [233] for details on fungal resistance to polyene antibiotics.

V. PRACTICAL ASPECTS

An important reason for investigating fungicide resistance is to find ways to prevent or delay its development or to successfully deal with it after it has developed. A solution to the problem, if possible at all, will have to come from a wide variety of information from biochemical and genetic studies and from epidemiological work on resistance patterns and attempts to control resistance by manipulation of fungal populations. In the latter field we do not have the experience of many years. The main problems encountered in the medical applications of antifungal compounds are the extremely rare occurrence of strains of Candida albicans resistant to amphotericin B and the resistance of many clinically important yeasts to 5-fluorocytosine [151]. Also in agriculture the use of fungicides was not confronted with practical resistance problems until very recently. In 1960 there had been almost no failures of chemical control of diseases of plants because of resistance development [234]. By that time, it must be pointed out, not only insecticide [235] and antibacterial antibiotic [236] resistance had created great difficulties for agriculture and public health, but also many plant varieties bred for resistance to a given fungal disease had been withdrawn because of genetic changes in the pathogen [237]. What made fungi appear less variable in their fungicide sensitivity than in their host-pathogen relationships? Two possible explanations were offered [234]: (a) in agricultural practice the environment in which fungi are exposed to chemicals might be unfavorable for resistance mechanism to be effective, and (b) the nonspecific mode of action of the agricultural fungicides available until then did not permit resistance development.

A. Resistance to Multisite and Specific-Site Fungicides

Before the recent introduction of new systemics, most antifungal compounds used in agriculture were of indiscriminate reactivity, each affecting a number of vital processes in the fungal cell. The dimethyl and diethyl dithiocarbamates, for example, are known to inhibit more than 20 different isolated enzymes [238], presumably by chelation of heavy metals. The importance of the multisite mode of action of the protectant fungicides in preventing resistance problems is now beyond dispute for the following reasons:

Firstly, even today stable resistance to many of the multisite fungicides has not been reported from laboratory studies. The main reason is, of course, that mutational modification of one cellular site would be without considerable effect on the toxicity of a multisite inhibitor. The lower the specificity, the more sites must be altered in order for resistance to develop. It also seems, however, that other mechanisms of resistance discussed in the previous section have not been important in the case of most protectant fungicides. The detoxification of organomercurials [136] and probably PCNB [144] and dodine [59] may be the exceptions that prove the rule.

Secondly, among the protectant fungicides, the members of the aromatic hydrocarbon group, resistance to which was easily obtained in the laboratory, did encounter the resistance problem in their agricultural applications, between 1960 and 1970 [239-242], in spite of their rather limited use. The same was true of dodine, resistance to which was first induced in the laboratory in 1967 [242] and was reported to cause failure of apple scab control two years later [243]. These examples supported the view that pathogenic fungi do develop field resistance to those fungicides for which a resistance mechanism is operative under laboratory conditions.

Thirdly, and most important, the introduction of the site-specific systemic fungicides in the second half of the 1960s was soon followed by a great increase in the number of reports of field-developed fungicide resistance. Table 2 lists some of the reports of resistant strains that were isolated after commercial application of the benzimidazole fungicides which became widely known in 1968 [244, 245] and were extensively used only after 1970. Large-scale failures of disease control due to the development of benzimidazole resistance by target fungi have been reported frequently in the last few years and undoubtedly there must be additional cases which have not appeared in the literature. Resistance of powdery mildew fungi to dimethirimol [265] and ethirimol [266], of *Pyricularia oryzae* to kasugamycin [203], and of

TABLE 2

Reports of Field-Developed Resistance to Benzimidazole Fungicides

Fungus	Crop	Country	Year	Reference
Botrytis cinerea	Cucumber	Japan	1975	326
	Cyclamen	Netherlands	1971	246
	Lettuce	England	1974	248
	Raspberry	Scotland	1973	247
	Strawberry	England	1974	327
		Scotland	1973	247
		U.S.	1976	328
	Tomato	England	1974	248
Ceratocystis ulmi	Elm	U.S.	1976	329
Cercospora apii	Celery	U.S.	1974	249
C. arachidicola	Peanut	U.S.	1974	250, 251
C. beticola	Sugarbeet	Greece	1973	252
		Italy	1974	343
		U.S.	1974	253
Cercosporidium personatum	Peanut	U.S.	1974	250
Colletotrichum coffeanum	Coffee	Kenya	1976	330
C. musae	Banana	West Indies	1973	254
Erysiphe cichoracearum	Eggplant	Japan	1975	326
E. graminis	Bluegrass	U.S.	1973	255

Fusarium oxysporum	Gladiolus	U.S.	1974	256
Penicillium brevicompactum	Cyclamen	Netherlands	1971	257
P. corymbiferum	Lily	Netherlands	1971	257
P. digitatum	Citrus	U.S.	1972	171
		Japan	1976	172
P. italicum	Citrus	U.S.	1972	171
		Israel	1974	260
		Japan	1976	172
		Australia	1974	173
Sclerotinia fructicola	Cherry	Australia	1976	258
S. homoeocarpa	Turfgrass	U.S.	1974	261
Sphaerotheca fuliginea	Cucurbits	U.S.	1969	262
		Israel	1970	263
		Japan	1975	326
		Greece	1976	259
S. humuli	Strawberry	Japan	1975	326
Venturia inaequalis	Apple	Australia	1974	264
		Japan	1975	326
V. pyrina	Pear	Israel	1976	331
Verticillium dahliae	Tomato	England	1976	332
V. malthousei	Mushroom	U.S.	1974	333
		Netherlands	1975	334

Alternaria kikuchiana to the site-specific polyoxins [222] have caused similar problems. It is fully justified to say that the introduction of the systemics, which certainly created new possibilities for the chemical control of plant diseases, at the same time placed the problem of fungicide resistance in its proper perspective.

The use of systemic fungicides in agriculture has shown that pathogen variability can challenge chemical control as it has challenged the use of resistant host varieties. From the point of view of the pathogen's chances to overcome control measures [267], the use of a site-specific fungicide is comparable to the cultivation of a race-specific resistant variety (vertical resistance) [267]. On the other hand, the use of multisite inhibitors as fungicides is analogous to the use of general resistance of host plants (horizontal resistance), according to the following scheme:

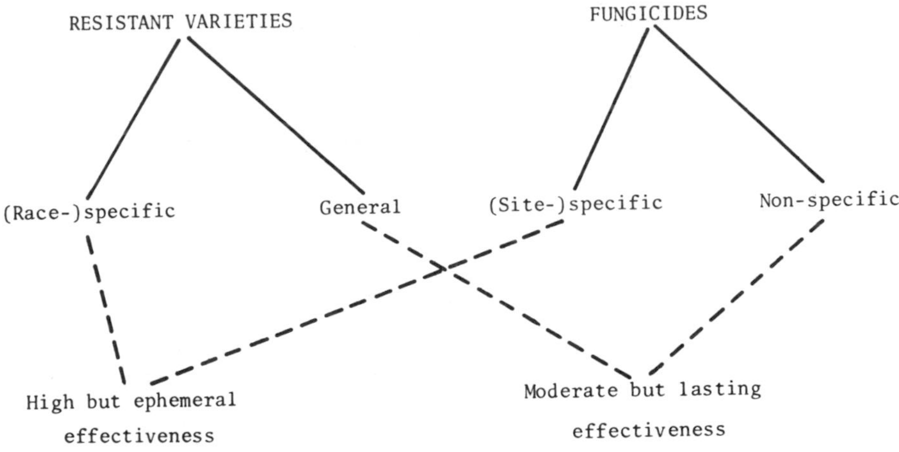

Although this may be an oversimplification, there is now good reason to accept this analogy. Some of the successes that we had so far with the use of systemics against particular fungi did not last longer than some race-specific resistant varieties [268, 269]. The dithiocarbamates, on the other hand, are still used successfully after 40 years. The fact that resistance to some multisite fungicides has developed, does not contradict the analogy since this development has taken very long (about 30 years for organomercurial resistance of P. avenae) and "erosion" of general resistance in plants cannot definitely be excluded [17].

The variety of fungal species which have developed resistance to benzimidazoles, following their application to various crops under different conditions (Table 2) indicates that whether resistance problems will arise or not depends mainly on the type of fungicide. If the type of fungicide permits, probably most fungi are variable enough to meet the challenge under most conditions. From what we know today, the most important single factor which describes the type of fungicide in this regard, is whether or not its action is site-specific. In fact, the ease with which resistant mutants are obtained seems to be a rather safe criterion for the biochemical specificity of fungicides in most cases.

However, the specific or nonspecific action of the fungicides is not sufficient to answer all questions that might be asked. Griseofulvin, for example, seems to act specifically on spindle function as explained in Section IV. Mutational modification of the site of action as in the case of MBC [169] should not be difficult. Still, genetically determined griseofulvin resistance not only has not been the cause of therapeutic failure [270], but it also does not seem to have been reported from laboratory studies.

With other site-specific antifungal compounds, mutational resistance may be easily obtained in the laboratory but it does not seem to have caused many problems in practice. The carboxamide fungicides may be a case in point. Modification of the site of carboxamide action leads to well defined resistance [33, 37, 92] which is expressed in treated plants [35], but we have only one report [335] of practical failure so far. Is resistance to carboxamides more difficult to develop in the field than resistance to the other groups of systemic fungicides and if so, why? We do not yet have the answers to such questions, but some possibilities may be examined.

Carboxamides should probably not be compared to the benzimidazoles because the former are not used against nearly as many types of diseases and have not even been cleared for use in some countries. The behavior of the benzimidazoles so far, however, gives the impression that these fungicides would have encountered resistance even if used on a small scale. We do not know if the ability of benzimidazoles to accelerate recombination [117] and, perhaps, to induce gene mutations [123, 336] has been a factor. Some people feel that it has been [271]. In experiments with _Cercospora musae_, the frequency of recovery of benomyl-resistant mutants by plating conidia on benomyl-containing medium was not appreciably increased by nitrosoguanidine pretreatment [272]. In any case, the pyrimidines and the polyoxins appear genetically inactive [119] but resistance limited their agricultural uses soon after their introduction [222, 265]. Genetic activity, therefore, if important at all, does not seem to be the only factor.

Another possibility to examine is whether the way of application is important. Unlike other systemics, the carboxamides are mostly used for seed treatment and not as foliar sprays. The type of environment may play some role. Davies and Savage [152], for example, found that addition of serum reduces the number of colonies of _Candida albicans_ which develop resistance to 5-fluorocytosine, apparently because it induces hyphal growth which is characterized by fewer mitoses than in blastospore multiplication. With the use of oxycarboxin as a foliar spray for the control of chrysanthemum rust, resistance was encountered [335]. Recently, pyracarbolide, also a carboxamide, has been recommended for foliage spray against coffee rust [273]. It will be interesting to see whether resistance develops following these applications. It must also be pointed out that the carboxamides are mostly used to control the dicaryotic phase of basidiomycetes. The requirement for two resistant homocaryons to meet before disease can develop in the presence of the fungicide may undoubtedly have delayed difficulties.

Finally, the assumption that systemic fungicides are site specific while protectants are not, may be an oversimplification. The trialkyltin compounds, for example, are not considered to be systemic fungicides but their action, at least in mammalian cells [209] and in S. cerevisiae [28], seems to be quite specific for oxidative phosphorylation. Do these fungicides act as phosphorylation inhibitors in those fungi against which they are used in agriculture and if so, why has no resistance developed after many years of use? In yeast a large number of mutants, exhibiting a 10-to-20-fold resistance to inhibition of growth were rather easily obtained [28]. Pyrazophos, on the other hand, shows considerable systemic activity [154], but attempts to induce resistance to its apparent fungitoxic product (PP) in U. maydis were unsuccessful [274]. Systemicity, of course, requires a somewhat discriminate reactivity but this may not necessarily mean interference with only one vital step in the fungal cell. Besides, what we call "systemicity" often implies only movement in the apoplast with limited exposure to reactive components of plant tissues [275].

B. Selection and Adaptability of Resistant Strains

The rate of change by selection depends on the amount of heritable variance in the population and on the intensity of selection. In the case of fungicide resistance, heritable variance is expressed by the number and mutability of the genes which can independently mutate to resistance to a fungicide (see Section III). Selection intensity depends on the number of applications relative to the reproductive cycle of the fungus, the persistence of the fungicide, and the sensitivity difference between wild type and mutant that determines the length of time during which a discriminating concentration of the chemical is maintained. Obviously, the higher the tolerance level, the wider the range of concentrations which will inhibit the wild type but not the mutant and the higher the intensity of the selection. These factors, together with the fitness of the mutant alleles will determine the useful life of a fungicide for the control of a particular disease.

The useful life of a resistant variety similarly depends on pathogen variability but the selection pressure imposed by the introduction of such a variety may be different for obligate and nonobligate parasites. With nonobligate plant diseases, effective control may be achieved for many years with the introduction of a major resistance gene if avirulent forms are more fit on nonhost substrates [17]. Such a factor may be important for the selection pressure imposed by a fungicide applied locally, i.e., with seed or furrow application. But if the chemical is sprayed over large areas, host and nonhost substrates are poisoned and all of the pathogen population is subjected to the selection pressure, independently of obligate or nonobligate parasitism. In fact, while the pressure of resistant varieties does not extend beyond the fungal species against which the varieties were bred, selective fungicides may affect saprophytic antagonists with the result of higher severity of diseases caused by insensitive species [276].

Little detailed information on rates of changes in the strain composition of a pathogen population following fungicidal applications is available. In the past, fungicides have been put into service without regard to the fact that they may be rendered ineffective by new strains of the fungi they were to control. Therefore, the effect of the fungicide on the population was not considered until resistance was suspected from failures of control in the field. By that time, however, the population may be resistant up to 100% [252, 253, 259, 265, 277]. Naturally, resistant strains must increase beyond a critical frequency in the population in order to cause a practical problem [13].

Some idea as to relative rates of changes can be obtained from the time interval between first application and first appearance of an economic problem, provided, of course, that the type and the frequency of applications is considered. In some areas of New York State, for example, intensive use of dodine against apple scab created a resistance problem within 10 years, while equally intensive use of captan for 20 years has given no indication of tolerant strains [278]. We do not know whether benzimidazoles have been used against scab in the same areas but from South Australia the first failures were reported [264] within 3 years of regular use of benomyl. The difference between captan and the other two fungicides can be explained on the basis of heritable variance in the fungus, since captan resistance has not been obtained even in the laboratory. Resistance to dodine, however, is not difficult to obtain in *Venturia inaequalis* [88] and the slow (compared to benomyl resistance) attainment of a critical frequency must be attributed to the intensity of the selection. Probably the low degree of dodine resistance, which may also allow for some control with increasing fungicide dosage [278], has been an important factor. Recently, dodine-resistant isolates of *V. inaequalis* were also reported from Michigan [61].

With pathogens such as *V. inaequalis* which do not have a means of spore dispersal assuring dissemination over large areas, the changes in the strain composition of the population are easy to follow in a relatively small field experiment. An attempt of this kind with regard to benomyl resistance in *Cercospora beticola* was made in Greece in the summer of 1973 [278]. In this experiment, laboratory-grown conidia were added to the plots and the original percentage of resistant conidia could be manipulated. As an example it is mentioned that two benomyl sprays, one on June 30 and one on July 13, were sufficient to increase the percentage of resistant conidia from an average of 3.5% on June 26 to 91.5% on July 30. Since infestation with outside inoculum was completely impossible, the benomyl sprays apparently created a very high selection pressure. This can explain the suddenness with which most benzimidazole-resistance problems have appeared.

Ethirimol resistance of barley powdery mildew, *Erysiphe graminis*, presents a quite different problem. Seed treatment with this fungicide is sufficient to delay the onset of the mildew epidemic so that the amount of disease which develops toward the end of the season has little effect on yield [280]. In this case, the chemical concentration in the plant decreases during the season with the result of a declining selection pressure. Tolerant forms of the pathogen are disseminated from treated to adjacent and even isolated areas and are found with decreasing frequency as the distance from a treated area increases [266]. On the other hand, sensitive spores may migrate from untreated to treated areas and may become established if the ethirimol activity has declined, thus confusing the picture of the effect the fungicide has on the population [281].

Changes in strain composition after the use of a fungicide has been discontinued are also very important. Such changes would provide information on the survival value or fitness of resistance genes in the absence of the agent responsible for selection [13]. One important factor to consider is the pathogenicity of fungicide-resistant mutants compared to wild-type strains. Studies with a few isolates, particularly if they have been in culture for some time, cannot give a measure of the possibilities which can be exploited by nature. Not only may one resistance mutation affect pathogenicity and others may not [282], but also reduced virulence in a fungicide-resistant mutant may be due to chance recombination between resistance alleles and alleles for low pathogenicity [283]. One would expect that with the great variability of fungal pathogens, forms able to successfully compete with fungicide-sensitive forms on untreated substrates would readily evolve in nature.

Other factors, in addition to pathogenicity, that are important in determining competitiveness include production of resting structures and overwintering, length of reproductive cycle, amount of conidiation, and ability for saprophytic colonization. A particular fungicide-resistance gene may well affect any of these adaptability characteristics. Artificial selection, in general, is believed to be counteracted by natural selection [284]. This is a consequence of the previous evolutionary history of the organism directed toward the maintenance of a phenotypic balance between fitness-determining characters (genetic homeostasis). This would be highly desirable but whether it will be found to be the case with all types of artificial selection is not known. Considering insecticide resistance, Crow [285] cited cases in which regression to susceptibility was observed after the selection for resistance was relaxed and others in which such regression had not been noticed. In 1961 Gordon [235] wrote: "The many vigorous multiresistant strains now known prove that R genes for many insecticides can coexist in one individual, without necessarily rendering it very susceptible to natural controls." In 1967 Keiding [286] also concluded that high level resistance may persist in insect populations for a long time and will usually decrease only partially when the selection pressure is relaxed. Whether the response in fungal populations is comparable is not known. In pathogenic fungi, some virulence genes seem to lower fitness while others do not. This question is discussed by Day [17] and by Wolfe [287].

The short time since the most important problems of resistance to fungicides were recognized has not permitted collection of large amounts of data on the survival value of resistance genes in the absence of fungicide. According to Bent [288], the dimethirimol-resistant strains of _Sphaerotheca fuliginea_ have been replaced by sensitive ones in glasshouses of Western Europe. The totally sensitive population of the fungus from the field-grown crops in the same areas [265] might have been responsible for this replacement. In England, avoidance of ethirimol treatment of winter barley has been recommended in order to avoid the "bridging" effect between two spring crops [266]. It appears that this recommendation has indeed been of value in prolonging the useful life of the fungicide because of selection against the resistant forms on untreated winter barley [281]. In Greece, on the other hand, after 3 years during which no benzimidazole fungicides were used against sugarbeet leaf spot we have no indication that the frequency of resistant conidia of _Cercospora beticola_ is declining [279]. Even in areas where this frequency was kept low, due to limited use of the fungicide, no regression of resistance has been observed. The experience with ethirimol resistance of _Erysiphe graminis_ f. sp. _hordei_ in England has been different with rapid shifts in sensitivity in both directions [337]. This can probably be explained by the rareness of strains of the pathogen with stable resistance to ethirimol [338].

Probably when selection for resistance is accompanied by a great deal of selection for general fitness, the only kind of resistance genes to become frequent in the population will be those that cause little or no reduction in fitness. On the other hand, selection under circumstances where there is very little natural competition may provide some very poor specimens with regard to natural survival. We are very much in need of data on the adaptability of fungicide-resistant mutants but the important information cannot be provided by studies of a few mutant and wild-type strains. Rather, the gross behavior of all representatives of each class which have evolved in the field, with both artificial and natural selection operating should be considered.

Entomologists know that stability in nature is influenced by the "age" of resistance, i.e., the length of time during which the population has been under pressure by the chemical; this time indicates the extent to which the resistant genotypes have been selected for fitness [339]. The first resistant population in an area may contain only a small proportion of strains with a good combination of genetic determinants for resistance and fitness. If the selection pressure is removed at that point, regression will follow but new applications of the chemical should then lead to a resistant population of high fitness. Undoubtedly a follow-up of some of the recently developed resistant populations of plant pathogenic fungi would deepen our insight into some of these problems.

C. Recommendations

The experience thus far with agricultural uses of multisite and specific-site antifungal compounds indicates that the resistance problem may be minimized if use of fungicides of the second category is avoided. As indicated earlier in this section, our understanding of the relation of specificity of action to the development of field resistance may be incomplete but, in general, it would be wise to limit the use of site-specific fungicides to those cases in which no effective multisite alternatives are available. Unfortunately, new fungicides are introduced for commercial use before their biochemical specificity is known. Some indication can, of course, be provided from data on systemic activity which are collected during the preliminary screening of compounds. However, information on the possible useful life of a fungicide in a given situation can easily be obtained directly within a few weeks in the laboratory. In rather simple experiments it can be determined whether mutations to resistance are possible and frequent, and what the levels of tolerance of the mutants are. From this information the type of selection pressure which is likely to be imposed may be deduced during the development stage of future fungicides and before large-scale applications are recommended.

It must be pointed out that elimination of fungicides which have been found or are suspected to be vulnerable to the development of fungal resistance in the field cannot easily be recommended. The systemic and curative action of such fungicides offers certain advantages [289] which agriculture can ill afford not to explore and in many cases it has done so with considerable success [290-293]. In addition, even in cases where only protectant action is required and a good multisite alternative is available, it cannot be taken for granted that this alternative will never be confronted with fungal resistance or encounter other problems. _Sclerotinia homoeocarpa_, for example, has developed tolerance to Dyrene [294], which indiscriminately reacts with many cell constituents [218]. Van der Kerk [295] gives the example of organomercurials which may have to be totally replaced as seed dressing agents because of government bans. Finally, a recent review [296], in which captan is given as a possible genetic toxicant of man and animals, should also be mentioned. It would be highly desirable, therefore, not to have to eliminate site-specific fungicides but to try to prolong the useful life of each such compound in the control of those diseases for which it is most suitable or indispensable. A number of possibilities have been examined by other authors [13, 14, 271, 297-300].

Undoubtedly, the sudden and dramatic prevalence of fungal strains resistant to some of the systemic fungicides in many areas during the last few years was not unrelated to their careless use. In the years since Horsfall's [18] recommendation "not to await the arrival of resistant strains in the field," there has been ample warning that the introduction of fungicides which depend on a single-site mechanism would be followed by development of resistance [9, 11, 234, 301, 302]. In spite of such warning, the mistake was made of applying the same selective compound to large acreages and many times during a growing season. Valuable antifungal compounds have thus lost a good place in the market.

There may, however, still be some time to avoid mistakes. The same compounds may be useful for the control of other diseases or of the same diseases in areas where resistant strains do not exist or are still very infrequent. In addition, new specific fungicides may be discovered. In all of these cases where a high selection pressure is expected, it is important to limit the use of the selective factor in space and time as much as possible, so that a background of competition by a sensitive population remains. This, together with the maintenance of a close watch on the pathogen population for any shifts in the occurrence of insensitivity should prevent the creation of an acute problem. Such an effort has been undertaken by the Plant Breeding Institute at Cambridge, Great Britain, with the purpose of monitoring on a regional basis the occurrence of resistance in the barley mildew pathogen to ethirimol and tridemorph [281]. Similar monitoring seems to be underway for other fungus/fungicide combinations [303, 304].

High frequencies of resistant strains may be delayed and perhaps prevented if the selective fungicide is applied to the minimum area necessary at a time when optimal disease control can be expected. This, particularly in the case of nonobligate parasites, will result in a limited selection pressure. Control may be supplemented by the cultivation of semiresistant varieties [287] or the use of a perhaps less effective, multisite fungicide or another specific-site fungicide which does not select for the same type of mutants. The question of cross-resistance is very important here. A positive correlation usually exists between compounds of related structure. However, the aromatic hydrocarbon fungicides, grouped together on the basis of cross-resistance experiments [9], include diverse chemical structures. Similarly, mutants of *Cladosporium cucumerinum* selected for resistance to triarimol are cross-resistant to the unrelated structure of triforine [226]. If only structural similarity were considered, one would expect to control mutants resistant to the former compound by the use of the latter. Both of these compounds inhibit the biosynthesis of ergosterol [225-228]. But positive correlation is possible even between compounds of basically different modes of action in case of pleiotropic effects of a single gene. The practical importance of laboratory studies on mechanisms of action and mechanisms of resistance can thus hardly be overemphasized.

It would be helpful to have negatively correlated fungicides at our disposal. This would permit the use of one fungicide against the mutants, the frequency of which increased following applications of another. Such a negative correlation has been reported to exist between carboxin and antimycin A [63] and between benomyl and TBZ [84]. Unfortunately, in both of these cases only one type of mutants selected for by the one compound are more sensitive than the wild type to the other compound. In the remaining mutants either no correlation [63] or a positive correlation [84] exists. Effectiveness of phosphoramidate and phosphorothiolate fungicides has also been reported to be negatively correlated [23] but in this case the genetics of the two strains of Piricularia oryzae that were used is not known. Although we have limited knowledge on this subject at present, it seems that some interesting possibilities exist. In a study of a large number of carboxin analogs, White [305] found that some are highly inhibitory to the succinic dehydrogenase complex of carboxin-resistant mutants in vitro and considerably less so to the wild-type preparation. Such negative correlations have been observed in vivo. The scientific interest of such observations from the point of view of the possible mutational alterations of receptor cell sites does not need to be stressed.

The main danger of exposing a population to more than one site-specific toxicant of different mechanisms of action, either simultaneously or alternatively, is the possibility of multiresistant individuals. Whether the risk can be counterbalanced by the benefit should be studied in each particular case. Multiresistant strains are known in bacteria and are one of the reasons for which avoidance of combinations of drugs is recommended [306]. In insects, multiple resistance is also common [235, 307] and seems to be responsible for the abandonment of the negative correlation approach to the resistance problem [235]. Entomologists seem to prefer successions of selective insecticides rather than mixtures or alternations [340]. There are, however, other factors which should be considered such as synergism between compounds. In several yeast-like fungi, a combination of amphotericin B and 5-fluorocytosine produces toxicity which is more than could be attributed to an additive effect of the two drugs [308]. Since resistance to the pyrimidine analog may result from impermeability [78] the damage to the membrane produced by amphotericin B might counteract resistance to 5-fluorocytosine in treatments of fungal infections.

The use of nonfungitoxic compounds which increase host resistance or affect the host-pathogen interaction has been suggested [302, 309] as a possible solution to the resistance problem. Such compounds are not available now for commercial use but they may be so in the future. It is true that our aim is not to kill fungi but to prevent losses. It is not at all certain, however, that such a compound would have no problems with pathogen variability. If host resistance were induced there is no reason why mutation to overcome this resistance should not be possible for the pathogen. The treatment might then select for a pathogen race as has happened with disease-resistant varieties. The same may be true of interference with the host-pathogen interaction. For example, if a compound blocks one of the steps in the biosynthesis of a toxin responsible for disease, the treatment may prevent the establishment and reproduction of the pathogen. In such a case any mutation modifying the sensitive enzyme system, so that inhibition is reduced, would have high survival value as long as the "fungicide" is present in the host. This does not mean that compounds with such mechanisms are condemned to failure. It is difficult to estimate the research effort needed in these areas. It cannot be excluded that the chemical may induce the plant to form a fungal inhibitor of multisite activity (in which case we may have a situation analogous to that of polygenic resistance to disease) or to produce more than one phytoalexin as is probably the case with the dichlorocyclopropanes [341, 342].

VI. CONCLUSIONS

A few fungal species have long been used in genetical and biochemical studies of modified sensitivity to toxic compounds but fungicide resistance is only now becoming a scientific field in its own right. Without question the number of papers in this field, particularly on epidemiological and pathological aspects, will be increasing in the years to come. It is important that in all cases resistance is described as accurately as possible.

In laboratory studies, addition of impure fungicides to poorly defined media and mass transfers of highly variable and heterocaryon-forming fungi are not likely to provide significant information. Attempts to "train" fungi to tolerate fungicides may lead to some ill-defined physiological adaptation which does not justify the effort. By contrast, selection from a large number of mutagenized or nonmutagenized cells on fungicide concentrations completely inhibitory to the wild type is a much better way to obtain resistant mutants. On the other hand, if the fungus chosen is not easily amenable to genetic analysis, every effort should be made to at least determine whether the material used is genetically homogeneous. Single-spore or hyphal-tip analysis may be helpful.

The availability of site-specific fungicides will undoubtedly be of great help in understanding biochemical mechanisms. The isolation of the system involved in fungitoxicity from cell-free preparations and the recognition of its possible mutational alterations which lead to resistance should be the main objective. In this way an increase in knowledge, not only of gene action but also of fungicidal mechanisms, may be anticipated.

Knowledge on field behavior of resistant strains may also be expected to accumulate as a result of the use of biochemically specific fungicides. Whether it will be possible to successfully meet the technological challenge presented by the variability of fungal pathogens to the chemical control of disease is not known. Judging from the experience of entomologists and medical bateriologists, no long-term successes can be expected. The main recommendation that can be made is to avoid high frequencies of resistant strains by intermittent selection which requires that any one selective fungicide be used only for a limited time. However, whether the increase in frequency of resistant genes during the use of a fungicide will be compensated by a decrease during a period of disuse remains to be determined. Increased knowledge of the nature of vulnerable cell sites may permit manipulation of fungal populations by the use of appropriate compounds.

ACKNOWLEDGMENT

I am indebted to Drs. P. R. Day of the Connecticut Agricultural Experiment Station, and M. S. Wolfe of the Plant Breeding Institute, University of Cambridge, England, for reading parts of the manuscript and making some useful suggestions.

REFERENCES

1. H. S. Moyed, Ann. Rev. Microbiol., 18, 347 (1964).
2. S. Falkow, E. M. Johnson, and L. S. Baron, Ann. Rev. Genet., 1, 87 (1967).
3. J. S. Kiser, G. O. Gale, and G. A. Kemp, Advan. Appl. Microbiol., 11, 77 (1969).
4. W. V. Shaw, Advan. Pharmacol. Chemotherap., 9, 131 (1971).
5. T. J. Franklin and G. A. Snow, Biochemistry of Antimicrobial Action, Chapman and Hall Ltd., London, 1971.
6. A. M. Reynard, in Drug Resistance and Selectivity (E. Mihich, ed.), Academic Press, New York, 1973, pp. 127-183.
7. R. Benveniste and J. Davies, Ann. Rev. Biochem., 42, 471 (1973).
8. J. Ashida, Ann. Rev. Phytopathol., 3, 153 (1965).
9. S. G. Georgopoulos and C. Zaracovitis, Ann. Rev. Phytopathol., 5, 109 (1967).
10. J. Dekker, World Rev. Pest Control, 8, 79 (1968).
11. S. G. Georgopoulos, Bioscience, 19, 971 (1969).
12. J. Dekker, Proc. 6th Brit. Insectic. Fungic. Conf. Brighton, 1971, Vol. 3, p. 715.
13. M. S. Wolfe, Proc. 6th Brit. Insectic. Fungic. Conf. Brighton, 1971, Vol. 3, p. 724.
14. J. Dekker, in Systemic Fungicides (R. W. Marsh, ed.), Wiley, New York, 1972, pp. 156-174.
15. M. G. Sevag, in Origins of Resistance to Toxic Agents (M. G. Sevag, R. D. Reid, and O. E. Reynolds, eds.), Academic Press, New York, 1955, pp. 370-408.

16. G. C. Ainsworth, Dictionary of the Fungi, 6th ed., Commonwealth Mycological Institute, Kew, Great Britain, 1971.
17. P. R. Day, Genetics of Host Parasite Interaction, Freeman, San Francisco, 1974.
18. J. G. Horsfall, Principles of Fungicidal Action, Waltham, Mass. Chronica Botanica, 1956.
19. W. Szybalski and V. Bryson, J. Bacteriol., 64, 489 (1952).
20. D. J. Hutchison and F. A. Schmid, in Drug Resistance and Selectivity (E. Mihich, ed.), Academic Press, New York, 1973, pp. 73-126.
21. A. W. A. Brown, Pest Control, 29, 24 (1961).
22. G. H. Rank and N. T. Bech-Hansen, Molec. Gen. Genet., 126, 93 (1973).
23. Y. Uesugi, M. Katagiri, and O. Noda, Agr. Biol. Chem., 38, 907 (1974).
24. Y. A. Chabbert, J. G. Baudens, and D. H. Bouanchaud, in Bacterial Episomes and Plasmids, Ciba Foundation Symposium (G. E. W. Wolstenholme and C. M. O'Connor, eds.), Churchill, London, 1969, pp. 227-243.
25. E. G. Diacumakos, L. Garnjobst, and E. L. Tatum, J. Cell Biol., 26, 427 (1965).
26. M. Rabinowitz and H. Swift, Physiol. Rev., 50, 376 (1970).
27. A. Kaars Sijpesteijn, World Rev. Pest Control, 9, 85 (1970).
28. W. E. Lancashire and D. E. Griffiths, FEBS Lett., 17, 209 (1971).
29. D. E. Mathre, Pestic. Biochem. Physiol., 1, 216 (1971).
30. G. A. White, Biochem. Biophys. Res. Commun., 44, 1212 (1971).
31. S. G. Georgopoulos and V. Vomvoyanni, in Proc. 2nd Intern. Congr. Pestic. Chem., Tel Aviv, 1971 (A. S. Tahori, ed.), Vol. 5, Gordon and Breach, London, 1972, pp. 337-346.
32. J. T. Ulrich and D. E. Mathre, J. Bacteriol., 110, 628 (1972).
33. S. G. Georgopoulos, E. Alexandri, and M. Chrysayi, J. Bacteriol., 110, 809 (1972).
34. G. A. White and G. D. Thorn, Pestic. Biochem. Physiol., 5, 380 (1975).
35. S. G. Georgopoulos, M. Chrysayi, and G. A. White, Pestic. Biochem. Physiol., 5, 543 (1975).
36. I. A. U. N. Gunatilleke, C. Scazzocchio, and H. N. Arst, Jr., Heredity, 32, 285 (1974).
37. I. A. U. N. Gunatilleke, H. N. Arst, Jr., and C. Scazzocchio, Genet. Res. Cambridge, 26, 297 (1976).
38. D. R. Mackenzie, R. R. Nelson, and H. Cole, Phytopathology, 61, 471 (1971).
39. D. Y. Thomas and D. Wilkie, Genet. Res. Cambridge, 11, 33 (1968).
40. A. W. Linnane, A. J. Lamb, C. Christodoulou, and H. B. Lukins, Proc. Natl. Acad. Sci. U.S., 59, 1288 (1968).
41. K. D. Stuart, Biochem. Biophys. Res. Commun., 39, 1045 (1970).

42. C. L. Bunn, C. H. Mitchell, H. B. Lukins, and A. W. Linnane, Proc. Natl. Acad. Sci. U.S., 67, 1233 (1970).
43. C. H. Mitchell, C. L. Bunn, H. B. Lukins, and A. W. Linnane, Proc. Australian Biochem. Soc., 4, 67, 1233 (1970).
44. L. A. Grivell, L. Reijnders, and H. de Vries, FEBS Lett., 16, 159 (1971).
45. R. A. Kleese, R. C. Grotbeck, and J. R. Snyder, Can. J. Genet. Cytol., 14, 713 (1972).
46. G. A. Rank, Can. J. Microbiol., 20, 9 (1974).
47. A. W. Linnane, J. M. Halsam, H. B. Lukins, and P. Nagley, Ann. Rev. Microbiol., 26, 163 (1972).
48. P. R. Avner and D. E. Griffiths, Eur. J. Biochem., 32, 312 (1973).
49. P. R. Avner and D. E. Griffiths, Eur. J. Biochem., 32, 301 (1973).
50. R. T. Rowlands and G. Turner, Molec. Gen. Genet., 126, 201 (1973).
51. R. T. Rowlands and G. Turner, Molec. Gen. Genet., 132, 72 (1974).
52. D. S. Shaw and I. A. Khaki, Genet. Res., Cambridge, 17, 165 (1971).
53. I. A. Khaki and D. S. Shaw, Genet. Res., Cambridge, 23, 165 (1974).
54. L. W. Timmer, J. Castro, D. C. Erwin, W. L. Belser, and Z. A. Zentmeyer, Amer. J. Bot., 57, 1211 (1970).
55. D. Lloyd, The Mitochondria of Microorganisms, Academic Press, London, 1974.
56. S. G. Georgopoulos, Phytopathology, 53, 1086 (1963).
57. S. G. Georgopoulos and N. J. Panopoulos, Can. J. Genet. Cytol., 8, 347 (1966).
58. R. J. Threlfall, J. Gen. Microbiol., 52, 35 (1968).
59. A. Kappas and S. G. Georgopoulos, Genetics, 66, 617 (1970).
60. F. J. Plach, Phytopathology, 63, 1189 (1973).
61. K. S. Yoder and E. J. Klos, Phytopathology, 62, 799, abstr. (1972).
62. A. C. Hastie and S. G. Georgopoulos, J. Gen. Microbiol., 67, 371 (1971).
63. S. G. Georgopoulos and H. D. Sisler, J. Bacteriol., 103, 745 (1970).
64. J. M. van Tuyl, Neth. J. Plant Pathol., 81, 122 (1975).
65. D. Wilkie and B. K. Lee, Genet. Res. Cambridge, 6, 130 (1965).
66. K. S. Hsu, J. Gen. Microbiol., 32, 341 (1963).
67. V. Vomvoyanni, private communication, 1975.
68. J. A. Roper and E. Käfer, J. Gen. Microbiol., 16, 660 (1957).
69. J. H. Parker, I. R. Trimble, Jr., and J. R. Mattoon, Biochem. Biophys. Res. Commun., 33, 590 (1968).
70. K. A. Ahmed and R. A. Woods, Genet. Res. Cambridge, 9, 179 (1967).
71. R. A. Woods, J. Bacteriol., 108, 69 (1971).
72. M. Grindle, Molec. Gen. Genet., 130, 81 (1974).
73. M. L. Teles Grilo and W. Klingmüller, Molec. Gen. Genet., 133, 123 (1974).
74. U. Sinha, Beitr. Biol. Pflanzen, 48, 171 (1972).
75. S. Srivastava and U. Sinha, Genet. Res. Cambridge, 25, 29 (1975).

76. K. K. Jha, Molec. Gen. Genet., 114, 156 (1971).
77. G. R. Hoffman and H. V. Malling, J. Gen. Microbiol., 83, 319 (1974).
78. R. Jund and F. Lacroute, J. Bacteriol., 102, 607 (1970).
79. M. V. Elorza and H. N. Arst, Jr., Molec. Gen. Genet., 111, 185 (1971).
80. W. Klingmüller, Z. Naturforschung, 22b, 227 (1967).
81. G. Balassa, Molec. Gen. Genet., 104, 73 (1969).
82. S. D. Martinelli and A. J. Glutterbuck, J. Gen. Microbiol., 69, 261 (1971).
83. A. C. Hastie, private communication, 1975.
84. J. M. van Tuyl, L. C. Davidse, and J. Dekker, Neth. J. Plant Pathol., 80, 165 (1974).
85. V. Vomvoyanni, Nature, 248, 508 (1974).
86. V. E. Vomvoyanni and S. G. Georgopoulos, Phytopathology, 56, 1330 (1966).
87. J. Bartels-Schooley and B. H. MacNeill, Phytopathology, 61, 816 (1971).
88. B. H. MacNeill and J. Schooley, Can. J. Bot., 51, 379 (1973).
89. P. V. Patel and J. R. Johnston, Appl. Microbiol., 16, 164 (1968).
90. B. B. Chatoo and U. Sinha, Mutation Res., 23, 41 (1974).
91. D. S. Shaw and C. G. Elliott, J. Gen. Microbiol., 51, 75 (1968).
92. Y. Ben-Yephet, Y. Henis, and A. Dinoor, Phytopathology, 64, 51 (1974).
93. D. Cooper, D. V. Banthorpe, and D. Wilkie, J. Mol. Biol., 26, 347 (1967).
94. J. R. Warr and J. A. Roper, J. Gen. Microbiol., 40, 273 (1965).
95. R. K. Webster, J. M. Ogawa, and C. J. Moore, Phytopathology, 58, 997 (1968).
96. O. F. Esuruoso and R. K. S. Wood, Ann. Appl. Biol., 68, 271 (1971).
97. W. Greenaway and J. W. Cowan, Trans. Brit. Mycol. Soc., 54, 127 (1970).
98. S. G. Georgopoulos, Nature, 194, 148 (1962).
99. R. W. Meyer and J. R. Parmeter, Jr., Phytopathology, 58, 472 (1968).
100. T. H. Pittenger and K. C. Atwood, Genetics, 41, 227 (1956).
101. K. Borck and H. D. Braymer, J. Gen. Microbiol., 85, 51 (1974).
102. D. Lewis, Nature, 200, 151 (1963).
103. B. R. Nebel, Nature, 142, 257 (1938).
104. K. D. Wuu and W. F. Grant, Can. J. Genet. Cytol., 8, 481 (1966).
105. I. Prasad and D. Pramer, Phytopathology, 58, 1188 (1968).
106. M. Ahmed and W. F. Grant, Mutation Res., 14, 391 (1972).
107. M. A. Pilinskaya, ГЕНЕТИКА, 7, 138 (1971), in Russian with English summary.

108. L. Larizza, G. Simoni, F. Tredici, and L. de Carmi, Mutation Res., 25, 123 (1974).
109. W. S. Boyle, J. Heredit., 64, 49 (1973).
110. J. A. Styles and R. Garner, Mutation Res., 26, 177 (1974).
111. T. R. Thyagarajan, O. P. Srivastava, and V. V. Vora, Naturwissenschaften, 50, 524 (1963).
112. S. H. B. Crackower, Can. J. Microbiol., 18, 683 (1972).
113. K. Gull and A. P. J. Trinci, Nature, 244, 292 (1973).
114. R. S. Hammerschlag and H. D. Sisler, Pestic. Biochem. Physiol., 3, 42 (1973).
115. L. C. Davidse, Pestic. Biochem. Physiol., 3, 317 (1973).
116. A. C. Hastie, Nature, 226, 771 (1970).
117. A. Kappas, S. G. Georgopoulos, and A. C. Hastie, Mutation Res., 26, 17 (1974).
118. A. Kappas and S. G. Georgopoulos, J. Bacteriol., 119, 334 (1974).
119. S. G. Georgopoulos, A. Kappas, and A. C. Hastie, Phytopathology, 66, 217 (1976).
120. G. Ficsor and G. M. Nii, Environ. Mutagen Soc. Newsletter, 3, 38 (1970).
121. C. H. Clarke, Mutation Res., 11, 247 (1971).
122. H. V. Malling and F. J. De Serres, Environ. Mutagen Soc. Newsletter, 3, 37 (1970).
123. B. Dassenoy and J. A. Meyer, Mutation Res., 21, 119 (1973).
124. J. P. Seiler, Mutation Res., 15, 273 (1972).
125. D. Siebert, F. K. Zimmermann, and E. Lemperle, Mutation Res., 10, 533 (1970).
126. S. E. Luria and M. Delbrück, Genetics, 28, 491 (1943).
127. A. Antoine, Exptl. Cell. Res., 37, 278 (1965).
128. B. D. Davis, in Drug Resistance in Microorganisms, Ciba Foundation Symposium (G. E. W. Wolstenholme and C. M. O'Connor, eds.), Churchill, London, 1957, pp. 165-182.
129. C. F. Heredia and A. Sols, Biochim. Biophys. Acta, 86, 224 (1964).
130. D. Moore and G. R. Stewart, Genet. Res. Cambridge, 18, 341 (1971).
131. A. Čapek and A. Šimek, Antimicrob. Agents, 10, 364 (1971).
132. A. Čapek and A. Šimek, Antimicrob. Agents, 16, 472 (1971).
133. M. Noble, Q. D. MacGavrie, A. F. Hams, and E. L. Leafe, Plant Pathol., 15, 23 (1966).
134. J. P. Malone, Plant Pathol., 17, 41 (1968).
135. J. E. Sheridan, J. H. Tickle, and Y. S. Chin, New Zealand J. Agr. Res., 11, 601 (1968).
136. W. Greenaway, Trans. Brit. Mycol. Soc., 56, 37 (1971).
137. W. Greenaway, J. Gen. Microbiol., 73, 251 (1972).
138. I. S. Ross and K. M. Old, Trans. Brit. Mycol. Soc., 60, 293 (1973).
139. I. S. Ross and K. M. Old, Trans. Brit. Mycol. Soc., 60, 301 (1973).
140. A. O. Summers and S. Silver, J. Bacteriol., 112, 1228 (1972).

141. A. O. Summers and E. Lewis, J. Bacteriol., 113, 1070 (1973).
142. T. Nakanishi and H. Oku, Phytopathology, 59, 1563 (1969).
143. T. Nakanishi and H. Oku, Phytopathology, 59, 1761 (1969).
144. T. Nakanishi and H. Oku, Ann. Phytopathol. Soc. Japan, 36, 67 (1970).
145. J. A. Bartz and J. E. Mitchell, Phytopathology, 60, 350 (1970).
146. R. W. Brockman and E. P. Anderson, Ann. Rev. Biochem., 32, 463 (1963).
147. J. Dekker, in Mechanisms of Action of Fungicides and Antibiotics (W. Girbardt, ed.), Akademie Verlag, Berlin, 1967, pp. 333-339.
148. J. Dekker, Neth. J. Plant Pathol., 74, 127 (1968).
149. J. Dekker and A. J. P. Oort, Phytopathology, 54, 815 (1964).
150. H. M. Dekhuijzen and J. Dekker, Pestic. Biochem. Physiol., 1, 11 (1971).
151. S. Shadomy, Appl. Microbiol., 17, 871 (1969).
152. R. R. Davies and M. A. Savage, Sabouraudia, 12, 302 (1974).
153. M. Grenson, Eur. J. Biochem., 11, 249 (1969).
154. M. A. de Waard, Ph.D. Thesis, Agricultural University, Wageningen, The Netherlands (1974).
155. D. M. Miller, World Rev. Pest Control, 10, 60 (1971).
156. K. T. Huang, T. Misato, and H. Asuyama, J. Antibiot. (Tokyo), 17A, 71 (1964).
157. M. Wagenman, W. Klingmüller, and W. Neupert, Arch. Microbiol., 100, 105 (1974).
158. U. Sinha, Genetics, 62, 495 (1969).
159. H. Cherest and H. de Robichon-Szulmajster, Genetics, 54, 981 (1966).
160. R. A. Butow and M. Zeydel, J. Biol. Chem., 243, 2545 (1968).
161. H. D. Sisler, Ann. Rev. Phytopathol., 7, 311 (1969).
162. R. W. Tillman and H. D. Sisler, Phytopathology, 63, 219 (1973).
163. H. Kersten and W. Kersten, in Inhibitors, Tools in Cell Research (Th. Bücher and H. Sies, eds.), Springer-Verlag, Berlin, 1969, pp. 11-31.
164. J. P. Lynch and H. D. Sisler, Phytopathology, 57, 367 (1967).
165. N. T. Bech-Hansen and G. H. Rank, Can. J. Genet. Cytol., 14, 681 (1972).
166. G. P. Clemons and H. D. Sisler, Pestic. Biochem. Physiol., 1, 32 (1971).
167. G. P. Clemons and H. D. Sisler, Phytopathology, 59, 705 (1969).
168. J. W. Vonk and A. Kaars Sijpesteijn, Pestic. Sci., 2, 160 (1971).
169. L. C. Davidse, in Microtubules and Microtubular Inhibitors (M. Borgers and M. de Brabander, eds.), North-Holland Publ. Co., Amsterdam, 1975, pp. 483-495.
170. L. C. Davidse and W. Flash, Neth. J. Plant Pathol., in press.
171. P. R. Harding, Jr., Plant Disease Reptr., 56, 256 (1972).

172. T. Kuramoto, Plant Disease Reptr., 60, 168 (1976).
173. I. F. Muirhead, Australian J. Exp. Agric. Anim. Husbandry, 14, 698 (1974).
174. S. G. Georgopoulos and V. Vomvoyanni, Can. J. Bot., 43, 165 (1965).
175. J. Kuiper, Australian J. Agric. Anim. Husbandry, 7, 275 (1967).
176. S. G. Georgopoulos, C. Zafiratos, and E. Georgiadis, Physiol. Plantarum, 20, 373 (1967).
177. D. J. Weber and J. M. Ogawa, Phytopathology, 55, 159 (1965).
178. B. Macris and S. G. Georgopoulos, Phytopathology, 59, 879 (1969).
179. R. J. Threlfall, J. Gen. Microbiol., 71, 173 (1972).
180. B. Macris and S. G. Georgopoulos, Z. Allgem. Mikrobiol., 13, 415
181. S. G. Georgopoulos, unpublished work, 1975.
182. A. Kappas, private communication, 1975.
183. K. J. Bent and R. H. Moore, in Biochemical Studies on Antimicrobial Drugs, 16th Symp. Soc. Gen. Microbiol. (B. A. Newton and P. E. Reynolds, eds.), Cambridge University Press, London, 1966, pp. 82-110.
184. L. M. Grisham, K. G. Bensch, and L. Wilson, J. Cell Biol., 59, 125a (1973).
185. R. E. Bauerle and H. R. Garner, Biochim. Biophys. Acta, 93, 316 (1964).
186. R. L. Metzenberg, M. S. Kappy, and J. W. Parson, Science, 145, 1434 (1964).
187. S. B. Galsworthy and R. L. Metzenberg, Biochem., 4, 1183 (1965).
188. M. R. Siegel and H. D. Sisler, Nature, 200, 675 (1963).
189. M. R. Siegel and H. D. Sisler, Biochim. Biophys. Acta, 87, 83 (1964).
190. M. R. Siegel and H. D. Sisler, Biochim. Biophys. Acta, 103, 558 (1965).
191. W. Mckeehan and B. Hardesty, Biochem. Biophys. Res. Commun., 36, 625 (1969).
192. M. Pongratz and W. Klingmüller, Molec. Gen. Genet., 124, 359 (1973).
193. A. Neuhäuser, W. Klingmüller, and F. Kaudewitz, Molec. Gen. Genet., 106, 180 (1970).
194. S. S. Rao and A. P. Grollman, Biochem. Biophys. Res. Commun., 29, 696 (1967).
195. A. Jimenez, B. Littlewood, and J. Davies, in Molecular Mechanisms of Antibiotic Action on Protein Biosynthesis and Membranes (E. Munoz, F. Garcia-Ferrandiz, and D. Vazquez, eds.), Elsevier Publishing Co., New York, 1972, pp. 292-306.
196. B. E. H. Maden, Progr. Biophys. Molec. Biol., 22, 127 (1971).
197. L. S. Kogerson, C. McLaughlin, and E. Wakatama, J. Bacteriol., 116, 818 (1973).
198. P. Grant, L. Sanchez, and A. Jimenez, J. Bacteriol., 120, 1308 (1974).

199. D. Schindler, P. Grant, and J. Davies, Nature, 248, 535 (1974).
200. T. Misato, in Antibiotics (D. Gottlieb and P. D. Shaw, eds.), Vol. I, Springer-Verlag, New York, 1967, pp. 434-439.
201. K. T. Huang, T. Misato, and H. Asuyama, J. Antibiot. (Tokyo), 17A, 71 (1964).
202. K. Ohmori, J. Antibiot. (Tokyo), 20A, 109 (1967).
203. H. Miura, M. Katagiri, T. Yamaguchi, Y. Uesugi, and H. Ito, Ann. Phytopathol. Soc. Japan, 42, 117 (1976).
204. T. L. Helser, J. E. Davies, and J. E. Dahlberg, Nature, New Biol., 235, 6 (1972).
205. A. P. Grollman, J. Biol. Chem., 242, 3226 (1967).
206. R. D. Tinline, Can. J. Bot., 39, 1695 (1961).
207. T. Kinoshita, G. Kawano, and N. Tanaka, Biochem. Biophys. Res. Commun., 33, 769 (1968).
208. E. Otaka, H. Teraoka, M. Tamaki, K. Tanaka, and S. Osawa, J. Molec. Biol., 48, 499 (1970).
209. W. N. Aldridge and B. W. Street, Biochem. J., 124, 221 (1971).
210. T. K. Sharples and R. A. Butow, J. Biol. Chem., 245, 58 (1970).
211. J. C. Mounolou, FEBS Lett., 29, 275 (1973).
212. C. Leben, D. M. Boone, and G. W. Keitt, Phytopathology, 45, 467 (1955).
213. C. J. P. Grimmelikhuijzen and E. C. Slater, Biochim. Biophys. Acta, 305, 67 (1973).
214. A. Tzagoloff and P. Meagher, J. Biol. Chem., 247, 594 (1972).
215. I. A. U. N. Gunatilleke, C. Scazzocchio, and H. N. Arst, Jr., Molec. Gen. Genet., 137, 269 (1975).
216. C. Shannon, R. Enns, L. Wheelis, K. Burchiel, and R. S. Griddle, J. Biol. Chem., 248, 3004 (1973).
217. A. Goffeau, Y. Landry, F. Foury, and M. Briquet, J. Biol. Chem., 248, 7097 (1973).
218. H. D. Sisler, in Perspectives of Biochemical Plant Pathology (S. Rich, ed.), Conn. Agr. Exptl. Sta. Bull. 663, 1963, pp. 116-136.
219. A. Endo, K. Kakiki, and T. Misato, J. Bacteriol., 104, 189 (1970).
220. M. Hori, K. Kakiki, and T. Misato, Agr. Biol. Chem., 38, 691 (1974).
221. M. Hori, K. Kakiki, and T. Misato, Agr. Biol. Chem., 38, 699 (1974).
222. S. Nishimura, K. Kohmoto, and H. Udagawa, Rept. Tottori. Mycol. Inst. (Japan), 10, 677 (1973).
223. M. Hori, J. Eguchi, K. Kakiki, and T. Misato, J. Antibiot. (Tokyo), 27, 260 (1974).
224. A. Kaars Sijpesteijn, A. Kerkenaar, and J. C. Overeem, Mededel. Fak. Landbouwwetenschap. Gent, 39, 1027 (1974).
225. N. N. Ragsdale and H. D. Sisler, Pestic. Biochem. Physiol., 3, 20 (1973).

226. J. L. Sherald, N. N. Ragsdale, and H. D. Sisler, Pestic. Sci., 4, 719 (1973).
227. N. N. Ragsdale, Biochim. Biophys. Acta, 380, 81 (1975).
228. J. L. Sherald and H. D. Sisler, Pestic. Biochem. Physiol., 5, 477 (1975).
229. D. Gottlieb and P. D. Shaw, Ann. Rev. Phytopathol., 8, 371 (1970).
230. E. K. Hebeka and M. Solotorovsky, J. Bacteriol., 84, 237 (1962).
231. J. M. T. Hamilton-Miller, Microbios, 8, 209 (1973).
232. M. Grindle, Molec. Gen. Genet., 120, 283 (1973).
233. J. M. T. Hamilton-Miller, Advan. Appl. Microbiol., 17, 109 (1974).
234. H. D. Sisler and C. E. Cox, in Plant Pathology: An Advanced Treatise (J. G. Horsfall and A. E. Dimond, eds.), Academic Press, New York, 1960, Vol. 2, pp. 507-552.
235. H. T. Gordon, Ann. Rev. Entomol., 6, 27 (1961).
236. S. Mudd, Scientific American, 200, 41 (1959).
237. E. C. Stackman and J. J. Christensen, in Plant Pathology: An Advanced Treatise (J. G. Horsfall and A. E. Dimond, eds.), Academic Press, New York, 1960, Vol. 3, pp. 567-624.
238. R. G. Owens, in Fungicides: An Advanced Treatise (D. L. Torgeson, ed.), Academic Press, New York, 1969, Vol. 2, pp. 147-301.
239. R. Duran and S. M. Norman, Plant Disease Reptr., 45, 475 (1961).
240. J. Kuiper, Nature, 206, 1219 (1965).
241. S. B. Locke, Phytopathology, 59, 13, abstr. (1969).
242. A. Kappas and S. G. Georgopoulos, Experientia, 24, 181 (1968).
243. M. Szkolnik and J. D. Gilpatrick, Plant Disease Reptr., 53, 861 (1969).
244. C. J. Delp and H. L. Klopping, Plant Disease Reptr., 52, 95 (1968).
245. D. C. Erwin, J. J. Sims, and J. Partridge, Phytopathology, 58, 860 (1968).
246. G. J. Bollen and G. Scholten, Neth. J. Plant Pathol., 77, 83 (1971).
247. W. R. Jarvis and A. J. Hargreaves, Plant Pathol., 22, 139 (1973).
248. M. W. Miller and J. T. Fletcher, Trans. Brit. Mycol. Soc., 62, 99 (1974).
249. R. D. Berger, Plant Disease Reptr., 57, 837 (1973).
250. E. M. Clark, P. A. Backman, and R. Rodriguez-Kabana, Phytopathology, 64, 1468 (1974).
251. R. H. Littrell, Phytopathology, 64, 1377 (1974).
252. S. G. Georgopoulos and C. Dovas, Plant Disease Reptr., 57, 321 (1973).
253. E. G. Ruppel and P. R. Scott, Plant Disease Reptr., 58, 434 (1974).
254. P. J. Griffee, Trans. Brit. Mycol. Soc., 60, 433 (1973).
255. J. M. Vargas, Jr., Phytopathology, 63, 1366 (1973).
256. R. O. Magie and G. J. Wilfret, Plant Disease Reptr., 58, 256 (1974).
257. G. J. Bollen, Neth. J. Plant Pathol., 77, 187 (1971).
258. J. H. Whan, Plant Disease Reptr., 60, 200 (1972).
259. N. Petsikos-Panayotarou, Neth. J. Plant Pathol., in press.

260. Y. Gutter, U. Yanko, M. Davidson, and M. Rahat, Phytopathology, 64, 1477 (1974).
261. C. G. Warren, P. Sanders, and H. Cole, Phytopathology, 64, 1139 (1974).
262. W. T. Schroeder and R. Provvidenti, Plant Disease Reptr., 53, 271 (1969).
263. D. Netzer and I. Dishon, Plant Disease Reptr., 54, 909 (1970).
264. T. Wicks, Plant Disease Reptr., 58, 886 (1974).
265. K. J. Bent, A. M. Cole, J. A. W. Turner, and M. Woolner, Proc. 6th Brit. Insectic. Fungic. Conf., Brighton, 1971, Vol. 1, p. 274.
266. M. S. Wolfe and A. Dinoor, Proc. 7th Brit. Insectic. Fungic. Conf., Brighton, 1973, Vol. 1, p. 11.
267. S. G. Georgopoulos, in Radiation and Radioisotopes for Industrial Microorganisms, International Atomic Energy Agency, Vienna, 1971, pp. 129-133.
268. C. Person, Can. J. Bot., 45, 1193 (1967).
269. M. E. Gallegly, Ann. Rev. Phytopathol., 6, 375 (1968).
270. R. R. Davies, private communication, 1975.
271. W. Greenaway and F. R. Whatley, Current Advan. Plant Sci., 7, 335 (1975).
272. I. Fourcade and E. Laville, Fruits, 28, 103 (1973).
273. H. Stingl, 2nd Intern. Plant Pathol. Congr., Minneapolis, 1973, Abstr. 128.
274. S. G. Georgopoulos, J. W. G. Geerligs, and J. Dekker, Neth. J. Plant Pathol., 81, 38 (1975).
275. S. H. Crowdy, in Systemic Fungicides (R. W. Marsh, ed.), Wiley, New York, 1972, pp. 92-115.
276. R. J. Williams and A. Ayanaba, Phytopathology, 65, 217 (1975).
277. P. R. Harding, Jr., Plant Disease Reptr., 46, 100 (1962).
278. M. Szkolnik and J. D. Gilpatrick, Plant Disease Reptr., 57, 817 (1973).
279. C. Dovas, G. Skylakakis, and S. G. Georgopoulos, Phytopathology, 66, 1452 (1976).
280. D. H. Brooks, Ann. Appl. Biol., 75, 136 (1973).
281. M. S. Wolfe, private communication, 1975.
282. S. G. Georgopoulos, Phytopathology, 53, 1081 (1963).
283. A. Kappas and S. G. Georgopoulos, Phytopathology, 61, 1093 (1971).
284. I. M. Lerner, Genetic Homeostasis, Wiley, New York, 1954.
285. J. F. Crow, Ann. Rev. Entomol., 2, 227 (1957).
286. J. Keiding, World Rev. Pest Control, 6, 115 (1967).
287. M. S. Wolfe, Ann. Appl. Biol., 75, 132 (1973).
288. K. J. Bent, private communication, 1975.
289. D. C. Erwin, FAO Plant Prot. Bull., 18, 73 (1970).
290. D. H. Brooks, in Systemic Fungicides (R. W. Marsh, ed.), Wiley, New York, 1972, pp. 186-205.
291. D. M. Spencer, in Systemic Fungicides (R. W. Marsh, ed.), Wiley, New York, 1972, pp. 206-224.

292. R. B. Maude, in Systemic Fungicides (R. W. Marsh, ed.), Wiley, New York, 1972, pp. 225-236.
293. R. J. W. Byrde, in Systemic Fungicides (R. W. Marsh, ed.), Wiley, New York, 1972, pp. 237-254.
294. J. F. Nicholson, W. A. Meyer, J. B. Sinclair, and J. D. Butler, Phytopathol. Z., 72, 169 (1971).
295. G. J. M. van der Kerk, Neth. J. Plant Pathol., 75, Suppl. 1, 5 (1969).
296. M. Legator and S. Zimmering, Ann. Rev. Pharmacol., 15, 387 (1975).
297. E. Evans, Pestic. Sci., 2, 192 (1971).
298. J. Dekker, Acta Phytopathol. Acad. Scient. Hungaricae, 6, 329 (1971).
299. A. H. M. Kirby, Pestic. Abstr. News Sum. Sect. B. Fungic. Herbic., 18, 1 (1972).
300. D. C. Erwin, Ann. Rev. Phytopathol., 11, 389 (1973).
301. E. Somers, Span, 6, 94 (1963).
302. G. J. M. van der Kerk, Acta Phytopathol. Acad. Scient. Hungaricae, 6, 311 (1971).
303. P. Chidambaram and G. W. Bruehl, Plant Disease Reptr., 57, 935 (1973).
304. K. G. Tate, J. M. Ogawa, B. T. Manji, and E. Bose, Plant Disease Reptr., 58, 663 (1974).
305. G. A. White, private communication, 1975.
306. E. Jawetz, Ann. Rev. Pharmacol., 8, 151 (1968).
307. J. M. Grayson and D. G. Cochran, World Rev. Pest Control, 7, 172 (1968).
308. G. Medoff, M. Comfort, and G. S. Kobayashi, Proc. Soc. Exptl. Biol. Med., 138, 571 (1971).
309. D. Woodcock, Chem. Brit., 7, 415 (1971).
310. J. M. van Tuyl, Neth. J. Plant Pathol., in press.
311. S. G. Georgopoulos and B. N. Ziogas, Neth. J. Plant Pathol., in press.
312. J. M. van Tuyl, Mededel. Fac. Landbouwhogeschool Rijksuniv. Gent, 40, 691 (1975).
313. A. Nachmias and I. Barash, J. Gen. Microbiol., 94, 167 (1976).
314. M. Hori, K. Kakiki, and T. Misato, J. Pestic. Sci., 1, 31 (1976).
315. D. H. Lambert and P. J. Wuest, Phytopathology, 66, 1144 (1976).
316. M. Yoshikawa, A. Okuyama, and N. Tanaka, J. Bacteriol., 122, 796 (1975).
317. Y. Ben-Yephet, A. Dinoor, and Y. Henis, Phytopathology, 65, 936 (1975).
318. Y. Ben-Yephet, Y. Henis, and A. Dinoor, Phytopathology, 65, 563 (1975).

319. T. P. Singer, H. Beinert, B. A. C. Ackrell, and E. B. Kearney, in Proc. 10th FEBS Meeting, Paris, 1974 (Y. Raoul, ed.), Elsevier, Amsterdam, 1975, pp. 173-185.
320. P. C. Mowery, B. A. C. Ackrell, T. P. Singer, G. A. White, and G. D. Thorn, Biochem. Biophys. Res. Commun., 71, 354 (1976).
321. H. Lyr and T. Schewe, Acta Biol. Med. Ger., 34, 1631 (1975).
322. M. F. Henry and E. J. Nyns, Sub-Cell. Biochem., 4, 1 (1975).
323. T. Kato, S. Tanaka, M. Ueda, and Y. Kawase, Agr. Biol. Chem., 38, 2377 (1974).
324. T. Kato, S. Tanaka, M. Ueda, and Y. Kawase, Agr. Biol. Chem., 39, 169 (1975).
325. H. Buchenauer and F. Grossman, Neth. J. Plant Pathol., in press.
326. W. Iida, Japan. Pestic. Inform., 23, 13 (1975).
327. V. W. L. Jordan and D. V. Richmond, Plant Pathol., 23, 81 (1974).
328. A. H. McCain, Calif. Plant Pathol., 32, 1 (1976).
329. L. R. Schreiber and A. M. Townsend, Phytopathology, 66, 225 (1976).
330. R. T. A. Cook and J. L. Pereira, Ann. Appl. Biol., 83, 365 (1976).
331. E. Shabi and Y. Ben-Yephet, Plant Disease Reptr., 60, 451-454 (1976).
332. T. Locke and I. G. Thorpe, Plant Pathol., 25, 59 (1976).
333. P. J. Wuest, H. Cole, and P. L. Sanders, Phytopathology, 64, 331 (1974).
334. G. J. Bollen and A. van Zaayen, Neth. J. Plant Pathol., 81, 157 (1975).
335. K. Abiko, K. Kishi, and A. Yoshioka, Ann. Phytopathol. Soc. Japan, 41, 100 (1975).
336. A. Kappas, M. H. L. Green, B. A. Bridges, A. M. Rogers, and W. J. Muriel, Mutation Res., 40, 379 (1976).
337. M. C. Shephard, K. J. Bent, M. Woolner, and A. M. Cole, Proc. 8th Brit. Insectic. Fungic. Conf., Brighton, 1975, Vol. 1, p. 59.
338. D. W. Holomon, Proc. 8th Brit. Insectic. Fungic. Conf., Brighton, 1975, Vol. 1, p. 51.
339. G. P. Georghiou, Ann. Rev. Ecol. Systematics, 3, 133 (1972).
340. A. W. A. Brown, Proc. Amer. Phytopathol. Soc., 3, in press.
341. P. Langcake and S. G. A. Wickins, Physiol. Plant Pathol., 7, 113 (1975).
342. P. Langcake, Neth. J. Plant Pathol., in press.
343. V. D'Ambra, S. Mutto, and G. Caruba, Ind. Saccar. Ital., 1, 11 (1974).

Chapter 13

ANTIFUNGAL COMPOUNDS ASSOCIATED WITH DISEASE RESISTANCE IN PLANTS*

Joseph Kuć and Louis Shain

Department of Plant Pathology
University of Kentucky
Lexington, Kentucky

*Published as Journal Paper No. 75-11 of the Kentucky Experiment Station, Lexington, Kentucky.

I. INTRODUCTION

Green plants represent the broadest spectrum of synthetic capabilities since the initial reactants are water, carbon dioxide, nitrogen (elemental and in inorganic salts), phosphorus compounds, and traces of inorganic salts. With the vast diversity of higher green plants and their chemical constituents, it is not surprising that many plants contain antifungal compounds. It is also not surprising that some of these compounds may participate in disease resistance mechanisms. Some would be effective against a broad spectrum of fungi and could effectively protect a broad spectrum of plants other than those in which they were produced. Plant products make a vital contribution to our list of medicinals. The widespread use of antibiotics of microbial origin does not preclude the existence of equally potent antibiotics in higher plants, and plant products may find application in the control of plant disease. The use of plant products as agents for disease control can be considered from at least two perspectives. One is the use of the chemical itself as a protectant. The second is the elicitation of the chemical's production in plants. The latter consideration includes the accumulation and retention of the antifungal compound throughout the life of the plant as well as the transitory, but rapid, accumulation of the compound to fungitoxic levels after exposure to the fungus. The activation of the potential to respond may be the key to protection rather than the maintenance of a high level of the compound.

The suggestion that plants respond biochemically to infectious agents and that this response is associated with disease resistance, was proposed as early as the late 19th century. The validity of a response mechanism in plants and its association with resistance was brought to public attention by the reviews of Chester [1]. The complex defense mechanism in animals, including the production of antibodies, is well accepted and is the basis for preventive medicine. A phenomenon similar to induced immunity in animals was reported by Muller and Borger [2]. They found that potato tubers developed localized resistance to a pathogenic race of Phytophthora infestans if they were first inoculated with a race of the fungus to which they were resistant. This observation led to the development of the "phytoalexin theory." The basic concept of this theory was that a chemical compound, a phytoalexin, was produced by plant cells as a result of metabolic interaction between host and infectious agent. This compound was not detectable in the plant before infection and it inhibited nonpathogens of the host. It was proposed that pathogens were either not sensitive to the phytoalexin or not able to cause its accumulation. In recent years, phytoalexins have been detected in trace quantities in healthy tissue and their accumulation is not dependent on infection only. Fungi and fungal metabolites, viruses, bacteria, mechanical injury, and numerous chemicals cause the accumulation of phytoalexins. It also seems unlikely that a single phytoalexin is responsible for resistance since a number of fungitoxic compounds have been demonstrated to accumulate after infection in many host-parasite interactions. Some of these compounds are structurally related, e.g., the derivatives of phaseollin in the green bean; others are not, e.g., chlorogenic acid and rishitin in the Irish potato. Some of the compounds are also present in tissue before infection and may serve as a "passive" chemical defense.

This chapter emphasizes only those antifungal compounds that have been associated with disease resistance mechanisms in plants. Some reference will be made to the potential application of plant antifungal compounds and their elicitation for the practical control of disease.

Most of the work on antifungal compounds in trees has been related to the decay resistance of forest products. Resistance to decay fungi was largely attributed to certain phenolic compounds occurring in the heartwood of many species. This subject was reviewed in 1966 by Scheffer and Cowling [3]. Since then interest has continued to develop in resistance mechanisms of the living tree. These large, perennial hosts offer unique opportunities, as well as some obvious difficulties, for the study of disease resistance. Such terms as discolored wood, false heartwood, protection wood, reaction zone, etc. [4] have been applied to what appears to be a dynamic response of living xylem to injury and infection. Inhibitory substances are being identified in these tissues as well as the phloem. The effect of oleoresin in the Pinaceae on disease resistance has received considerable attention since the late 1960s. It is these areas that will be emphasized in Section III of this chapter.

The efficacy of the disease resistance mechanisms of plants is evident in their survival on earth, and the mechanisms in plants are as important for survival as those in animals. Our first line of defense against plant disease is the development of disease resistant plants.

II. ANTIFUNGAL COMPOUNDS IN HERBACEOUS PLANTS

A. Leguminosae

1. Garden Pea

Early workers studying the role of phytoalexins in disease resistance of plants were without the benefit of chemically characterized phytoalexins. Gauman and Kern [5], Uritani and Maramatsu [6], Kuć et al. [7], Condon and Kuć [8], Cruickshank and Perrin [9], and their colleagues were among the first to work with chemically characterized phytoalexins and to introduce biochemical techniques to studies of their role in disease resistance.

Using the drop-diffusate technique, Cruickshank and Perrin [9, 10] and Perrin and Bottomley [11] isolated pisatin (1) from the pods of the garden pea inoculated with fungi. Pisatin is a weak antibiotic with a broad biological

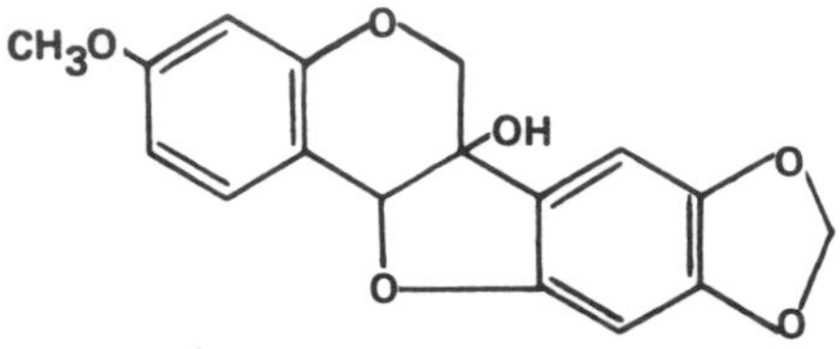

(1)

spectrum [12]. Mycelial growth of Monilinia fructicola is three times more sensitive to pisatin than is spore germination and this is characteristic of the response of many other fungi to phytoalexins. Fungi pathogenic to pea are generally insensitive to the amounts of pisatin accumulating after infection, whereas nonpathogens of pea are generally sensitive.

The antifungal spectrum of pisatin includes Phycomycetes, Ascomycetes, Basidiomycetes, and Fungi Imperfecti with representatives of many phytopathogens [12]. The ED_{50} of pisatin for mycelial growth of 38 out of 44 nonpathogens of pea was less than 50 μg/ml and only one of them had an ED_{50} in excess of 75 μg/ml. The ED_{50} for 11 of the fungi was less than 25 μg/ml and 38 were inhibited more than 90% at 100 μg/ml. Pisatin accumulation is stimulated by many fungi, metabolic inhibitors, spore-free germination fluids, psoralens, ultraviolet (UV) radiation and ethylene [13-18]. In addition, many medicinals elicit pisatin accumulation in pods, including antihistaminic, antiviral, antimalarial, and tranquilizing drugs [19, 20]. With few exceptions, the elicitors of pisatin accumulation also increase the activity of phenylalanine ammonia lyase (EC 4.3.1.5) in treated tissues. Pisatin is degraded by a number of pathogens of pea and since pisatin is not a stable end product in plant tissues, degradation by the plant itself is also a consideration [21-24].

Heath and Wood [25] reported the accumulation of pisatin in infected pea leaflets, and work with pods and leaflets clearly indicates that pisatin accumulation is not restricted to resistant interactions. Schwochau and Hadwiger [26] presented an induction hypothesis for resistance based on the Jacob-Monod model for gene activation in bacteria. They proposed that microbial metabolites in low concentration induce high metabolic activity, protein synthesis, and phytoalexin accumulation. The accumulation of phytoalexins, such as pisatin, may be a response to stress and part of a general tissue-repair mechanism since UV radiation and many compounds of diverse chemical structure cause their accumulation.

Pisatin appears to be synthesized jointly by the acetate-malonate and shikimate pathways [27].

2. Green Bean

Cruickshank and Perrin [10, 28] and Perrin [29] identified phaseollin (2) in diffusates from seed cavities of green bean pods inoculated with M. fructicola, a nonpathogen of bean. A number of compounds structurally related to phaseollin, phaseollidin (3), phaseollinisoflavan (4), kievitone (5), and coumestrol (6) have recently been isolated from diffusates or tissues

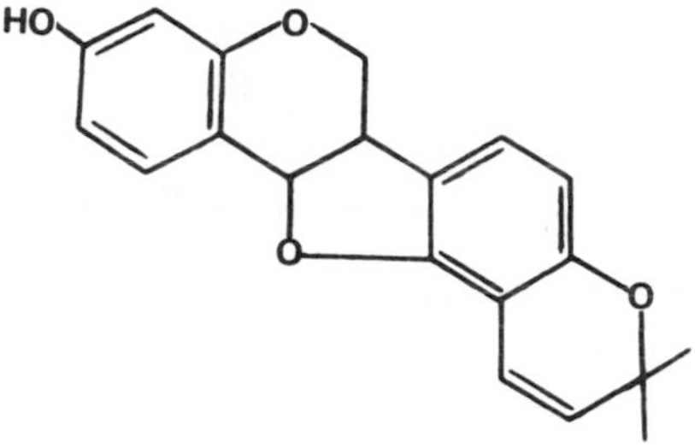

(2)

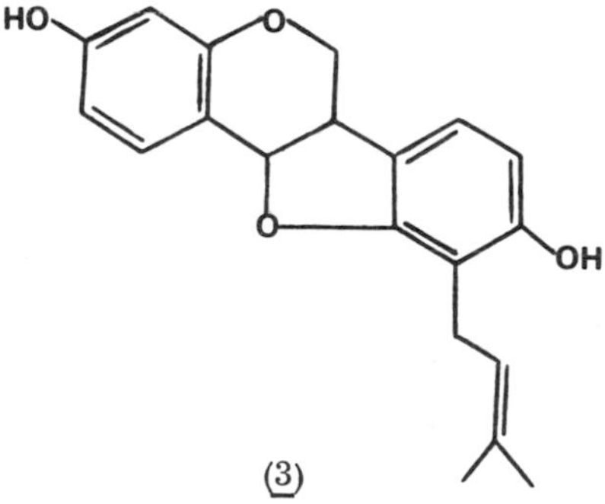

(3)

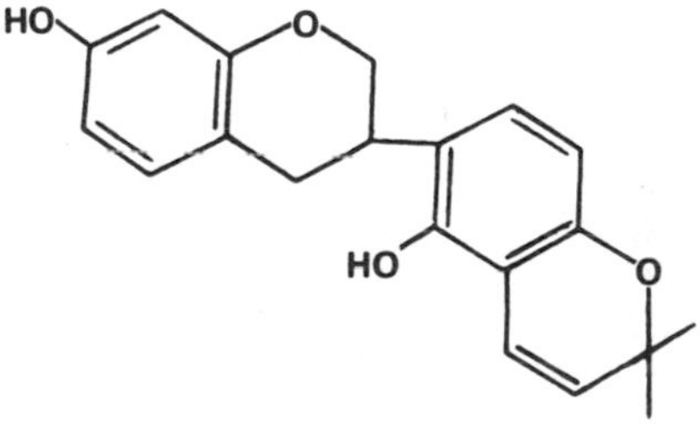

(4)

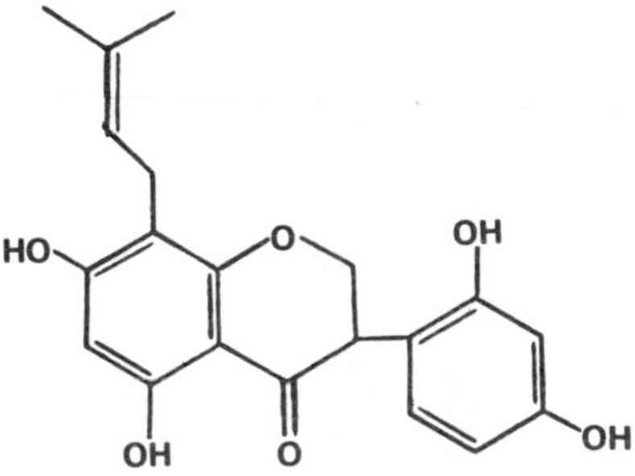

(5)

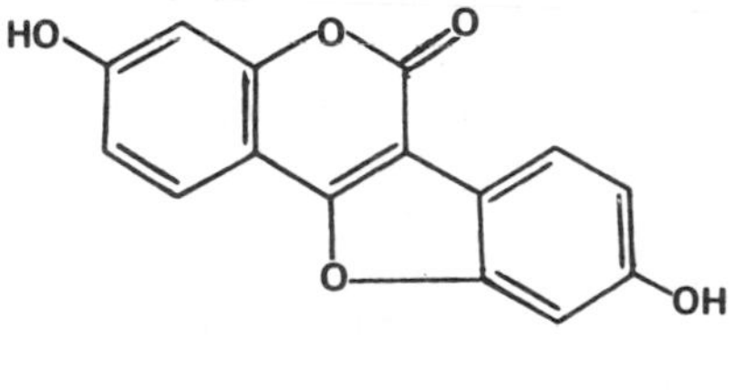

(6)

inoculated with fungi, bacteria, or viruses [30-35]. In addition, a number of degradation products of phaseollin have been reported in infected tissue, culture filtrates of fungi, and diffusates from infected tissue [36-39]. Burden and Bailey [40] reported 6a-hydroxyphaseollin, formerly incorrectly thought to be the major phytoalexin in soybean, as a metabolite arising from the action of Colletotrichum lindemuthianum on phaseollin.

Phaseollin, like pisatin, is a weak antibiotic [28]. It is inhibitory to a broad spectrum of fungi, but not to bacteria, and is fungistatic rather than fungicidal [28, 41]. The ED_{50} of phaseollin for mycelial growth is less than 10 μg/ml for 14 of 27 nonpathogens of bean tested and 10-20 μg/ml for seven others. Three of six pathogens of bean tested had an ED_{50} in excess of 50 μg/ml. At 50 μg/ml the growth of 21 of the 27 fungi tested was inhibited 85-100% [28]. The ED_{50} of phaseollin, phaseollidin, phaseollinisoflavan, and kievitone for mycelial growth of Rhizoctonia solani ranged from 18 to 36 μg/ml [35].

Monilicolin A, a peptide from the mycelium of M. fructicola, stimulates phaseollin accumulation in bean pod endocarp at 2.5×10^{-9} M [42]. The peptide is not fungitoxic or phytotoxic and it does not cause the accumulation of pisatin or viciatin in pea and broad bean, respectively. It dissolves readily in water and has a molecular weight of about 8,000. Necrosis of plant tissue is not necessary for the accumulation of phaseollin which is also caused by low concentrations of heavy metal ions, metabolic inhibitors, and antibiotics [28]. Rathmell and Bendall [33] suggest phaseollin accumulation arises because of a specific stimulation of isoflavonoid metabolism, which is separate from a general increase in phenols associated with necrosis.

The first report associating the accumulation of phaseollin and phaseollin-like compounds with varietal resistance was by Rahe et al. [43]. They reported that the appearance of the hypersensitive reaction of bean to Colletotrichum lindemuthianum corresponded to the time of rapid phaseollin accumulation in tissues. Phaseollin accumulated in susceptible varieties at the time large lesions became apparent. Thus, the ability to accumulate phaseollin or its derivatives is not restricted to resistant plants. The time of accumulation and its magnitude, rather than the ability to accumulate the phytoalexins, appears to distinguish susceptibility from resistance. In subsequent studies, both the development of the fungus in infected bean hypocotyls and accumulation of phaseollin and phaseollin-like compounds were followed [34, 44, 45]. It appears that the concentrations of these compounds is sufficient to markedly inhibit elongation of hyphae and is associated with restricted growth of the fungus. Still unknown, however, are the critical factors for resistance or susceptibility, which determine when phaseollin accumulates, the magnitude of its accumulation, and the causes of the hypersensitive death of resistant cells. It has been demonstrated that bean hypocotyls inoculated with an incompatible (nonpathogenic) race or heat-attenuated compatible (pathogenic) race of C. lindemuthianum are protected from disease when subsequently inoculated with a compatible race [46-50]. It was demonstrated further that hypocotyls were protected at a distance from the site of inoculation with an incompatible race [50]. Phaseollin accumulation was apparent in protected tissue only after the compatible race had penetrated [51]. Induced resistance at a distance involves effects upon the cells of the plant before challenge, but spore germination and penetration of the challenge are not affected on protected tissue [51]. The concept that resistance appears dependent on the expression of a potential rather than its presence or absence was strengthened by the observation that protection was induced in varieties of bean, apparently susceptible to all races of C. lindemuthianum, by Colletotrichum spp. nonpathogenic on green bean [51]. One interpretation of the data is that protected tissue reacts to the compatible race as if it were an incompatible race and this leads to phaseollin accumulcation, necrosis, and restricted development of the pathogen. Skipp and Deverall [52] reported that culture filtrates of compatible and incompatible races caused protected tissue to produce an inhibitor of spore germination. Berard et al. [53, 54] demonstrated the presence of a nonfungitoxic substance in cell-free diffusates from incompatible, but not compatible, interactions which caused protection. Thus, factors produced by the fungus or the interaction may influence resistance even though they do not directly result in the accumulation of phaseollin. The evidence presented by Kuć [55] and Cruickshank et al. [36] that a number of phytoalexins contribute to resistance and that their concentration is influenced by the infectious agent, host, and phase of infection, strengthens the role of phytoalexins in resistance. It is difficult to envision a single phytoalexin being responsible for the resistance of a plant to all pathogens. Phaseollin appears to be synthesized by the joint participation of the shikimate and acetate-malonate pathways [56].

3. Soybean

Fungitoxic compounds were reported to be produced in open soybean pods inoculated with nonpathogens of soybean [57, 58] and were leached from infected soybean plants by means of strings through inoculated wounds in the hypocotyl [59-61]. The compounds were found in fungitoxic concentrations in leachates from hypocotyls inoculated with nonpathogenic races of Phytophthora megasperma var. sojae, but not in uninoculated tissue or tissue inoculated with pathogenic races. They accumulated in soybeans inoculated with nonpathogens regardless of varietal resistance or susceptibility to the pathogen. Biehn et al. [62] found a number of phenols accumulated in etiolated seedlings of soybean, green bean, and lima bean inoculated with the pathogen of corn, Helminthosporium carbonum. The same phenols accumulated when soybean seedlings were inoculated with any one of five nonpathogens of soybean, but the quantity accumulated depended upon the fungus. Total phenol content increased 4-5-fold 24 hr after soybean seedlings were inoculated with H. carbonum, and the increase was accompanied by a marked increase in phenylalanine ammonia-lyase. One of the major phenols, probably the soybean phytoalexin subsequently isolated by Keen and his colleagues, accumulated to levels which markedly inhibited fungal growth [63].

A pterocarpan, characterized as 6a-hydroxyphaseollin, was isolated from soybean hypocotyls infected with Phytophthora megasperma var. sojae [64-66]. Burden and Bailey [40] subsequently corrected the structure and the compound has been named glyceollin (7). It is closely related to pisatin

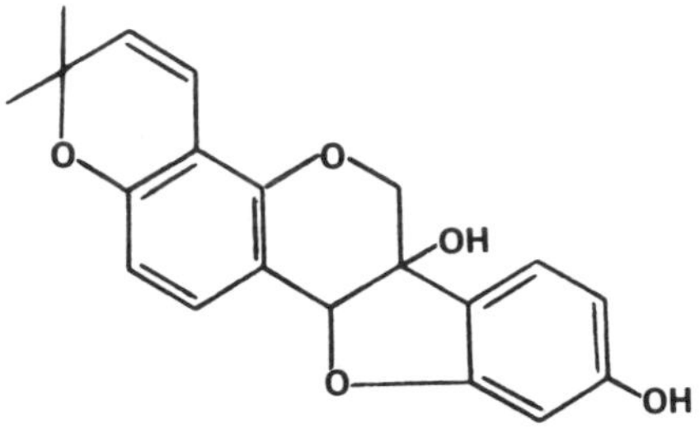

(7)

and phaseollin, and it accumulated 10-100 times faster in hypocotyls infected with an incompatible as compared to a compatible race of the fungus. In incompatible and compatible reactions, the concentrations of the phytoalexin is 100-400 and 1-4 times, respectively, the ED_{50} concentration for inhibition of mycelial growth of the fungus. The ED_{50} for mycelial growth of P. megasperma var. sojae is approximately 25 μg/ml. Like pisatin and phaseollin, glyceollin is fungistatic rather than fungicidal. The work of Keen and his colleagues strongly supports the role of glyceollin in disease resistance [64-66]. It is among the few and most thorough investigations of varietal resistance. Protection against pathogens by nonpathogenic races of pathogens or nonpathogens has been demonstrated [67, 68] and several fungicides, UV radiation, and tobacco necrosis virus elicit accumulation of the phytoalexin [69-71].

A recent paper by Keen [72] reports that both specific and nonspecific elicitors of phytoalexin accumulation are produced by P. megasperma var. sojae. A specific elicitor from culture filtrates of race 1 of the fungus caused greater accumulation of glyceollin in a resistant than a near-isogenic susceptible cultivar of soybean. It appears that the soybean phytoalexins daidzein, coumestrol, and sojagol accumulate in infected hypocotyls due to a general activation of isoflavonoid biosynthesis, with metabolites directed to the biosynthesis of pterocarpans [73]. Several phytoalexins, pterocarpan and nonpterocarpan, probably contribute to resistance in soybean.

4. Alfalfa, Red Clover, and Broad Bean

Coumestrol (6) accumulates in alfalfa inoculated with the pathogenic fungi Ascochyta imperfecta, Cylindrocladium scoparium, Colletotrichum trifolii, and *Uromyces striatus* [74]. It was not translocated from the infected roots to foliage or vice versa and very little accumulated in leaves inoculated with nonpathogenic fungi or a pathogenic bacterium. Coumestrol did accumulate, however, in detached leaves inoculated with nonpathogenic fungi. The concentration of coumestrol appeared directly related to the degree of infection. In addition to coumestrol, a broad spectrum of flavonoid aglycones and glycosides increased markedly during infection [75]. Infection of leaves did not alter the flavonoid content of roots and some of the flavonoid aglycones may have arisen by the action of fungal glycosidases. The phytoalexin medicarpin (8) was isolated from alfalfa leaves inoculated with spores of the corn pathogen, *Helminthosporium turcicum* [76]. *Stemphylium botryosum*, a pathogen of alfalfa, but not *H. turcicum*, a nonpathogen of alfalfa, degraded medicarpin to noninhibitory compounds [77, 78]. *Colletotrichum phomoides*, a nonpathogen, and *Stemphylium loti*, a weak pathogen of alfalfa, also degraded medicarpin, but the degradation products were inhibitory to these fungi [79]. The compounds accumulating in susceptible host-parasite interactions may be degradation products of phytoalexins. These compounds are synthesized by the host or are reaction products formed from metabolites accumulating after infection, e.g., amino acid addition products of quinones. Two phytoalexins, medicarpin and maackiain (9), were isolated from red clover foliage infected with H. turcicum [80]. The ED_{50} of medicarpin and maackiain for mycelial growth of *H. turcicum* is 45 μg/ml. The nonpterocarpan phytoalexins, wyerone and wyerone acid, were isolated from broad bean leaves infected by *Botrytis fabae* [81]. The ED_{50} of wyerone acid for mycelial growth of *Botrytis fabae*, *B. cinerea*, and *B. allii* are 13.5, 2.0, and 0.7 μg/ml, respectively. A concentration of 9 μg/ml completely prevented germination of *B. allii* spores.

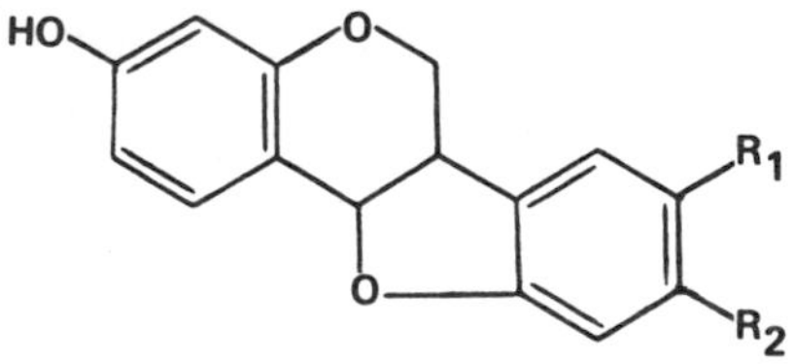

(8) $R_1 = H$, $R_2 = OCH_3$

(9) $R_1 = R_2 = OCH_2O$

B. Solanaceae

1. The Irish Potato

The foliage and tuber of the potato contains many preformed fungitoxic chemicals and both can accumulate phytoalexins. Chlorgenic (10) and caffeic acids, scopolin (11) , α-solanine, and α-chaconine, are the major fungitoxic

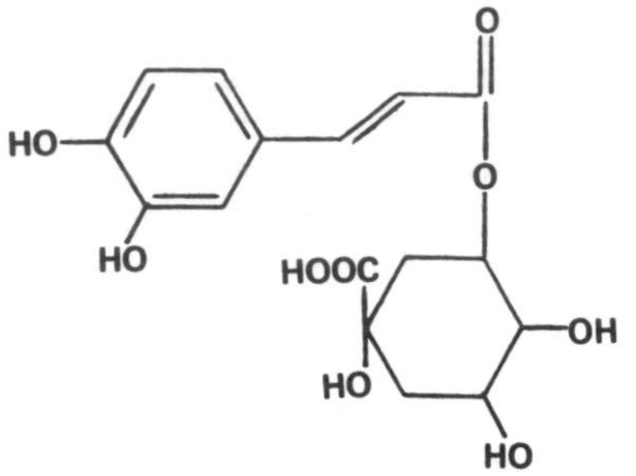

(10)

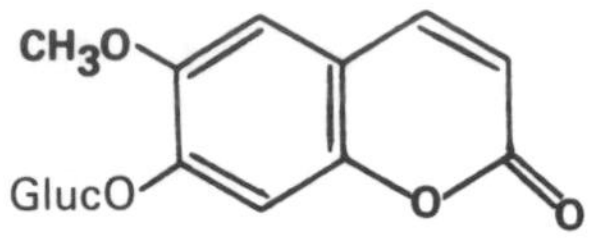

(11)

compounds found in healthy peel and foliage and they accumulate in injured and infected tubers. Rishitin (12), rishitinol, phytuberin (13), lubimin (14) and several spirovetiva derivatives, though possibly present in trace quantities in healthy tubers, accumulate markedly after infection.

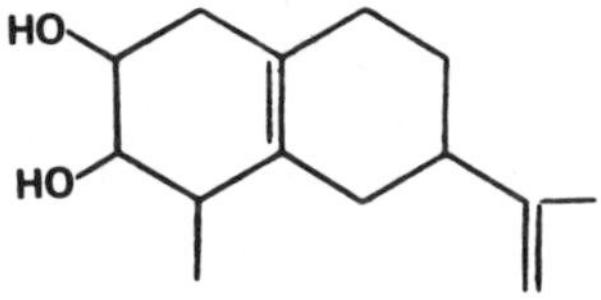

(12)

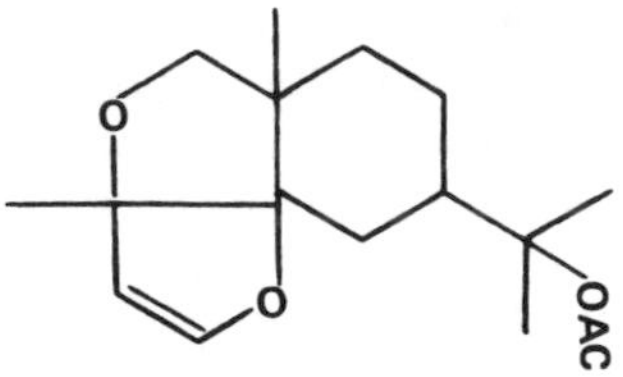

(13)

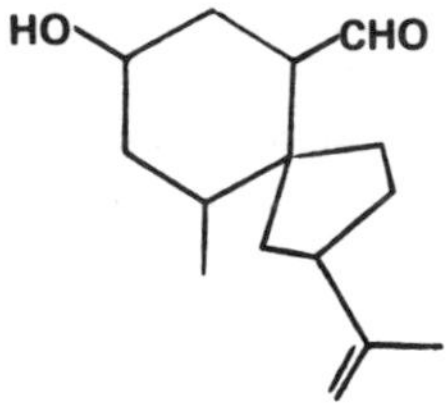

(14)

Chlorogenic and caffeic acids are weakly fungistatic and the ED_{50} values for mycelial growth of Fusarium solani are 0.75 mg and 0.30 mg/ml, respectively [82]. It is unlikely that they play a direct role in resistance as inhibitors of fungal development. These acids may affect metabolism of the host or pathogen and thereby indirectly influence resistance, or they may be converted to fungitoxic agents by the action of host and pathogen. Many investigators [82-86] suggest that quinones of chlorogenic and caffeic acids are phytoalexins. The o- and p-quinones are highly reactive and they take part in 1,4 additions to many amines and sulfhydryl compounds including proteins. They are nonspecific inhibitors of many enzymes and rapidly polymerize to lignin-like compounds in or on plant cell walls. Such cell walls may be more resistant to extracellular hydrolases, including pectinases and cellulases. Reactive quinones are toxic to plant cells as well as infectious agents, perhaps by altering cellular membranes, and their production may explain the necrosis often associated with hypersensitive resistance. The polymerization of quinones may function as a detoxication mechanism for plants and microorganisms, in the latter case, if it occurs in advance of the microorganism. Oxidation products of chlorogenic and ascorbic acid inhibit aldolase and the inhibition may shift respiration to the pentose pathway [87]. This can lead to an increase in erythrose phosphate and nicotinamide adenine dinucleotide phosphate (NADPH) which are necessary for the synthesis of phenols by the shikimate pathway. NADPH is also required for phenol synthesis in the acetate-malonate pathway and terpenoid synthesis in the acetate-mevalonate pathway. Phenols and quinones may also influence hormonal levels in plants, including the action and level of indole acetic acid [88, 89]. The role of quinones in disease resistance has largely been speculative. The compounds are formed in host-parasite interactions and they have properties which suggest their role in resistance. The danger of this approach is illustrated in work with the resistance of immature apples by a rot caused by _Botryosphaeria ribis_ [90]. Phenolics are oxidized more readily in green than mature apples, but it does not appear that the oxidation of phenols is responsible for the resistance of the green fruits.

Recently, Clarke [91, 92] reported that scopolin accumulates in potato tissue in response to infection by viruses, fungi, and an actinomycete. The greatest accumulation occurs in tissue infected by virulent isolates of pathogens, including _Phytophthora infestans_. The compound does not accumulate in response to wounding, avirulent isolates, or nonpathogens. The author suggests that virulent pathogens inhibit the development of lignin-like polymeric barriers, which form part of the host's resistance mechanism, by diverting precursors into other metabolic pathways. One such pathway may be involved in the synthesis of scopolin, a compound resistant to oxidation by peroxidases or phenoloxidases, instead of a readily oxidizable phenol, such as chlorogenic acid. Its synthesis would reduce the accumulation of toxic quinones and retard development of the hypersensitive reaction as well as the formation of protective polymers. Scopolin is an in vitro inhibitor of peroxidases which are involved in the oxidation of phenols and their subsequent polymerization.

The sites within the host cell in which a particular compound accumulates must be considered in determining whether or not it can inhibit the growth of an invading parasite. Scopolin accumulates within the protoplasts of host cells, probably within vacuoles, and thus may not participate directly in the host-parasite interaction as an enzyme inhibitor. The deposition of scopolin in the vacuole, however, is appropriate if it acts as a sink for phenolic metabolites which might otherwise be converted to oxidizable phenols. The accumulation of conjugates of cinnamic acid derivatives other than those mentioned, as well as the binding of phytoalexins to cell walls, may also be important in determining the nature of a host-parasite interaction [91, 92].

Another group of phytoalexins accumulating in potato after infection are isoprenoid derivatives. Rishitin, a bicyclic norsesquiterpene alcohol, was isolated by Tomiyama et al. [93] from resistant tubers infected with P. infestans, and its structure was established by Katsui et al. [94]. Other fungi [93, 95] as well as bacteria [96] also induce its accumulation. The development of P. infestans in susceptible and resistant tuber or leaf petiole tissue and the rate of rishitin accumulation suggests that rishitin has a role in disease resistance [85, 97]. Rishitin is first detected when inhibition of fungal growth begins and rapidly accumulates to levels many times that necessary to completely prevent it. Browning and restricted cell death induced by many chemicals and physiological stimuli do not cause rishitin accumulation. However, tuber slices treated with boiled cell-free sonicates of P. infestans accumulate high levels of rishitin with little browning in all cultivars tested, including those susceptible to all known races of the fungus [95].

Sato et al. [98] demonstrated that rishitin accumulated in incompatible interactions of four cultivars containing one of four "R" genes for resistance. A variety susceptible to all known races of the fungus accumulated rishitin when dipped into a cell-free homogenate of the fungus. Working with 11 cultivars and three races of the fungus, Varns et al. [99] demonstrated that the incompatible interaction is associated with rapid necrosis and the accumulation of 16-18 terpenoids, including rishitin and phytuberin. The latter is an aliphatic unsaturated, sesquiterpene acetate, $C_{17}H_{26}O_4$, isolated and partially characterized by Varns [100] and recently completely characterized by Coxon et al. [101] and Hughes and Coxon [102]. Rishitinol [103], lubimin [104, 105], spirovetiva-1(10),11-diene-2-one and spirovetiva-1(10),3,11-trien-2-one [106] have also been isolated from tubers infected by P. infestans. The ED_{50} values of rishitin, lubimin, and phytuberin for mycelial growth of P. infestans are 50, 40, and 60 μg/ml, respectively. Higher concentrations of phytuberin appear to lyse spores of P. infestans [100]. Fungal growth is completely inhibited at a concentration of 100 μg/ml rishitin. The potential for synthesis and accumulation of these phytoalexins exists in completely susceptible cultivars, but this potential is not readily expressed when cultivars are challenged by compatible races of P. infestans [95, 98]. This is consistent with other phytoalexins studied and disease resistance mechanisms found in animals. Tomiyama [107] indicated that a compatible race of the fungus delayed the hypersensitive collapse of cells. Varns and Kuć [108] demonstrated that a compatible race suppressed both necrosis and the accumulation of rishitin and phytuberin in tubers subsequently inoculated with an incompatible race. The suppression of the response to the incompatible race, or inoculation with the compatible race alone, was accompanied by the accumulation of unidentified nonfungitoxic terpenoids which were not detected in incompatible reactions [109]. Once the host was inoculated with an incompatible race, suppression from a subsequent inoculation with a compatible race did not occur. An incubation of 12 hr was sufficient either to establish suppression or elicit the hypersensitive response. A compatible interaction also suppressed the hypersensitive host response to cell-free sonicates of the fungus. In the interaction of potato and P. infestans, susceptibility may be determined by the suppression of a general resistance mechanism. The work of Clarke [92] supports this hypothesis. However, other workers [110, 111] suggest that necrosis of host tissue and subsequent accumulation of the terpenoid phytoalexins occurs in incompatible reactions after the fungus has been contained. They offer evidence that rishitin accumulation does not have a primary role in resistance.

At least two other isoprenoids, the steroid glycoalkaloids α-solanine and α-chaconine, may also be associated with resistance. They have been reported in potato tubers and foliage and appear localized around sites of injury in tubers [112]. The steroid glycoalkaloids are largely restricted to the peel of whole tubers, and they are the major antifungal compounds in potato peel [113]. The ED_{50} of α-solanine for mycelial growth of H. carbonum, a nonpathogen of potato, is 35 μg/ml and α-chaconine is approximately twice as toxic. This value is for potato dextrose agar at pH 5.6. The compounds are 100 times more active at pH 6.3-7.0 and the free base appears to be the active form. The accumulation of α-solanine and α-chaconine at the surface of cut tissue slices is markedly suppressed by inoculation with P. infestans, and the suppression is most marked after inoculation with incompatible races of the fungus [114-116].

The primary event determining a compatible or incompatible reaction of potato to P. infestans appears dependent upon an interaction (recognition) which occurs within hours and probably seconds after penetration and is responsible for establishing the nature of the reaction. All metabolic alterations, including the accumulation of phytoalexins, are the result of the initial interaction.

2. Pepper

The sesquiterpene phytoalexin capsidiol (15) was isolated from the fruit of sweet pepper inoculated with a number of fungi [117, 118]. It accumulates rapidly in some interactions but in others it is rapidly oxidized to capsenone [119]. Capsenone is less fungitoxic than capsidiol but pathogenicity may not solely depend upon the ability to detoxicate capsidiol. The ED_{50} of capsidiol for mycelial growth of M. fructicola is 15 μg/ml and that for Alternaria alternata is 60 μg/ml. Recently, the control of late blight of tomato under

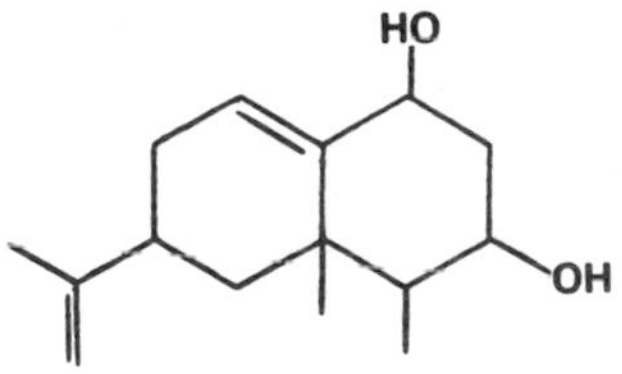

HO

OH

(15)

growth-room conditions was demonstrated with capsidiol applied as a spray at concentrations of 2.5×10^{-3} to 5×10^{-1} M [120].

C. Malvaceae

1. Cotton

The rate of accumulation of gossypol (16) and gossypol-like compounds in boll cavities, stem sections, and intact plants of species of *Gossypium* appears directly related to host resistance and inversely related to virulence

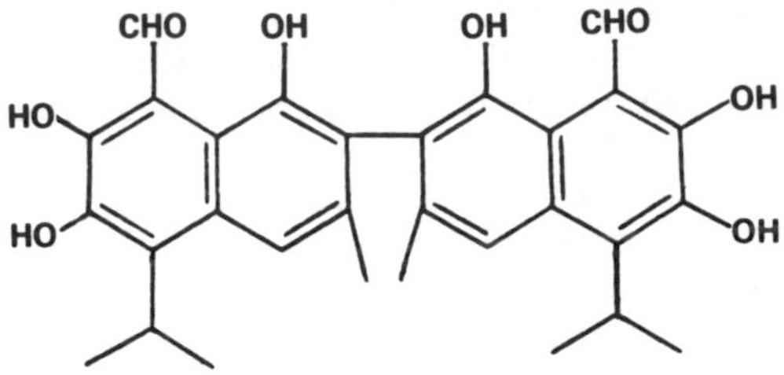

(16)

of the pathogen [121, 122]. The speed of colonization of host by *Verticillium albo-atrum* relative to the speed of gossypol accumulation appears to be an important factor in determining wilt resistance in cotton [123]. The accumulation of gossypol and gossypol-like compounds and the disease reaction are sensitive to temperature. The optimum temperature for accumulation of the compounds corresponds to the temperature at which cotton is resistant. Susceptibility and resistance are obviously not absolute conditions but are expressions within environmental limits. This is supported further by the observation that heat-killed conidia or conidia attenuated by heat in stem sections cause gossypol accumulation and increase resistance [124]. Avirulent strains of the fungus also protect against damage from subsequent inoculation with virulent strains. Gossypol accumulates more rapidly in resistant as compared to susceptible or tolerant varieties infiltrated with heat-killed conidia of *V. albo-atrum* or with solutions of cupric chloride. The ability to differentiate resistant from susceptible varieties on the basis of phytoalexin accumulation caused by chemicals or attenuated fungus has not been reported for other phytoalexins. Gossypol accumulation is activated by low concentrations of many toxic chemicals, wounding, and chilling. The gossypol-like compounds, vergosin (17) and hemigossypol (18) also accumulate in resistant reactions. These compounds, rather than gossypol, appear to be the major fungitoxic agents in extracts of infected cotton [125, 126].

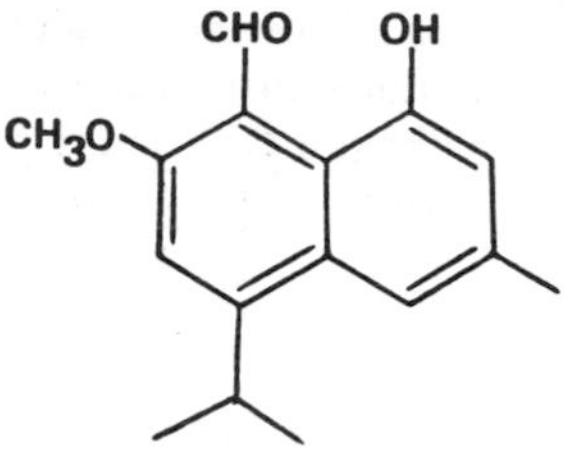

(17)

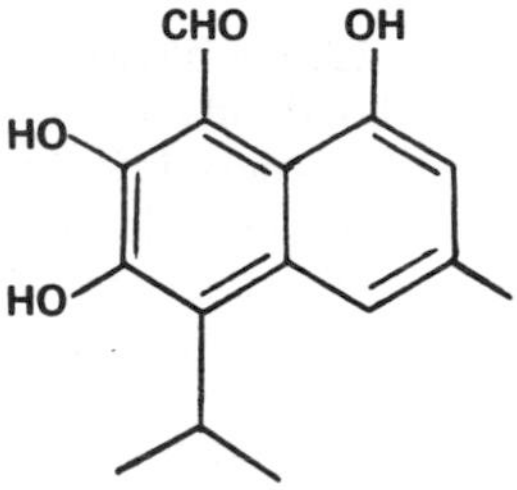

(18)

The ED_{50} of gossypol and vergosin for the inhibition of mycelial growth of V. albo-atrum are 100 and 60 μg/ml, respectively. Vergosin at 100 μg/ml completely inhibited growth of V. albo-atrum. Hemigossypol appears to be as fungitoxic as vergosin, although the ED_{50} was not determined because of its reactivity. A protein-lipopolysaccharide complex (PLP) isolated from culture filtrates of V. albo-atrum elicits the accumulation of gossypol derivatives in cotton [123]. The sensitivity of 16 varieties of cotton to the complex was positively correlated with susceptibility [127], even though PLP elicited the accumulation of gossypol derivatives [126].

D. Convolvulaceae

1. Sweet Potato

Sweet potato root resembles potato tuber in the number of antifungal compounds which accumulate after infection, injury, or treatment with chemicals. Chlorogenic and isochlorogenic acid, caffeic acid, umbelliferone, scopoletin, esculetin, ipomeamarone (19) and at least 10 furanoterpenoids accumulate in sweet potato root [128-132]. The peel of sweet potato also contains

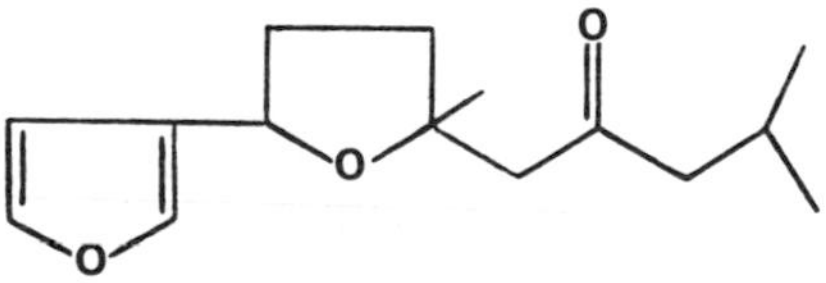

(19)

all of the above. Ipomeamarone inhibits the mycelial growth of a number of nonpathogens of sweet potato, with an ED_{50} of 0.003-0.03% [133]. A concentration of 0.1% ipomeamarone inhibits the growth of the pathogen Ceratocystis fimbriata 25%, but the level of the compound in sweet potato tissue infected with the pathogen often exceeds 1% of the fresh weight. Isochlorogenic acid and chlorogenic acid accumulate more rapidly in resistant than in susceptible roots infected with the pathogen C. fimbriata, but the acids are not appreciably inhibitory to the growth of C. fimbriata at levels considerably greater than that in infected tissue [129]. The acids, as in potato, may indirectly influence disease reaction by their effect on plant metabolism, including plant hormones.

Umbelliferone, scopoletin, esculetin, and glycosides of umbelliferone and scopoletin accumulate more rapidly in resistant as compared to susceptible sweet potato roots infected with C. fimbriata [130]. Umbelliferone and scopoletin markedly inhibit growth of C. fimbriata in vitro at a concentration of 0.1% but this level may be higher than that found in diseased tissue. Ipomeamarone was characterized by Kubota and Matsuura [131], and compounds of similar structure isolated from diseased tissue are ipomeanine, ipomeanic acid, and ipomeamaroneol. More than 10 furanoterpenes were separated by Akazawa [132]. Ipomeamarone appears to be synthesized by the acetate-mevalonate pathway [134, 135]. Uritani and co-workers have suggested that ipomeamarone may be associated with the disease resistance of sweet potato to C. fimbriata. It accumulates more rapidly in several infected resistant varieties, as compared with susceptible ones, and markedly inhibits growth of the fungus in vitro at concentrations found in infected tissue [128, 129].

Weber and Stahmann [136] reported that a nonpathogenic isolate of C. fimbriata protected sweet potato root from damage by a pathogenic isolate. Ipomeamarone, however, was not detected in the protected tissue. Further work by Hyodo et al. [137] also suggested ipomeamarone does not have a primary role in disease resistance. The development of prune and coffee isolates of C. fimbriata, nonpathogenic on sweet potato, was severely restricted in sweet potato roots, although the accumulation of furanoterpenes was less than in tissue infected by a pathogenic isolate. It appears that several mechanisms for resistance exist in sweet potato and ipomeamarone accumulation is, at best, one of them. The isolation of β-1-cinnamoyl-D-glucose from sweet potato root suggests its possible role in disease resistance [138].

E. Umbelliferae

1. Carrot

6-Methoxymellein (20) is an isocoumarin which reaches fungitoxic levels in carrot root after infection by fungi [8, 139-141]. The accumulation of 6-methoxymellein can also be induced by chemicals [137], cold treatment [142, 143], and ethylene [144-146]. The compound is responsible for the condition of bitter carrot, a physiological disorder of carrots resulting from storage at suboptimal temperatures. Biosynthesis of 6-methoxymellein appears to

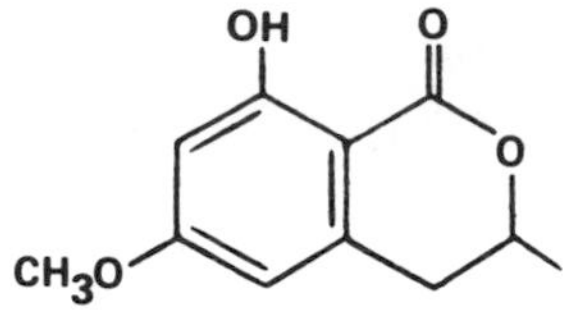

OH O CH3O O

(20)

proceed by a modification of the acetate-malonate pathway [140, 146]. The compound has an ED_{50} of 40 μg/ml for the mycelial growth of C. fimbriata.

2. Parsnip

Xanthotoxin (21) was isolated from the top mm of parsnip root discs inoculated with C. fimbriata, a nonpathogen of parsnip [147]. The compound was characterized by thin-layer (TLC) and gas-liquid (GLC) chromatography,

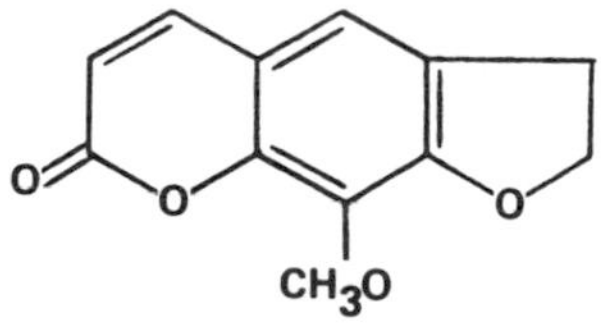

O O O CH3O

(21)

UV, nuclear magnetic resonance (NMR), infrared (IR), and mass spectroscopy (MS), and comparison to authentic xanthotoxin. The content of xanthotoxin in the top mm of inoculated tissue and control tissue treated with water 72 hr after inoculation or treatment was 1.0 and 0.05 mg/g fresh wt., respectively. Inoculation with other nonpathogens of parsnip, H. carbonum, Alternaria sp. or C. lindemuthianum β-race also resulted in accumulation of xanthotoxin. The ED_{50} of xanthotoxin for the mycelial growth of C. fimbriata is 22 μg/ml. Parsnip leaves and roots of hogweek (Heracleum sphordylium) and wild angelica (Angelica sylvestris) also contain strongly antifungal agents, probably furocoumarins [148].

F. Compositae

1. Safflower

Safynol (22) is a phytoalexin which accumulates in safflower leaves infected with Phytophthora drechsleri [149-151]. It rapidly accumulates to fungitoxic levels in resistant tissue after infection [152]. Safynol is not the only

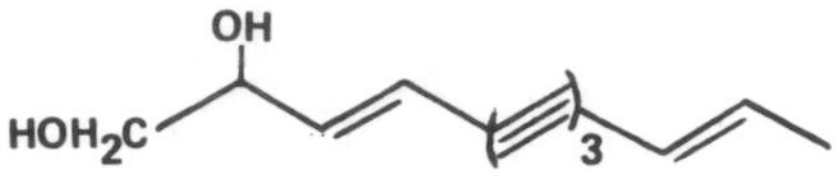

OH

HOH_2C 3

(22)

antifungal compound which accumulates in the infected hypocotyls and it appears, as with other phytoalexins studied, that series of polyacetylenic compounds accumulate [153]. Fungitoxic polyacetylenes appear to be widespread throughout the *Compositae*. The ED_{50} of safynol for mycelial growth of *P. drechsleri* is 12 μg/ml and growth is completely inhibited at 30 μg/ml [152].

G. Orchidaceae

Three dihydrophenanthrenes, orchinol, hircinol, and loroglossol have been isolated from bulbs, roots, and stalks of infected orchids [5, 154-156]. Orchinol and hircinol inhibit the growth of *Candida lipolytica* at 50-100 ppm, but the inhibition is not permanent.

III. ANTIFUNGAL COMPOUNDS IN TREES

A. Pinaceae

1. Oleoresin

Oleoresin, which occurs in the family Pinaceae is a hydrophobic mixture composed largely of resin and fatty acids in turpentine or a volatile oil, consisting mainly of a variety of mono and sesquiterpenes and a few alkanes. Resin acids, which comprise the major portion of the nonvolatile portion of oleoresin, are diterpenes with the basic formula of $C_{20}H_{30}O_2$ [157].

Oleoresin is produced and maintained under pressure in resin canals. Hence, oleoresin exudation and soaking of tissues occurs when resin canals are severed or when the epithelial parenchyma surrounding functioning resin canals is killed. Genera which feature resin canals in their normal anatomy are *Pinus*, *Picea*, *Larix*, and *Pseudotsuga* [158]. The most highly developed system of resin canals occurs in *Pinus* where the horizontal and vertical systems are interconnected. A parallel has been drawn between the decreasing effectiveness of the canal system from *Pinus* to *Picea* to *Pseudotsuga* and the increasing infection of these genera by *Fomes annosus* [159]. The above genera, as well as those that normally do not produce resin canals (e.g., *Abies*, *Sequoia*, *Tsuga*), may produce "traumatic" resin canals in the vicinity of wounds.

The accumulation of oleoresin in the vicinity of wounds and infected tissue may protect trees from pathogenic fungi by serving as a mechanical barrier, a water deficient barrier, and a chemical barrier.

Little direct evidence is available to implicate resin-soaked tissue as a mechanical barrier to fungal penetration. This notion, however, seems plausible, particularly when hardened crusts of oleoresin cover wounds. F. annosus grew around, but not on, large droplets of spruce (Picea abies) oleoresin on an agar surface during a test over three weeks [160].

Verrall [161] concluded that resin-soaked fire scars of red pine (Pinus resinosa) were protected from decay fungi because of a water-proofing, and not a toxic effect of oleoresin. These conclusions were based upon the similar decay rates of resin-soaked and ether-extracted cores taken from fire wounds and incubated with the decay fungus Lentinus lepideus for 90 days. The resin-soaked tissues contained about 35% ether-soluble extractives (considered to be oleoresin) and only about 20% moisture in vivo. Decay fungi require a moisture content above the fiber saturation point (usually ca. 25%) to actively degrade woody tissue; hence, sufficient moisture was made available for decay under the conditions of this test.

It appears from the foregoing that the replacement of moisture with oleoresin could render resin-soaked tissue resistant to decay. It should be noted, however, that resin-soaked wood decayed significantly less than wood not soaked in resin in similar tests with other fungi, e.g., F. annosus [162], Polyporus versicolor, and Poria monticola [163].

Although there are some conflicting reports, the weight of evidence strongly favors the presence of antifungal components in oleoresin. Some of the volatile components of oleoresin were shown to have a fungistatic and even fungicidal effect on a number of conifer pathogens and nonpathogens. Verrall's [161] inability to demonstrate differences in decay in the extracted and unextracted cores could be due to the escape of these volatiles during the long incubation period. Escape of volatiles, as well as the low solubility of oleoresin in aqueous media, could also explain another report [164] where F. annosus was not inhibited by pine oleoresin in an antibiotic disc assay.

Cobb et al. [165] tested the effects of nine volatile components of oleoresin from Ponderosa pine (Pinus ponderosa) on the linear growth of F. annosus and four Ceratocystis species. A saturated atmosphere of n-heptane completely inhibited the growth of F. annosus on potato dextrose agar, although its effect was fungistatic rather than fungitoxic. Saturated atmospheres of other volatile compounds inhibited this fungus to a lesser degree, such as myrcene (23), limonene (24), 72% β-phellandrene (25), β-pinene, 3-carene (26), 60 % α-pinene (27), and 25% camphene (28). Growth was not significantly reduced by undecane. Greater inhibition was obtained when these compounds were incorporated into the agar medium at the rate of 2.5%, the only concentration tested. The Ceratocystis sp. were affected similarly by these compounds but generally to a lesser degree than F. annosus.

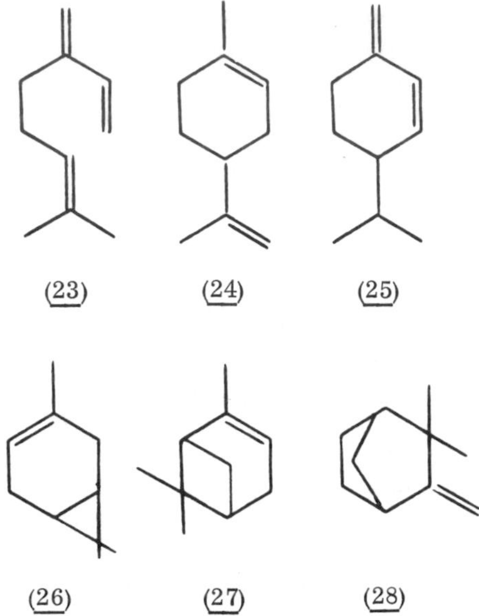

Under similar assay conditions (i.e., measurement of linear growth in atmospheres saturated with test terpenes), Trichoderma viride, Lenzites saepiaria, and Schizophyllum commune were inhibited more than 50% by l-α-pinene, d-α-pinene, l-β-pinene, l-limonene, d-limonene, p-cymene, dl-camphene, myrcene, and terpinolene. The blue-stain fungus Ceratocystis minor, the least sensitive fungus in this series of tests [166], was inhibited more than 50% only by d-α-pinene, p-cymene, and terpinolene. Because the pinenes are the predominant terpenes in southern pine, it was suggested that d-α-pinene was the principal terpene which limits invasion of southern pine by this fungus. T. viride was the most sensitive fungus tested; this could explain the delayed entry of this fungus in beetle-killed Southern pine.

However, under different assay conditions (measurement of mycelial weight after incubation in terpene-containing media), T. viride as well as Fomes pini, F. pinicola, and Ceratocystis pilifera were not inhibited by α- and β-pinene at concentrations up to 1% (v/v) [167]. The conflicting results for T. viride, and possibly other fungi, could be a reflection of the different assay techniques employed. In the absence of in vivo tests, the interpretation of in vitro tests based upon mycelial weight vs. linear growth remains open to question.

Hintikka [168] studied the effects of α- and β-pinene, camphene, 3-carene, and limonene on 16 hymenomycetes that degrade confierous wood and 22 that degrade hardwood. Under his assay conditions limonene was the most inhibitory. A concentration of 0.005% (volume of terpene per volume of enclosed air) completely inhibited the growth of half of the conifer-degrading fungi and all of the hardwood-degrading fungi that were tested. This author suggests that a terpene saturated atmosphere exists in the vicinity of wounds in pine and spruce and that these compounds could play a role in resistance as well as host selectivity among decay fungi.

The role of volatile terpenes in the ecology of ectomycorrhizae formation in pine is not as clear. Krupa et al. [169] found that mycorrhizal roots of shortleaf pine (Pinus echinata) formed in association with Pisolithus tinctorus contained a 43-fold increase in 3-carene as compared to nonmycorrhizal roots. Yet this mycorrhizal fungus was inhibited to a much greater degree by 3-carene (86% at 20 μl per vial per petri dish culture and fungitoxicity at 40 μl) than the root pathogens Phytophthora cinnamomi, Pythium irregulare (no inhibition at 40 μl), and F. annosus (47% inhibition at 40 μl). In an earlier paper [170] it was proposed that volatile inhibitory substances were involved in restricting mycorrhizal fungi within the host, resulting in a symbiotic rather than a pathogenic relationship.

Comparatively little has been done to determine the biological activity of resin acids. These compounds are the major fraction of oleoresin and resin-soaked tissues which can contain more than 25% of their dry weight in resin acids [157, 162]. Consequently, resin acids need not be inhibitory at low concentrations to be biologically active in vivo. Abietic acid (29) inhibited the linear growth of F. annosus by 60% at a concentration of 0.25% in malt agar (w/v) [160]. Wood blocks impregnated to contain about 15% of their dry weight in dehydroabietic acid or a mixture of resin acids (35% dehydroabietic, 35% abietic, and smaller amounts of other resin acids and oxidized materials) were decayed significantly less by P. versicolor and P. monticola than were nonimpregnated blocks [163].

Levopimaric acid (30), however, was not inhibitory to F. pinicola and a species of Haplosporella at concentrations up to 1% (v/v). This compound, a major resin acid of white pine (Pinus strobus), also served as a sole source of carbon for these test fungi under assay conditions based upon mycelial weight [171]. Evidence for the degradation of resin acids by F. annosus in vivo has also been provided [162].

Numerous workers have linked the accumulation of oleoresin or some of its components with resistance to a variety of pathogens. Gibbs [159]

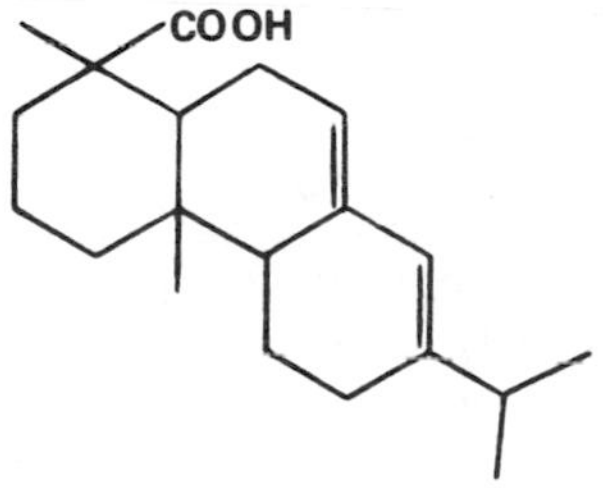

COOH

(29)

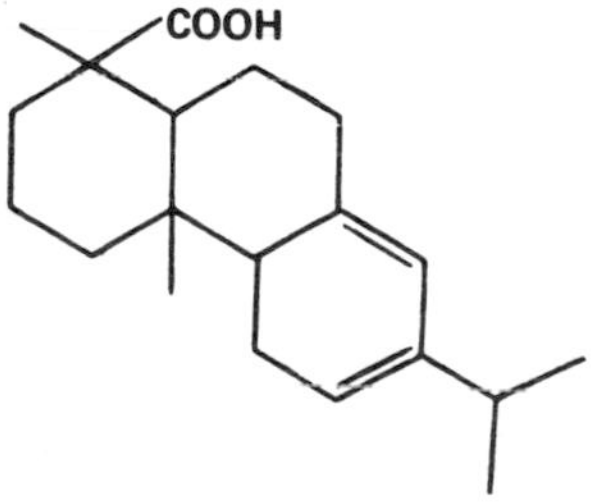

COOH

(30)

concluded that resistance to F. annosus was correlated with the ability of Scotch pine (Pinus sylvestris) and Corsican pine (Pinus nigra var. calabrica) to mobilize oleoresin in response to attack. Scotch pine with a high content of 3-carene and a relatively low content of α-pinene were immune to artificial inoculation with F. annosus [172]. The accumulation of large quantities of oleoresin in response to attack was reported to drown the mountain pine beetle (Dendroctonus ponderosae) and localize its associated blue stain fungi in lodgepole pine (Pinus contorta var. latifola) considered resistant to these pests [173].

In a comparison of the relative susceptibility of loblolly (Pinus taeda), longleaf (Pinus palustris), and slash (Pinus elliottii var. elliottii) pines to artificial inoculation with F. annosus, it was found that more oleoresin accumulated in affected tissue of loblolly pine, the most susceptible, than in longleaf pine, the most resistant [174]. It would have been most interesting if this brief report indicated whether this negative correlation between oleoresin accumulation and disease resistance applied also within the species studied.

Monoterpene content also may serve as a marker for disease resistance. A relatively high concentration of β-phellandrene in the bark of slash pine was correlated with resistance to Cronartium fusiforme [175], whereas a low concentration of this compound in the bark of loblolly pine was correlated with resistance to the same pathogen [176].

2. Pinus

The stilbenes, pinosylvin, PS (31) and pinosylvin monomethyl ether, PSM (32), are fungal inhibitors that occur as normal constituents of pine heartwood, generally at a ratio of 3 PSM to 1 PS in the hard pines investigated

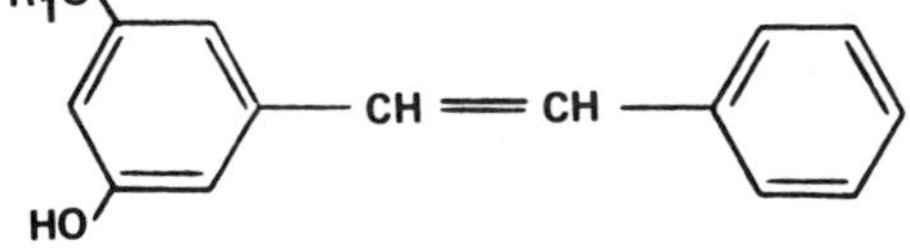

R_1O ... CH ═ CH ... HO

(31) $R_1 = H$

(32) $R_1 = CH_3$

[162, 177, 178]. Heartwood is composed of dead xylem, as the formation of this tissue is accompanied by the senescence and death of sapwood parenchyma. The outermost, or most recently formed, heartwood usually contains the highest amount of these inhibitory extractives, up to 1.5% dry weight [177]. This pattern is typical of the heartwood of numerous species; hence the outer heartwood tends to be more decay resistant than its older, inner portions [179, 180]. A logical explanation for this is that heartwood extractives, which are largely phenolic, are oxidized or otherwise degraded with increasing time [181]. Phenol-oxidizing enzymes, which could assist in this process, were found in the heartwood of Monterey pine (P. radiata) in the apparent absence of microorganisms [182].

The pinosylvins also are produced in the sapwood of living pines in response to injury and infection. The sapwood of loblolly pine reacted to infection by F. annosus with the formation of an oleoresin and PS-enriched zone in advance of the pathogen. Several points of evidence supported the view that this necrotic, extractive-enriched zone, termed a "reaction zone," functioned to restrain the pathogen from invading additional sapwood. The inhibitory constituents of the reaction zone, therefore, were considered phytoalexins produced during a hypersensitive response [162]. Similar responses were obtained with different injurious agents, demonstrating that this phenomenon is a nonspecific reaction to injury and infection [162, 183-185].

The production of PS in the sapwood of Monterey pine was stimulated by ethylene [186]. Substantially greater quantities of this gas were produced in a dry metabolically active zone surrounding the reaction zone produced in response to Sirex-Amylostereum attack, than in unaffected sapwood. A relationship among ethylene production, PS production, and disease resistance was suggested [187, 188].

Bioassays in malt agar generally indicate that decay fungi are more sensitive to PS than to PSM. Lethal dosages for both compounds in this medium were in the order to 100-200 ppm [189]. Malt agar containing 10% heartwood meal (lodgepole pine) and 400 ppm of PS, however, was strongly inhibiting but not lethal to F. pini and Peniophora pseudopini [190]. Decayed heartwood of pine further illustrates that some fungi can withstand much higher concentrations of these compounds in vivo than in standard in vitro tests. The detoxification [191] and degradation [162, 192, 193] of these compounds by fungi have been reported. In an attempt to explain the decreased tolerance of some fungi to these compounds in malt agar as compared to media containing heartwood, it was suggested that malt agar is an impoverished media lacking the enzyme inducers, cofactors, etc., that are present in heartwood and are required for these fungi to detoxify or degrade these inhibitors [190, 193]. Another plausible explanation is that possibilities for the compartmentalization of these inhibitors and adsorption of their active sites are greater in wood than in an agar medium. The pinosylvins, therefore, appear to retard, rather than stop, the rate of spread of fungal pathogens in heartwood and, perhaps more dramatically, in sapwood.

Uncoupling and inhibition of some sulfhydryl-containing enzymes are the reported modes of action of PSM [194, 195].

3. Picea

Five phenolic compounds isolated from the inner bark of Norway spruce (Picea abies) inhibited the growth of F. annosus in vitro. These were the flavonoids taxifolin or dihydroquercetin [see (36)], D-catechin, and quercetin, and the glucosides of the stilbenes piceatannol (33) and isorhapontigenin (34). The ED_{50} was in the order of 0.2% for the aglycones and 2% for the

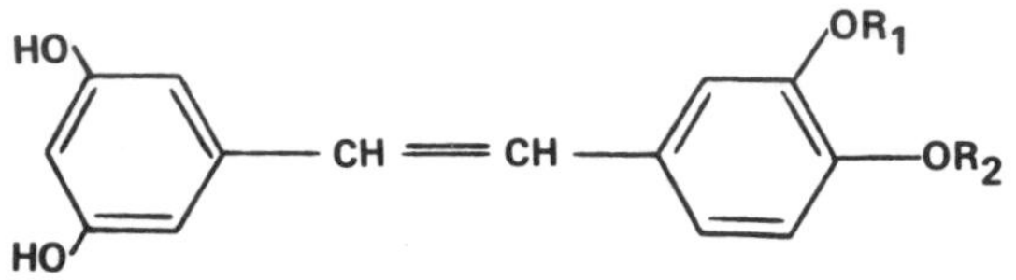

(33) $R_1 = R_2 = H$

(34) $R_1 = CH_3$, $R_2 = H$

glucosides and catechin [196]. Drought-stressed, as well as *F. annosus*-infected trees had lower concentrations of these preformed inhibitors than their healthy counterparts [196, 197].

Sapwood of living Norway spruce appears to resist infection by *F. annosus* and other decay fungi despite extensive decay of neighboring heartwood. This seems to be due to the formation of an alkaline (about pH 8.0) reaction zone containing relatively high amounts of the lignan hydroxymatairesinol, HMR (35) (up to 6% dry wt.) as compared to uninfected heartwood (less than

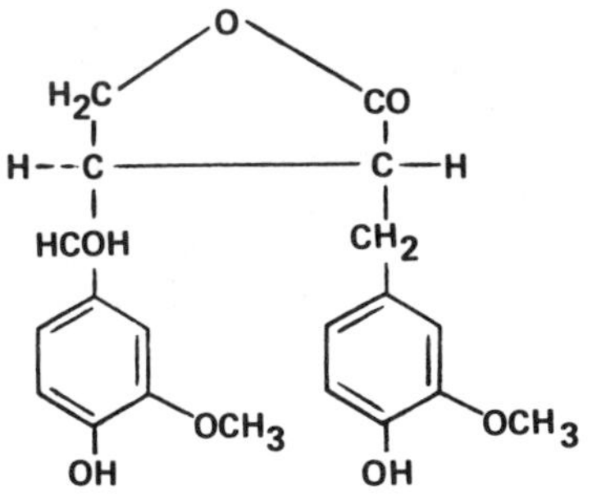

(35)

0.5%) [160, 198]. The significance of the elevated pH, which probably is due to the accumulation of inorganic carbonates, is that *F. annosus* appears to be inhibited strongly at levels exceeding pH 7.0 [199]. Preliminary tests suggested that calcium levels in the reaction zone were sufficient to reduce significantly the concentration of ATP in *F. annosus* at the pH of this tissue [200]. Bioassays in malt agar of HMR and two other lignans that occur in the heartwood and reaction zone (conidenrin and matairesinol) revealed that HMR was the most inhibitory to F. annosus (40% inhibition at 0.4% w/v) [198]. Butanone and water extracts of the reaction zone inhibited the laccase produced by this pathogen [201]. A common feature of the sapwood response to injury and infection in Picea and Pinus, therefore, is the disproportionate accumulation of a normal, antifungal, heartwood constituent.

3. Pseudotsuga

Although taxifolin (36) occurs in several genera, e.g., Cedrus, Larix, Picea, Thuja [202], and Prunus [203], its relationship with disease resistance has been studied most intensively in Douglas fir (Pseudotsuga menziesii). The heartwood of Douglas fir contains from 0.2-1.5% dry wt. of

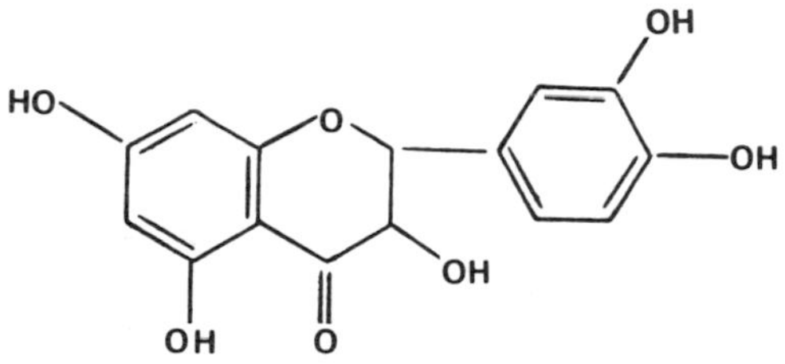

(36)

this compound, with the maximum concentration occurring usually at the heartwood periphery [204]. An exception to this pattern occurs in trees with target ring or alternating bands of included sapwood within the heartwood. In such a tree, the taxifolin content rose to about 1% in the heartwood and decreased to about 0.25% in the bands of included sapwood. Observations suggested that included sapwood was more susceptible to decay than heartwood [205].

Bioassays in malt agar revealed that taxifolin was more inhibitory to F. annosus and L. lepideus than other heartwood extractives of Douglas fir, requiring 0.45% and 0.7% for total inhibition, respectively [206]. In another test using liquid media, the growth of four white-rot fungi was inhibited by taxifolin (ED_{50} about 0.02%), while four brown-rot fungi were generally unaffected. The recovery of taxifolin from cultures containing the brown-rot fungi and its degradation by the white-rot fungi suggested that the degradation products may be more inhibitory than the parent compound [207]. Decay of wood blocks or sawdust containing taxifolin at concentrations up to 1.2% was from 40-92% of controls for a number of brown-rot fungi and 100% of controls for the white-rot fungus P. versicolor [3, 208].

The concentration of taxifolin increased in Douglas fir roots attacked, but not thoroughly decayed, by Poria weirii as compared to unaffected roots of the same tree [209]. This observation, combined with assay results indicating an ED_{50} of 0.02% for P. weirii [207] suggest a possible role for this compound in the resistance to this important root pathogen. It seems unlikely that an unidentified compound associated with roots in the advanced stages of decay would be involved, as suggested [209], in disease resistance.

B. Cupressaceae

Wood products composed of cedar heartwood are known for their decay resistance. This has been attributed to the variety of inhibitory heartwood extractives that occur in this family. Of particular interest are the tropolones, e.g., nootkatin (37) and γ-thujaplicin (38), characterized by a seven-carbon ring. These compounds were particularly active against numerous

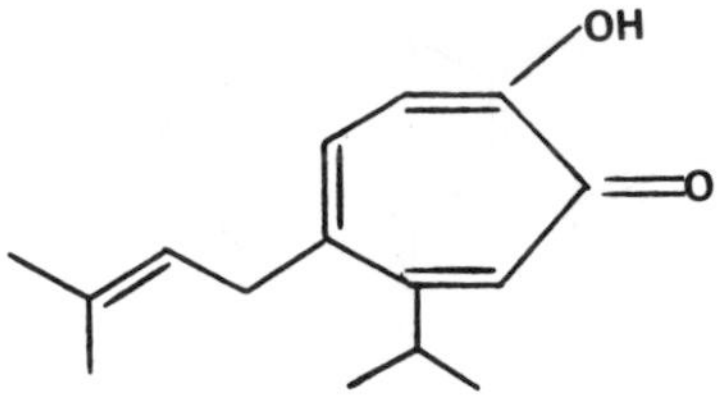

OH
O

(37)

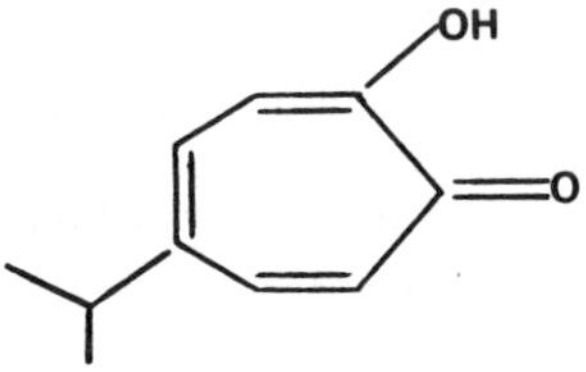

OH
O

(38)

decay fungi, with lethal dosages from 10-20 ppm in malt agar [189] and significant reductions in decay of impregnated wood blocks [3, 181]. The cedars also contain antifungal terpenoids in their heartwood. Based upon standard decay tests, the decreasing order of effectiveness of the terpenoids in incense cedar (Libocedrus decurrens) heartwood was hydrothymoquinone (39), p-methoxythymol, carvacrol, p-methoxycarvacrol (40), and thymoquinone [181].

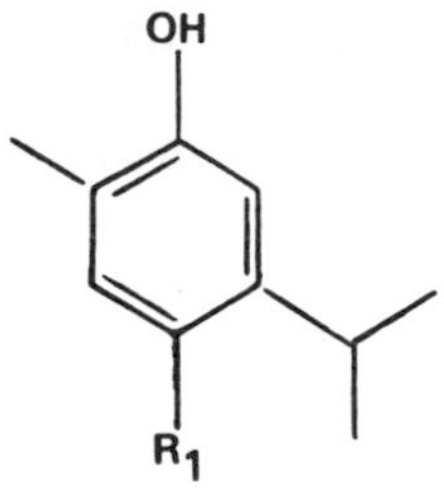

(39) R_1 = OH

(40) R_1 = H

Despite this arsenal of inhibitory extractives it seems strange that heartrot could be a major problem in standing trees of some species. Yet, heartrot has claimed more than 36% of the volume of incense cedar in California [210]. Additional information is required to explain this apparent anomaly. The fungi used for bioassays have been largely those associated with decay of wood products, rather than with decay in living trees. It is possible that the fungi responsible for decay in trees are more tolerant to these extractives than those that attack wood products. Bioassays indeed indicate that *Polyporus amarus*, which is largely responsible for the losses cited above, is more tolerant to the heartwood extractives of incense cedar than the wood products destroyed *P. monticola* and *L. lepideus* [211]. However, if increased tolerance to inhibitors were the complete explanation, then one would expect that the fungi causing heartrot in trees would be involved to a similar degree in the degradation of wood products. That this is not the case suggests that heartwood is changed in some manner during its conversion into products. For example, the concentrations of carbon dioxide and oxygen, which affect the growth of decay fungi [212] would be expected to differ considerably in the heartwood of trees and its products.

Another contributing factor could be that other fungi capable of degrading these inhibitory extractives precede the decay fungi. Evidence for this was obtained in yellow cedar (*Chamaecyparis nootkatensis*) where black-stain fungi, commonly found in the heartwood of this species, significantly reduced the concentration of nootkatin [213, 214].

It is also interesting to question why the decay fungi in the living tree are confined to the seemingly hostile environment of the heartwood. Are they inhibited from penetrating the sapwood by the higher concentration of extractives that usually occurs at the heartwood periphery? Is there a dynamic response in the sapwood resulting in the production of an even more inhibitory reaction zone, as observed in spruce and pine [160, 162]? These and other possibilities await investigation.

C. Aceraceae

Gallic acid was detected in the clear sapwood of red and sugar maple (Acer rubrum and A. saccharum, respectively), at a concentration of about 1%. This compound, however, was not detected in wood decayed by Fomes connatus or in the discolored tissue separating decayed and clear wood [215]. Phialophora melinii, a nonhymenomycete commonly associated with discolored wood, was capable of growing in the presence of gallic acid under certain conditions in vitro, while F. connatus, under similar conditions, was not. F. connatus, however, was capable of growing on media containing gallic acid after this media was altered by P. melinii. These results were taken to suggest that sapwood invasion by F. connatus is facilitated by the removal of an inhibitory phenol by P. melinii [216].

In contrast to the above, substituted coumarins were isolated recently from mineral-stained sugar maple in yields up to 1%. These compounds, however, were not found in clear wood. Media containing 1% of some of these compounds prevented the growth of L. trabea [217].

D. Betulaceae

Red alder (Alnus rubra), an important hardwood of the Pacific northwest, sometimes occurs in mixed stands with Douglas fir (Pseudotsuga menziesii). Douglas fir is susceptible to root rot caused by Poria weirii, whereas red alder is not [218]. Moreover, this pathogen did not persist as long in soils of forest stands containing red alder as compared to soils of conifer stands [219]. Attempts are being made to determine the basis for these results.

Soils in stands with red alder contained more lipids than soils in conifer stands. Linoleic acid was selected for study because this compound was isolated from red alder [220] and it was previously reported to have some antifungal activity [221]. Bioassays showed that the potassium salt of linoleic acid inhibited P. weirii linearly at concentrations up to 1 mg/ml with no growth occurring at 1-4 mg/ml [222].

The phenolic acids ferulic, syringic, p-coumaric, and vanillic were tested alone and in all combinations on the growth of two isolates of P. weirii. All four compounds were found in the roots of red alder, while only the last two were found in the roots of Douglas fir. The concentrations tested were based upon the concentrations found in hydrolized root extracts. The two isolates responded differently to a number of these tests, indicating the variability of the pathogen. Both isolates were significantly inhibited by media containing all four phenolic acids (red alder hydrolysate). The media containing p-coumaric and vanillic acids (Douglas fir hydrolysate), however, stimulated one isolate and inhibited the other [223].

E. Fagaceae

Tannins [180] and more recently ellagitannins [224] have been implicated in the resistance of heartwood of white oak (Quercus alba) to decay. Extracts containing these compounds inhibited the brown-rot fungus L. trabea [180] and P. monticola [224] but not the white-rot fungus P. versicolor [224]. Extracts treated with hide powder or containing polyvinylpyrrolidone or Tween 80 were not inhibitory, suggesting that free phenols were required for the observed inhibition.

Discolored sapwood of white oak, induced by increment core wounds, was intermediate in decay resistance to P. versicolor and P. hirsutus as compared to the more susceptible sapwood and the more resistant heartwood. Discolored sapwood, however, was more resistant to decay by P. monticola than its neighboring tissues. Wounded sapwood of black locust (Robinia pseudoacacia) and osage orange (Maclura pomifera) also was more resistant to decay than nonwounded sapwood [225] indicating that the sapwood of other angiosperms can also react to produce protective tissues.

F. Magnoliaceae

Glaucine (41), a nonphenolic aporphine alkaloid, was found in discolored sapwood of yellow poplar (Lirodendron tulipifera) at a concentration of 0.9% as compared to 0.2% of normal heartwood. This compound totally inhibited

OCH3
CH3O
CH3O
CH3O
N
CH3

(41)

five of six test fungi that commonly attack this species. Fourteen alkaloids were identified in the discolored sapwood and heartwood. The nonphenolic alkaloids tended to be more inhibitory than the phenolic alkaloids [226].

G. Myrtaceae

The decay resistance of Eucalyptus heartwood has been attributed to its methanol-soluble extractives [227]. In a recent study of the decay-resistant heartwood of red iron bark (E. sideroxylon), fractions of the methanol extract were partially characterized and assayed for antifungal activity [228]. The ellagitannins comprised a major portion of the water-soluble fraction, while the ether-soluble fraction was rich in resveratrol (3,4',5-trihydroxystilbene) (42). Both fractions were inhibitory to P. monticola and P. versicolor when incorporated into liquid media at concentrations less than

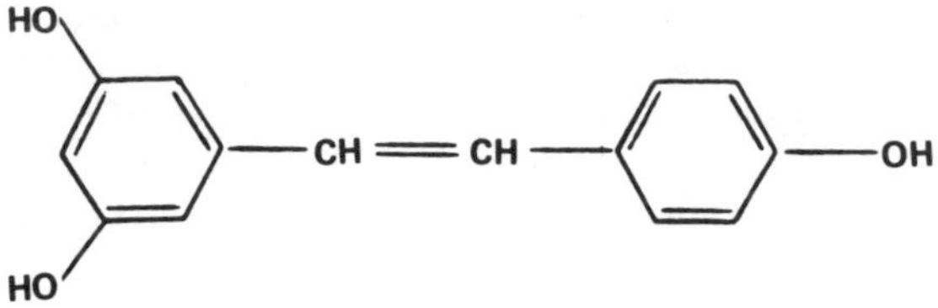

HO, HO — C₆H₃ — CH ═ CH — C₆H₄ — OH

(42)

those occurring in heartwood. However, wood blocks impregnated with these extracts and with resveratrol, were not particularly protected from decay by these test fungi. These results again point up the difficulty in interpreting bioassays of wood extracts with regard to their possible role in disease resistance.

Kino, an exudate produced by many eucalypts after cambial injury during the growing season, is composed largely of polymerized leuco-anthocyanins [229]. It would be of interest to determine if kino plays a role in disease resistance.

H. Salicaceae

The inhibitory effect of bark extracts of quaking aspen (Populus tremuloides) on the important canker pathogen Hypoxylon pruinatum was attributed to pyrocatechol and the aglycones benzoic and salicylic acid [230-232]. Total inhibition of this pathogen was attained in vitro by 5×10^{-3} M pyrocatechol [231] and salicylic acid and by 4×10^{-3} M benzoic acid [232]. Media containing benzoic and salicylic acids each at a concentration of 2×10^{-3} M also totally inhibited H. pruinatum, indicating a slight synergism between these two compounds [232].

However, H. pruinatum is a major pathogen of quaking aspen despite these inhibitors. This was explained partially by evidence that the pathogen invades sapwood, which lacks these inhibitors, beyond the margin of the canker. Further evidence suggested that bark is killed by a toxic substance produced by the pathogen in the underlying sapwood. Invasion of necrotic bark is then thought to occur after the oxidation of the inhibitory phenols [233].

IV. CONCLUSIONS

This chapter is not intended to present an exhaustive literature review. A number of reviews of antifungal compounds and their role in disease resistance are available [3, 55, 234-240]. The phytoalexins are generally not notably antifungal, e.g., ED_{50} ca. 40 μg/ml, and are not produced as a specific response to infection. They may occur in trace quantities in apparently healthy plants and they often accumulate to antifungal concentrations in infected tissues, in tissues treated with some microbial metabolites, antibiotics, low concentrations of metabolic inhibitors, UV radiation, fungicides, or in tissues subjected to mechanical injury or temperature stress. They may be part of a general tissue-repair mechanism which may have a role in resistance. In most host-parasite interactions a number of phytoalexins accumulate. Phytoalexin accumulation after infection is not restricted to a resistant or immune plant. The time and magnitude of accumulation may be vital in determining the nature of an interaction. Phytoalexins do not appear to be stable end products of the plant's biosynthetic process. They generally accumulate for 6-96 hr after infection and then decline to levels often found in uninfected plants.

Induced-protection studies indicate that protection is not dependent on the maintenance of a high level of phytoalexin in protected tissues. The rapidity of many incompatible interactions, e.g., hypersensitive response of potato tuber to incompatible races of P. infestans, suggests a surface phenomenon based on components on the surfaces of cell walls or membranes. As in disease resistance in animals, resistance in plants is a complex of interacting factors. The ability to synthesize antibodies is not the sole determinant of resistance, and the inability to induce antibody production is not the sole determinant of pathogenicity in the interaction of animals and infectious agents. Induced protection in plants, a phenomenon similar to immunization in animals, shows promise as a method for plant disease control in the future. Induced protection may depend upon attainment and maintenance of high levels of antifungal agents in plants; however, it appears the activation of the potential to accumulate phytoalexins is more important than the maintenance of a high level.

There appears to be a growing interest in resistance mechanisms in trees to fungal pathogens. With the current emphasis on tree improvement through selection and breeding, this trend is expected to continue. Antifungal compounds produced before or in response to attack constitute one of probably several mechanisms involved in the survival of these large perennial hosts. Fractions of extracts inhibitory to pathogens have been reported in numerous studies. Characterizations of these extracts will add to the growing list of antifungal compounds. Additional information is needed on the physiology and biochemistry of the elicitation of such compounds as well as their modes of action.

REFERENCES

1. K. Chester, Quart. Rev. Biol., 8, 129, 275 (1933).
2. K. Muller and H. Borger, Arb. Biol. Reichsanstalt Landwirtsch. Forstwissenschaft Berlin, 23, 189 (1940).
3. T. C. Scheffer and E. B. Cowling, Ann. Rev. Phytopathol., 4, 147 (1966).
4. A. L. Shigo and W. E. Hillis, Ann. Rev. Phytopathol., 11, 197 (1973).
5. E. Gaumann and H. Kern, Phytopathol. Z., 35, 347 (1959).
6. I. Uritani and M. Maramatsu, J. Agr. Chem. Soc. Japan, 27, 161 (1953).
7. J. Kuć, R. E. Henze, A. J. Ullstrup, and F. W. Quackenbush, J. Amer. Chem. Soc., 78, 3123 (1956).
8. P. Condon and J. Kuć, Phytopathology, 52, 182 (1962).
9. I. Cruickshank and D. Perrin, Nature, 187, 799 (1960).
10. I. Cruickshank and D. Perrin, Life Sci., 2, 680 (1963).
11. D. Perrin and W. Bottomley, J. Amer. Chem. Soc., 84, 1919 (1962).
12. I. Cruickshank, Australian J. Biol. Sci., 15, 147 (1962).
13. M. Schwochau and L. Hadwiger, Arch. Biochem. Biophys., 134, 34 (1969).
14. L. Hadwiger, Plant Physiol., 47, 346 (1971).
15. L. Hadwiger and M. Schwochau, Plant Physiol., 47, 588 (1971).
16. I. Cruickshank and D. Perrin, Australian J. Biol. Sci., 16, 111 (1963).
17. E. Chalutz and M. Stahmann, Phytopathology, 59, 1972 (1969).
18. L. Hadwiger, Plant Physiol., 49, 779 (1972).
19. L. Hadwiger, Biochem. Biophys. Res. Commun., 46, 71 (1972).
20. L. Hadwiger and A. Martin, Biochem. Pharmacol., 20, 3255 (1971).
21. A. De Witt-Elshove, Neth. J. Plant Pathol., 74, 44 (1968).
22. A. De Witt-Elshove, Neth. J. Plant Pathol., 75, 164 (1969).
23. A. De Witt-Elshove and A. Fuchs, Physiol. Plant Pathol., 1, 17 (1971).
24. F. Nonaka and K. Kawakami, Proc. Assoc. Plant Protect. (Kyushu), 16, 114 (1970).
25. M. Heath and R. Wood, Ann. Bot., 35, 475 (1971).
26. M. Schwochau and L. Hadwiger, Rec. Advan. Phytochem., 3, 181 (1970).
27. L. Hadwiger, Phytochemistry, 5, 523 (1966).
28. I. Cruickshank and D. Perrin, Phytopathol. Z., 70, 209 (1971).
29. D. Perrin, Tetrahedron Lett., 1964, No. 1, p. 29.
30. D. Perrin and C. Whittle, Tetrahedron Lett., 1972, No. 17, p. 1673.

31. R. Burden, J. Bailey, and G. Dawson, Tetrahedron Lett., 1972, No. 41, p. 4175.
32. D. Smith, H. Van Etten, J. Serum, T. Jones, D. Bateman, T. Williams, and D. Coffen, Physiol. Plant Pathol., 3, 293 (1973).
33. W. Rathmell and D. Bendall, Physiol. Plant Pathol., 1, 351 (1971).
34. J. Bailey, Physiol. Plant Pathol., 4, 477 (1974).
35. D. Smith, H. Van Etten, and D. Bateman, Physiol. Plant Pathol., 5, 51 (1975).
36. I. Cruickshank, D. Biggs, D. Perrin, and C. Whittle, Physiol. Plant Pathol., 4, 261 (1974).
37. M. Heath and V. Higgins, Physiol. Plant Pathol., 3, 107 (1973).
38. J. Van Den Heuvel and H. Van Etten, Physiol. Plant Pathol., 3, 327 (1973).
39. J. Van Den Heuvel, H. Van Etten, H. Serum, J. Coffin, and T. Williams, Phytochemistry, in press.
40. R. Burden and J. Bailey, Phytochemistry, 14, 1389 (1975).
41. J. Van Etten and D. Bateman, Phytopathology, 61, 1363 (1971).
42. I. Cruickshank and D. Perrin, Life Sci., 7, 449 (1968).
43. J. Rahe, J. Kuć, C. Chuang, and E. Williams, Neth. J. Plant Pathol., 75, 58 (1969).
44. J. Bailey and B. Deverall, Physiol. Plant Pathol., 1, 435 (1971).
45. R. Skipp and B. Deverall, Physiol. Plant Pathol., 2, 357 (1972).
46. J. Rahe, J. Kuć, C. Chuang, and E. Williams, Phytopathology, 59, 1641 (1969).
47. J. Rahe and J. Kuć, Phytopathology, 60, 1006 (1970).
48. J. Rahe, Phytopathology, 63, 572 (1973).
49. J. Rahe, Can. J. Bot., 52, 1339 (1973).
50. J. Elliston, J. Kuć, and E. Williams, Phytopathology, 61, 1110 (1971).
51. J. Elliston, Ph.D. Thesis, Purdue University, Lafayette, Indiana, 1975.
52. R. Skipp and B. Deverall, Physiol. Plant Pathol., 3, 299 (1973).
53. D. Berard, J. Kuć, and E. Williams, Physiol. Plant Pathol., 2, 123 (1972).
54. D. Berard, J. Kuć, and E. Williams, Physiol. Plant Pathol., 3, 51 (1973).
55. J. Kuć, in The Dynamic Role of Molecular Constituents in Plant Parasite Interaction (C. Mirocha and I. Uritani, eds.), Bruce Publ., St. Paul, 1967, pp. 183-202.
56. S. Hess, L. Hadwiger, and M. Schwochau, Phytopathology, 61, 79 (1971).
57. K. Uehara, Ann. Phytopathol. Soc. Japan, 23, 225 (1958).
58. F. Nonaka, S. Isayama, and H. Furukawa, Agr. Bull. Saga Univ. Japan, 22, 51 (1966).
59. W. Klarman and J. Gerdemann, Phytopathology, 53, 863 (1963).
60. W. Karlman and J. Gerdemann, Phytopathology, 53, 1318 (1963).

61. G. Gray, W. Klarman and M. Bridge, Can. J. Bot., 45, 285 (1968).
62. W. Biehn, J. Kuć, and E. Williams, Phytopathology, 58, 1255 (1968).
63. W. Biehn, E. Williams, and J. Kuć, Phytopathology, 58, 1261 (1968).
64. J. Sims, N. Keen, and V. Honward, Phytochemistry, 11, 827 (1972).
65. N. Keen, Physiol. Plant Pathol., 1, 265 (1971).
66. N. Keen, J. Sims, D. Erwin, E. Rice, and J. Partridge, Phytopathology, 61, 1084 (1971).
67. J. Paxton and D. Chamberlain, Phytopathology, 57, 352 (1967).
68. W. Svoboda and J. Paxton, Phytopathology, 62, 1457 (1972).
69. J. Reilly and W. Klarman, Phytopathology, 62, 1113 (1972).
70. W. Klarman and F. Hammerschlag, Phytopathology, 62, 719 (1972).
71. M. Bridge and W. Klarman, Phytopathology, 63, 606 (1973).
72. N. Keen, Science, 187, 74 (1975).
73. N. Keen, A. Zaki, and J. Sims, Phytochemistry, 11, 1031 (1972).
74. R. Sherwood, A. Olah, W. Oleson, and E. Jones, Phytopathology, 60, 684 (1970).
75. A. Olah and R. Sherwood, Phytopathology, 61, 65 (1971).
76. D. Smith, A. McInnes, V. Higgins, and R. Millar, Physiol. Plant Pathol., 1, 41 (1971).
77. V. Higgins and R. Millar, Phytopathology, 59, 1493 (1969).
78. V. Higgins and R. Millar, Phytopathology, 59, 1500 (1969).
79. V. Higgins and R. Millar, Phytopathology, 60, 296 (1970).
80. V. Higgins and D. Smith, Phytopathology, 62, 235 (1972).
81. R. Letcher, B. Widdowson, B. Deverall, and J. Mansfield, Phytochemistry, 9, 249 (1970).
82. L. Metlitskii and O. Ozeretskovskaya, Plant Immunity, Plenum, New York, 1968, pp. 67-79.
83. G. Johnson and L. Schaal, Phytopathology, 45, 626 (1955).
84. B. Rubin and V. Aksenova, Biokhimiya (Engl. Transl.), 22, 191 (1957).
85. N. Sato, K. Kitazawa, and K. Tomiyama, Physiol. Plant Pathol., 1, 289 (1971).
86. J. Kuć, in Microbial Toxins (S. Ajl, G. Weinbaum, and S. Kadis, eds.), Vol. 8, Academic Press, New York, 1972, pp. 211-247.
87. K. Tomiyama, R. Sakai, Y. Otani, and T. Takemori, Plant Cell Physiol., 8, 1 (1967).
88. M. Zucker, K. Hanson, and E. Sondheimer, in Phenolic Compounds and Metabolic Regulation (B. Finkle and V. Runeckles, eds.), Appleton, New York, 1967, pp. 68-93.
89. L. Greasy and M. Zucker, in Metabolic Regulation of Secondary Plant Products (V. Runeckles and E. Conn, eds.), Academic Press, New York, 1974, pp. 1-19.
90. J. Kuć, E. Williams, M. Maconkin, J. Ginzel, F. Ross, and L. Freedman, Phytopathology, 57, 38 (1967).
91. D. Clarke, Proc. Roy. Soc. London B., 181, 303 (1972).
92. D. Clarke, Physiol. Plant Pathol., 3, 347 (1973).

93. K. Tomiyama, T. Sakuma, V. Ishizaka, N. Sato, J. Katsui, M. Takasugi, and T. Masamune, Phytopathology, 58, 115 (1968).
94. N. Katsui, A. Murai, M. Takasugi, K. Imaizumi, and T. Masamune, Chem. Commun., 1968, p. 43.
95. J. Varns, W. Currier, and J. Kuć, Phytopathology, 61, 968 (1971).
96. G. Lyon, Physiol. Plant Pathol., 2, 411 (1972).
97. N. Sato and K. Tomiyama, Ann. Phytopathol. Soc. Japan, 35, 202 (1969).
98. N. Sato, K. Tomiyama, N. Katsui, and T. Masamune, Ann. Phytopathol. Soc. Japan, 34, 140 (1968).
99. J. Varns, J. Kuć, and E. Williams, Phytopathology, 61, 174 (1971).
100. J. Varns, Ph.D. Thesis, Purdue University, Lafayette, Ind., 1970.
101. D. Coxon, R. Curtis, K. Price, and B. Howard, Tetrahedron Lett., 1974, No. 27, p. 2363.
102. D. Hughes and D. Coxon, Chem. Commun., 1974, p. 822.
103. N. Katsui, A. Matsunaga, K. Imaizumi, T. Masamune, and K. Tomiyama, Tetrahedron Lett., 1971, No. 2, p. 83.
104. A. Stoessl, J. Stothers, and E. Ward, Chem. Commun., 1974, p. 709.
105. A. Masamune, Tetrahedron Lett., 1974, No. 51/52, p. 4483.
106. D. Coxon, K. Price, B. Howard, S. Osman, E. Kalan, and M. Zacharuis, Tetrahedron Lett., 1974, No. 34, p. 2921.
107. K. Tomiyama, Ann. Phytopathol. Soc. Japan, 32, 181 (1966).
108. J. Varns and J. Kuć, Phytopathology, 61, 178 (1971).
109. S. Subramanian, J. Varns, and J. Kuć, Proc. Indiana Acad. Sci., 80, 367 (1970).
110. Z. Kiraly, B. Barna, and T. Ersek, Nature, 239, 456 (1972).
111. T. Ersek, B. Barna, and Z. Kiraly, Acta Phytopathol. Acad. Sci. Hungaricae, 8, 3 (1973).
112. R. McKee, Ann. Appl. Biol., 43, 147 (1955).
113. E. Allen and J. Kuć, Phytopathology, 58, 776 (1968).
114. K. Tomiyama, N. Ishizaka, N. Sato, T. Masamune, and N. Katsui, in Biochemical Regulation in Diseased Plants and Injury, Phytopathological Society of Japan, Tokyo, 1968, pp. 287-292.
115. M. Shih, J. Kuć, and E. Williams, Phytopathology, 63, 821 (1973).
116. M. Shih and J. Kuć, Phytopathology, 63, 826 (1973).
117. A. Stoessl, C. Unwin, and W. Ward, Phytopathol. Z., 74, 141 (1972).
118. M. Gordon, A. Stoessl, and J. Stothers, Can. J. Chem., 51, 748 (1973).
119. A. Stoessl, C. Unwin, and W. Ward, Phytopathology, 63, 1225 (1973).
120. E. Ward, C. Unwin, and A. Stoessl, Phytopathology, 65, 168 (1975).
121. A. Bell, Phytopathology, 57, 759 (1967).
122. A. Bell, Phytopathology, 59, 1119 (1969).
123. A. Bell and J. Presley, Phytopathology, 59, 1141 (1969).
124. A. Bell and J. Presley, Phytopathology, 59, 1147 (1969).

125. A. Zaki, N. Keen, J. Sims, and D. Erwin, Phytopathology, 62, 1398 (1972).
126. A. Zaki, N. Keen, and D. Erwin, Phytopathology, 62, 1402 (1972).
127. N. Keen, M. Long, and D. Erwin, Physiol. Plant Pathol., 2, 317 (1972).
128. I. Uritani, Conn. Agr. Expt. Sta. Bull., 663, 4 (1963).
129. T. Akazawa and K. Wada, Plant Physiol., 36, 139 (1961).
130. T. Minamikawa, T. Akazawa, and I. Uritani, Plant Physiol., 38, 493 (1963).
131. T. Kubota and T. Matsuwra, J. Chem. Soc. Japan, 74, 248 (1953).
132. T. Akazawa, Arch. Biochem. Biophys., 90, 82 (1960).
133. F. Nonaka and K. Yasui, Agr. Bull. Saga Univ., 22, 39 (1966).
134. T. Akazawa, I. Uritani, and Y. Akazawa, Arch. Biochem. Biophys., 99, 52 (1962).
135. H. Imaseki and I. Uritani, Plant Cell Physiol., 5, 133 (1964).
136. D. Weber and M. Stahmann, Phytopathology, 56, 1066 (1966).
137. H. Hyodo, I. Uritani, and S. Akai, Phytopathol. Z., 65, 332 (1969).
138. M. Kojuma and I. Uritani, Plant Cell Physiol., 13, 1075 (1972).
139. P. Condon and J. Kuć, Phytopathology, 50, 267 (1960).
140. P. Condon, J. Kuć, and H. Draudt, Phytopathology, 53, 1244 (1963).
141. B. Herndon, J. Kuć, and E. Williams, Phytopathology, 56, 187 (1966).
142. A. Dodson, N. Fukui, C. Ball, R. Carolus, and H. Sell, Science, 124, 984 (1956).
143. E. Sondheimer, J. Amer. Chem. Soc., 79, 5036 (1957).
144. E. Chalutz, J. Devay, and E. Maxie, Plant Physiol., 44, 235 (1969).
145. J. Jaworski, J. Kuć, and E. Williams, Phytopathology, 63, 408 (1973).
146. J. Jaworski and J. Kuć, Plant Physiol., 53, 331 (1974).
147. C. Johnson, D. Brannon, and J. Kuć, Phytochemistry, 12, 2961 (1973).
148. J. Martin, E. Baker, and R. Byrde, Ann. Appl. Biol., 57, 501 (1966).
149. H. Aldwinckle, Phytopathology, 59, 1015 (1969).
150. F. Bohlman, S. Kohn, and C. Arndt, Chem. Ber., 99, 3433 (1966).
151. C. Thomas and E. Allen, Phytopathology, 60, 261 (1970).
152. C. Thomas and E. Allen, Phytopathology, 60, 1153 (1970).
153. E. Allen and C. Thomas, Phytopathology, 61, 1107 (1971).
154. E. Gaumann and E. Kern, Phytopathol. Z., 36, 1 (1959).
155. E. Gaumann, J. Neusch, and R. Rimpaw, Phytopathol. Z., 38, 274 (1960).
156. M. Fisch, H. Brigitta, H. Flick, and J. Arditti, Phytochemistry, 12, 437 (1973).
157. D. B. Mutton, in Wood Extractives (W. E. Hillis, ed.), Academic Press, New York, 1962, pp. 331-363.

158. H. P. Brown, A. J. Panshin, and C. C. Forsaith, Textbook of Wood Technology, Vol. I, McGraw-Hill, New York, 1949, p. 50.
159. J. N. Gibbs, Ann. Bot., 32, 649 (1968).
160. L. Shain, Phytopathology, 61, 301 (1971).
161. A. F. Verrall, J. Forest., 36, 1231 (1938).
162. L. Shain, Phytopathology, 57, 1034 (1967).
163. J. H. Hart, J. F. Wardell, and R. W. Hemingway, Phytopathology, 65, 412 (1975).
164. R. V. Bega and J. Tarry, Phytopathology, 56, 870 (1966).
165. F. W. Cobb, Jr., M. Kristic, E. Zavarin, and H. W. Barber, Jr., Phytopathology, 58, 1327 (1968).
166. R. C. DeGroot, Mycologia, 64, 863 (1972).
167. C. R. Keyes, Phytopathology, 59, 400 (1969).
168. V. Hintikka, Karstenia, 11, 28 (1970).
169. S. Krupa, J. Andersson, and D. H. Marx, Eur. J. Forest Pathol., 3, 194 (1973).
170. S. Krupa and N. Fries, Can. J. Bot., 49, 1425 (1971).
171. C. R. Shriner and W. Merrill, Phytopathology, 60, 578 (1970).
172. A. V. Chudnyi, R. A. Krangauz, and E. I. Gundaeva, Lesnoe Khozyaistvo, 7, 60 (1972) (in Russian); Forest Abstr., 34, 4654 (1973).
173. D. M. Shrimpton, Can. J. Bot., 51, 1155 (1973).
174. C. S. Hodges, Phytopathology, 59, 1031 (1969).
175. D. L. Rockwood, Phytopathology, 64, 976 (1974).
176. D. L. Rockwood, Phytopathology, 63, 551 (1973).
177. H. Erdtman and A. Misiorny, Svensk Papperstid., 55, 608 (1952).
178. T. Kondo and Y. Kitamura, J. Jap. Forest. Soc., 37, 157 (1955).
179. T. C. Scheffer and H. Hopp, U.S. Dept. Agr. Tech. Bull. No. 984, 1949, p. 37.
180. R. A. Zabel, N.Y. State Coll. Forestry Tech. Publ. No. 68, 1948.
181. A. B. Anderson, T. C. Scheffer, and C. G. Duncan, Holzforschung, 17, 1 (1963).
182. L. Shain and J. F. G. Mackay, Forest Sci., 19, 153 (1973).
183. E. Jorgensen, Can. J. Bot., 39, 1765 (1961).
184. W. E. Hillis and T. Inoue, Phytochemistry, 7, 13 (1968).
185. M. P. Coutts, Australian Forest. Res., 4, 15 (1970).
186. L. Shain and W. E. Hillis, Can. J. Bot., 51, 1331 (1973).
187. L. Shain and W. E. Hillis, Phytopathology, 62, 1407 (1972).
188. L. Shain and W. E. Hillis, 2nd Intern. Congr. Plant Pathol., Minneapolis, 1973, Abstr. 0696.
189. E. Rennerfelt and G. Nacht, Svensk Bot. Tidsskr., 49, 419 (1955).
190. A. A. Loman, Can. J. Bot., 48, 1303 (1970).
191. H. Lyr, Flora, 152, 570 (1962).
192. E. Rennerfelt, Svensk Bot. Tidsskr., 39, 311 (1945).
193. A. A. Loman, Can. J. Bot., 48, 737 (1970).
194. H. Lyr, Flora, 150, 227 (1961).

195. H. Lyr, Enzymologia, 23, 231 (1961).
196. M. Alcubilla, M. P. Diaz-Palacio, K. Kreutzer, W. Laatsch, K. E. Rehfuess, and G. Wenzel, Eur. J. Forest Pathol., 2, 100 (1971).
197. G. Wenzel, K. Kreutzer, and M. Alcubilla, Forstwissenschaft Zentr., 89, 372 (1970).
198. L. Shain and W. E. Hillis, Phytopathology, 61, 841 (1971).
199. E. Rennerfelt and S. K. Paris, Oikos, 4, 58 (1953).
200. M. Johansson and O. Theander, Physiol. Plantarum, 30, 218 (1974).
201. M. Johansson and O. Theander, Proc. 4th Intern. Conf. Fomes Annosus, Athens, Georgia, 1974, p. 133.
202. H. L. Hergert and O. Goldschmid, J. Org. Chem., 23, 700 (1958).
203. M. Hasegawa and T. Shirato, J. Amer. Chem. Soc., 68, 700 (1954).
204. J. A. F. Gardner and G. M. Barton, Forest Prod. J., 10, 171 (1960).
205. R. W. Kennedy and J. W. Wilson, Forest Prod. J., 6, 230 (1956).
206. R. W. Kennedy, Forest Prod. J., 6, 80 (1956).
207. A. J. Cserjesi, Can. J. Microbiol., 15, 1137 (1969).
208. P. Rudman, Holzforschung, 16, 74 (1962).
209. G. M. Barton, Can. J. Bot., 45, 1545 (1967).
210. W. W. Wagener and R. V. Bega, U.S. Dept. Agr. Forest Serv. Forest Pest Leaflet, 30, 1958, p. 7.
211. W. W. Wilcox, Phytopathology, 60, 919 (1970).
212. K. F. Jensen, Forest Sci., 13, 384 (1967).
213. R. S. Smith, Can. J. Bot., 48, 1731 (1970).
214. R. S. Smith and A. J. Cserjesi, Can. J. Bot., 48, 1727 (1970).
215. T. A. Tattar and A. E. Rich, Phytopathology, 63, 167 (1973).
216. W. C. Shortle, T. A. Tattar, and A. E. Rich, Phytopathology, 61, 552 (1971).
217. J. F. Manville and N. Levitin, Bimonthly Res. Notes, Can. Forest Serv., 30, 3 (1974).
218. G. Wallis, in Biology of Alder (J. M. Trappe, J. F. Franklin, R. F. Tarrent, and G. M. Hansen, eds.), U.S. Dept. Agr. Forest Serv. Pacif. N.W. Forest Range Expt. Sta., Portland, Ore., 1968, p. 195.
219. E. E. Nelson, U.S. Dept. Agr. Forest Serv. Res. Note PNW-83, 1968.
220. E. F. Kurth and E. L. Becker, Tappi, 36, 461 (1953).
221. E. Honkanen and A. I. Virtanen, Suomen Kemistilehti B, 33, 171 (1960).
222. C. Y. Li, K. C. Lu, J. M. Trappe, and W. B. Bollen, Forest Sci., 16, 329 (1970).
223. J. M. Trappe, C. Y. Li, K. C. Lu, and W. B. Bollen, Forest Sci., 19, 191 (1973).
224. J. H. Hart and W. E. Hillis, Phytopathology, 62, 620 (1972).
225. J. H. Hart and K. C. Johnson, Wood Sci. Technol., 4, 267 (1970).
226. C. Huang, H. Chang, C. Chen, and M. P. Levi, private communication, 1975.

227. P. Rudman, Holzforschung, 15, 56 (1962).
228. J. H. Hart and W. E. Hillis, Phytopathology, 64, 939 (1974).
229. W. E. Hillis, in Wood Extractives (W. E. Hillis, ed.), Academic Press, New York, 1962, pp. 60-131.
230. M. Hubbes, Science, 136, 156 (1962).
231. M. Hubbes, Can. J. Bot., 44, 365 (1966).
232. M. Hubbes, Can. J. Bot., 47, 1295 (1969).
233. M. Hubbes, Can. J. Bot., 42, 1489 (1964).
234. J. Kuć, Ann. Rev. Microbiol., 20, 334 (1966).
235. J. Kuć, World Rev. Pest Control, 7, 42 (1968).
236. J. Kuć, Ann. Rev. Phytopathol., 10, 207 (1972).
237. J. Ingham, Bot. Rev., 38, 343 (1972).
238. I. Cruickshank, D. Biggs, and D. Perrin, J. Indian Bot. Soc., 50A, 1 (1971).
239. A Stoessl, Rec. Advan. Phytochem., 3, 142 (1970).
240. T. Kosuge, Ann. Rev. Phytopathol., 7, 195 (1969).

Chapter 14

TOXICOLOGICAL ASPECTS OF FUNGICIDES

Lawrence Fishbein

Chemistry Division
National Center for Toxicological Research
Jefferson, Arkansas

I. INTRODUCTION

There is increasing recognized concern over the possible toxicological hazards posed by a spectrum of environmental chemicals, food additives, and drugs, either considered singly as a class or specific agent or in combination.

A number of fungicides, based on aspects of their biological and toxicological properties per se (as well as those of their trace impurities or degradation products), their ubiquity, persistence, presence, and concentration in the food chain, constitute a significant source of potential environmental hazard to mammalian species including man.

The major objectives of this chapter are to highlight the salient toxicological features (primarily teratogenic, carcinogenic, and mutagenic) of a number of the major fungicidal agents including their important trace impurities and degradation products and concomitantly consider germane aspects of their acute toxicity and effects on growth.

It should be noted that there exists considerable disagreement concerning the scientific assessment of teratogenic, mutagenic, and carcinogenic effects of these agents, principally due to the lack of definitive characterization of the agent per se, and the nature of its trace impurities or occluded degradation products. The variety of testing procedures employed further complicates comparisons within any one toxicological area as well as hinders the unambiguous recognition and interpretation of the toxic event and the extrapolation and relevance of experimental findings to man.

Definitive descriptions of the general considerations, interrelationships, and most salient respective teratogenicity, carcinogenicity, and mutagenicity testing procedures employed are beyond the scope of this chapter. However, the reader is referred to excellent compendia and articles covering these aspects [1-19].

II. ETHYLENEBISDITHIOCARBAMATES

A. Maneb and Zineb

The metal derivatives of the ethylenebisdithiocarbamates constitute one of the most important classes of fungicides currently used in agricultural practice. The major agents in this group are the manganese and zinc derivatives maneb and zineb, respectively, and the disodium derivative nabam. More recently introduced related agents include mancozeb, manganese-zinc ethylenebis(dithiocarbamate) and propineb, zinc propylenebis(dithiocarbamate). Until recently, this class has been regarded as relatively harmless because of generally low mammalian toxicity, their biodegradability, as well as their continuous use for more than 30 years.

The acute oral LD_{50} in rats (mg/kg) for maneb and zineb is 6,750 and >5,200, respectively [20]. The acute dermal LD_{50} in rats for zineb (25% suspension in propylene glycol) is >2,500 mg/kg [20].

The effects of zineb and maneb on the blood and hematopoietic organs of animals has been noted by Kurbat [21]. Administration of zineb and maneb in the form of 2-20% suspensions in starch given to male and female rabbits internally in daily doses of 0.05-0.5 g/kg resulted in leukopenia in all rabbits. These chantes were caused by maneb at a 0.05 g/kg dose and by zineb at a 0.1 g/kg dose. Zineb and maneb caused a decrease in weight and appetite and decline in the general condition of the animals. With the development of leukopenia, the rabbits died between day 12 and day 78 of the study. Multiple petechial hemorrhages were observed in the mucosa of the gastrointestinal (GI) tract, external layers of myocardium, and under the visceral pleura.

Aspects of the subchronic toxicity in female Wistar rats maintained on a diet containing 117.5 and 163 mg/kg body weight of zineb and maneb, respectively, for 7 months, have been reported by Przezdziecki et al. [22]. Generally a defined drop in food intake, body weight, and growth was observed in all intoxicated animals, the most pronounced being in the group fed maneb. Zineb and maneb caused statistically significant increases in the weight of the thyroid and a decrease in the weight of the kidneys, adrenal glands, and ovaries. No pathological changes were observed in the parenchymatous organs. While all groups of rats showed decreased blood glucose level and liver glycogen, no effect was noted on total serum proteins and its fractions, or on the content of Na^+, K^+, and Ca^{++}.

Changes in the permeability of blood-tissue barriers in rats treated with 1/50, 1/500, and 1/5,000 LD_{50} doses of zineb were reported by Yudina and Novikov [23]. The kidney was the most permeable organ, while the brain was the least permeable. The decrease in permeability of the spleen, kidney, adrenals, pancreas, thyroid, and liver was directly proportional to the concentration of zineb. When zineb was administered in combination with chlorophos (trichlorfon), a decrease in dose-dependent blood permeability was found in all organs studied which was believed to be associated with the hormonal impairment caused by these compounds.

Changes in the immunological structures of the spleen have been reported [24] in albino rats 30-35 days old treated daily with 125 mg/kg of zineb for 6 months. Plethora of the spleen and retardation of the development of the lymphatic follicles were observed in animals treated with large doses of zineb. The changes were similar to those produced in the control groups treated with daily doses of 2 mg/kg of cyclophosphamide. The cytostatic effect of zineb primarily involved the T-lymphocytes while the B-lymphocytes were rather resistant. In contrast to the cyclophosphamide-treated group, the pesticide-treated groups (zineb, carbaryl, linuron, and tribufon) showed few changes in the structure and weight of the thymus.

Engst et al. [25] compared the toxicity of maneb, zineb, and nabam in mice and rats with a number of their degradation products. Changes in the weight of organs and body were influenced more significantly by feeding pure maneb than by feeding degradation products, while an increase in monosulfide content caused an increased inactivation of hydrosulfuric enzymes.

Blackwell-Smith et al. [26] and Clayton et al. (quoted in Evaluation of the Toxicity of Pesticide Residues in Food, 1965) [27], reported thyroid hyperplasia in animals treated with maneb, zineb, thiram, and mancozeb. Reduced ^{131}I concentration rates were found in dogs after 2 years of administration of mancozeb.

In short-term feeding of male albino rats [26], zineb was found to have less than one-tenth the goitrogenic potency and toxicity of maneb. This appeared to be due to the fact that only 11-17% of the ingested dose was absorbed by the gastrointestinal tract. When ingested chronically over periods up to 2 years, zineb produced mild thyroid hyperplasia in about 40% of the rats studied at the lowest dietary level of 500 ppm [26]. However, in the dog hyperplastic changes in the thyroid were not produced by levels as high as 2,000 ppm when fed for 1 year [26].

In the rat, growth and mortality were unaffected by the chronic ingestion of zineb (incorporated into Purina Dog Chow as the basic diet) at levels of 2,500 ppm and below in the female, and at levels of 5,000 ppm and below in the male. In the dog, neither body weight, nor mortality were affected by the ingestion of 10,000 ppm of zineb in the diet for 1 year [26]. The blood picture in dogs and rats was apparently not affected by the chronic ingestion of diets containing as much as 10,000 ppm of zineb, and furthermore, there appeared to be no relationship between zineb and the occurrence of malignancies in rats on these diets [26]. It was emphasized in the studies of Blackwell-Smith et al. [26] that the goitrogenic effect appeared to be the only specific pathological change induced by zineb and "it would appear unlikely that the one ppm or less which might appear in the total human diet would afford a hazard to the consumer."

Innes et al. [29] reported the nontumorigenicity of maneb, zineb, and nabam in chronic feeding studies on three strains of mice.

In subacute experiments [30, 31], decreased concentration rates of ^{131}I were found in experimental animals treated with zineb, maneb, and mancozeb. Ivanova-Chemishanska et al. [30] studied rat thyroid glands after 30 days of application of 0.1 LD_{50} of maneb, zineb, and mancozeb by means of histology, histochemistry, and electron microscopy and reported evidence of increased thyroid function as a result of increased thyrotropin stimulation. The intrafollicular hyperplasia and formation of microfollicles that lead to markedly increased thyroid weight in animals treated with maneb and mancozeb were suggested to be processes that occur later, after the functional limits of the gland are reached [32].

An assessment of the teratogenic activity of technical-grade maneb and zineb was provided by Petrova-Vergieva and Ivanova-Chemishanska [33]. These fungicides were given in a single oral dose to groups of pregnant rats on day 11 or 13 of organogenesis. Single doses of 1-4 g maneb per kg and 2-8 g zineb per kg induced congenital anomalies in 12-100% of the fetuses. The malformations were located mainly in the fetal brain, facial part of the skull, limbs, and tail. The no-effect levels for a single rat dose were 0.5 g/kg for maneb and 1 g/kg for zineb, respectively. No adverse effect on the intraciterial development of the progeny was observed when the latter dose levels were given daily to groups of rats from day 2 to day 21 of pregnancy or when the rats were exposed in a dynamic inhalation chamber to a concentration of 100 mg zineb per m^3 for 4 hr per day from day 4 of pregnancy. This exposure was 100 times higher than the maximum allowable concentration (MAC) of 1 mg/m^3. It was estimated by Petrova-Vergieva and Ivanova-Chemishanska [33] that effective teratogenic doses of maneb and zineb in their study were at least 1,000 times higher than the daily human intake that could result from the consumption of foods containing the maximum permitted residues of these compounds. It was thus concluded that the present exposure to dithiocarbamate fungicides in agriculture is unlikely to present a hazard to the normal development of the human embryo [33].

Maneb injected into the stomach of rats every other day at 50 mg/kg (0.001 LD_{50}) during pregnancy caused resorption and death of the embryos and had an unfavorable effect in 21% of the pregnancies compared with 12% in the controls [34]. Maneb administered for 1 month in the same dose to sexually immature male and female rats reversible reduced fertility.

The blastomogenic properties of zineb as well as ziram (zinc dimethyldithiocarbamate) in mice was described by Chernov and Khitsenko [35]. Blastomogenesis in the mouse lungs initiated by these two compounds was qualitatively similar to that induced by urethane, but quantitatively less significant.

A paucity of information exists regarding the mutagenicity of the ethylenebisdithiocarbamates. In limited studies, zineb has been found to be nonmutagenic when tested in <u>Drosophila melanogaster</u> [36] while maneb was reported to be of doubtful mutagenicity when tested in TA1532 and TA1534 Salmonella strains and nonmutagenic to his G46, TA1530, and TA1531 Salmonella strains [37].

B. Nabam

Nabam, disodium ethylene-1,2-bis(dithiocarbamate), has been largely superseded as a protective fungicide by maneb and zineb, though the latter may be prepared in the field by mixing solutions of nabam and zinc sulfate.

The acute oral LD_{50} for rats was 395 mg/kg [26], when nabam was administered as a 19% solution via stomach tube. Death in most cases occurred within 12 to 60 hr after dosing and was preceded by diarrhea, general weakness, and prostration. The goitrogenic activity of nabam in rats has been reported by Seifter and Ehrich [28] and Blackwell-Smith et al. [26]. The skin irritating properties of nabam have also been noted [26, 38] in animals and man [26]. Nabam was found to be about 1/70 as toxic as the dithiocarbamates ziram, ferbam, and thiram when tested by the chick embryo test, e.g., the LD_{50} (μg per egg) of nabam was 140.0 compared to 2.1 and 2.2 for ziram and ferbam, respectively [39].

Deleterious effects of nabam on the cells and tissues of Xenopus laevis embryos were noted by Prahlad et al. [40]. Nabam decreased the survival time of the embryos as well as induced definitive ultrastructural alterations in tissues such as the pigmented retina, notochord, and skin.

C. Impurities and Metabolic and Degradation Products of Ethylenebisdithiocarbamates

Ethylenethiourea (ETU, 2-imidazolidinethione) occurs as an impurity in technical ethylenebis(dithiocarbamate) fungicides [41-43], as well as from their metabolic [44-49] and nonbiological alterations [42, 50-63].

The acute oral LD_{50} for ETU in rats is 1,832 mg/kg (confidence interval from 1,379 to 2,562) [64]. Graham and Hansen [64] in a short-term (90 day) feeding study with ETU (of unstated purity), using Osborne-Mendel rats, reported increases in thyroid weight, decreases in ^{131}I uptake, decreases in body weight, and thyroid hyperplasia at high doses (500 and 750 ppm). Adenomas were also observed in rats fed at these levels.

Seifter and Ehrich [28] had previously noted decreased growth, increased thyroid weight, and marked thyroid hyperplasia when ETU was fed to weanling rats at 0.1% of the diet (1,000 ppm) for 8 days.

Innes et al. [29] reported the induction of liver tumors (hepatomas) in mice fed ETU (the thyroids were not examined). Ulland et al. [65] reported a study in which Charles River rats were fed 350 or 175 ppm of technical-grade ETU in their diets for 18 months, followed by a control diet for 6 months. Each test group consisted of 26 males and 26 females. In the group given 350 ppm ETU, thyroid carcinoma was found in 17 males and eight females, with pulmonary in metastasis in two males. The first tumor was found at week 68 in the 350-ppm group. Hyperplastic goitre with extreme thyroid enlargement was noted in 17 males and 13 females of the same 350-ppm group, as well as in nine males and six females of the 175-ppm group. (It is possible that serial sections of these glands might have revealed further carcinomas.) In addition, there was a scattered incidence of solid-cell adenoma, and also of hyperplastic liver nodules, but no liver tumors were noted. Apart from one simple goitre, none of these lesions occurred in a large group of control animals. Ulland et al. [65] concluded from the above study that ETU has an action like that of a number of other thio compounds which cause thyroid carcinomas and indirectly affect the liver.

Statistically significant decreases in body weight and increases in thyroid-to-body weight ratios in male and female Charles River rats exposed to levels of 125, 250, and 500 ppm in their diets for up to one year have been noted by Graham et al. [66]. Uptake of ^{131}I was significantly decreased in male rats after 12 months at 500 ppm, but was increased in females.

After 6 months, the thyroids of males fed the 500-ppm level were all hyperplastic with eight or more adenomatous nodules present among the thyroid sections. Female rats fed ETU at 500 ppm also had hyperplastic thyroids, two of these had low-grade carcinomas. After 12 months, the vascularity of the thyroid was increased at all test levels (5, 25, 125, 250, and 500 ppm) in both male and female rats. Nodular hyperplasia developed in males at 125 ppm and carcinomas were evident at 250 ppm. At the 500-ppm level, carcinomas were found in 77% of the male rats and 42% of the female rats.

Increased vascularity and hyperplasia give some indications of an overactive thyroid gland, effects seen even at 5 ppm. In these studies, all thyroids exhibited a series of changes which included diffuse microfollicular hyperplasia, diffuse and nodular hyperplasia, nodular hyperplasia with papillary and cystic deviations, and finally the development of adenocarcinoma. It was suggested that ETU initially reduces thyroid activity after which compensation occurs by an increased release of thyrotropin (TSH) and that this increase in TSH stimulates thyroid weight in an attempt to overcome the blocking effect of ETU [65].

The teratogenicity of ETU has been demonstrated in rats and rabbits by Khera [67] who administered ETU orally in daily doses of 5, 10, 20, 40, or 80 mg/kg. The rats were treated from 21 to 42 days before conception to day 15 of pregnancy or on days 6-15 or 7-20 of pregnancy, whereas the rabbits received ETU only on days 7-20 of pregnancy. No overt signs of toxicity in the mothers of any test group were noted with the exception of the rats given 80 mg/kg per day, 82% of which died after treatment for 7 or 8 days. In this species, the numbers of corpora lutea and live fetuses were unaffected by the treatments, but fetal weight was reduced at 40 and 80 mg/kg.

Irrespective of the time at which ETU was administered to the rats, lesions were produced in the central nervous and skeletal systems, their severity depending on the dose.

The severe effects seen at the two highest dose levels include meningo-encephalocele, meningorrhagia, meningorrhoea, hydrocephalus, obliterated neural canal, abnormal pelvic limb posture with equinovarus, micrognathia, oligodactyly, and absent, short, or kinky tail. Less serious defects were induced by 20 mg/kg, and at 10 mg/kg there was only a retardation of parietal ossification and of cerebellar Purkinje-cell migration. Retarded parietal ossification was the sole abnormality seen at 5 mg/kg, its incidence being limited to small areas and to a few large litters. Rabbits were far less affected by ETU treatment, showing only an increase in resorption sites and a reduction in brain weight at the highest dose level. Renal lesions, characterized by degeneration of the proximal convoluted tubules, were noted microscopically, but there were no skeletal abnormalities that could be attributed to ETU.

Ethylenethiourea in moderate concentrations has been found to induce small but reproducible changes in mutation frequency (through base substitutions) when tested in cultures of Salmonella typhinurium strain, his G46 [68] at concentrations of 100 and 1,000 ppm. ETU had a relative mutagenicity activity of 2.51 and 2.44, respectively ($P < 0.001$).

Although ethylenethiourea, in terms of toxicological significance is by far the most important impurity or metabolic and degradative substance associated with the ethylenebisdithiocarbamates, it should be noted that this class of fungicides can give rise to a variety of other degradative and metabolic products, most of which have not been as extensively studied toxicologically. These compounds include ethylenethiuram monosulfide (ETM [41, 47, 51, 53-56, 63, 69]), ethylenethiuram monosulfide polymer (ETM polymer [63, 69]), ethylenethiuram disulfide (ETD [41, 53-56, 76]), ethylenethiuram disulfide polymer (ETD-polymer [56, 63]), ethyleneurea [56, 70-76], ethylene-1,2-bis(thiocarbamoyl dimethylthiocarbamoyl disulfide) [73], hydrogen sulfide [56, 75], and sulfur [41, 53, 56, 63]. Ethylene-1,2-bis(thiocarbamoyl dimethylthiocarbamoyl disulfide) (Tecoram) is an oxidation product of nabam. Van Steenis and van Logten [73] reported that Tecoram, when administered at doses of 0.01, 0.1, 1 or 10 mg per egg in propylene glycol or in saline to chick embryos caused paralysis, shortening of the extremities, muscular atrophy, dwarfing, and death. Microscopically, the neurotoxic effects included signs of peripheral neuropathy, mainly confined to the distal parts of the peripheral nerves and muscular atrophy.

III. DITHIOCARBAMATES

A. Ferbam and Ziram

The dithiocarbamates are the half amides of dithiocarbamic acid with strong strong metal-binding characteristics; they interact with compounds containing sulfhydryl groups and have been widely used as fungicides and as vulcanization accelerators in the rubber industry. The principal fungicidal dithiocarbamates are ferbam (ferric dimethyldithiocarbamate) and ziram (zinc dimethyldithiocarbamate).

The acute oral LD_{50} for rats for ferbam is over 4,000 mg/kg and that for ziram 1,400 mg/kg [77]. Given as a single oral dose, ferbam had a much lower acute toxicity than ziram, e.g., 17 g/kg of ferbam killed only one of ten rats, whereas the LD_{50} for ziram was 1.4 g/kg. However, when repeated doses were given to rats by mixing ferbam or ziram in the diets, ferbam was more toxic. For example, the daily ingestion of as much as 1/5 of the LD_{50} dose of ziram or 1/50 of the LD_{50} of ferbam produced death within 30 days in some of the test animals. No mortality followed daily doses for 1 month of 100 mg/kg body weight and no growth retardation was observed with this dose of ziram or with three times this dose of ferbam [77]. Normal thyroid tissues were found in rats that had ingested diets containing 0.25% of ferbam or ziram for one month.

When administered a diet for a period of 2 years, growth and mortality in albino rats of the Rochester strain (ex-Wistar 1923) were not affected by 0.025 or 0.0025% of ferbam or ziram, respectively [78]. (The source and purity of the dithiocarbamates tested was not specified.) Growth was reduced by diets containing 0.25% and life span was shortened by this amount of ferbam but not by ziram. Neurological changes appeared after 2 months or more in the rats given the 0.25% diets of ferbam or ziram. Cystic brain lesions were found in many of the female rats and a few of the male rats given the 0.25% ferbam diet, and a number of rats on the higher percentage diets (0.025 and 0.25%) exhibited atrophic testes. A number of tumors developed during the 2-year test period, six in the ferbam-fed rats and seven in the ziram-fed animals. Three neoplasms of the pituitary were found in rats fed ziram; these were considered probably malignant and were described as angioinvasive adenomata. The administration of ziram was not considered a causal factor as morphologically similar tumors were found in 2-year animals [78]. Samples of thyroid tissue were normal, except in three rats given ziram in the 0.0025% group; one thyroid was hyperplastic and two rats in the 0.25% group had adenomata of the thyroid. Neither ferbam or ziram were considered by Hodge et al. to be carcinogenic in these studies [78].

When ferbam and ziram were administered in doses of 0.5, 5, and 25 mg/kg per day for 1 year to groups of two young adult beagle dogs, ill effects were observed only in the dogs given 25 mg/kg [78]. With one exception, these suffered convulsive seizures, and two dogs given this dose of each compound died during this period. The failure of cystic lesions to develop may have depended on the relative doses, e.g., the highest dog dose was only 1/10 the highest rat dose.

The toxicity of a number of dithiocarbamates (including ferbam and ziram) in young and domestic fowl was noted by Rasul and Howell [79]. These compounds had an adverse effect on body weight, retarded testicular development and produced degeneration in the seminiferous epithelium of mature birds. Adult fowl were found to be more susceptible to the toxic action of dithiocarbamates than were young chicks. Ziram was found to be the most toxic of the five dithiocarbamates tested.

Reproductive and teratogenic effects of ferbam in rodents have been reported by Minor et al. [80]. Daily feeding, equivalent to 23, 66, or 109 mg/kg per day, to male rats for 13 weeks caused deaths and weight loss at the highest dosage without any effect on reproduction. Daily feeding, equivalent to 15 or 51 mg/kg per day, to females for 2 weeks caused a severe weight loss at the high dose and a slight, but not significant, decrease in neonatal survival. Ferbam, administered to rats on days 6-15 of gestation at dosage of 150 mg/kg per day resulted in some deaths, increased resorption, decreased fetal weight, and caused a slight increase in soft and skeletal tissue anomalies. Ferbam, administered to mice on days 6-14 of gestation at 30 or 300 mg/kg per day, failed to show any teratological effect. These studies suggest that high doses of ferbam have mild teratological effects in rats [80].

The mutagenicity of ziram appears to be equivocal. For example, ziram has been reported to be mutagenic when tested against *Salmonella typhinurium* strains TA1534 and TA1530 but nonmutagenic against *his* G46, TA1531, and TA1532 *Salmonella* strains [37]; in the Miller-5 test in *Drosophila melanogaster* [36]; in the liquid-holding test, i.e., forward mutation in *E. coli* to 5-methyltryptophan (5-MT) resistance [81]; spot test, i.e., back-mutation to prototrophy in two auxotrophic strains a21 and a742 of *Serratia marcescens* and forward mutations to galactose prototrophy in a phenotypic galactose-negative (Gal R^5) strain of *E. coli* [81]; liquid-holding test, i.e., forward mutation to streptomycin-resistance in *E. coli* [81]; and liquid-holding test, i.e., mitotic gene conversion in *S. cerevisiae* [81].

Ziram has been reported to induce chromosome breaks in cultured peripheral human lymphocytes [82]. These breaks were nonrandom, most of them confined to chromosome 2. The main quantitative characteristic of the cytogenic effect of ziram at 0.003 μg/ml concentration were similar to those shown by the analysis of blood of individuals having a professional contact with this fungicide [82].

Antonovich et al. [83] reported a comparative toxicological assessment of a number of dithiocarbamates and related fungicides. Based on tests with experimental animals (oral administration), and tissue culture, ziram, thiram (TMTD), and maneb were generally more toxic than were zineb and polycarbacine. Ziram possessed greater cumulative teratogenic, gonadotoxic, and blastogenic effects on rats, mice, and rabbits, than did the other pesticides, and possessed high cytogenic activity as well. Maneb exhibited the strongest embryotoxic activity while thiram had the greatest mutagenic effect and the greatest LD_{50} in chronic poisoning experiments.

It is also of importance to note aspects of the metabolism of ziram in terms of potential toxicological consideration of the metabolites per se.

For example, Vekshtein and Khitsenko [84] reported that orally administered ziram was retained for prolonged periods in rats. After 6 days only part of the ingested ziram had been excreted through exhalation or in the urine. The largest concentration of ziram metabolites, e.g., tetramethylthiuram disulfide (thiram), tetramethylthiourea, and dimethyldithiocarbamic acid dimethylamine salt was observed in the spleen; carbon disulfide and dimethyldithiocarbamic acid salt were found in the lungs.

The recent finding by Eisenbrand et al. [85], that the carcinogenic nitrosamine, dimethylnitrosamine (DMN) could be formed rapidly by interaction of nitrite with ziram and ferbam in vitro under simulated gastric conditions and in vivo in rat stomach, adds another dimension to the toxicological potential of this class of fungicides. Residual levels of such fungicides in the human diet represent potential precursors for the formation of carcinogenic nitrosamines when they are ingested with nitrite. Under the selected conditions, the optimum pH for the formation of DMN was 1.5-2.0. At pH 2.0, more than 1 mg DMN was produced after a 10-min incubation of 10^{-4} moles ziram with a twenty-fold molar excess of nitrite. This corresponded to about 8% of the theoretical yield, assuming that two molecules of the carcinogen were formed from one molecule of ziram.

IV. THIRAM

Thiram, tetramethylthiuram disulfides, bis(dimethylthiocarbamoyl) disulfide, TMTD, is an extensively employed fungicide for both foliage application and seed treatment. At high doses, it is repellent to mice in the field. The mammalian toxicity (acute oral LD_{50}s) range from 350 to 2,500 mg/kg. The acute oral LD_{50} in both male and female Sherman strain rats if 620-640 mg/kg when thiram is administered as a 20% suspension in propylene glycol [20]. The acute oral LD_{50} to rabbits is 350 mg/kg. In female virgin mice of the NMRI strain, oral LD_{50} of thiram was 2,300 mg/kg (2,050 to 2,500 mg/kg) [86].

Thiram was classified by Agaeva [87] as having medium toxicity. With intragastric administration, rats were more sensitive than mice or cats and the LD_{50} for rats was found to be 740 mg/kg.

A single application of thiram to the skin of rats and rabbits in the form of an oily solution in doses of 1,000-2,000 mg/kg and 500-1,000 mg/kg did not produce an irritating and general toxic effect. Repeated application of thiram to the skin of rabbits in a dose of 50 mg/kg did not have an irritating effect right up to the time when the animals died. However, thiram has cumulative properties with the cumulation coefficient at 1/10-1/20 LD_{50} established at 2.1-2.85 [86]. Poisoning with thiram is characterized by intensely pronounced affections of the nervous system. This is manifested in the depression of the animals, adynamia, disturbances in coordination of movements, and in clonicotonic convulsions. An early manifestation of thiram poisoning is pronounced eosinopenia [86].

Contact dermatitis of the allergic type, following intracutaneous administration of 0.1 ml of 0.5% thiram to the skin of albino rabbits has been reported [88]. Humoral antihapten antibody was found to participate in the development of the contact dermatitis, and the antibody level was dependent on the intensity of the allergic process.

Thiram has been found teratogenic to the hamster (Golden Syrian strain) when administered on days 7 or 8 at 250 mg/kg suspended in carboxymethyl cellulose (CMC), and when it was dissolved in dimethyl sulfoxide, DMSO (31 mg/kg), the teratogenicity was additive or possibly more than additive [89]. The various defects found in hamsters included exencephaly, cranial pimple, fused rib, cleft palate, and short or curved tail. The effects from thiram in DMSO must be compared with results from administration of an equal level of solvent, since DMSO is teratogenic as well as embyotoxic in the hamster [90].

Roll [91] found thiram to be teratogenic to two strains of mice (MMRI and SW) when administered between day 6 and day 17 of pregnancy (as well as during certain other periods of gestation) in oral doses that were nontoxic (e.g., 50-30 mg per adult animal). In addition to increased resorption of embryos and clearly retarded fetal development, a syndrome of skeletal malformations, including cleft palate, "wavy" ribs, curved long bones of the extremities, and micrognathia, were observed to be dose related in both strains. It was also found that varying times of application allowed for a determination of stage-specific effects. Day 12 and day 13 of embryonic development proved to be the most susceptible phase. There was a considerably higher susceptibility of the NMRI than SW strain with regard to the induction of cleft palates. The teratogenically ineffective dose of thiram for mice was considered to be about 250 mg/kg [91]. (The thiram used in the study of Roll was from Farbenfabriken Bayer A.G.-Leverkusen, melting at 149-152° C).

Matthiaschk [86] reported that oral doses of 10, 20, and 30 mg thiram per animal given to MMRI-mice from day 5 to day 15 of their pregnancy, resulted in increased resorption during the intermediate and late stages of fetogenesis, as well as a typical syndrome of malformations (with an almost linear dose-effect relationship), including cleft palates, micrognathy, hyphoses, "wavy" ribs, and distorted wavy and block-shaped bones of the extremities and their suspension apparatus.

Simultaneous oral administration of 2.5 mg thiram and intraperitoneal injection of 5 mg of L-cysteine per animal from day 5 to day 15 of pregnancy resulted in an obvious reduction in the severity of individual malformations, while the incidence of malformations remained constant.

Following administration of 30 mg thiram per day on days 12 and 13 of pregnancy, both the overall rate of malformations and the number of individual malformations were increased as compared with the results of administration from day 6 to day 15 of pregnancy [86].

In a three-generation study in rats, 100 ppm of thiram was reported to have caused no toxic or teratogenic effects [92].

Davydova [93] described the effects of thiram on the estrous cycle and the reproductive function of female albino rats. When animals were exposed to thiram concentrations of 3.8 ± 0.058 mg/m^3 for 6 hr per day, 5 days per week, for 4.5 months, extension of the estrous cycle at the expense of the resting cycle was observed during the initial period of the experiments. Although the estrous cycle normalized again by the end of the experiment, impairment of the reproductive functions of the animals, e.g., reduced rate of conception, reduced fertility, and underweight fetuses, was observed.

Thiram given to young and adult domestic fowl caused an adverse effect on body weight and retarded testicular development, and produced degeneration in the semineferous epithelium of mature birds [79]. Thiram (99.9% pure) was found effective in inhibiting ovulation in bobwhite quail (Colinus virginianus) [94]. Recovery occurred about 14 days after its removal from the diet. A dose level of 8.8 mg/kg per day caused a 50% reduction in egg production. It was noted that this level of thiram in the diet of bobwhites could be reached readily when many seed species are treated according to manufacturer's recommendations. The general effects of thiram poisoning observed during the reproductive period suggest an alteration of hormone levels. These effects include (a) significant weight loss of the ovary and oviduct, (b) decrease in serum calcium level, which is estrogen controlled, (c) lower egg production, (d) alteration in the normal maturation pattern of the ova in treated birds, and (e) an increase in relative activity [94].

Treatment with 1/100 LD_{50} of thiram has been found to totally suppress egg production in chickens [95]. Following a total intake equalling 0.20 LD_{50} or 168 mg/kg, increased infertility of the eggs (34.2%), a hatching rate of 50%, reduced hatching weight, and considerable slowdown of the postembryonal development, as well as malformations of the extremities in 11.3% of the test embryos was observed [95]. Antonovich and Vekshtein [96] reported that thiram administered orally to rats at 0.5, 1 2, or 4 mg/kg daily for 8-18 months, impaired growth, but did not affect life expectancy. At the larger chronic doses, thiram increased the hemoglobin, sulfhydryl, and erythrocyte levels of the blood, disrupted hippuric acid metabolism and caused fatty and protein dystrophy of the liver and kidneys. After 8 months of chronic administration, even at the lowest dose, thiram was detected in the blood and all the major organs and glands of the body. The degree and pathological damage and morphological changes induced by thiram was found to be dose related.

The literature on the cytogenetic and mutagenic effects of thiram is scant. Mice given thiram intraperitoneally (i.p.) at 100 mg/kg developed 2.5 times more chromosomal aberrations in their bone marrow cells than did mice given the same dose introgastrically [97]. Ziram was less effective than thiram in producing aberrations, while maneb and zineb were still less effective. All the aberrations produced by these compounds were of the chromatid type [97].

A number of studies involving human exposure to thiram have been reported. Sivitskaya [98] described ophthalmological changes in 50 workers, 20-58 years of age, having prolonged occupational contact with thiram. Some of the test subjects were being treated for astheno-vegetative syndrome and vegetative vascular dystonia. Lachrymation and photophobia were the initial, temporary symptoms of exposure after prolonged interruption of contact with thiram. Chronic conjunctivitis was seen in 14% of the cases, along with reduced visual acuity, delayed dark adaptation, reduced corneal sensitivity, change in the diameter of the retinal vessels (in 34%), increased pressure in the retinal artery, and increased tonometric pressure.

Lyubchenko et al. [99] described the effects of 5-20 years of exposure of a benzene hexachloride-thiram combination in 22 female and male workers in the 35-to-54-year age bracket.

With enhanced detection techniques contact dermatitis involving the fungicides thiram and Difolatan (captafol) is apparently becoming increasingly recognized in the tropical developing countries [100].

V. BENZIMIDAZOLE DERIVATIVES

Benomyl, methyl 1-(butylcarbamoyl)-2-benzimidazolyl-carbamate (Benlate), is one of the more recently developed derivatives of this group of substituted benzimidazole compounds that exhibit a broad spectrum of fungicidal as well as nematocidal and mite ovicidal activity [101].

Benomyl is degraded by nucleophilic attack or UV radiation [102] to methyl 2-benzimidazolecarbamate (benzimidazolecarbamate methyl ester, carbendazim, MBC) which apparently is the fungitoxic principle [102-106]. Benomyl decomposes in aqueous solutions to MBC as well as butylisocyanate [107].

The acute oral LD_{50} of benomyl in rats is 9.5 g/kg [108]. The low mammalian toxicity of benomyl may be due to its rapid metabolism and excretion of hydroxylated products [109, 110], e.g., in mice, rabbits, and sheep [109] and rats [110].

The teratogenic and embryotoxic effects of benomyl in Wistar rats (as affected by the protein level in the diet) were described by Torchinskiy [111]. The animals were kept on diets with 5, 10, 19, and 35% protein, either for 1 month prior to mating and during the entire pregnancy or only during pregnancy. Benomyl was administered in doses of 250 and 145 mg/kg body weight on day 12 of pregnancy. The teratogenic and embryotoxic action of benomyl did not change in groups where the females received the diet containing 5, 10, or 19% protein from the very beginning of prenancy.

An increased spontaneous postimplantational lethality of the embryos was noted in females kept on a diet with 5% protein since 1 month before conception. The teratogenic action of benomyl was weaker in animals which had received a diet with 10% protein since 1 month before conception, than in the groups kept on diets with 19 and 35% protein.

The cytogenetic and mutagenic activity of benomyl and a number of benzimidazole derivatives have been recently assessed. In eukaryotes, benomyl may affect ploidy since it induces segregation in the diploid strain of *Aspergillus nidulans*, although having no apparent effect on the frequency of point mutations [112]. Kappas et al. [113] reported that benomyl and MBC greatly increased the number of diploid and particularly haploid segregants in diploid *A. nidulans*. The increased numbers of diploid sectors were not the result of chromosome breakage and deletion. Primary genetic changes were induced and although increased frequency of mitotic crossing over was a possibility, nondisjunction was considered to be probably the main effect [113].

Chromosomal aberrations have been induced in *Allium cepa* roots in 15 hr in solutions containing 96 and 48 ppm benomyl. The cells were affected by benomyl in the prometaphasic and metaphasic stages, developing elongated chromosomes, while the chromosomes exposed to 12 ppm of benomyl tended toward fragmentation [114].

Benomyl and MBC have been reported to exhibit mutagenic effects when tested in the *Salmonella*, his G46 and TA1530 strains, but to be nonmutagenic to the *Salmonella* strains, TA1531, TA1532, and TA1534 [37, 115].

Styles and Garner [116] showed that benomyl and MBC were cytoxins causing mitotic arrest, mitotic delay, and a low incidence of chromosome damage when tested on specific-pathogen-free Wistar rats and various mammalian cell lines (human adult liver, HeLa, mouse connective tissue, Syrian Hamster kidney, and Chinese hamster lung).

In the tissue culture experiments, benomyl and MBC caused pyknosis at concentrations of 10^{-2} and 10^{-3} M, arrested metaphases, pyknotic nuclei, and morphologically abnormal nuclei (10^{-4} to 10^{-5} M) after 24 hr incubation. After 4 hr incubation, benomyl and MBC (10^{-5} M and above) caused metaphase arrest.

A host-mediated assay, using serum from animals dosed with MBC or related compounds, revealed that the cytotoxicity of sera, following intraperitoneal administration of MBC, benomyl, and PP010 (methylbenzothiadiazine), were similar. However, following oral dosing, the cytotoxicity of serum from benomyl-treated rats was 10-fold lower. Since MBC is probably the degradation product responsible for the cytotoxicity of both PP010 and benomyl [102, 106] administration may be due to differences in stability, rate of absorption, or metabolism of the parent compound.

In man, benomyl has been reported to induce contact dermatitis and bilateral conjunctivitis [117].

VI. THIOPHANATES

Thiophanate, 4,4'-*o*-phenylenebis(3-thioallophanate), 1,2-bis(3-ethoxy carbonyl-2-thioureido)benzene, and its methyl homolog, thiophanate-methyl, dimethyl-4,4'-*o*-phenylenebis(3-thioallophanate); 1,2-bis(3-methoxy carbonyl-2-thioureido)benzene, are recently introduced systemic fungicides whose antifungal spectrum is similar to that of benomyl. This similarity is essentially due to the common degradation and metabolic products, e.g., benzimidazolecarbamic acid ethyl ester (MBC) and benzimidazolecarbamic acid methyl ester (BCE). The former arising from benomyl and thiophanate and the latter from thiophanate-methyl, respectively. In the case of the thiophanates, MBC and BCE formation is due to degradative cyclization [118, 119]. The acute oral and dermal LD_{50} for rats is more than 15,000 mg/kg [120]. The acute oral LD_{50} (g/kg) of thiophanate-methyl in several species [121] is listed below:

	male	female
dd Y mice	3.51	3.40
Wistar rats	7.50	6.64
Guinea pig	3.64	6.70
Rabbit	2.27	2.50

The minimal lethal dose of thiophanate-methyl was 5.0 g/kg for both male and female dogs (beagle). The acute dermal LD_{50} of thiophanate-methyl for dd Y mice, Wistar rats, guinea pig, rabbit, and dog (beagle) was greater than 10 g/kg [121]. The acute intraperitoneal LD_{50} (g/kg) of thiophanate-methyl for dd Y mice and Wistar rats was 0.79 and 1.64 for the males and 1.11 and 1.14 for the females of the species, respectively [121]. Technical-grade thiophanate-methyl, used in the above acute toxicity determinations, contained 94% active material with impurities consisting of approx. 2.0% sulfur, 2.0% inorganic chloride, 2.0% unknown volatile substances, and less than 0.5% unknown aromatic amines. When large amounts of thiophanate-methyl were given, animals exhibited tremors 1-2 hr after administration, became sensitive to touch and had tonic or clonic convulsions (the degrees and duration of signs varied with dose). No noteworthy toxic signs were observed in animals treated by dermal administration.

Inhalation studies involving the administration of Topsin-M (a hydrophilic powder containing 75% thiophanate-methyl and 25% fine diamorphous powder and surfactants) to male dd Y mice at an aerosol concentration of 100 mg/liter for periods of 30, 60, and 120 min failed to reveal any toxicity [121].

A decrease of about 20% in the white cell count of rabbits, 1-3 hr after treatment with 1,000 mg/kg thiophanate-methyl (one-half of LD_{50}) orally (po), as well as a slight hemolytic and anticoagulant activity of the fungicide was reported by Hashimoto et al. [122].

Daily oral administration of thiophanate at levels ranging from 5-500 mg/kg per day to beagle dogs for 180 days did not produce significant changes in body and organ weights. No significant morphological changes in organs were observed [122]. Graded single intraperitoneal injections of technical-grade thiophanate-methyl (94% active ingredient) suspended in 1% gum arabic at levels ranging from 8-500 mg/kg to male mice or oral administration (as a suspension in a 5% gum arabic aqueous solution) of 40-500 mg/day to pregnant mice from day 1 to day 15 of gestation, did not produce significant mutagenic, cytogenic, or teratogenic effects [123]. The male mice were mated with untreated females in a subsequent 8-week period. The animals produced no mutagenic effect in the dominant lethal assay [124-126] as measured by an increase in early fetal deaths. The orally treated females and controls presented no differences in the number of dead fetuses and the body weight of the living ones.

Additional in vivo cytogenetic studies were described in Wistar strain albino male rats, which received daily i.p. administrations for 5 days of technical-grade thiophanate-methyl (94% active component) in 5% gum arabic (in total amounts of 312.5 to 5,000 mg/kg) [123]. Cytogenetic analysis on both somatic and germinal cells failed to reveal any chromosomal aberrations.

Thiophanate as well did not exert significant mutagenic, cytogenetic, and teratogenic effects when tested under analogous experimental conditions [127].

The mutagenic effects of thiophanate and thiophanate-methyl, when tested for their ability to induce genetic segregation in a heterozygous green diploid strain of *Aspergillus nidulans*, have been reported by Kappas et al. [113].

Analogous studies with benomyl and MBC led Kappas et al. to the hypothesis that "the toxicity of the new benzimidazole and thiophanate fungicides is directly related to their genetic activity; the mechanism of genetic activity may itself be the cause of toxicity" [113]. MBC was suggested to be the compound responsible for genetic activity as it is for fungitoxicity [113]. In an analogous fashion, ethyl-2-benzimidazole carbamate (BCE) may be responsible for the genetic activity of thiophanate [128].

VII. PHTHALIMIDES

A. Captan and Folpet

The phthalimide fungicides, e.g., captan, folpet, and captafol (difolatan) are almost as widely used as the dithiocarbamates. Captan, *N*-(trichloromethylthio)-3a, 4, 7, 7a-tetrahydrophthalimide, *N*-(trichloromethylthio)-cyclohex-4-en-1, 2, -dicarboximide, Orthocide, is the most important of these fungicides, mainly used for foliar, soil, and seed applications [129].

Folpet, *N*-(trichloromethylthio)phthalimide, controls certain powdery mildews against which captan has little effect. Captan and folpet have apparent low acute oral toxicity for laboratory and farm animals [130-134].

The acute oral LD_{50} of captan to rats and rabbits is 10,000 mg/kg and 3,000 mg/kg, respectively. Oral and i.p. doses of captan are not uniformly toxic [135]. For example, acute oral LD_{50} and acute intraperitoneal LD_{50} have been reported to be 15,000 mg/kg and 50-100 mg/kg, respectively. This difference may arise from the fact that captan is susceptible to degradation in the GI tract [136]. For example, captan has been found to be rapidly excreted via urine, feces, and expired air from orally dosed rats [136]; 50% of the oral dose was cleared within 9 hr. In comparison, 50% excretion of an i.p. dose was not achieved until 2 days after injection [136]. Urinary metabolites of orally administered captan included thiazolidine-2-thione-4-carboxylic acid, a salt of dithiobis(methane sulfonic acid), and the disulfide monoxide derivative of dithiobis(methane sulfonic acid). The latter two metabolites were not detected in urine of rats treated i.p. with captan [136]. Metabolism of captan appears to involve evolution of thiophosgene derived from the trichloromethylthio moiety. The formation of thiophosgene by the reaction of captan with thiol compounds has been shown in vitro [137-140]. Thiophosgene, as well as tetrahydrophthalimide can arise from the reaction of captan with cysteine [141]. Captan is also susceptible to hydrolytic cleavage, e.g., half-life of 2.5 hr at pH 7.0 [142], which is accelerated as the pH is increased [143]. Solutions of inorganic sulfites and thiosulfates have been shown to decompose captan by scission of the N-S bond [144].

Female Wistar rats maintained on a diet containing 158.6 mg/kg body weight of captan for 7 months suffered a defined drop in food intake, body weight, and growth. Captan caused an increase in weight of kidneys, adrenal glands, and ovaries and no pathological changes were observed in the parenchymatous organ [22].

The acute oral LD_{50} of folpet in rats is more than 10 g/kg [131]. Acute parenteral LD_{50} values of 7,540 ± 876.3 mg/kg in rats, 1,546 ± 120 mg/kg in mice, and 1,115 mg/kg in rabbits have been reported [145]. The threshold dose of folpet was 200 mg/kg. Doses up to 300 mg/kg had no gonadotoxic effect in these species [145]. Doses of 35, 25, 20, and 10 mg/kg, corresponding to the respective total doses of 280, 200, 160, and 80 mg/kg in a subacute experiment, retarded the embryonic development, but caused no fetal death or teratogenesis when administered to rats during the first or second week of pregnancy [145].

In a 6-month chronic experiment in rats, 50, 35, and 25 mg/kg doses caused reduced hemoglobin level in the blood, leukopenia, reduced hippuric acid excretion, and reduction of the prothrombin time. The cumulation coefficient was determined as 13.5 in rats and 8.2 in mice [145]. The above findings of Mandzhgaladze et al. [145] indicated that folpet is a low-toxicity, embryotoxic fungicide. No evidence of chronic toxicity to dogs fed for 66 weeks with technical-grade captan at rates of 300 mg/kg per day has been noted [131]. The "no-effects" level from 2-year dietary tests on rats was 1,000 mg/kg of captan [131].

The phthalimide fungicides appear to be the most intensively investigated fungicides primarily in regard to their potential teratogenic and mutagenic activity. This is in part due to the earlier demonstrated embryotoxicity of thalidomide [146-149] and the possibility that this activity could be associated with the N-substituted phthalimide portion of the molecule.

Preliminary studies in 1962 by Marliac and co-workers [150, 151] indicated that captan and folpet were teratogens in the developing chick embryo. A more definitive study was reported in 1969 by Verrett et al. [152], who compared the effects of captan, folpet, captafol (N-tetrachloroethylthio-4-cyclohexene-1,2-dicarboximide), captan epoxide, captafol epoxide, tetrahydrophthalimide, phthalimide, phthalic acid, and tetrahydrophthalic acid in the developing chick embryo.

Technical-grade captan, folpet, captafol, and captafol epoxide (in the solvent DMSO) injected into either the yolk or air cells of White Leghorn eggs before incubation, demonstrated a high and specific teratogenic activity with the nature of the malformations of the wings and legs in the chicken embryo being particularly significant. However, the studies were not extensive enough to allow a conclusive determination of the specific metabolites or portions of the molecules responsible for the teratogenic activity of captan, folpet, and captafol in the chicken embryo [152].

Fabro et al. [153] were unable to demonstrate embryologic malformations with captan and folpet in rabbits.

Captan and folpet were nonteratogenic in two strains of rabbits (New Zealand white and Dutch Belted) when administered daily via gelatin capsules from day 6 through gestation at levels of 18.75, 37.5, and 75.0 mg/kg [154]. However, New Zealand albino rabbits showed an increased number of resorption sites and a corresponding decrease in the number of young per litter with captan (75 mg/kg) and folpet 37.5 mg/kg. Progeny size was reduced by 15-28% among young derived from New Zealand dogs treated at 75 mg/kg with captan or folpet.

Offspring of albino rats given captan and folpet, either during gestation only or over three consecutive generations, were free of abnormalities [154]. Corn oil suspension of the test compounds were administered orally via gavage from gestation day 6 through 15 inclusive. In the multigeneration study, the fungicides were incorporated into the basal breeder ration at dietary levels up to 0.1%. Offspring from Golden Syrian hamsters fed captan at levels adjusted to maintain an average daily intake of up to 1,000 mg/kg body weight during the entire gestation period did not indicate any significant impairment of embryonic development [154]. Although gross malformations such as microphthalmia, exencephaly or limb and tail abnormalities were observed infrequently, no relation could be shown between treatment and occurrence of abnormalities.

At levels of captan corresponding to 250 mg/kg and above, marked reduction in food intake was observed. The maternal animals in these groups of hamsters either lost weight or failed to show the weight gain expected during gestation. Resorption became significant, and the number of young per litter was reduced only in the group fed an equivalent of 1,000 mg/kg of captan.

Treatment of pregnant rhesus monkeys and stump-tailed macaques with technical-grade captan and folpet (89.5 and 90% pure, respectively) at doses of 10.0, 25.0, and 75.0 mg/kg (as suspensions in a solution of cream of coconut) from day 21 to day 34 of gestation caused neither fetal deformities nor systemic toxic effects in the mothers [155]. Abnormal fetuses were observed following administration of thalidomide in nonhuman primates during the above periods of fetal limb development. The only abnormality noted among any of the fetuses receiving captan or folpet was the presence of a 13th pair of ribs in two of the rhesus fetuses from mothers dosed with folpet, e.g., one at 25.0 and one at 75.0 mg/kg, while administration of thalidomide resulted in only two normal fetuses among 11 rhesus monkeys and no normal fetuses among nine stump-tailed macaques.

Teratogenic effects of captan on the embryo of the common snapping turtle, *Chelydra serpentina*, following administration of 1 mg of the fungicide injected into the yok sac of each egg away from the embryo, have also been noted [156].

Despite the fact that captan has been intensively investigated in regard to its genetic effects, the results are conflicting. The mutagenic activity of captan was first noted by Legator et al. in 1969 [157] in both streptomycin and thymine-dependent strains of *E. coli*, as well as in a heteroploid human embryonic lung cell line and in a cell line derived from the kidney of the rat-kangaroo (*Protorus tridactylis*) [157]. Captan at a concentration of 1,000 μg per assay disc showed a 10-fold increase in production of mutants over control in both strains of *E. coli*. At concentrations of captan less than 5 μg/ml, both growth and mitosis in the heteroploid embryonic lung cells was demonstrated as well as chromosome aberrations within 2-5 hr after exposure. In the rat-kangaroo cell line, a disproportionate number of breaks was found in the sex chromosome (which accounts for less than 10% of the total chromosome mass) with the chromosome breaks and mitotic inhibition being proportional to the concentration of captan.

Bridges et al. [158], using a spot-test technique with selected repair-deficient strains of *E. coli*, showed that a substantial part of the mutagenic activity of captan was due to excisable DNA damage mediated by a volatile breakdown product. The production of this volatile mutagen was greater at alkaline pH and was believed to result from hydrolysis. Unlike other excisable DNA damage, that produced by the volatile mutagen did not depend on the $ExrA^+$ and $RecA^+$ repair functions for its mutagenicity. In addition, a diffusible mutagen was also found (following exposure of the bacteria to filter paper containing 0.1 ml of a solution of 10 mg recrystallized "Orthocide" dissolved in 1 ml chloroform), that produced excisable damage the mutagenic action of which was Exr^+-dependent (in contrast to that of the volatile mutagen). Bridges et al. [158] stated that "it is now indisputable that captan is a potent mutagen at the cellular level, with both bacteria and mammalian cells." The hazard of captan exposure to the gonodal cells of man, however, appeared to be unlikely via transport of captan through the blood stream, except after massive doses, since it has a half-life of 10 sec in serum, and reacts quickly with thiols. However, it was suggested that a more likely hazard might be carcinogenesis via the inhalation route, since most, if not all, mutagens producing excisable damage are also carcinogens [158], and it was postulated that once a particle of captan in the dry state reached the lung, it could remain for some time, being poorly soluble, and slowly releasing both volatile and diffusible mutagens into the tissues.

Captan induced mutations in the Uvr^+ strain of E. coli at levels as low as 2 μg/ml and in the Uvr^+ strain of E. coli at levels as low as 2 μg/ml and in the Uvr^- strain at 0.1 μg/ml [159].

Captan has been classified as a base-change mutagen in the rec-assay using B. subtillus strains [160]. The rec-assay system is capable of detecting mutagenicity indirectly by analyzing differences in growth sensitivities of rec^+ (recombination) and rec^- mutant cells of Bacillus subtillus.

Shirasu [161] reported the mutagenicity of captan and folpet in a microbial test system consisting of rec-assay in combination with reverse mutation systems. The reverse mutation systems included four histidine-requiring mutants of S. typhinurium (TA1535, 1536, 1537, and 1538) and two tryptophane-requiring mutants of E. coli (WP2 hcr^+ and WP2 hcr^-).

Seiler [37] reported captan and folpet to be mutagenic against his G-46 and TA1531, TA1532, and TA1534 Salmonella strains.

Captan and folpet induced back mutation to prototrophy in two auxotrophic strains (a21 and a742) of Serratia marcescens and forward mutations to galactose prototrophy in a phenotypic galactose-negative (Gal R) strain of E. coli [162].

Captan has been shown to induce forward mutations in Neurospora crassa [163], and mitotic gene conversion in S. cerevisiae [164].

The nonmutagenicity of captan and folpet in Drosophila melanogaster [165-167] has been attributed to inactivation of the compounds before reaching the germ cells [167].

A relatively low mutagenic index for captan was reported by Epstein and Shafner [168] after a single-dose (i.p. or oral) dominant lethal assay. More definitive dominant lethal assays of captan [169] and folpet [170] have been described. Technical-grade captan was given either i.p. in doses of 2.5, 5.0, or 10 mg/kg per day for 5 days or by oral intubation in doses of 50, 100, or 200 mg/kg per day for 5 days to groups of 15 male, Osborne-Mendel rats and 15 male mice (strain CBA-J). Carboxymethyl cellulose, the suspending medium for captan, provided the regular control and triethylene melamine (TEM) administered to similar groups as the positive control for 5 days i.p. at a dose of 0.05 mg/kg per day or orally at 0.2 mg/kg per day.

Male rats, 9-10 weeks old, were treated and were then mated with one virgin female of the same age weekly for the following 10 weeks. Male mice were treated at about 8 weeks of age and were then mated with two virgin females weekly for the following 12 weeks. Captan produced negligible antifertility effects and a minimal reduction in the mean number of total implants per pregnancy. Increase in the mean number of early fetal deaths per pregnancy, a measure of mutagenic effect, were seen in both species after i.p. and oral administration. Significant linear trends were also observed for the numbers of litters with two or more early deaths in both rats and mice. Captan at the highest i.p. dose (10 mg/kg) produced effects on mouse spermatocytes (week 4-5), but when given by intubation, it affected mainly epididymal sperm and late spermatids (week 1-2).

The increase in mean early deaths observed suggests that captan may exhibit mutagenic properties when administered in repeated doses. Mutagenic effects appeared more closely dose-related after oral than i.p. administration, suggesting possible metabolic activation during the digestive processes. In the studies of Collins [169], the no-effect level of orally intubated captan in rats and mice was found to be 50 mg/kg per day or about 500 ppm. On the basis of 1970 tolerances, the maximum human dietary exposure to captan was calculated to be about 13 ppm. It was suggested by Collins [169] that there appeared to be no mutagenic danger from captan from the maximum levels encountered in the human diet in 1970.

Folpet was studied by Collins [170] utilizing analogous dominant lethal assay in rats as described previously for captan [169]. Technical-grade folpet was given either i.p. in doses of 2.5, 5.0, or 10.0 mg/kg per day for 5 days or by oral intubation in doses of 50, 100, or 200 mg/kg per day; also for 5 days to groups of 15 male rats of the Osborne-Mendel strain, which were subsequently mated with untreated females for the following 10 weeks. Neither degrees of fertility, nor mean total implants, were affected. Mean early deaths per pregnancy were consistently higher than those of the controls after either i.p. or oral administration of folpet. The percentages of litters with two or more early deaths showed statistically significant increases after treatment, and the increases appeared to be dose-related. The following six stages of spermatogenesis can be sampled in male rats after treatment: sperm in the vas deferens, 3-4 days; epidymal sperm, 3-4 days to 2 weeks; late spermatids, 3 weeks; mid-to-early spermatids, 4-5 weeks; spermatocytes, 6-8 weeks; spermatogonia, from 9 weeks [171]. Whereas i.p. treatment with folpet had negligible effects on fertility, treatment by oral intubation affected sperm in the vas deferens and epididymis.

Antifertility effects and mutagenic effects, as reflected in mean early deaths per pregnancy and number of litters with multiple early deaths, appeared greater after oral intubation than after i.p. dosage, suggesting (as in the case of captan) possible metabolic activation during the digestive process. The no-effect level for mutagenicity of folpet, when given by oral intubation, was 50 mg/kg per day or about 500 ppm.

Scant information exists as to the carcinogenic potential of the phthalimide fungicides. Captan and folpet were found to be nontumorigenic when tested in several strains of mouse by Innes et al. [29].

B. Captafol (Difolatan)

Captafol, N-(1,1,2,2-tetrachloroethylthio)-3a,4,7,7a-tetrahydrophthalimide, N-(1,1,2,2-tetrachloroethylthio)-cyclohex-4-en-1,2-dicarboximide, is used both for commercial application to raw agricultural products as well as on vegetable gardens, flower beds, and lawns.

The acute oral LD_{50} for rats is 5,000-6,200 mg/kg [131]. When administered as an aqueous suspension of the 80% wettable powder, the acute oral LD_{50} for rats is 2,500 mg/kg. The acute dermal LD_{50} for rabbits is more than 15,400 mg/kg.

In 2-year feeding tests with captafol, no apparent effects resulted at the 500-ppm level on rats or at the 10-mg/kg per day level on dogs [131]. Captafol, as well as captafol epoxide, have been shown to possess a high and specific teratogenic activity in the chicken embryo [157]. Captafol was found to be nonteratogenic in two strains of rabbit (New Zealand white and the Dutch belted) that were particularly sensitive to the embryotoxic effects of thalidomide [154].

Offspring of Sprague-Dawley derived albino rats treated with captafol (incorporated in the basal breeder ration at dietary levels up to 0.10%), either during gestation only, or over three consecutive generations, were free of abnormalities [154].

Captafol (99% pure) was found to be nonteratogenic when tested in the rhesus monkey and stump-tailed macaque at doses of 6.25, 12.5, and 25.0 mg/kg body weight (administered po as suspensions in a solution of cream of coconut) from day 22 to day 32 of gestation [155].

Captafol was classified as a weak base-change mutagen [160, 172] when examined by rec-assay procedures utilizing *E. coli* WP2 strain and Ames' *Salmonella* system [173]. Captafol was found to be mutagenic in the *Salmonella*, *his* G46 and TA1530 strains, and nonmutagenic when tested in the TA1531, TA1532, and TA1534 strains of *Salmonella* [37].

Captafol (100% pure, Chevron Chemical Co.) was tested in the dominant lethal assay, like captan [169] and folpet [170], e.g., given either i.p. in doses of 2.5, 5.0, or 10.0 mg/kg per day for 5 days, or by oral intubation in doses of 50, 100, or 200 mg/kg per day, or for 5 days to groups of 15 male rats of Osborne-Mendel strain. The animals were mated and neither degrees of fertility nor mean total implants were affected. The number of mean early deaths per pregnancy reflected effects on epididymal sperm and late spermatids in a dose-related manner, when captafol was administered either i.p. or orally. In addition, mid-spermatids appeared to be affected by intubation of captafol.

The maximum human dietary level of captafol was estimated as 22 ppm and the no-effect level for mutagenicity was about 500 ppm (analogous to captan and folpet) [169, 170]. It was concluded that, as in the case of folpet and captan, there is no mutagenic danger to man from captafol at the current (1970) levels of intake in the diet [170].

VIII. HEXACHLOROBENZENE

Hexachlorobenzene (HCB) is a selective fungicide used primarily to control bunt of wheat. It is encountered as a waste product from the manufacture of chlorinated hydrocarbons [174, 175], e.g., perchloroethylene via the chlorination of hydrocarbons. Hexachlorobenzene has a low acute toxicity to mammals. For example, the acute oral LD_{50} for male rats is 10,000 mg/kg, and for guinea pigs greater than 3 g/kg [176, 177]; 500 mg/kg i.p. is nonlethal in rats [178] and the oral lethal dose of a 15% suspension of HCB in female quail is greater than 1 g/kg [178]. A minimum lethal dose of about 2 g/kg for both rats and mice has been reported by Savitskii [179].

Although HCB has a low single-dose toxicity, repeated daily exposures to small doses appear to be more hazardous than the single-dose oral toxicity would suggest. Feeding of a diet containing 0.2% HCB resulted within a month in the death of 13 of 33 rats [180]. Rats fed a diet of 0.2% HCB showed marked enlargements of the hepatocytes after the third week [181].

Gurfein and Pavlova [182] reported a possible disruption of conditioned reflexes in rats that were maintained on drinking water containing 0.1 mg/liter HCB, providing a daily dose of 0.025 mg/kg for 4-8 months.

In a preliminary subacute toxicity study in Japanese quail (*Coturnix japonica*), it was found that feeding of 20 ppm HCB (lowest level) was still toxic [183].

Vos et al. [178] reported the effects of dietary concentrations of 0, 1, 5, 20, and 80 ppm HCB (99.5% purity) given Japanese quail for 90 days. Tremors and mortality occurred in birds fed 80 ppm. Red fluorescence of tissues, liver damage (enlargement of nuclei and nucleoli, proliferation of bile ductules, necrosis of hepatocytes), erythrophagocytosis in the spleen, ceroid granules in the tubules of the kidney, reduced reproduction, and reduced volume of eggs were also found in birds fed at this level. Increased liver weight, slight liver damage, and enlarged fecal excretion of coporporphyria occurred in birds fed at the 5 ppm level and the no-effect level was established at 1 ppm. Significantly higher brain residues of HCB were found in male, compared to female birds.

Schwetz et al. [184] described reproduction studies in Japanese quail fed 20 ppm HCB for 90 days. No effect on the general health of the birds was noted, nor was egg production or eggshell thickness altered, as found previously by Vos et al. [178]. However, the 7-day survival of hatched chicks was found to be significantly decreased, and an increase in the liver weight of adult birds at the end of the study was noted, confirming the observations by Vos et al. [178, 183]. Differences between the studies of Schwetz et al. [184] and Vos et al. [178, 183] in certain parameters, could well have resulted from the differences in sample purity, the small number of animals used in each study, and the expected amounts of biological variation which limits the reproducibility of any animal experiment.

The toxicity of technical hexachlorobenzene in the Sherman strain rat has been elaborated by Kimbrough and Linder following administration of the fungicide at the dietary levels of 0, 100, 500, and 1,000 ppm for 4 months [185]. Two of 10 males and 14 of 20 females died at the dietary level of 500 ppm and three of the 10 males, as well as 19 of 20 females, died at the dietary level of 1,000 ppm for 4 months [185]. Various morphological changes were observed in the livers of the dosed animals. In addition, hyperplasia of the adrenal cortex, as well as large numbers of macrophages in the pulmonary alveoli and lung fibrosis were noted. UV fluorescence of the GI tract and the liver indicated that HCB also induced porphyria. These preliminary findings indicate that the primary organs affected in the rat by HCB are the adrenals, liver, heart, and lungs.

It should be noted that the technical-grade HCB used in the above study of Kimbrough and Linder [185] was about 93% purity and contained trace amounts of a number of toxicants including carbon tetrachloride, perchloroethylene, hexachloroethane, hexachlorobutadiene, pentachlorobenzene, octachlorodibenzofuran, decachlorobiphenyl, octachlorobiphenylene, and a few others [186]. It would appear obvious that additional studies are necessary to elicit whether the reported pathology of HCB is due to the effect of HCB per se or its trace impurities.

Grant et al. [187] studied the effects on liver porphyria levels and microsomal enzymes in male and female Sprague-Dawley rats fed diets containing 0, 10, 20, 40, 80, or 160 ppm of HCB (analytical standard grade, BDH Chemicals) for 9 and 10 months. Hexachlorobenzene residues in the liver were dose related and levels were similar in males and females. The liver/body weight ratios were increased in both sexes fed the 80- and 160-ppm diets. Female rats acquired chemical porphyria readily, whereas male rats were resistant. No neurological effects (tremors) were observed in any of the test animals.

The effects of subacute feeding of diets containing HCB in the Charles River (COBS) strain rat were described by Kuiper-Goodman et al. [188]. Seven hundred male and female rats were fed a diet consisting of Master Fox cubes with 4% corn oil to which was added HCB to give 0, 0.5, 2.0, 8.0, and 32.0 mg/kg body weight. Groups of four rats of both sexes were sacrificed at 3, 6, 9, and 12 weeks of feeding with HCB. Females were more sensitive than males to the lethal effects of HCB, e.g., in the highest dose group, 26% of the females died and no males died. Females developed a more severe porphyria with very high porphyrin concentrations in the liver. Males showed a significant increase in liver weight, which was found to be mainly due to an increase in smooth endoplasmic reticulum correlated to increased amounts of drug-metabolizing enzymes at the two highest dose levels.

Aspects of the placental transfer of HCB in the rabbit have been noted by Villeneuve et al. [189]. White New Zealand rabbits were mated (day 0) and then orally administered HCB (99.5% pure, BDH Chemicals) in corn oil from day 1 to day 27 with subtoxic doses of 0, 0.1, 1.0, or 10 mg/kg.

Hexachlorobenzene was found to cross the placenta and accumulate in the fetus in a dose-dependent manner. In the dams, the tissue with the highest HCB concentration was fat, followed in decreasing order of concentration, by liver, heart, kidney, brain, lung, spleen, and plasma. In the fetus, liver concentrations were higher than the corresponding maternal organ and comprised from 70-80% of the fetal body burden. No fetotoxic effects, e.g., increase in fetal deaths or deformities, resorptions or abortions, were noted at the dose levels of HCB employed when compared to control animals.

Khera [190] described teratogenicity and mutagenicity studies with HCB in rats. Rats were given single daily po doses of 0, 10, 20, 40, 60, 80, or 120 mg HCB/kg during days 6-9, 10-13, 6-16, or 6-21 of gestation; 80 and 120 mg/kg doses resulted in maternal neurotoxicity and reduction in fetal weight. When HCB was administered from days 10-13, 6-16, and 6-21 of gestation, the incidence of uni- and bilateral 14th rib among fetuses was significantly increased above control values. (The incidence was related to the duration of treatment and the dose.) However, parameters such as numbers of live and dead fetuses, resorption sites, fetal weight, and visceral and other skeletal anomalies were within control limits.

For the dominant lethal tests, male rats distributed in four groups each of 15 males were dosed orally with 0, 20, 40, or 60 mg/kg HCB for 10 consecutive days, followed by 14 sequential mating trials. No significant differences in the incidence of pregnancies, corpora lutea, live implants, and decidaomas in the test and control groups were observed [190].

It is extremely important to note the effects of HCB in man. Hexachlorobenzene has been implicated in more than 3,000 cases of human poisoning in Turkey in the years 1955-1959, as a result of the consumption of wheat treated with 2 g of HCB per 100 kg of seed [191]. The main symptoms of this poisoning include porphyria cutanea tarda with hypersensitivity of the skin to sunlight, hyperpigmentation and hypertrichosis, hepatomegaly, weight loss, osteoporosis, and enlargement of the thyroid and lymph nodes [191-193].

It should be noted (aside from the tragic poisoning episodes in Turkey) that residues of HCB have also been found in human adipose tissue [194-195], and blood [196-198] in samples from areas in the United States, Australia, and Japan.

A recent study by Burns and Miller [198] disclosed that plasma HCB residues in a Louisiana population, exposed through transportation and disposal of chemical waste containing HCB, averaged 3.6 ± 4.3 parts per billion (ppb) in a sample of 86 people. The highest level in the general population was 23 ppb and in a waste disposal facility workers, 345 ppb. Perchloroethylene and carbon tetrachloride production workers had HCB levels of up to 233 ppb. No evidence of cutaneous porphyria by history or skin examination in any of the exposed subjects was found. The detection of HCB in the blood of 52 nonoccupationally exposed persons in Australia [197] indicates that the level of contamination in the Louisiana study may not be exceptional [198].

Hexachlorobenzene residues have also been reported in fish [199-201], cattle [199], and wildlife [178], poultry products such as eggs [202], milk [203], and other dairy products [204].

Hexachlorobenzene arising as a trace contaminant in other fungicides, e.g., pentachloronitrobenzene (PCNB) [185, 203] should be noted.

Aspects of trace contaminants in commercial HCB bears particular emphasis and restressing. For example, Villaneuva et al. [186] found the following impurities in trace amounts in samples of commercial HCB hepta- and octadichlorodibenzofurans, octachlorodibenzo-p-dioxin, octa-, nona- and decachlorobiphenyls, 1-pentachlorophenyl-2,2-dichloroethylene, hexachlorocyclopentadiene, pentachloroiodobenzene, and heptachlorotropilium.

The toxicities of some of the chlorinated dibenzo-p-dioxins and chlorodibenzofurans have been reported [205-209].

IX. CHLOROPHENOLS

Chlorinated phenols are used widely, not only as fungicides, but also as herbicides, insecticides, bactericides, molluscacides, slimicides, and wood preservatives, as well as in the production of phenoxy acids such as 2,4-D and 2,4,5-T.

A. Pentachlorophenol

Pentachlorophenol (PCP) and sodium pentachlorophenate (PCP-Na) are the most widely employed of the chlorophenols with a broad spectrum of applications in agriculture and industry. Their utility as fungicides and bactericides includes the processing of cellulosic products, adhesives, starches, proteins, leather, oils, paints, rubber, textiles, and use in food processing plants to control mold and slime and in construction and lumber industries to control mold, termite infestation, and wood-boring insects. Aspects of the properties of PCP (and its salts) in terms of its occurrence as a residue in human and animal tissues and toxicology have been reviewed (up to 1967) by Bevenue and Beckman [210].

Acute toxicity studies have been performed in several species [211-216]. The LD_{50} values of PCP per kg body weight, following oral administration are: 120-140 mg for the mouse; 27-100 mg for the rat; about 100 mg for the guinea pig; 100-130 mg for the rabbit; 150-200 mg for the dog; about 120 mg for the sheep; and about 140 mg for the calf. The acute symptoms of intoxication are vomiting, hyperpyrexia, elevated blood pressure, increased respiration rate and amplitude, and tachycardia and hyperglycaemia; later frequent defecation, weakened eye reflex, and developing motor reflexes are noted [217]. Deaths have occurred in mammals not only after oral or parenteral administration, but also after cutaneous absorption of PCP [218, 219]. Mean lethal doses ranging from about 40 to above 200 mg/kg, varying with sex, species, mode of administration, compound purity, and solvent have been reported [210-220].

The effects of acute and chronic intoxications of PCP-Na in the rabbit have been noted by McGavack et al. [215]. The minimum lethal doses found after cutaneous, subcutaneous, intraperitoneal, and oral administration were 512.5, 275.0, 135.5, and 550.0, average mg/kg, respectively. Cerebral changes seen in chronic intoxications in rabbits with PCP-Na were quite similar to those observed in long-standing cases of hypoglycemia and could readily result from the associated tissue anoxemia. A chemical peritonitis was produced in all animals receiving PCP-Na, i.p., while repeated injections by both the intramuscular and subcutaneous routes were associated with several necrotic and ulcerative lesions, even when only 1/20 MLD (mean lethal dose) was used.

A number of subacute toxicity studies with PCP have been reported [211-214, 217]. Deichmann et al. [212] reported no tissue damage in rats fed 3.9 mg PCP daily for 28 weeks and a reduction of food intake at a dosage of 558 ppm in the food for 26 weeks.

Knudsen et al. [217] recently described a short-term toxicity study in Wistar strain rats designed to establish a no-toxic effect level. Rats were fed PCP, m.p. 159-176° C, (Dynamite-Nobel, Germany) containing 200 ppm octachlorodibenzo-p-dioxin and 82 ppm pre-octachlorodibenzo-p-dioxin, but no tetrachlorodioxins), in a semisynthetic diet at levels of 0, 25, 50, and 200 ppm for 12 weeks. Growth was decreased in the group of female rats fed 200 ppm. The treatment had no effect on food intake and behavior. Liver weight (in the females) was increased at the 50- and 200-ppm dose levels, accompanied by an increased activity of microsomal liver enzymes. Higher hemoglobin and hematocrit values were found in males (in week 6) fed 50 and 200 ppm. However, by week 11, the hemoglobin values and number of erythrocytes decreased which suggested a decreased ATP content in the cells. Centrilobular vacuolization in the livers was found in animals fed 50 and 200 ppm PCP. This phenomenon is also found in cases of human fatal intoxications caused by PCP [219, 221, 222]. A striking dose-related decrease of calcium deposits at the corticomedullary junction of the kidneys was also found and it was suggested that the Ca^{2+} blood levels could be lowered through an indirect action of PCP on calcium metabolism.

Schwetz et al. [223] further studied the effect of purified and commercial grade PCP on rat embryonal and fetal development. The commercial grade contained 88.4% PCP, 4.4% tetrachlorophenol, less than 0.1% trichlorophenol, and 6.2% higher chlorinated phenoxyphenols; the nonphenolics (ppm) consisted of dibenzo-p-dioxins: 2,3,7,8-tetrachloro (<0.05), hexachloro (4), heptachloro (125), and octachloro (2,500); dibenzofurans: hexachloro (30), heptachloro (80), and octachloro (80). The purified PCP contained 98% PCP, 0.27% tetrachlorophenol, 0.05% trichlorophenol, and 0.5% higher chlorinated phenoxyphenols; the nonphenolics (ppm) consisted of dibenzo-p-dioxins: 2,3,7,8-tetrachloro (<0.05), hexachloro (<0.5), heptachloro (<0.5), octachloro (<1.0); dibenzofurans: hexachloro (<0.5); heptachloro (<0.5), and octachloro (<0.5).

Dose levels up to and including a maximum tolerated dose were administered po (in 2.0 mg/kg corn oil) to pregnant Sprague-Dawley rats on days 6 through 15, 8 through 11, and 12 through 15 of gestation. In the teratological study, doses of 0, 5, 15, 30, and 50 mg/kg per day of commercial grade or purified PCP were administered by gavage to groups of 20-40 bred rats on days 6-15 inclusive of gestation. Additional groups of rats were given 0 or 30 mg PCP/kg per day on days 8-11 or 12-15 of gestation. Following the administration of PCP, signs of embryotoxicity and fetoxicity, such as resorptions, subcutaneous edema, dilated ureters, and anomalies of the skull, ribs, vertebrae, and sternebrae were observed at an incidence which increased with increasing the dose. Purified PCP, with its low nonphenolic content was slightly more toxic than the commercial-grade PCP containing a much higher level of nonphenolics. The developing embryo is most susceptible to the toxic effects of a given dose of PCP during the period of early organogenesis. The estimated maternal dose that would be lethal to one-half of the embryos in utero was 16 mg of purified PCP/kg per day compared to 44 mg of the commercial grade/kg per day (estimated from a plot of probability of percent resorptions versus maternal log dose of test material). The no-effect dose level of the commercial-grade PCP was 5 mg/kg per day when administered on days 6 through 15 of gestation. At the same dose level, purified PCP caused a statistically significant increase in the incidence of delayed ossification of skull bones, but had no other effect on embryonal and fetal development.

Schwetz et al. [223] suggested that the amounts of the nonphenolics contained in commercial-grade PCP do not significantly contribute to the effects of this material on the developing rat embryo and fetus. The two grades of PCP differed only in the sense that they caused slightly different degrees of the same forms of toxicity.

The results reported in the above study of Schwetz et al. [223] are similar to those of Hinkle [224] who administered 1.25 to 20 mg of PCP/kg per day to pregnant Golden Syrian hamsters from days 5 to 10 of gestation. Courtney [225] observed that the administration of 75 mg/kg of PCP to rats had no significant effect on survival or embryonal or fetal development.

In regard to carcinogenicity, PCP was found not to cause a significant increase in the number of tumors (at 0.01 level of significance) in a chronic feeding study involving one strain of mouse [29].

Information as to the mutagenic effects of the chlorophenols is scant and ambiguous. No evidence for a genetic activity of PCP has been found in *Drosophila melanogaster* [166]. It was effective in inducing mitotic gene conversion in *S. cerevisiae* [81], but displayed only weak activity in inducing back-mutation in bacteria incubated in the peritoneal cavity of a mouse [81, 226].

The common use of PCP (and its salts) has created opportunities for human intoxication and some of the episodes have led to fatalities [210, 219, 227-233]. The general clinical symptoms include loss of appetite and body weight, high temperature, excessive respiration, headache, congestion of ocular and nasal mucosal, irritation of respiratory organs, thorax constriction, weakness, abdominal pain, nausea, and vomiting. Cases of neuralgia following exposure to PCP have been described by Campbell [234] and Barnes [235]. After dermal exposure to PCP (and its salts) local skin irritation and acne are found. One case of fatal aplastic anemia, following continued dermal exposure for about 1 year, has been reported [236].

At least 24 fatalities in industry have been attributed to PCP exposure [217]. At autopsy, the following changes were noted: edematous brain, heart dilatation, tubular degeneration of the kidneys, edematous lungs with congestion and intraalveolar hemorrhages, and a slightly congested liver with centrilobular degeneration [219, 221, 222, 228].

The misuse of PCP-Na in a laundry product in a hospital nursery in St. Louis, Mo., has resulted in 20 of about 80 newborn infants developing signs of intoxication leading to seven deaths [237, 238] due to percutaneous absorption of PCP which had been used in the laundering of the diapers and the infants bed linen. The clinical features of the affected infants included profuse, generalized diaphoresis, fever, tachycardia, tachypnea, hepatomegaly, and metabolic acidosis. Autopsy revealed fatty infiltration of the liver and fatty degeneration of renal tubular cells. The levels of PCP (mg/100 gm) in autopsy tissue from one exposed infant were: kidney, 2.8; adrenal, 2.7; heart and blood vessel, 2.1; fat, 3.4; and connective tissue, 2.7.

B. Tetrachlorophenol

The acute oral LD_{50} of 2,3,4,6-tetrachlorophenol in the rat is 140 mg/kg. The effects of purified and commercial-grade 2,3,4,6-tetrachlorophenol on rat embryonal and fetal development were studied by Schwetz et al. [239]. The commercial grade contained 73% tetrachlorophenol and 27% pentachlorophenol; the nonphenolics (ppm) consisted of dibenzo-*p*-dioxins: 2,3,7,8-tetrachloro (<0.05), hexachloro (28), octachloro (30; dibenzofurans: hexachloro (55), heptachloro (100), and octachloro (25). The purified 2,3,4,6-tetrachlorophenol contained 99.6% tetrachlorophenol, 0.1% pentachlorophenol, and the nonphenolics (ppm) consisted of dibenzo-*p*-dioxins: 2,3,7,8-tetrachloro (<0.05) and less than 0.5 ppm each of hexachloro-, heptachloro- and octachlorodibenzo-*p*-dioxins; dibenzofurans: less than 0.5 ppm each of hexachloro-, heptachloro-, and octachlorodibenzofurans.

Dose levels up to and including a maximum tolerated dose were administered po (in 2.0 ml/kg corn oil) to pregnant Sprague-Dawley rats on days 6 through 15 of gestation. The maximum tolerated dose of commercial tetrachlorophenol following 10 days of administration in a preliminary tolerance study was 30 mg/kg per day. Signs of toxicity and death were observed at 100 and 300 mg/kg per day. In the teratological study, maternal weight gain during gestation was not affected by the administration of 10 or 30 mg of commercial-grade or purified tetrachlorophenol per kg per day on days 6-15 of gestation. No signs of maternal toxicity were observed among dams treated with any dose of either sample during gestation or at the time of cesarean section and autopsy on day 21 of gestation.

The only fetal skeletal anomaly associated with the administration of tetrachlorophenol was delayed ossification of the skull bones. At 30 mg commercial-grade tetrachlorophenol per kg per day, the incidence was significantly increased among litters (35%, 7/20). The incidence of delayed ossification of the skull bones among the control animals was 8% (14/173) of the fetal population and 19% (6/31) of the litters. The no-effect dose level of both commercial and purified tetrachlorophenol for embryotoxicity was 10 mg/kg per day. The only soft tissue anomaly which occurred at an increased incidence among the fetal population and litters was subcutaneous edema, which was not dose related but associated with the administration of 10 mg of either commercial grade or purified tetrachlorophenol/kg per day.

The results of the studies of Schwetz et al. [239] indicate that commercial-grade and purified tetrachlorophenol induce embryotoxic and fetotoxic but not teratogenic responses at the maximum tolerated dose level. There was essentially no difference in the degree of toxicity associated with the two grades of tetrachlorophenol and the amounts of nonphenolics, e.g., dibenzo-p-dioxins and dibenzofurans contained in commercial-grade tetrachlorophenol do not contribute significantly to the effects of this material on the developing rat embryo and fetus. Relative to the maximum tolerated dose for the maternal animal, pentachlorophenol had a greater effect on embryonal and fetal development [213] than did tetrachlorophenol.

C. Trace Impurities in Chlorophenols

Technical formulations of 2,3,4,6-tetrachlorophenol and pentachlorophenol can contain a spectrum of impurities, e.g., lower chlorophenols, higher chlorinated diphenyl ethers (1) and non-phenols, e.g., chlorinated dibenzo-p-dioxins (2) and dibenzofurans (3) as previously shown in the work of Schwetz et al. [223, 239].

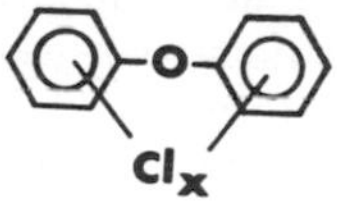

Cl_x (1) Cl_x (2) Cl_x (3)

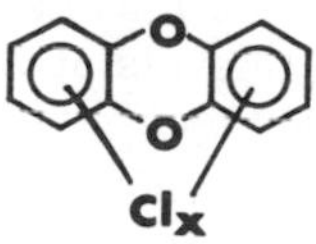

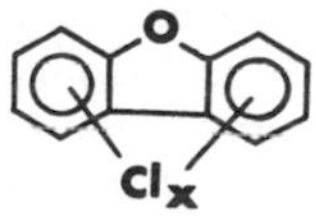

Pentachlorophenols are usually manufactured by alkaline treatment of hexachlorobenzene [240, 241] or by direct liquid-phase chlorination of phenol [240]. The vigorous thermal conditions employed may favor many side reactions, e.g., condensation products such as structures (1-3). Technical formulations of 2,4,6-tri-, 2,3,4,6-tetra-, and pentachlorophenol have also been found to contain 2-phenoxyphenols with four to nine chlorine atoms (4) [241, 244-247], also called predioxins [246, 247], which are converted into dioxins when heated [241, 246-248]. Isomers of predioxins, chlorinated 4- or 3-hydroxydiphenyl ethers (isopredioxins) (5) have also been identified [246]. These do not undergo ring closure to form dioxins [241, 246]. Other isopredioxins have been found, e.g., 2-hydroxydiphenyl ethers (with no chlorine in the 2' position) which do not undergo ring closure to form dioxins [241, 246-247].

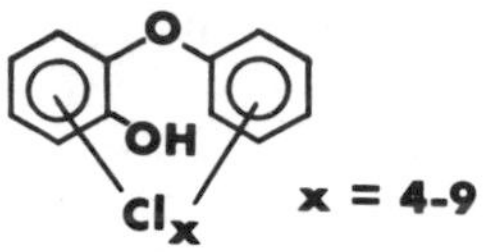

$x = 4-9$

(4) 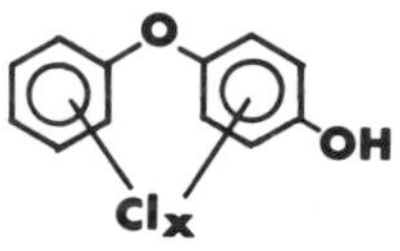 (5)

Many dibenzofurans with four to eight chlorine atoms and ethers with four to ten chlorine atoms have been found in tetra- and pentachlorophenol samples [242]. There was evidence for the presence of tetrachlorodibenzofuran and pentachloroanisole in some PCP samples. Levels found by Firestone et al. [242] were 0.17-39 ppm of hexachlorodioxin in all eight PCP samples examined, while 2,3,7,8-tetrachlorodibenzo-p-dioxin (TCDD) was not found in any of the 11 tetra- and pentachlorophenol samples.

Of 11 samples of PCP examined by Woolson et al. [243], 10 contained chlorodioxins. Although no TCDD was found in any samples examined, seven samples had hexachlorodibenzo-p-dioxin contents in the range of 10-100 ppm. Heptachlorodibenzo-p-dioxin was found in four samples at levels of 10-100 ppm and in six samples at levels between 100 and 1,000 ppm. Octachlorodibenzo-p-dioxin levels in four samples were between 10 and 100 ppm and in six samples, the levels ranged from 100 and 1,000 ppm. Three tetrachlorophenol samples examined contained hexa-, hepta-, and octachlorodibenzo-p-dioxins at levels below 100 ppm [243].

Swedish samples of PCP have been found [246, 247] to contain 870 ppm of heptachlorodibenzo-p-dioxin and 1,300 and 1,200 ppm, respectively, of its predioxin and isopredioxin. Levels of octachlorodibenzo-p-dioxin in six samples ranged from 50 to 3,300 ppm, while that of its predioxin and isopredioxin ranged from 0.6 to 1,600 ppm and from a trace to 1,600 ppm, respectively [246]. The total dimeric content of some PCP samples was on the order of 10 g/kg [247].

Johnson et al. [249] have assessed a number of commercial PCP compositions, which are used as antimicrobial agents for the preservation of wood. The composition was as follows: pentachlorophenol, 85-90%; tetrachlorophenol, 4-8%; trichlorophenol, <0.1%; higher chlorophenols, 2-6%; and caustic insolubles (maximum), 1%. The caustic insolubles sometimes referred to as the "nonphenolic or neutral impurities," include chlorinated dibenzo-p-dioxins and chlorinated dibenzofurans [240, 250]. The concentration range of some chlorinated dioxins found by Johnson et al. [249] in commercial PCP were as follows: 2,3,7,8-tetrachlorodibenzo-p-dioxin, none; hexachlorodibenzo-p-dioxins, 9-27 ppm; and octachlorodibenzo-p-dioxin, 575-2,510 ppm. (The limits of detection of 2,3,7,8-tetrachloro-p-dioxin was 0.05 ppm.)

It is also important to note some of the reactions of pentachlorophenol that can occur under environmental conditions. For example, it has been shown that transformation of PCP by light results in the formation of dioxins [250, 251] or their precursors [240, 250, 251].

Chlorinated dibenzo-p-dioxins may also arise from the combustion products of wood [247, 252] and paper treated with PCP [252].

Studies dealing with the occurrence of PCP in sewage and water [253, 254], its accumulation in fish tissues following industrial discharge [256, 257], and the environmental significance of chlorodioxins [255, 258, 259], including aspects of their photodecomposition [260] and persistence and metabolism in soils [258, 261], have all been reported.

Severe toxicological responses have been associated with certain chlorodibenzo-p-dioxins as well as chlorodibenzofurans. TCDD has been associated with occupational chloroacne in workers engaged in the manufacture of technical chlorophenols and their derivatives [262-265]. The chloroacnegenic potency of TCDD in experimental animals was also demonstrated [266, 267]. TCDD is the most toxic of the chlorodibenzo-p-dioxins having an LD_{50} in the μg/kg range [270-274].

Tetrachlorodibenzo-p-dioxin is prophyrogenic in the mouse [271], but not in rats and guinea pigs [273, 275]. The hepatic porphyria evoked by TCDD, probably associated with liver damage is a uroporphyria [271, 276].

In the mouse, the most severe lesions seen are in the thymus and liver with a no-effect level for their weights being 0.2 μg/kg dose [271].

Severe TCDD-induced liver damage has been reported in rats [273], rabbits [262], and chickens [274], but not in guinea pigs [273]. Progressive proliferation of bile ducts and bile-duct epithelial cells combined with bizane pleomorphism of hepatocytes have been noted in rat liver following a single administration of TCDD [269, 273]. In the mouse, degenerative and necrotic changes in the liver were essentially centrilobular and were accompanied by cellular infiltrates and ceroid pigment deposition [271].

A depression in growth rate or a loss of weight after single doses of 1 μg to 10 mg TCDD given i.p., or orally has been frequently noted [269, 271, 277-279].

Although TCDD is an extremely toxic agent, it has a very prolonged action and the time of death in rats can be over 40 days after a single oral dose [269]. The delayed deaths in rats (and guinea pigs) cannot be explained by any consistent pathological change in the dead or dying animals [269]. Although all rats given a single oral dose of 200 μg/kg, for example, had liver damage; the extent of this was such that it was difficult to attribute the death of the animals to a gross hepatic failure [266]. The histological appearance of the liver was unusual in that there was a progressive failure with the formation of multinucleate cells and some fibrosis. It was suggested that TCDD interferes with the capacity of the liver to maintain their correct organization; in some cells this leads to death and in others to disorganization of structure [269]. TCDD has also been shown to be fetotoxic (embryotoxic) and teratogenic in rats and mice [274, 280-286]. Subcutaneous administration of TCDD at a dose level of 3 μg/kg in mice on days 6 through 15 of gestation produced pups with cleft palates and kidney anomalies [282]. The C57B1/6 mouse was the most sensitive of three strains tested to the TCDD-induced kidney effects in that almost 100% of the fetuses developed kidney anomalies [282]. Hydronephrotic kidneys have been found in mouse pups that nursed a mother treated with TCDD during pregnancy or at time of parturition [284]. Variations in kidney development, consistent with hydronephrosis were observed in fetuses examined at gestation day 18. It was hypothesized that the prenatal and postnatal kidney anomaly are of the common etiology and that the incidence and degree of hydronephrosis is a function of dose and length of target organ exposure [284].

Oral treatment of pregnant Wistar rats with 0.25 μg/kg per day (or more) of TCDD for 10 days during gestation resulted in adverse effects on rat development, while no adverse effects were noted at the 0.125 μg/kg per day level [285].

The no-effect level for embryotoxicity in Sprague-Dawley rats has been reported to be 0.03 μg/kg per day [274].

Information as to the mutagenicity of the chlorodibenzo-p-dioxins is sparse. TCDD has been reported to be mutagenic when tested in E. coli [37] and S. typhinurium [37, 287], with its mutagenic effect being attributed to intercalation with DNA [287]. In the dominant lethal test in male Wistar rats, TCDD induced hyperplastic changes and sperm granulomas in the epidymis, but no apparent lethal mutations were noted during postmeiotic phases of spermatogenesis [285]. Inhibition of mitosis has been observed [288] in cultures of endosperm cells of the African blood lily exposed to TCDD at concentrations of 0.2 μg/liter.

Although it is recognized that TCDD is by far the most toxic of the chlorinated dibenzo-p-dioxins, it is also important to briefly consider aspects of two other chlorinated dioxins frequently found in PCP samples, e.g., the hexa- and octachloro derivatives. Hexachlorodibenzo-p-dioxin (HCDD) is known to be positive for the chick edema factor, a condition characterized by hydropericardium, ascites, and anascara [289, 290].

The chlorodibenzo-p-dioxins differ in their toxicological properties [268, 274], e.g., HCDD is highly toxic but much less so than TCDD. Limited data suggest that oral doses of about 100 mg/kg of HCDD are needed to cause death in male Sprague-Dawley rats [268, 274]. In a teratology study, no deaths occurred following administration of 100 μg/kg of HCDD to female Sprague-Dawley rats for 10 consecutive days. At this level, HCDD is teratogenic when given orally on days 6 through 15. Embryotoxicity was evidenced by a dose-related decrease in fetal body weight and crown-rump length, and an increase in the incidence of fetal resorptions. Similarly, the incidence of certain soft tissue and skeletal anomalies increased in a manner related to the dose level of HCDD. A 0.1 μg/kg per day dosage of HCDD had no effect on embryonal or fetal development [268, 274]. Octachlorodibenzo-p-dioxin (OCDD) and 2,7-dichlorodibenzo-p-dioxin (2.7-DCDD) have low acute toxicities [268, 274]. OCDD caused embryotoxicity but was not teratogenic in Wistar strain rats at the 500 mg/kg per day level [274]. OCDD and 2,7-DCDD caused neither teratogenicity nor embryotoxicity at 100 mg/kg per day [274].

Both 2,3,7,8-TCDD and HCDD were acnegenic and gave positive results in chick edema [289]. OCDD and 2.7-DCDD have been found to be nonchloroacnegenic [268, 274], and OCDD was negative for the chick edema factor [268, 274].

Pathological changes observed in animals treated with chlorodibenzo-p-dioxins were inconsistent from animal to animal and species to species. Although, hepatic lesions were observed consistently, their nature, degree, and distribution were variable, and changes in organs other than the liver were sporadic and unpredictable [274].

Chlorinated dibenzofurans, such as the tetra- and pentachloro isomers have been shown to cause hydropericardium in chicks and hyperkeratosis of the rabbit's ear [291-294].

X. CHLORONITROBENZENES

Pentachloronitrobenzene, Quintozene, Terrachlor, PCNB, is used as a seed and soil fungicide, and is effective against bunt of wheat, and species of Botrytis, Rhizoctonia, and Sclerotinia. The impurities of technical-grade PCNB can include 2,3,5,6-tetrachloronitrobenzene, tecnazene, TCNB [291-293]; 2,3,4,5-tetrachloronitrobenzene [294], pentachlorobenzene [291-293], and hexachlorobenzene [291-293]. Pentachlorobenzene, PCA, is both a soil degradation product formed by microorganisms in moist soils [295] and a biological metabolite of PCNB [291]. Pentachlorothioanisole, methylpentachlorophenyl sulfide, PCTA, has been identified in PCNB treated soils [293] and is a metabolite of PCNB in dogs, rats, and plants [294]. Other related compounds isolated from PCNB treated soil are tetrachloroaniline, TCA, and tetrachlorothioanisole, TCTA [293]. The acute oral LD_{50} of PCNB for rats is greater than 12,000 mg/kg [296].

Pentachloronitrobenzene produced hepatomas in (C57B1/6 XC_3H/Amf) F_1 mice when fed initially at 464 mg/kg in 0.5% gelatin intragastrically (daily from days 7-28), followed by its incorporation in the diet at 1,206 ppm and fed ad-lib for 24 days [29].

Jordan and Borzelleca [297] reported that PCNB given in oral dosages up to 1,563 ppm daily to albino rats of the Charles River strain on days 6 through 15 of gestation was nonteratogenic. The most common defects, e.g., dilated renal pelvis, hydronephrosis and hydroureter, were found in control and treated groups and appeared unrelated to treatment.

When PCNB was administered orally to C57B1/6 mice at levels of 500 mg/kg, from day 7 to 11 of gestation, kidney formation was inhibited in 80% of the litters [298]. Renal agenesis occurred unilaterally about twice as often as bilaterally.

Pentachloronitrobenzene was nonmutagenic when tested in _Drosophila_ [166], _his_ G46, TA1531, and TA1532 _S. typhinurium_ strains [37], in _E. coli_ and _S. cerevisiae_ [81], and host-mediated assay back mutation in _S. typhinurium_ and _S. carscescens_ [226], and possessed equivocal mutagenicity in TA1530 and TA1534 _Salmonella_ strains [237].

2,3,5,6-Tetrachloronitrobenzene, in addition to being an impurity in technical PCNB, is a selective fungicide effective for the control of the dry rot fungus of potato tubers _Fusarium caerulum_, and is also applied in smokes as a fungicide for lettuce. Its mammalian toxicity is considered to be low. Rats fed 6.8 mg TCNB daily (or a dose of about 57 mg/kg per day for 10 weeks) exhibited no reduction of growth or toxic symptoms. A dose level of 34 mg per day or 400 mg/kg per day caused a diminished rate of growth, but no other toxic symptoms and no deaths. A dose level of 1,111 mg/kg per day produced a rapid loss of weight and death of all animals within 5 weeks.

Mice fed a daily dose of TCNB (215 mg/kg per day) suffered no ill effects. However, a dose of 1,750 mg/kg per day caused an inhibition of growth rate in mice. In rabbits, no untoward reactions or symptoms of irritation were found as a result of eye or skin exposure to TCNB [299], and no adverse effects were noted in pigs fed TCNB-treated potatoes [300].

XI. ORGANOMERCURY COMPOUNDS

A large variety of organomercury compounds have been used extensively as fungicides in agriculture, horticulture, and forestry, as well as in the paper-pulp, textile, and paint industries. The organomercurial fungicides all have the general structure R—Hg—X, where R is usually an organic radical (aryl, aryloxy, alkyl, or alkoxyethyl) with a carbon-mercury bond, and X denotes a more or less dissociable anion, organic, or inorganic, e.g., halide, hydroxide, nitrate, acetate, urea, carbamate, etc.

The properties of the organomercury compounds (especially the effects on animal organisms), depend primarily on the organic radical. The anion portion is less important for the biological effect, but to some extent influences physical properties such as solubility and volatility.

The toxicity of mercury residues, from any source, depends primarily on the chemical nature of the mercury compounds [301-311]. Organomercury compounds (especially alkyl derivatives) are more toxic to man than inorganic mercury salts [301-323]. However, organomercurials have different degrees of toxicity. Published data on the toxicity vary, mainly because of differences in technique and animal strains used. Alkoxyalkyl mercury compounds have realtively low toxicities, whereas alkyl mercury, particularly methyl mercury compounds, are highly toxic [302-311, 320-324].

The major danger from mercury compounds, arises from the presence of alkyl mercury compounds (primarily methyl mercury) in the environment, due to the fact that the different forms of mercury that get into the environment can be converted by soil and sediment microorganisms to methyl mercury [325-328], which is then taken up by aquatic organisms and concentrated in the food chains. It should be noted that fish, plants, or animals cannot themselves convert other mercury compounds into methyl mercury [329].

Methyl mercury and other alkyl mercury compounds are used in agriculture as seed dressings (either as dusts or slurries, generally containg 1.5-3.2% metallic mercury). Accidental poisoning from this use can occur from eating treated grain, e.g., bread or meat from swine fed such seed [322, 324, 330-333] or from careless handling or exposure to the chemicals [311, 334-335].

Several major incidents of poisoning have occurred in Iraq, Pakistan, and Guatemala due to the ingestion of flour and wheat seed, treated with methyl and ethyl mercury compounds [322, 324, 330-331]. The fungicide ethylmercury-p-toluenesulfonamide was claimed to be responsible for two outbreaks in Iraq in 1956 and 1960. In 1960, an estimated 1,000 patients were affected by methyl mercury poisoning and 370 were admitted to hospital. In Guatemala, cases that were originally believed to be viral encephalities were reported during the wheat growing seasons of 1963 to 1965. Forty-five people were affected and 20 died. Methyl mercurydicyandiamide, used to treat the seed wheat before distribution to farmers, was established as the causative agent. By far the worst catastrophe of this type ever recorded in terms of its extent and the ensuing morbidity and mortality, occurred in Iraq in 1972 [324]. A total of 6,530 cases were admitted to hospitals and there were 459 hospital deaths attributed to methyl mercury poisoning as a result of ingestion of home made bread prepared from seed wheat treated with methyl mercurial fungicides. Mercury in treated grain may enter the body as a result of inhalation or skin contact in addition to oral ingestion. The primary signs and symptoms of methyl mercury poisoning result from damage to the nervous system and are characterized by loss of sensation at the extremities of the hands and feet and in areas around the mouth (paresthesia), loss of coordination in gait (ataxia), slurred speech, concentric constriction of the visual fields, impairment of heating, symptoms from the autonomic nervous and the extrapyrimidal system, as well as "mental disturbances" [319, 323, 324, 336-338]. Severe poisoning can cause blindness, coma, and death. (The illness is often called the Minamata disease.) There is a latent period of weeks or months between exposure to methyl mercury and the development of poisoning symptoms [318, 324]. Prenatal exposure to methyl mercury has resulted in mental retardation with cerebral palsy [324]. Humans are believed to be most sensitive to methyl mercury during the early stages of the life cycle, including pre- and the postnatal periods [324, 339].

Although 250 cases of human poisoning by ethyl mercury compounds have been reported, most of them have not been described in detail [302]. The majority of those affected had consumed seed dressed with various ethyl mercury compounds [322, 331, 340-342]. The consumption of seed wheat with Agrosan GN (a mixture of ethylmercury chloride and phenylmercury acetate) has been implicated in several deaths in Pakistan in 1961 [331]. Most recently the deaths of 20 persons (out of 144 cases) who had ingested maize, which was dressed with ethylmercuric chloride and intended for sowing, was reported in Ghana [342]. Ethyl mercury exposure has been associated with symptoms from the GI tract including abdominal pain, vomiting, and diarrhea [302, 342], dehydration, oliguria, and general weakness [342]. In the Ghanian poisoning episode [342], there appeared to be a latent period of 5 to 14 days after the first meal before the toxic effects were manifest. These effects appeared earlier and were more severe in children than in adults, with no apparent sex differences noted. The central nervous system involvement resembled that of methyl mercury poisoning. Death was usually due to dehydration, shock and renal failure.

Ten cases of prenatal intoxication with ethyl mercury have been reported [343]. The mothers had shown symptoms of poisoning to Granosan (ethylmercuric chloride) up to 3 years before pregnancy. In three of the prenatal cases, severe mental and neurological symptoms in accordance with those seen in Minamata, Japan, were described [302-304].

Phenylmercuric acetate, PMA, is the most widely used of the aryl mercurial fungicides, although the benzoate, lactate, nitrate, oleate, and propionates are also employed and with PMA find additional application as herbicides and slimicides. There appears to be no significant difference in human toxicity of the various phenyl mercurials [344]. Although phenyl mercurials may be absorbed through the intact skin or mucous membranes, reports of poisoning due to these compounds are apparently rare and chronic occupational poisoning is virtually unknown [308]. However, phenyl mercury compounds have been reported to cause dermatitis in skin exposure [344-346]. It is also important to consider the toxicity of a number of alkyl, alkoxyalkyl, and aryl mercurials in animals. Methyl mercury is absorbed most quickly, retained the longest in animal tissue [305, 310], and is the most toxic to all animals studied [302, 304, 305]. The risk is mainly due to slow elimination and a biological half-life, which varies between 20 and 70 days in different species (in man it is 70 days). Methyl mercury accumulates in the central nervous system and concentrations of about 10 ppm or over in the brain can cause poisoning [347]. The fetal brain appears to be more sensitive than the adult brain to methyl mercury.

In most species, anorexia and weight loss were the first signs of intoxication. However, neurological symptoms dominated the clinical picture in all of the species studied, with most species exhibiting a definite similarity. The most prominent sign first observed was ataxia. Morphological changes in methyl mercury intoxication were reported in mice, rats, ferrets, cats, and monkeys [302, 304]. Damage was noted in cerebral and cerebellar cortex and in peripheral nerves with their dorsal roots, separately or in combination. The pattern is similar to that observed in humans.

Ethyl mercury poisoning has been described in rats [348], rabbits [335], cats [349], sheep [350], swine [351], and calves [352], with the symptoms being similar to those in methyl mercury poisoning [302].

Nose [353] reported that the chemical structure of the lower alkyl mercuric compounds bore a close relationship to the manifestation of symptoms of poisoning in the rat, and that the increase in the number of carbon atoms correlated with a diminution in the symptoms observed. The greater portion of methyl, ethyl, n-propyl, and n-butyl mercuric compounds which accumulated in various organs, were metabolized by breakage of the C–Hg bond and were excreted as inorganic mercury.

Few reports have described the toxicity of phenyl mercurials. The acute intravenous LD_{50} of PMA for rats and mice has been reported to range from 20-23 mg/kg [307, 354] and 8-20 mg/kg [307, 354] for rats and mice, respectively.

A decreased weight gain, histopathological kidney damage, and reduced survival period in rats exposed to PMA in the diet at varying levels for 12 to 24 months has been reported by Fitzhugh et al. [355]. Oral administration of PMA to swine has been found to result in anorexia, diarrhea, weight loss, and renal damage [356]. Wien [357] reported gastrointestinal lesions after parenteral administration of phenyl mercuric nitrate in rats and mice. Gage and Swan [358] found PMA to be much less toxic than certain methyl mercury compounds and no more toxic than inorganic mercury salts. The central nervous system lesions produced by methyl mercury salts were not produced by PMA.

The discovery that lower alkyl mercury compounds can pass through the placenta [359-362] has caused concern regarding the possible embryopathic effects of mercury compounds in man and as a consequence has spurred a variety of teratogenic studies in animals. Studies have been reported in mice [363, 364], chicks [365], and hamsters [366] involving the use of single high-dose exposures to organomercury compounds during the period of organogenesis. The malformations observed have included hypoplasia of the cerebellum, exencephaly, encephalocele, hydrocephalus, vaulted cranium, cleft palate and lip micrognathia, microglossia, anophthalmia, rib fusion, syndactyly, and clubfoot.

Oral administration of methylmercury chloride to pregnant rats on days 7 to 20 was reported to have resulted in decreased weight of offspring when 6 mg/kg per day was given. Marked reduction of litter size occurred when a dose of 8 mg/kg per day was administered [367].

Effects of chronic low-dose subcutaneous exposure of Sprague-Dawley rats near term to methylmercury hydroxide were reported by Mottet [368]. The near-term maternal liver content of mercury served as an index of body content. At liver contents below 15 μg/g, maternal and fetal mortality and fetal size were unaffected. The threshold effect occurred at 15 μg/g maternal liver. From the threshold level to 75 μg/g (maternal toxic level) there was a generally dose-response relation characterized by increasing maternal death, prenatal resorption and fetal death, and decreasing fetal weight. Generalized fetal edema occurred at the maternal toxic dose level and no gross or microscopic malformations were found in the fetuses. The development of behavioral characteristics in offspring of rats exposed to comparable levels of methyl mercury was similar to the controls throughout 21 days of observation. The results of these behavioral studies suggest that antenatal exposure of rats to methyl mercury at the threshold level did not alter brain development sufficiently to produce grossly defective behavioral responses.

Khera and Tabacova [369] reported similar results in mice exposed to 1 mg/kg per day from day 6 to 17 of pregnancy. Although some changes in the migration of granular cells of the cerebellum were found, no adverse behavioral effects were noted.

Gale and Ferm [370] elaborated embryopathic effects of mercuric acetate and phenylmercuric acetate administered intravenously (i.v.) to Golden Syrian hamsters on day 8 of gestation. All doses of mercuric acetate and PMA at 7.5, 8, and 10 mg/kg induced resorption rates significantly higher than the control level of 4%. Both mercury compounds induced identical defects in hamster fetuses removed on either day 12 or 14 of gestation. The defects included exencephaly, encephalocele, anophthalmia, microphthalmia, cleft lip and palate rib fusions, and syndactylia. Both mercurials also produced weight loss, kidney lesions, diarrhea, slight tremor, and somnolence in the maternal system.

Piechocka [371] reported a reduction in litter size of rats given food containing 8 mg Hg/kg as PMA for 6 months compared to control. No clinical symptoms of poisoning were apparent in the exposed rats.

Khera [372] described the reproductive capability of Wistar rats and Swiss Webster mice. Male rats were administered single daily doses of 0, 1, 2.5, and 5 mg Hg^{2+} (as methylmercuric chloride) per kg po for 7 consecutive days, then mated after dosing was discontinued. A dose-related reduction in mean litter size per pregnancy attributed to preimplantation losses occurred at all doses during days 5 to 20 of posttreatment. In a long-term study, male rats were dosed po throughout a maximum of 21 mating trials, each of 5 days duration. From 30 days onward at 1 mg/kg per day and from 90 days onward at 0.5 mg/kg per day, reduction in average litter size due to preimplantation losses was noted, an effect not observed at 0.1 mg/kg per day. Male rats dosed for 7 days had a dose-related reduction in incidence of fertile matings, although this effect was not apparent during chronic treatment.

Male mice administered daily doses po of up to 5 mg Hg^{2+} (as methylmercuric chloride) per kg for 5 or 7 consecutive days and then sequentially mated for five to seven trials produced no postimplantation losses or reduction of fertility [372]. The results in the two species appear to reflect differences in susceptibility of the sperm-cell population or in the distribution of mercury in the male genital system.

The genetic effects of mercury have been reviewed by Ramel in 1972 [373]. A broad spectrum of mercury compounds exhibits various effects on genetic material. Colchicine-mitosis (C-mitosis), an inactivation of the spindle fiber mechanism at cell division, has been demonstrated in cells of *Allium cepa* roots exposed to concentrations of alkyl and phenyl mercury compounds at levels as low as 15×10^{-7} M (about 0.05 ppm). Alkoxyl compounds are less active and require levels as high as 30×10^{-7} M (about 0.7 ppm) for comparable effects. Although all the organomercurials tested produce C-mitosis, the effects were not identical. Phenyl compounds give rise to more bridges and fragments than do other mercury compounds [374]. The organomercurial fungicides found to induce C-mitosis in *Allium cepa* were methyl mercuric dicyandiamide, methylmercuric hydroxide, phenylmercuric hydroxide, methoxyethylmercuric chloride and Panogen, a fungicidal mixture containing methyl mercuricdicyandiamide [375, 376].

Methyl and phenyl mercury are among the most active C-mitotic agents known, 200-1,000 times more effective than colchicine. Inorganic mercury is about 200 times less effective than are the organic mercury compounds in the production of C-mitosis [376, 377].

Colchicine-mitotic effects similar to the ones found in plants have been observed in animal cells, e.g., tissue cultures of HeLa cells, when treated with phenyl and ethyl mercury chloride [378]. Similar effects on human leucocytes, treated in vitro with methyl mercury chloride were noted by Fiskesjö [379]. The lowest concentration level for C-mitosis found for methyl mercury chloride was between 1 and 2×10^{-6} M, a similar order of magnitude to that producing C-mitosis in plant cells.

Okada and Oharazawa [380] noted significantly increased frequency of polyploidy in tissue cultures from mice treated in vivo with subcutaneous injections of ethyl mercury phosphate.

Methyl and phenyl mercury compounds have also been shown to cause genetic effects in *D. melanogaster* [374, 381-382], while methoxy mercuric chloride and inorganic mercury were not active even in higher concentrations in this system. The effect of methyl mercury on nondisjunction of chromosomes in *Drosophila* confirms at a genetic level the cytological observation of the C-mitotic effect of mercurials. The effect of methyl mercury on the distribution of the sex chromosomes was restricted almost entirely to treatments of females. A substantial effect after treatment of males could only be obtained with the rise of chromosomal aberrations, which cause a high spontaneous nondisjunction [383]. It should be noted that the direct mutagenic effect of methyl mercury, as revealed by the recessive lethal tests in Drosophila is of comparatively small magnitude [373]. A reason for this may be that the majority of the mercury molecules entering the cell gets "trapped" and inactivated by various proteins and polypeptides before reaching the nucleus and the chromosomes [373].

Khera [384] described a dominant lethal experiment with male rats treated orally with 0, 1, 2.5, and 5 mg/Hg per kg as methylmercury chloride for 7 days. For 5 days each, 14 sequential matings to untreated females were performed. From matings during the first 20 days following the treatment, a reduction in the average number of viable embryos per litter was found in all the mercury-treated groups, and it was suggested that the highest sensitivity must have involved gametes treated as spermatozoa or late spermatids [373]. When males were treated continuously with 0.1, 0.5, and 1 mg/Hg per day, the highest dose caused a significant depression in the average litter size after 25 days.

Skerving et al. [382] reported a statistical increase in chromosomal breaks in blood cells (lymphocytes) of human consumers of large amounts of fish containing methyl mercury. Their data would appear to indicate that the mercury pollution has reached a level at which genetic effects on humans actually do occur, although little is known of the medical significance of chromosomal defects in blood cells [373] in the absence of evidence of clinical poisoning [382].

XII. ORGANOTIN COMPOUNDS

The organotin fungicides of importance are triphenyltin hydroxide (fentin hydroxide, Du-ter) and triphenyltin acetate (fentinacetate, Brestan). These fungicides are effective against many of the fungi susceptible to copper fungicides, but at smaller dosages. Triaryltin compounds are general biocides and are inhibiters of oxidative phosphorylation in mammals, as well as in plants and fungi.

In regard to the toxicity of organotin compounds, distinction must be noted between the effects of the triaryl, trialkyl, and dialkyl compounds. Compared to both the trialkyl (especially the lower members), which have a specific effect on the central nervous system [385-387], and the dialkyl compounds, which are severe skin irritants and exhibit an inflammatory action on the bile ducts [385-388], the triphenyltin compounds have a relatively low order of acute and chronic mammalian toxicity [385, 386].

The acute oral LD_{50} of triphenyltin hydroxide in rats is 171 mg/kg (range 100-295) in males and 208 mg/kg (range 205-344) in females [389], and 27.1 mg/kg in female guinea pigs [390].

The acute oral LD_{50} of triphenyltin acetate for rats is 127 mg/kg and the acute dermal LD_{50} for rats is 500 mg/kg [390].

Abnormality was observed in the spleen of rats in the form of extramedullary hematoporesis following administration of triphenyltin hydroxide [389]. Acute 25-hr and subacute 14-day repeated application of triphenyltin hydroxide to rabbit skin was without gross effect. However, it was an extreme eye irritant, even when the eyes were washed 2 sec after application, producing corneal opacities in all animals.

The chronic effects of triphenyltin hydroxide, given to rats in the diet at the rate of 400 ppm, were elaborated by Gaines [391]. This amount reduced the food consumption, and within 7 to 34 days caused death from starvation and hemorrhage in all tested animals. Dietary levels of 100 and 200 ppm of triphenyltin hydroxide reduced food consumption and weight gain, particularly during the first week. By week 7, the animals had adapted to these concentrations and the food consumption was the same as the controls. The food intake or weight gain was not affected in rats fed 50 ppm of triphenyltin hydroxide for 276 days. After 64 days of dietary exposure at 200 or 100 ppm, reduced fertility was observed in male rats. However, fertility improved later, and was comparable to that of the control rats by day 113. The difference was explained by the partial starvation which the animals suffered, particularly during the first week of the experiment. Gross and macroscopic examination of rats fed triphenyltin hydroxide at the rate of 200, 100, or 50 ppm for 276 days did not reveal changes in any organ that could be related to the compound.

A teratogenic study in Sprague-Dawley rats at a dosage of 15 mg/kg showed that triphenyltin hydroxide prevented implanation when given from day 1 through 7. When given from day 8 or beyond, the compound was feticidal [389].

The effects of triphenyltin acetate (TPTA) and triphenyltin chloride (TPTC) on C_3H strain mice were reported by Brown and Clemson [392]. TPTA and TPTC were given to adult female C_3H mice in a 14-month feeding tumor-growth rate study (10 mg/kg per day for 6 weeks and 15 mg/kg per day from week 6 to 60. Administering TPTA and TPTC po for 14 months significantly decreased animal weights by week 8 and 10, respectively. No weight differences were seen when compounds were given for less than 8 weeks. The above dosage schedule for 60 days significantly decreased kidney weight of both TPTA and TPTC treated mice, while the spleens of TPTC treated mice weighed significantly less and the livers weighed significantly more than the same organs of the controls. TPTA and TPTC for 60 days gave no difference in the amount of tin in the spleen, fat, or brain tissue, but the feces and kidneys of treated mice contained significantly greater amounts of tin than the controls. Higher amounts of tin were also found in the livers of mice treated with TPTA (but not TPTC).

XIII. GRISEOFULVIN

Griseofulvin, 7-chloro-4,6-dimethoxycoumaran-3-one-spiro-1'(2'-methoxy-6'-methylcyclohex-2'-en-4'-one), is a fermentation product of three species of *Penicillium*, e.g., *P. patulum*, *P. griseofulvum*, and *P. janczewski*, which in very low concentration suppresses growth and induces striking changes in the hyphae of a number of fungal species. Griseofulvin has also been extensively used in man in the systemic treatment of dermatophyte infection. It has been reported that orally administered griseofulvin inhibits the growth of dermatophytes in vivo without inducing significant adverse effects on either man or experimental animals [393-397].

In laboratory animals, the lethal dose of griseofulvin in high. Wistar rats of either sex tolerate a daily dose of 2 g/kg given i.p. in aqueous suspension for several days [398]. However, following dosage of this level for 4 days, the seminal epithelium was found to be severely damaged and in many places necrotic. The intestinal epithelium also showed abnormalities, but these were less marked.

The median lethal dose of griseofulvin to rats by intravenous injection is of the order of 400 mg/kg. Approximately 6 hr after a single intravenous dose of griseofulvin (100 or 200 mg/kg) was administered, a striking arrest of mitosis in metaphase was evident in sites of mitotic activity in the rat [398]. It was very marked in the bone marrow and intestine, and in rats bearing transplanted tumor (Walker Carcinoma 256 or a lymphosarcoma) in the growing tumor tissue. Cells arrested in metaphase show, in varying degree, disorientation and scattering of the chromosomes through the cytoplasm. Hence, in this respect, griseofulvin resembles colchicine [398].

Similar effects by griseofulvin have also been observed in plant cells by Paget and Walpole [398] and in the induction of multinuclearity in mammalian cells cultured in vitro [399-401].

Larizza et al. [401] reported a marked effect on the dynamics of chromosomal complements both in diploid and heteroploid cell lines (including hybrids produced by fusion) after treatment with griseofulvin for 3 days with doses ranging from 40 to 60 μg/ml. According to the cell type, a tendency to a doubling of the chromosomal set was evident.

The fungistatic action of griseofulvin had been suggested to be possibly due to an interference of nucleic acid synthesis [402]. Since compounds which block such synthesis have caused teratogenic changes in mammals [403], griseofulvin may have teratogenic potential [398, 402-406].

Klein and Beall [406] administered oral doses of 125 to 1,500 mg/kg per day to pregnant rats of the CAW:CFE SD spf strain during organogenesis from day 6 through 15. Evaluation of the offspring from dams treated with the largest doses, e.g., 1,250 or 1,500 mg/kg per day (63 and 75 times a therapeutic dose in man) indicated decreased survival rates and a syndrome of malformations. The incidence of resorptions, as a percentage of nidation sites, was 16 and 64% in the dams given 1,250 and 1,500 mg/kg, respectively, compared to 5% in the controls. The incidence of malformations between the 1,250 and 1,500 mg/kg treated groups increased only 2%, suggesting that at 1,500 mg/kg griseofulvin approaches the limit of fetal tolerance. It was also shown that griseofulvin given to male rats daily for 63 days in oral doses up to 1,500 mg/kg did not adversely affect spermatogenesis [406].

The results of Klein and Beall [406] could apparently be correlated with those obtained in humans by MacLeod and Nelson [407]. In their studies, 14 men were given daily doses of 2,000 mg of microsize griseofulvin for 3 months. Sperm counts and histological examination of testicular biopsies indicated no adverse effects on spermatogenesis [407].

The enhancement by the pesticidal synergist piperonyl butoxide of acute subcutaneous toxicity of griseofulvin in infant mice (random-bred Swiss 1 CR/Ha strain) has been reported by Epstein et al. [408]. The increase of 21 day mortalities from 43 to 100% for griseofulvin (at an initial dose of 0.25 mg) was highly significant.

Fuji and Epstein [409] reported increase of about 5-fold (compared to controls) of hepatomas in Swiss mice after 1 year following initial treatment at ages 1, 7, 14, and 21 days after birth with repeated subcutaneous injections of griseofulvin.

XIV. MISCELLANEOUS FUNGICIDES

Carboxin (2,3-dihydro-6-methyl-5-phenylcarbamoyl-1,4-oxathiin; 5,6-dihydro-2-methyl-1,4-oxathiin-3-carboxanilide; Vitavax) is a systemic fungicide used primarily for the seed treatment of cereals against smuts and bunts and of cotton, peanuts, and vegetables against Rhizoctonia, the activity possibly being due to inhibition of respiration at or near the site of succinate oxidation [410].

The acute oral LD_{50} of carboxin for rats if 3,820 mg/kg and the acute dermal LD_{50} for rabbits is more than 8,000 mg/kg [411]. Albino rats fed diets containing 600 ppm of carboxin for 2 years, suffered no detectable symptoms [411].

Stefan et al. [412] reported decreased erythrocyte and increased leukcyte count in rats treated for 30-45 days with 0.015 mg of dietary carboxin per day. Carboxin decreased the serum total protein and albumin and increased the α_1- and α_2-globulin levels. The serum glutamic-oxalacetic transaminase activity was decreased and glutamic-pyruvic transaminase and alkaline phosphatase increased.

A major soluble metabolite of carboxin in barley seedlings and mature plants has been recently identified as the p-hydroxyphenyl derivative [413]. Production of a phenol could increase the toxicity and perhaps explain earlier findings that carboxin inhibits respiration, photosynthesis [414], and the tricarboxylic acid cycle [415]. Elaboration of the mammalian toxicity of this metabolite, as well as the sulfoxide of carboxin [416] would be of importance since the metabolites of carboxin can possibly enter the food chain as trace residues.

XV. SUMMARY

A number of major fungicides from a broad spectrum of structural categories, such as metallic ethylenebisdithiocarbamates, dithiocarbamates, thiocarbamoyl disulfides, phthalimides, benzimidazoles, thiophanates, chlorobenzenes, chlorophenols, chloronitrobenzenes, organomercury, organotin compounds and spirolactones were examined, primarily in regard to the major areas of toxicity, e.g., carcinogenicity, teratogenicity, and mutagenicity, and secondarily to their acute, subacute and chronic toxicity, as well as their effects on growth.

The importance of elaborating the identity and toxicological role of trace impurities and degradation and metabolic products cannot be overstressed. This is particularly manifest in such examples as ethylenethiourea in ethylenebisdithiocarbamates and chlorinated dibenzo-*p*-dioxins and their precursors, as well as chlorinated dibenzofurans in chlorophenol fungicides.

The difficulty of extrapolating mammalian toxicity findings to man is apparent. This is so not only from the reported conflicting data per se, arising from the differences in homogeneity and purity of the test material, but also from differences in testing protocol, e.g., choice of strain, species, dosage, age of animal, duration of test, nutritional and immunological aspects of host, etc. Compounding the dilemma is the lack of unanimity concerning interpretation and significance of findings to man, the ultimate concern.

REFERENCES

1. Health and Welfare Board of Canada, *The Testing of Chemicals for Carcinogenicity, Mutagenicity, and Teratogenicity,* Health and Welfare Board of Canada, Ottawa, 1973.
2. WHO Scientific Group, *World Health Organ. Tech. Rept. Ser.*, 1969, No. 426.
3. WHO Scientific Group, *World Health Organ. Tech. Rept. Ser.*, 1967, No. 364.
4. WHO Scientific Group, *World Health Organ. Tech. Rept. Ser.*, 1971, No. 482.
5. WHO Scientific Group, *World Health Organ. Tech. Rept. Ser.*, 1974, No. 546.
6. U.S. Department of Health, Education, and Welfare, *Secretary's Commission on Pesticides and Their Relationship to Environmental Health Report*, U.S. Govt. Print. Off., Washington, D.C., 1969.

7. H. Kalter, in Chemical Mutagens: Principles and Methods for Their Detection (A. Hollaender, ed.), Plenum, New York, 1971, p. 57.
8. E. C. Miller and J. A. Miller, in Chemical Mutagens: Principles and Methods for Their Detection (A. Hollaender, ed.), Plenum, New York, 1971, p. 83.
9. L. Fishbein, W. G. Flamm, and H. L. Falk, Chemical Mutagens: Environmental Effects on Biological Systems, Academic Press, New York-London, 1970.
10. M. S. Legator and H. V. Malling, in Chemical Mutagens: Principles and Methods for Their Detection (A. Hollaender, ed.), Plenum, New York, Vol. 2, 1971, p. 569.
11. World Health Organization, Health Hazards of the Human Environment, World Health Organization, Geneva, 1972, p. 213.
12. H. E. Sutton and M. I. Harris, Mutagenic Effects of Environmental Contaminants, Academic Press, New York, 1972.
13. U.S. Congress, Senate, Government Operations Committee, Chemicals and the Future of Man: Hearings Before the Subcommittee on Executive Reorganization and Government Research, 92nd Congress, 1st Session, April 6 and 7, U.S. Govt. Print. Off., Washington, D.C., 1971.
14. S. S. Epstein, Environ. Health Persp., 6, 23 (1973).
15. S. Green and J. A. Springer, Envion. Health Persp., 6, 37 (1973).
16. J. G. Brewen and R. J. Preston, Environ. Health Persp., 6, 157 (1973).
17. E. Freese, Environ. Health Persp., 6, 171 (1973).
18. W. G. Flamm, Environ. Health Persp., 6, 215 (1973).
19. M. S. Legator and W. G. Flamm, Ann. Rev. Biochem., 42, 683 (1973).
20. T. B. Gaines, Toxicol. Appl. Pharmacol., 14, 515 (1969).
21. N. M. Kurbat, Zh. Dravookhr. Beloruss., 2, 49 (1968) (in Russian); Chem. Abstr., 71, 48683T (1969).
22. Z. Przezdziecki, J. Bankowska, W. Komorowsk-Malewska, and T. Janicka, Rocz. Panstw. Zakl. Hlg., 20, 133 (1969) (in Polish); Chem. Abstr., 71, 48674R (1969).
23. T. V. Yudina and Y. V. Novikov, Gig. Sanit., 36, 14 (1971) (in Russian); Pestic. Abstr., 5, 1946 (1972).
24. S. K. Dinoyeva, Gig. Sanit., 39, 85 (1974) (in Russian); Pestic. Abstr., 1956 (1974).
25. R. Engst, W. Schnaak, and H. J. Lewerenz, Z. Lebensm.-Unters. Forsch., 146, 91 (1971).
26. R. Blackwell-Smith, J. K. Finnegan, P. S. Larson, and M. L. Dreyfuss, J. Pharmacol. Exptl. Therap., 109, 159 (1953).
27. Food and Agriculture Organization (FAO) and World Health Organization (WHO), FAO Meeting Rept., Geneva, No. PL/1965/10/1, WHO/Food Add./27.65, p. 142.
28. J. Seifter and W. E. Ehrich, J. Pharmacol. Exptl. Therap., 92, 303 (1948).

29. J. R. M. Innes, B. M. Ulland, M. G. Valerio, L. Petrucelli, L. Fishbein, E. R. Hart, A. J. Pallotta, R. R. Bates, H. L. Falk, J. J. Gart, M. Klein, I. Mitchel, and J. Peters, J. Natl. Cancer Inst., 42, 1101 (1969).
30. L. Ivanova-Chemishanska, G. Dimov, and N. Mosheva-Izmirova, Hyg. Ziraveopazvane, 11, 554 (1968) (in Bulgarian).
31. L. Ivanova-Chemishanska, Letopisi Na Hei, 9, 46 (1969) (in Bulgarian); L. Ivanova-Chemishanska, D. V. Markov, and G. Dashev, Environ. Res., 4, 201 (1972).
32. L. Ivanova-Chemishanska, D. V. Markov, and G. Dashev, Environ. Res., 4, 201 (1971).
33. T. Petrova-Vergieva and L. Ivanova-Chemishanska, Food Cosmet. Toxicol., 11, 239 (1973).
34. L. V. Marston, Farmakol. Toksikol. (Moscow), 32, 731 (1969) (in Russian); Chem. Abstr., 72, 658700 (1970).
35. O. V. Chernov and I. I. Khitsenko, Vop. Onkol., 15, 71 (1969) (in Russian); Chem. Abstr., 71, 21105V (1969).
36. V. Benes and R. Sram, Ind. Med. Surg., 38, 50 (1969).
37. J. P. Seiler, Experientia, 29, 622 (1973).
38. A. M. Kligman and W. Rosensweig, J. Invest. Dermatol., 10, 59 (1948).
39. D. O. E. Gebhardt and M. J. Van Logten, Toxicol. Appl. Pharmacol., 13, 316 (1968).
40. K. V. Prahlad, R. Bancroft, and L. Hanzely, Cytobios, 9, 121 (1974).
41. C. J. Czeglédi-Jankó and A. Hallo, J. Chromatog., 31, 89 (1967).
42. J. W. Vonk and A. Kaars Sijpesteijn, Ann. Appl. Biol., 65, 489 (1970).
43. W. R. Bontoyan, L. B. Looker, T. E. Kaiser, P. G. Ang, and B. M. Olive, J. Assoc. Offic. Anal. Chemists, 55, 923 (1972).
44. R. Engst and W. Schnaak, Z. Lebensm.-Unters. Forsch., 134, 216 (1967).
45. R. Engst, W. Schnaak, and H. J. Lewerenz, Z. Lebensm.-Unters. Forsch., 146, 91 (1971).
46. R. Kato, Y. Odanaka, S. Teramoto, and O. Matano, Bull. Environ. Contam. Toxicol., 5, 546 (1976).
47. H. Seidler, M. Haertig, W. Schnaak, and R. Engst, Nahrung, 14, 363 (1970).
48. J. W. Vonk, Mededel. Landbouwhoogeschool Opzoekinsstat. Staat Gent, 36, 109 (1971).
49. R. Truhaut, M. Fujita, N. P. Lich, and M. Chaigneau, Compt. Rend., 276, 229 (1973).
50. H. L. Klöpping and G. J. M. van der Kerk, Rec. Trav. Chim., 70, 949 (1951).
51. G. Petrosini, F. Tafuri, and M. Businelli, Notiz. Mal. Piante, 65, 9 (1963); Chem. Abstr., 59, 8066 (1963).
52. G. Viel and M. Chancogne, Phytiatr. Phytopharm., 13, 109 (1971).
53. L. Fishbein and J. Fawkes, J. Chromatog., 19, 364 (1965).
54. L. E. Lopatecki and W. Newton, Can. J. Bot., 30, 131 (1952).

55. J. W. Hylin, Bull. Environ. Contam. Toxicol., 10, 227 (1973).
56. G. D. Thorn and R. A. Ludwig, The Dithiocarbamates and Related Compounds, Elsevier, Amsterdam, 1962.
57. R. Engst and W. Schnaak, Z. Lebensm.-Unters. Forsch., 143, 99 (1970).
58. R. A. Ludwig, G. D. Thorn, and D. M. Miller, Can. J. Bot., 32, 48 (1954).
59. W. H. Newsome and G. W. Laver, Bull. Environ. Contam. Toxicol., 10, 151 (1973).
60. C. H. Blazquez, J. Agr. Food Chem., 21, 330 (1973).
61. R. R. Watts, R. W. Stroherr, and J. H. Onley, Bull. Environ. Contam. Toxicol., 12, 224 (1974).
62. R. A. Ludwig and G. D. Thorn, Can. J. Bot., 36 (1958).
63. G. D. Thorn and R. A. Ludwig, Can. J. Chem., 32, 872 (1954).
64. S. L. Graham and W. H. Hansen, Bull. Environ. Contam. Toxicol., 7, 19 (1972).
65. B. M. Ulland, J. H. Weisburger, E. K. Weisburger, J. M. Rice, and R. Cypher, J. Natl. Cancer Inst., 49, 583 (1972).
66. S. L. Graham, W. H. Hansen, K. J. Davis, and C. H. Perry, J. Agr. Food Chem., 21, 324 (1973).
67. K. S. Khera, Teratology, 7, 243 (1973).
68. J. P. Seiler, Mutation Res., 26, 189 (1974).
69. R. A. Ludwig, Can. J. Bot., 33, 42 (1955).
70. R. D. Ross and D. G. Crosby, J. Agr. Food Chem., 21, 335 (1973).
71. P. A. Cruickshank and H. C. Farrow, J. Agr. Food Chem., 21, 333 (1973).
72. J. W. Vonk and A. Kaars Sijpesteijn, Pestic. Biochem. Physioc., 1, 163 (1971).
73. G. van Steenis and M. J. van Logten, Toxicol. Appl. Pharmacol., 19, 675 (1971).
74. C. E. Cox, H. D. Sisler, and R. A. Spurr, Science, 114, 643 (1951).
75. R. W. Barratt and J. G. Horsfall, Conn. Agr. Expt. Sta., New Haven Bull., No. 508, 1947.
76. H. L. Klöpping and G. J. M. van der Kerk, Rec. Trav. Chim., 70, 917 (1951).
77. H. C. Hodge, E. A. Maynard, W. Downs, H. J. Blanchet, Jr., and C. K. Jones, J. Amer. Pharm. Assoc., Sci. Ed., 41, 662 (1952).
78. H. C. Hodge, E. A. Maynard, W. L. Downs, R. D. Coye, Jr., and L. T. Steadman, J. Pharmacol. Exptl. Therap., 118, 174 (1956).
79. A. R. Rasul and J. M. Howell, Toxicol. Appl. Pharmacol., 30, 63 (1974).
80. J. L. Minor, J. Q. Russell, and C. C. Lee, Toxicol. Appl. Pharmacol., 29, 120 (1974).
81. R. Fahrig, Chemical Carcinogenesis Assays (R. Montesano and L. Tomatis, eds.), Intern. Agency for Research on Cancer, Publ. No. 10, Lyon, France, 1974, p. 161.

82. M. A. Pilinskaya, Genetika, 7, 138 (1971).
83. E. A. Antonovich, O. V. Chernov, L. V. Samosh, L. V. Marston, M. A. Pilinskaya, L. I. Kurinnyi, M. Sh. Vekshtein, V. S. Marston, P. N. Balin, and I. I. Khitsenko, Gig. Sanit., 50, 25 (1972) (in Russian); Chem. Abstr., 78, 24925S (1973).
84. M. Sh. Vekshtein and I. I. Khitsenko, Gig. Sanit., 36, 23 (1971) (in Russian).
85. G. Eisenbrand, O. Ungerer, and R. Preussmann, Food Cosmet. Toxicol., 12, 229 (1974).
86. G. Matthiaschk, Arch. Toxikol., 30, 251 (1973).
87. T. M. Agaeva, Gig. Primen Toksikol. Pestits. Klin. Otravceni., 4, 163 (1966) (in Russian); Chem. Abstr., 69, 34906Q (1968).
88. Y. S. Brusilovskiy and A. M. Fial'kovskiy, Vestn. Dermatol. Venerol., 47, 28 (1973) (in Russian); Pestic. Abstr., 7, 74-2688 (1974).
89. J. F. Robens, Toxicol. Appl. Pharmacol., 15, 152 (1969).
90. V. H. Ferm, Lancet, 1, 208 (1966).
91. R. Roll, Arch. Toxikol., 27, 173 (1971).
92. K. S. Khera, Can. Med. Assoc. J., 100, 167 (1969).
93. T. B. Davydova, Gig. Sanit., 38, 108 (1973) (in Russian); Pestic. Abstr., 6, 73-2929 (1973).
94. J. Wedig, A. Cowan, and R. Hartung, Toxicol. Appl. Pharmacol., 12, 293 (1968).
95. N. I. Zhavoronkov and S. D. Antsiferov, Veterinariya (Moscow), 8, 92 (1972) (in Russian); Pestic. Abstr., 7, 74-1451 (1974).
96. E. A. Antonovich and M. S. Vekshtein, Gig. Primen. Toksikol. Pestits. Klin. Otravlenii., 8, 221 (1970) (in Russian); Chem. Abstr., 77, 110262J (1972).
97. A. I. Kurinnyi and T. I. Kendratenko, Tsitol. Genet., 6, 225 (1972) (in Russian); Chem. Abstr., 77, 110245 F (1972).
98. I. I. Sivitskaya, Oftalmol. Zh., 28, 286 (1974) (in Russian); Pestic. Abstr., 7, 74-2847 (1974).
99. P. N. Lyubchenko, A. B. Chemnyy, Z. I. Boyarchuk, D. A. Ginzburg, and V. M. Sukova, Gig. Truda Prof. Zabolevaniya, 17, 50 (1973) (in Russian); Pestic. Abstr., 7, 74-1366 (1974).
100. A. R. H. B. Verhagen, Trans. St. John's Hosp. Dermatol. Soc., 60, 86 (1974).
101. W. W. Kilgore and E. R. White, Bull. Environ. Contam. Toxicol., 5, 67 (1970).
102. J. J. Sims, H. Mee, and D. C. Erwin, Phytopathology, 59, 1775 (1969).
103. G. P. Clemons and H. D. Sisler, Phytopathology, 59, 705 (1969).
104. G. P. Clemons and H. D. Sisler, Pestic. Biochem. Physiol., 1, 32 (1971).
105. R. S. Hammerschlag and H. D. Sisler, Pestic. Biochem. Physiol., 2, 123 (1972).

106. C. A. Peterson and L. V. Edgington, J. Agr. Food Chem., 17, 898 (1969).
107. R. M. Krupka, Pestic. Sci., 5, 211 (1974).
108. E. Lemperle and E. Kerner, Z. Anal. Chem., 254, 117 (1971).
109. P. G. C. Douch, Xenobiotica, 3, 367 (1973).
110. J. A. Gardiner, R. K. Brantley, and H. Sherman, J. Agr. Food Chem., 16, 1050 (1968).
111. A. M. Torchinskiy, Vop. Pitaniya, 3, 76 (1973) (in Russian); Pestic. Abstr., 7, 74-2948 (1974).
112. A. C. Hastie, Nature, 226, 771 (1970).
113. A. Kappas, S. B. Georgopoulos, and A. C. Hastie, Mutation Res., 26, 17 (1974).
114. H. M. Moll, M. C. Vila, and C. Heras, Rev. Acad. Cienc., 28, 373 (1973) (in Spanish); Pestic. Abstr., 7, 74-2682 (1974).
115. J. P. Seiler, Mutation Res., 15, 273 (1972).
116. J. A. Styles and R. Garner, Mutation Res., 26, 177 (1974).
117. L. E. Savitt, Arch. Dermatol., 105, 926 (1972).
118. T. Noguchi, K. Ohkuma, and S. Kosaka, in Proc. 2nd Intern. IUPAC Congr. Pestic. Chem. (A. S. Tahori, ed.), Vol. V, Gordon & Breach Science Publishers, New York, 1971, p. 263.
119. T. Noguchi, K. Ohkuma, and S. Kosaka, 2nd Intern. Congr. Pestic. Chem., Tel-Aviv, Israel, 1971.
120. Y. Hashimoto, T. Makita, T. Mori, T. Nishibe, T. Noguchi, S. Tsubol, and G. Ohta, Pharmacometrics, 4, 5 (1970).
121. Y. Hashimoto, T. Makita, N. Ohnuma, and T. Noguchi, Toxicol. Appl. Pharmacol., 23, 606 (1972).
122. Y. Hashimoto, T. Mori, N. Ohnuma, and T. Noguchi, Toxicol. Appl. Pharmacol., 23, 616 (1972).
123. Y. Hashimoto, T. Makita, T. Nishibe, T. Mori, N. Ohnuma, T. Noguchi, and Y. Fukuda, Oyo Yakurl, 6, 873 (1972) (in Japanese); Chem. Abstr., 78, 93337P (1973).
124. T. Makita, Y. Hashimoto, and T. Noguchi, Toxicol. Appl. Pharmacol., 24, 206 (1973).
125. A. J. Bateman, Nature, 210, 205 (1966).
126. S. S. Epstein and H. Shafner, Nature, 219, 385 (1968).
127. E. Vogel and G. Röhrborn, Chemical Mutagenesis in Mammals and Man, Springer-Verlag, Heidelberg-New York, 1970, p. 148.
128. T. Makita, Y. Hashimoto, and T. Noguchi, Pharmacometrics, 4, 23 (1970).
129. J. W. Vonk and A. Kaars Sijpesteijn, Pestic. Sci., 2, 160 (1971).
130. E. F. Davis, B. L. Tuma, and L. C. Lee, Handbook of Toxicology, Vol. 5 (D. S. Dittmer, ed.), W. B. Saunders, Philadelphia, 1959, p. 37.
131. H. Martin, Pesticide Manual, 3rd ed., British Crop Protection Council, Droitwich, England, 1972, pp. 81, 82, 271.

132. T. M. Dowe, J. Matsushima, and V. H. Arthoud, J. Animal Sci., 16, 93 (1957).
133. D. F. Johnson, Southwestern Vet., 8, 30 (1954).
134. R. P. Link, J. C. Smith, and C. C. Morrill, J. Amer. Vet. Med. Assoc., 128, 614 (1956).
135. W. S. Spector, Handbook of Toxicology, Vol. I, W. B. Saunders Co., Philadelphia and London, 1956, p. 58.
136. J. R. De Baun, J. B. Miaullis, J. Knarr, A. Mihailovski, and J. J. Menn, Xenobiotica, 4, 101 (1974).
137. R. J. Lukens and H. D. Sisler, Science, 127, 650 (1958).
138. R. J. Lukens and H. D. Sisler, Phytopathology, 48, 235 (1958).
139. S. S. Block and J. P. Weidner, Develop. Ind. Microbiol., 3, 213 (1963).
140. R. G. Owens and G. Blaak, Contrib. Boyce Thompson Inst., 20, 475 (1960).
141. R. J. Lukens and H. D. Sisler, Science, 127, 650 (1958).
142. H. P. Burchfield and J. Schechtman, Contrib. Boyce Thompson Inst., 19, 411 (1958).
143. R. H. Daines, R. J. Lukens, E. Brennan, and I. A. Leone, Phytopathology, 47, 567 (1957).
144. O. K. Kohn, U.S. Pat. 3,305,441 (1967).
145. R. N. Mandzhgaladze, V. I. Geradze, and I. S. Gvineriya, Gig. Truda Prof. Zabolevaniya, 5, 50 (1974) (in Russian); Pest. Abstr., 7, 74-2640 (1974).
146. M. J. Seller, Lancet, 1962, II, p. 249.
147. G. F. Somers, Lancet, 1962, I, p. 912.
148. I. D. Fratta, E. B. Sigg, and K. Maiorana, Toxicol. Appl. Pharmacol., 7, 268 (1965).
149. S. Fabro, H. Schumacher, R. L. Smith, R. B. L. Stagg, and R. T. Williams, Brit. J. Pharmacol., 25, 352 (1965).
150. J. P. Marliac, Federation Proc., 23, 105 (1964).
151. J. P. Marliac, M. J. Verrett, J. McLaughlin, and O. G. Fitzhugh, Toxicol. Appl. Pharmacol., 7, 25 (1965).
152. M. J. Verrett, M. K. Mutchler, W. F. Scott, E. F. Reynaldo, and J. McLaughlin, Ann. N.Y. Acad. Sci., 160, 334 (1969).
153. S. Fabro, R. L. Smith, and R. T. Williams, Food Cosmet. Toxicol., 3, 587 (1966).
154. G. Kennedy, O. E. Fancher, and J. C. Calandra, Toxicol. Appl. Pharmacol., 13, 420 (1968).
155. J. F. Vondruska, O. E. Fancher, and J. C. Calandra, Toxicol. Appl. Pharmacol., 13, 420 (1968).
156. J. T. Mitchell and C. L. Yntema, Anat. Record, 175, 390 (1973).
157. M. S. Legator, F. J. Kelly, S. Green, and E. J. Oswald, Ann. N.Y. Acad. Sci., 160, 344 (1969).
158. B. A. Bridges, R. P. Mottershead, M. A. Rothwell, and M. H. L. Green, Chem. Biol. Interact., 5, 77 (1972).

159. B. A. Bridges, R. P. Mottershead, and C. Colella, Mutation Res., 21, 303 (1973).
160. T. Kada, New Methods in Environmental Chemistry, presented at Intern. Symp. on Ecological Chemistry, Susono, Japan, Nov. 1973 (F. Coulston, F. Korte, and M. Goto, eds.), p. 127.
161. Y. Shirasu, 3rd Intern. Symp. on Chemical and Toxicological Aspects of Environmental Quality, Tokyo, Japan, Nov. 19-22, 1973, p. 98.
162. G. Mohn, Arch. Toxikol., 28, 93 (1971).
163. H. V. Malling and F. J. De Serres, Environ. Mutation Soc. News Letter, 3, 41 (1970).
164. D. Siebert, F. K. Zimmermann, and D. Lemperle, Mutation Res., 10, 533 (1970).
165. P. G. N. Kramers and G. A. C. Knapp, Mutation Res., 21, 149 (1973).
166. E. Vogel and J. L. R. Chandler, Experientia, 30, 621 (1974).
167. P. Mollet, Mutation Res., 21, 137 (1973).
168. S. S. Epstein and H. Shafner, Nature, 219, 385 (1968).
169. T. F. X. Collins, Food Cosmet. Toxicol., 10, 353 (1972).
170. T. F. X. Collins, Food Cosmet. Toxicol., 10, 363 (1972).
171. A. J. Bateman, Genet. Res., 1, 381 (1960).
172. T. Kada, M. Moriya, and Y. Shirasu, Mutation Res., 26, 243 (1974).
173. B. N. Ames, F. D. Lee, and W. E. Durston, Proc. Natl. Acad. Sci. U.S., 70, 782 (1973).
174. Anon., Chem. Week, 112, 21 (1973).
175. EPA, Environmental Contamination From Hexachlorobenzene, Environmental Protection Agency, Washington, D.C., 1973.
176. E. F. Davis, B. L. Tama, and C. C. Lee, Handbook of Toxicology, Vol. 5 (D. S. Ditter, ed.), W. B. Saunders, Philadelphia, 1959, p. 19.
177. H. Martin, Pesticide Manual, 3rd ed., British Crop Protection Council, Droitwich, England, 1972, p. 283.
178. J. G. Vos, H. L. Van Der Maas, A. Musch, and E. Ram, Toxicol. Appl. Pharmacol., 18, 944 (1971).
179. I. U. Savitskii, Kievsk. Med. Inst., 158 (1964) (in Russian); Chem. Abstr., 63, 8952D (1965).
180. R. K. Ockner and R. Schmid, Nature, 189, 499 (1961).
181. A. Medline, E. Bain, A. I. Menon, and H. F. Haberman, Arch. Pathol., 96, 61 (1973).
182. L. N. Gurfein and Z. K. Pavlova, Sanit. Okhrana Vodoemov ot Zagryazneniya Promy. Stochnymi Vodami, 4, 117 (1960) (in Russian); Chem. Abstr., 56, 7060i (1962).
183. J. G. Vos, H. A. Breeman, and H. Benschop, Meded. Rijksfac. Landbouwwetenschap. Gent, 33, 1263 (1968).
184. B. A. Schwetz, J. M. Norris, R. J. Kociba, P. A. Keeler, R. F. Cornier, and P. J. Gehring, Toxicol. Appl. Pharmacol., 30, 255 (1974).

185. R. D. Kimbrough and R. E. Linder, Res. Commun. Chem. Pathol. Pharmacol., 8, 653 (1974).
186. E. C. Villaneuva, R. W. Jennings, V. W. Burse, and R. D. Kimbrough, J. Agr. Food Chem., 22, 916 (1974).
187. D. L. Grant, F. Iverson, G. V. Hatina, and D. L. Villeneuve, Environ. Physiol. Biochem., 4, 159 (1974).
188. T. Kuiper-Goodman, D. Grant, G. Korsrud, C. A. Moodie, and I. C. Munro, Toxicol. Appl. Pharmacol., 29, 101 (1974).
189. D. C. Villeneuve, L. G. Panopio, and D. L. Grant, Environ. Physiol. Biochem., 4, 112 (1974).
190. K. S. Khera, Toxicol. Appl. Pharmacol., 29, 109 (1974).
191. C. Cam and G. Nigogosyan, J. Amer. Med. Assoc., 183, 88 (1963).
192. R. Schmid, New Engl. J. Med., 263, 397 (1960).
193. F. DeMatteis, B. E. Prior, and C. Rimington, Nature, 191, 363 (1961).
194. A. Curley, V. W. Burse, R. W. Jennings, E. C. Villanueva, L. Tomatis, and K. Akazaki, Nature, 242, 338 (1973).
195. M. N. Brady and D. S. Siyali, Med. J. Australia, 1, 158 (1972).
196. L. Acker and E. Schulte, Naturwissenschaft, 57, 497 (1970).
197. D. S. Siyali, Med. J. Australia, 2, 1063 (1973).
198. J. E. Burns and F. M. Miller, Arch. Environ. Health, 30, 44 (1975).
199. J. L. Johnson, D. L. Stalling, and J. W. Hogan, Bull. Environ. Contam. Toxicol., 11, 393 (1974).
200. A. V. Holden, Pestic. Monit. J., 4, 117 (1970).
201. V. Zitko, Bull. Environ. Contam. Toxicol., 6, 464 (1971).
202. W. Stanhope, J. Agr. Victoria Australia, 67, 245 (1969).
203. J. Goursaud, F. M. Luquet, J. F. Boudier, and J. Casalis, Ind. Aliment. Agr. Paris, 89, 31 (1972).
204. L. G. M. T. Tuinstra and J. B. Roos, Jr., Ned. Melk Zuiveltijdschr., 24, 65 (1970).
205. D. T. Williams, H. M. Cunningham, and B. J. Blanchfield, Bull. Environ. Contam. Toxicol., 7, 57 (1972).
206. G. R. Higginbotham, A. Huang, D. Firestone, M. J. Verrett, J. Ress, and A. D. Campbell, Nature, 200, 702 (1968).
207. J. E. Huff and J. S. Wassom, Intern. J. Environ. Studies, 6, 13 (1974).
208. B. A. Schwetz, J. M. Norris, G. L. Sparschu, V. K. Rowe, P. J. Gehring, J. L. Emerson, and C. G. Gerbig, Environ. Health Persp., 5, 88 (1973).
209. M. E. King, A. M. Shefner, and R. R. Bates, Environ. Health Persp., 5, 163 (1973).
210. A. Bevenue and H. Beckman, Residue Rev., 19, 83 (1967).
211. P. Bayle, Sur La Toxicologie Du Pentachlorophenol, Ph.D. Thesis, University of Lyon, France, 1924.
212. W. Deichmann, W. Machle, K. V. Kitzmiller, and G. Thomas, J. Pharmacol. Exptl. Therap., 76, 104 (1942).

213. D. L. Harrison, New Zealand Vet. J., 7, 89 (1959).
214. E. F. Stohlman, U.S. Public Health Rept., 66, 1303 (1951).
215. T. H. McGavack, L. J. Boyd, F. V. Piccione, and R. Terranova, J. Ind. Hyg. Toxicol., 23, 239 (1941).
216. A. Pleskova and K. Bencze, Pracovni Lekarstvi, 7X1, 348 (1959) (in Czech.); Chem. Abstr., 54, 16661A (1960).
217. I. Knudsen, H. G. Verschuuren, E. M. den Tonkelaar, R. Kroes, and P. F. W. Helleman, Toxicology, 2, 141 (1974).
218. T. S. Carswell and H. K. Nason, Ind. Eng. Chem., 30, 622 (1938).
219. R. Truhaut, P. L. Epee, and E. Boussemart, Arch. Maladies Prof., 13, 567 (1952).
220. R. H. Kehoe, K. Deichmann-Gruebler, and K. V. Kitzmiller, J. Ind. Hyg. Toxicol., 21, 160 (1939).
221. M. F. Mason, S. M. Wallace, E. Foerster, and W. Drummond, J. Forensic Sci., 10, 136 (1965).
222. H. Bergner, P. Constantinidis, and J. H. Martin, Can. Med. Assoc. J., 92, 448 (1965).
223. B. A. Schwetz, P. A. Keeler, and P. J. Gehring, Toxicol. Appl. Pharmacol., 28, 151 (1974).
224. D. K. Hinkle, Toxicol. Appl. Pharmacol., 25, 454 (1973).
225. K. D. Courtney, personal communication in Ref. 223.
226. W. Buselmaier, G. Röhrborn, and P. Propping, Biol. Zentr., 91, 311 (1972).
227. H. C. E. W. Baader and H. J. Bauer, Ind. Med. Surg., 20, 286 (1951).
228. J. A. Menon, Brit. Med. J., 1, 1156 (1958).
229. M. Shirakawa, H. Nagatoshi, and M. Hirakawa, Kurume Med. J., 6, 24 (1959).
230. G. E. M. M. Viniegra, R. E. Marquez, and R. S. Saraboso, Salud Publica Med., 5, 915 (1963).
231. W. J. Hayes, Jr., Clinical Handbook of Economic Poisons, U.S. Dept. of Health, Education, and Welfare, Communicable Diseases Center, Atlanta, Ga., 1963.
232. D. Gordon, Med. J. Australia, 43, 485 (1956).
233. S. Nomura, Bull. Hyg., 29, 294 (1954).
234. A. M. G. Campbell, Brit. Med. J., 23, 415 (1952).
235. J. M. Barnes, WHO Monograph Series No. 16, 1953, p. 67.
236. H. J. Roberts, Southern Med. J., 56, 632 (1963).
237. A. M. Robson, J. M. Kissane, N. H. Elvick, and L. Pundavela, J. Pediat., 75, 309 (1969).
238. R. W. Armstrong, E. R. Eichner, D. E. Klein, W. F. Barthel, J. V. Bennett, V. Jonsson, H. Bruce, and L. E. Loveless, J. Pediat., 75, 317 (1969).
239. B. A. Schwetz, P. A. Keeler, and P. J. Gehring, Toxicol. Appl. Pharmacol., 28, 146 (1974).
240. J. R. Plimmer, Environ. Health Persp., 5, 41 (1973).

241. C. A. Nilsson and L. Renberg, J. Chromatogr., 89, 325 (1974).
242. D. Firestone, J. Ress, N. L. Brown, R. P. Barron, and J. N. Damico, J. Assoc. Offic. Anal. Chemists, 55, 85 (1972).
243. E. A. Woolson, R. F. Thomas, and P. D. J. Ensor, J. Agr. Food Chem., 20, 351 (1972).
244. C. Rappe and C. A. Nilsson, J. Chromatog., 67, 247 (1972).
245. K. Andersson, C. A. Nilsson, and C. Rappe, 166th Natl. Meet. Amer. Chem. Soc., Chicago, August, 1973.
246. S. Jensen and L. Renberg, Ambio, 1, 1 (1972).
247. S. Jensen and L. Renberg, Environ. Health Persp., 5, 37 (1973).
248. C. A. Nilsson, K. Andersson, C. Rappe, and S. O. Westermark, J. Chromatog., 96, 137 (1974).
249. R. L. Johnson, P. J. Gehring, R. J. Kociba, and B. A. Swetz, Environ. Health Persp., 5, 171 (1973).
250. J. R. Plimmer, J. M. Ruth, and E. A. Woolson, J. Agr. Food Chem., 21, 90 (1973).
251. K. Munakata and M. Kuwahaja, Residue Rev., 25, 13 (1969).
252. R. H. Stehl, R. R. Papenfuss, R. A. Bredeweg, and R. W. Roberts, Chlorodioxins—Origin and Fate (E. H. Blair, ed.), Advan. Chem. Ser., 120, 119 (1973).
253. D. R. Buhler, M. E. Rasmusson, and H. S. Nakaue, Environ. Sci. Technol., 7, 919 (1973).
254. D. Bandt and D. Nehring, Z. Fischind., 10, 543 (1962).
255. V. Zitko, O. Hutzinger, and P. M. K. Choi, Bull. Environ. Contam. Toxicol., 12, 649 (1974).
256. A. Stark, J. Agr. Food Chem., 17, 871 (1969).
257. L. Rudling, Water Res., 4, 533 (1970).
258. P. C. Kearney, A. R. Isensee, C. S. Helling, E. A. Woolson, and J. R. Plimmer, Chlorodioxins—Origin and Fate (E. H. Blair, ed.), Advan. Chem. Ser., 120, 105 (1973).
259. P. C. Kearney, E. A. Woolson, A. R. Isensee, and C. S. Helling, Environ. Health Persp., 5, 273 (1973).
260. D. G. Crosby, A. S. Wong, J. R. Plimmer, and E. A. Woolson, Science, 173, 748 (1971).
261. P. C. Kearney, E. A. Woolson, and C. P. Ellington, Environ. Sci. Technol., 6, 1017 (1972).
262. J. Kimmig and K. H. Schulz, Dermatologica, 115, 540 (1957).
263. H. Bauer, K. H. Schulz, and W. Spiegelberg, Arch. Gewerbepathol. Gewerbehyg., 18, 538 (1961).
264. A. P. Poland and D. Smith, Arch. Environ. Health, 22, 316 (1971).
265. H. T. Hofman, Arch. Exptl. Pathol. Pharmakol., 232, 228 (1957).
266. E. L. Jones and H. Krizek, J. Invest. Dermatol., 39, 511 (1962).
267. M. H. Milnes, Nature, 232, 395 (1971).
268. B. A. Schwetz, J. M. Norris, G. L. Sparschu, V. K. Rowe, P. J. Gehring, J. L. Emerson, and C. G. Gerbig, Chlorodioxins—Origin and Fate (E. H. Blair, ed.), Advan. Chem. Series, 120, 55 (1973).

269. J. B. Greig, G. Jones, W. H. Butler, and J. M. Barnes, Food Cosmet. Toxicol., 11, 585 (1973).
270. K. H. Schulz, Arbeitsmed. Sozialmed. Arbeitshyg., 3, 25 (1968).
271. J. G. Vos, J. A. Moore, and J. G. Zinkl, Toxicol. Appl. Pharmacol., 29, 229 (1974).
272. M. W. Harris, J. A. Moore, J. G. Vos, and B. N. Gupta, Environ. Health Persp., 5, 101 (1973).
273. B. N. Gupta, J. G. Vos, J. A. Moore, J. G. Zinkl, and B. C. Bullock, Environ. Health Persp., 5, 125 (1973).
274. B. A. Schwetz, J. M. Norris, G. L. Sparschu, V. K. Rowe, P. J. Gehring, J. L. Emerson, and C. G. Gerbig, Environ. Health Persp., 5, 87 (1973).
275. J. S. Woods, Environ. Health Persp., 5, 221 (1973).
276. J. A. Goldstein, P. Hickman, H. Bergman, and J. G. Vos, Res. Commun. Chem. Pathol. Pharmacol., 6, 919 (1973).
277. N. P. Buu-Hoi, C. Pham-Huu, G. Sesque, M. C. Azum-Gelade, and G. Saint-Ruf, Naturwissenschaft, 59, 174 (1972).
278. H. M. Cunningham and D. T. Williams, Bull. Environ. Contam. Toxicol., 7, 45 (1972).
279. G. Jones and W. H. Butler, J. Pathol., 109, 15 (1973).
280. G. L. Sparschu, F. L. Dunn, and V. K. Rowe, Toxicol. Appl. Pharmacol., 17, 317 (1970).
281. G. L. Sparschu, F. L. Dunn, and V. K. Rowe, Food Cosmet. Toxicol., 9, 405 (1971).
282. K. D. Courtney and J. A. Moore, Toxicol. Appl. Pharmacol., 20, 396 (1971).
283. D. Neubert and I. Dillmann, Arch. Pharmacol., 272, 243 (1972).
284. J. A. Moore, B. N. Gupta, J. G. Zinkl, and J. G. Vos, Environ. Health Persp., 5, 81 (1973).
285. K. S. Khera and J. A. Ruddick, Advan. Chem. Series, 120, 70 (1973).
286. D. Neubert, P. Zens, A. Rothenwallner, and H. J. Merker, Environ. Health Persp., 5, 67 (1973).
287. S. Hussain, L. Ehrenberg, G. Lofroth, and T. Gejvall, Ambio, 1, 32 (1972).
288. W. T. Jackson, J. Cell Sci., 10, 15 (1972).
289. G. R. Higginbotham, A. Huang, D. Firestone, M. J. Verrett, J. Ress, and A. D. Campbell, Nature, 200, 702 (1968).
290. Anon., Chem. Eng. News, 45, 10 (1967).
291. J. Beck and K. E. Hansen, Pestic. Sci., 5, 41 (1974).
292. FAO/WHO, 1969, Evaluations of Some Pesticide Residues in Food, Food and Agriculture Organization, Rome, 1970, p. 199.
293. R. H. deVos, M. C. ten Noever de Brauw, and P. D. A. Olthof, Bull. Environ. Contam. Toxicol., 11, 567 (1974).
294. E. J. Kuchar, F. O. Geenty, W. P. Griffith, and R. J. Thomas, J. Agr. Food Chem., 17, 1237 (1969).
295. W. H. Ko and J. D. Farley, Phytopathology, 59, 64 (1969).

296. H. Martin, Pesticide Manual, 2nd ed., British Crop Protection Council, Droitwich, England, 1971, p. 427.
297. R. L. Jordan and J. F. Borzelleca, Toxicol. Appl. Pharmacol., 25, 454 (1973).
298. D. Courtney, Toxicol. Appl. Pharmacol., 25, 454 (1973).
299. G. A. H. Buttle and F. J. Dyer, J. Pharm. Pharmacol., 2, 371 (1950).
300. J. T. Abrams, Brit. Vet. J., 106, 413 (1950).
301. P. L. Bistrup, Toxicity of Mercury and Its Compounds, Elsevier, Amsterdam, 1964.
302. S. Skerving and J. Vostal, in Mercury in the Environment (L. Friberg and J. Vostal, eds.), CRC Press, Cleveland, 1972, pp. 93-107.
303. L. Friberg and G. F. Nordberg, in Mercury in the Environment (L. Friberg and J. Vostal, eds.), CRC Press, Cleveland, 1972, pp. 113-139.
304. S. Skerving, in Mercury in the Environment (L. Friberg and J. Vostal, eds.), CRC Press, Cleveland, 1972, pp. 141-168.
305. Secretary's Pesticide Advisory Committee of Environ. Res., 4, 1 (1971).
306. F. Berglund and M. Berlin, in Chemical Fallout (M. W. Miller and G. G. Berg, eds.), Charles C. Thomas, Springfield, Ill., 1969, pp. 258-273.
307. A. Swensson, Arhiv. Higi. Rada i Toksikol., 24, 307 (1973).
308. L. J. Goldwater, Environ. Control, 4, 305 (1974).
309. L. J. Goldwater, Sci. Amer., 224, 15 (1971).
310. Anon., Arch. Environ. Health, 19, 891 (1969).
311. J. G. Saha and K. S. McKinley, Toxicol. Environ. Chem. Rev., 1, 271 (1973).
312. H. L. Friedman, Ann. N.Y. Acad. Sci., 65, 461 (1957).
313. N. L. Hughes, Ann. N.Y. Acad. Sci., 65, 54 (1957).
314. A. A. Swensson, Acta Med. Scand., 143, 365 (1952).
315. I. M. Weiner, R. I. Levy, and G. H. Mudge, J. Pharmacol. Exptl. Therap., 138, 96 (1962).
316. O. Hoshino, K. Tanzawa, T. Terao, T. Ukita, and A. Ohuchi, J. Hyg. Chem., 12, 94 (1966).
317. A. Ahlmark, Brit. J. Ind. Med., 5, 117 (1948).
318. O. Hook, K. D. Lundgren, and A. Swensson, Acta Med. Scand., 150, 131 (1954).
319. K. D. Lundgren and A. Swensson, Nord. Hyg. Tidskr., 29, 1 (1948).
320. T. Takewchi, Proc. 7th Intern. Congr. Neurology, Rome, 1961, p. 1.
321. Report of Expert Committee, Nord. Hyg. Tidskr., Suppl. 4 (1971).
322. M. A. Jalili and A. H. Abbasi, Brit. J. Ind. Med., 18, 303 (1961).
323. Study Group of Minamata Disease, Minamata Disease (M. Kutsuna, ed.), Kumamoto Univ., Japan (1968).

324. F. Bakir, S. F. Damluji, L. Amin-Zakl, M. Murtadha, A. Khalidi, N. Y. Al-Rawi, S. Tikriti, H. I. Dhahir, T. W. Clarkson, J. L. Smith, and R. A. Doherty, Science, 181, 230 (1973).
325. S. Jensen and A. Jernelov, Nature, 223, 753 (1969).
326. A. G. Johnels and T. Westermark, in Chemical Fallout (M. W. Miller and G. G. Berg, eds.), Charles C. Thomas, Springfield, 1969, p. 221.
327. A. Jernelov, in Chemical Fallout (M. M. Miller and G. G. Berg, eds.), Charles C. Thomas, Springfield, 1969, p. 68.
328. J. Wood, Environment, 14, 33 (1972).
329. J. G. Saha, Residue Rev., 42, 1 (1972).
330. J. V. Ordonez, J. A. Carrillo, and C. M. Miranda, Bull. Pan Amer. Sanit. Bur., 60, 510 (1966).
331. I. U. Haq, Brit. Med. J., 1963, p. 1579.
332. W. H. Likosky, P. E. Pierce, and A. H. Hinman, Proc. Amer. Acad. Neurology, Bal Harbour, Fla., April 27-30, 1970.
333. A. Curley, V. A. Sedlak, E. F. Girling, R. E. Hawk, W. F. Barthel, P. E. Pierce, and W. H. Likosky, Science, 172, 65 (1971).
334. L. T. Kurland, S. N. Faro, and H. Siedler, World Neurol., 1, 370 (1960).
335. H. Schmidt and R. Harzmann, Intern. Arch. Arbeits Med., 26, 71 (1970).
336. B. Storrs, J. Thompson, G. Fair, M. S. Dickerson, L. Nickey, W. Barthel, and J. E. Spaulding, Morbidity Mortality, 19, 25 (1970).
337. B. Storrs, J. Thompson, L. Nickey, W. Barthel, and J. E. Spaulding, Morbidity Mortality, 19, 169 (1970).
338. K. D. Lundgren and A. Swensson, J. Ind. Hyg., 31, 190 (1949).
339. R. D. Snyder, New Engl. J. Med., 284, 1014 (1971).
340. S. S. Dahhan and H. Orfaly, Amer. J. Cardiol., 14, 178 (1964).
341. A. D. Kantarjian, Neurology, 11, 639 (1961).
342. L. K. A. Derban, Arch. Environ. Health, 28, 49 (1974).
343. A. V. Bakulina, Soviet. Med., 31, 60 (1968).
344. A. C. Ladd, L. J. Goldwater, and M. B. Jacobs, Arch. Environ. Health, 9, 43 (1964).
345. J. Hartung, Berufsdermatosen, 13, 116 (1965).
346. G. E. Morris, Arch. Environ. Health, 1, 53 (1960).
347. T. Suzuki and T. Miyama, Ind. Health, 9, 51 (1971).
348. Y. Itsuno, in Minamata Disease (M. Katsuna, ed.), Kumamoto University, Japan, 1968, p. 267.
349. M. Yamashita, J. Japan Soc. Internal Med., 53, 539 (1964).
350. J. S. Palmer, J. Amer. Vet. Assoc., 142, 1385 (1963).
351. E. L. Taylor, J. Amer. Vet. Assoc., 111, 46 (1947).
352. W. T. Oliver and N. Platonow, Amer. J. Vet. Res., 21, 906 (1960).
353. K. Nose, Japan J. Hyg., 24, 359 (1969).

354. N. J. Eastman and A. B. Scott, Human Fertil., 9, 33 (1944).
355. O. G. Fitzhugh, A. A. Nelson, E. P. Laug, and F. M. Kunze, Ind. Hyg. Occup. Med., 2, 433 (1950).
356. L. Tryphonas and N. O. Nielsen, Can. J. Comp. Med., 34, 181 (1970).
357. R. Wien, Quart. J. Pharm., 12, 212 (1939).
358. J. C. Gage and W. Swan, Biochem. Pharmacol., 8, 77 (1961).
359. M. Berlin and S. Ullberg, Arch. Environ. Health, 6, 610 (1963).
360. E. J. Fujita, J. Kumamoto Med. Soc., 43, 47 (1969).
361. T. N. Suzuki, N. Matsumoto, and H. Katsunuma, Ind. Health, 5, 149 (1967).
362. N. E. Garrett, R. J. B. Garrett, and J. W. Arch Deacon, Toxicol. Appl. Pharmacol., 22, 649 (1972).
363. J. M. Spyker, S. B. Sparber, and R. M. Goldberg, Science, 177, 621 (1972).
364. M. Inouye, K. Hoshimo, and U. Murakami, Ann. Rept. Res. Inst. Environ. Med., Nagoya Univ., 19, 69 (1972).
365. E. Rosenthal and S. B. Sparber, Life Sci., 11, 883 (1972).
366. S. B. Harris, J. G. Wilson, and R. H. Printz, Teratology, 6, 139 (1972).
367. D. J. Clegg, Proc. Roy. Soc. Can. Symp., Toronto, Feb. 15-16, 1971, p. 141.
368. N. K. Mottet, Teratology, 10, 173 (1974).
369. K. S. Khera and S. A. Tabacova, Food Cosmet. Toxicol., 11, 245 (1973).
370. T. F. Gale and V. H. Ferm, Life Sci., 10, 1341 (1971).
371. J. Piechocka, Rocz. Panstw. Zakl. Hig., 19, 385 (1968) (in Polish); Chem. Abstr., 70, 19197M (1969).
372. K. S. Khera, Toxicol. Appl. Pharmacol., 24, 167 (1973).
373. C. Ramel, in Mercury in the Environment (L. Friberg and J. Vostal, eds.), CRC Press, Cleveland, 1972, pp. 169-181.
374. C. Ramel, Hereditas, 57, 445 (1967).
375. C. Ramel, Hereditas, 57, 448 (1967).
376. C. Ramel, Hereditas, 61, 208 (1969).
377. F. Y. Fahmy, Cytogenic Analysis of the Action of Some Fungicide Materials, Ph.D. Thesis, University of Lund, Sweden, 1951.
378. M. Umeda, K. Saitu, K. Hirose, and M. Salto, J. Exptl. Med., 39, 47 (1969).
379. G. Fiskesjö, Hereditas, 64, 142 (1970).
380. M. Okada and H. Oharazawa, in Report on the Cases of Mercury Poisoning in Niigata, Ministry of Health and Welfare, Tokyo, 1967, p. 63.
381. C. Ramel and J. Magnusson, Hereditas, 61, 231 (1969).
382. S. Skerving, K. Hansson, and J. Lindsten, Arch. Environ. Health, 21, 133 (1970).
383. C. Ramel, Proc. 1st Ann. Meet. Environ. Mutagen Soc., Washington, D.C., 1970, p. 22.

384. K. S. Khera, Proc. 2nd Ann. Meet. Environ. Mutagen Soc., Washington, D.C., March 21-24, 1971.
385. J. M. Barnes and H. B. Stoner, Pharmacol. Rev., 11, 211 (1959).
386. E. Browning, Toxicity of Industrial Metals, 2nd ed., Butterworths, London, 1969, pp. 323-330.
387. J. M. Barnes and H. B. Stoner, Brit. J. Ind. Med., 15, 15 (1958).
388. J. M. Barnes and P. N. Magee, J. Pathol. Bacteriol., 75, 267 (1958).
389. M. J. Marks, C. L. Winek, and S. P. Shanor, Toxicol. Appl. Pharmacol., 14, 627 (1969).
390. H. Martin, Pesticide Manual, 3rd ed., Brit. Crop Protection Council, Droitwich, England, 1972, p. 263.
391. T. B. Gaines, Toxicol. Appl. Pharmacol., 12, 397 (1968).
392. N. M. Brown and S. C. Clemson, Dissertation Abstr., 33, 5356 (1973).
393. Anon., Lancet, 2, 1216 (1958).
394. D. I. Williams, R. H. Marten, and I. Sarkany, Lancet, 2, 1212 (1958).
395. J. C. Gentles, Nature, 182, 476 (1958).
396. H. Blank and F. J. Roth, Jr., Arch. Dermatol., 79, 259 (1959).
397. G. Riehl, Hautarzt, 10, 136 (1959).
398. G. E. Paget and A. C. Walpole, Nature, 182, 1320 (1958).
399. L. Larizza, F. Tredici, C. Rossi, and L. De Carli, Atti. Assoc. Genet. Ital., 17, 19 (1972).
400. C. Rossi, F. Tredici, and L. De Carli, Atti. Assoc. Genet. Ital., 16, 45 (1971).
401. L. Larizza, G. Simoni, F. Tredici, and L. De Carli, Mutation Res., 25, 123 (1974).
402. E. McNall, Arch. Dermatol., 81, 657 (1960).
403. D. Karnofsky, in Teratology Principles and Techniques (J. G. Wilson and J. Warkany, eds.), University of Chicago Press, Chicago, 1965, p. 185.
404. N. Sloniskaya, Antibiotiki, 14, 44 (1969).
405. P. Koller, Trans. St. John's Hosp. Dermatol. Soc., 45, 38 (1960).
406. M. F. Klein and J. R. Beall, Science, 175, 1483 (1972).
407. J. MacLeod and W. Nelson, Proc. Soc. Exptl. Biol. Med., 102, 259 (1959).
408. S. S. Epstein, J. Andrea, P. Clapp, and D. Mackintosh, Toxicol. Appl. Pharmacol., 11, 442 (1967).
409. K. Fuji and S. S. Epstein, Toxicol. Appl. Pharmacol., 14, 613 (1969).
410. D. E. Mathre, Pest. Biochem. Physiol., 1, 216 (1971).
411. H. Martin, Pesticide Manual, 3rd ed., Brit. Crop Protection Council, Droitwich, England, 1972, p. 89.
412. A. Stefan, N. Stancioiu, and S. Ceausescu, Analele Univ. Bucuresti, Biol. Anim., 22, 125 (1973) (in Rumanian); Chem. Abstr., 81, 131359U (1974).

413. D. E. Briggs, R. H. Waring, and A. M. Hackett, Pestic. Sci., 5, 599 (1974).
414. L. W. Carlson, Can. J. Plant. Sci., 50, 627 (1970).
415. N. N. Ragsdale and H. D. Sisler, Phytopathology, 60, 1422 (1970).
416. W. T. Chin, G. M. Stone, A. E. Smith, Proc. 5th Brit. Insect. Fungic. Conf. Brighton, 1969, Vol. 2, p. 322.

AUTHOR INDEX

Numbers in brackets are reference numbers and indicate that an author's work is referred to although his name is not cited in the text. Underlined numbers give the page on which the complete reference is listed.

C

D

E

F

G

K

L

O

U

SUBJECT INDEX

G

Q

R

S

T